Graduate Texts in Physics

Graduate Texts in Physics publishes core learning/teaching material for graduate- and advanced-level undergraduate courses on topics of current and emerging fields within physics, both pure and applied. These textbooks serve students at the MS- or PhD-level and their instructors as comprehensive sources of principles, definitions, derivations, experiments and applications (as relevant) for their mastery and teaching, respectively. International in scope and relevance, the textbooks correspond to course syllabi sufficiently to serve as required reading. Their didactic style, comprehensiveness and coverage of fundamental material also make them suitable as introductions or references for scientists entering, or requiring timely knowledge of, a research field.

More information about this series at http://www.springer.com/series/8431

Udo W. Pohl

Epitaxy of Semiconductors

Physics and Fabrication of Heterostructures

Second Edition

Udo W. Pohl
Institute of Solid State Physics
Technische Universität Berlin
Berlin, Germany

ISSN 1868-4513 ISSN 1868-4521 (electronic)
Graduate Texts in Physics
ISBN 978-3-030-43871-5 ISBN 978-3-030-43869-2 (eBook)
https://doi.org/10.1007/978-3-030-43869-2

This Springer imprint is published by the registered company Springer Nature Switzerland AG
The registered company address is: Gewerbestrasse 11, 6330 Cham, Switzerland

Preface to the Second Edition

The book *Epitaxy of Semiconductors* has been well received by students and the scientific community. The first edition focused concisely on the very basics and omitted quite a number of important aspects. In the second edition, I therefore added a few new chapters and also amended the initial text. The new parts comprise the emerging field of organic semiconductors with their specific peculiarities and the important in situ analysis of epitaxial growth, which is today routinely implemented in commercial growth systems for online monitoring and control. A chapter on the application of surfactants is added, illustrating an additional means to control growth and pointing out the effect of kinetics. One new chapter is dedicated to special techniques, which are applied by all major epitaxy methods MOVPE, MBE, and LPE. Selective area growth is widely employed in nanostructure fabrication and epitaxial lateral overgrowth to create layers with areas of low defect density; migration-enhanced epitaxy is used to improve the performance of MBE, and vapor-liquid-solid growth is a popular and versatile method for fabricating nanowires with high quality.

I would like to thank my colleagues for critically reading parts of the new manuscript: Armin Dadgar (Univ. Magdeburg), Mario Dähne (TU Berlin), Norbert Esser (ISAS Berlin), and Kolja Haberland (LayTec Berlin). I am indebted to Nigel Mason (Oxford, UK) for stimulating the inclusion of in situ sensing, and I appreciate encouraging comments I received from Ganesh Balakrishnan (Univ. Albuquerque, NM).

Berlin
September 2019

Udo W. Pohl

Preface to the First Edition

Epitaxy—the growth of a crystalline layer on a crystalline substrate—represents the basis for the fabrication of semiconductor heterostructures and devices. Textbooks on semiconductor physics and devices usually describe the design of a heterostructure and subsequently measured data of a respective realization. The chain between these end points requires to solve many basic problems related to physics and technology. Such steps are generally described in specialized literature focusing on diverse aspects. Students and researchers starting in the field need to study papers and books on quite specific problems in a wide field. This textbook attempts to bridge the gap between well-established books on semiconductor physics on one side and texts on completed heterostructures like semiconductor devices on the other.

The book is based on a one-semester course held at Technical University of Berlin for undergraduate and graduate students in physics and engineering physics. It is primarily addressed to the non-specialist with some basic knowledge in solid state and semiconductor physics. The field of epitaxy is rapidly evolving and includes many materials and growth techniques. The text, therefore, focuses on basics and important aspects of epitaxy, emphasizing particularly the physical principles. Problems are illustrated for important semiconductors with zincblende or wurtzite structure.

The subject matter first covers the properties of heterostructures. Structural aspects implying elasticity and strain relaxation by dislocations are addressed as well as electronic properties including band alignment and electronic states in low-dimensional structures. Then, the thermodynamics and kinetics of epitaxial layer growth are considered, introducing the driving force of crystallization and paying special attention to nucleation and surface structures. Instructive examples are given for self-organized growth of quantum dots and wires. Afterward, aspects of doping, diffusion, and contacts are discussed. Eventually, the most important methods used for epitaxial growth are introduced: metalorganic vapor-phase epitaxy, molecular-beam epitaxy, and liquid-phase epitaxy.

I am grateful to my students, who consistently engaged me in discussions about fundamentals of epitaxy and stimulated an active motivation to write this book. I am also indebted to A. Krost for critical manuscript reading, and I appreciated the cooperation with C. Ascheron, Springer Science+Business Media.

Berlin
July 2012

Udo W. Pohl

Contents

About the Author

Dr. rer. nat. Udo W. Pohl, professor of experimental physics, studied physics in Aachen and Berlin, Germany, and received his Ph.D. degree in 1988 from the Technical University of Berlin, where he is currently principal investigator in the Institute of Solid State Physics. He is a member of the German Physical Society, participated in international conference committees, and has given keynote addresses at international conferences. Since the early 1990s he has given lectures on semiconductor physics and epitaxy. In 2009, he was appointed Adjunct Professor of Physics at Technical University of Berlin. He is series editor of Materials Sciences, has authored over 200 journal articles and conference papers, the books *Epitaxy of Semiconductors* and—together with Prof. Dr. Karl W. Böer—*Semiconductor Physics*, thirteen book contributions, and patents. His current research interests include physics and epitaxy of semiconductor nanostructures and devices.

Abbreviations

0D	zero-dimensional
1D	one-dimensional
2D	two-dimensional
3D	three-dimensional
ACS	American Chemical Society
AIP	American Institute of Physics
APS	American Physical Society
AVS	American Vacuum Society—The Science & Technology Society
bcc	body-centered cubic
BEP	beam equivalent pressure
CB	conduction band
CBE	chemical beam epitaxy
CVD	chemical vapor deposition
DAS	dimer adatom stacking-fault (model)
DBR	distributed Bragg reflector
DOS	density of states
ELO	epitaxial lateral overgrowth (also: ELOG)
EMA	effective mass approximation
EXAFS	extended X-ray absorption fine-structure
fcc	face-centered cubic
FET	field-effect transistor
FWHM	full width at half maximum
GSMBE	gas-source molecular beam epitaxy
hcp	hexagonally closed packed
hh	heavy hole
HRTEM	high-resolution transmission electron microscopy
HRXRD	high-resolution X-ray diffraction
HUC	half unit cell
LEED	low-energy electron diffraction
LED	light-emitting diode

lh	light hole
LPE	liquid phase epitaxy
MBE	molecular beam epitaxy
ML	monolayer
MOCVD	metal-organic chemical vapor deposition
MOMBE	metal-organic molecular beam epitaxy
MOVPE	metal-organic vapor-phase epitaxy
MRS	Materials Research Society
MWQ	multiple quantum well
PL	photoluminescence
PLE	photoluminescence excitation (spectroscopy)
PVD	physical vapor deposition
QD	quantum dot
QW	quantum well
QWR	quantum wire
RHEED	reflection high-energy electron diffraction
rms	root mean square
RSC	Royal Society of Chemistry
SEM	scanning electron microscopy
si	semi-insulating
SIMS	secondary ion mass spectrometry
SL	superlattice
slm	standard liters per minute
so	spin-orbit, also split-off
STM	scanning tunneling microscopy
TEC	thermal expansion coefficient
TEM	transmission electron microscopy
TLK	terrace-ledge-kink (model)
TSK	terrace-step-kink (model)
UHV	ultra-high vacuum
VB	valence band
VCA	virtual crystal approximation
VCSEL	vertical-cavity surface-emitting laser
ViGS	virtual gap states
VPE	vapor phase epitaxy
Wiley	John Wiley and Sons
XPS	X-ray photoelectron spectroscopy
XRD	X-ray diffraction
XSTM	cross-sectional scanning tunneling microscopy
ZB	zincblende

Chapter 1
Introduction

Abstract This introductory chapter provides a brief survey on the development of epitaxial growth techniques and points out tasks for the epitaxy of device structures. Starting from early studies of alkali-halide overgrowth in the beginning of the 20th century, basic concepts for lattice match between layer and substrate were developed in the late 1920s, followed by the theory of misfit dislocations introduced about 1950. Major progress in epitaxy was achieved by technical improvements of the growth techniques, namely liquid phase epitaxy in the early, and molecular beam epitaxy and metalorganic vapor phase epitaxy in the late 1960s. Current tasks for epitaxial growth are often motivated by needs for the fabrication of advanced devices, aiming to control carriers and photons.

Most semiconductor devices fabricated today are made out of a thin stack of layers with a typical total thickness of only some μm. Layers in such a stack differ in material composition and may be as thin as a single atomic layer. All layers are to be grown with high perfection and composition control on a bulk crystal used as a substrate. The growth technique employed for coping with this task is termed epitaxy. In the following we briefly consider the historical development and illustrate typical issues accomplished using epitaxy.

1.1 Epitaxy

1.1.1 Roots of Epitaxy

Crystalline solids found in nature show regularly shaped as-grown faces. The faces of a zinc selenide crystal grown from the vapor phase in the lab are shown in Fig. 1.1a. Mineralogists found that the angles between corresponding faces are always the same for different samples of the same type of crystal. They concluded already in the 18th century that such regular shapes originate from a regular assembly of identical building blocks forming the crystal as indicated in Fig. 1.1b. In the early 19th century mineralogists noticed that naturally occurring crystals sometimes grew together with

U. W. Pohl, *Epitaxy of Semiconductors*, Graduate Texts in Physics,
https://doi.org/10.1007/978-3-030-43869-2_1

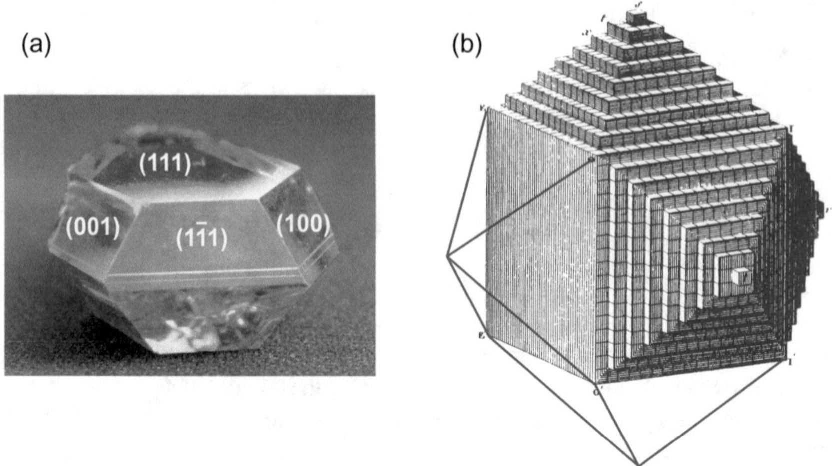

Fig. 1.1 **a** ZnSe bulk crystal with {100} and {111} growth faces. **b** Faces formed from regularly repeating building blocks described around 1800 [3]. The complete shape of the displayed rhomb-dodecahedron is indicated by *red lines*

a unique relationship of their orientations [1]. A first successful artificial reproduction of this effect in a laboratory was reported by Moritz L. Frankenheim in 1836 [2]. He demonstrated a parallel oriented growth of sodium nitrate $NaNO_3$ from solution on a freshly cleaved calcite crystal $CaCO_3$.

First systematic studies on the crystal growth on top of another crystalline material were reported by Thomas V. Barker starting 1906 [4]. At that time growth from solution and optical microscopy were the only readily developed techniques for growth and characterization of samples, respectively. Baker investigated a large number of NaCl-type alkali halides like chlorides, bromides, iodides, and cyanides of, e.g., rubidium and cesium. He placed a drop of saturated solution of one halide onto a freshly cleaved surface of another halide and observed the nucleation of a crystalline structure under a microscope. Crystals of the solute appeared as a rule in a few seconds, but sometimes nucleation was too rapid to be observed or difficulties arose due to a greater solubility for the crystal than for the dilute dropped on top. He concluded that crystalline growth of alkali halides was more likely to occur if the molecular volumes of the two inter-growing materials were nearly equal. We note that such conditions often imply a similar size of the building blocks mentioned above and consequently a small misfit of the lattice constants of the two materials.

The discovery of X-ray diffraction (1912) and electron diffraction (1927) by crystals had a strong impact on the knowledge about crystal structure. When Louis Royer made his seminal comprehensive studies with a wide variety of layers and substrate materials in 1928, he could precisely report on the effect of substrate crystal structure on the crystalline orientation of the layer [5]. Royer introduced the term *epi-*

taxy from the Greek επι (*epi*, upon, attached to)—ταξισ (*taxis*, arrangement, order) and concluded general rules for epitaxy. He noted that oriented growth occurs only when it involves the parallelism of two lattice planes which have lattice networks of identical or quasi-identical form and closely similar spacing. More precisely, he found that the differences between the lattice-network spacing (lattice parameters) for growth of alkali halides upon other alkali halides or mica (minerals of the form $XY_{2\text{-}3}Z_4O_{10}(OH, F)_2$) should be no more than 15%. Such geometrical considerations are still prominent today, though it was established later that epitaxy may also occur for much larger misfits.

In the 1930s, G. I. Finch and A. G. Quarrell concluded from a study of zinc oxide on sputtered zinc that the initial layer is strained in order to attain lattice matching parallel to the interface [6]. The layer lattice-parameter vertical to the interface was also considered to be changed to maintain approximately the bulk density. They named this phenomenon *pseudomorphism*. Though later the experimental evidence was pointed out to be by no means conclusive [1], the concept of pseudomorphic layers proved to be of basic importance for epitaxial structures.

The introduction of the theory of misfit dislocations at the interface between the substrate and the layer by F. C. Frank and Jan H. van der Merve in 1949 [7–9] extended the pseudomorphism approach and predicted a limit for the misfit of pseudomorphic growth. The incorporation of edge dislocations allows to accommodate misfit strain and consequently makes epitaxy possible also for structures with larger misfit. Conclusive experimental evidence for this was provided by John W. Matthews and co-workers in the 1960s [10, 11]. Thin metal layers on single-crystal metals with low misfit proved to grow pseudomorphic, while misfit dislocations were generated in thicker layers with a density depending on layer thickness. Later also conditions not included in the mentioned concepts like, e.g., interface alloying or surface energies were found to significantly affect epitaxial growth.

The emerging semiconductor industry in the early 1960s had a strong impact on the interest in epitaxy. In addition, advancements in the technique of producing high vacuum and pure materials, and progress in experimental techniques like electron microscopy and X-ray diffraction allowed to efficiently develop methods for epitaxial growth. Liquid-phase epitaxy enabled epitaxial growth of multilayered device structures of high complexity like separate-confinement semiconductor lasers. The advancement of this technology was facilitated by its similarity to the well-studied growth of bulk single-crystals from seeded solution. In the 70s the more sophisticated methods of molecular-beam epitaxy [12] and metalorganic vapor-phase epitaxy emerged. These techniques opened up epitaxy far from thermodynamic equilibrium and hence fabrication of structures with atomically sharp interfaces, which cannot be produced near equilibrium. Understanding and control of the epitaxial growth techniques was significantly advanced by the application of in-situ studies of the nucleation and growth process, and by the development of computational techniques. In the late 80s the modern techniques attained a maturity to get in the lead of device mass-production. Today a large variety of electronic, optoelectronic, magnetic, and superconducting layer structures are fabricated using epitaxial techniques, including structures of reduced dimensionality on a nanometer scale.

1.1.2 Epitaxy and Bulk-Crystal Growth

Crystal growth in an epitaxial process proceeds basically in the same way as conventional growth of bulk crystals. In epitaxy, however, layer and substrate differ in the nature and strength of the chemical bond. Moreover, both materials may have a different crystal structure and generally have an unequal lattice parameter—at least if the temperature is varied. We may therefore say that crystals differ *energetically and geometrically* in epitaxy [13]. Crystalline deposition of different materials on each other is also termed *heteroepitaxy*.

Such a definition does obviously not apply for the epitaxial deposition on a substrate of the same material. The process is usually referred to as *homoepitaxy* (sometimes also *autoepitaxy*). Let us consider a simple example to illustrate that there may still be a difference to conventional crystal growth. Epitaxy of electronic devices is usually performed on doped substrates and often starts by depositing a layer of the same material termed homoepitaxial buffer layer. Doping often alters the lattice parameter without significantly changing the chemical bond in bulk material. A p-type doping of, e.g., silicon with boron to a level of 2×10^{19} cm^{-3} induces a change of the lattice parameter by -1%. An undoped layer on a doped substrate of the same material will hence not differ energetically but geometrically. This provides a clear distinction of homoepitaxy from conventional crystal growth. According to this differentiation deposition of a layer of the same kind and doping like the substrate underneath should be termed crystal growth instead of epitaxy. Following the general usage we will, however, use the term homoepitaxy less strictly. Deposition of a layer of the same material as the substrate is usually just one of many layers to follow in the growth process of a device structure. Furthermore, fabrication of bulk crystals is usually performed applying a different growth regime than in epitaxy, allowing for much higher growth rates. Accordingly also different experimental setups are employed. The term homoepitaxy will be used here for deposition of the same material as the substrate, just to distinguish from heteroepitaxy.

1.2 Issues of Epitaxy

1.2.1 Convention on Use of the Term "Atom"

Solids are composed of atoms, which may be charged due to the character of the chemical bond. When used in a general way in this book, the word "atom" denotes both, an atom or an ion. Uncharged atoms are hence for simplicity usually not distinguished from charged atom cores in, e.g., ionic crystals or metals unless explicitly pointed out.

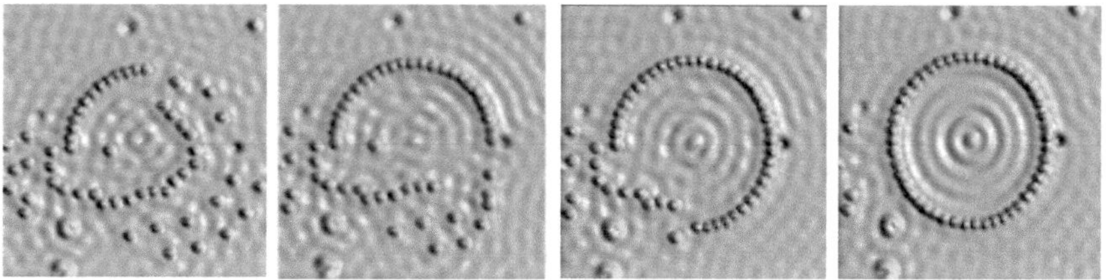

Fig. 1.2 Gradual circular arrangement of 48 iron adatoms on a copper surface assembled using a low-temperature scanning tunnelling microscope. Interference ripples originate from electron surface states. From [15]

1.2.2 Assembly of Atoms

Epitaxy denotes the regular assembly of atoms on a crystalline substrate. By inventing the scanning tunneling microscope (STM) a particular means has been developed to control the assembly of single atoms for forming an ordered structure. The method uses the finite force an STM tip always exerts on an adsorbed atom (adatom) attached to the surface of a solid [14]. The magnitude of the force can be tuned by adjusting the voltage and the position of the tip. Since generally less force is required to move an atom across a surface than to pull it away the tip parameters can be set to allow for positioning an individual adatom while it remains bound to the surface. The example given in Fig. 1.2 shows iron atoms on a (111) copper surface. The initially disordered Fe atoms were carefully positioned at chosen locations. May such procedure also be applied to deposit an epitaxial layer onto a substrate? Imagine a skillful operator placing one atom per second exactly on the correct site of a layer. For typically about 10^{15} sites per cm^2 such procedure requires 31 Mio. years for a single atomic layer being deposited on one cm^2. The fabrication of epitaxial structures hence requires other methods.

The problem of growing a layer by the assembly of single atoms is comparable to the issue of describing the behavior of a gas by formulating the equation of motion for each atom. This cannot be accomplished for exceedingly large ensembles like a considerable fraction of 6×10^{23} atoms present in one mole. The approach of kinetic theory of gases hence solely describes averages of certain quantities of the vast number of atoms in the ensemble, concluded from the behavior of one single atom. These averages correspond to macroscopic variables. We will basically follow a comparable approach. In epitaxial growth we seek to establish favorable conditions for atoms of a nutrient phase to finding proper lattice sites in the solid phase. The macroscopic control parameters are governed by both thermodynamics and kinetics, and their effect depends on the materials and the applied growth method.

1.2.3 Tasks for Epitaxial Growth

Semiconductor devices control the flow and confinement of charge carriers and photons. To fulfill its task a device is composed of crystalline layers and corresponding interfaces with different physical properties. Epitaxy is employed to assemble such layer structure. The precise control of the growth process in epitaxy requires the accomplishment of issues with quite different nature. We consider the example of a semiconductor device-structure to illustrate a number of tasks addressed during fabrication and indicate the connection to respective chapters of this book. The addressed concepts basically apply also for insulators and metals. In this book we focus on semiconductor materials.

The demand for increasing data-rate capacity in data-communication networks raised the need for suitable optical interconnects, particularly for light sources. Vertical-cavity surface-emitting lasers (VCSELs) have characteristics meeting ambitious requirements of fiber communication and meanwhile emerged from the laboratory to the marketplace. VCSEL devices are also widely used in computer mice due to a good shape of their optical radiation field. A VCSEL is a semiconductor laser, which emits the radiation vertically via its surface—in contrast to the more common edge-emitting lasers. Like any laser it consists of an active zone where the light is generated, overlapping with a region where the optical wave is guided. Light is generated by recombination of electrons and holes which are confined in quantum wells (MQW, multiple quantum well). The generated photons contribute to the light wave which travels back and forth in an optical Fabry-Pérot resonator built by two mirrors, and a small fraction representing the laser radiation is allowed to emerge from the top mirror. Since the resonator in a VCSEL is very short, the reflectivity of the mirrors must be very high ($R > 99\%$) to maintain lasing oscillation. They hence are made from distributed Bragg reflectors (DBR) of many pairwise $\frac{1}{4}\lambda/n$ thick layers with a difference in the respective refractive index n, λ being the operation wavelength. If the index step Δn is low many pairs are required (a few tens). The basic design of a VCSEL is given in Fig. 1.3.

The realization of a semiconductor device is a complex process. The basic design layout is determined by a number of operation parameters like, e.g., the emission wavelength for an optoelectronic device. Already at this stage materials aspects play an important role. Obviously a wide-bandgap semiconductor material, e.g., must be used in the active region if the device is to radiate at high photon energy. The design stage comprises simulation work on electrical, optical, and other properties of the device depending on the employed materials and the specific purpose of the device. The design eventually yields a list of materials composition, thickness, and doping for each individual layer in the entire layer stack to be epitaxially grown.

The crystalline epitaxial growth on a single-crystalline substrate requires a well-defined relationship of the substrate structure with respect to that of the layers grown on top. For this purpose the spacing of the atoms parallel to the interface between substrate and the layers of the device structure on top has to accommodate. Since the lateral lattice constants never match perfectly and the total layer thickness of the

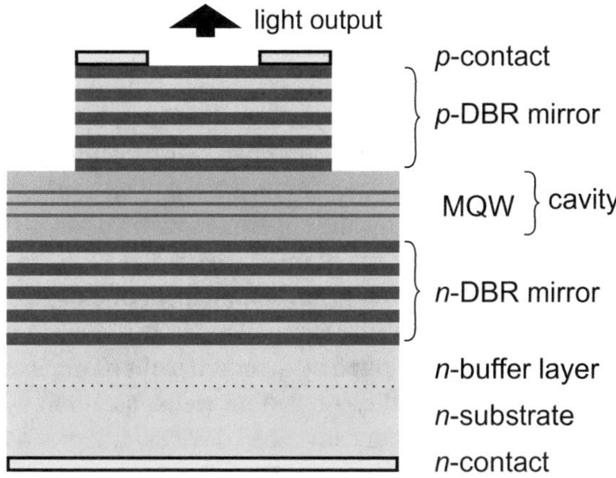

Fig. 1.3 Schematic of a vertical-cavity surface-emitting laser indicating the sequence of differently alloyed and doped semiconductor layers

device structure is generally much below the substrate thickness, epitaxial layers are elastically strained. Chapter 2 introduces into the structural and elastic properties of epitaxial layers and points out a critical limit for such strain. As a consequence of overcritical stress, plastic relaxation occurs by the introduction of misfit dislocations. Prominent species of such dislocations for the important crystal structures zincblende and wurtzite are treated. Prior to the growth of the device layers usually a buffer layer is grown on the substrate. This layer is introduced to keep defects located at the interface to the substrate and dislocations originating from large misfits away from the device layers. A further challenge occurs from a large change of composition within a layer sequence. The device depicted in Fig. 1.3 comprises Bragg mirrors, which require a large step of the refractive index between consecutive quarter-lambda thick layers. A large index step of the layer pairs in the mirror stack is connected not only to a large difference of the fundamental bandgap of the layers, but usually also to a large difference of lattice constants. To keep the strain below the critical limit, layers with a composition mix of materials are used to maintain the lattice constant while changing the refractive index. The change of lattice constant in mixed layers and means to compensate the total strain in a layer stack are also considered in Chap. 2.

Organic semiconductors are increasingly applied in (opto-) electronic devices. Their structure differs substantially from that of their inorganic counterparts and is discussed in Chap. 3. Organic semiconductors comprise small-molecule crystals and polymers; they both feature weak intermolecular forces, yielding usually highly defective crystals with a close molecule packing. The structure of a thin small-molecule film often deviates from the bulk structure and depends on the film thickness and on the interaction to the substrate. The exact register of substrate and layer lattices commonly observed for inorganic heterostructures is usually not found for organic layers. The new epitaxial coincidence modes occuring in such structures and their relation to intra- and intermolecular elasticity are outlined in Chap. 3. Various kinds of defects are described, including the prevailing vacancies and grain boundaries between single-crystal domains.

Strain and interfaces between layers with different bandgap affect the electronic properties of the layer structure. In Chap. 4 the effect of strain on valence and conduction-band states is outlined. Furthermore, the consequence of alloying on the fundamental bandgap is considered. The contact of two semiconducting layers raises the question how the uppermost valence bands mutually align. Models treating this problem and effects of interface composition are reviewed in Chap. 4. The band discontinuities determine the confinement of charge carriers in a sandwich structure and are also affected by strain. This is particularly important for the active layers which are usually formed by quantum wells as depicted in Fig. 1.3. Structures with a reduced dimensionality—quantum wells, quantum wires, and quantum dots—form the active core of many advanced devices. Chapter 4 points out the basic electronic properties of such quantum structures to indicate the required dimensions, which have to be realized in the epitaxial growth process.

In organic semiconductors the highest occupied und lowest unoccupied molecular orbitals (HOMO and LUMO) form the valence and conduction bands. Chapter 5 considers the special properties of these bands originating from the molecular structure, and the consequences for the electronic properties. The environment of a charge carrier in an organic crystal experiences a strong structural and electronic relaxation. The resulting strongly bound Frenkel excitons and the pronounced polaron character of mobile carriers are pointed out, and the transport properties are discussed. The band alignment at interfaces with organic semiconductors depends sensitively on molecule charge and orientation. In many cases the classical electron-affinity rule known from Chap. 4 yields a good description of the heterojunction between two organic solids. The effect of molecule orientation is discussed in Chap. 5, and also the band alignment for a contact to a metal is treated.

Growth occurs at some deviation from thermodynamic equilibrium. Epitaxy is a controlled transition from the gas or liquid phase to a crystalline solid. The nature of the driving force depends on the particular material system and growth conditions. Chapter 6 introduces into the thermodynamics of growth for simple one- and two-component systems. Nucleation of a layer, epitaxial growth modes, and a thermodynamic approach to surface energies are described. Even though epitaxial growth processes may occur under conditions far from equilibrium, thermalization times of atoms arriving from the nutrition phase on the surface can be much less than the time required to grow a single monoatomic layer. In such cases thermodynamic descriptions are often successfully applied to model the growth process. For the device structure illustrated in Fig. 1.3 thermodynamics may constitute limits for the stability of mixed crystals used to meet the requirements for lattice constants, bandgaps, band alignments, and doping. In such cases epitaxy may still be possible in a restricted temperature range.

The fabrication of atomically sharp interfaces between two dissimilar solids requires a significant deviation from equilibrium to suppress interdiffusion. This is particularly important for quantum structures like the active layers in the VCSEL structure shown in Fig. 1.3, usually realized using multiple quantum wells. Under such nonequilibrium conditions growth is strongly affected by kinetic influences. Kinetics and atomistic aspects of epitaxial growth are addressed in Chap. 7. A kinetic description of nucleation and layer growth accounts for the detailed steps atoms experience on the growing surface. They depend on the structure of the surface, where the arrangement of atoms may differ substantially from that found in the solid bulk underneath. Such surface reconstructions are specific for the given material and change with growth conditions—they are pointed out for specific examples. Growth modes depending on strain or specific surface states are often employed in epitaxy to fabricate low-dimensional structures by self-organized processes. The basics of such self-organized formation of quantum dots and quantum wires is considered in Chap. 7. VCSEL devices like that depicted in Fig. 1.3 are also fabricated using quantum dots in the active region, formed in the self-organized Stranski-Krastanow growth mode.

The online analysis of a layer structure already during its epitaxial growth is a prominent tool for the development of device structures and for ensuring run-to-run reproducibility. Sensors for the in situ measurement of sample temperature, growth rate, strain, and other parameters of the growth process are considered in Chap. 8. The data are deduced by probing the sample surface, or from the analysis of the ambient in the vicinity of the growing sample using mass spectrometry or optical spectroscopy. Most online tools discussed in Chap. 8 are applicable to all major growth techniques, and some of them can even resolve the growth of a single monolayer. Prominent techniques for the in situ analysis of epitaxy comprise structural analysis by diffraction methods and various optical probes; besides pyrometry techniques such as deflectometry, reflectometry, ellipsometry, and others are addressed providing both chemical and structural information.

The surface energy can be modified during epitaxy by surface-active species referred to as surfactants. Chapter 9 discusses how the three-dimensional island growth occurring during mismatched heteroepitaxy can be suppressed, yielding smooth two-dimensional growth for an appropriate combination of surfactant and semiconductor material. During growth the surfactant layer floates on the semiconductor layer; altered adatom diffusivity and nucleation rate are considered, and processes for strain relaxation in the epitaxial layer are pointed out. Chapter 9 also introduces models treating both kinetic and thermodynamic effects, and highlights the prominent role of the passivation of step edges at nucleating monolayer islands. For specific examples exchange pathways of adatoms and surfactant atoms are discussed.

Electronic and optoelectronic devices require control of charge carriers in semiconductors and a contact of the semiconductor structure to a metal for a connection to the electric circuit. Chapter 10 gives a brief introduction to problems in epitaxy connected to doping and contact fabrication. For a given semiconductor material thermodynamics may impose limits in the doping level originating from restricted solubility of the dopants, an amphoteric behavior of the dopants, or compensation by native defects. Nonequilibrium epitaxial growth may relieve some restrictions. Growth far from equilibrium also enables delta-like doping profiles, used to fabricate devices employing a two-dimensional electron gas with a high mobility. Doping and heterostructure composition profiles may be affected by redistribution of atoms due to diffusion phenomena. Mechanisms of diffusion in semiconductors are considered and examples are given for dependences on the ambient atmosphere and on the kind of doping. Ohmic contacts between semiconductor and metal are a classical subject, but epitaxy also enables the growth of specific contact structures to achieve a low contact resistance. Some examples are outlined in Chap. 10.

Various techniques for epitaxial growth have been established and are described in Chap. 11. The first method which attained maturity to produce complex devices was liquid-phase epitaxy. It operates close to thermodynamic equilibrium and may hence be well described by thermodynamics. More versatile control is achieved using the more sophisticated methods of molecular-beam epitaxy and metalorganic vapor-phase epitaxy, which both operate far from thermal equilibrium. They hence both allow for a control of layer thickness down to a fraction of a single atomic layer and are used to fabricate devices with quantum structures in the active region. The two methods are also employed in VCSEL production on a large scale.

Chapter 12 addresses special techniques applied using common epitaxial methods. In selective area growth an epitaxial layer is grown only in selected parts of a substrate, defined by windows in a mask. The conditions for selectivity are pointed out, and the effects of the mask and the crystallographic orientation on the growth rate are discussed. Epitaxial lateral overgrowth of the mask is highlighted as an important application. In the following section vapor-liquid-solid (VLS) growth is introduced as a prominent method to fabricate semiconductor nanowires with high performance. Controlled axial and radial growth enabling the fabrication of axial and core-shell heterostructures is pointed out, and the very large critical thicknesses for coherent epitaxy is discussed. Techniques with an alternately pulsed supply of gaseous precursors are considered next, comprising atomic layer epitaxy (ALE) or deposition (ALD), and the related migration enhanced epitaxy (MEE). The MEE mode of MBE leads to an increased lifetime of adatoms, enabling particularly sharp interfaces and doping profiles. ALD of amorphous or polycrystalline layers is widely used in semiconductor technology due to an exceptional conformality and thickness control on nonplanar surfaces. Chapter 12 concludes considering the epitaxial deposition of organic crystals using organic molecular beam deposition or organic vapor-phase deposition. Differences with respect to inorganic layer deposition is discussed, and the demanding nucleation issues are emphasized.

References

1. D.W. Pashley, A historical review of epitaxy, in *Epitaxial Growth (Part B)*, ed. by J.W. Matthews (Academic Press, New York, 1975), pp. 1–27
2. M.L. Frankenheim, Über die Verbindung verschiedenartiger Krystalle. Ann. Phys. **113**, 516 (1836). (On the interface of different crystals, in German)
3. M. L'Abbé, R.J. Haüy, *Traité Élémentaire de Physique*, 3rd edn. (Mme. V. Courcier, Libraire Pour Les Sciences, Paris, 1821) (in French)

4. T.V. Barker, Contributions to the theory of isomorphism based on experiments on the regular growths of crystals on one substance on those of another. J. Chem. Soc. Trans. **89**, 1120 (1906)
5. L. Royer, Recherches expérimentales sur l'epitaxie ou orientation mutuelle de cristaux d'espèces différentes. Bull. Soc. Fr. Minéral. Cristallogr. **51**, 7 (1928). (Experimental investigation on the epitaxy or mutual orientation of crystals of different species, in French)
6. G.I. Finch, A.G. Quarrell, The structure of magnesium, zinc and aluminium films. Proc. R. Soc. Lond. A **141**, 398 (1933)
7. F.C. Frank, J.H. van der Merve, One-dimensional dislocations. I. Static theory. Proc. R. Soc. Lond. A **198**, 205 (1949)
8. F.C. Frank, J.H. van der Merve, One-dimensional dislocations. II. Misfitting monolayers and oriented overgrowth. Proc. R. Soc. Lond. A **198**, 216 (1949)
9. F.C. Frank, J.H. van der Merve, One-dimensional dislocations. III. Influence of the second harmonic term in the potential representation, on the properties of the model. Proc. R. Soc. Lond. A **200**, 125 (1949)
10. W.A. Jesser, J.W. Matthews, Evidence for pseudomorphic growth of iron on copper. Philos. Mag. **15**, 1097 (1967)
11. W.A. Jesser, J.W. Matthews, Pseudomorphic growth of iron on hot copper. Philos. Mag. **17**, 595 (1968)
12. J. Orton, T. Foxon, *Molecular Beam Epitaxy: A Short History* (Oxford University Press, Oxford, UK, 2015)
13. I.V. Markov, *Crystal Growth for Beginners* (World Scientific, Singapore, 2003)
14. D.M. Eigler, E.K. Schweizer, Positioning single atoms with a scanning tunnelling microscope. Nature **344**, 524 (1990)
15. Details and the last STM image are reported in M.F. Crommie, C.P. Lutz, D.M. Eigler, Confinement of electrons to quantum corrals on a metal surface. Science **262**, 218 (1993). Image originally created by IBM Corporation, http://www.almaden.ibm.com/vis/stm/corral.html

Chapter 2
Structural Properties of Heterostructures

Abstract Structural properties of epitaxial layers are pointed out in this chapter with some emphasis on zincblende and wurtzite crystals. After a brief review on perfect, polytype, and mixed bulk crystals we focus on elastic properties of pseudomorphic strained-layer structures. Then the concept of critical layer thickness is introduced, and dislocations relieving the strain in epitaxial layers are presented. X-ray diffraction—the standard tool for structural characterization—is outlined at the end of the chapter.

2.1 Basic Crystal Structures

Atoms in an ideal solid have a regular periodic arrangement, which represents a minimum of total energy. The structure of such a crystalline solid is described by a *lattice*, which constitutes the translational periodicity, and a *basis*, which represents the atomic details of the recurring unit cell. In 1848 Auguste Bravais showed that there are only 14 lattices in three-dimensional space. These 14 *Bravais lattices* may be divided into 7 crystal systems, which differ in shape of their unit cell as shown in Table 2.1. The given unit cells reflect the symmetry of the structures. In most cases they are not primitive, i.e., they are larger than the smallest possible unit cell and contain more than one basis. Some important crystal structures are pointed out explicitly in this chapter.

2.1.1 Notation of Planes and Directions

Lattice planes and directions in crystals are generally denoted by triplets of integer numbers h, k, l called *Miller indices*. To determine the Miller indices of a given plane, first a coordinate system is constructed using base vectors of the unit cell (regardless of being primitive or not). Then the intersection points of the plane with the axes are determined in units of the lattice constants, i.e., as multiples of the base vectors. Finally the reciprocal of these three values are expanded to the smallest set of integer

U. W. Pohl, *Epitaxy of Semiconductors*, Graduate Texts in Physics,
https://doi.org/10.1007/978-3-030-43869-2_2

Table 2.1 The seven crystal systems and the 14 Bravais lattices. **a**, **b**, and **c** are lattice vectors spanning the unit cell; α, β, and γ are angles between these vectors

System	Unit cell	Bravais lattices	Symmetry axes
Cubic	$\mathbf{a} = \mathbf{b} = \mathbf{c}$, $\alpha = \beta = \gamma = 90°$	Simple cubic, body-centered cubic, face-centered cubic	4 threefold axes parallel to the diagonals of the unit cell
Tetragonal	$\mathbf{a} = \mathbf{b} \neq \mathbf{c}$, $\alpha = \beta = \gamma = 90°$	Simple tetragonal, body-centered tetragonal	1 fourfold axes of rotation or inversion parallel to **c**
Rhombohedral	$\mathbf{a} = \mathbf{b} = \mathbf{c}$, $\alpha = \beta = \gamma \neq 90°$	Rhombohedral	1 threefold axes of rotation or inversion parallel to $\mathbf{a} + \mathbf{b} + \mathbf{c}$
Hexagonal	$\mathbf{a} = \mathbf{b} \neq \mathbf{c}$, $\alpha = \beta = 90°$, $\gamma = 120°$	Hexagonal	1 sixfold axes of rotation or inversion parallel to **c**
Orthorhombic	$\mathbf{a} \neq \mathbf{b} \neq \mathbf{c}$, $\alpha = \beta = \gamma = 90°$	Simple orthorhombic, base-centered orthorhombic, body-centered orthorhombic, face-centered orthorhombic	3 mutually perpendicular twofold axes of rotation or inversion parallel to **a**, **b**, and **c**
Monoclinic	$\mathbf{a} \neq \mathbf{b} \neq \mathbf{c}$, $\alpha = \gamma = 90°$, $\beta \neq 90°$	Simple monoclinic, face-centered monoclinic	1 twofold axes of rotation or inversion, e.g. parallel to **b**
Triclinic	$\mathbf{a} \neq \mathbf{b} \neq \mathbf{c}$, $\alpha \neq \beta \neq \gamma$, $\alpha, \beta, \gamma \neq 90°$	Triclinic	None

values. Example: A plane intersects the three axes of a cubic coordinate system at (3, 1, 2) yielding the reciprocals (1/3, 1, 1/2), and consequently the Miller indices (2, 6, 3). Different kind of brackets are used to distinguish between various planes, directions, and lattice points:

(hkl) a *specific plane*, normal parentheses
$\{hkl\}$ a *set of* crystallographically *equivalent planes*, curly brackets
$[hkl]$ a *specific direction*, square brackets
$\langle hkl \rangle$ a *set of equivalent directions*, cuspid brackets
hkl the *position* of a specific lattice point.

Finally, negative values are denoted by a bar on top, e.g., $[\bar{h}\bar{k}l]$ means $[-h\ -k\ l]$. Position and nomenclature of some important low-index planes in cubic and hexagonal lattices are depicted in Fig. 2.1.

In the hexagonal system usually the *four Miller-Bravais* indices are used, i.e., $hktl$. The index t refers to a third vector $\mathbf{a}_3$ in the base plane of the unit cell and

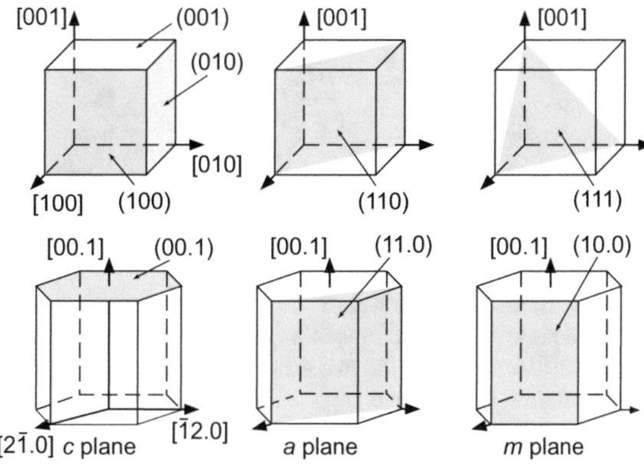

Fig. 2.1 Position of important planes and their Miller indices for cubic and hexagonal lattices

depends on the indices of the vectors along $\mathbf{a}_1$ and $\mathbf{a}_2$ shown in Fig. 2.4d by the relation $t = -(h + k)$, i.e., $\mathbf{a}_3 = -(\mathbf{a}_1 + \mathbf{a}_2)$. Miller-Bravais indices of a plane are defined by the primes referring to the reciprocals of intercepts of the plane with *all four* axes. For indexing *planes* the index t is redundant and commonly replaced by a dot. Sometimes this dot is dropped and the Miller indexing system is used, so that indices of cubic and hexagonal lattices look similar. For *directions* the relation between the four-index and three-index systems is more complicate, because the third index *cannot* just be left out. The conversion of any direction

$$u\,\mathbf{a}_1 + v\,\mathbf{a}_2 + t\,\mathbf{a}_3 + w\,\mathbf{c} = u'\,\mathbf{a}_1 + v'\,\mathbf{a}_2 + w'\,\mathbf{c}$$

between the four-index Miller-Bravais system and the (primed) three-index Miller system is given by the relations

$$\begin{array}{ll} u' = 2u + v & u = \frac{1}{3}(2u' - v') \\ v' = 2v + u & v = \frac{1}{3}(2v' - u') \\ & t = -\frac{1}{3}\left(u' + v'\right) = -(u + v) \\ w' = w & w = w'. \end{array}$$

After conversion all indices must be reduced to relative primes. The Miller-Bravais-indexed $\left[2\bar{1}\bar{1}0\right]$, $\left[\bar{1}2\bar{1}0\right]$, and $\left[11\bar{2}0\right]$ directions hence correspond to the Miller-indexed [100], [010], and [110] directions, respectively. The directions of the hexagonal system given in Fig. 2.1 refer to the commonly used convention.

2.1.2 *Wafer Orientation*

Thin slices of bulk semiconductor crystals referred to as *wafers* are used for growth of epitaxial layers. Wafers made of Si, GaAs, and InP with surfaces vicinal to {100},

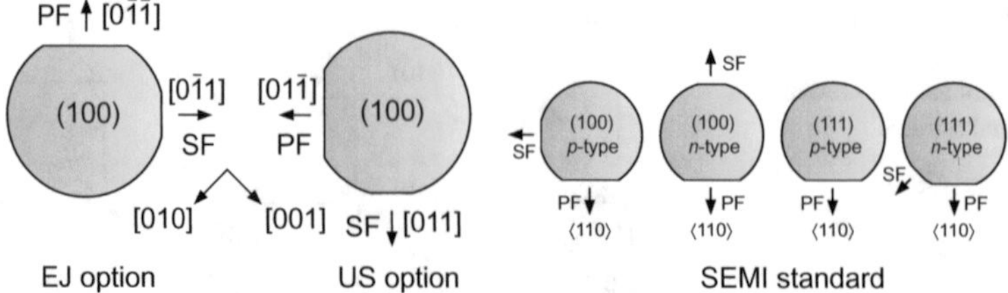

Fig. 2.2 Semiconductor wafers with standard flat orientations according to various options. *PF* and *SF* denote primary and secondary flats, respectively. The *arrows* along [010] and [001] refer to the coordinate system of the two wafers on the *left hand side*, the primary flat in the SEMI standard is located on a {110} plane

{111}, and {110} are most common and produced by cutting an on-axis crystal at the appropriate angle. Their cleavage planes are discussed in Sect. 2.1.4. The wafers are usually of circular shape with 2–6 in. diameter (silicon up to 300 mm) and 300 up to 1000 μm thick (depending on diameter). Information on the orientation of the crystallographic axes and the dopant type (n-type, p-type, or undoped) are indicated by location and number of *flats*, features machined at the perimeter of the wafer. Primary or major (orientation) flats and shorter secondary or minor (identification/index) flats are used according to different standards. Common standards are the EJ (Europe/Japan), US, and SEMI options. Figure 2.2 shows some wafer geometries of semiconductors with cubic structure along with crystallographic orientations.

2.1.3 Face-Centered Cubic and Hexagonal Close-Packed Structures

The face-centered cubic and the hexagonal structure are often found in metals, and the most important semiconductors crystallize in the related diamond, zincblende, or wurtzite structures. Let us consider the atoms as spheres, which do not prefer any direction of bonding. If such atoms are closely arranged in a plane, each one has contact to six next neighbors, cf. Fig. 2.3a. The atoms form a hexagonal net plane, which we label plane B. To obtain a closely stacked three-dimensional arrangement, atoms within a plane A on top of plane B must be arranged similarly, starting with an atom above the interstice of three B-atoms. Note that only 3 of the 6 hexagonally arranged interstices of B are occupied by atoms in plane A. Hence there are two alternatives for a plane C below plane B: Their atoms either use the interstices not used above in plane A, or they are arranged similar to plane A. These two different structures are referred to as *face-centered cubic* (fcc, stacking sequence $ABCABC$ along the [111] stacking direction) and *hexagonal close-packed* (hcp, sequence $ABAB$). Note that both structures have the *same* stacking density with 74.1% space filling.

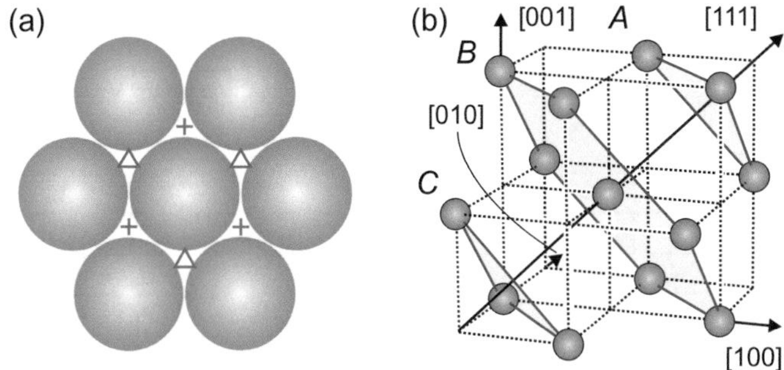

Fig. 2.3 **a** Hexagonal net planes in a closed-packed crystal structure. Δ and + denote locations of atoms in the planes above and below, respectively, forming the fcc structure shown in (**b**). **b** Face-centered cubic structure. *A*, *B*, and *C* label different hexagonal lattice planes in a face-centered cubic lattice

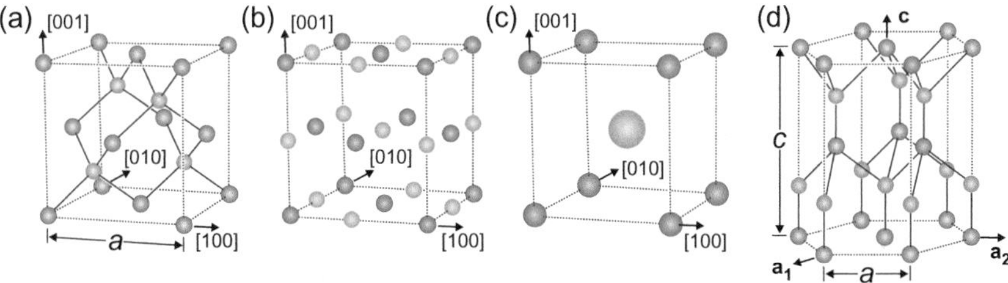

Fig. 2.4 The crystal structures of **a** zincblende, **b** rocksalt (NaCl), **c** cesium chloride (CsCl), and **d** wurtzite

Figure 2.3b shows that the *ABC*-stacked atoms are located on a *cubic* lattice, with additional atoms centered on the six faces of the cube (the contour of the front half of the fcc unit cell is only partially drawn for clarity). This is not the case in the unit cell of the hcp structure, represented by the dark spheres in Fig. 2.4d.

2.1.4 Zincblende and Diamond Structures

Many solids are composed of more than one kind of atoms. Crystal structures of some important binary bulk compounds with an at least partial ionic bond are given in Table 2.2. Some of these solids may crystallize also in another structure as noted in Table 2.3. ZnS and CdSe, e.g., have also a stable high temperature phase (here wurtzite), which may be preserved at room temperature by suitable growth conditions. Using epitaxy also structures which differ from that of the stable bulk crystal may be stabilized.

Table 2.2 Stable crystal structures of AB compounds composed of either group II and VI, or group III and V elements. W, ZB, and Gr denote wurtzite, zincblende, and graphite structure, the compound marked by a star does not exist. From [1]

Cation group II	Anion group VI				Cation group III	Anion group V			
	O	S	Se	Te		N	P	As	Sb
Be	W	ZB	ZB	ZB	B	Gr	ZB	ZB	*
Mg	NaCl	NaCl	NaCl	W	Al	W	ZB	ZB	ZB
Zn	W	ZB	ZB	ZB	Ga	W	ZB	ZB	ZB
Cd	NaCl	W	ZB	ZB	In	W	ZB	ZB	ZB

In the *zincblende structure*, the basis of the primitive unit cell consist of *two* atoms, which generally have different polarity. Their location is at $(0, 0, 0)$ and $(1, 1, 1) \times a/4$, a being the lattice constant (cf. Fig. 2.4a). We will refer to these atoms as *cations* and *anions* (like zinc and sulfur in ZnS). The zincblende structure is composed of an fcc lattice of cations (represented by, e.g., dark spheres in Fig. 2.4a) and an fcc lattice of anions. These two sublattices are displaced by a quarter of the cube diagonal, i.e. by $(\sqrt{3}/4) \times a$. As a result, each ion is tetrahedrally surrounded by four ions of opposite polarity as depicted in Fig. 2.4a. The zincblende structure has no center of inversion, in contrast to the related diamond structure introduced below. $\langle 110 \rangle$ planes of the zincblende structure contain an equal number of cations and anions, arranged along zig-zag chains. Since their charges balance, $\langle 110 \rangle$ crystal faces are non-polar. Crystals may usually be well cleaved parallel to non-polar planes, a fact used, e.g., in laser devices made from zincblende semiconductors to fabricate resonator facets by cleaving along two parallel $\langle 110 \rangle$ planes: Zincblende wafers with (100) orientation scribed along $[011]$ and $[01\bar{1}]$ form side facets perpendicular to the wafer surface and perpendicular to each other. In contrast, $\langle 100 \rangle$ and $\langle 111 \rangle$ planes contain only one kind of ions and hence form polar faces. The $[111]$ direction is not equivalent to the $[\bar{1}\bar{1}\bar{1}]$ direction. In AB compounds like zincblende ZnS (A an B generally denoting cations and anions, respectively, not to be confused with layer labels), wafers with (111) surface have a cation-terminated A-face and an opposing anion-terminated B-face, which can be distinguished, e.g., by different chemical etching behavior. Typical compounds with zincblende structure are ZnS, GaAs, InP, and CuCl.

If the atoms on the two fcc sublattices are identical, we obtain the *diamond structure* (consider *all* atoms in Fig. 2.4a to be blue). The diamond structure has a center of inversion at the midpoint of each connection line between two next neighbors. Crystal faces of this structure are non-polar, crystals may also be cleaved parallel to planes differing from $\langle 110 \rangle$. Elements crystallizing in diamond structure are C (diamond phase), Ge, Si, and Sn (α phase). In Si and Ge primary cleaving planes are $\langle 111 \rangle$ planes: They produce less dangling bonds than $\langle 110 \rangle$ planes upon cleaving.

Si wafers with (100) orientation scribed along [011] or $[01\bar{1}]$ form {111} side facets inclined to the wafer surface by an angle of 54.7°.

2.1.5 Rocksalt and Cesium-Chloride Structures

The cubic *rocksalt structure* (also termed sodium-chloride structure) shown in Fig. 2.4b is composed of two fcc lattices for cations and anions, respectively, displaced by half of the cube diagonal, i.e., by $(\sqrt{3}/2) \times a$. The two ions of the basis are located at (0, 0, 0) and $(1, 1, 1) \times a/2$. As a result, each ion is octahedrally surrounded by six ions of opposite polarity. Their charges balance on ⟨100⟩ crystal faces, which are hence non-polar and build cleavage planes of this structure. Compounds which crystallize in rocksalt structure are, e.g., NaCl, KCl, AgBr, FeO, PbS, MgO, and TiN.

The *cesium-chloride structure* depicted in Fig. 2.4c consists of a simple cubic lattice with a diatomic base. The two ions are located at (0, 0, 0) and $(1, 1, 1) \times a/2$, similar to the rocksalt structure. Typical compounds with cesium-chloride structure are CsCl, NH_4Br, TlCl, AgZn, AlNi, and CuZn.

2.1.6 Wurtzite Structure

The wurtzite structure shown in Fig. 2.4d is composed of two hcp sublattices, one for cations and one for anions. In the ideal structure these sublattices are displaced by $\mathbf{u} = 3/8\mathbf{c}$ along the [00.1] direction, c being the vertical lattice parameter. The ideal ratio of vertical and lateral lattice parameters is $c/a = (8/3)^{1/2} = 1.633$. The primitive unit cell contains 1/3 of the volume of the usually applied unit cell of Fig. 2.4d and is spanned by the vectors $\mathbf{a}_1$, $\mathbf{a}_2$, $\mathbf{c}$; its basis comprises *four* atoms with two anions and two cations. The wurtzite structure has no center of inversion.

The u/c ratio of wurtzite-type crystals often deviates from the ideal value 0.375, and likewise the c/a ratio. The deviation gives rise to a macroscopic spontaneous polarization (Sect. 4.1.3), which can cause strong internal electric fields—e.g. in group-III nitrides up to 3 MV/cm [2]. Lattice parameters for some crystal structures are listed in Tables 2.3 and 2.4. In wurtzite crystals the zig-zag chains of bonding lie within a non-polar ⟨11.0⟩ plane (Fig. 2.6, a plane Fig. 2.1). It is a cleavage plane perpendicular to the (00.1) c plane. The top face of the unit cell represents the polar (00.1) plane. It is often referred to as *basal* plane, i.e., a plane which is perpendicular to the principal axis (c axis) in a hexagonal or tetragonal structure. The [00.1] direction is not equivalent to $[00.\bar{1}]$; the (00.1) face is made up of cations and, e.g., in GaN referred to as (00.1) Ga face, while anions build the $(00.\bar{1})$ face ($(00.\bar{1})$ N in GaN).

Table 2.3 Lattice parameters of existing and hypothetical tetrahedrally coordinated crystals [1]. Italicized numbers are theoretical results from pseudopotential calculations, $\Delta E_{\text{W-ZB}}$ is the calculated energy difference between wurtzite and zincblende structures at $T = 0$

Solid	Wurtzite				Zincblende a (Å)	$\Delta E_{\text{W-ZB}}$ (meV/atom)
	a (Å)	c (Å)	c/a	u/c		
GaN	3.192	5.196	1.628		4.531	
	3.095	*5.000*	*1.633*	*0.378*	*4.364*	*−9.9*
InN	3.545	5.703	1.609			
	3.536	*5.709*	*1.615*	*0.380*	*4.983*	*−11.4*
AlN	3.112	4.980	1.600		5.431	
	3.099	*4.997*	*1.612*	*0.381*	*4.365*	*−18.4*
GaAs					5.653	
	3.912	*6.441*	*1.647*	*0.374*	*5.654*	*12.0*
AlAs					5.660	
	3.979	*6.497*	*1.633*	*0.376*	*5.620*	*5.8*
ZnS	3.823	6.261	1.638		5.410	
	3.777	*6.188*	*1.638*	*0.375*	*5.345*	*3.1*
CdS	4.137	6.716	1.624		5.818	
	4.121	*6.682*	*1.621*	*0.377*	*5.811*	*−1.1*
Si					5.431	
	3.800	*6.269*	*1.650*	*0.374*	*5.392*	*11.7*
C	2.51	4.12	1.641			
	2.490	*4.144*	*1.665*	*0.374*	*3.539*	*25.3*

2.1.7 Thermal Expansion

Epitaxy is performed at temperatures far above room temperature to provide sufficient mobility of adatoms on the growing crystal surface. Typical growth temperatures are around 600 °C, but materials may require substantially lower or higher temperatures (e.g., compound semiconductors containing Hg below 200 °C or nitrides well above 1000 °C). Cooling after epitaxy over a wide range to room temperature or cryogenic temperatures is accompanied by a significant diminution of the lattice constant due to anharmonic terms of the crystal potential. The difference in the thermal expansion of substrate and layer materials may induce large strain in the structure, leading to bending of the substrate and, in case of tensile strain, to cracks in the epitaxial layer. Large differences in the thermal expansion of layers within a heteroepitaxial structure may hence seriously affect structural properties and lead to the requirement of inserting suitable buffer layers to accommodate the thermal mismatch.

Within the thermal range of interest usually the description of a temperature-dependent lattice parameter $a(T)$ by a *linear* thermal expansion coefficient (TEC) α is applied according to

Table 2.4 Linear thermal expansion coefficients and lattice parameters of some cubic and hexagonal semiconductors at 300 K. Data reported in literature have significant scatter, particularly for hexagonal structures. Listed values represent some mean data

Semiconductor	Lattice parameter (Å)		Thermal expansion coefficient (10^{-6} K^{-1})	
	a	c	α_a	α_c
GaAs	5.6535	–	5.7	–
AlAs	5.661	–	5.2	–
InAs	6.058	–	5.1	–
GaP	5.4509	–	4.8	–
AlP	5.4635	–	~5	–
InP	5.8690	–	4.7	–
ZnS	5.410	–	7.1	–
ZnSe	5.668	–	~7.4	–
CdTe	6.484	–	4.9	–
Si	5.4310	–	2.6	–
Ge	5.6576	–	5.7	–
SiC (3C)	4.3596	–	2.8	–
SiC (6H)	3.0806	15.1173	4.2	4.7
GaN	3.189	5.186	~4.5	~4.0
AlN	3.112	4.982	~4.2	~5.3
InN	3.533	5.693	~3.8	~2.7
ZnO	3.249	5.207	~4.7	~2.9
Al_2O_3	4.758	12.991	~8.1	~7.3

$$\frac{a(T) - a(T_0)}{a(T_0)} = \alpha \times (T - T_0).$$

Usually room temperature is taken as reference temperature T_0. The thermal change of volume is in general anisotropic and α is consequently a second rank tensor. It is described by three components α_i the principal axes of which are along those of the strain tensor (Sect. 2.2.2). Along a principal axis the respective component α_i is defined by

$$\alpha_i = \frac{1}{a}\left(\frac{\partial a}{\partial T}\right)_P.$$

For cubic crystals the three components α_i are identical and the expansion coefficient becomes a scalar. Crystals with a hexagonal structure have two independent components, namely $\alpha_a = \alpha_\perp$ perpendicular to the c axis and $\alpha_c = \alpha_\parallel$ along the c axis. Linear expansion coefficients of semiconductors are typically of the order mid 10^{-6} K^{-1}. Data for some semiconductors are given in Table 2.4.

The thermal expansion is actually not a linear function of temperature. The expansion coefficients depend on temperature and are usually positive quantities. Below

Table 2.5 Coefficients for a polynomial fit of the temperature dependent lattice parameter $a(T)$ of some cubic semiconductors; the reference temperature is $T_0 = 300$ K

Semiconductor	A	B (10^{-6} K^{-1})	C (10^{-9} K^{-2})	D (10^{-12} K^{-3})	Range (K)
GaAs	-1.47×10^{-3}	4.239	2.916	−0.936	200–1000
Si	-0.71×10^{-3}	1.887	1.934	−0.4544	293–1600
Ge	-1.533×10^{-3}	4.636	2.169	−0.4562	293–1200

100 K semiconductors with zincblende or diamond structure show a commonly not observed negative thermal expansion [3, 4]. Above 300 K coefficient values slightly increase if the temperature is raised. An empirical relation by a polynomial fit is sometimes applied to account for the temperature dependence of the thermal expansion. In a specified temperature range the lattice parameter is then described by

$$a(T) = a(T_0)\left(1 + A + BT + CT^2 + DT^3\right),$$

where T_0 is the reference temperature (generally 300 K) and T is the absolute temperature in K. Values for some cubic semiconductors are given in Table 2.5.

2.1.8 *Structural Stability Map*

Rules to predict crystal structures of solids from properties of atoms require first-principle calculations with a high degree of accuracy, because the difference in equilibrium energies of a given compound in two closely related structures is often less than 0.1% of the cohesive energy. In the corresponding problem of the regularities of the periodic table of the elements, a scheme with the two integral coordinates principal and orbital quantum number was found. Starting from classical approaches using atomic radii and Pauling's electronegativity, today some largely universal schemes for predicting the structural stability of many intermetallic compounds of the form A_xB_y, of sp-bonded A^nB^{p-n} ($p = 8$ or $2, \ldots, 6$) semiconductors and insulators, and of high-T_c superconductors have been suggested. The success rate generally exceeds 95%, and domains of different overlapping structure types could be related to polymorphic structural forms [1, 5–8]. Most schemes use orbital radii coordinates R, which are linear combinations of the s- and p-orbital radial functions R^A and R^B of the atoms building the AB compound:

$$R_\sigma(A, B) = \left|\left(R_p^A + R_s^A\right) - \left(R_p^B + R_s^B\right)\right|, \tag{2.1a}$$

$$R_\pi(A, B) = \left|\left(R_p^A - R_s^A\right) + \left(R_p^B - R_s^B\right)\right|. \tag{2.1b}$$

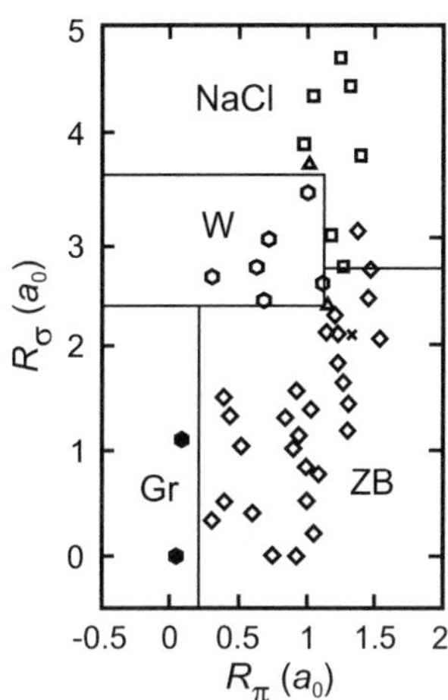

Fig. 2.5 Structural stability map for binary AB compounds, calculated using atomic orbital radial functions in units of the Bohr radius. NaCl, W, ZB, and Gr denote sodium-chloride, wurtzite, zincblende, and graphite structures, respectively. The misaligned cross in the zincblende range marks a NiAs structure, the triangles cinnabar (HgS) structure. From [1]

R_σ provides a measure of the size difference between atoms A and B, and roughly scales with ionicity. R_π scales with the atomic s and p energy difference and measures to some extent sp-hybridization. Figure 2.5 shows a structural stability diagram for binary compounds of the form $A^n B^{8-n}$ like $Ga^{III}As^{V}$, using radii of (2.1a)–(2.1b) at which the all-electron atomic radial orbitals $r \times R_{nl}(r)$ have their outer maxima.

Zincblende and wurtzite are the most common crystal structures of binary semiconductors. For these structures the orbital radii coordinates given in (2.1a)–(2.1b) were used to calculate the equilibrium energy difference $\Delta E_{\text{W}-\text{ZB}}$ between the ideal wurtzite and the zincblende structure, cf. Table 2.3. Calculated structural deviations from ideal wurtzite are small, the energy gain due to such relaxations is generally smaller than 1 meV per atom (except for AlN, -2.7 meV) [1]. The study confirms the phenomenological correlation, that wurtzite-stable compounds have smaller c/a ratios than the ideal value, while zincblende and diamond-stable compounds have larger ratios. As a further rule of thumb, a large difference in electronegativity and atomic radii of atoms A and B favor the wurtzite structure.

2.1.9 Polytypism

Small values of equilibrium energy differences between wurtzite and zincblende structure $\Delta E_{\text{W-ZB}}$ given in Table 2.3 indicate an intrinsic low stability of the crystal structure of some solids. In fact many solids may crystallize in multiple crystal structures having identical stoichiometries, a phenomenon called polymorphism. Wurtzite and zincblende are two structures of a one-dimensional polymorphism referred to as *polytypism*. For a graphic representation of such polytypes of tetrahedrally coordinated AB compounds a plot of the cation-anion zigzag chains, which lie in the (11.0) plane of the hexagonal lattice and in the (110) plane of the zincblende lattice,

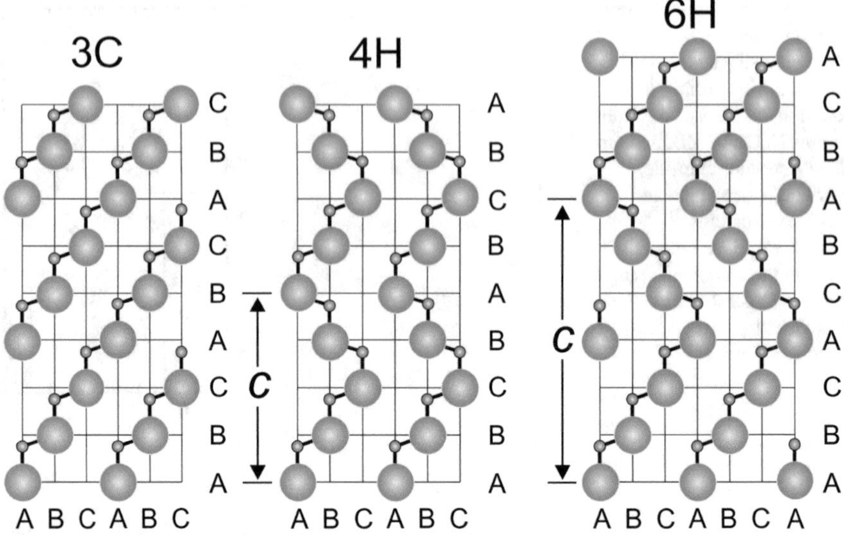

Fig. 2.6 Stacking sequences of binary AB compound polytypes depicted in the (11.0) plane of the hexagonal lattice. Labels A, B, and C on the horizontal axis refer to atom sites, while such labels on the vertical axis denote net planes as shown in Fig. 2.3b

is useful. Figure 2.6 shows three prominent polytypes, the zincblende 3C structure, and the two structures 4H and 6H. The number in the notation refers to the periodicity, while C and H designate a cubic and hexagonal structure, respectively. Using this notation, the wurtzite structure is labeled 2H.

There are numerous such polytypes known from some compounds, more than, e.g., 100 for ZnS and about 170 for SiC; they build cubic (3C), hexagonal (H), and rhombohedral (R) structures. All these modifications have the same lateral lattice constant a in the representation chosen above, while c is an integer multiple of the layer thickness. All polytypes may be considered as mixtures of 3C and 2H. The 4H modification is then composed equally of cubic (ABC next neighborhood) and hexagonal (ABA) bonds. The notation of a given polytype by periodicity is not unambiguous, e.g., periodic stacks of $ABCACB$ and $ABCBAB$ form inequivalent hexagonal 6H modifications.

The different polytypes of a given compound have widely ranging physical properties. The bandgap energy generally becomes greater as the wurtzite component increases, see Table 2.6. The number of atoms within a unit cell and that of inequivalent atom sites increases with the size of the unit cell, and consequently also the number of phonon branches. The electron mobility of the high symmetry 3C polytype may be higher due to reduced phonon scattering. Furthermore, a particular doping impurity on inequivalent sites has different electronic properties due to a more cubic or hexagonal environment.

Table 2.6 Physical properties of some polytypes, arranged in increasing order of wurtzite character. The number of atoms refers to the primitive unit cell, E_g denotes the bandgap energy at 2 K. Data from [9–11]

Polytype		3C	15R	6H	4H	2H
Space group symmetry		T_d	C_{3v}	C_{6V}	C_{6V}	C_{6V}
Atoms per unit cell		2	10	12	8	4
Inequivalent sites		1	5	3	2	1
SiC	E_g^X (eV)	2.390	2.906	3.023	3.265	3.330
	a (Å)	4.349	3.08	3.081	3.073	3.076
	c (Å)		37.70	15.117	10.053	5.048
ZnS	E_g^Γ (eV)	3.85				3.91
	a (Å)	5.410	3.83	3.82	3.81	3.823
	c (Å)	–	46.88	18.72	12.46	6.260
GaN	E_g^Γ (eV)	3.21				3.503
	a (Å)	4.50				3.189
	c (Å)	–				5.185

A self-contained theory explaining the origin of polytypes with up to more than 100 planes within a period does not yet exist. The occurrence of long periods is often associated with dislocations (Sect. 2.3), e.g., screw dislocations which induce spiral growth. In such case the periodicity is determined by the step height of the growth spiral. Short-period polytypes may be stabilized by growth conditions given by temperature, pressure or gas-phase composition, due to different minima in the formation energy.

2.1.10 Random Alloys and Vegard's Rule

Mixing two or more solids to an *alloy*, i.e., a solid solution, is an old technique to modify properties of materials. Alloys with two, three, or four components are called *binary*, *ternary* or *quaternary* alloys, respectively. In substitution alloys atoms of comparable size are simply substituted for one another in the crystal structure. Such alloys are routinely made from semiconductors to engineer properties like lattice parameter or bandgap. Metals may also form interstitial alloys, where atoms of one component are substantially smaller than the other and fit into the interstices between the larger atoms.

Usually a random mixing of the atoms on the semiconductor lattice-sites of the alloy is intended. Limits in the miscibility of the components or ordering effects in the alloy may, however, lead to significant deviations from a random distribution and consequently to altered physical properties of the alloy. There are three extreme cases occurring in alloy formation.

- *Random alloy*: The probability of an atom next to a given atom in an $A_x B_{1-x}$ alloy is x for an A-atom and $(1-x)$ for a B-atom. This is the case usually wanted for applications.
- *Ordered alloy*: Atoms in the alloy have a regular periodic structure. The crystal structure is then given by placing at each site of the Bravais lattice a multiatomic base, yielding translational symmetry of the Bravais lattice in the alloy. Examples are β-brass with alternating Cu and Zn atoms along [111] or CuPt structure of $In_{1-x}Ga_xP$ at $x \approx 0.5$. Since ordering requires a specific ratio in the number of the atom types forming such alloys, they are also referred to as *stoichiometric alloys*.
- *Phase separation*: Atoms of type A and B do not mix and are located at different regions in the solid.

We consider the technologically important *random alloy* in more detail. We assume two compound semiconductors AC and BC having the same crystal structure, A and B representing cations and C an anion. Alloying leads to a semiconductor $A_x B_{1-x} C$ with a mixture of A and B atoms on the cation sublattice, while all sites on the anion sublattice remain occupied by C atoms. Since the alloying forms a binary *sub*lattice (the unmixed materials AB and AC are already binary compounds themselves) such alloys are called *pseudobinary* (often also referred to as ternary, though in a true ternary all components mix on the same lattice). The concept may be extended to compounds $A_x B_{1-x} C_y D_{1-y}$ usually termed quaternary compound.

The composition parameter x in the pseudobinary alloy $A_x B_{1-x} C$ denotes that on the *average* an anion C has x neighbors of type A and a fraction of $1-x$ neighbors of type B. A number of properties of true alloys are expected to scale by a smooth interpolation between the two endpoint materials. This is usually well fulfilled for the lattice constant. The empirical rule called *Vegard's rule* [12] states that a *linear* interpolation exists, at constant temperature, between the crystal lattice constant of an alloy and the concentrations of the constituent elements. The lattice constant a_{alloy} from two materials A and B with the same crystal structure and lattice constants a_A and a_B, respectively, is hence given by

$$a_{\mathrm{alloy}} = x a_A + (1-x) a_B. \tag{2.2a}$$

The lattice constant of a (pseudobinary) ternary alloy with a mixture of compounds AC and BC is given by the same relation putting a_A and a_B to a_{AC} and a_{BC}, respectively. The linear relationship also holds for quaternary alloys. For compounds of the type $A_x B_y C_{1-x-y} D$ the interpolation yields

$$a_{\mathrm{alloy}} = x a_{AD} + y a_{BD} + (1-x-y) a_{CD}. \tag{2.2b}$$

The lattice parameter for alloys of the type $A_x B_{1-x} C_y D_{1-y}$, where atoms A and B mix on one sublattice and atoms C and D on another sublattice, is calculated from the ternary parameters a_{ABC}, a_{ABD}, a_{ACD}, and a_{BCD},

$$a_{\text{alloy}} = \frac{x(1-x)[y a_{ABC}(x) + (1-y) a_{ABD}(x)] + y(1-y)[x a_{ACD}(y) + (1-x) a_{BCD}(y)]}{x(1-x) + y(1-y)},$$

$$a_{ABC}(x) = x a_{AC} + (1-x) a_{BC}, \tag{2.2c}$$

$a_{ABD}(x)$, $a_{ACD}(y)$, and $a_{BCD}(y)$ accordingly.

Using Vegard's rule (2.2a)–(2.2c) lattice matching to a substrate lattice parameter is achieved using an appropriate composition x (and y) of materials with a larger and a smaller lattice parameter. Example: $In_{0.53}Ga_{0.47}As$ for epitaxy on InP ($a_{\text{InAs}} = 6.0584$ Å, $a_{\text{GaAs}} = 5.6533$ Å, $a_{\text{InP}} = 5.8688$ Å). It must be noted that the adjustment of lattice parameters by alloying is accompanied by a change of the bandgap energy and other properties. Often the composition is rather chosen for obtaining a desired band gap in the alloy. In a *quaternary* alloy lattice parameter and bandgap energy can be adjusted independently (Sect. 4.1.5). Note that exact lattice matching is generally met only for a given temperature, because the lattice parameter varies as the temperature is changed (e.g., from growth temperature to room temperature) and different materials like substrate and layer have usually different thermal expansion coefficients (cf. Sect. 2.1.7).

For stable two- and three-dimensional central-force networks (e.g. triangular net or fcc lattice) Vegard's rule was derived from quite general assumptions [13]. Deviations from the linear relation are always found in metallic alloys [14], but they are in practice usually negligible in miscible semiconductor alloys. Vegard's rule is therefore routinely used to measure the composition of a studied alloy from its lattice parameter a_{alloy} using, e.g., X-ray diffraction.

Vegard's rule may more generally apply for the average interatomic distances in solids if structural changes without change of the coordination number occur in an alloy. The example given in Fig. 2.7 shows the next-neighbor distance on the cation sublattice in the diluted magnetic semiconductor $Zn_{1-x}Mn_xSe$ with magnetic Mn^{2+} ions substituting the nonmagnetic Zn^{2+} cations. For a fraction exceeding $x \approx 0.3$ the crystal structure changes from zincblende to wurtzite. The c/a ratio of wurtzite ZnMnSe is close to the ideal value of $\sqrt{8/3}$, indicating a close relationship to an ideal close packing arrangement [15]. The type of coordination does not change across

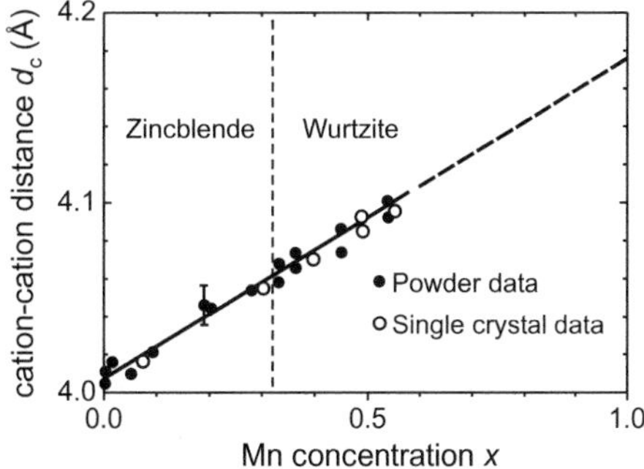

Fig. 2.7 Mean cation-cation distance d_c in $Zn_{1-x}Mn_xSe$ for varying Mn concentration x. The distance d_c increases linearly with x according to Vegard's rule also across a change of crystal structure. From [15]

the region in which the alloy changes between the crystal structures, which both are based on a closest packing of spheres. Such factors favor compliance with Vegard's rule.

The simple linear relation of the concentration-weighted average bond length expressed by Vegard's rule (2.2a), (2.2b) suggests that the chemical bond of atoms in an alloy smoothly changes between the values of the end-point materials. Such assumption is the premise of the virtual-crystal approximation (VCA) discussed below (Sect. 2.1.11). In real semiconductors this is not compelling. Experiments like extended X-ray absorption fine structure measurements (EXAFS) [16, 17] show that bond lengths in a semiconductor alloy are actually much closer to those of the end-point materials. The bond length calculated using Vegard's rule originates essentially from the weighted average of alternating bonds [18].

The difference between Vegard's rule and actual bond lengths is illustrated for the pseudobinary alloy $Ga_{1-x}In_xAs$. Both GaAs and InAs crystallize in zincblende structure. They are miscible over the entire range and form a random alloy. In X-ray diffraction (XRD) measurements the lattice constant of this alloy is found to vary linearly with the In composition x according to Vegard's rule. This behavior is shown in terms of the measured cation-anion bond length $\sqrt{3} \times a/4$ in Fig. 2.8a (squares) [19]. In contrast, a measurement of the local bond length using EXAFS demonstrates that actually two different types of bonds exist in the alloy: shorter Ga–As bonds and longer In–As bonds, yielding *in average* a bond length varying

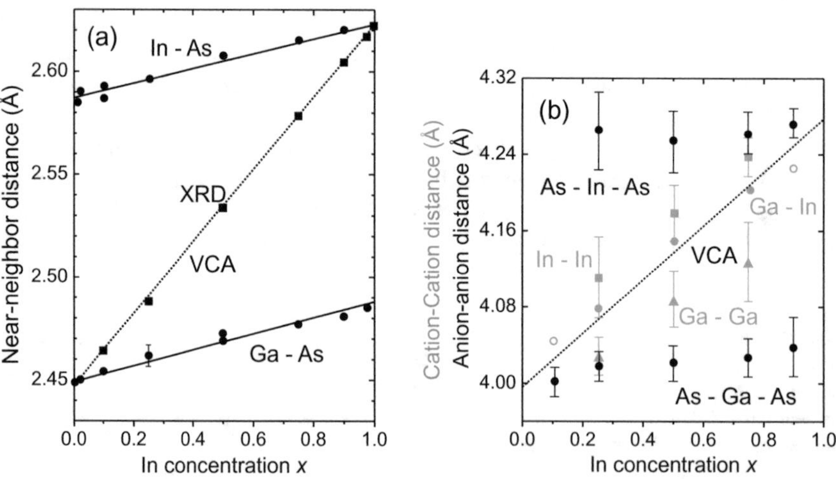

Fig. 2.8 **a** Cation-anion near-neighbor distance as a function of In mole fraction in $Ga_{1-x}In_xAs$ alloys. *Circles* and *squares* refer to EXAFS and XRD data, respectively, the *dotted line* is the cation-anion bond length ($\sqrt{3} \times a/4$) according to the virtual-crystal approximation (VCA), calculated from the X-ray lattice constant. **b** As–As second-neighbor distance corresponding to As–Ga–As and As–In–As bonds (*black circles*) and Ga–Ga, In–In, and Ga–In second-neighbor distances, denoted by *gray triangles*, *squares*, and *circles*, respectively. The *dotted line* represents values expected from the VCA. Reproduced with permission from [16], © 1983 APS

linearly with composition. Figure 2.8a shows that the actual individual bond lengths are close to those of GaAs and InAs, and that they vary only little. Consequently the two second-next neighbor distances of anions As–Ga–As and As–In–As given in Fig. 2.8b differ and remain almost constant for all compositions x. This is not the case for the second-next neighbor distances of *cations* depicted with gray symbols in the figure. The distances of atoms on the cation sublattice are all within ~0.05 Å of the linear interpolation suggested by Vegard's rule and the VCA considered below. Distances of In–In are above, Ga–Ga below, and Ga–In very close to the VCA line given in Fig. 2.8b. It should be noted that the bimodal near-neighbor distance shows a quite small width of the distribution, whereas all second-nearest neighbor distributions are rather broad [16]. The data demonstrate that the atomic scale structure of $Ga_{1-x}In_xAs$ alloys features a near-neighbor distribution which consists of two well defined distances. In contrast, the second-nearest neighbor mixed cation distances exhibit a single broadened distribution, and the corresponding common anion distribution is bimodal. The cation sublattice hence approaches a linearly interpolated VCA lattice, while the anions suffer a local displacement from the average position in the lattice. The origin of this behavior was explained in terms of differences in the distortion energy and the consequential mixing enthalpy [20].

2.1.11 Virtual-Crystal Approximation

The virtual-crystal approximation (VCA) assumes an ideal hypothetical crystal to model the properties of a random alloy formed by atoms A and B on a (sub-) lattice. In a VCA crystal atoms A and B are replaced by a single kind of atoms C whose properties are assumed as a linear average of those of A and B. The virtual alloy made of pseudo-atoms C has hence a crystal structure common to the crystals made of either A or B atoms with a linearly averaged crystal potential. The assumption strongly reduces computational complexity and leads to a great popularity of the approach. The VCA has been applied to semiconductor alloys within various schemes like the semiempirical pseudopotential method or the empirical tight-binding method.

A VCA crystal contains solely a single type of bonds and yields the linear relationship of lattice parameters in the alloy expressed by Vegard's rule (2.2a)–(2.2c). The approximation succeeded in calculating a number of problems like the fact of an optical bandgap bowing discussed in Sect. 4.1.5. It must be noted that properties depending on local differences in the potentials of individual atoms like charge redistribution or polarization are precluded in the VCA approach. Effects arising from such differences as atom-A-like and atom-B-like features, also referred to as chemical disorder, are not correctly described by the VCA approach.

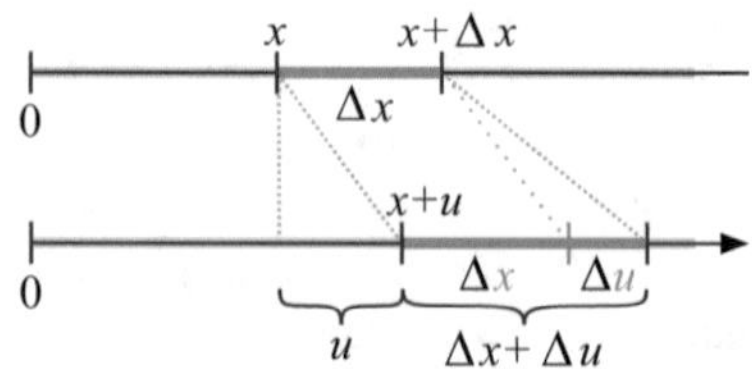

Fig. 2.9 Deformation of a dilatable string in an unstrained (*top*) and strained state (*bottom*). The origin is assumed to be fixed

2.2 Elastic Properties of Heterostructures

An epitaxial layer with a chemically different composition and potentially also a different structure compared to the substrate represents a heterostructure. If the interface does not contain structural defects, the layer—and to some minor extent also the substrate—are *strained*, because the lattice parameters of layer and substrate will generally differ due to differences in bond lengths and thermal expansion coefficients. The elastic properties of the heterostructure can, to a very good approximation, be determined using continuum mechanics, i.e., a macroscopic theory. To assess the effect of the structural and thermal misfits, we first consider the macroscopic effect of stress on the deformation of solids in general. An introduction into this classical subject may be found in, e.g., [21, 22].

2.2.1 *Strain in One and Two Dimensions*

Stress leads to a deformation (strain) of a solid. For a *one-dimensional* problem the effect of strain on a dilatable string is illustrated in Fig. 2.9. A point at an arbitrary position x is displaced by u. In *homogeneous* stretching u is a *linear* function of x, and the section Δx is strained to $\Delta x + \Delta u$. The *strain* of the section Δx can now be defined as $e = du/dx$. It is a dimensionless quantity and small compared to 1.

We now extend the problem to the *two-dimensional strain* of a plane sheet. Here, the strain is specified by the four quantities

$$e_{ij} = du_i/dx_j \quad \text{for } i, j = 1, 2. \tag{2.3}$$

The quantities e_{ij} can be regarded as components of a second-rank tensor described by a 2×2 matrix. Without any distortion all components are expected to vanish. This is, however, not fulfilled for components defined in such a way; a simple rigid-body rotation leads to non-zero off-diagonal components. To meet the condition of zero components at zero distortion, the tensor is expressed as a sum of a symmetrical and an antisymmetrical tensor, yielding components $e_{ij} = \varepsilon_{ij} + \overline{\varepsilon_{ij}}$. Only the *symmetrical* part

$$\varepsilon_{ij} = \frac{1}{2}(e_{ij} + e_{ji}) = \varepsilon_{ji} \tag{2.4}$$

is defined as the strain (the antisymmetrical components are $\overline{\varepsilon_{ij}} = \frac{1}{2}(e_{ij} - e_{ji}) = -\overline{\varepsilon_{ji}}$). The two-dimensional strain tensor then reads

$$\boldsymbol{\varepsilon} = \begin{bmatrix} e_{11} & \frac{1}{2}(e_{12} + e_{21}) \\ \frac{1}{2}(e_{12} + e_{21}) & e_{22} \end{bmatrix}. \tag{2.5}$$

In absence of a rotation, the diagonal components ε_{11} and ε_{22} directly measure the extensions or compressions per length along the x- and y-axes, respectively. The off-diagonals $\varepsilon_{12} = \varepsilon_{21}$ measure the shear strain. The *anti*symmetrical tensor contains only the two off-diagonal elements $\frac{1}{2}(e_{12} - e_{21})$ and describes a pure solid-body rotation potentially connected to the deformation of the area.

The deformation of an area $(\Delta x_1, \Delta x_2)$ in the plane sheet is given by $(\Delta x_1 + \Delta u_1, \Delta x_2 + \Delta u_2)$ with the two displacement components

$$\Delta u_i = \frac{\partial u_i}{\partial x_1}\Delta x_1 + \frac{\partial u_i}{\partial x_2}\Delta x_2 = e_{i1}\Delta x_1 + e_{i2}\Delta x_2 \quad \text{for } i = 1, 2.$$

2.2.2 Three-Dimensional Strain

The description of *three-dimensional strain* is a generalization of the former cases. The three diagonal components $\varepsilon_{ii} = e_{ii}$ $(i = 1, 2, 3) = \varepsilon_{xx}, \varepsilon_{yy}, \varepsilon_{zz}$, are the tensile or compressive strains. They occur along the x, y, and z-axis, respectively, if no rotation is connected to the deformation. The sum of the diagonal components yields the relative change of volume originating from the strain,

$$\frac{\Delta V}{V} = \sum_{i=1}^{3} \varepsilon_{ii}. \tag{2.6}$$

In the off-diagonal elements the quantities e_{ij} describe rotations similar to the two-dimensional case. e_{12} describes an anticlockwise rotation about the z-axis, $e_{21} = -e_{12}$. Similarly, e_{13} and e_{23} describe rotations about the y and x-axes, respectively. The off-diagonal terms of $\boldsymbol{\varepsilon}$ are again symmetrized and represent shear strains,

$$\varepsilon_{ij} = \frac{1}{2}\left(\frac{\partial u_i}{\partial x_j} + \frac{\partial u_j}{\partial x_i}\right), \tag{2.7}$$

yielding the three-dimensional strain tensor

$$\boldsymbol{\varepsilon} = \begin{bmatrix} e_{11} & \frac{1}{2}(e_{12}+e_{21}) & \frac{1}{2}(e_{13}+e_{31}) \\ \frac{1}{2}(e_{12}+e_{21}) & e_{22} & \frac{1}{2}(e_{23}+e_{32}) \\ \frac{1}{2}(e_{13}+e_{31}) & \frac{1}{2}(e_{23}+e_{32}) & e_{33} \end{bmatrix}. \tag{2.8}$$

Since the strain tensor is symmetric, only 6 of the 9 components are independent. The 6 independent components are often written according to the Voigt notation in form of a 6×1 matrix for short. This symbolic vector is formed by the index substitution $11 \to 1$, $22 \to 2$, $33 \to 3$, 23 and $32 \to 4$, 31 and $13 \to 5$, and 12 and $21 \to 6$, yielding

$$\boldsymbol{\varepsilon} = \begin{bmatrix} \varepsilon_{11} & \varepsilon_{12} & \varepsilon_{13} \\ \varepsilon_{12} & \varepsilon_{22} & \varepsilon_{23} \\ \varepsilon_{13} & \varepsilon_{23} & \varepsilon_{33} \end{bmatrix} \to \boldsymbol{\varepsilon} = \begin{pmatrix} \varepsilon_1 \\ \varepsilon_2 \\ \varepsilon_3 \\ \varepsilon_4 \\ \varepsilon_5 \\ \varepsilon_6 \end{pmatrix}. \tag{2.9}$$

In the Voigt notation the shear components of the strain tensor are usually defined in terms of an *engineering convention* instead of the physical convention used above. The engineering quantities are $\varepsilon_{ij}^{\text{engin.}} = 2\varepsilon_{ij}, i \neq j$. This allows for writing Hooke's law in the simple notation with reduced indices (2.10). A 3×3 matrix composed of the off-diagonals $\varepsilon_{ij}^{\text{engin.}}$ and the diagonals ε_{ii} of $\boldsymbol{\varepsilon}$ does, however, *not* form a tensor, because such array does not transform according to the rules of a second-rank tensor [21]. This must be considered if the strain occurs not along the principal directions.

We briefly consider the three-dimensional *stress*, which gives rise to the strain treated above. Similar to the strain $\boldsymbol{\varepsilon}$ all stress components are combined in a symmetric second order stress tensor $\boldsymbol{\sigma}$, which often is likewise written in form of a vector. Any stress may be composed of three components: uniaxial stress, shear stress, and hydrostatic stress. Uniaxial tensile or compressive stresses σ_{xx}, σ_{yy}, and σ_{zz}, act along the axes x, y, and z, respectively. They are built by force pairs acting normal to the surfaces (see Fig. 2.10) and have the unit of a force per area, i.e., N/m^2 or Pa. If the forces act tangentially, they create a shear stress and are labeled σ_{xy}, σ_{yz}, and σ_{xz}. The first and second index denote the axis of force direction and the normal of the surface where the force acts, respectively. In the Voigt notation the same index substitution is used as that applied for the strain, but no factor of 2 is put into the off-diagonal elements.

2.2.3 Hooke's Law

If the strain is not too large, the interaction potential of the atoms in the solid is well described in the harmonic approximation, and we obtain a linear relationship between stress and resulting strain. For the analogous problem of a spring, Robert Hooke stated in the 17th century that the force exerted by a mass attached to a spring is

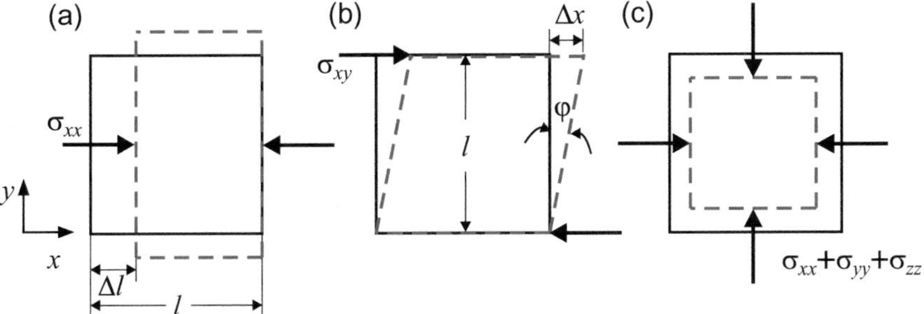

Fig. 2.10 Deformation of a solid by **a** uniaxial stress, **b** shear stress, and **c** hydrostatic stress

proportional to the amount the spring is stretched. In the generalized Hooke's law the constant of proportionality in the one-dimensional case is replaced by the elasticity stiffness tensor **C**. **C** is a fourth-order tensor comprising $3^4 = 81$ components C_{ijkl}. By considering all symmetries present in crystalline solids, the number of non-zero components is reduced to 36. **C** is usually written according to Voigt's notation in form of a 6×6 matrix by putting $C_{ijkl} = C_{mn}$ $(i, j, k, l = 1, 2, 3; m, n = 1, \ldots, 6)$. Using the Voigt notation for stress and strain defined in (2.9) and accordingly the 6×6 matrix representation for **C**, Hooke's law reads

$$\begin{pmatrix} \sigma_1 \\ \sigma_2 \\ \sigma_3 \\ \sigma_4 \\ \sigma_5 \\ \sigma_6 \end{pmatrix} = \begin{pmatrix} C_{11} & C_{12} & C_{13} & C_{14} & C_{15} & C_{16} \\ C_{21} & C_{22} & C_{23} & C_{24} & C_{25} & C_{26} \\ C_{31} & C_{32} & C_{33} & C_{34} & C_{35} & C_{36} \\ C_{41} & C_{42} & C_{43} & C_{44} & C_{45} & C_{46} \\ C_{51} & C_{52} & C_{53} & C_{54} & C_{55} & C_{56} \\ C_{61} & C_{62} & C_{63} & C_{64} & C_{65} & C_{66} \end{pmatrix} \begin{pmatrix} \varepsilon_1 \\ \varepsilon_2 \\ \varepsilon_3 \\ \varepsilon_4^{\text{engin.}} \\ \varepsilon_5^{\text{engin.}} \\ \varepsilon_6^{\text{engin.}} \end{pmatrix}, \tag{2.10}$$

where $\varepsilon_i^{\text{engin.}}$ denotes shear components of the strain according to the engineering convention. Equation (2.10) may be written $\boldsymbol{\sigma} = \mathbf{C}\boldsymbol{\varepsilon}$ for short. The inverse relation is given by the compliance matrix $\mathbf{S} = \mathbf{C}^{-1}$, yielding $\boldsymbol{\varepsilon} = \mathbf{S}\boldsymbol{\sigma}$. In Voigt's notation of **S** some factors appear, which are not necessary in **C** [21]: $S_{mn} = S_{ijkl}$ when m and n are 1, 2, or 3; $S_{mn} = 2S_{ijkl}$ for either m or n being 4, 5, or 6; $S_{mn} = 4S_{ijkl}$ when both m and n are 4, 5, or 6.

A particular stress component is given by

$$\sigma_i = \sum_{k=1}^{6} C_{ik}\varepsilon_k, \tag{2.11}$$

where ε_k denotes the engineering quantities when $k > 3$. The components of the elasticity matrix **C** are called (second order) elastic stiffness constants. **C** is a symmetric matrix, which has at most 21 independent coefficients. The factor of 2 in $\varepsilon_k^{\text{engin.}}$ with respect to ε_k had to be introduced in order to write Hooke's law in this simple form with reduced indices. If we write (2.11) in the initial full tensor notation

$$\sigma_{ij} = \sum_{k=1}^{3} \sum_{l=1}^{3} C_{ijkl} \varepsilon_{kl}, \tag{2.11a}$$

we obtain for the off-diagonals the summands $C_{ijkl}\varepsilon_{kl} + C_{ijlk}\varepsilon_{lk}$ $(k \neq l)$. In the shorter matrix notation (2.11) the corresponding summands $C_{mn}\varepsilon_n$ $(n = 4, 5, 6)$ only appear once. The missing factor of 2 is therefore put into the off-diagonal elements by introducing the engineering convention. The procedure has the disadvantage that it applies only for strain along the principal axes for the usually listed constants C_{mn}, e.g., along the [001] direction. For other orientations (2.11a) needs to be rotated to obtain a transformed elasticity tensor C'_{ijkl} [23].

We return to the simplified **C** matrix of (2.10) with 21 independent coefficients. For crystal structures with high symmetry, many of these coefficients are 0. In addition, some coefficients are related to others. Any cubic crystal structure has only 3 independent elastic stiffness coefficients, namely C_{11}, C_{12}, and C_{44}. This strong reduction originates basically from the equivalence of the three cubic axes. Any hexagonal crystal structure has 5 independent coefficients. As an example, the matrix notations of the elasticity tensors of a cubic and a hexagonal structure are given by

$$\begin{aligned}
\mathbf{C}_{\text{cubic}} &= \begin{pmatrix} C_{11} & C_{12} & C_{12} & 0 & 0 & 0 \\ C_{12} & C_{11} & C_{12} & 0 & 0 & 0 \\ C_{12} & C_{12} & C_{11} & 0 & 0 & 0 \\ 0 & 0 & 0 & C_{44} & 0 & 0 \\ 0 & 0 & 0 & 0 & C_{44} & 0 \\ 0 & 0 & 0 & 0 & 0 & C_{44} \end{pmatrix}, \\
\mathbf{C}_{\text{hexagonal}} &= \begin{pmatrix} C_{11} & C_{12} & C_{13} & 0 & 0 & 0 \\ C_{12} & C_{11} & C_{13} & 0 & 0 & 0 \\ C_{13} & C_{13} & C_{33} & 0 & 0 & 0 \\ 0 & 0 & 0 & C_{44} & 0 & 0 \\ 0 & 0 & 0 & 0 & C_{44} & 0 \\ 0 & 0 & 0 & 0 & 0 & \frac{C_{11}-C_{12}}{2} \end{pmatrix}.
\end{aligned} \tag{2.12}$$

Values of C_{mn} for some solids are listed according to the engineering notation in Table 2.7.

For isotropic solids only 2 independent stiffness constants exist, describing the response on axial and shear stress. For such materials **C** is given by the cubic matrix in (2.12), putting $C_{44} = \frac{1}{2}(C_{11} - C_{12})$. The two independent elastic constants are called Lamé constants $\mu = G = C_{44}$ and $\lambda = C_{12}$. G is the shear modulus. Often any other two constants are used, and a number of relations exist to express their dependences.

Table 2.7 Elastic stiffness constants of some cubic and hexagonal solids at room temperature, given in units of 10^{10} Pa. Column 2 denotes structures as discussed in Sect. 2.1, values for carbon (C) refer to the diamond modification

Material	Structure	C_{11}	C_{12}	C_{13}	C_{33}	C_{44}
Au [24]	fcc	19.0	16.1	–	–	4.2
Al [25]	fcc	10.9	6.3	–	–	2.8
W [25]	bcc	51.5	20.4	–	–	15.6
Mg [24]	hcp	5.9	2.6	2.1	6.2	1.6
NaCl [24]	NaCl	4.9	1.3	–	–	1.3
CsCl [24]	CsCl	3.7	0.9	–	–	0.8
C [9]	D	107.6	12.5	–	–	57.7
Si [9]	D	16.6	6.4	–	–	8.0
ZnS [9]	ZB	9.8	6.3	–	–	4.5
ZnSe [9]	ZB	9.0	5.3	–	–	4.0
GaAs [9]	ZB	11.9	5.4	–	–	6.0
AlAs [9]	ZB	11.9	5.7	–	–	5.7
InAs [9]	ZB	8.3	4.5	–	–	4.0
GaN [26]	W	39.0	14.5	10.6	39.8	10.5
AlN [26]	W	39.6	13.7	10.8	37.3	11.6
InN [26]	W	22.3	11.5	9.2	22.4	4.8
ZnO [27]	W	20.6	11.7	11.8	21.1	4.4

If the elastic properties of isotropic solids are expressed in terms of components of the compliance matrix **S**, the two independent components are $S_{11} = 1/E$ and $S_{12} = -\nu/E$; $S_{44} = S_{55} = S_{66}$ then equals $(2 + 2\nu)/E$. E is Young's modulus (also referred to as elastic modulus) and represents the ratio stress/strain, and ν is Poisson's ratio considered below. The two quantities are related to the Lamé constants by

$$E = \frac{\mu(2\mu + 3\lambda)}{\mu + \lambda}, \qquad \nu = \frac{\lambda}{2(\mu + \lambda)}.$$

2.2.4 *Poisson's Ratio*

Virtually all common materials undergo a transverse contraction when longitudinally stretched, and a transverse expansion when longitudinally compressed. The quotient of such deformations is a material property termed *Poisson's ratio* ν: $\nu = -$(transverse strain/longitudinal strain) for a uniaxial tensile stress applied in longitudinal direction. It is usually a positive quantity, though also negative Poisson's ratios were reported for novel foam structures called anti-rubber which have inter-atomic bonds realigning with deformation [28]; for more information see [29].

For isotropic solids comprising polycrystalline materials ν is given by the ratio of the two independent stiffness constants C_{11}/C_{12}. For non-isotropic materials like any crystalline solid ν depends on the stress direction. For cubic materials and stress along an axis of the unit cell, the ratio is $\nu = C_{12}/(C_{11} + C_{12})$, for other axes and symmetries the relation gets more bulky [30]. Values for Poisson's ratios range between 0.5 (incompressible medium) and -1 (perfect compressibility), values for ordinary materials are between 0.25 and 0.3 for common semiconductors, 0.45 for Pb, and 0.33 for Al.

2.2.5 *Pseudomorphic Heterostructures*

We consider a free standing crystalline structure consisting of two layers, which have a common interface as delineated in Fig. 2.11. We assume the layers to have the same cubic crystal structure, but—in absence of the common interface—different *unstrained* lattice constants a and thicknesses t labeled a_1, t_1, a_2, and t_2, respectively. If the difference in lattice constants is not too large (say, below 1%), the layers may form an interface without structural defects and adopt a common in-plane lattice constant $a_{\|}$ parallel to the interface, with an intermediate value $a_1 > a_{\|} > a_2$. Since $a_{\|}$ is smaller than the relaxed (unstrained) lattice parameter a_1, layer 1 is compressively strained in lateral direction (i.e., parallel to the interface) by the contact to layer 2. Layer 1 consequently experiences a distortion also in the vertical direction to approximately maintain its bulk density. The vertical lattice constant $a_{1\perp}$ of the strained layer 1 is hence larger than the unstrained value a_1. Vice versa layer 1 exerts a laterally tensile stress on layer 2, leading likewise to a vertical strain $a_{2\perp} < a_2$. Such a heterostructure is called *pseudomorphic*, and the layers are designated *coherently* strained.

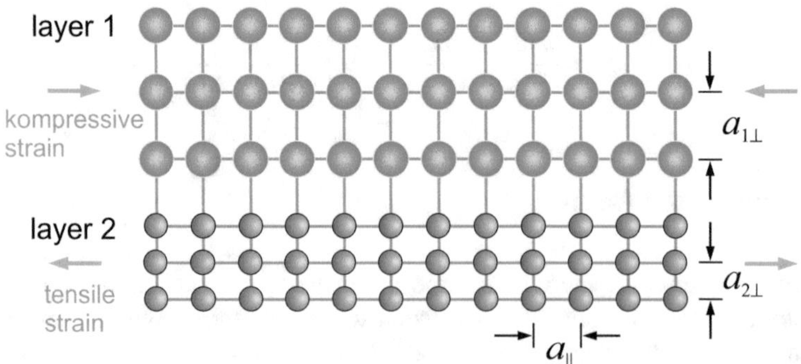

Fig. 2.11 Schematic of a heterostructure consisting of two layers with a common interface. $a_{\|}$ is the common lateral lattice constant, $a_{1\perp}$ and $a_{2\perp}$ denote the vertical lattice constants of the strained layers 1 and 2

In the strained cubic heterostructure illustrated in Fig. 2.11, the strain in the two lateral directions, say x and y, is equal, $\varepsilon_{xx} = \varepsilon_{yy} = \varepsilon_{\|}$. Such a biaxial strain results in a tetragonal distortion. Since the structure is assumed to be free-standing, no force is applied along the z-direction. The stress tensor component σ_{zz} is thus always zero. Inserting the cubic elasticity tensor (2.12) into Hooke's law (2.10) and setting up (2.11) for σ_{zz}, we obtain a relation between the diagonal strain components and hence between $\varepsilon_{\|}$ and $\varepsilon_{\perp}$:

$$\begin{aligned} \sigma_{zz} &= 0 = C_{12}\varepsilon_{xx} + C_{12}\varepsilon_{yy} + C_{11}\varepsilon_{zz}, \\ \varepsilon_{zz} &= -\frac{C_{12}\varepsilon_{xx} + C_{12}\varepsilon_{yy}}{C_{11}} = -\frac{2C_{12}}{C_{11}}\varepsilon_{xx}, \\ \varepsilon_{\perp} &= -D_{001}\varepsilon_{\|}. \end{aligned} \tag{2.13}$$

The distortion factor D_{001} is given in (2.17) also for other orientations of the interface. Vertical strain $\varepsilon_{zz} = \varepsilon_{\perp}$ and lateral strain $\varepsilon_{\|}$ virtually always have opposite sign, since common stiffness constants C_{ij} are positive quantities. The volume of the unit cell is usually not preserved during deformation: The effect of the counteracting strain perpendicular to the interface does not fully compensate the in-plane strain. In the case of a tetragonal distortion no in-plane shear strain occurs, the respective off-diagonal elements of the strain tensor are hence zero. According to (2.6) the volume change is given by the trace of the strain tensor,

$$\Delta V/V = \varepsilon_{xx} + \varepsilon_{yy} + \varepsilon_{zz}. \tag{2.14}$$

The in-plane lattice constant $a_{\|}$ of the pseudomorphic heterostructure (Fig. 2.11) is given by a balance of the elastic strain minimizing the strain energy [31],

$$\begin{aligned} a_{\|}G_1t_1 + a_{\|}G_2t_2 &= a_1G_1t_1 + a_2G_2t_2, \\ a_{\|} &= \frac{a_1G_1t_1 + a_2G_2t_2}{G_1t_1 + G_2t_2}. \end{aligned} \tag{2.15}$$

G denotes the shear modulus and depends on the crystal structure and the crystallographic orientation of the interface plane. a and t are the unstrained lattice parameter and the thickness of the respective layer. For a cubic structure, like e.g. zincblende, the shear moduli G_i of layers 1 and 2 are given by

$$G_i = 2\left(C_{11}^i + 2C_{12}^i\right)(1 - D_i/2), \tag{2.16}$$

with the distortion factors D of the respective orientations

$$D_{001} = \frac{2C_{12}}{C_{11}},$$
$$D_{110} = \frac{C_{11} + 3C_{12} - 2C_{44}}{C_{11} + C_{12} + 2C_{44}}, \tag{2.17}$$
$$D_{111} = \frac{2C_{11} + 4C_{12} - 4C_{44}}{C_{11} + 2C_{12} + 4C_{44}}.$$

The factors D represent the ratios $-\varepsilon_{i\perp}/\varepsilon_{i\parallel}$ for the respective orientations of the interface, i.e., the ratios of normal and lateral strains for a biaxial deformation. The assumption of a *constant* ratio in strained layers is based on the assumed validity of Hooke's law (2.10).

It should be noted that distortions along $\langle 110\rangle$ or $\langle 111\rangle$ reduce the crystal symmetry in a way that the atomic positions in the unit cell are not uniquely determined by strain. The shifts originate from changes in the bond strengths: Strain along [111] in zincblende or diamond structures, e.g., makes the [111] bond inequivalent to the other $[\bar{1}\bar{1}1]$, $[\bar{1}1\bar{1}]$, and $[1\bar{1}\bar{1}]$ bonds, leading to a static displacement of the sublattices. Introducing an internal strain parameter ξ [32], the [111] bond is elongated by $(1 - \xi)\varepsilon_{44}a\sqrt{3}/4$. $\xi = 0$ and 1 correspond to perfectly strained positions and rigid bond lengths, respectively [33].

Using the constants D also the vertical change of the unit cell in layer i can be calculated,

$$a_{i\perp} = a_i\big(1 - D_i(a_\parallel/a_i) - 1\big). \tag{2.18}$$

It should be noted that $a_\parallel$ and $a_\perp$ represent the actual lattice constants only for the (001) orientation of the interface plane. For other orientations these quantities express the change of the unit cell dimensions under strain. The lateral and vertical strains of layer i are related to the respective strained lattice parameters by

$$\varepsilon_{i\parallel} = \frac{a_\parallel}{a_i} - 1, \tag{2.19a}$$
$$\varepsilon_{i\perp} = \frac{a_{i\perp}}{a_i} - 1, \tag{2.19b}$$

where the quantity a_i denotes the unstrained lattice constant of layer i. In a corresponding *hexagonal* structure with a basal-plane interface, the lattice constants a and c of the considered layer are inserted into $a_\parallel$ and $a_\perp$ of (2.19a), (2.19b), respectively. Using analogous insertions made to obtain (2.13), the relation reads for hexagonal structures

$$\varepsilon_\perp = -\frac{2C_{13}}{C_{33}}\varepsilon_\parallel. \tag{2.19c}$$

Pseudomorphic Layer

A thin epitaxial layer grown on a much thicker substrate is described by considering $t_1/t_2 \to 0$ in (2.15). We see that then $a_\parallel = a_2$. The unstrained lattice constant a_L of the layer (a_1 in (2.15)) laterally adopts the lattice constant of the substrate a_S (a_2 in (2.15)) at epitaxial growth and can be varied using different substrates. The substrate remains virtually unstrained due to its large thickness, and the *misfit* (or, *lattice mismatch*) f between the two crystals is usually expressed by

$$f = \frac{a_\mathrm{S} - a_\mathrm{L}}{a_\mathrm{L}}. \tag{2.20a}$$

It must be noted that also other definitions for the lattice mismatch are used in literature, particularly the relations

$$f_\mathrm{alternative1} = \frac{a_\mathrm{L} - a_\mathrm{S}}{a_\mathrm{S}} \tag{2.20b}$$

and

$$f_\mathrm{alternative2} = \frac{a_\mathrm{L} - a_\mathrm{S}}{a_\mathrm{L}}. \tag{2.20c}$$

Note the change in sign of the alternative relations with respect to (2.20a).

If we compare (2.20a) with (2.19a), we notice that the misfit is equal to the lateral strain of the epitaxial layer: $f = \varepsilon_\parallel$. This applies for *coherent* growth, i.e., growth with an *elastic* relaxation of the strain without formation of defects. Such a layer is called *pseudomorphic*. f may have either sign as illustrated in Fig. 2.12 by comparing a compressively and a tensely strained layer.

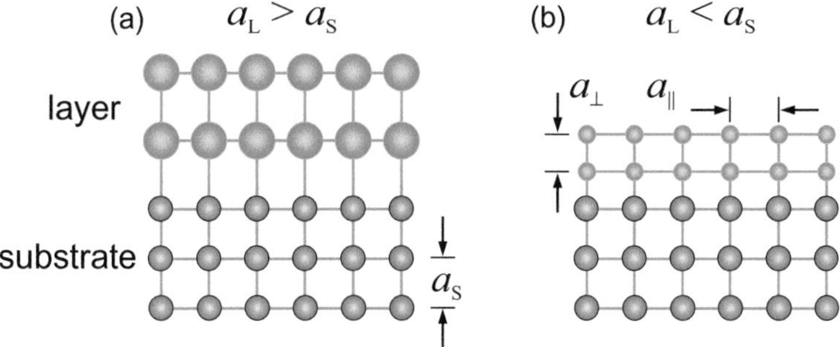

Fig. 2.12 Biaxially strained layers (*yellow atoms*) on substrates (*blue atoms*) with another lattice constant a_S. In **a** the unstrained lattice constant of the layer a_L is larger than a_S, and the layer is compressively strained in lateral direction; in **b** the layer is tensely strained

2.2.6 Critical Layer Thickness

A strained solid like the layers considered in Fig. 2.12 contains a strain energy per volume E/V. The differential work of deformation for a small increment of strain is given by

$$\frac{1}{V}dE = \sum_{i=1}^{6} \sigma_i d\varepsilon_i. \tag{2.21}$$

Integration of (2.21) using (2.11) yields the elastic energy density of a strained solid,

$$\frac{E}{V} = \frac{1}{2}\sum_{i=1}^{6}\sum_{k=1}^{6} C_{ik}\varepsilon_i\varepsilon_k. \tag{2.22}$$

For a cubic material (2.22) reads

$$\begin{aligned}\frac{E}{V} = \frac{1}{2}\big(&C_{11}\left(\varepsilon_{xx}^2 + \varepsilon_{yy}^2 + \varepsilon_{zz}^2\right) + 2C_{12}(\varepsilon_{yy}\varepsilon_{zz} + \varepsilon_{zz}\varepsilon_{xx} + \varepsilon_{xx}\varepsilon_{yy})\\ &+ C_{44}\left(\varepsilon_{yz}^2 + \varepsilon_{zx}^2 + \varepsilon_{xy}^2\right)\big),\end{aligned} \tag{2.23}$$

with the three respective terms of hydrostatic, uniaxial, and shear strains. We see that the homogeneous energy density increases quadratically with strain in the harmonic approximation of Hooke's law. If a cubic, biaxially strained pseudomorphic layer is considered, which is allowed to elastically relax according to Poisson's ratio, the elastic energy density is given by [34]

$$\frac{E}{At_{\mathrm{L}}} = 2G\varepsilon_{\parallel}^2\frac{1+\nu}{1-\nu}, \tag{2.24}$$

A, t_{L}, G, $\varepsilon_{\parallel}$, and ν being the area, the *strained* layer thickness, the shear modulus, the in-plane strain parallel to the interface, and Poisson's ratio of the layer, respectively. According to (2.24) the areal density of the elastic strain energy in the layer E/A increases linearly with the layer thickness t_{L} and quadratically with $\varepsilon_{\parallel}$, i.e., the misfit f for a pseudomorphic layer. Areal strain energy is hence accumulated as the layer thickness increases. At some critical thickness t_{c} this energy is larger than the energy required to form structural defects which *plastically* relax the strain. Such defects are dislocations, the nature of which is studied in more detail in Sect. 2.3. The plastic relaxation reduces the overall strain, but at the same time the dislocation energy increases from zero to a value determined by the particular dislocation.

Quite a few models were developed to calculate the critical thickness t_{c} of a pseudomorphic epitaxial layer depending on the misfit f with respect to the substrate. They all consider some energy balance or, equivalently, a balance of forces or stresses, by comparing the amount of homogeneous strain energy relaxed by the introduction of a particular defect with the energy cost associated with the formation of this defect.

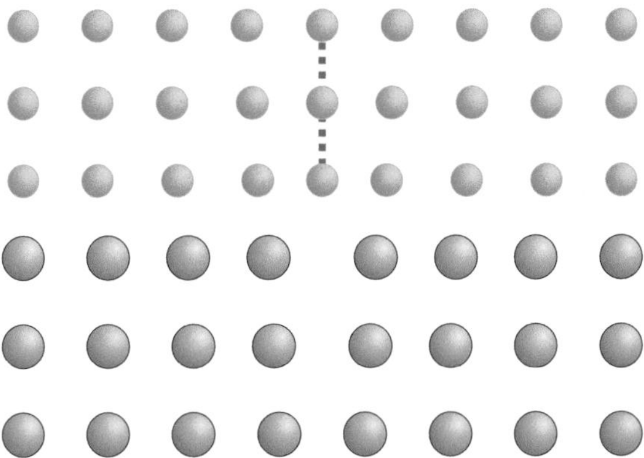

Fig. 2.13 Scheme of a dislocation introduced into a layer, that plastically relaxes the strain. Layer and substrate are represented by *yellow* and *blue atoms*, respectively. The inserted extra plane is shown in cross section and represented by the *dashed red line*

Basic early work by Frank, Van der Merve, and co-workers described the plastic relaxation by an array of parallel, equally spaced misfit dislocations (Sect. 2.3.6) [35–37], or by a two-dimensional grid of such dislocations at the interface [38–41]. A more simple approach for predicting the critical thickness was developed by Matthews and Blakeslee [42, 43]. The model considers the force on a threading dislocation penetrating from the substrate through the epitaxial layer, which creates a dislocation at the interface and gives rise to a comparable scenario as considered by Van der Merve and co-workers.

The general picture is outlined below, assuming a structural lattice defect that can relax the strain of the layer by a plastic deformation. Such a dislocation formed by, e.g., the insertion of an extra lattice plane in the layer is depicted in Fig. 2.13. The end of the additional half-plane near the interface of the heterostructure builds a dislocation line with a locally highly strained region. The formation of this dislocation requires a formation energy E_D. This energy cost is balanced by the elastic relief of homogeneous strain energy in the layer lattice outside the core region of the dislocation line.

Whether or not plastic relaxation occurs in the epitaxial layer depends on the minimum of the elastic energy E_I at the interface. E_I is given by the sum of homogeneous strain energy E_H in the layer (denoted E in (2.24)) and dislocation energy E_D. The dislocation energy E_D depends on the particular geometry and the amount of plastic relaxation as expressed by the Burgers vector (Sect. 2.3). The remaining mismatch f of the (partially) relaxed layer refers to an *average* lattice constant of the layer. In the presence of dislocations f is less than the natural misfit, which is defined by the *unstrained* lattice constants in (2.20a)–(2.20c) and denoted f_0 in the following. To a good approximation f is given by the sum of f_0 and the lateral strain

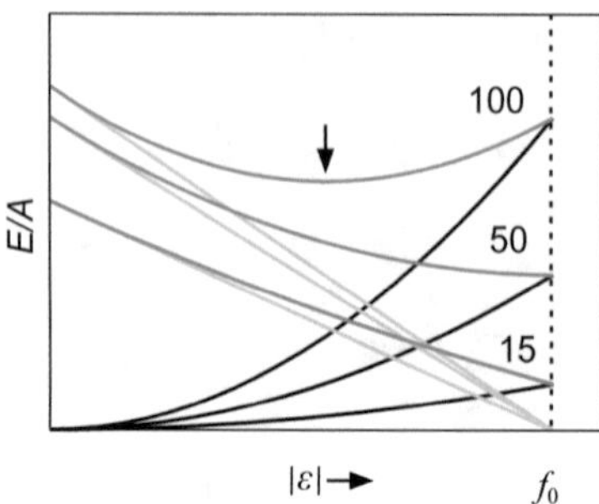

Fig. 2.14 Areal energy densities occurring at a biaxially strained layer with a thickness of 15, 50 and 100 times the substrate lattice constant. ε denotes lateral strain, $f_0 = -0.36$ % is the assumed natural misfit. The *black*, *gray*, and *light gray curves* are homogeneous strain, strain at the interface, and dislocation energy, respectively. The *arrow* denotes an energy minimum of a thick layer attained by plastic strain relaxation

$$\varepsilon = \frac{a_{\rm L} - a_{\rm L,0}}{a_{\rm L,0}}$$

which normally has opposite sign; $a_{\rm L,0}$ denotes the unstrained lattice parameter of the layer. The dependence of $E_{\rm I}$ from ε is then given by

$$\begin{aligned} E_{\rm I}/A &= E_{\rm H}/A + E_{\rm D}/A, \\ E_{\rm H}/A &\propto t_{\rm L}\varepsilon^2, \\ E_{\rm D}/A &\propto \varepsilon + f_0. \end{aligned} \tag{2.25}$$

The relation for the interface energy given by (2.25) is illustrated in Fig. 2.14 for a $GaAs_{0.9}P_{0.1}$ layer on GaAs substrate with a natural misfit $f_0 = -0.36\%$ and a thickness below, just at, and above the critical value for plastic relaxation [40]. The homogeneous strain-energy density disappears at zero strain and increases quadratically with ε (black curves), while the dislocation energy density gets zero at $f = \varepsilon + f_0 = 0$ (light gray curves). In any case the energy density at the interface $E_{\rm I}/A$ (gray curves) tends to attain a minimum. The criterion for the critical thickness $t_{\rm c}$ is [44]

$$\partial(E_{\rm I}/A)/\partial(|\varepsilon|) = 0, \tag{2.26}$$

evaluated at $|\varepsilon| = |f_0|$. For $t_{\rm L} > t_{\rm c}$ the homogeneous strain of the layer gets larger than the natural misfit $|f_0|$, and dislocations introduce a strain of opposite sign, thereby reducing $|\varepsilon|$ and hence $E_{\rm I}/A$. The critical thickness is usually inverse to the natural misfit in a wide range, cf. Fig. 2.15.

For the evaluation of the critical thickness in a given heterostructure the geometry of the strain-relaxing dislocations must be specified. We consider the example of a semiconductor layer with diamond (or zincblende) structure and accommodating

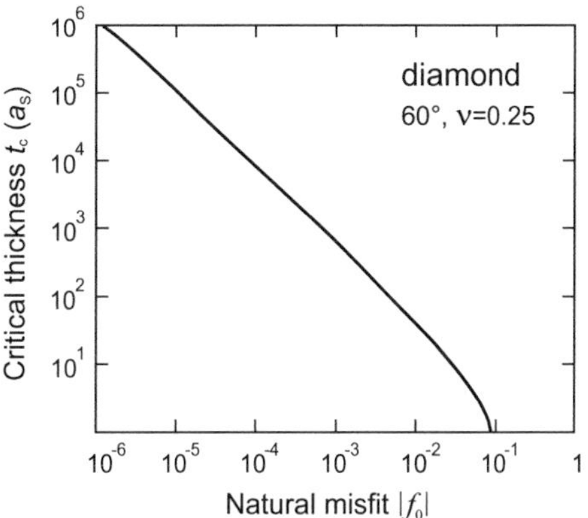

Fig. 2.15 Critical thickness of a biaxially strained layer with diamond or zincblende structure in units of the substrate lattice constant a_s for the introduction of accommodating 60° dislocations. From [40]

60° dislocations, which are most prominent in such solids and are treated in detail in Sect. 2.3. We assume a biaxially strained layer with a given Poisson ratio ν. Taking the dislocation energy E_D from (2.36), the energy at the interface E_I (2.25) reads

$$\begin{aligned} E_I/A &= E_H/A + E_D/A \\ &= 2Gt\varepsilon^2\frac{1+\nu}{1-\nu} + bG\frac{|\varepsilon + f_0|(1-\nu\cos^2\alpha)}{2\pi(1-\nu)\sin\alpha\cos\beta}\ln\left(\frac{\rho R}{b}\right). \end{aligned} \tag{2.27}$$

The strain denotes the in-plane components $\varepsilon = \varepsilon_{xx} = \varepsilon_{yy} = \varepsilon_{\parallel}$, and the geometry of the dislocation is expressed in terms of the absolute value b of its Burgers vector **b**, the angle α between the Burgers vector and the dislocation line vector, and the angle β between the glide plane of the dislocation and the interface. R is the cut-off radius defining the boundary of the strain field produced by the dislocation core, and the factor ρ accounts for the strain energy of the dislocation core and is on the order of unity. We now apply condition (2.26) to the interface energy (2.27) and put $R = t_c$, because the first dislocations are expected to appear if the cut-off radius is equal to the critical layer thickness. This gives the relation [40]

$$t_c = b\frac{(1-\nu\cos^2\alpha)}{8\pi|f_0|(1+\nu)\sin\alpha\cos\beta}\ln\left(\frac{\rho t_c}{b}\right). \tag{2.28}$$

The critical thickness t_c can be calculated numerically from this transcendent equation for a given natural misfit f_0. If the geometrical parameters of the 60° dislocation are inserted for the considered geometry, $b = a/2 \times [101] = a/\sqrt{2}$, $\alpha = 60°$, $\beta = 54.7°$, and assuming $\rho = 4$ for the dislocations and $\nu = 0.25$ for layer, we obtain the dependence of t_c shown in Fig. 2.15 [40]. The geometry of the 60° dislocation in zincblende crystals is depicted in Fig. 2.25b. Consideration of another geometry yields a different prefactor of *ln* in (2.28), producing a vertical shift of the curve in Fig. 2.15.

The dependence given in Fig. 2.15 follows approximately the relation $t_c \sim 1/|f_0|$. It should be noted that the actual degree of plastic relaxation does not solely depend on the misfit, but also on the growth procedure: Kinetic barriers for the generation and movements of misfit dislocations may lead to significantly larger values of the critical thickness than expected from equilibrium calculations. For metal layers, the experimentally determined values agree reasonably well with the predictions of equilibrium theory. For semiconductor layers, however, much larger values were observed. Equilibrium considerations are still useful to provide a reasonable measure for the lower limit of the critical layer thickness.

2.2.7 Approaches to Extend the Critical Thickness

Epitaxy of heterostructures free of dislocations (at least within the diffusion length of charge carriers) is of particular importance for electronic and optoelectronic applications. Dislocations introduce local inhomogeneities, giving rise to a short lifetime and non-radiative recombination of charge carriers in semiconductors. Furthermore, piezoelectric effects associated with local strains change electronic properties. Dopants may precipitate at dislocations or strongly change diffusion characteristics. Fast degradation of devices often is connected to the action of dislocations. In device fabrication the reduction of the dislocation density in the active region below a level usually defined by the diffusion length of charge carriers is therefore an important issue. Typical values in semiconductor industry are in the range below 10^2 cm^{-2} for Si and 10^2–10^3 cm^{-2} for III–V arsenides. Nitride semiconductors have much higher values in the range 10^4–10^6 cm^{-2} for lasers and 10^7–10^9 cm^{-2} for LEDs and electronic devices.

A number of concepts is applied to extend the critical thickness beyond the limit given by the natural misfit. In many cases *alloying* of the layer materials by mixing crystals of different lattice constants is used to closely match the substrate lattice constant. The lattice constants of alloys are generally found to vary linearly with composition, a relation expressed by Vegard's rule (Sect. 2.1.10). Since simultaneously the bandgap of the alloy varies, this may require mixing of more than two components with counteracting effects on bandgap and lattice constant (Sect. 4.1.5). Growth of such ternaries or quaternaries sometimes encounters thermodynamic limits, stimulating the search for alternatives.

Buffer Layer

A widely applied method to reduce the dislocation density of lattice-mismatched heterostructures is the introduction of a *buffer* between the epitaxial structure and the substrate [45]. Such layers or layer stacks confine misfit dislocations in a region below the active part and should in particular suppress the *penetration* of dislocation lines, i.e., the *threading dislocations*. A simple approach for a buffer is a *uniform buffer layer* (thick, μm range) with a high defect density near the interface to the substrate. The density of threading dislocations in a mismatched layer is usually found to decrease inverse to its thickness due to reactions of dislocations with opposite sign. Such buffer layers are referred to as *metamorphic*. They are to a large extend plastically relaxed and their lattice parameter approaches the unstrained value at the surface. Due to the thermal mismatch with respect to the substrate there will always remain some strain below growth temperature. A thick uniform buffer layer is also called *virtual substrate*.

The approach of a uniform buffer grown at low temperature is widely applied in nitride growth on sapphire [46, 47]. This systems is strongly structurally and thermally mismatched, giving rise to peeling off and cracking of epitaxial GaN layers grown without buffer. The basic idea is the deposition of a nitride layer with poor crystallinity at a low temperature where adatom mobility is low, and a subsequent crystallization by an annealing step. The interface to the substrate then contains numerous defects making the buffer compliant to accommodate misfits, while the surface consists of well oriented crystallites which can coalesce during overgrowth.

It should be noted that the epitaxial layer may have a different crystallographic orientation or a different crystal structure than the substrate. A prominent example is the prevalent growth of wurtzite (hexagonal) GaN on basal plane corundum (trigonal) α-sapphire Al_2O_3. The [00.1] directions (or (00.1) planes, respectively) of GaN and Al_2O_3 coincide, but the [10.0] direction of GaN is parallel to the $[1\bar{1}.0]$ direction of the Al_2O_3 substrate [48, 49]. This corresponds to a rotation of the epitaxial unit cell by 30° about the [00.1] c-axis as compared to that of the substrate. *General matching conditions* for any pair of crystal lattices and any orientation may be found by considering interface translational symmetry [50]. The two lattices match if the two two-dimensional lattices, formed by the crystal translations of the paired lattices parallel to the common interface, have a common superlattice. Such geometrical lattice match may induce epitaxial growth, but interface chemistry will always play a major role.

A drawback of the uniform relaxed buffer layer is the large thickness required to obtain a threading-dislocation density suitable for device-grade material of the epitaxial structure grown on top. In zincblende semiconductors a typical buffer-layer thickness on the order of some μm is required to achieve a high quality for device growth. In wurtzite layers grown along the common [0001] direction the reduction of dislocation density is less efficient, and thicker uniform buffers are needed [51].

Another approach is the *graded buffer* consisting of a continuously (or stepwise) change of the lattice constant by alloying with consecutively changing alloy parameters. Usually a *linear* grading of the composition—and consequently also of the lattice mismatch—is applied. Such a buffer is expected to have a uniform volume density of misfit dislocations throughout its thickness; this is in contrast to the high dislocation density of a uniform buffer (or layer) near the interface to the substrate, decreasing with increasing separation to the interface, see Fig. 2.16. The critical thickness t_c of a

Fig. 2.16 Transmission electron micrograph of a 1.8 μm thick ZnTe layer grown on (001)-oriented GaAs, imaged along the [110] direction. *Inset*: Diffraction pattern from the ZnTe/GaAs interface region. Reproduced with permission from [62], © 1993 Elsevier

graded buffer exceeds that of a uniform buffer, and above t_c the density of threading dislocations at the surface is lower. An empirical model indicates that this density is proportional to the concentration gradient and also to the applied growth rate, if the gliding process of dislocations is not significantly impeded [52]. Growth conditions should hence favor the glide of threading dislocations to create the longest possible strain-relieving misfit segments, e.g., by a high growth temperature. For a review on various graded buffer layers see [53].

A more sophisticated approach is the so-called dislocation filtering using a *strained-layer superlattice*. Multilayer structures with either alternating strain or alternating elastic stiffness were employed. Strains in the superlattice may cause bending of dislocation lines, leading to a meandering in alternating strain fields. Bending is also promoted when a dislocation line enters an elastically softer layer. The bending of dislocation lines increases the probability for defect reactions and favors annihilation. The thickness of the individual layers in the superlattice must be below the critical value for that particular layer, because the layers should induce a bending of dislocation lines without introducing new dislocations.

Epitaxial Lateral Overgrowth

A conceptual different approach is *epitaxial lateral overgrowth* (ELO or ELOG) [54, 55]. In the ELO process the substrate is first covered with an amorphous mask layer (e.g., SiO_2) which contains windows. In subsequent epitaxy, growth is controlled to occur only in the windows. When the layer thickness exceeds that of the mask, the mask is laterally overgrown. Since defects from the interface between substrate and layer cannot penetrate through the mask, the overgrowth region has a very low defect density. ELO growth requires a growth anisotropy with the lateral growth rate well exceeding the vertical rate; furthermore, nucleation and growth on the mask must be prevented. ELO is an application of selective area growth and discussed in detail in Sect. 12.1.4.

Compliant Substrate

Concepts discussed so far use thick substrates which define the lateral lattice constant of the heterostructure in pseudomorphic growth. Considering (2.15), the critical layer thickness can be largely extended by using a thin and therefore *compliant substrate* [56–58]. The strain is then distributed between epilayer and substrate. For a sufficiently thin, free-standing substrate the epitaxial layer can virtually be arbitrarily thick.

Compliant substrates were implemented using various approaches. The compliant substrate is generally composed of a thin compliant layer, which is mechanically decoupled from a thicker substrate required for handling. On a small area thin free-standing membranes with a support at the side were realized. Approaches for larger area use various methods for bonding the thin layer to the mechanically stable support, in particular oxide bonding, borsilicate-glass-bonding, and twist-bonding. Reviews on this topic are given in [59, 60].

2.2.8 Partially Relaxed Layers and Thermal Mismatch

The increase of elastic strain energy in a lattice-mismatched epitaxial layer up to a critical thickness t_c was pointed out in Sect. 2.2.6. If the layer thickness exceeds t_c misfit dislocations are formed in the layer and the strain is partially or, in sufficiently thick layers, fully relieved. We consider a cubic crystal. Usually strain relief of a biaxially strained and partially relaxed layer is symmetric and the two independent lateral lattice constants remain equal. The relief of strain may then be described by a strain parameter γ,

$$\gamma = 1 - \frac{a_{\parallel} - a_S}{a_L - a_S}, \tag{2.29}$$

a_S and a_L being the unstrained (natural) lattice constants of substrate and layer, respectively, and $a_{\parallel}$ the lattice constant of the layer in its strained state parallel to the interface. The fully strained pseudomorpic layer then corresponds to $\gamma = 1$ ($a_{\parallel} = a_S$), and $\gamma = 0$ to the fully relaxed layer ($a_{\parallel} = a_L$). Usually partial strain relief leads to a graded strain profile, i.e., γ is not constant throughout the layer. The density of threading dislocations is approximately invers to the layer thickness and is modeled for semiconductors with fcc (and related zincblende) structure in [61] and for wurtzite structure in [51]. An example is given in Fig. 2.16. The image shows a strongly mismatched zincblende ZnTe layer ($a_L = 6.123$ Å at $T_{growth} = 350$ °C) grown on GaAs substrate ($a_S = 5.665$ Å at 350 °C, $f = -7.5\%$) [62]. Dark lines indicate dislocation lines which form a dense network extending approximately 300 nm from the interface. Most of them are inclined by about 55° to the interface, equal to the angle of the intersection line of {111} planes with the (110) image plane. The angle indicates the occurrence of 60° dislocations (Sect. 2.3.5). The density of dislocations decreases as the ZnTe layer thickness grows, leading to regions with much lower

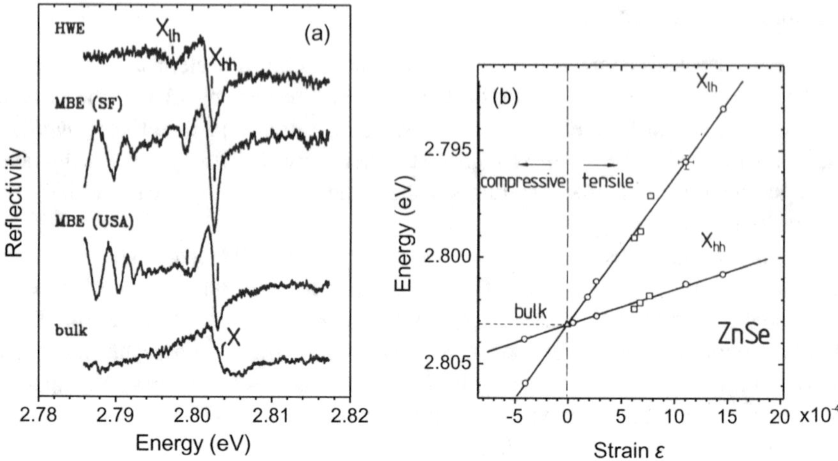

Fig. 2.17 **a** Reflectivity spectra of free excitons in ZnSe/GaAs epitaxial layers which were grown at different temperatures. The lowest spectrum refers to a ZnSe bulk crystal. From [64]. **b** Energies of free light-hole (X_{lh}) and heavy-hole (X_{hh}) excitons in differently prepared ZnSe samples on GaAs substrate (*squares*) and ZnSe layers released from the substrate (*circles*). The triangle at the crossing point marks the exciton energy measured with a bulk crystal. Reproduced with permission from [63], © 1992 Elsevier

defect density above 600 nm. The inset in Fig. 2.16 shows a diffraction pattern from the interface region, consisting of the superimposed response from the two zincblende crystals GaAs (outer points of each reflection due to a smaller lattice constant) and ZnTe. The pattern proves the epitaxial relation of layer and substrate.

The decreasing defect density for increased layer thickness is accompanied by a progressive strain relief ($\gamma \to 0$) and consequently by a gradual approach of the layer lattice constants $a_\parallel$ and $a_\perp$ toward the unstrained value a_L.

It must be noted that the unstrained lattice constant varies as the temperature is changed (e.g., from growth temperature to room temperature, Sect. 2.1.7), and that different materials (e.g., substrate and layer) have different thermal expansion coefficients.

An example of different built-in strains in ZnSe epilayers grown at different temperatures on GaAs is given in Fig. 2.17. The lattice mismatch (2.20a) at room temperature is $f = -0.27\%$, inducing laterally a biaxial compressive strain in the layer. The studied ZnSe layers were thicker than the critical thickness t_c ($\cong$150 nm) considered in Sect. 2.2.6 and were thus partially relaxed. Since the thermal expansion coefficient of GaAs is substantially smaller than that of ZnSe ($\sim 5.7 \times 10^{-6}$ K^{-1} compared to $\sim 7 \times 10^{-6}$ K^{-1} at RT, respectively), the ZnSe lattice contracts more pronounced when the heterostructure is cooled from growth temperature to the applied measurement temperature. The strain may be evaluated from the splitting of the valence band, reflected in the splitting of the free exciton (Sect. 4.1.2). The measurement of the free-exciton states in Fig. 2.17a shows that different strain states exist in samples which were prepared at different growth temperatures between 320 °C (molecular beam epitaxy) and 450 °C (hot wall epitaxy) [63, 64]. The exciton reflection-loops progressively split and shift to lower energy as the growth temperature is increased, indicating increasing *tensile* strain. The strain state was calculated from the difference of thermal expansion coefficients times the difference between growth and measurement temperature. The measurement of the near-surface strain state Fig. 2.17b derived from reflectivity spectra recorded at 1.6 K reveals a large range of values including also negative strain, i.e., still compressive in-plane strain of the ZnSe layer at 1.6 K. The crossing point agrees well with the energy of the unsplit free exciton of a ZnSe bulk crystal.

2.3 Dislocations

A crystalline defect is any region where the microscopic arrangement of atoms differs from that of a perfect crystal. Defects are classified into point defects, line defects, and surface defects, depending on whether the imperfect region is bounded in three, two or one dimensions, respectively. There is a very large variety of defects in solids, and for many of such imperfections their presence is a general thermal equilibrium phenomenon. We will focus on only two important kinds of defects, namely dislocations and substitutional impurities. The latter are point defects, which particularly govern electronic and optical properties of semiconductors and are studied in Chap. 10. In the following some basic properties of dislocations are outlined. More details are found in, e.g., [65, 66]. Examples are particularly considered for the fcc and hcp structures and their related important zincblende and wurtzite structures.

2.3.1 Edge and Screw Dislocations

Dislocations are line defects which are present in virtually any real crystal. Dislocation densities in actual crystals range typically from 10^2 cm^{-2} in good semiconductors to 10^9 cm^{-2} in metals or highly defective semiconductors. The vector along the dislocation line, i.e., along the core of the dislocation, is called *line vector* **l**. A dislocation is characterized by **l** and the *Burgers vector* **b**. **b** is the dislocation-displacement vector and determined by a closed path around the dislocation core, the Burgers circuit. The procedure to find **b** is illustrated in Fig. 2.18. The Burgers vector completes the path around the dislocation line with respect to a similar path within a perfect reference crystal. According to the *right-hand screw convention*, the Burgers circuit is formed *clockwise* around the dislocation line, when looking in the positive sense of the line vector (in Fig. 2.18 **l** points into the plane of projection; in many structures an arbitrariness in the sense of **l** cannot be avoided). If the path incloses more than one dislocation, the resulting Burgers vector is given by the sum of the Burgers vectors of all single dislocations.

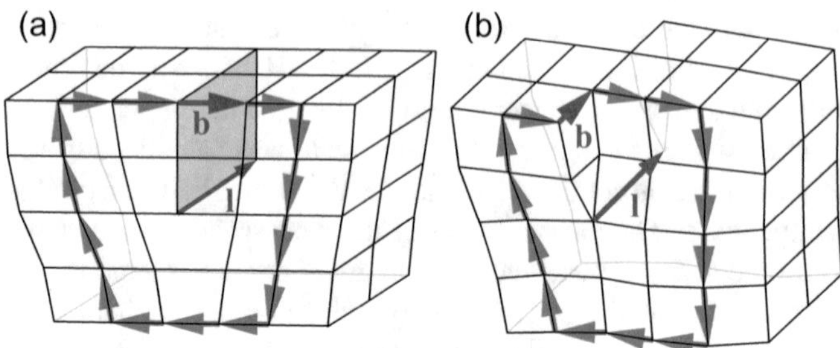

Fig. 2.18 **a** Edge dislocation and **b** screw dislocation. **b** and **l** denote the Burgers vector and the dislocation-line vector, respectively

There are two basic types of dislocations, edge dislocations and screw dislocations. In an *edge dislocation* **b** is perpendicular to **l**. Such a dislocation may be formed by the insertion of an extra half-plane spanned by **l** and $\mathbf{b} \times \mathbf{l}$, see Fig. 2.18a. In a *screw dislocation* **b** is parallel to **l**. This kind of dislocation is built by a shift of one part of the solid by an amount **b** as shown in Fig. 2.18b. Most dislocations occurring in solids are of mixed character with an edge component and a screw component. They are generally denoted by specifying the angle between **b** and **l**. Pure edge dislocations obey $\mathbf{b} \cdot \mathbf{l} = 0$, for pure screw dislocations $\mathbf{b} \cdot \mathbf{l} = b$ (for a right-handed screw, $-b$ otherwise). The Burgers vector along a dislocation line is constant, if the dislocation line is a straight line; consequently the type of dislocation does not change in this case. However, the type of dislocation and hence also the angle between Burgers vector and line vector change if the dislocation line bends.

2.3.2 *Motion of Dislocations*

Geometrical considerations show that a dislocation line can neither begin nor end within the crystalline solid [65]. A dislocation line therefore either forms a closed loop within the crystal, or it begins and ends at an interface or grain boundary of the crystal. If a force acts on the crystal, the dislocation line can move along specific slip planes through the crystal. This process requires the breaking of bonds and is hence thermally activated. Since the position of individual atoms changes only a fraction of the lattice constant during such movements, the required energy is quite small. The presence of dislocations thereby accounts for the fact that real solids withstand much smaller shear strengths against slipping of atomic planes than perfect crystals would do (factor $\sim 10^{-4}$).

Two kinds of dislocation movements are distinguished: Glide and climb processes. They are illustrated in Fig. 2.19. During *gliding* the dislocation moves in the direction of the Burgers vector by turning crystal planes as illustrated in Fig. 2.19a. Such shear displacement is termed glide when it is produced by a single dislocation, and slip

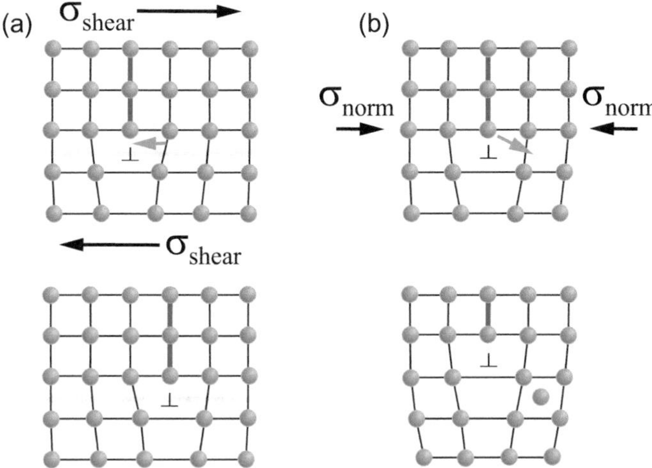

Fig. 2.19 Edge dislocation performing **a** a gliding process and **b** a climbing process. σ_{shear} and σ_{norm} indicate shear and normal stresses acting on the solid, the *yellow line* in (**a**) indicates the glide plane

when it is produced by a number of dislocations. The total number of atoms and lattice sites is conserved in such motions. For pure edge dislocations the process can only occur along *slip planes* which contain both the Burgers vector and the dislocation line. Pure screw dislocations can glide along any plane, since **l** and **b** are parallel. *Climbing* is a motion out of the glide plane; it occurs within a plane, which contains the dislocation line but is perpendicular to the Burgers vector. Climbing is accompanied by a material transport, i.e., emission or absorption of interstitials or vacancies (point defects) moving to or from the dislocation core, as indicated in Fig. 2.19b. The symbol to represent a general dislocation is $\perp$. For an edge dislocation the upwards pointing arm of the inverted T points to the direction of the added or removed material.

The total geometry of a dislocation is specified by the glide plane in addition to the Burgers vector and the line vector; this set is referred to as *slip system*. Slip planes in a crystal are usually planes with closest packing of atoms; they have largest separation. In zincblende semiconductors these are $\{111\}$ planes, see Fig. 2.3.

Dislocation lines in real crystals are not straight lines; they contain short segments, where the dislocation line is displaced on an atomic scale. A *kink* is formed when part of the dislocation is shifted within the same glide plane by one lattice plane from its original position (Fig. 2.20a, b). Such a kink does not impede glide of the dislocation. If the dislocation line moves from one atomic slip plane to another a *jog* is formed (Fig. 2.20c, d). Jogs are segments of the dislocation line that have a component normal to the glide plane. If this sement extends over more than one interplanar spacing it is termed *superjog*. The jog on a screw dislocation has an edge component; it can only glide in the plane containing the dislocation line before and

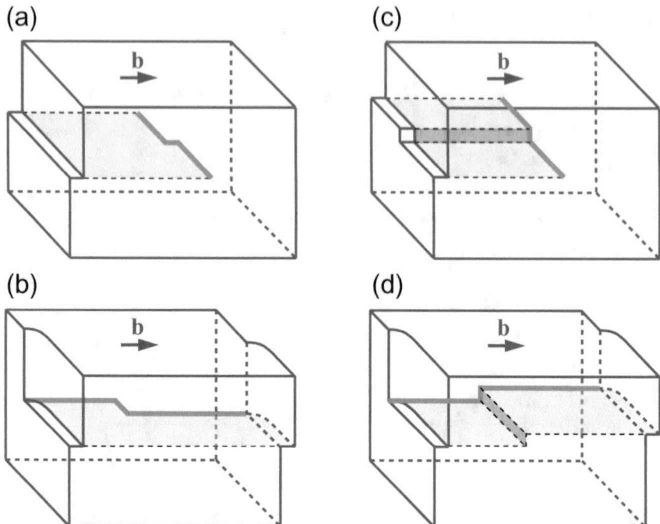

Fig. 2.20 Dislocation line with a kink (**a**, **b**) and a jog segment (**c**, **d**) on a scew (**b**, **d**) and an edge dislocation (**a**, **c**). The glide plane is marked in *yellow*

after the jog (paper plane in Fig. 2.20d). Both kinks and jogs can nucleate in pairs on an initially straight dislocations line. Such pairs (of opposite sign) are formed by thermal fluctuations in the crystal and can initiate movement of the dislocation line.

If the last line of atoms along an edge dislocation is removed, e.g., by diffusion to the crystal surface, then the dislocation has climbed by one atomic spacing. A climb is a motion out of the glide plane. However, the diffusion of only a few atoms from the edge is more likely. This results in a climb of only a fraction of the dislocation, with a jog to the undisturbed part. Climb by nucleation of jog pairs is equivalent to dislocation motion in perpendicular direction by nucleation of kink pairs. Widening of the distance between jog pairs usually requires diffusion of the interspacing line of atoms to the surface.

2.3.3 Dislocation Network

Section 2.2 outlined how a strained epitaxial layer of a pseudomorphic heterostructure may relax strain energy by introducing a misfit dislocation. The process of this plastic strain relaxation is now considered in more detail. To reduce the tensile or compressive strain in an overcritical epilayer with thickness $t > t_c$, the introduction or omission of a lattice plane is favorable, respectively, creating a dislocation line at the layer/substrate interface. Since the dislocation line can neither begin nor end within a crystal, its ends must lie at the surface. There exist two possibilities which fulfill this condition [42]. One mechanism is based upon a dislocation line with a suitable

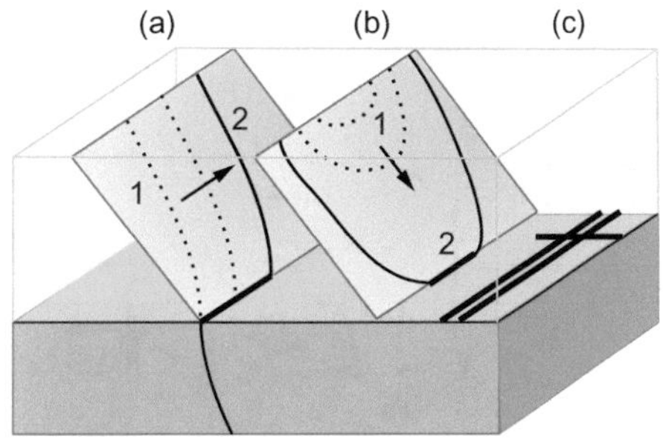

Fig. 2.21 Generation of a misfit dislocation network (**c**) at the interface between layer (*upper part*) and substrate (*lower part, blue*) from (**a**) a preexisting threading dislocation of the substrate and (**b**) from the nucleation of a dislocation half loop

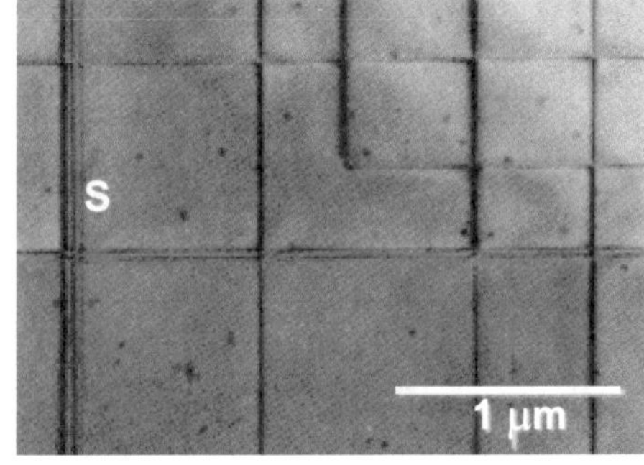

Fig. 2.22 Plan view transmission-electron micrograph of a 2 μm thick Si layer on (001)-oriented GaP substrate. The parallel and closely spaced fringes denoted *S* originate from a stacking fault. Reproduced with permission from [67], © 1987 AIP

Burgers vector already existing in the substrate and terminating at its surface. As illustrated in Fig. 2.21a the dislocation is replicated in the layer and forms a *threading dislocation*, i.e. a dislocation penetrating the layer. Under the action of coherency strain the dislocation line bends and glides along the interface (from position 1 to 2). Thereby a strain-relaxing misfit segment of the dislocation line is created at the interface. Another mechanism is the nucleation of a *dislocation half loop* at the layer surface, illustrated in Fig. 2.21b. The half loop represents the border of an extra plane (in tensely strained layers), which expands and glides towards the interface over slip planes. Both mechanisms lead to the formation of a *dislocation network* at the interface (Fig. 2.21c). The average spacing p of the parallel line segments at the interface is given by (2.33).

The formation of a dislocation network at the layer/substrate interface of incoherent (plastically relaxed) epilayers may be observed in plan view micrographs due to the accompanying strain field at the dislocation core. An example for an epitaxial layer with 0.5% natural misfit at growth temperature is given in Fig. 2.22.

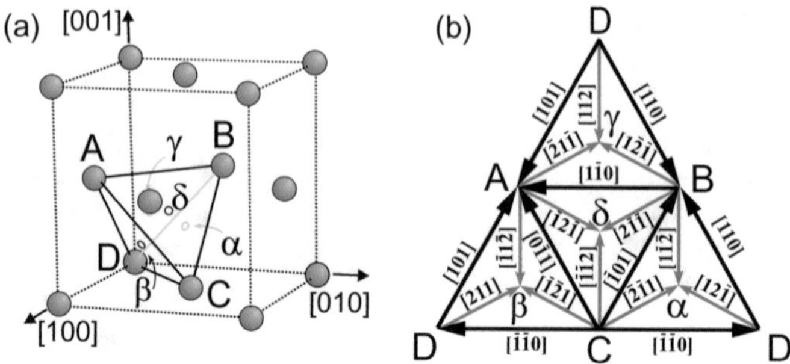

Fig. 2.23 Thompson's reference tetrahedron **a** in the fcc unit cell and **b** in a planar development showing the different crystallographic directions

2.3.4 *Dislocations in the fcc Structure*

The face-centered cubic and hexagonally close-packed structures are the two most common structures for metallic elements and—as constituents of the related zincblende and wurtzite structures—also for important semiconductors (Sect. 2.1). For these structures we consider simple reference schemes to denote Burgers vectors of possible dislocations and stacking faults. Figure 2.23 represents *Thompson's tetrahedron* used for fcc structures [68]. The side faces of the tetrahedron are the four equivalent $\{111\}$ faces, which lie parallel to the corresponding close-packed planes of the fcc structure and represent the possible glide planes. The edges point towards $\langle 110\rangle$ directions and correspond to the six glide directions denoted AB, etc. They represent vectors of the type $\frac{1}{2}a\ \langle 110\rangle$ with length $a/\sqrt{2}$, which are primitive translations of the fcc structure. Dislocations with such a Burgers vector are termed *perfect dislocations*. Vectors from a corner to the midpoint of a side facet are of the type $\frac{a}{6}\langle 112\rangle$ and are labeled $A\beta$, etc. Dislocations with a Burgers vector which does *not* correspond to primitive translations (like $A\beta$) are referred to as *partial dislocations*.

Partial dislocations lead to a change in the ABC stacking order of the $\{111\}$ planes of the fcc structure. The resulting planar defects are referred to as *stacking faults*. Possible Burgers vectors of dislocations in the fcc structure and the notation of the corresponding dislocation type are given in Table 2.8. Notations with two pairs represent the sum vector, e.g., $DB/CA \equiv DC + BA$ yields [100].

The energy of a dislocation is generally proportional to the square of the absolute value of its Burgers vector. A dislocation with Burgers vector $\mathbf{b}$ can, therefore, lower its strain energy and divide into two (or more) partial dislocations with Burgers vectors $\mathbf{b}_1$ and $\mathbf{b}_2$, if *Frank's rule* is satisfied:

$$|\mathbf{b}_1|^2 + |\mathbf{b}_2|^2 < |\mathbf{b}|^2. \tag{2.30}$$

The common validity of Frank's rule originates from the weak dependence of the line energy of a dislocation on its character (edge or screw type or mixed). In any case

Table 2.8 Burgers vectors and types of dislocations in the fcc structure, a is the lattice constant of the unit cell

Burgers vector (Thompson's notation)	Burgers vector (crystallographic notation)	Type of dislocation	$\|\mathbf{b}\|^2$ ($\times 36/a^2$)
AB	$\frac{1}{2}a\langle 110\rangle$	Perfect	18
$A\alpha$	$\frac{1}{3}a\langle 111\rangle$	Frank partial	12
$A\beta$	$\frac{1}{6}a\langle 112\rangle$	Shockley partial	6
$\alpha\beta$	$\frac{1}{6}a\langle 110\rangle$	Stair-Rod partial	2
$\delta\alpha/CB$	$\frac{1}{3}a\langle 100\rangle$	Stair-Rod partial	4
$\delta D/C\gamma$	$\frac{1}{3}a\langle 110\rangle$	Stair-Rod partial	8
$\delta\gamma/BD$	$\frac{1}{6}a\langle 013\rangle$	Stair-Rod partial	10
$\delta B/D\gamma$	$\frac{1}{6}a\langle 123\rangle$	Stair-Rod partial	14

$E_{\text{dislocation}}/\text{length} \propto \text{const} \times Gb^2$, where G is the shear modulus and the constant differs by less than a factor of 2.

Only dislocations with the shortest Burgers vectors are stable. Perfect dislocations with shortest Burgers vectors in the fcc structure are of the type $\frac{1}{2}a\langle 011\rangle$. Such a dislocation tends to dissociate into two (stable) partial dislocations with shorter Burgers vectors. From Thompson's tetrahedron Fig. 2.23 we read, e.g.,

$$AD = A\beta + \beta D, \quad \text{or}$$
$$\frac{a}{2}[\bar{1}0\bar{1}] = \frac{a}{6}[\bar{1}0\bar{2}] + \frac{a}{6}[\bar{2}\bar{1}\bar{1}].$$

This dislocation reaction describes the dissociation of a perfect dislocation into two energetically favorable Shockley partial dislocations, cf. Table 2.8. The process is illustrated in Fig. 2.24. The dislocation line of the perfect dislocation AD splits into two lines of the partial dislocations. Since their Burgers vectors are not lattice vectors, they lead from a lattice site to a crystallographic not equivalent site: They produce a stacking fault. The two dislocation lines border a strip of stacking faults which keeps

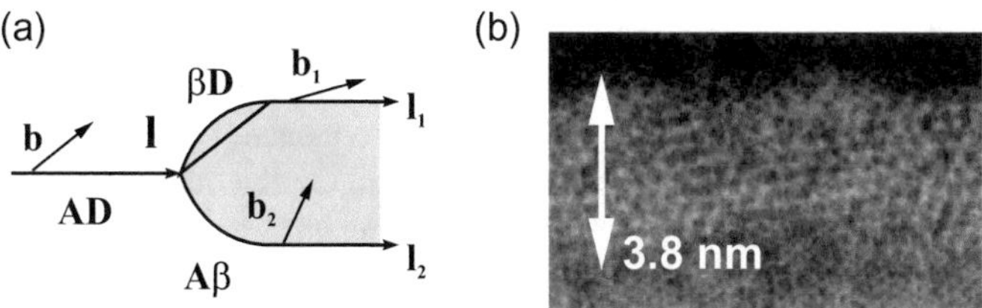

Fig. 2.24 **a** Dissociation of a perfect dislocation into two Shockley partial dislocations. **b** High-resolution transmission electron micrograph of a stacking-fault ribbon in Ge bounded by two Shockley partial dislocations, the *double arrow* indicates the stacking fault width expected from the stacking fault energy. Reproduced with permission from [69], © 1998 Elsevier

the partial dislocations together at some equilibrium distance and forms a so-called extended dislocation. In the stacking-faulted region the regular $ABCABC$ stacking order is changed by removal or insertion of one layer, e.g., to $ABABC$. The three equivalent $\langle 110 \rangle$ directions and four equivalent $\{111\}$ planes of a cubic semiconductor crystal yield 12 slip systems, collectively denoted $\frac{a}{2}\langle 110 \rangle \{111\}$. Dislocations on these slip systems comprise pure edge, pure screw, and mixed 60° dislocations.

2.3.5 Dislocations in Diamond and Zincblende Structures

The diamond and zincblende structures result from the fcc structure by adding to each atom on the ABC-stacked fcc lattice one atom of another type in the distance $\frac{a}{4}[111]$, i.e., a corresponding abc-stacking. This yields a total $AaBbCc$ stacking, with different kind of atoms (charges) on the two sublattices in zincblende (Sect. 2.1). Dislocations in semiconductors which crystallize in diamond or zincblende structures are therefore also described using Thompson's tetrahedron. Mismatched layers of such semiconductors generally show larger values of the critical thickness for plastic relaxation and slower relaxation of the elastic strain than metal films. This is due to a lower mobility of dislocations in semiconductors and a consequently reduced length of misfit segments along the interface. Furthermore, the larger perfection of semiconductor substrates necessitates the nucleation of new dislocations, instead of, as in metal films, glide of preexisting ones or heterogeneous nucleation at precipitates.

The three main types of perfect dislocations in diamond and zincblende semiconductors with $\frac{a}{2}\langle 110 \rangle$ Burgers vectors are edge dislocations, screw dislocations, and 60°-mixed dislocations. The 60°-mixed type is the most common perfect dislocation, because the edge type has a high core energy and the screw type cannot relax tetragonal mismatch which arises from (001)-oriented heterostructure growth. The screw dislocation and 60°-mixed dislocation in zincblende structure are illustrated in Fig. 2.25. The respective dislocations of the diamond structure are obtained when all atoms are considered to be equal.

There are two different sets of dislocations due to the $AaBbCc$ double-layer atomic arrangement. Dislocations of the *shuffle set* (or type I) have glide planes lying between layers of the same index, e.g., Aa. Gliding of shuffle-set dislocations requires breaking of *one* bond per atom. Dissociation of perfect dislocations of such set into partials is not favorable, because it would produce a stacking fault of the type $AaBbC|bCcAa$ with a high energy of the CbC sequence. Furthermore, the partial dislocations cannot glide like a perfect dislocation. On the other hand, dislocations of the *glide set* (or type II) dissociate into partials which are glissile by, e.g., forming a stacking fault of the type $AaBb|AaBbCc$. Gliding of perfect glide-set dislocations requires breaking of *three* bonds per atom; they are much more mobile than shuffle-set dislocations, though. Dissociation and interactions of these dislocations are identical to those of the fcc structure. A glide-set dislocation can transform to the shuffle set and vice versa by climb. Dislocations in the zincblende structure require a further distinction with respect to those in diamond structure, because the dislocation line

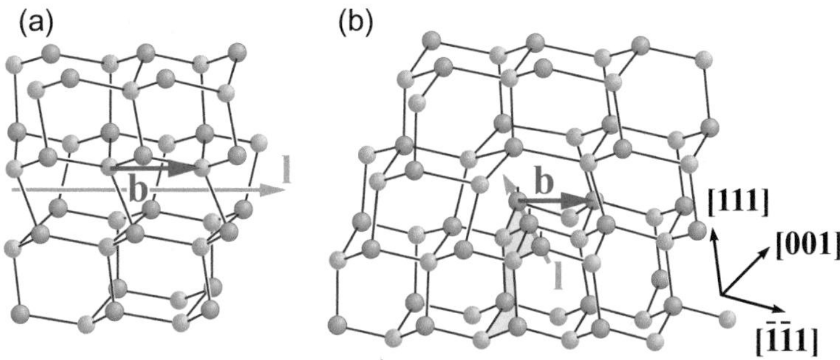

Fig. 2.25 Perfect dislocations in the zincblende structure: **a** Screw dislocation, **b** 60° dislocation. The [001] direction shown in (**b**) lies in the plane spanned by [111] and [$\bar{1}\bar{1}1$]

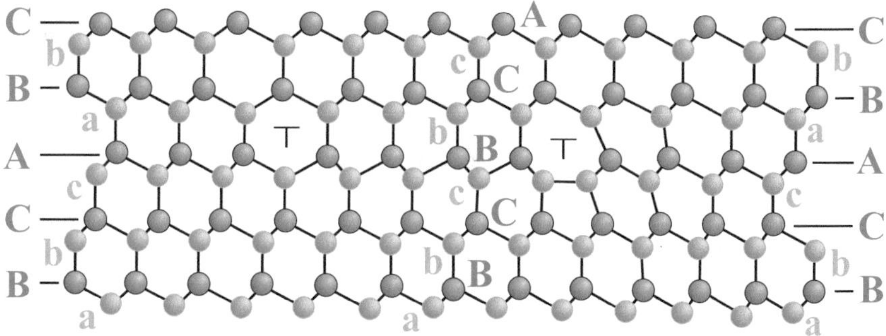

Fig. 2.26 Dissociated perfect dislocation with a bounded stacking fault in zincblende structure viewed along [$1\bar{1}0$]

may either lie on a row of anions or cations. Dislocations with l parallel to the [$1\bar{1}0$] direction are termed α type. Their core comprises cations in the shuffle set and anions in the glide set. Accordingly, β type dislocations have a dislocation line along the [110] direction with anions in the shuffle set and cations in the glide set.

The dissociation of a perfect zincblende 60° dislocation with type II glide plane into an extended dislocation of a stacking fault bounded by two Shockley partial dislocations is shown in Fig. 2.26. This scheme corresponds to the stacking fault ribbon imaged in Fig. 2.24, if all atoms are considered to be of the same kind to obtain the diamond structure.

2.3.6 Dislocation Energy

Near a dislocation atoms are displaced from their equilibrium positions which they occupy in a perfect crystal. A corresponding cost of elastic energy is required to produce the dislocation and is referred to as dislocation energy E_D. To obtain an estimate for E_D we first consider the simple case of a *screw dislocation* in an isotropic continuum forming a coaxial cylinder as depicted in Fig. 2.27 [65]. The dislocation is produced by a shear displacement in the z direction across the xz glide plane and assumed to increase uniformly with the angle φ.

Displacements of this dislocation are given by $u_x = u_y = 0$, and $u_z = b \times \varphi/(2\pi) = b/(2\pi) \times \tan^{-1}(y/x)$. The only nonzero stresses of such displacement are $\sigma_{xz} = -Gb/(2\pi) \times y/(x^2 + y^2)$, and σ_{yz} with y in the enumerator being replaced by $-x$. G is the shear modulus and b the length of the Burgers vector. In polar coordinates the stress reads $\sigma_{\varphi z} = Gb/(2\pi r)$, and the strain resulting from Hooke's law is $\varepsilon_{\varphi z} = \varepsilon_{z\varphi} = \sigma_{\varphi z}/(2G) = b/(4\pi r)$. The strain field has a pure shear character and decays with $1/r$ from the dislocation line **l**. The elastic energy per length L follows from the integration of the elastic energy density $E/V = 2G\varepsilon_{\varphi z}^2$, yielding

$$\frac{E_{\mathrm{screw}}^{\mathrm{isotropic}}}{L} = \int_{r_c}^{R} 2G\varepsilon_{\varphi z}^2 r dr d\varphi = \frac{Gb^2}{4\pi} \ln(R/r_c). \tag{2.31}$$

The lower bound r_c denotes the radius of the dislocation core. At a distance r from the center of the dislocation below a value r_c the linear elastic theory cannot be applied. A somewhat arbitrary cutoff parameter ρ is often used to account for the non-linear elastic energy and dangling bonds in the core, where $r_c = b/\rho$ with $\rho \approx 2, \ldots, 4$. Sometimes in literature instead of using ρ a constant is added to the logarithm term of (2.31), yielding $(\ln(R/b) + \mathrm{const})$ instead of $\ln(\rho R/b)$; const $= 1 (= \ln(e))$ is then usually applied. The divergence of $E_{\mathrm{screw}}^{\mathrm{isotropic}}/L$ for an infinite upper bound R indicates that the dislocation does not have a specific characteristic energy, but depends on the size of the solid. An appropriate choice for R is the shortest distance of the dislocation line to the surface or, in case of a solid with many dislocations of both signs, roughly their mutual average distance.

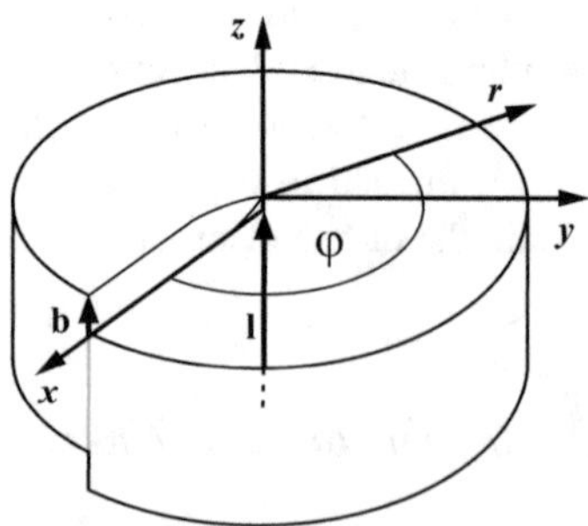

Fig. 2.27 Screw dislocation along the axis of a cylinder

The strain field of an *edge dislocation* comprises all kind of strains. Considering the edge dislocation Fig. 2.18a with an inserted lattice plane above the line vector **l**, the strain field has dominant compressive and tensile hydrostatic components above and below **l**, respectively. At the right and left hand side of **l** we find predominantly shear strains of opposite sign, and on the diagonals compressive and tensile axial components. In isotropic media the tensile stress in radial direction σ_{rr} and that along the circumference $\sigma_{r\varphi}$ should be proportional $1/r$ and change sign by exchanging x and y. A continuum-mechanical deduction of the stress field of the edge dislocation yields $\sigma_{rr} = \sigma_{\varphi\varphi} = -Gb/(2\pi(1-\nu)) \times \sin\varphi/r$, and $\sigma_{r\varphi} = Gb/(2\pi(1-\nu)) \times \cos\varphi/r$ for the shear stress [65]. Integration leads to the elastic energy per length $E_{\text{edge}}^{\text{isotropic}}/L$, which, apart from an additional factor $1/(1-\nu)$, has the same form as that of the screw dislocation (2.31). The considerations likewise apply for anisotropic media, where the particular geometry examined introduces additional terms.

In a dislocation with *mixed character* the strain fields of the screw and edge components superimpose. We consider the example of the prominent 60° dislocation in crystals with diamond or zincblende structure, introduced at a (001) interface between a substrate and a layer to accommodate misfit strain. The angle α between line vector and Burgers vector enters the elastic energy per unit length, yielding [40]

$$\frac{E_\alpha}{L} = \frac{Gb^2}{4\pi}\left(\cos^2\alpha + \frac{\sin^2\alpha}{(1-\nu)}\right)\ln\left(\frac{\rho R}{b}\right). \tag{2.32}$$

To evaluate the strain energy E_{D} of *all* dislocations in a layer, their total number has to be considered. Figures 2.21 and 2.22 illustrate that the strain in a layer which exceeds the critical thickness is actually accommodated by a *network* of dislocations, which in many cases form a more or less regular grid at the interface. The average spacing p in such an array of parallel misfit dislocations is related to the actual misfit f. We consider the strongly mismatched, relaxed heterostructure depicted in Fig. 2.28 to obtain an expression for p. In Fig. 2.28 the lateral distance p comprises n planes of the substrate and $(n+1)$ planes of the layer (here $a_{\text{L}} < a_{\text{S}}$). The lateral spacing between the depicted planes of the substrate is then given by $a_{\text{S}} = p/n$, and that of the layer planes is $a_{\text{L}} = p/(n+1)$. Taking the definition of the misfit f from (2.20a) we obtain for $a_{\text{L}} < a_{\text{S}}$

$$f = \frac{a_{\text{S}} - a_{\text{L}}}{a_{\text{L}}} = \frac{\frac{p}{n} - \frac{p}{n+1}}{\frac{p}{n+1}} = \frac{1}{n}.$$

If $a_{\text{L}} > a_{\text{S}}$ there are only $n-1$ layer planes for n substrate planes, yielding $a_{\text{L}} = p/(n-1)$. Insertion into (2.20a) then leads to $f = -1/n$ $(f < 0)$. The spacing p of the parallel misfit dislocations is hence

$$p = n a_{\text{S}} = \frac{a_{\text{S}}}{|f|}. \tag{2.33}$$

The inverse of p is the (linear) density of the dislocations illustrated in Fig. 2.28.

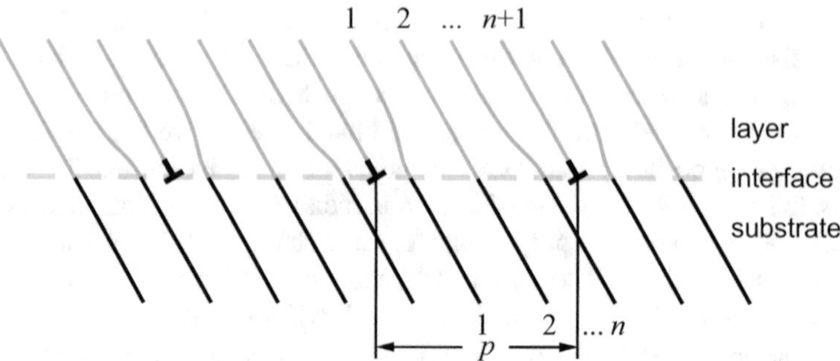

Fig. 2.28 Equally spaced dislocations accommodating the strain in the layer of a mismatched, relaxed heterostructure. The *lines* represent lattice planes which lie perpendicular to the image plane and contain the dislocation line. Lattice planes in substrate and layer are drawn *black* and *gray*, respectively

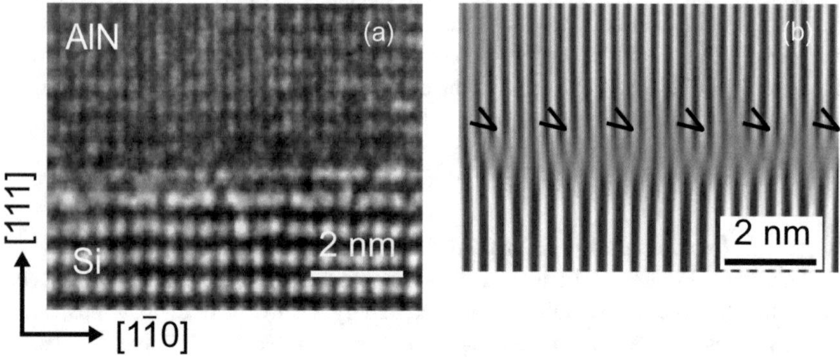

Fig. 2.29 Equally spaced step dislocations at the interface between an AlN(0001) epilayer and the Si(111) substrate. **a** High-resolution transmission-electron micrograph taken along the [11 −2] axis of Si. **b** The fast-Fourier filtered image shows the Si(−110) and the AlN(−2110) net planes, respectively. Reproduced with permission from [70], © 1999 Elsevier

A dense *regular* spacing of dislocations may occur, if a large lattice mismatch f between epilayer (lattice spacing d_L) and substrate (d_S) is accommodated by a *coincidence lattice* fulfilling the condition $m \times d_S \approx n \times d_L$ with small integer numbers m and n. The remaining mismatch is in this case $f = (m \times d_S - n \times d_L)/n \times d_L$. An example is given in Fig. 2.29. The lateral lattice constant of the AlN epilayer ($d_L = 3.11$ Å) is much smaller than that of the (111)-oriented Si substrate ($d_S = 5.43$ Å$/\sqrt{2} = 3.84$ Å), yielding a natural misfit $(d_S - d_L)/d_L = 0.23$. By introducing one edge dislocation into each unit cell of the coincidence lattice on the epilayer side with a ratio $m : n = 4 : 5$ the mismatch is reduced to $f = -0.01$ [70]. The edge dislocation was more recently considered being composed of the edge components of two 60° dislocations with $\frac{1}{2}\langle 10\bar{1}0\rangle$ edge component each and opposite screw components [71]; there are three such equivalent 60° dislocations with Burgers vectors of type $\frac{1}{3}\langle 11\bar{2}0\rangle$ lying in the (0001) basal plane. These dislocations create a regular triangular network of misfit dislocations at the AlN/Si(111) interface. Coincidence relations play a prominent role in organic epitaxy discussed in Sect. 3.2.1.

The elastic dislocation energy E_D per area A of a network of dislocations at the interface is

$$\frac{E_D}{A} = \frac{2}{p}\frac{E_{disloc}}{L}. \tag{2.34}$$

The factor 2 accounts for the two independent lateral directions assumed here to be equivalent, and E_{disloc}/L is the energy per length of one dislocation, like, e.g., (2.32). If the degree of relaxation is anisotropic, as was observed for, e.g., 60° dislocations in zincblende type layers along the [110] and $[1\bar{1}0]$ directions, (2.34) splits into a sum with *two* periods p_i and two related $E_{disloc,i}/L_i$. An explicit relation of (2.34) also comprises the angle β between the glide plane containing the Burgers vector and the interface. β is related to the angle α between line vector $\mathbf{l}$ and Burgers vector $\mathbf{b}$, and also to the substrate plane spacing a_S according to

$$a_S = b \sin\alpha \cos\beta. \tag{2.35}$$

Using (2.32), (2.33), and (2.35) yields a general relation for the dislocation energy of a layer containing a dislocation network [40],

$$\frac{E_D}{A} = bG\frac{|\varepsilon + f_0|(1-\nu\cos^2\alpha)}{2\pi(1-\nu)\sin\alpha\cos\beta}\ln\left(\frac{\rho R}{b}\right). \tag{2.36}$$

The geometrical parameters for the 60° dislocation are the length of the Burgers vector $b = a/2 \times [101]$, the angle between Burgers vector and line vector $\alpha = 60°$, and the angle between the {111} glide planes and the considered (001) interface plane $\beta = 54.7°$. The critical thickness of a strained epitaxial layer accommodated by such misfits is given by the implicit relation (2.28), the solution of which is depicted in Fig. 2.15.

2.3.7 Dislocations in the hcp and Wurtzite Structures

In the *hcp structure* the triangular bipyramid given in Fig. 2.30 is a widely used reference construction for classifying dislocations [77]. In this representation basal-plane lattice vectors are labeled $AB = \mathbf{a}_1$ $(= \frac{1}{3}[\bar{1}2\bar{1}0])$, $BC = \mathbf{a}_2$, and $CA = \mathbf{a}_3$. The lattice vector $\mathbf{c}$ (= [0001]) is given by TS, the line $\sigma S = T\sigma = \frac{1}{2}\mathbf{c}$ does not represent a lattice vector. Further vectors of interest are $TS + AB = \mathbf{c} + \mathbf{a}_1 = \frac{1}{3}[\bar{1}2\bar{1}3]$, and $A\sigma = \frac{1}{3}[01\bar{1}0]$.

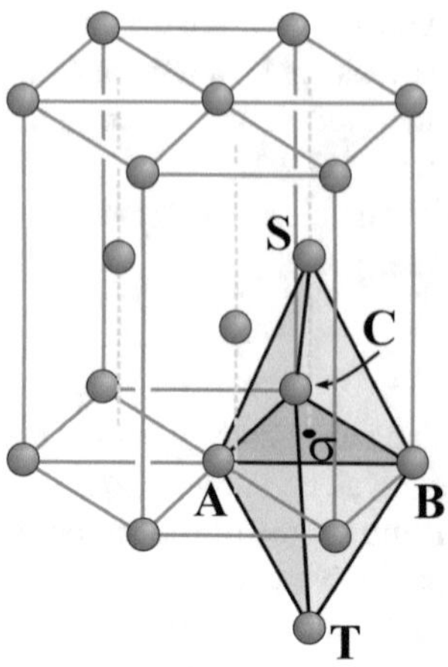

Fig. 2.30 Triangular bipyramid used as reference system to describe dislocations in the hcp structure

In crystals with hcp structure the close-packed basal planes are the most frequently observed glide planes. Dislocations with a slip vector of the type AB operating on these (0001) planes are therefore of particular importance. For vectors lying in the basal plane it is sufficient to use the base of the bipyramid as a reference system. As an example we consider the dissociation of a perfect glide dislocation with vector AB into Shockley partial dislocations. From the reference bipyramid we read

$$AB = A\sigma + \sigma B. \tag{2.37}$$

The partial dislocation $A\sigma$ changes the layer sequence from the ordinary *abab* stacking to a *bcbc* sequence as illustrated in Fig. 2.31. The second partial σB in (2.37) restores the initial order. Thereby a stacking-fault ribbon bound by two partials is created, comparable to the case shown in Figs. 2.24 and 2.26. Another kind of such a ribbon is formed from the same perfect glide dislocation if the order of the partial dislocations is reversed, i.e.

$$AB = \sigma B + A\sigma. \tag{2.38}$$

In this case the ribbon contains layers stacked in an *acac* sequence, cf. Fig. 2.31b. Both kinds of ribbons have the same energy, although they are crystallographically distinguishable.

The *wurtzite structure* is formed from the hcp structure by adding to each atom a further atom of another kind, shifted by $\frac{3}{8}c$ along [0001], cf. Sect. 2.1. Dislocations in the resulting $AaBbAaBb$ stacking are described in the same framework as those of the hcp structure. The basic edge and screw dislocations in a wurtzite crystal are shown in Fig. 2.32. Panel (a) shows a high-resolution transmission-electron micrograph of a pure edge dislocation with a Burgers vector of $\frac{1}{3}[11\bar{2}0]$, determined from the Burgers cycle marked by open circles. The inserted plane is perceived by viewing the image at a glancing angle from the lower left corner. This type of threading

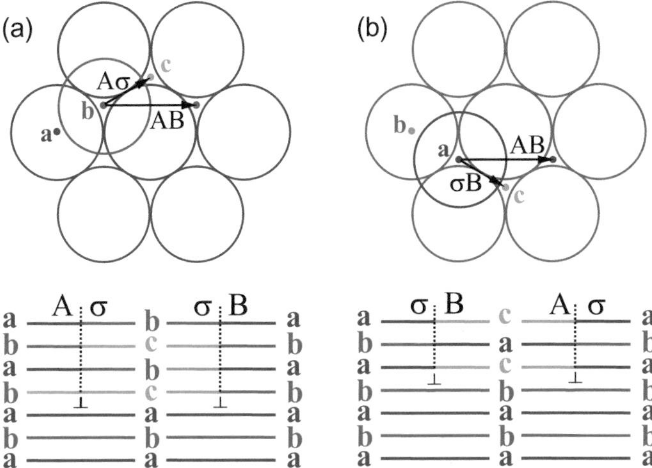

Fig. 2.31 Dissociation of a perfect dislocation AB in a hcp structure into two partial dislocations $A\sigma$ and σB. **a** Displacement of an atom in the b layer (*red*) on *top* of the a layer (*blue circles*) by $A\sigma$. **b** Displacement of an atom in the a layer on *top* of the b layer by σB. The stacking sequence of the ribbons bounded by the two partial dislocations is illustrated in the *bottom*

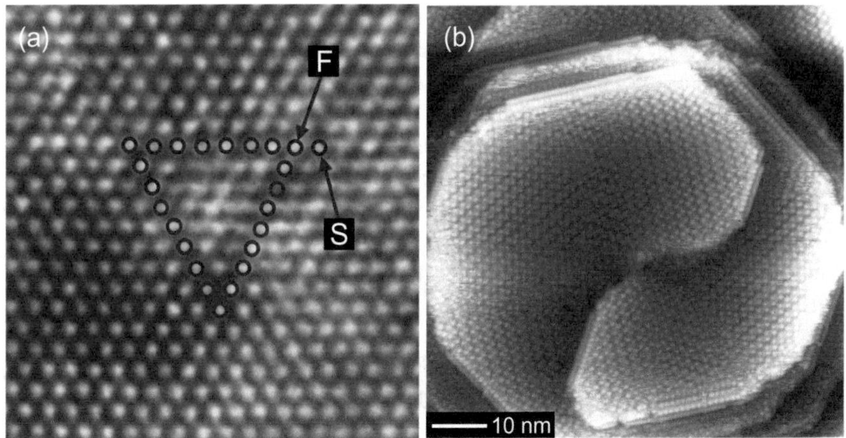

Fig. 2.32 **a** Plan view transmission-electron micrograph of a pure edge dislocation in wurtzite GaN. S and F mark begin and end of a Burgers cycle which is indicated by *open circles*. Reprinted with permission from [78], © 2000 APS. **b** Scanning tunneling micrograph of a screw dislocation on the N-face of GaN. Reprinted with permission from [79], © 1998 AVS

dislocation is most common in GaN layers epitaxially grown on (0001) sapphire or on (111) Si.

Screw dislocations can provide a steady source of kinks on a growing crystal surface by producing kink sites and may thereby lead to high growth rates. Figure 2.32b shows an STM image of a pure screw dislocation at a GaN surface (N-face). The measured step heights at the two spiral growth fronts are each one GaN bilayer, yielding a Burgers vector of $c[000\bar{1}]$ for this dislocation.

2.3.8 Antiphase Domains

An antiphase domain, also referred to as *inversion domain*, is a region in a polar crystal where the atoms are located on regular lattice sites but in the opposite order with respect to the undisturbed crystal. In the domains of zincblende or wurtzite crystals anions occupy cation sites and vice versa. Such a disorder nucleates particularly at the heterointerface of a polar semiconductor grown on a nonpolar semiconductor. Here the antiphase domain is created during growth at monoatomic steps on the nonpolar semiconductor [72]: if growth starts with an anion layer, initial growth on a one monolayer higher region yields an inequivalent crystal orientation as illustrated in Fig. 2.33. Boundaries between domains of reversed orientation correspond to stacking faults and are also termed inversion-domain boundaries (IDBs). At these boundaries bonds among equally charged ions occur, see the bonds across the IDBs marked in green in Fig. 2.33. Inversion-domain boundaries (IDBs) generally create states in the bandgap. They introduce nonradiative carrier recombination and degrade the performance of electronic and optoelectronic devices. The epitaxy of III–V semiconductors on Si is exceedingly interesting, since the integration of a direct-bandgap III–V device on the indirect-bandgap Si combines Si electronics with efficient photonics applications. Growth issues of this demanding interface are reviewed in [73, 74].

In the epitaxy of a zincblende semiconductor on group-IV semiconductor (e.g., InP on (001)-oriented Si) the IDBs are inclined to the interface; a second monoatomic step may then induce a second IDB and lead to an annihilation of the inversion domain as illustrated in Fig. 2.33 for IDBs on {111} planes; such an annihilation may also occur with IBDs on {110} planes. IBDs have, however, usually irregular shapes and lie only partly on low-index planes, making the annihilation process less efficient. The creation of an antiphase domain can be avoided by double steps on the nonpolar semiconductor [72]; this is illustrated at the right in Fig. 2.33, where no disorder in the layer results from the double step of the substrate. On Si(001) substrate such double steps can be formed by applying a $\sim 6°$ offcut orientation toward a {110} direction for creating single-step terraces and a subsequent thermal treatment for double-step formation [75].

Antiphase domains occurring in epitaxial GaN or AlN on $Al_2O_3(0001)$ do not have such a favorable inclination of their boundaries. They have a hexagonal shape

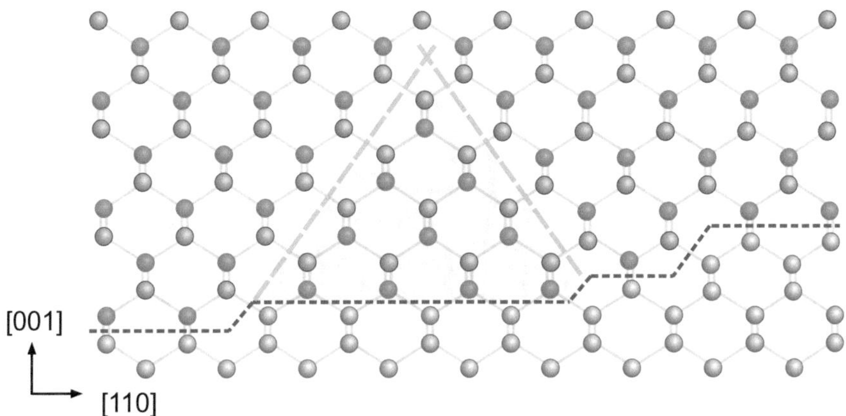

Fig. 2.33 Antiphase domain (*shaded region*) created at a monoatomic step at the interface of a nonpolar to a polar zincblende semiconductor. The *green dashed lines* mark inversion-domain boundaries. No such boundary forms at a double step depicted right. The crystal orientations in the non-shaded and shaded regions are invers

with $\{10\bar{1}0\}$ side walls being parallel to the (0001) growth direction; they hence persist in the entire epitaxial layer [76].

2.3.9 Mosaic Crystal

Faults like the line defects considered in the last few pages exist in virtually any crystal. A periodic pattern of dislocations may form an interface between undistorted regions called low-angle grain boundary as depicted in Fig. 2.34. Such a boundary leads to local tilts or twists of crystal planes and separate regions with perfect crystal structure. The tilt angle Θ of a tilt boundary formed from a linear sequence of edge dislocations as shown in Fig. 2.34 is given by their average distance d and the absolute value of their Burgers vector b. From the figure follows $\tan(\Theta/2) = b/(2d)$ or, since Θ is small, $\Theta = b/d$. There exist also twist boundaries formed from a sequence of screw dislocations. Low-angle grain boundaries are normally composed of a mixture of tilt and twist boundaries.

In practice crystals usually consist of a mosaic of small blocks with undistorted structure. The single-crystalline blocks are typically on the order of a few microns in vertical and lateral dimensions (with respect to the growth plane), and they are randomly slightly misoriented with respect to each other. The finite size of the crystallites limits the coherence of scattered X-ray radiation. The dimensions are therefore called vertical and lateral coherence length, respectively. Out-of-plane rotation perpendicular to the surface leads to a mosaic tilt. In-plane rotation around the surface normal is referred to as mosaic twist. The characteristic parameters of mosaicity are

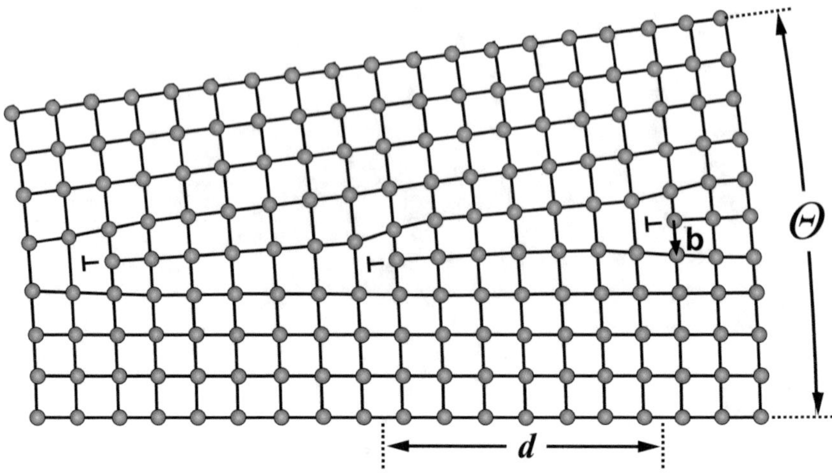

Fig. 2.34 Scheme of a low-angle grain tilt boundary inducing a mutual tilt of adjacent crystalline regions by an angle Θ

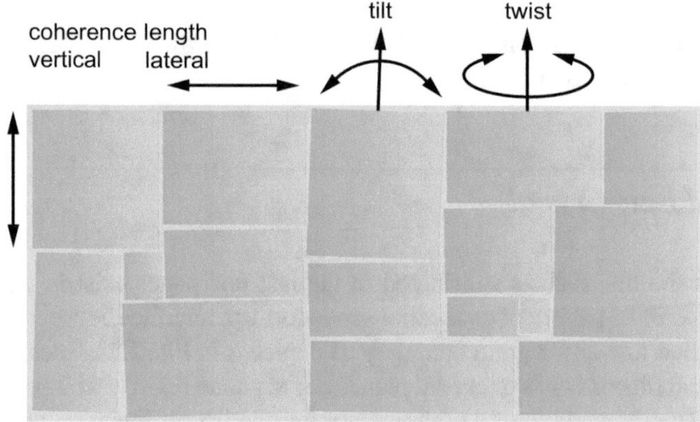

Fig. 2.35 Characteristic parameters of mosaicity in single crystals

illustrated in Fig. 2.35. It should be noted that the distortions are small and the crystal is still considered a single crystal, albeit with some mosaicity. In epitaxial layers with line defects mainly running parallel to the surface normal, the vertical coherence length is commonly related to the layer thickness. The mosaicity observed for ZnSe/GaAs(001) layers is discussed in Sect. 2.4.6. A model of dislocations accounting for the twist found in highly mismatched epilayers such as AlN/Si(111) is reported in [71].

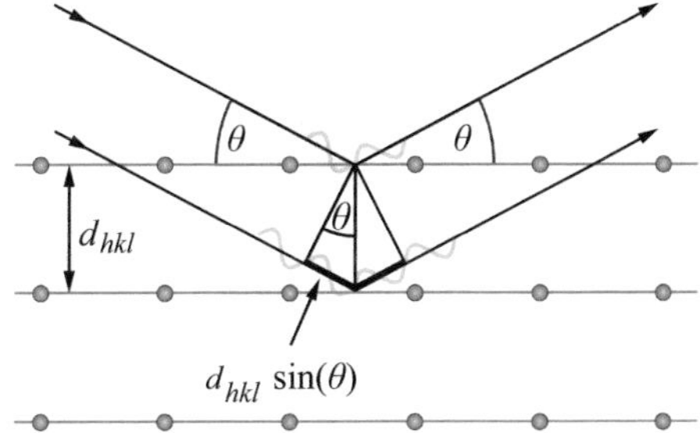

Fig. 2.36 X-ray diffraction of two rays at neighboring lattice planes. For constructive interference the path difference of the diffracted waves $2d_{hkl} \sin \Theta$ must be equal an integral number of wavelengths

2.4 Structural Characterization Using X-Ray Diffraction

X-ray diffraction is a powerful tool for investigating the structural properties of epitaxial layers and is employed in virtually any growth laboratory. Interatomic distances in a solid are typically on the order of some angstrom. A diffraction probe of the microscopic structure must hence have a wavelength at least this short. For an electromagnetic wave 1 Å corresponds to a photon energy $E = hc/\lambda = 12.40$ keV, a characteristical X-ray energy. High resolution X-ray diffraction (HRXRD) is the most commonly applied technique for analyzing lattice parameters, strains, and defects in epitaxial heterostructures. The technique is outlined in this section.

2.4.1 Bragg's Law

The periodic arrangement of atoms in the solid leads to an elastic scattering of impinging X-ray waves. Diffracted waves have definite phase relations among each other. They interfere constructively, if the phase difference is an integral number n of wavelengths λ. This is depicted in Fig. 2.36 for two rays which are diffracted at two parallel lattice planes spaced a distance d_{hkl} apart. According to this picture the waves behave as being reflected at the lattice planes.

The path of the lower ray is $2 \times d_{hkl} \times \sin \Theta$ longer, Θ being the angle of incidence (measured from the diffracting lattice plane). The condition for X-ray reflections was concluded by William L. Bragg and his father William H. Bragg in 1913 from characteristic patterns of X-rays reflected from crystalline solids and reads

$$2d_{hkl} \sin \Theta = n\lambda. \tag{2.39}$$

n is called the order of the reflection. Note that *Bragg's law* (2.39) only defines the *direction* of a diffracted beam (not the intensity), and that it allows reflections only for $\lambda \le 2d_{hkl}$. The lattice-plane distance d_{hkl} and hence Θ depend on shape and size

Table 2.9 Bragg's equations for the seven crystal systems

Crystal system	Angles $\sin(\Theta)$ of Bragg reflections
Cubic	$\sin(\theta) = \frac{\lambda}{2a}\sqrt{h^2+k^2+l^2}$
Tetragonal	$\sin(\theta) = \frac{\lambda}{2a}\sqrt{h^2+k^2+(\frac{a}{c})^2 l^2}$
Rhombohedral	$\sin(\theta) = \frac{\lambda}{2a}\sqrt{\frac{(h^2+k^2+l^2)\sin^2\alpha+2(kl+lh+hk)(\cos^2\alpha-\cos\alpha)}{1-3\cos^2\alpha+2\cos^3\alpha}}$
Hexagonal	$\sin(\theta) = \frac{\lambda}{2a}\sqrt{\frac{4}{3}(h^2+k^2+hk)+(\frac{a}{c})^2 l^2}$
Orthorhombic	$\sin(\theta) = \frac{\lambda}{2a}\sqrt{h^2+(\frac{a}{b})^2 k^2+(\frac{a}{c})^2 l^2}$
Monoclinic	$\sin(\theta) = \frac{\lambda}{2}\sqrt{\frac{h^2}{a^2\sin^2\beta}+\frac{k^2}{b^2}+\frac{l^2}{c^2\sin^2\beta}-\frac{2hl\cos\beta}{ac\sin^2\beta}}$
Triclinic	$\sin(\theta) = \frac{\lambda}{2}\sqrt{h^2a^{*2}+k^2b^{*2}+l^2c^{*2}+2klb^*c^*\cos\alpha^*+2lhc^*a^*\cos\beta^*+2hka^*b^*\cos\gamma^*}$, $a^* = \frac{1}{D}bc\sin\alpha,\ \cos\alpha^* = \frac{\cos\beta\cos\gamma-\cos\alpha}{\sin\beta\sin\gamma}$ $b^*,\ c^*,\ \cos\beta^*,\ \cos\gamma^*$ correspondingly with cyclic change of $a,\ b,\ c,\ \alpha,\ \beta,\ \gamma,$ and $D = abc\sqrt{1+2\cos\alpha\cos\beta\cos\gamma-\cos^2\alpha-\cos^2\beta-\cos^2\gamma}$

of the unit cell and on the position of the planes with respect to the axes of the crystal system. They can be calculated for the seven crystal systems listed in Table 2.1 from the lattice vectors **a**, **b**, **c**, the angles α, β, and γ included by them, and the Miller or Miller-Bravais indices of the lattice planes h, k, l. The result is given in Table 2.9 in terms of Bragg angles $\sin(\Theta)$ of allowed scattering directions.

2.4.2 The Structure Factor

The *intensity* of the diffracted beam depends on the spatial arrangement of the atoms in the unit cell and their scattering power. The intensity of the wave scattered by all N atoms of the unit cell results from the sum of their individual contributions and their respective phases. This is expressed by the *structure factor* F_{hkl},

$$F_{hkl} = \sum_{n=1}^{N} f_n \exp\big(i2\pi(hu_n + kv_n + lw_n)\big). \tag{2.40}$$

The *atomic scattering factor* f_n expresses the scattering power of the nth atom in the unit cell and is determined by its inner electronic charge distribution. The exponential factor gives the phase relation of the N contributions and depends on the positions of the atoms in the unit cell, expressed by the coordinates u_n, v_n, and w_n. The absolute value of the structure factor $|F_{hkl}|$ yields the amplitude of the diffracted wave, i.e., the amplitude of the wave scattered by all atoms of the unit cell with respect to the amplitude of a wave scattered by a free electron. The intensity I_{hkl} of the scattered wave is eventually given by its square,

$$I_{hkl} \propto |F_{hkl}|^2 = F^*_{hkl} F_{hkl}. \tag{2.41}$$

We consider for example the structure factor of a solid with body centered cubic (bcc) structure. The unit cell contains 2 atoms located at 000 and $\frac{1}{2}\frac{1}{2}\frac{1}{2}$ (in units of the lattice constant a), which are identical ($f_1 = f_2 = f$). From (2.40) we obtain

$$\begin{aligned} F &= f\exp\big(i2\pi(0h + 0k + 0l)\big) + f\exp\big(i2\pi\big(\tfrac{1}{2}h + \tfrac{1}{2}k + \tfrac{1}{2}l\big)\big) \\ &= f\big(1 + \exp\big(i\pi(h + k + l)\big)\big), \end{aligned}$$

i.e., $F = 2f$ when $h + k + l$ is even, and $F = 0$ when $h + k + l$ is odd. For fcc structure we similarly obtain $F = 4f$ when h, k, l are all even or all odd, and $F = 0$ when h, k, l are mixed even or odd.

In *zincblende structure* we have two kinds of atoms with atomic form factors f_1 and f_2 ($> f_1$). In this case we find for the structure factor (with n being an integer):

$$\begin{array}{ll} h, k, l \text{ all even and } h + k + l = 4n, & F = 4(f_1 + f_2), \\ h, k, l \text{ all even and } h + k + l = 4n + 2, & F = 4(f_1 - f_2), \\ h, k, l \text{ all odd,} & F = 4\big(f_1^2 - f_2^2\big)^{1/2}, \\ h, k, l \text{ are mixed even and odd,} & F = 0. \end{array}$$

Note that all these structures belong to the cubic crystal system.

In the hexagonal *wurtzite structure* the structure factor has the following form:

$$\begin{array}{ll} l = 2n \text{ and } h + 2k = (3m \text{ or } 3m \pm 1), & F = 2\big(f_1^2 + 2f_1 f_2 \cos(2\pi ul) + f_2^2\big)^{1/2}, \\ l = 2n + 1 \text{ and } h + 2k = 3m, & F = 0, \\ l = 2n + 1 \text{ and } h + 2k = 3m \pm 1, & F = \sqrt{3}\big(f_1^2 + 2f_1 f_2 \cos(2\pi ul) + f_2^2\big)^{1/2}. \end{array}$$

n and m are integers and u ($\approx \frac{3}{8}c$) is the mutual displacement of the anion and cation sublattices.

A number of further factors determines the intensity of the scattered wave in addition to the structure factor F which solely considers the geometric structure of the scattering unit cells. They account for the polarization of the scattered wave (polarization factor $P \propto (1 + \cos^2 2\Theta)$), deviations from a perfect parallel and monochromatic incident radiation (Lorentz factor $L \propto \lambda^2/\sin^2\Theta$), the effect of temperature (temperature factor or Debye-Waller factor $f_T \propto \exp(\sin^2\Theta/\lambda)$), and absorption corrections depending on the measurement geometry (reflection or transmission setup). They are not included in this brief introduction.

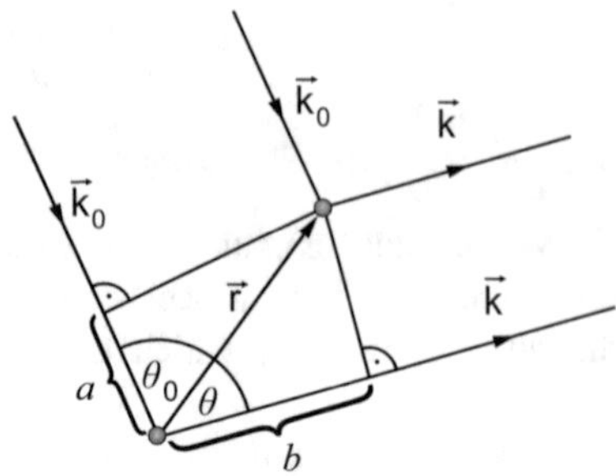

Fig. 2.37 Path difference $a + b$ of X-rays scattered from two atoms separated by $\mathbf{r}$

2.4.3 The Reciprocal Lattice

For a detailed description of the scattering geometry the approach formulated by Max von Laue in 1911 is more appropriate than the simplified picture presented in Fig. 2.36. In this framework we regard the crystal as being composed of atoms which reradiate the incident radiation in all directions. Sharp reflections are obtained only in directions for which the rays from all lattice sites interfere constructively. We first consider scattering of only two atoms separated by a vector $\mathbf{r}$, see Fig. 2.37.

The incident X-ray beam along the direction $\mathbf{s}_0$ with wavelength λ has the wave vector $\mathbf{k}_0 = (2\pi/\lambda)\mathbf{s}_0$, $\mathbf{s}_0$ being the unit vector $\mathbf{k}_0/k_0$. Constructive interference is obtained along a scattered direction $\mathbf{s} = \mathbf{k}/k$, if the path difference of the rays scattered by the two atoms is an integral number of wavelengths. From Fig. 2.37 we read that the path difference is composed by a sum of $a = r\cos\Theta_0 = -\mathbf{s}_0\mathbf{r}$ and $b = r\cos\Theta = \mathbf{sr}$, yielding

$$(\mathbf{s} - \mathbf{s}_0)\mathbf{r} = n\lambda. \tag{2.42}$$

$\lambda = \lambda_0$ because the scattering is assumed to be elastic. Multiplying (2.42) by $2\pi/\lambda$ we obtain

$$(\mathbf{k} - \mathbf{k}_0)\mathbf{r} = n2\pi. \tag{2.43}$$

We now consider many scattering atoms placed at the sites of a Bravais lattice. All lattice sites are displaced from one another by a Bravais lattice vector $\mathbf{R}$. Constructive interference of all scattered rays occurs if condition (2.43) holds simultaneously for all values of $\mathbf{r}$, i.e.

$$(\mathbf{k} - \mathbf{k}_0)\mathbf{R} = n2\pi. \tag{2.44}$$

Equation (2.44) may be rewritten by introducing the *reciprocal lattice*. A unique reciprocal lattice can be constructed for each particular Bravais lattice. It is a Bravais lattice for itself, and its reciprocal lattice is the original direct lattice. The base vectors $\mathbf{g}_i$ of the reciprocal lattice are defined by the condition

$$\mathbf{g}_i \mathbf{a}_j = 2\pi \delta_{ij}, \tag{2.45}$$

$\mathbf{a}_i$ being the base vectors of the corresponding Bravais lattice and Kronnecker's symbol $\delta_{ij} = 1$ for $i = j$ and 0 otherwise. The vector $\mathbf{g}_1$ is thus normal to $\mathbf{a}_2$ and $\mathbf{a}_3$, and can be calculated from

$$\mathbf{g}_1 = \frac{\mathbf{a}_2 \times \mathbf{a}_3}{\mathbf{a}_1 (\mathbf{a}_2 \times \mathbf{a}_3)}. \tag{2.46}$$

The vectors $\mathbf{g}_2$ and $\mathbf{g}_3$ are obtained from (2.46) by a cyclic permutation of the indices. From the construction of the reciprocal lattice we conclude that each vector $\mathbf{G}_{hkl}$ is normal to a lattice plane of the real lattice with Miller indices (hkl). Furthermore, the distance d_{hkl} of two neighboring parallel lattice planes with Miller indices (hkl) is inverse to the absolute value of the corresponding reciprocal lattice vector $\mathbf{G}_{hkl}$,

$$d_{hkl} = \frac{2\pi}{|\mathbf{G}_{hkl}|}. \tag{2.47}$$

Using the reciprocal lattice vectors defined in (2.45) and a reciprocal lattice vector $\mathbf{G}$ built from the base vectors $\mathbf{g}_i$ we represent (2.44) by the *von Laue condition*

$$(\mathbf{k} - \mathbf{k}_0) = \mathbf{G}. \tag{2.48}$$

Equation (2.48) means that constructive interference of scattered X-rays occurs if the change in wave vector (referred to as diffraction vector) $\Delta\mathbf{k} = \mathbf{k} - \mathbf{k}_0$ is a vector of the reciprocal lattice. Conditions (2.48) and (2.39) are equivalent criteria for constructive interference of X-rays. We may derive Bragg's law (2.39) from (2.48) by noting that $\mathbf{G}$ is an integral multiple of the shortest parallel reciprocal lattice vector, the length of which is $2\pi/d_{hkl}$, i.e., $G = n2\pi/d_{hkl}$. G is also related to the Bragg angle Θ. From (2.48) and the triangle built by the vectors $\mathbf{k}_0$, $\mathbf{k}$, and $\Delta\mathbf{k}$ in Fig. 2.38 we find the relation $G = \Delta k = 2 \times k \sin\Theta$. Putting this together we find

$$k \sin\theta = \frac{n\pi}{d_{hkl}}. \tag{2.49}$$

By expressing k in terms of $2\pi/\lambda$ we obtain Bragg's law from (2.49). We use here the convention, which is common in solid state physics. It should be noted that *in crystallography* the length of the wave vector is usually defined by $k = 1/\lambda$, i.e., without the factor 2π; the reciprocal lattice spacing is then inverse to the associated lattice-plane spacing in real space.

The reciprocal lattice defined by (2.45) and (2.46) helps to find allowed Bragg reflections for a given crystal and the corresponding Bravais lattice. Such reflections fulfill the Bragg and von Laue conditions for a *set* of parameters consisting of lattice-plane distance d_{hkl}, angle Θ, and X-ray wavelength λ. Methods of X-ray

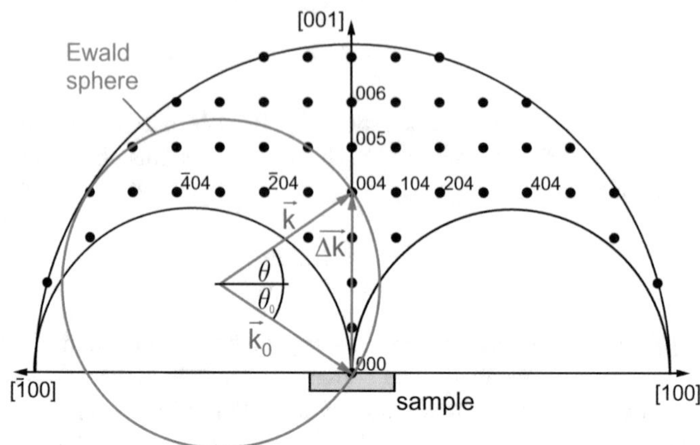

Fig. 2.38 Reciprocal space map of all Bragg reflections accessible for an incident X-ray radiation with wave vector of length k_0. The *horizontal axis* denotes a diffraction vector $\Delta\mathbf{k}$ along a [100] direction lying in the sample surface, the *vertical axis* signifies a $\Delta\mathbf{k}$ normal to the surface. The *outer sphere* with radius $2k_0$ refers to a deflection angle $2\Theta = 180°$

characterization differ in the parameters varied. They may be well represented using a simple geometric construction introduced by Paul P. Ewald.

2.4.4 The Ewald Construction

To represent X-ray diffraction using the Ewald construction we first draw the reciprocal lattice, i.e., a collection of points derived from the real (direct) lattice of the crystal structure to be studied using (2.46). Now the incident wave vector $\mathbf{k}_0$ is added with the tip pointing to the origin $hkl = 000$, see Fig. 2.38. An *Ewald sphere* of radius k_0 $(= 2\pi/\lambda)$ centered on the origin of $\mathbf{k}_0$ is drawn. Scattered wave vectors $\mathbf{k}$ (with same length as $\mathbf{k}_0$) also starting at the center of the sphere satisfy the von Laue condition if and only if their tips end on a reciprocal lattice point, i.e., the lattice points lie on the surface of the sphere. In this case the diffraction vectors $\Delta\mathbf{k}$ are vectors $\mathbf{G}_{hkl}$ of the reciprocal lattice. Note that each vector $\mathbf{G}_{hkl}$ represents a *set* of parallel lattice planes of the real crystal with Miller indices hkl.

The reciprocal space map shown in Fig. 2.38 is a representation of the real lattice in the reciprocal space. It is *fixed* with respect to the studied sample and its diffracting Bragg planes. Ewald's construction therefore provides a good survey of possible reflections for varied angles or wavelengths. For fixed λ and consequently fixed lengths of $\mathbf{k}_0$ and $\mathbf{k}$ all possible reciprocal lattice points lie within a sphere of radius $2k$ (corresponding to Bragg scattering with $2\Theta = 180°$). Some points in the lower part of the map cannot be accessed since either $\Theta_0 < 0$ or $\Theta < 0$, i.e., $\mathbf{k}_0$ cannot

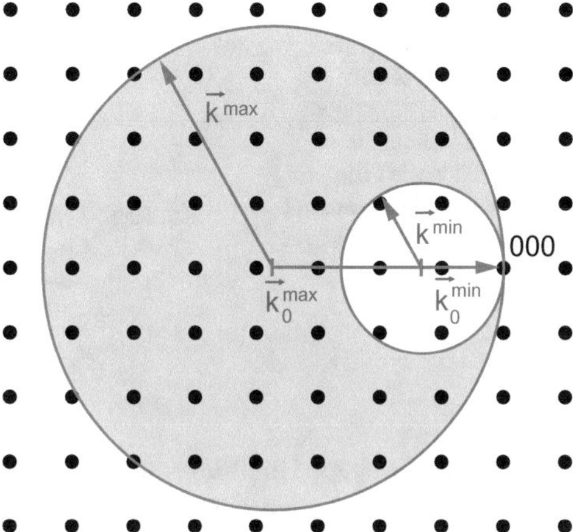

Fig. 2.39 Ewald's construction of the von Laue method. The crystal is irradiated with a polychromatic X-ray beam

enter the sample (the incident beam is below the surface) or **k** cannot emerge from the sample (the diffracted beam is below the surface).

If all angles are kept fixed and λ is varied over all possible values we directly obtain an image of the reciprocal lattice. This technique was applied by von Laue and co-workers in 1912 to prove the wave nature of X-rays and is mainly used today to determine the crystallographic orientation of samples with a known structure. Ewald's construction of the *Laue method* provides spheres for all wavelengths between λ_{min} (corresponding to k_{max}) and λ_{max}. They all have a common touch point at the origin because the crystal is kept stationary and the direction of incidence is hence fixed during data collection, see Fig. 2.39. Variation of λ is performed by using white radiation, i.e., continuum Bremsstrahlung radiation of an X-ray tube or not monochromatized synchrotron radiation. The fraction of the reciprocal lattice filled by the shaded area then diffracts simultaneously.

2.4.5 High-Resolution Scans in the Reciprocal Space

X-ray diffractometers are widely employed for non-destructive structural analyses of crystalline samples. Epitaxial structures are usually characterized applying *high-resolution X-ray diffraction* (HRXRD). The large natural line width of the generally used $K_{\alpha 1}$ radiation ($\Delta\lambda/\lambda = 3 \times 10^{-4}$ for Cu) limits the resolution of a simple single-crystal diffractometer to $\Delta\Theta/\Theta \cong 10^{-2}$. Furthermore, the close vicinity to the $K_{\alpha 2}$ line (1.5443 Å for Cu, with half the intensity of $K_{\alpha 1}$ at 1.5405 Å) complicates the analysis particularly for multilayer structures. In a modern HRXRD setup the incident X-ray radiation is therefore monochromized using Bragg reflections of a primary crystal (commonly a highly perfect Ge or Si crystal with a channel cut for multiple reflections). Such a monochromator improves the resolution to typically 10^{-5}. Often

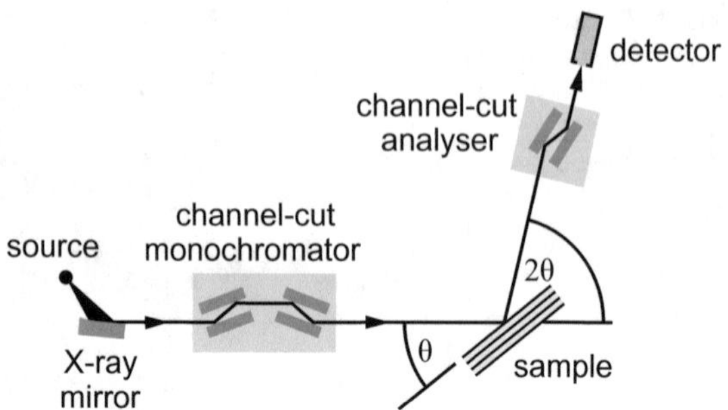

Fig. 2.40 High-resolution X-ray diffractometer for structural analyses of heteroepitaxial structures. The studied sample is symbolized by its diffracting Bragg planes. These are not parallel to the sample surface in case of asymmetric reflections

also a parabolic shaped multilayer mirror is used to convert the divergent radiation of the X-ray tube to an intense quasi-parallel beam. Such a beam conditioning reduces the required measuring time significantly. A high-resolution X-ray setup is illustrated in Fig. 2.40.

The setup shown in Fig. 2.40 is also referred to as triple-axis diffractometer, the three rotation axes being the Bragg angles of the monochromator crystal, of the investigated sample, and of the analyzer crystal. Many other configurations of X-ray optics are used, depending on the required resolution. X-ray measurements are conveniently described in the reciprocal space illustrated in Fig. 2.38. Various types of scans are used to characterize or map a Bragg reflection. Commonly applied scan directions are depicted in Fig. 2.41.

The incident and diffracted wave vectors are denoted $\mathbf{k}_\mathrm{i}$ and $\mathbf{k}_\mathrm{f}$, respectively. Due to the elastic scattering their absolute values are equal and both given by the applied radiation wavelength, $k = 2\pi/\lambda$. The components of the diffraction vector $\Delta\mathbf{k} = \mathbf{k}_\mathrm{f} - \mathbf{k}_\mathrm{i}$ are

$$q_{\parallel} = k(\cos\alpha_\mathrm{f} - \cos\alpha_\mathrm{i}),$$
$$q_{\perp} = k(\sin\alpha_\mathrm{f} + \sin\alpha_\mathrm{i}).$$

In the X-ray diffractometer the angle of rotation of the sample surface is usually labeled ω, and the angle of the detector rotation is labeled 2Θ. The goniometer angle ω represents the angle of incidence with respect to the sample surface α_i, and 2Θ equals the sum of α_i and the angle of the diffracted beam with respect to the sample surface α_f. The components of the diffraction vector are thus directly measured by scanning ω and Θ.

In reciprocal space the $\omega - 2\Theta$ scan is oriented along the direction of the origin to the studied reciprocal-lattice point. The direction of $\Delta\mathbf{k}$ remains fixed, while the length changes. For such a radial scan the sample is rotated by an amount $\Delta\omega$ and the detector is rotated by $\Delta 2\Theta = 2\Delta\omega$. In the case of a symmetric reflection, i.e., when the Bragg planes are parallel to the sample surface and $\alpha_i = \alpha_f$, the relation $\omega = 2\Theta/2$ applies. The diffraction vector $\Delta\mathbf{k}$ is then always normal to the surface as depicted in Fig. 2.38.

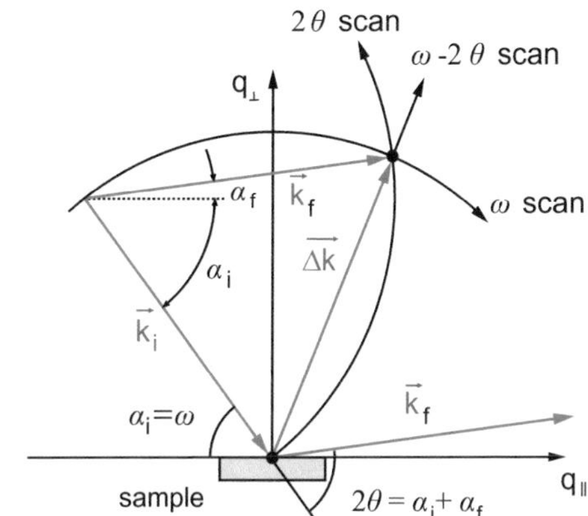

Fig. 2.41 Scan directions in the reciprocal space commonly applied in high-resolution X-ray diffraction. $\mathbf{k}_i$ and $\mathbf{k}_f$ denote initial and final wave vectors of the monochromatic X-ray beams, respectively. Angles α represent angles of the incident and diffracted (final) beams with respect to the sample surface

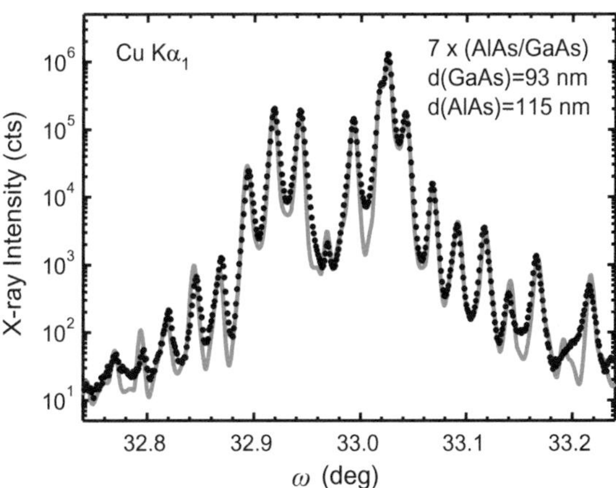

Fig. 2.42 ω scan of the symmetric (004) reflection of a 7 period AlAs/GaAs superlattice structure grown on (001)-oriented GaAs. *Data points* and the *gray curve* are measured and simulated diffraction patterns, respectively

The ω scan traces a circle of radius $|\Delta\mathbf{k}|$ about the origin and is nearly oriented perpendicular to the $\omega - 2\Theta$ scan. The sample is rotated ('rocked') about the ω axis. The angle 2Θ and hence the length of $\Delta\mathbf{k}$ remains fixed, while the direction of $\Delta\mathbf{k}$ changes. Usually no analyzer and no slit is attached to the detector for recording all intensity reflected from the sample. This scan is usually referred to as 'rocking curve' measurement, although also the $\omega - 2\Theta$ scan requires sample rotation.

The 2Θ scan traces an arc on the Ewald sphere. The incident angle α_i and hence ω is kept fixed, and α_f is changed by varying the detector angle 2Θ. Both length and direction of the diffraction vector $\Delta\mathbf{k}$ change.

The ω scan is widely applied for the measurement of composition and layer thickness in coherent heterostructures. Data evaluation then applies Vegard's law (2.2a)–(2.2c) and a simulation of the intensity based on the dynamical theory of X-ray diffraction. In (001)-oriented structures of a cubic crystal system often the intense symmetric (004) reflection is used. A measured and simulated diffraction pattern of an AlAs/GaAs superlattice structure is shown in Fig. 2.42.

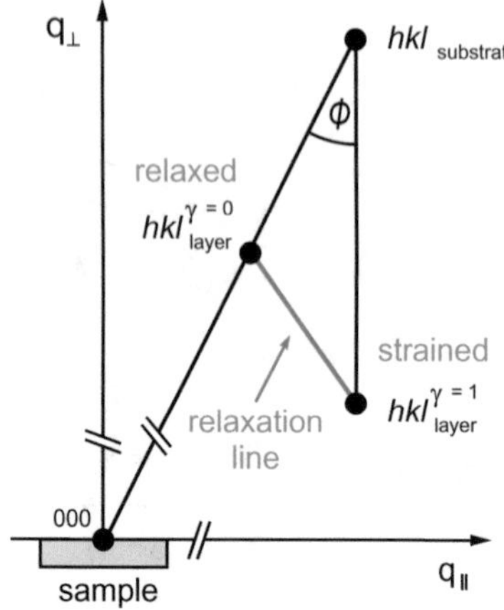

Fig. 2.43 Positions of reciprocal-lattice points *hkl* of a fully strained ($\gamma = 1$) and fully relaxed ($\gamma = 0$) epitaxial layer

2.4.6 Reciprocal-Space Map

The analysis of elastically relaxed heteroepitaxial structures requires the measurement of both, vertical and lateral lattice parameters. For the evaluation of lateral components asymmetric reflections must be used. Strain relief in the structure may be quite inhomogeneous as shown in Fig. 2.16. Furthermore, Bragg planes of an epitaxial structure may be tilted with respect to those of the substrate, and mosaicity of layers with tilted and twisted crystallites as illustrated in Sect. 2.3.9 may occur. The analysis of such imperfections is conveniently performed by combining a series of scans in reciprocal space to a map.

An analysis of the strain state in a partially relaxed epitaxial layer from reciprocal-space maps is illustrated in Figs. 2.43 and 2.44. For simplicity we assume an equal crystal structure for layer and substrate. Partial relaxation may be described by the relaxation parameter γ (2.29) introduced in Sect. 2.2.8. The reflection hkl_{layer} of a *coherently strained* pseudomorpic layer ($\gamma = 1$) has the same lateral component of the scattering vector as the corresponding reflection of the substrate, because the lateral lattice constant of the layer $a_{\parallel}$ adopts the value of the substrate a_{S}. In Fig. 2.43 a layer with a natural (unstrained) lattice constant a_{L} larger than the substrate lattice constant a_{S} is assumed. hkl_{layer} therefore lies below $hkl_{\text{substrate}}$ in the reciprocal space map, i.e. at a *smaller* $q_{\perp}$ value. For a completely *relaxed* layer ($\gamma = 0$) the Bragg planes lie parallel to that of the substrate. Consequently the scattering vector $\Delta\mathbf{k}$ has the same direction for layer and substrate, i.e., hkl_{layer} and $hkl_{\text{substrate}}$ lie on a common line starting at the reciprocal origin 000. Intermediate strain states of the layer lie on a *relaxation line* bounded by the points for $\gamma = 0$ and 1, cf. Fig. 2.43. This line is straight due to the assumed validity of Hook's law during the relaxation process.

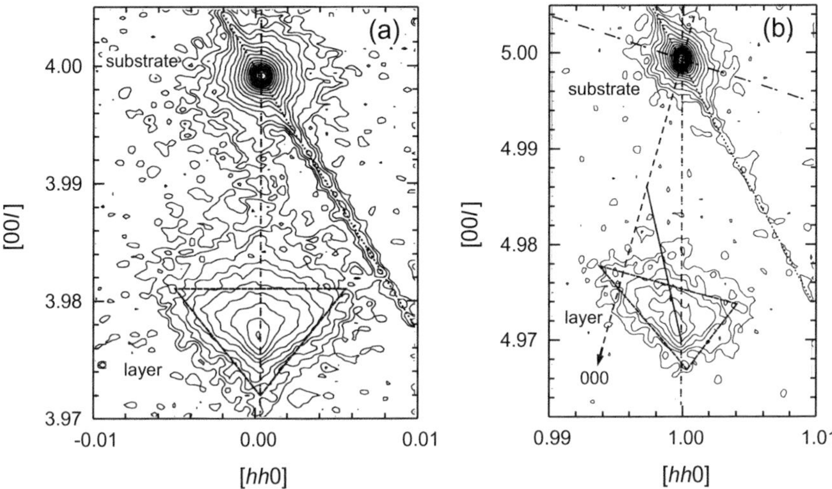

Fig. 2.44 Reciprocal-space maps of the symmetrical (004) reflection (*left*) and the (115) reflection (*right*) of a 310 nm thick epitaxial ZnSe layer on (001) GaAs substrate. *Contour lines* refer to scattered X-ray intensities on a logarithmic scale with steps of 0.2, coordinates are given in terms of non-integer Miller indices. The *straight line* in the *right* map indicates the relaxation line. The *dotted lines* in the maps indicate sections of Ewald spheres, the *streaks* along these lines are analyzer artefacts. Reproduced with permission from [80], © 1994 Elsevier

The experimental result given in Fig. 2.44 represents a cut of the reciprocal space using the [110] axis of the cubic crystal lattice as azimuth reference: $q_{\|} = [hh0] \times \sqrt{2}a_S$, i.e., h always equals k. The normal reference is $q_{\perp} = [00l] \times a_S$. The measurements show the scattered X-ray intensity near the (004) and (115) reflections of an epitaxial ZnSe layer on (001)-oriented GaAs substrate [80]. Both materials have zincblende structure.

The reciprocal-lattice points of the layer show a clear broadening with a non-circular shape indicated by triangles. The broadening along the *vertical* dash-dotted line (along the surface normal) demonstrates the coexistence of various strain states in the layer. Small $00l$ values refer to highly strained parts of the layer ($\gamma \to 1$), large values to relaxed parts. The distribution is asymmetric. Perpendicular to the scattering vector (i.e., the vertical line) another broadening mechanism originating from mosaicity occurs. The triangular shape of the reciprocal-lattice point indicates that this mosaicity is related to the strain state: small for large strains (lower triangle corner) and large for relaxed parts. The triangles referring to the (004) and (115) reflections are geometrically similar. The different shapes result from a transformation which depends on the inclination angle ϕ of the Bragg planes [80].

In the next example the mosaicity of an epitaxial layer is considered. Perturbations in the periodicity of the crystal lattice lead to incoherent scattering of X-rays and consequently to a broadening of reciprocal lattice points. From the broadening the average extension of coherently scattering regions in the crystal (i.e., regions free of

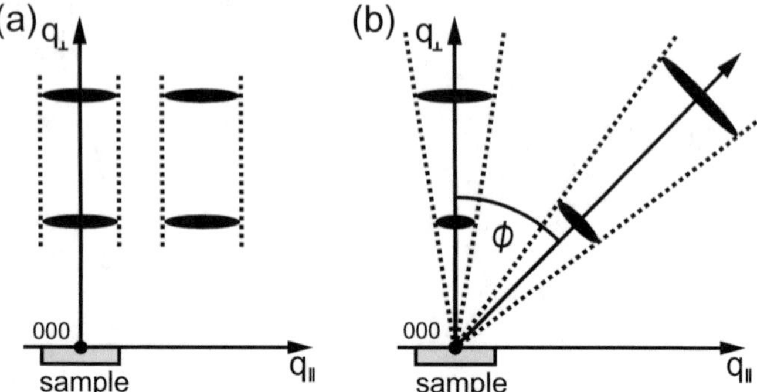

Fig. 2.45 Scheme of broadened reciprocal-lattice points due to mosaicity in an epitaxial layer originating predominantly from **a** finite lateral coherence length and **b** tilts of mosaic blocks

defects) can be derived. The relation between this so-called coherence length d and the broadening of a reciprocal lattice point due to the finite size of perfect crystallites is expressed by an equation formulated by P. Scherrer [81],

$$FWHM = \frac{K\lambda}{d \times \cos\Theta}. \tag{2.50}$$

In the *Scherrer equation* (2.50) $FWHM$ denotes the full width at half maximum of an X-ray reflection scattered from crystallites with average dimension d, and K is the Scherrer constant. K is on the order of unity and usually put to 0.9 for a Gaussian function of the reflection shape. λ and Θ are the X-ray wavelength and the Bragg angle, respectively. In addition to finite-size effects in lateral and vertical direction expressed by Scherrer's formula mutual tilts and twists of the small crystalline blocks building the mosaic of a real crystal (Sect. 2.3.9) contribute to the broadening of reciprocal lattice points. The action of tilts differs from that of a finite size as illustrated in Fig. 2.45.

Finite sizes of mosaic blocks along the lateral and vertical directions lead to broadenings along $q_{\|}$ and $q_{\perp}$, respectively. This holds for symmetric and asymmetric reflections of any order in the same way as illustrated in Fig. 2.45a for a limited lateral size. Tilts induce a broadening which linearly increases with the order of the reflection. In symmetric reflections the superimposed broadening contributions due to lateral finite size and tilt can be separated by plotting the $FWHM$ of rocking curves over the respective reflection order, yielding the tilt from the slope and the lateral coherence length from the inverse of the ordinate interception-point. Asymmetric reflections are broadened by pure tilts along an axis inclined by the angle ϕ which represents the angle between the diffracting Bragg planes and the sample surface. Superimposed finite size effects modify the angle.

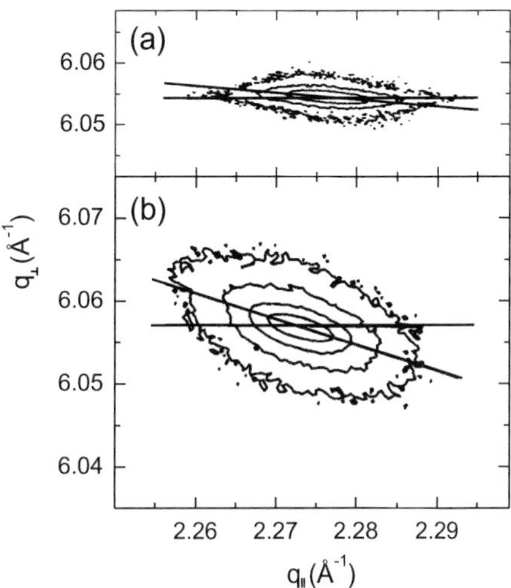

Fig. 2.46 Reciprocal-space maps of the asymmetric (10.5) reflection of GaN layers with **a** 350 nm and **b** 4200 nm mosaic grains, epitaxially grown on (0001) sapphire substrate. *Contour lines* refer to equal X-ray intensities. Reproduced with permission from [82], © 2003 AIP

Twists lead to a broadening of reciprocal lattice points in the plane spanned by $q_{\|}$ and a second independent lateral direction. An evaluation can be obtained from grazing incidence diffraction which uses scattering from Bragg planes lying perpendicular to the sample surface or from measuring the edge of the sample. Also selected-area electron diffraction of plan view scanning TEM was used to image tilt by broadened diffraction spots [71].

The effect of finite mosaic grain size and tilt on asymmetric reflections is illustrated in Fig. 2.46. The reciprocal-space maps show (10.5) reflections of wurtzite GaN grown on basal-plane sapphire [82]. The large lattice mismatch leads to very high dislocation densities in the range of typically 10^9 cm^{-2}. The measured reflections have nearly elliptical shape with an inclination of the main axis with respect to the azimuth. The angle depends on the average size of mosaic grains in the epitaxial layer. Small angles are due to a dominant effect of lateral finite size. Pure tilts reproduce the inclination ϕ with respect to the surface (20.5° for the given reflection).

2.5 Problems Chapter 2

2.1 The stable modification of GaN is the wurtzite (α) structure. Epitaxial layers can also be grown in the metastable cubic zincblende (β) structure.

(a) Determine the lattice mismatch f of β-GaN on (001)-oriented zincblende GaAs at room temperature.
(b) Find a suitable ratio of small integers for a coincidence lattice and determine the respective coincidence-lattice mismatch.

(c) On (111)-oriented GaAs the GaN layer tends to grow in the α phase. What is the wurtzite a lattice parameter of the GaAs(111) plane? Prove that the lattice mismatch of α-GaN/GaAs(111) is similar to that of β-GaN/GaAs(001), if α-GaN is assumed to have the same bond length as β-GaN.

2.2 GaN is grown on various substrates due to a lack of well lattice-matched materials.

(a) Calculate the lattice mismatch for growth on the most commonly used basal-plane sapphire Al_2O_3 in case of an epitaxial relation $[0001]_{GaN} \parallel [0001]_{Al_2O_3}$ and $[100]_{GaN} \parallel [100]_{Al_2O_3}$ (or $\left[2\bar{1}\bar{1}0\right]_{GaN} \parallel \left[2\bar{1}\bar{1}0\right]_{Al_2O_3}$, i.e., a and c axes of substrate and layer are parallel). Compare this value to that often derived from the alternative definition $f_{\text{alternative1}}$ given in the text. The lattice parameters of Al_2O_3 are $a_{Al_2O_3} = 4.758$ Å, $c_{Al_2O_3} = 12.991$ Å.

(b) The a axes of the GaN layer are actually rotated by 30° about the c axis, yielding the relation $[100]_{GaN} \parallel [1\bar{1}0]_{Al_2O_3}$ (cf. figure below). What is the lattice mismatch for this epitaxial relation?

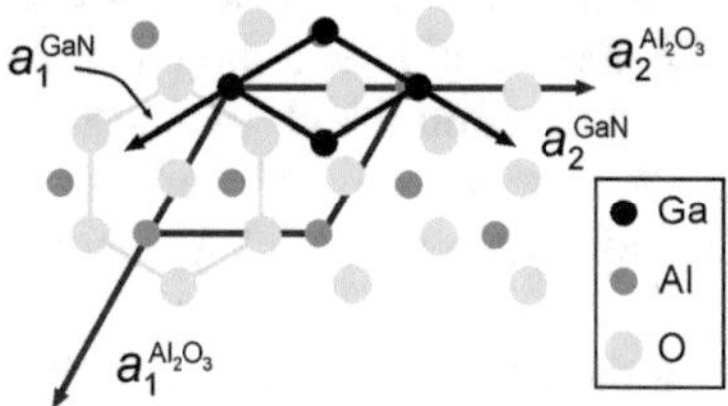

(c) The growth of m-plane wurtzite GaN (with a nonpolar surface) is often performed on γ-$LiAlO_2$, which has a tetragonal structure with lattice parameters $a = b = 5.169$ Å and $c = 6.268$ Å. Compute the lattice mismatch of the two edges of the GaN m-plane aligned along the two different axes of γ-$LiAlO_2$, when every second Ga atom along the a axis of the GaN sublattice matches an Al atom along the c axis of the γ-$LiAlO_2$ sublattice.

(d) An alternative m-plane substrate for m-plane wurtzite GaN is the readily available 6H-SiC polytype with $ABCACB$ stacking order. What is the lattice mismatch for an equal orientation of the unit cells? It should be noted that the GaN layers tend to grow in a 2H sequence and only partially adopt the 6H sequence (in contrast to AlN layers), yielding a high density of stacking faults.

2.3 (a) How does the Al composition parameter x in the quaternary compound $Al_xGa_yIn_{1-x-y}As$ depend on the Ga composition parameter y for a layer lattice-matched to InP according to Vegard's law?

(b) What are the maximum of the Ga composition parameter y and the maximum of the Al composition parameter x?

(c) Write the relation for a quaternary AlGaInAs layer lattice-matched to InP in terms of two ternary alloys X_z and Y_{z-1} which are both lattice-matched to InP.

2.4 Apply in the following problem *linear* expansion coefficients (which actually underestimate the thermal expansion at higher temperatures) and a linearly weighted quantity for the alloy.

(a) Calculate the lattice mismatch f of a ZnS layer on Si substrate at a growth temperature of 360 °C.
(b) Determine the composition parameter x of a $ZnS_{1-x}Se_x$ layer for lattice-matched conditions on Si at growth temperature.
(c) Calculate the thermally induced lateral strain of the $ZnS_{1-x}Se_x$ layer after cooling the lattice-matched $ZnS_{1-x}Se_x$ layer of (b) from 360 °C to room temperature (23 °C).
(d) Which composition parameter x produces a strain-free $ZnS_{1-x}Se_x$ layer *at room temperature* for growth at 360 °C? How large is then the mismatch at growth temperature?

2.5 Consider a pseudomorphic ZnSe layer on (001)-oriented GaAs substrate at room temperature.

(a) Determine the strain of the ZnSe layer perpendicular to the interface.
(b) Calculate the relative change of the unit-cell volume induced by the strain.
(c) The strain in the layer changes the distance d_{hkl} between nearest lattice planes. In crystal systems with orthogonal axes and lattice vectors of lengths a, b, and c the distance can be expressed by $d_{hkl} = ((h/a)^2 + (k/b)^2 + (l/c)^2)^{-1/2}$. Use this relation to calculate the distances between nearest (111) planes in unstrained ZnSe and the pseudomorphically strained ZnSe/GaAs layer. Relate the factor $\sqrt{3}$ in the unstrained material to the stacking sequence of the zincblende structure.
(d) Express the thickness of a 285 Å thick layer in units of monolayers.
(e) Calculate the strain energy per unit area of a 285 Å thick layer. Which strain energy per unit area has a 114 nm thick layer?
(f) Which strain energy is stored in one unit cell of the pseudomorpic ZnSe layer?

2.6 The strain in pseudomorphic lattice-mismatched layers is sometimes compensated by inserting additional layers with opposite lattice mismatch, yielding layer stacks which are *in total* lattice-matched to a substrate.

(a) Determine the thickness of an $In_{0.15}Ga_{0.85}As$ layer, which is to be pseudomorphically grown on a 1 monolayer thick (001)-oriented GaP layer to obtain a total lateral lattice constant coinciding with that of GaAs. How many monolayers of $In_{0.15}Ga_{0.85}As$ correspond to this thickness? The stiffness coefficients of GaP are $C_{11} = 141$ GPa, $C_{12} = 62$ GPa; use a linearly weighted shear modulus for the $In_{0.15}Ga_{0.85}As$ layer determined similar to the unstrained lattice parameter of this layer.

(b) The strain in a pseudomorphic superlattice with 10 $In_xGa_{1-x}As$ quantum wells separated by GaAs barriers is to be compensated by the additional insertion of a counteracting $GaAs_{1-y}P_y$ layer into the center of each of these barriers. The entire layer stack of the strain-compensated superlattice should adopt the same lateral lattice parameter as the (001)-oriented GaAs substrate. Calculate the composition parameter y for the case that in each of the 20 nm thick barrier layers 12 nm of GaAs is replaced by $GaAs_{1-y}P_y$. The quantum wells have a thickness of 10 nm and a composition of 22% indium. Apply a linearly weighted shear modulus for the wells, but approximate the value for the barriers by that of GaAs (check if such a simplification is justified).

2.7 Determine the energy per unit length for the following perfect misfit dislocations in GaAs within a radius of 300 nm around the dislocation line. A single dislocation with a dislocation-core radius of $\frac{1}{3}$ of the Burgers vector is assumed.

(a) Pure screw dislocation.
(b) Pure edge dislocation.
(c) 60° dislocation.

2.8 The composition parameter x of a (001)-oriented alloy layer A_xB_{1-x} with diamond structure was approximately adjusted for achieving lattice matching to a substrate with lattice parameter $a_S = 5.5$ Å, yielding a misfit $f = 10^{-3}$.

(a) Estimate the critical thickness for the introduction of 60° misfit dislocations for a Poisson ratio of the layer $\nu = 0.25$, using the graphical result given in the text.
(b) By which factor changes the layer thickness for $\nu = 0.35$?
(c) Which misfit must be achieved for extending the critical layer thickness by a factor of 5 for a Poisson ratio $\nu = 0.25$?

2.9 A ZnSe/GaAs(001) layer is characterized by X-ray diffraction using $CuK_{\alpha 1}$ radiation.

(a) Calculate the separation between the Bragg angles for the (004) reflections of the substrate and the layer for a completely relaxed layer and a pseudomorphically strained layer.
(b) Repeat (a) for the asymmetric (115) reflection.
(c) The intensity of the scattered radiation is largely determined by the square of the structure factor. Calculate the approximate intensity ratio $I(004)/I(115)$ of the two reflections, if the ratio of the atomic scattering factors $f_{Se}/f_{Zn} = 1.17$.

2.10 An $In_xGa_{1-x}N$ layer is to be grown lattice-matched on the basal plane of a ZnO substrate.

(a) Find the In composition x_1 and the Bragg angle of the (00.2) reflection of a lattice-matched $In_xGa_{1-x}N$ layer for $CuK_{\alpha 1}$ radiation.

(b) Calculate the In composition x_2 of a *relaxed* (not lattice-matched) $In_x Ga_{1-x}N$ layer producing the (00.2) reflection at a Bragg angle 400 s below the value of a lattice-matched layer. Compare this value to the composition x_3 of a *pseudomorphic* $In_xGa_{1-x}N$ layer, whose (00.2) reflection appears at the same Bragg angle (find x_3 by using x_2 and interpolation; use for simplicity elastic constants of pure GaN).

2.6 General Reading Chapter 2

J.F. Nye, *Physical Properties of Crystals* (Clarendon Press, Oxford, 1972)
A.S. Saada, *Elasticity Theory and Applications* (Pergamon Press, New York, 1974)
J.P. Hirth, J. Lothe, *Theory of Dislocations*, 2nd edn. (Wiley, New York, 1982)
D. Hull, D.J. Bacon, *Introduction to Dislocations*, 4th edn. (Butterworth-Heinemann, Oxford, 2001)
S. Amelinckx, Dislocations in particular structures, in *Dislocations in Solids*, vol. 2, ed. by F.R.N. Nabarro (North-Holland, Amsterdam, 1979)
J.E. Ayers, T. Kujofsa, P. Rago, J.E. Raphael, *Heteroepitaxy of Semiconductors: Theory, Growth, and Characterization*, 2nd edn. (CRC Press, Boca Raton, 2017)

References

1. C.-Y. Yeh, Z.W. Lu, S. Froyen, A. Zunger, Zinc-blende-wurtzite polytypism in semiconductors. Phys. Rev. B **46**, 10086 (1992)
2. O. Ambacher, J. Smart, J.R. Shealy, N.G. Weimann, K. Chu, M. Murphy, W.J. Schaff, L.F. Eastman, R. Dimitrov, L. Wittmer, M. Stutzmann, W. Rieger, J. Hilsenbeck, Two-dimensional electron gases induced by spontaneous and piezoelectric polarization charges in N- and Ga-face AlGaN/GaN heterostructures. J. Appl. Phys. **85**, 3222 (1999)
3. H. Ibach, Thermal expansion of silicon and zinc oxide. Phys. Status Solidi **31**, 625 (1969)
4. K. Haruna, H. Maeta, K. Ohashi, T. Koike, The negative thermal expansion coefficient of GaP crystal at low temperatures. J. Phys. C **19**, 5149 (1986)
5. J.St. John, A.N. Bloch, Quantum-defect electronegativity scale for nontransition elements. Phys. Rev. Lett. **33**, 1095 (1974)
6. J.R. Chelikowsky, J.C. Phillips, Quantum-defect theory of heats of formation and structural transition energies of liquid and solid simple metal alloys and compounds. Phys. Rev. B **17**, 2453 (1978)
7. S.B. Zhang, M.L. Marvin, Determination of AB crystal structures from atomic properties. Phys. Rev. B **39**, 1077 (1989)
8. P. Villars, K. Mathis, F. Hulliger, Environment classification and structural stability maps, in *The Structure of Binary Compounds*, ed. by F.R. de Boer, D.G. Pettifor (North-Holland, Amsterdam, 1989)
9. O. Madelung (ed.), *Semiconductors-Basic Data*, 2nd revised edn. (Springer, Berlin, 1996)

10. G.L. Harris (ed.), *Properties of Silicon Carbide*. EMIS Datareview Series, vol. 13 (INSPEC, London, 1995)
11. J.H. Edgar (ed.), *Properties of Group III Nitrides*. EMIS Datareview Series, vol. 11 (INSPEC, London, 1994)
12. L. Vegard, [De]Die Konstitution der Mischkristalle und die Raumfüllung der Atome. Z. Phys. **5**, 17 (1921). (in German)
13. M.F. Thorpe, E.J. Garboczi, Elastic properties of central-force networks with bond-length mismatch. Phys. Rev. B **42**, 8405 (1990)
14. W.B. Pearson, *A Handbook of Lattice Spacing and Structures of Metals and Alloys*, vol. 1 (Pergamon Press, London, 1958)
15. D.R. Yoder-Short, U. Debska, J.K. Furdyna, Lattice parameters of $Zn_{1-x}Mn_xSe$ and tetrahedral bond lengths in $A^{II}_{1-x}Mn_xB^{VI}$ alloys. J. Appl. Phys. **58**, 4056 (1985)
16. J.C. Mikkelsen Jr., J.B. Boyce, Extended X-ray absorption fine-structure study of $Ga_{1-x}In_xAs$ random solid solutions. Phys. Rev. B **28**, 7130 (1983)
17. A. Balzarotti, N. Motta, A. Kisiel, M. Zimnal-Starnawska, M.T. Czyzyk, Model of the local structure of random ternary alloys: experiment versus theory. Phys. Rev. B **31**, 7526 (1985)
18. J.L. Martins, A. Zunger, Bond lengths around isovalent impurities and in semiconductor solid solutions. Phys. Rev. B **30**, 6217 (1984)
19. J.C. Mikkelsen Jr., J.B. Boyce, Atomic scale structure of random solid solutions: extended X-ray absorption fine-structure study of $Ga_{1-x}In_xAs$. Phys. Rev. Lett. **49**, 1412 (1982)
20. T. Fukui, Atomic structure model for $Ga_{1-x}In_xAs$ solid solutions. J. Appl. Phys. **57**, 5188 (1985)
21. J.F. Nye, *Physical Properties of Crystals* (Clarendon Press, Oxford, 1972)
22. A.S. Saada, *Elasticity Theory and Applications* (Pergamon Press, New York, 1974)
23. D.J. Dunstan, Strain and strain relaxation in semiconductors. J. Mater. Sci., Mater. Electron. **8**, 337 (1997)
24. K.-H. Hellwege, A.M. Hellwege, *Landolt-Börnstein New Series Group III, vol. 2, Elastic, Piezoelectric, Piezooptic, Electrooptic Constants, and Nonlinear Dielectric Susceptibilities of Crystals*, 6th edn. (Springer, Berlin, 1966)
25. K.-H. Hellwege, A.M. Hellwege, *Landolt-Börnstein New Series Group III, vol. 1, Elastic, Piezoelectric, Piezooptic and Electrooptic Constants of Crystals*, 6th edn. (Springer, Berlin, 1966)
26. I. Vurgaftman, J.R. Meyer, L.R. Ram-Mohan, Band parameters for III-V compound semiconductors and their alloys. J. Appl. Phys. **89**, 5815 (2001)
27. G. Carlotti, D. Fioretto, G. Socino, E. Verona, Brillouin scattering determination of the whole set of elastic constants of a single transparent film of hexagonal symmetry. J. Phys. Condens. Matter **7**, 9147 (1995)
28. R.S. Lakes, Foam structures with a negative Poisson's ratio. Science **235**, 1038 (1987)
29. E. Pasternak, A.V. Dyskin, Architectured materials with inclusions having negative Poisson's ratio or negative stiffness, in *Architectured Materials in Nature and Engineering*, ed. by Y. Estrin, Y. Bréchet, J. Dunlop, P. Fratzl (Springer, Cham, 2019)
30. A. Ballato, Poisson's ratio for tetragonal, hexagonal, and cubic crystals. IEEE Trans. Ultrason. Ferroelectr. Freq. Control **43**, 56 (1996)
31. C.G. Van de Walle, R.M. Martin, Theoretical calculation of heterojunction discontinuities in the Si/Ge system. Phys. Rev. B **34**, 5621 (1986)
32. L. Kleinman, Deformation potentials in silicon. I. Uniaxial strain. Phys. Rev. **128**, 2614 (1962)
33. O.H. Nielsen, R.M. Martin, Stress in semiconductors: ab initio calculations on Si, Ge, and GaAs. Phys. Rev. B **32**, 3792 (1985)
34. W.A. Jesser, D. Kuhlmann-Wilsdorf, On the theory of interfacial energy and elastic strain of epitaxial overgrowth in parallel alignment on single crystal substrates. Phys. Status Solidi **19**, 95 (1967)
35. F.C. Frank, J.H. van der Merve, Proc. R. Soc. Lond. A **198**, 205 (1949)
36. F.C. Frank, J.H. van der Merve, Proc. R. Soc. Lond. A **198**, 216 (1949)
37. F.C. Frank, J.H. van der Merve, Proc. R. Soc. Lond. A **200**, 125 (1949)

38. J.H. Van der Merve, Crystal interfaces. Part I. Semi-infinite crystals. J. Appl. Phys. **34**, 117 (1963)
39. J.H. Van der Merve, C.A.B. Ball, Energy of interfaces between crystals, in *Epitaxial Growth. Part B*, ed. by J.W. Matthews (Academic Press, New York, 1975), pp. 493–528
40. C.A.B. Ball, J.H. Van der Merve, The growth of dislocation-free layers, in *Dislocations in Solids*, vol. 6, ed. by F.R.N. Nabarro (North-Holland, Amsterdam, 1983)
41. W.A. Jesser, J.H. Van der Merve, An exactly solvable model for calculating critical misfit and thickness in epitaxial superlattices. II. Layers of unequal elastic constants and thicknesses. J. Appl. Phys. **63**, 1928 (1988)
42. J.W. Matthews, A.E. Blakeslee, Defects in epitaxial multilayers I. Misfit dislocations. J. Cryst. Growth **27**, 118 (1974)
43. J.W. Matthews, S. Mader, T.B. Light, Accommodation of misfit across the interface between crystals of semiconducting elements or compounds. J. Appl. Phys. **41**, 3800 (1970)
44. C.A.B. Ball, On bonding and structure of epitaxial bicrystals. Phys. Status Solidi **42**, 357 (1970)
45. R. Beanland, D.J. Dunstan, P.J. Goodhew, Plasic relaxation and relaxed buffer layers for semiconductor epitaxy. Adv. Phys. **45**, 87 (1996)
46. I. Akasaki, H. Amano, Y. Koide, K. Hiramatsu, N. Sawaki, Effects of a buffer layer on crystallographic structure and on electrical and optical properties of GaN and $Ga_{1-x}Al_xN$ ($0<x=0.4$) films grown on sapphire substrate by MOVPE. J. Cryst. Growth **98**, 209 (1989)
47. J.N. Kuznia, M. Asif Khan, D.T. Olson, R. Kaplan, J. Freitas, Influence of buffer layers on the deposition of high quality single crystal GaN over sapphire substrates. J. Appl. Phys. **73**, 4700 (1993)
48. R.C. Powell, N.-E. Lee, Y.-W. Kim, J.E. Greene, Heteroepitaxial wurtzite and zincblende structure GaN grown by reactive-ion molecular-beam epitaxy: growth kinetics, microstructure, and properties. J. Appl. Phys. **73**, 189 (1993)
49. P. Kung, C.J. Sun, A. Saxler, H. Ohsato, M. Razeghi, Crystallography of epitaxial growth of wurtzite-type thin films on sapphire substrates. J. Appl. Phys. **75**, 4515 (1994)
50. A. Zur, T.C. McGill, Lattice match: an application to heteroepitaxy. J. Appl. Phys. **55**, 378 (1984)
51. S.K. Mathis, A.E. Romanov, L.F. Chen, G.E. Beltz, W. Pompe, J.S. Speck, Modeling of threading dislocation reduction in growing GaN layers. J. Crystal Growth **231**, 371 (2001)
52. E.A. Fitzgerald, A.Y. Kim, M.T. Currie, T.A. Langdo, G. Taraschi, M.T. Bulsara, Dislocation dynamics in relaxed graded composition semiconductors. Mat. Sci. Eng. B **67**, 53 (1999)
53. J.E. Ayers, Low-temperature and metamorphic buffer layers, in *Handbook of Crystal Growth—Thin Films and Epitaxy: Basic Techniques, vol. 3 Part A*, 2nd edn., ed. by T.F. Kuech (Elsevier, Amsterdam, 2015), pp. 1007–1056
54. R.W. McClelland, C.O. Bozler, J.C.C. Fan, A technique for producing epitaxial films on reusable substrates. Appl. Phys. Lett. **37**, 560 (1980)
55. P. Gibart, Metal organic vapour phase epitaxy of GaN and lateral overgrowth. Rep. Prog. Phys. **67**, 667 (2004)
56. Y.H. Lo, New approach to grow pseudomorphic structures over the critical thickness. Appl. Phys. Lett. **59**, 2311 (1991)
57. D. Teng, Y.H. Lo, Dynamic model for pseudomorphic structures grown on compliant substrates: an approach to extend the critical thickness. Appl. Phys. Lett. **62**, 43 (1993)
58. W.A. Jesser, J.H. van der Merve, P.M. Stoop, Misfit accommodation by compliant substrates. J. Appl. Phys. **85**, 2129 (1999)
59. K. Vanhollebeke, I. Moerman, P. Van Daele, P. Demeester, Compliant substrate technology: integration of mismatched materials for opto-electronic applications. Prog. Cryst. Growth Charact. Mater. **41**, 1 (2000)
60. J.E. Ayers, Compliant substrates for heteroepitaxial semiconductor devices: theory, experiment, and current directions. J. Electron. Mater. **37**, 1151 (2008)
61. A.E. Romanov, W. Pompe, G. Beltz, J.S. Speck, Modeling of threading dislocation density reduction in heteroepitaxial layers. Phys. Stat. Sol. B **198**, 599 (1996)

62. S. Bauer, A. Rosenauer, P. Link, W. Kuhn, J. Zweck, W. Gebhardt, Misfit dislocations in epitaxial ZnTe/GaAs (001) studied by HRTEM. Ultramicroscopy **51**, 221 (1993)
63. G. Kudlek, N. Presser, U.W. Pohl, J. Gutowski, J. Lilja, E. Kuusisto, K. Imai, M. Pessa, K. Hingerl, H. Sitter, Exciton complexes in ZnSe layers: a tool for probing the strain distribution. J. Cryst. Growth **117**, 309 (1992)
64. G. Kudlek, Struktur und Dynamik exzitonischer Komplexe in verspannten ZnSe- und ZnTe-Heteroschichten, PhD Thesis, Technische Universität Berlin, D83, Berlin, 1992 (in German)
65. J.P. Hirth, J. Lothe, *Theory of Dislocations*, 2nd edn. (Wiley, New York, 1982)
66. S. Amelinckx, Dislocations in particular structures, in *Dislocations in Solids*, vol. 2, ed. by F.R.N. Nabarro (North-Holland, Amsterdam, 1979)
67. P.M. Marée, J.C. Barbour, J.F. Van der Veen, K.L. Kavanagh, C.W.T. Bulle-Lieuwma, M.P.A. Viegers, Generation of misfit dislocations in semiconductors. J. Appl. Phys. **62**, 4413 (1987)
68. N. Thompson, Dislocation nodes in face-centred cubic lattices. Proc. Phys. Soc. B **66**, 481 (1953)
69. M. Inoue, K. Suzuki, H. Amasuga, M. Nakamura, Y. Mera, S. Takeuchi, K. Maeda, Reliable image processing that can extract an atomically-resolved line shape of partial dislocations in semiconductors from plan-view high-resolution electron microscopic images. Ultramicroscopy **75**, 5 (1998)
70. H.P.D. Schenk, G.D. Kipshidze, U. Kaiser, A. Fissel, J. Kräußlich, J. Schulze, W. Richter, Investigation of two-dimensional growth of AlN(0001) on Si(111) by plasma-assisted molecular beam epitaxy. J. Cryst. Growth **200**, 45 (1999)
71. N. Mante, S. Rennesson, E. Frayssinet, L. Largeau, F. Semond, J.L. Rouvière, G. Feuillet, P. Vennéguès, Proposition of a model elucidating the AlN-on-Si (111) microstructure. J. Appl. Phys. **123**, 215701 (2018)
72. H. Kroemer, Polar-on-nonpolar epitaxy. J. Cryst. Growth **81**, 193 (1987)
73. O. Supplie, O. Romanyuk, C. Koppka, M. Steidl, A. Nägelein, A. Paszuk, L. Winterfeld, A. Dobrich, P. Kleinschmid, E. Runge, T. Hannappel, Metalorganic vapor phase epitaxy of III-V-on-silicon: experiment and theory. Prog. Cryst. Growth Charact. Mater. **64**, 103 (2018)
74. B. Kunert, Y. Mols, M. Baryshniskova, N. Waldron, A. Schulze, R. Langer, How to control defect formation in monolithic III/V hetero-epitaxy on (100) Si? A critical review on current approaches. Semicond. Sci. Technol. **33**, 093002 (2018)
75. T. Sakamoto, G. Hashiguchi, Si (001)-2 $\times$ 1 single-domain structure obtained by high temperature annealing. Jpn. J. Appl. Phys. **25**, L78 (1986)
76. P. Ruterana, Convergent beam electron diffraction investigationof inversion domains in GaN. J. Alloys Compounds **401**, 199 (2005)
77. A. Berghezan, A. Fourdeux, S. Amelinckx, Transmission electron microscopy studies of dislocations and stacking faults in a hexagonal metal-zinc. Acta Met. **9**, 464 (1961)
78. S. Vézian, J. Massies, F. Semond, N. Grandjean, P. Vennéguès, In situ imaging of threading dislocation terminations at the surface of GaN(0001) epitaxialy grown on Si(111). Phys. Rev. B **61**, 7618 (2000)
79. A.R. Smith, V. Ramachandran, R.M. Feenstra, D.W. Grewe, M.-S. Shin, M. Skowronski, J. Neugebauer, J.E. Northrup, Wurtzite GaN surface structures studied by scanning tunneling microscopy and reflection high energy electron diffraction. J. Vac. Sci. Technol. A **16**, 1641 (1998)
80. H. Heinke, M.O. Möller, D. Hommel, G. Landwehr, Relaxation and mosaicity profiles in epitaxial layers studied by high resolution X-ray diffraction. J. Cryst. Growth **135**, 41 (1994)
81. P. Scherrer, Bestimmung der Grösse und der inneren Struktur von Kolloidteilchen mittels Röntgenstrahlen. Nachr. Ges. Wiss. Gött. **26**, 98 (1918). (Determination of size and inner structure of colloidal particles by X-rays, in German)
82. R. Chierchia, T. Böttcher, H. Heinke, S. Einfeldt, S. Figge, D. Hommel, Microstructure of heteroepitaxial GaN revealed by high resolution X-ray diffraction. J. Appl. Phys. **93**, 8918 (2003)

Chapter 3
Structure of Organic Crystals

Abstract Organic semiconductors comprise small-molecule crystals and polymers, both featuring strong intramolecular bonding of carbon atoms by a system of conjugated π electrons. By contrast only weak intermolecular Van der Waals forces act between neutral nonpolar molecules, yielding usually defective crystals with a low stability. In two-component crystals additional longer-ranging dipole and Coulomb interactions occur. The weak intermolecular bonding favors a close molecule packing with large coordination numbers, often leading to a herringbone stacking in small-molecule crystals. Structures of thin films often deviate from the bulk structure and depend on the interaction to the substrate and on film thickness; organic crystal layers usually show epitaxial coincidence modes not observed in inorganic heterostructures. Grain boundaries between single-crystal domains and vacancies are prevailing defects in organic crystals. Polymers consist of long chain-like molecules packed largely uniformly in lamella of crystalline domains, which are separated by amorphous regions with tangled polymer chains. Crystallinity usually increases with the molecular length, get maximum at a critical molecular weight, and decreases beyond.

3.1 Building Blocks in Organic Crystals

Semiconductors made from organic materials gained recently much advertence; organic light-emitting diodes (OLEDs) and field-effect transistors (OFETs, processed as thin-film transistors TFT), are devices made of organic semiconductors which already entered the market [1, 2]. Also organic photovoltaic cells (OPVs) have achieved high conversion efficiencies [3]. Prominent molecules used in organic (opto-) electronics are listed in Sect. 3.1.2. Organic semconductors comprise two classes: *small-molecule crystals* (low molecular weight crystals) and *polymers*. A characteristic common feature of both classes is the bonding of carbon atoms by a system of *conjugated π electrons*. We first consider this intramolecular bonding, and then the intermolecular bonding acting between the building blocks of small-molecule crystals. Section 3.1 concludes with an outline of structures in polymers.

U. W. Pohl, *Epitaxy of Semiconductors*, Graduate Texts in Physics,
https://doi.org/10.1007/978-3-030-43869-2_3

3.1.1 Bonding in Organic Crystals

Intramolecular Bonding

In a system of conjugated π electrons two adjacent C atoms are not only bond by σ bonds, i.e., single-bonds, but in addition by multiple (usually double) bonds. A simple example is the ethen molecule C_2H_4 illustrated in Fig. 3.1. Three of the four valence electrons ($2s^2 2p^2$) of each C atom form σ bonds from sp^2 hybride orbitals: two to H atoms and one to the other C atom; all these bonds lie in one plane. The two remaining p_z electrons of the C atoms have their density distribution above and below this plane; they form an additional π bond, which is weaker than the strong σ bond because the overlap of the p_z wave functions of the adjacent C atoms is small.

Larger molecules of organic semiconductors have delocalized conjugated π electrons in alternating single and double bonds. As illustrated in Fig. 3.2 the molecules may be linear, like the long chains found in polymers, or cyclic. An introduction to organic crystals is given [4].

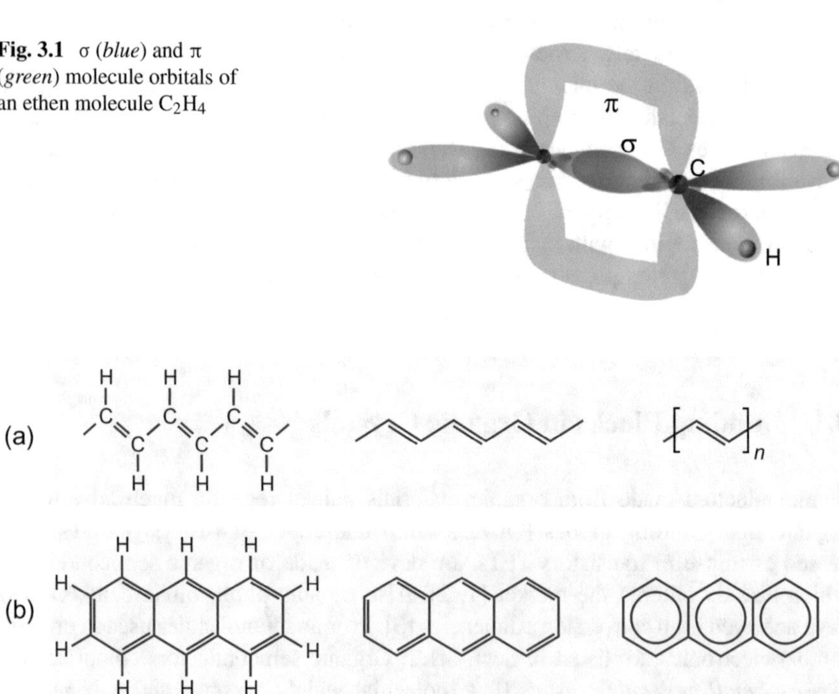

Fig. 3.1 σ (*blue*) and π (*green*) molecule orbitals of an ethen molecule C_2H_4

Fig. 3.2 Various representations of **a** polyethylene and **b** anthracene. Carbon atoms are generally left out, and usually also the hydrogen atoms. The π electrons are indicated by alternating single and double valence lines or, in cyclic molecules, also by a circle

Intermolecular Bonding

The structure of small-molecule crystals is determined by their intermolecular forces. Intermolecular bonding forces of electrically neutral and nonpolar molecules are pure Van der Waals interactions; nonpolar molecules comprise aliphatic hydrocarbons like the alcanes $CH_3(CH_2)_nCH_3$ and aromates like the oligoacenes $C_{4n+2}H_{2n+4}$ described in Sect. 3.1.2. In addition to the weak attracting force their bonding potential $V(r)$ comprises also a short-range repulsive component (Born forces) originating from the Coulomb repulsion of the core electrons and nuclei; this component can be approximated by an exponential term, yielding the Buckingham potential [5]

$$V = -\frac{\alpha}{r^6} + \beta\, e^{-\gamma\, r} \tag{3.1}$$

with empirical parameters α, β, and γ. If the molecules have a permanent dipole moment or polar substituents (see the charge-transfer complexes of the two-component semiconductors discussed below), static dipolar or ionic bonding are superimposed. In the presence of charges a term $+q_1 q_2/(4\pi\varepsilon_0 r)$ is added to the potential (3.1) [6].

For organic small-molecule crystals the distance r in (3.1) is not well defined; the distance between the molecules is of the same size as the extension of the molecules. A good description of experimental data is obtained, if the attractive potential between two neighboring molecules is calculated as the sum of all individual atom–atom potentials according to the Buckingham potential; the distance r in (3.1) is then the distance r_{ij} from an atom i of one molecule to an atom j of the other molecule [7, 8], yielding

$$V = \sum_{ij} -\frac{\alpha_{ij}}{r_{ij}^6} + \beta_{ij} e^{-\gamma_{ij}\, r_{ij}}. \tag{3.2}$$

Values for the parameters of (3.2) are given in Table 3.1.

Table 3.1 Parameters of the potential (3.2) for atom–atom pairs of neighboring hydrocarbon molecules [9]

Pair	α (kJ mol^{-1} Å^6)	β (kJ mol^{-1})	γ (Å^{-1})
C–C	568	83,630	3.60
C–H	125	8766	3.67
H–H	27.3	2654	3.74

Table 3.2 Melting points (m.p.) of crystals bonded by Van der Waals interaction

Molecule	Formula	m.p. (°C)
Anthracene	$C_{14}H_{10}$	216
Tetracene	$C_{18}H_{12}$	357
Pentacene	$C_{22}H_{14}$	300
Perylene	$C_{20}H_{12}$	274
Quaterthiophene	$C_{16}H_{10}S_4$	216
Quinquethiophene	$C_{20}H_{12}S_5$	253
Hexathiophene	$C_{24}H_{14}S_6$	290
Rubrene	$C_{42}H_{28}$	315–330

Since the intermolecular Van der Waals interaction is very weak, molecule crystals are usually soft and melt at comparably low temperatures. The low stability imposes limits for the epitaxy of these materials. Melting points of prominent organic crystals are listed in Table 3.2.

3.1.2 *Small-Molecule Crystals*

Growth Units in Small-Molecule Crystals

The growth units in organic semiconductors are bulky molecules with a lower symmetry than that of single atoms, which are the building blocks of *inorganic* semiconductors. Along with the weak intermolecular bonding organic crystals therefore show only little tendency to form stable solids and crystallize usually with low-symmetry unit cells. All physical properties have consequently tensor character with often large anisotropies. The versatile ability for synthesizing organic molecules leads to a huge and steadily increasing number of organic crystals. Most of them are insulating, but quite a few show conductive or semiconducting properties.

We distinguish *single-component* and *two-component* (charge-transfer) *semiconductors*. The first group contains the classical organic semiconductors, such as the acenes described below, which show comparatively high mobilities; the second group includes highly conductive compounds, some of which show semiconductor-metal transitions and even superconductivity; two-component semiconductors are described at the end of Sect. 3.1.2.

Single-Component Semiconductors are usually insulators, but may become photoconductive with sufficient optical excitation. The class of aromatic hydrocarbons like the acenes has been more thoroughly investigated. They have bandgap energies between 2 and 5 eV (see Sect. 5.1.2). The carrier mobility is low compared to inorganic semiconductors, but comparatively high for organic semiconductors: typically in the 10^{-2} to 10 cm^2/(Vs) range for electrons and holes at 300 K, falling with increasing temperature. The prominent organic semiconductors listed in in Table 3.3 are widely used particularly due to their high carrier mobility and stability. Highest hole mobilities at room temperature were reported for pentacene (35 cm^2/Vs [10]) and rubrene crystals (40 cm^2/Vs [11]); CuPc, known as blue dye in artificial organic pigments, is used in organic FETs, and Alq3 is commonly applied in organic LEDs.

Oligomers are the building blocks in many important small-molecule crystals; an oligomer (from Greek *oligos*, "a few", and *meros* "part") is a molecule consisting of a small number of the repeat units of a polymer, while a polymer (from Greek *poly* "many", and *meros* "part") is a large molecule composed of many repeated subunits. The family of *acenes* is formed from polycyclic aromatic hydrocarbons fused in a linear chain of conjugated benzene rings (Fig. 3.3a). The polycyclic aromate *rubrene* (5,6,11,12-tetraphenylnaphthacene, the numbers indicate where four phenyl groups are attached to tetracene) is built on a tetracene backbone with two phenyl rings on either side that lie a in a plane which is perpendicular to the plane of the backbone (Fig. 3.3b). The *perylene* molecule shown in Fig. 3.3e consists of two naphthalene molecules (each being similar to anthracene Fig. 3.3a with only *two* benzole rings, i.e., $n = 0$), connected by a carbon–carbon bond; all of the carbon atoms in perylene are sp^2 hybridized. A popular derivative of perylene is *PTCDA* (3,4,9,10-perylene-tetracarboxylytic dianhydride) shown in Fig. 3.3f. The heterocyclic *thiophenes* Fig. 3.3d include a sulfur atom in their ring structure. Examples for more complex compounds used in organic devices are copper phthalocyanine (CuPc) shown in Fig. 3.3c and Tris(8-hydroxyquinolinato)aluminum (Alq3, Fig. 3.3g). There are numerous derivatives of all these compounds obtained from substituting one or several hydrogen atoms (which are not drawn in Fig. 3.3) for organic groups like, e.g., methyl (CH_3), or a halogen like Cl, or a cyclic phenyl ring (C_6H_5) as those shown in rubrene Fig. 3.3b; for the representation of chemical structures see Fig. 3.2.

Unit Cells and Molecule Ordering

The *molecules* introduced above represent the building blocks of small-molecule crystals. They may either be arranged in a regular order of a crystal or irregularly; while both forms are applied for organic electronics, generally best performance is obtained with crystalline modifications. The molecules of organic semiconductors do not have unsaturated bonds and hence no free valences like the atoms of inorganic semiconductors.

Due to the weak attractive intermolecular interaction the molecules tend to crystallize in lattices with closest packing for maximizing the number of intermolecular contacts. The packing density is described by the *molecular packing coefficient* [7]

$$k = Z\,V_{\text{molecule}}/\,V_{\text{uc}}, \tag{3.3}$$

where V_{uc} is the volume of the unit cell and V_{molecule} the volume of one of the Z molecules of the unit cell. V_{molecule} can basically be computed from the geometric molecule structure and the atomic radii [7]; however, in the calculation of the charge density the volume is found to depend on the environment [12]. The *Van der Waals volume* is substantially larger than V_{molecule} used in (3.3) [13, 14]; in the bulk of an organic crystal it is given by V_{uc}/Z. Stable crystals have packing coefficients between 0.65 and 0.77; this is on the same order as the close packing of spheres

Table 3.3 Crystallographic data of some organic semiconductors. Z denotes the number of molecules per unit cell, values of vectors a, b, c are given in Å, volume V of the unit cell in Å^3, and angles α, β, γ in degrees; α is the angle between **b** and **c**, β and γ are defined correspondingly. Most molecules form various polymorphs, only data of a few are given

Organic crystal	Formula	Crystal system	Z	a	b	c	α	β	γ	V
Anthracene	$C_{14}H_{10}$	Monoclinic	2	8.6	6.0	11.2	90	125	90	474
Tetracene	$C_{18}H_{12}$	Triclinic	2	7.9	6.0	13.5	100	113	86	583
Pentacene	$C_{22}H_{14}$	Triclinic	2	7.9	6.1	16.0	102	113	86	692
Rubrene	$C_{42}H_{28}$	Monoclinic	2	8.7	10.1	15.6	90	91	90	1383
		Triclinic	1	7.0	8.5	11.9	93	106	96	684
		Orthorhombic	4	26.9	7.2	14.4	90	90	90	2736
Perylene (α phase)	$C_{20}H_{12}$	Monoclinic	4	11.4	10.9	10.3	90	101	90	1249
Perylene (β phase)	$C_{20}H_{12}$	Monoclinic	2	11.3	5.9	9.7	90	92	90	394
Quaterthiophene (α-4T)	$C_{16}H_{10}S_4$	Monoclinic	4	30.5	7.9	6.1	90	92	90	1471
Hexathiophene (α-6T)	$C_{24}H_{14}S_6$	Monoclinic	4	44.7	7.9	6.0	90	91	90	2117
CuPc (β phase)	$CuN_8C_{32}H_{16}$	Monoclinic	2	14.6	4.8	17.3	90	105	90	1171
Alq3 single crystal	$Al(C_9H_6NO)_3$	Triclinic	2	8.4	10.3	13.2	109	97	90	1072

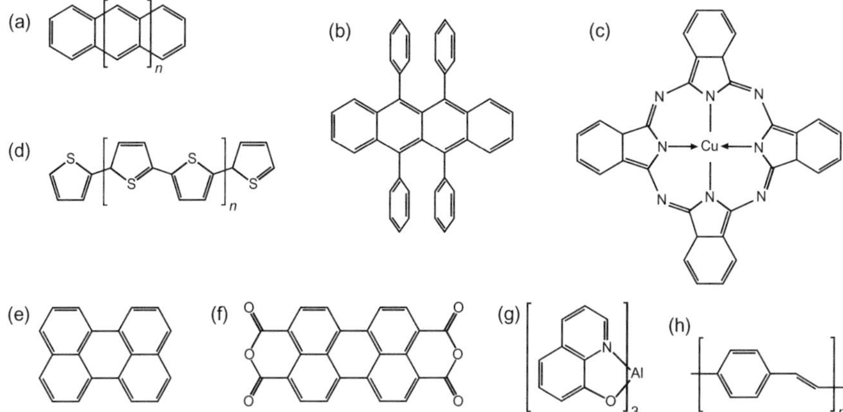

Fig. 3.3 Molecules of prominent organic semiconductors: **a** Oligoacenes anthracene (n = 1), tetracene (n = 2), pentacene (n = 3); **b** rubrene, the small width of the side rings indicates a twist by 85° out of the plane of projection; **c** copper phthalocyanine (CuPc), **d** oligothiophenes quarterthiophene (n = 1), hexathiophene (n = 2); **e** perylene; **f** PTCDA; **g** tris(8-hydroxyquinolinato)aluminum (Alq_3); **h** The repetition unit of the polymer poly(p-phenylene vinylene) PPV

(0.74, see *hcp* structure Fig. 2.3a). If the shape of a molecule does not allow for a coefficient exceeding 0.6 the respective solid tends to vitrification. Usually the packing coefficient is inverse to the compressibility.

The close packing is related to large coordination numbers. Usually organic crystals crystallize with a 12-fold coordination; in some cases the specific shape of molecules may lead to packing with 10-fold or 14-fold coordination.

The mutual arrangement of the molecules follows the trend of close spacing: planar molecules prefer a parallel alignment. Furthermore, atoms tend to locate at interstices between atoms of the adjacent molecule; the ensuing minimum distance of neighboring molecules is typically 3.4–3.5 Å, similar to the spacing of layers in graphite (3.35 Å). This favors a crystallization in a herringbone packing with an angle between adjacent columns of the planar molecules, observed, e.g., for oligoacenes and oligothiophenes. The rule of thumb for interstitial alignment does not apply if the molecules have a permanent dipole moment or polar substituents; even small dipolar or ionic contributions to the intermolecular bonding have a respective long-ranging $1/r^3$ or $1/r$ dependence and thus a significant effect on the crystal structure.

The frequently observed *herringbone alignment* of neighboring molecules is illustrated in Fig. 3.4a for a pentacene crystal. The nonpolar acene molecules are planar, and a similar crystalline arrangement of the molecules is found for the other members of this family. The angle enclosed by the normals to the molecular planes of the two molecules belonging to a unit cell is referred to as *herringbone angle*; all acenes feature herringbone angles around 50°. The respective shapes of monoclinic and triclinic unit cells do hence not differ much (except for the different molecule lengths and respective c values), as indicated by comparable angles α and γ near 90° listed in Table 3.3. The unit cell of an anthracene crystal is shown in Fig. 3.4b.

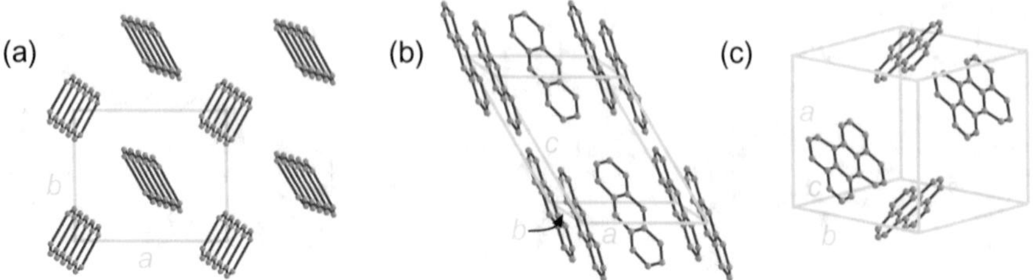

Fig. 3.4 Crystal structure of organic semiconductors; orange circles represent C atoms, H atoms are not shown. **a** The frequently observed herringbone packing of organic crystals, demonstrated for a top view on the *a*–*b* plane of a pentacene crystal. **b** Anthracene crystal with two anthracene molecules per unit cell. **c** Unit cell of the α phase of a perylene crystal comprising two pairs of perylene molecules

The plane of the molecule does not coincide with a face of unit cell. In pentacene the long axes of the two differently aligned molecules form respective angles of 22° and 20° to the *c* axis, their short axes form angles of 31° and 39° to the *b* axis, and their normal axes form angles of 27° and 32° to the *a* axis of the unit cell; comparable values are found for the other acene crystals. It should be noted that the prevailing bulk structure differs from the structure predominately found in thin-film growth. In thin films the tilt of the long molecule axis with respect to the *a*–*b* plane is much smaller than in bulk crystals (3° instead of 22° for pentacene, see [15]).

The herringbone packing is also realized in a variety of crystal structures of rubrene and perylene crystals. In the α phase of perylene shown in Fig. 3.4c the pattern is built by molecule *pairs*, while it is formed by single molecules in the β phase (not shown). The pairing leads to a roughly doubled *b* value of the α-phase unit cell, while the other parameters are similar.

Two-Component Semiconductors consist of pairs of complementary molecules with large differences in their redox properties: the organic molecules with a low ionization energy acts as electron donors *D*, and the other molecules with a high electron affinity acts as acceptors *A*. Such a combination produces organic crystals which may show very low or vanishing activation energies for free carriers and hence comparatively high conductivities. The crystals are formed by a sandwich-like stacking of planar molecules, where donors and acceptors form $D^{\delta+}\,A^{\delta-}\,D^{\delta+}\,A^{\delta-}$ structures, or they are located in separated stacks, i.e., the stacking contains $D^{\delta+}\,D^{\delta+}$... and $A^{\delta-}\,A^{\delta-}$ ... complexes for segregated stacking with face-to-face stacks. δ denotes the transferred charge per molecule in units of elementary charges.

The charge transfer (CT) may be incomplete, yielding two limiting cases: weak CT complexes and the strong CT complexes, which are also referred to as *radical ion salts* [16]. In radical ion salts, an organic radical cation (such as perylene$^+$) is combined with a counter anion (such as a halogen or PF_6^-), or an organic radical anion (such as $TCNQ^-$) is combined with a counter cation. Strong CT complexes consist of donors with relatively low ionization energy (below 7 eV) and acceptors with high

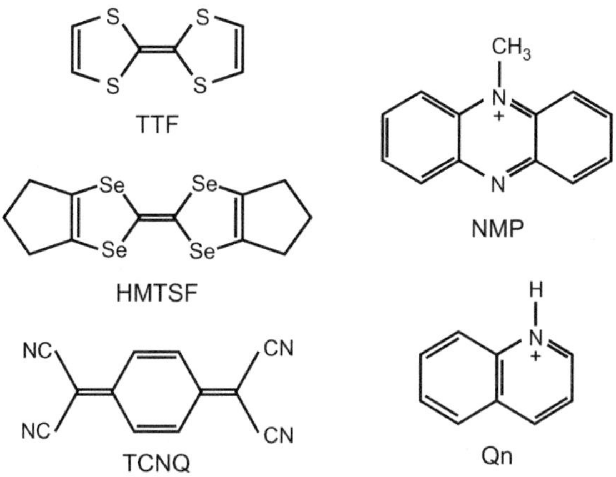

Fig. 3.5 Structure of some typical organic donor and acceptor molecules in charge-transfer crystals: Tetrathiofulvalene (TTF), Hexamethylenetetraselenofulvalene (HMTSF), Tetracyanoquinodimethan (TCNQ), N-methylphenazinium (NMP), Quinolinium (Qn)

electron affinity (above 2 eV). The solids have a pronounced ionic character, i.e., δ is often close to 1; usually $\delta < 1$ for conductive radical ion salts.

Charge-transfer complexes with δ significantly smaller than unity can show high conductivities, caused by the incomplete charge transfer between D and A, which results for the ground state in partially filled bands. Typical donor and acceptor molecules that form such charge-transfer crystals are given in Fig. 3.5. Many such semiconductors have low-lying electronically excited states in which an electron is transferred from D to A. The charge-transfer transition in the excited state (which has essentially ionic charge character) may be written as

$$DA \rightarrow D^+A^- \text{ with } E_{\mathrm{CT}} = I_D - A_A - C, \tag{3.4}$$

where I_D is the ionization energy of the donor, A_A is the electron affinity of the acceptor (both in the gas phase), and C is a Coulomb binding-energy of the excited state. E_{CT} is the "energy gap" between the ground state and the excited charge-transfer state [17]. The resulting structures are termed *neutral charge-transfer crystals*, typically with stacks of alternating D and A molecules: $DADADA$ .... The lowest excited state is $DADAD^+A^-DADA$ ...; the respective activation energy for semiconductivity is typically $E_s \cong \frac{1}{2}E_{\mathrm{CT}}$ [18].

The incomplete electron exchange and consequently partially filled bands result in a rather large semiconductivity—or even metallic conductivity. The resistivity of these semiconductors lies between 10^2 and 10^6 Ω cm at room temperature with a transfer-energy gap in the 0.1–0.4 eV range [19]. The conductivity is usually highly anisotropic, with the electron transfer-integral in the stacking direction typically a factor of 10 larger than in the direction perpendicular to the stacks [20]. Trapping is of minor importance in these semiconductors with a high carrier density [21]. The charge-transfer crystals provide an opportunity for fine tuning of the semiconductive properties by replacing TTF-type donors and TCNQ-type acceptors with other similar molecules [22], which could render such materials attractive for various technical applications.

3.1.3 Polymers

Polymers are organic compounds consisting of long, chain-like molecules with typically 10^2 to 10^4-fold repeated molecular units. In semiconducting polymers with π bonding in a conjugated chain the bonds have alternating lengths along the backbone. In the double (or triple) bonds π electrons of the p_z orbitals are localized and form an occupied valence band (or HOMO, highest occupied molecular orbital), separated by a bandgap from the conduction band (LUMO, lowest unoccupied molecular orbital).

Polymers with conjugated π electrons are interesting for semiconductor applications due to their favorable large-scale processing ability and their robustness. A simple example is poly(p-phenylene vinylene), PPV, shown in Fig. 3.3h. Since this polymer does not dissolve in common solvents, more conveniently prepared derivatives of PPV are widely applied. Thin films of polymers are often formed by solution processing such as spin casting, resulting in polycrystalline or amorphous solids with entangled long polymer chains. These films are more robust than the crystalline films prepared from small molecules; their electrical properties are, however, inferior to crystalline solids.

The structure of polymers is generally more distorted than that of small-molecule crystals; consequently the mobility of carriers is usually significantly lower. In a solid the long polymer molecules are generally packed together non-uniformly, building both crystalline and highly disordered amorphous domains. The amorphous regions are composed of coiled and tangled chains, whereas in crystalline (albeit still distorted) regions linear polymer chains are oriented in a three-dimensional matrix. Polythiophene (Fig. 3.3d) and its derivatives are interesting polymers due to their very high carrier mobility (up to ~10 cm^2/Vs). Figure 3.6 shows a schematic of the polymer structure for polythiophene; in crystalline regions the molecules are arranged in long sheets, which are oriented along the conjugated backbone direction labelled [001] and packed along the π–π stacking direction [010]. The sheets are additionally stacked along the [100] lamella-stacking direction. Each crystallite represents a grain in the polymer material. Grain boundaries are anisotropic regions of strong disorder with bend chains; low-angle grain boundaries are somewhat less distorted than high-angle grain boundaries.

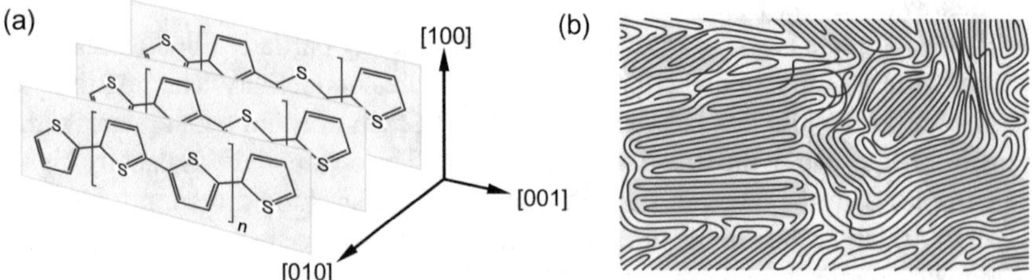

Fig. 3.6 **a** Packing motiv of a polymer semiconductor in crystalline regions. The *index at the bracket* indicates the *n*-fold repetition of the unit along the polymer chain. **b** Tangled polymer chains in the bulk with crystalline (*shaded*) and amorphous regions

3.2 Substrate-Layer Relation and Defects of Organic Crystals

Organic crystals generally suffer from their limited perfection. Crystal growth is hampered by various factors: the orientational degree of freedom of their building blocks in *small-molecule crystals* favors disorder-induced defects, crystal properties vary sensitively with the introduction of contaminants, and the rigid molecule structure combined with a weak intermolecular bonding make organic crystals fragile. Structural imperfections imply the frequently observed formation of polymorphs, which differ, e.g., in the herringbone angle or even in the number Z of molecules per unit cell. In *polymers* the strong disorder in the alignment of the polymer chains generally leads to low carrier mobility. Furthermore, defects in the bond alternation sequence yield unpaired electrons (Sect. 3.2.3). In the following the two classes of organic semiconductors are discussed separately.

3.2.1 Substrate-Layer Relation of Organic Crystals

In the well-established heteroepitaxy of *inorganic* crystals the atoms at the interface between substrate and layer are laterally usually in exact register, i.e., commensurate conditions exist. Since the elastic constants of substrate and layer are large compared to those of organic crystals, no large deformations can be accommodated by layer or substrate; in case of nonmatching conditions therefore only azimuthal rotation of the layer with respect to the substrate may occur to minimize interfacial energy—apart from the introduction of misfit dislocations.

In *organic epitaxy* the layers have small elastic constants and generally low-symmetry oblique unit cells, preventing to find matching conditions. A sensitive balance between noncovalent intralayer interactions among the organic molecules in the layer and the interaction of the layer to the substrate lead to epitaxial modes not observed in inorganic heterostructures [23, 24]. These modes can be classified by the relation of the coefficients m_{ij} in the matrix (3.5), which transforms the lateral substrate lattice-vectors ($\mathbf{a}_S, \mathbf{b}_S$) to the respective vectors ($\mathbf{a}_L, \mathbf{b}_L$) of the layer:

$$\begin{pmatrix} \mathbf{a}_L \\ \mathbf{b}_L \end{pmatrix} = \begin{pmatrix} m_{11} & m_{12} \\ m_{21} & m_{22} \end{pmatrix} \begin{pmatrix} \mathbf{a}_S \\ \mathbf{b}_S \end{pmatrix}. \tag{3.5}$$

The values of m_{ij} depend on the choice of the vectors describing the primitive unit cell of the layer in the interface plane, but all choices have equal area and thus identical matrix determinants.

Types of Substrate-Layer Relations

Point on point (POP) commensurism (or simply ***commensurism***) exists if all lattice points of the layer lie on a lattice point of the substrate. This can also be regarded as all layer lattice-points lying simultaneously on *two* primitive lattice *lines* of the substrate. All coefficients m_{ij} in (3.5) are integers; the matrix implies that each lateral primitive lattice vector of the layer is an integer multiple of a (primitive or non-primitive) substrate lattice vector with the same orientation [23].

Point on line (POL) coincidence (also referred to as ***coincidence I***) exists if all lattice points of the layer lie on primitive lattice *lines* of the substrate surface-lattice; the lattice lines are defined by the intersection of substrate lattice planes with the surface. At least two integers are confined to one column of the matrix in (3.5). There are two classes of POL I coincidence. In class IA both remaining matrix coefficients are rational numbers, being typical for coincidence conditions in inorganic heteroepitaxy; a layer supercell with matching corner points can then be found. In class IB at least one coefficient is an irrational number; no matching condition in the corresponding direction can be found.

POL coincidence II exists if only *some* lattice points of the layer lie on primitive lattice lines. Some lattices lines of the layer then coincide with primitive lattice lines of the substrate. The matrix in (3.5) contains only rationals and no integer column. Since the coefficients are rational, a layer supercell can be constructed with matching corner points.

Line on line (LOL) coincidence exists if all lattice lines of the layer lie on *non-primitive* lines of the substrate surface-lattice [24]. This means a coincidence between nonprimitive reciprocal lattice vectors of layer and substrate.

Incommensurism exists if no distinct registry between layer lattice and substrate lattice occurs; the layer lattice does not coincide with lattice points or primitive lattice lines of the substrate. At least one matrix coefficient is an irrational number and no column of integers exists.

Different types of substrate-layer relations are illustrated in Fig. 3.7.

Molecule-Substrate Interaction

Any type of epitaxial alignment is related to a minimum of the interaction potential V_{inter} at the interface between substrate and layer. The prediction of epitaxial order is hampered by the fine balance of the weak interaction forces and the large sizes of the molecular domains, which need to be taken into account. The total interface potential V_{tot} is generally given by a sum over atom–atom potentials that can be regrouped to express the distinction between the molecule–molecule pair interaction within the layer V_{intra} and the molecule-substrate interaction V_{inter},

$$V_{\text{tot}} = \sum_i \sum_j V_{\text{atom}}(\mathbf{r}_i - \mathbf{r}_j) = V_{\text{intra}} + V_{\text{inter}}, \tag{3.6}$$

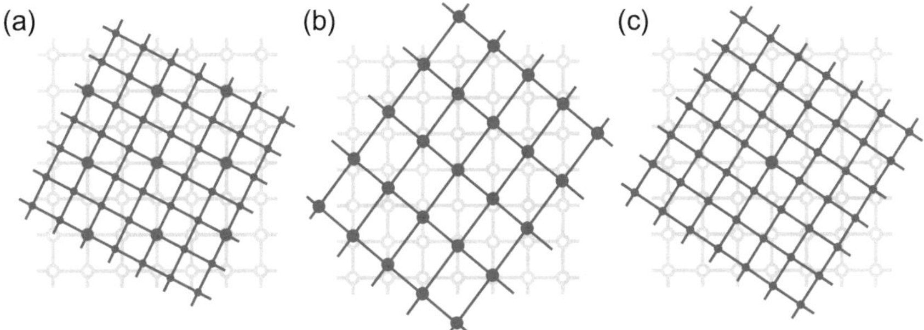

Fig. 3.7 Schematic of different types of epitaxial relations between substrate (symbolized by *open circles*) and layer (*full circles*). Coincident lattice points are marked by *large full circles*; **a** commensurate (point-on-point) relation, **b** point-on-line coincidence, **c** incommensurate relation

where $\mathbf{r}_i$, $\mathbf{r}_i$ are the coordinates of atoms i and j. The sum can be simplified by considering only additive two-body interactions, by treating the molecules rigid, and by assuming the substrate unperturbed by the presence of the layer on top. By introducing parametrized potentials, the minimum in the total interface potential V_{tot}, which has quite a complex structure for the large organic unit cells [25], was demonstrated for POL [26] and LOL [24] coincidence.

There is basically a decrease of the layer-substrate interaction from commensurism to point on line coincidence (with class IA being larger than class IB) and even further to line on line coincidence. The interaction strength depends essentially on the ratio of coinciding to non-coinciding lattice points and hence on the degree of the registry of layer and substrate lattices.

Since inorganic substrates have much smaller unit cells than organic crystals, a large organic molecule attaches during nucleation at docking groups to the substrate—if it is sufficiently reactive to form chemical bonds—or it is bonded by weaker electrostatic or Van der Waals interactions. While the binding energy of a single covalent hybridization is in the eV range and much stronger than the weak Van der Waals bond, which yields only 1–10 meV per atom, the respective interaction of the entire molecule may also be in the eV range; an intermediate bonding strength may result from electron transfer between π electrons of the molecule and the substrate.

The favored alignment of an organic layer can be estimated from geometric considerations like those pointed out above, when the elastic constants of the layer and the layer-substrate interface are taken into account [27, 28]. The elastic constants within the layer C_{intra} and of the interface C_{inter} are given by the second derivatives $e\,\partial^2 V/\partial x^2$ at the respective minima of the potentials V_{intra} and V_{inter} in (3.6), where x is a lateral coordinate parallel to the interface. The magnitudes of the elastic constants are hence proportional to the curvature of the respective potentials. If a lattice point of the layer does not coincide with a lattice point of the substrate, it must be displaced from its energetically most favored position and the elastic constants control the ease for this process. The interface energy $E_{\text{inter}} = eV_{\text{inter}}$ will thereby increase with the magnitude of the displacement and the number of non-coinciding lattice points of the layer. In addition, the ratio of C_{intra} and C_{inter} affects the interface energy and hence the preferred mode of epitaxial relation.

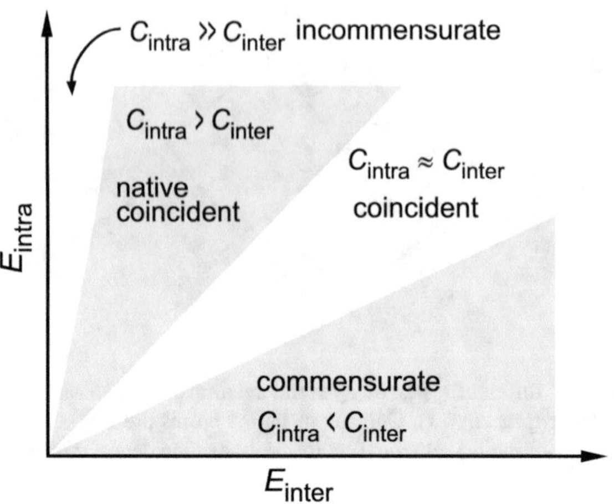

Fig. 3.8 Schematic of regimes for different epitaxial substrate-layer relations depending on the elastic constants within the layer C_{intra} and between layer and substrate C_{inter}. The energies E_{intra} and E_{inter} are given by the respective potentials V and related to the elastic constants C by the potential curvatures at minimum. *Native coincident* denotes a crystalline organic layer with its native unperturbed bulk structure

As a rule of thumb derived from a large number of interfaces, commensurate overlayers are favored when $C_{intra} < C_{inter}$; the molecule–substrate interactions are then relative strong, and the commensurate ratio is established at the expense of a dense packing within the compliant layer. On the other hand, if $C_{intra} \gg C_{inter}$ the layer tends to preserve its native structure; the azimuthal alignment of the layer unit-cell will then either adapt a commensurate configuration, if this is possible without too much increasing the interface energy, or otherwise one of the coincident configurations. An incommensurate alignment is much less favorable [23]; the preferred regimes of epitaxial modes for different ratios of the elastic constants C_{intra} and C_{inter} are illustrated in Fig. 3.8. In a coincident configuration the organic layer may adopt various polymorphic structures besides its native bulk structure to achieve some matching conditions.

3.2.2 *Ordered Pentacene Layers*

We consider the widely studied *pentacene* to illustrate some general features of ordered organic layers. This organic semiconductor is attractive due to the very high mobility of carriers found in bulk crystals and represents a model and benchmark system for organic small-molecule semiconductors. Pentacene ($C_{22}H_{14}$) is a polycyclic aromatic hydrocarbon composed of 5 fused benzene rings in a linear planar arrangement (Fig. 3.4a). Bulk crystals grow in the herringbone structure discussed in Sect. 3.1.2; the *thin-film phases* deviate from the bulk structure, representing largely modifications of the herringbone packing. Thin pentacene films were studied on many substrate materials.

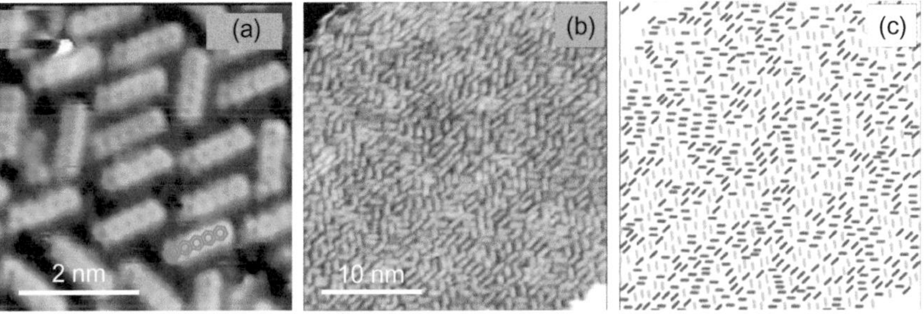

Fig. 3.9 Scanning-tunneling micrographs of pentacene molecules adsorbed on Ni(111). **a** Individual molecules with a superimposed structural model at the bottom. **b** Overviev image showing orientational disorder in the molecule alignment. **c** Same as (**b**) with color-coded orientations. Reproduced with permission from [29], © 2015 RSC Publishing

On metals the interaction between substrate and the delocalized electrons of the layer molecules is generally strong. The contact of individual molecules with the surface is hence large, leading mostly to a *lying-down configuration* of the first molecule layer(s): the molecular long axes are then parallel to the surface. A lateral ordering aligned to the periodically arranged metal atoms may occur at suitable growth conditions; the respective molecular spacing in the organic layer may substantially differ from that in an organic bulk crystal. For sufficiently strong intermolecular interactions often a continuous variation toward the bulk value with increasing distance to the interface is observed, including various structural phases.

The structure of a monolayer thick pentacene layer deposited at room temperature on a Ni(111) surface is shown in Fig. 3.9. Nickel crystallizes in the *fcc* structure, which has hexagonal net planes along the $\langle 111 \rangle$ direction (Fig. 2.3b); pentacene molecules are strongly adsorbed on the metal surface and aligned along the $\langle 1\bar{1}0 \rangle$ directions of the Ni substrate, with an approximately equal fraction for the three possible orientations [29]. The arrangement is referred to as "random tiling" phase, since the alignment of a given molecule cannot be concluded from the directional alignment of its neighboring molecules. Pentacene layers subsequently deposited on the flat-lying molecules align in a tilted arrangement yielding a pronounced islanding [30].

To reduce the strong metal-layer interaction a thin (1 monolayer thick) organic interlayer may be inserted, see below (*On an organic interlayer*).

On insulating substrates such as ionic crystals the interaction between the molecules of an organic layer and the atoms of the insulating substrate is much weaker. Still the effect of substrate atoms can lead to crystal structures distinct from the bulk structure. Usually the lateral alignment of layer molecules with respect to the substrate orientation is poor, due to the weak interaction and a much larger, normally incommensurate unit cell. Independent locations of nucleation with different orientations lead to the creation of extended defects during coalescence. Typically organic thin films exhibit a texture structure with a preferred crystallographic plane parallel to the interface.

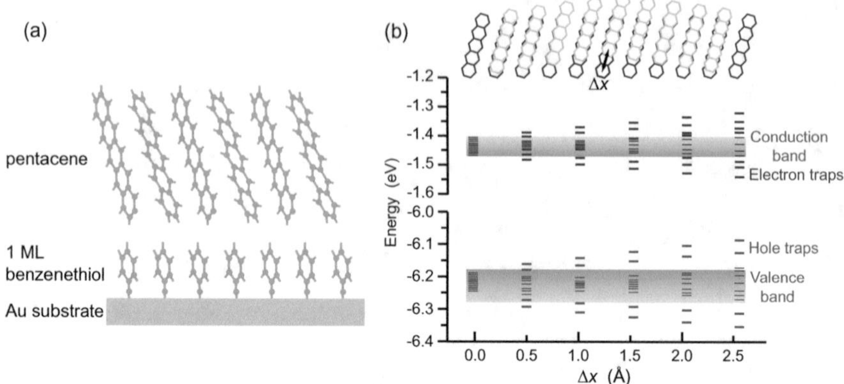

Fig. 3.10 **a** Orientation of pentacene molecules deposited at 295 K on a Au(111) surface passivated with a monolayer of benzenethiol. **b** Creation of trap states in the bandgap for holes and electrons due to the displacement of pentacene molecules by an amount Δx along the long molecular axis in a pentacene crystal. Reproduced from [38] with permission of AIP Publishing

The insertion of a thin organic interlayer between the substrate and the organic semiconductor similar to the passivation layer on a metal shown in Fig. 3.10a proved also beneficial for layers on insulating substrates. The achievement of a low density of electronic states in the bandgap for organic semiconductors with a high degree of crystallinity is discussed in Sect. 5.2.2.

Pentacene on silicon dioxide is a particularly interesting system for organic thin-film transistors; since the charge transport in such devices predominantly occurs in the first few monolayers above the dielectric [31, 32], carrier mobility largely depends on the molecular packing in the first monolayer that determines the overlap of orbitals between neighboring molecules. The two molecules in the thin-film unit cell of pentacene are more vertically oriented on SiO_2 (also on other substrates) with more rectangular in-plane vectors **a** and **b** ($\gamma = 89.8°$) than in the bulk unit cell ($\gamma = 86°$, see Table 3.3); molecules of the first monolayer adsorb exactly vertically on SiO_2 ($\alpha = \beta = 90°$ [33]). The molecules in the unit cell exhibit a herringbone packing (viewed along the interface normal) like the bulk structure. The upright alignment on the substrate indicates a somewhat stronger molecular–substrate interaction than the intermolecular interaction in the layer and yields in the first monolayer a slightly compressed in-plane unit cell with shorter a and b axes.

On an organic interlayer between an inorganic substrate and the organic semiconductor the initially formed layer structure can be controlled; this applies for both substrates of inorganic semiconductors and metals. Pentacene molecules form strong multiple covalent bonds on Si(001) [34] or Si(111) surfaces [35]. The pentacene growth hence commences with a disordered layer of strongly bonded molecules which form coalescing crystalline islands in a textured organic solid. Passivating the Si surface prior to pentacen growth by an exposure to organic molecules such as styrene, cyclopentane, or hexane allows to yield monolayer-thick pentacene films with low defect densities [36, 37].

The strong interaction of organic molecules to a metal substrate shown above may also be reduced by inserting an organic interlayer. A monolayer of benzenethiol on an Au(111) surface was, e.g., applied for growing a pentacene film, yielding a crystalline bulk-like *standing-up phase* with the *ab* plane (Fig. 3.4a) parallel to the interface as illustrated in Fig. 3.10a [38]. Using such layers it was shown that defect states may occur despite maintaining the two-dimensional crystalline packing with the herringbone motiv. Scanning-tunneling microscopy images of 2 monolayer thick pentacene layers showed a broad Gaussian distribution of slight thickness variations in the range of 1–2 Å, well *below* a thickness of one monolayer (~15 Å). A corresponding displacement of pentacene molecules along the long molecular axis illustrated in Fig. 3.10 breaks the translation symmetry in the crystal; calculations show that such a disorder is sufficient to create defect states in the bandgap as shown in Fig. 3.10b.

Another interesting substrate is graphite; its lattice on the basal plane is nearly identical to the carbon frame of polycyclic aromatic hydrocarbons like pentacene, and avoids strong molecule bonding. The good match of its basal plane yields highly ordered monolayers with a commensurate superstructure; subsequent molecular layers are gradually vertically tilted with increasing coverage [39]. In the first pentacene monolayer the molecules are adsorbed in closely spaced rows with their plane parallel to the surface; the long axis is oriented along one of the azimuth $\langle\bar{1}2\bar{1}0\rangle$ directions, and neighboring molecules are slightly shifted along the long axis. This yields an oblique surface unit-cell and a commensurate superstructure described by the coefficient matrix $\begin{pmatrix} 7 & 0 \\ -1 & 3 \end{pmatrix}$.

3.2.3 *Defects in Small-Molecule Crystals*

The defect types classified for inorganic crystals in Sect. 2.3 also occur in organic crystals: planar defects like stacking faults and grain boundaries, line defects like screw and edge dislocations, and point defects [37, 40]. A screw dislocation in a perylene single crystal (α phase) is shown in Fig. 3.11a; the step height of about 1 nm indicates the monomolecular increment. Figure 3.11b shows a stacking fault in pentacene. All these defects affect carrier transport; carrier mobilities are particularly high in some organic bulk single-crystals such as Rubrene and Pentacene, but orders of magnitude lower in respective thin films due to the very high defect density. The π electrons are delocalized over the individual molecules, but conductivity of the *crystal* largely depends on the transfer integral between neighboring molecules that

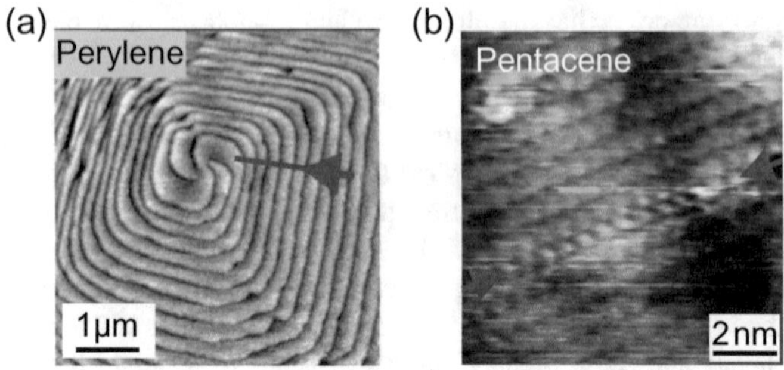

Fig. 3.11 **a** AFM migrograph of a screw dislocation in a perylene crystal with monomolecular steps measured along the *marked line*. **b** Scanning tunneling micrograph of a stacking fault in pentacene. Reproduced with permission from [41], © 2015 ACS, and from [37], © 2015 AIP Publishing

is strongly affected by the defects. Moreover, the numerous defects introduce a high density of trap states. In the following we discuss point defects and extended defects.

Prominent Point Defects

Point defects comprise intrinsic defects like vacancies or interstitials, and extrinsic impurities. Intrinsic point defects are mostly vacancies, as shown by self-diffusion experiments [42]. In two-component crystals these imperfections are easily introduced when the stoichiometry is not balanced. Extrinsic defects occur since most organic crystals are not grown under ultraclean conditions like those applied for inorganic semiconductors. Oxygen, hydrogen, and water are prominent contaminants known to affect also organic crystals. We focus on pentacene as a model system for the effect of such impurities.

Oxygen is easily introduced into a pentacene crystal due to a calculated energy drop of 0.13 eV per O_2 molecule [43]. Studies on pentacene thin-film transistors clearly prove that oxygen only creates defects states if the Fermi energy is high in the bandgap [44]. Total-energy calculations within the local DFT scheme yield two prominent oxygen-related defects forming trap states in the gap (Fig. 3.12). Breakup of the O_2 molecule leads to the energetic favorable complex Pn–2O illustrated in Fig. 3.12b, where the two dangling O atoms are bond to C atoms at positions 6 and 13 (depicted in panel *a*) of the same molecule (pentacenequinone). Oxygen removes a p_z orbital of the respective C atom from the planar π system of the molecule by forming a double bond and creates a localized gap state [45]; filled with 2 electrons, the energy of the Pn–2O state is computed to lie 0.36 eV above the valence-band edge [transition level (–/2–)]. The creation of the twofold charged defect yields a formation energy of $2(E_g - 0.36\text{ eV})$ with a Pn bandgap energy of 2.0 eV, greatly favoring defect formation at high Fermi level. A shallow (0/–) acceptor energy of 0.08 eV was determined for this complex as a possible source of p-type doping. The dangling oxygen at position C6 may build an intermolecular bridge to a C13 atom of a neighboring Pn molecule (Fig. 3.12c), thereby forming the Pn–$2O_{II}$ complex, gaining 0.61 eV formation energy, and creating a deeper acceptor level at $E(0/–) =$

0.29 eV above the valence-band edge. Such findings agree with the broad distribution (0.16 eV FWHM) of trap states peaking 0.28 eV above valence-band maximum (mobility edge) measured with pentacene thin-film transistors which were exposed to photo-oxidation [46]. Since the π electron system of pentacenequinone is smaller than that of Pn, it has a larger bandgap energy; the molecule is hence also believed to act as a scattering center.

Hydrogen added to a pentacene molecule acts quite similar to an added oxygen atom [45]. Adding H_2 to a pentacene molecule is exothermic; again the center ring of pentacene is most reactive, and bonding of two additional hydrogen atoms at the 6 and 13 positions (Fig. 3.12a) yielding dihydropentacene is favorable [47]. The calculated (+/0) level for *one* additional H atom at C6 position lies ~0.34 eV above the valence-band maximum and the (0/–) level occurs at ~0.80 eV, yielding a 0.5 eV-wide range of the Fermi energy for the stable neutral defect; adding *two* H atoms at the 6 and 13 positions yields a decreased valence-band edge and an increased conduction-band edge at the dihydropentacene molecule, so that no gap states are expected in this case [45]. Such a molecule will therefore also act as a scattering center.

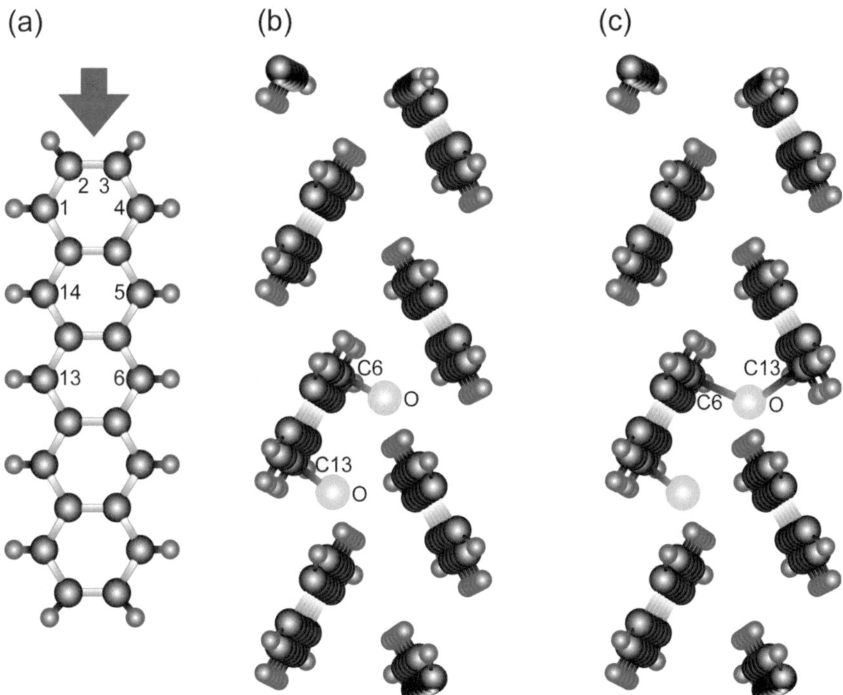

Fig. 3.12 **a** Numbering of the 14 outer C atoms in a pentacene (Pn) molecule; the arrow on top indicates the viewing directions in (**b**) and (**c**). **b** The Pn–2O complex with 2 dangling O atoms bonded to a single Pn molecule. (**c**) The Pn–$2O_{II}$ complex with an intermolecular O bridge and a dangling O atom. Adapted with permission from [44], © 2009 Wiley

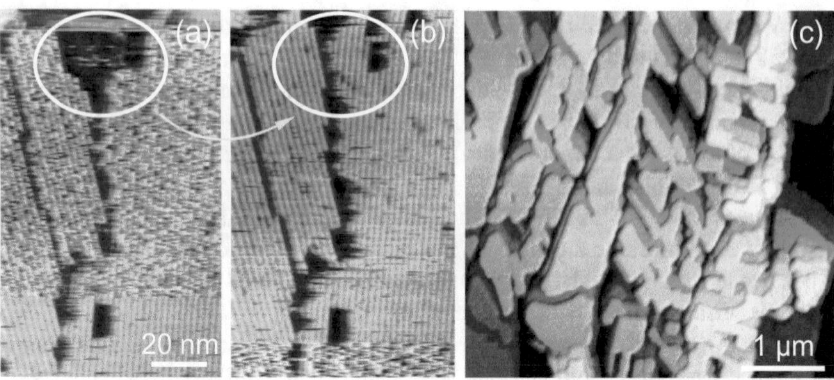

Fig. 3.13 **a** Scanning-tunneling micrographs of a monolayer-thick pentacene layer adsorbed on (0001) graphite; **b** same as (**a**) recorded some minutes later. Changes to a grainy part of the image indicate sudden tip switches due to molecule pick-up and release at the scanning tip. **c** Morphology of a layer with 35 nm nominal thickness. Reproduced with permission from [39], © 2010 APS

Water is also easily introduced into a pentacene crystal; an energy gain of 0.55 eV was calculated for the incorporation of a H_2O molecule at a favorable position between pentacene layers [43]; the study indicates that the molecule does not dissociate. Furthermore, neither the structure of the crystal nor the electronic density of states shows significant changes for intercalated water. Scattering of carriers and hence reduction of carrier mobility are the main effects expected from the incorporation of water into pentacene.

Extended Defects

A grain boundary between single-crystalline domains in a monolayer-thick pentacene layer grown on graphite is shown in Fig. 3.13. Pentacene molecules are quite mobile at room temperature due to the weak adsorption to this substrate: the ovals in panels (a) and (b) mark changes occurring by detachment from and attachment to existing islands on a time scale of minutes. Still the boundary persists and cannot be closed even in thicker layers and after thermal treatment. Figure 3.13c shows an atomic force micrograph of a thicker layer. The organic film consists of flat islands separated by narrow, deep crevices.

The conductivity of the crystalline organic layers is particularly strongly deteriorated by a high density of grain boundaries like those visible in Fig. 3.13. The effect of such boundaries on electronic properties was theoretically studied in a self-consistent polarization-field approach [48]; the optimum arrangement of the molecules at the boundary was calculated using molecular-dynamic force fields. Results shown in Fig. 3.14 display the energy of a free hole at a boundary-near molecule with respect to its energy in a perfect crystal; at increased energy the molecule acts as scattering center, at decreased energy as a trap. The calculation gives evidence that both, traps and scattering centers for free holes appear at the boundary. The hole energy is determined by the particular environment of a molecule near the boundary; the boundary

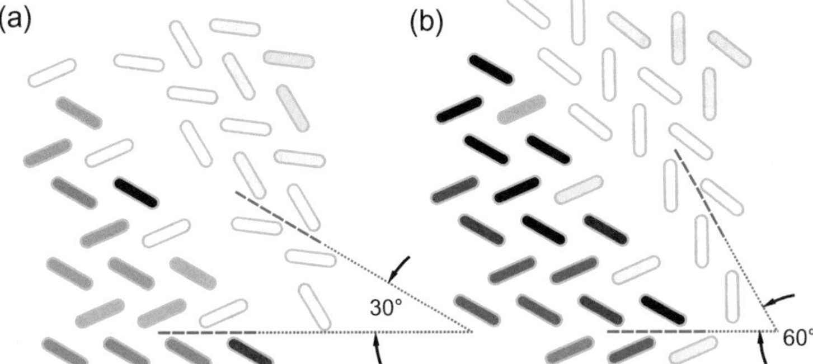

Fig. 3.14 Effect of boundaries between grains in a pentacene crystal misaligned by (**a**) 30° and (**b**) 60°. The molecules are viewed along their long axis, their plane is perpendicular to the image plane. The shading marks the energy of a hole at a molecule: states with energies below −0.25 eV (*dark shading*) are traps, states with energies above 0.25 eV (*bright shading*) act as scattering centers. Reproduced with permission from [48], © 2007 APS

has no periodicity on a short range, leading to a substantial variation in hole energies along the interface. At larger angles the energies shift gradually from trap states at the edge of the left grain and scattering centers at the right grain to the inverse case with trap states at the right grain (not shown in the figure); this is traced back to the interaction between the strongly polarizing hole and the quadrupole moments of neighboring molecules (polycyclic aromatic hydrocarbons like pentacene have a positively charged planar backbone of atom cores and negatively charged π electrons in front and at the back of this plane, yielding a permanent quadrupole moment) [48]. The study indicates that the potential landscape at a grain boundary in organic crystals is complicated and has a granular structure on the molecular level; charge transport across such extended defects is not described by a simple model.

3.2.4 *Defects in Polymers*

In the long conjugated chains of semiconducting polymers with π bonding an irregularity may occur due to the alternating bond lengths. This *bond-alternation defect* is illustrated for polyacetylene in Fig. 3.15. The terminal repeat units must have single bonds; if thus a finite polymer chain does not have the correct length, a mismatch in bond-alternation occurs, yielding the backbone structure shown in Fig. 3.15c. The C atom at this bond-alternation defect has two single C–C bonds and one unpaired π electron. The polymer chain remains electrically neutral, but the electron at the defect has an unpaired spin, giving rise to a finite paramagnetism of the otherwise diamagnetic chain. Due to similarities in the theoretical description the unpaired electron, which may move along the chain, is also referred to as *soliton*. The soliton

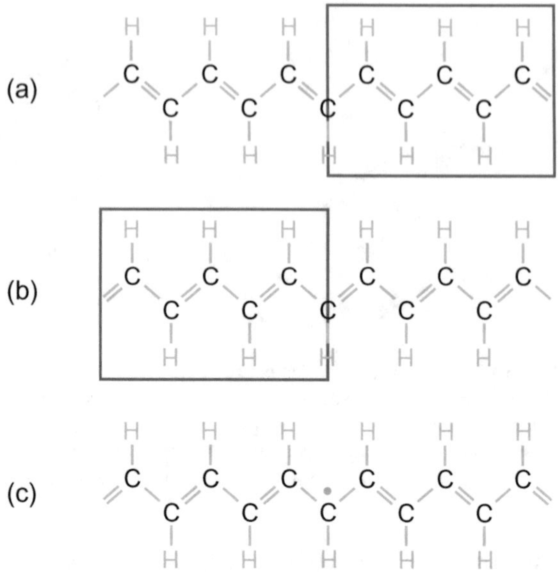

Fig. 3.15 Different bond-alternation order in sequences (**a**) and (**b**) of a polyacetylene chain; (**c**) bond-alternation defect occurring when sequences of (**a**) and (**b**) marked by red boxes meet. The dot on top of the central C atom represents an unpaired electron created at the defect

represents a boundary which separates domains in the phase of the π bonds in the polymer backbone.

Trap states appear in polymers predominantly in the most disordered regions; in most cases the structural nature of such states is unknown. Still the distortions in the crystalline regions are important, since these crystallites determine the carrier transport in a high-carrier-mobility polymer, where percolation effects allow for a dominating conduction in the ordered regions. Polymer segments extending from one ordered domain into another can enhance the mobility significantly if the π–π coupling is sufficiently large [49]. If no perculation by crystalline regions occurs, conduction is primarily carried through amorphous regions with very low mobility [50].

Significant distortions also occur in the crystalline regions of polymers; these regions are hence referred to as *paracrystalline*, denoting an intermediate state between crystalline and amorphous [51]. Such states can be characterized by a crystallographic *paracrystalline distortion parameter* g, defined by the average separation $\langle d_{hkl}\rangle$ of adjacent repetition units along a specific $[hkl]$ direction (considered ideal) and the relative statistical deviations of actual separations:

$$g^2 = \left(\left\langle d_{hkl}^2\right\rangle - \langle d_{hkl}\rangle^2\right)/\langle d_{hkl}\rangle^2 = \left\langle d_{hkl}^2\right\rangle/\langle d_{hkl}\rangle^2 - 1. \tag{3.7}$$

The paracrystallinity parameter g allows to classify materials with different degrees of distortions as listed in Table 3.4 [52]; values below 1% are indicative for crystalline properties, the range between 1 and 10% characterizes paracrystalline materials, and values of 10–15% represent a glass or a melt.

Table 3.4 Values of the paracrystalline distortion parameter g in various micro-paracrystals and other substances

Substance	g (%)
Single crystal	0
Crystallite in polymer	2
Bulk polymer	3
Graphite, coal tar	6
SiO_2 glass	12
Molten metals	15
Boltzmann gas	100

Distortions in the (para-) crystalline regions of a polymer consist, e.g., of random variations in the spacings between the backbone lamella illustrated in Fig. 3.6a. Such variations create band tailing into the bandgap comparable to the distortions shown in Fig. 3.10 for molecule shifts in a pentacene crystal. Such effects are discussed in Sect. 5.2.2.

Properties of polymers related to the macromolecular order (such as crystallinity) and consequential characteristics depend also on the length of the molecular chains, in addition to the chemical structure and the applied processing. With increasing length of the backbone chain (and hence molecular weight) most polymers transform from a highly ordered chain alignment to a structure composed of crystalline and amorphous regions. All structures have lamellae as those shown in Fig. 3.6a as building blocks, but their size, perfection, and molecular interconnectivity differ. A qualitative picture for the development of various material properties with increasing molecular length is given in Fig. 3.16 [53]. The material properties addressed in the black curve refer to crystallinity, density, melting temperature, or Young's modulus, which is related to the stiffness of the solid. Values of these quantities increase as the molecular chains gets longer, and get maximum at a critical molecular weight (i.e., length) when chains start to entangle. At this length a two-phase morphology with crystalline and amorphous regions appears, and often the considered values eventually decrease. For a respective study on charge transport see [54]. Other mechanical quantities such as tensile strength and elongation at break increase when the chains entangle (red curve). For a review on the connection between polymer conformation and materials properties see [53].

3.3 Problems Chapter 3

3.1 The intermolecular bonding of neutral and nonpolar molecules in organic crystals is reasonably described by the Buckingham potential. The empirical parameters α, β, and γ of this potential for a bond between a C atom of a molecule and an H atom of a neighboring molecule are 5.42 eV $\times$ Å^6, 380 eV, and 3.67 Å^{-1}, respectively.

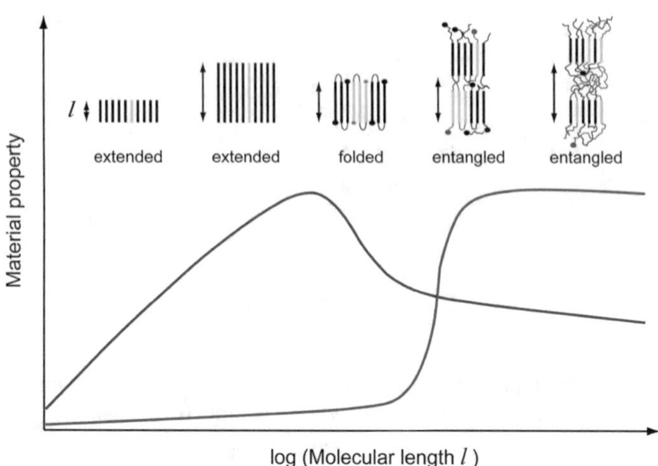

Fig. 3.16 Qualitative dependence of properties like crystallinity or melting temperature (*black curve*) and mechanical strength (*red curve*) on the length of the molecular chains in a polymer. The inset on top illustrates the respective macromolecular structure. Reproduced with permission from [53], © 2010 Wiley

(a) The parameters yield a C–H equilibrium distance of about 3.30 Å. Determine the corresponding bond energy.

(b) In a molecule crystal each atom is bond to all atoms of neighboring molecules with a hence varying distance. Calculate the energy change per varied bond distance at equilibrium, and at the distances (3.3 + 0.5) Å and (3.3 − 0.5) Å.

3.2 Pentacene $C_{22}H_{14}$ has a molecular weight of 278.3 g/mol. In the Campbell phase of the bulk it forms a triclinic crystal with 2 molecules per unit cell and a molecular packing coefficient of 0.743.

(a) Calculate the volume of a single pentacene molecule and the density of the crystal using data from Table 3.3.

(b) The unit cell of monolayer-thick films on SiO_2 substrate were reported to deviate from the bulk phase by different angles α, β, and γ (all 90°), and by shorter in-plane vectors a and b (7.6 and 5.9 Å, respectively). Determine the relative change $\Delta V/V$ of the unit cell assuming the c parameter listed in Table 3.3 unchanged. What is the resulting density of such thin films?

3.3 The carbon atoms in the basal plane of graphite are arranged in layers of interconnected hexagonal rings, such that each carbon atom is bonded to three neighboring atoms. The distance of 1.42 Å between adjacent carbon atoms in the plane matches well the C–C bond lengths of the five fused carbon rings of pentacene ranging from 1.35 (outermost ring) to 1.45 Å (central ring); a monolayer of pentacene on graphite is hence arranged in a lay-down configuration with its C atoms sticking to the graphite C atoms underneath. Using for the substrate the hexagonal basal-plane vectors $\mathbf{a}_1$, $\mathbf{a}_2$, $\mathbf{a}_3$ ($\mathbf{a}_1 = \mathbf{u}$ and $\mathbf{a}_2 = \mathbf{v}$ in Fig. 2.4d) we define the more appropriate substrate vectors $\mathbf{a}_S = \mathbf{a}_1 - \mathbf{a}_3$, $\mathbf{b}_S = \mathbf{a}_2 - \mathbf{a}_3$; the two vectors $\mathbf{a}_L$ and $\mathbf{b}_L$ of the pentacene monolayer on graphite are then given by the coefficient matrix $\begin{pmatrix} 7 & 0 \\ -1 & 3 \end{pmatrix}$; the long axis of all pentacene molecules are oriented along $\mathbf{a}_L$.

(a) How long is the fused chain of the 5 pentacene rings, if their C atoms match exactly the positions of the substrate atoms beneath? How many equivalent orientations exist?
(b) Calculate the lengths of the vectors $\mathbf{a}_L$ and $\mathbf{b}_L$ of the 2D unit cell and the angle between them.
(c) What is the nearest distance between C atoms of two neighboring pentacene molecules?

3.4 In the paracrystalline region of a polymer a mean distance of 4.10 Å in the π–π stacking direction between the lamellae, and a distortion parameter of 2.2% are observed.

(a) Compute the lower and upper values of the stacking distance corresponding to the standard deviation, if the stacking has a Gaussian distribution.
(b) Compute the binding energy from all C atoms between two adjacent 80 nm long lamella with 4.10 Å mutual distance and 1.26 Å average spacing of C atoms along the direction of the chain; the Buckingham parameters for intermolecular C bonds are 24.6 eV × Å^6, 3.63 keV, and 3.60 Å^{-1}. By which factor does the binding energy change if the average lamella spacing is reduced by 3%?

3.4 General Reading Chapter 3

A.I. Kitaigorodsky, *Molecular Crystals and Molecules* (Academic Press, New York and London, 1973)

J.D. Wright, *Molecular Crystals*, 2nd edn. (Cambridge University Press, Cambridge, 1995)

References

1. H. Sirringhaus, Organic field-effect transistors: the path beyond amorphous silicon. Adv. Mater. **6**, 1319 (2014)
2. M. Muccini, S. Toffanin, *Organic Light-Emitting Transistors: Towards the Next Generation Display Technology* (Wiley, Hoboken, New Jersey, 2016)

3. W. Zhao, D. Qian, S. Zhang, S. Li, O. Inganäs, F. Gao, J. Hou, Fullerene-free polymer solar cells with over 11% efficiency and excellent thermal stability. Adv. Mater. **28**, 4734 (2016)
4. M. Schwoerer, H.C. Wolf, *Organic Molecular Solids* (Wiley-VCH Verlag, Weinheim, 2007)
5. R.A. Buckingham, The classical equation of state of gaseous Helium, Neon and Argon. Proc. Roy. Soc. Lon. Ser. A, Math. Phys. Sci. **168**, 264 (1938)
6. T.L. Starr, D.E. Williams, Coulombic nonbonded interatomic potential functions derived from crystal-lattice vibrational frequencies in hydrocarbons. Acta Cryst. A **33**, 771 (1977)
7. A.I. Kitaigorodsky, *Molecular Crystals and Molecules* (Academic Press, New York and London, 1973)
8. S.R. Forrest, Y. Zhang, Ultrahigh-vacuum quasiepitaxial growth of model van der Waals thin films. I. Theory. Phys. Rev. B **49**, 11297 (1994)
9. D.E. Williams, Nonbonded potential parameters derived from crystalline aromatic hydrocarbons. J. Chem. Phys. **45**, 3770 (1966)
10. O.D. Jurchescu, J. Baas, T.T.M. Palstra, Effect of impurities on the mobility of single crystal pentacene. Appl. Phys. Lett. **84**, 3061 (2004)
11. J. Takeya, M. Yamagishi, Y. Tominari, R. Hirahara, Y. Nakazawa, T. Nishikawa, T. Kawase, T. Shimoda, S. Ogawa, Very high-mobility organic single-crystal transistors with in-crystal conduction channels. Appl. Phys. Lett. **90**, 102120 (2007)
12. R.F.W. Bader, M.T. Carroll, J.R. Cheeseman, C. Chang, Properties of atoms in molecules: atomic volumes. J. Am. Chem. Soc. **109**, 7968 (1987)
13. F.L. Hirshfeld, Bonded-atom fragments for describing molecular charge densities. Theoret. Chim. Acta **44**, 129 (1977)
14. M.A. Spackman, D. Jayatilaka, Hirshfeld surface analysis. Cryst. Eng. Comm. **11**, 19 (2009)
15. C. Ambrosch-Draxl, D. Nabok, P. Puschnig, C. Meisenbichler, The role of polymorphism in organic thin films: oligoacenes investigated from first principles. New J. Phys. **11**, 125010 (2009)
16. Z.G. Soos, Theory of π-molecular charge-transfer crystals. Ann. Rev. Phys. Chem. **25**, 121 (1974)
17. R.S. Mulliken, A comparative survey of approximate ground state wave functions of helium atom and hydrogen molecule. Proc. Natl. Akad. Sci. **38**, 160 (1952)
18. H. Kuroda, M. Kobayashi, M. Kinoshita, S. Takemoto, Semiconductive properties of tetracyanoethylene complexes and their absorption spectra. J. Chem. Phys. **36**, 45 (1962)
19. C.L. Braun, Organic semiconductors, in *Handbook on Semiconductors vol. 3 Materials Properties and Preparation*, ed. by T.S. Moss, S.P. Keller (North Holland Publication, Amsterdam 1980), pp. 857–873
20. H.J. Keller (ed.), *Chemistry and Physics in One-Dimensional Metals* (Plenum Press, New York, 1977)
21. N. Karl, *Organic Semiconductors*. Landoldt-Börnstein (Springer, Heidelberg, 1984)
22. A.N. Bloch, T.F. Carruthers, T.O. Poehler, D.O. Cowan, The organic metallic state: some physical aspects and chemical trends, in *Chemistry and Physics in One-Dimensional Metals*, ed. by H.J. Keller (Plenum Press, New York, 1977), pp. 47–85
23. D.E. Hooks, T. Fritz, M.D. Ward, Epitaxy and molecular organization on solid substrates. Adv. Mater. **13**, 227 (2001)
24. S.C.B. Mannsfeld, K. Leo, T. Fritz, Line-on-line coincidence: a new type of epitaxy found in organic-organic heterolayers. Phys. Rev. Lett. **94**, 056104 (2005)
25. C. Wagner, R. Forker, T. Fritz, On the origin of the energy gain in epitaxial growth of molecular films. Phys. Chem. Lett. **3**, 419 (2012)
26. S.C.B. Mannsfeld, T. Fritz, Understanding organic–inorganic heteroepitaxial growth of molecules on crystalline substrates: experiment and theory. Phys. Rev. B **71**, 235405 (2005)
27. A.C. Hillier, M.D. Ward, Epitaxial interactions between molecular overlayers and ordered substrates. Phys. Rev. B **54**, 14037 (1996)
28. S.R. Forrest, Ultrathin organic films grown by organic molecular beam deposition and related techniques. Chem. Rev. **97**, 1793 (1997)

29. L.E. Dinca, F. De Marchi, J.M. MacLeod, J. Lipton-Duffin, R. Gatti, D. Ma, D.F. Perepichkab, F. Rosei, Pentacene on Ni(111): room-temperature molecular packing and temperature-activated conversion to graphene. Nanoscale **7**, 3263 (2015)
30. D. Käfer, L. Ruppel, G. Witte, Growth of pentacene on clean and modified gold surfaces. Phys. Rev. B **75**, 085309 (2007)
31. D.V. Lang, X. Chi, T. Siegrist, A.M. Sergent, A.P. Ramirez, Amorphouslike density of gap states in single-crystal pentacene. Phys. Rev. Lett. **93**, 086802 (2004)
32. B.-N. Park, S. Seo, P. Evans, Channel formation in single-monolayer pentacene thin film transistors. J. Phys. D **40**, 3506 (2007)
33. S.C.B. Mannsfeld, A. Virkar, C. Reese, M.F. Toney, Z. Bao, Precise structure of pentacene monolayers on amorphous silicon oxide and relation to charge transport. Adv. Mater. **21**, 2294 (2009)
34. F.-J. Meyer zu Heringdorf, M.C. Reuter, R.M. Tromp, Growth dynamics of pentacene thin films. Nature **412**, 7 (2001)
35. A. Al-Mahboob, J.T. Sadowski, Y. Fujikawa, K. Nakajima, T. Sakurai, Kinetics-driven anisotropic growth of pentacene thin films. Phys. Rev. B **77**, 035426 (2008)
36. R.J. Hamers, Formation and characterization of organic monolayers on semiconductor surfaces. Annu. Rev. Anal. Chem. **1**, 707 (2008)
37. S. Seo, P.G. Evans, Molecular structure of extended defects in monolayer-scale pentacene thin films. J. Appl. Phys. **106**, 103521 (2009)
38. J.H. Kang, D. da Silva Filho, J.-L. Bredas, X.-Y. Zhu, Shallow trap states in pentacene thin films from molecular sliding. Appl. Phys. Lett. **86**, 152115 (2005)
39. J. Götzen, D. Käfer, C. Wöll, G. Witte, Growth and structure of pentacene films on graphite: Weak adhesion as a key for epitaxial film growth. Phys. Rev. B **81**, 085440 (2010)
40. T. Maeda, T. Kobayashi, T. Nemoto, S. Isoda, Lattice defects in organic crystals revealed by direct molecular imaging. Philos. Mag. B **81**, 1659 (2001)
41. A. Pick, M. Klues, A. Rinn, K. Harms, S. Chatterjee, G. Witte, Polymorph-selective preparation and structural characterization of perylene single crystals. Cryst. Growth Des. **15**, 5495 (2015)
42. J.N. Sherwood, Lattice defects in organic crystals. Mol. Cryst. Liquid Cryst. **9**, 37 (1969)
43. L. Tsetseris, S.T. Pantelides, Intercalation of oxygen and water molecules in pentacene crystals: first-principles calculations. Phys. Rev. B **75**, 153202 (2007)
44. D. Knipp, J.E. Northrup, Electric-field-induced gap states in pentacene. Adv. Mater. **21**, 2511 (2009)
45. J.E. Northrup, M.L. Chabinyc, Gap states in organic semiconductors: Hydrogen- and oxygen-induced states in pentacene. Phys. Rev. B **68**, 041202 (2003)
46. W.L. Kalb, K. Mattenberger, B. Batlogg, Oxygen-related traps in pentacene thin films: energetic position and implications for transistor performance. Phys. Rev. B **78**, 035334 (2008)
47. C. Mattheus, J. Baas, A. Meetsma, J.L. de Boer, C. Kloc, T. Siegrist, T.T.M. Palstra, A 2:1 cocrystal of 6,13-dihydropentacene and pentacene. Acta Crystallogr. E **58**, o1229 (2002)
48. S. Verlaak, P. Heremans, Molecular microelectrostatic view on electronic states near pentacene grain boundaries. Phys. Rev. B **75**, 115127 (2007)
49. J.E. Northrup, Mobility enhancement in polymer organic semiconductors arising from increased interconnectivity at the level of polymer segments. Appl. Phys. Lett. **106**, 023303 (2015)
50. R.A. Street, J.E. Northrup, A. Salleo, Transport in polycrystalline polymer thin-film transistors. Phys. Rev. B **71**, 165202 (2005)
51. A.M. Hindeleh, R. Hosemann, Microparacrystals: the intermediate stage between crystalline and amorphous. J. Mater. Sci. **26**, 5127 (1991)
52. R. Hosemann, A.M. Hindeleh, Structure of crystalline and paracrystalline condensed matter. J. Macromol. Sci. Phys. B **34**, 327 (1995)
53. A.A. Virkar, S. Mannsfeld, Z. Bao, N. Stingelin, Organic semiconductor growth and morphology considerations for organic thin-film transistor. Adv. Mater. **22**, 3857 (2010)

54. F.P.V. Koch, J. Rivnay, S. Foster, C. Müller, J.M. Downing, E. Buchaca-Domingo, P. Westacott, L. Yu, M. Yuan, M. Baklar, Z. Fei, C. Luscombe, M.A. McLachlan, M. Heeney, G. Rumbles, C. Silva, A. Salleo, J. Nelson, P. Smith, N. Stingelin, The impact of molecular weight on microstructure and charge transport in semicrystalline polymer semiconductors–poly(3-hexylthiophene), a model study. Prog. Poly. Sci. **38**, 1978 (2013)

Chapter 4
Electronic Properties of Heterostructures

Abstract This chapter presents electronic properties of a junction between two semiconductors and electronic states in low-dimensional structures. First, we consider the valence and conduction bands of zincblende and wurtzite bulk semiconductors and illustrate the effects of strain and alloying. Then, models describing the band lineup of heterostructures are introduced and the effect of interface stoichiometry is illustrated. The characteristic scale for the occurrence of size quantization is discussed, and electronic states in quantum wells, quantum wires, and quantum dots are described.

4.1 Bulk Properties

A heterostructure is formed by a junction between two dissimilar solids. Before considering the electronic properties of such an interface in more detail (Sect. 4.2) we briefly compile some electronic properties of a single constituent. We focus on the uppermost valence bands and the fundamental bandgap of bulk crystals with zincblende or wurtzite structure. Bulk denotes a size well above the limit of size-quantization effects (Sect. 4.3.2). In this sense a 100 nm thick epitaxial layer may be considered as a bulk crystal.

4.1.1 Electronic Bands of Zincblende and Wurtzite Crystals

Energy bands in semiconductors may fittingly be described using the effective mass approximation in the framework of a multiband **kp** method, which requires only a small set of experimentally determined parameters [cf., e.g., [1, 2]]. Tetrahedrally coordinated crystals with zincblende or wurtzite structure form three p-like valence bands and an s-like conduction band, leading to an 8 band **kp** Hamilton operator (4 bands $\times 2$ spin orientations) with terms linear and quadratic in k. The valence dispersions comprise bands for the heavy hole, the light hole, and the split-off hole (zincblende) or crystal hole (wurtzite). The effective mass m^* at the edges of valence

U. W. Pohl, *Epitaxy of Semiconductors*, Graduate Texts in Physics,
https://doi.org/10.1007/978-3-030-43869-2_4

and conduction bands near the center of the Brillouin zone is generally a tensor. Its components m_{ij}^* are related to the energy dispersion $E(\mathbf{k})$ according to

$$m_{ij}^* = \hbar^2 \left(\frac{\partial^2}{\partial k_i \partial k_j} E(\mathbf{k})|_{k_i,k_j=0} \right)^{-1} . \tag{4.1}$$

In *zincblende semiconductors* the anisotropic effective mass of the heavy hole is usually expressed in terms of band parameters A, B, C, or the related Luttinger (or: Kohn-Luttinger) parameters γ_1, γ_2, γ_3. We apply the widely used Luttinger parameters, yielding along the different crystallographic directions for the effective heavy-hole mass the relations

$$\begin{aligned} \left(\frac{m_0}{m_{\text{hh}}^*}\right)^{[100]} &= \gamma_1 - 2\gamma_2, \quad \left(\frac{m_0}{m_{\text{hh}}^*}\right)^{[110]} = \frac{1}{2}(2\gamma_1 - \gamma_2 - 3\gamma_3), \\ \left(\frac{m_0}{m_{\text{hh}}^*}\right)^{[111]} &= \gamma_1 - 2\gamma_3. \end{aligned} \tag{4.2a}$$

The corresponding effective light-hole mass is given by similar relations,

$$\begin{aligned} \left(\frac{m_0}{m_{\text{lh}}^*}\right)^{[100]} &= \gamma_1 + 2\gamma_2, \quad \left(\frac{m_0}{m_{\text{lh}}^*}\right)^{[110]} = \frac{1}{2}(2\gamma_1 + \gamma_2 + 3\gamma_3), \\ \left(\frac{m_0}{m_{\text{lh}}^*}\right)^{[111]} &= \gamma_1 + 2\gamma_3. \end{aligned} \tag{4.2b}$$

The split-off hole mass is given by

$$\frac{m_0}{m_{\text{so}}^*} = \gamma_1 - \frac{E_P \Delta_0}{3E_{\text{g}}(E_P + \Delta_0)}, \tag{4.2c}$$

where E_P is the momentum matrix-element between the p-like valence bands and the s-like conduction band. E_{g} and Δ_0 are the direct bandgap energy and the spin-orbit splitting, respectively. Parameters for some technologically important zincblende semiconductors are given in [3]. The 8 band **kp** approximation provides a good description up to about a quarter of the way from the center to the boundary of the Brillouin zone. Additional bands may be included to improve the description or to describe also indirect-bandgap semiconductors.

The valence-band structure near the center of the Brillouin zone (Γ point at $k = 0$) for a typical zincblende semiconductor (GaAs) is shown in Fig. 4.1a. The bands of heavy hole and light hole are degenerated at $k = 0$ in absence of symmetry-reducing strain. Away from the Γ point non-parabolicity occurs due to an anti-crossing behavior of these holes with the split-off hole, which lies below the other two bands at $k = 0$ by the amount of the spin-orbit energy Δ_0.

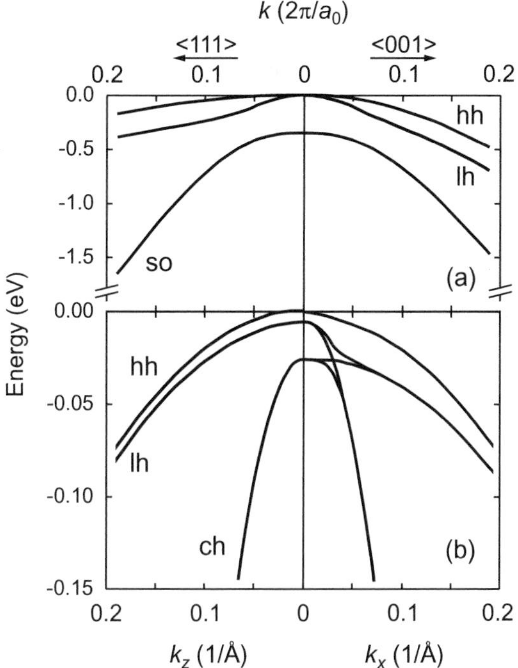

Fig. 4.1 Valence-band structure of **a** a zincblende and **b** a wurtzite semiconductor near the center of the Brillouin zone. The labels *hh*, *lh*, *so*, and *ch* denote valence bands of heavy hole, light hole, split-off hole, and crystal hole, respectively

Crystals with *wurtzite structure* have a different dispersion in the basal plane and perpendicular, i.e., along the c axis parallel [0001]. The valence band-structure of a typical wurtzite semiconductor (GaN) near $k = 0$ is shown in Fig. 4.1b. The degeneracy of the valence bands is lifted at the Γ point. The splittings between heavy hole and light hole, and that between heavy hole and crystal hole essentially reflect the effects of spin-orbit and crystal-field interactions, respectively. A set of seven Luttinger-like parameters which accounts for the non-cubic symmetry was introduced to describe the bands of wurtzite semiconductors [2, 4].

Wurtzite is the structure with highest symmetry compatible with the effect of spontaneous polarization (zincblende structure has no intrinsic bulk polarization) [5]. In epitaxial layers this effect is always accompanied by strain-induced piezoelectricity discussed Sect. 4.1.3.

4.1.2 Strain Effects

Virtually any heteroepitaxial structure is strained as pointed out in Sect. 2.2. Strain and the related atomic positions are determined by minimizing the elastic energy, under the constraint of a common lattice constant parallel to the interface $a_{\|}$ throughout the structure for pseudomorphic conditions. The resulting strain described by (2.13) to (2.19a)–(2.19c) affects the energy of the electronic bands. We first focus on *zincblende* semiconductors.

Zincblende Crystals

Shear components of the strain lead to a splitting of degenerate cubic valence bands and indirect conduction bands, but they do not affect the *average* valence-band energy $E_{\mathrm{v,av}}$. The hydrostatic component of the strain changes the volume and leads to a shift of the bands with respect to $E_{\mathrm{v,av}}$ and also affects the average electrostatic potential itself.

The effect of strain in cubic semiconductors is expressed in terms of deformation potentials a, b, and d [6, 7]. The total effect of hydrostatic strain on the valence band is described by the hydrostatic deformation potential a_{v} for the valence band,

$$a_{\mathrm{v}} = \frac{dE_{\mathrm{v,av}}}{d\ln V} = \frac{dE_{\mathrm{v,av}}}{\frac{1}{V}dV}. \tag{4.3}$$

The quantity a_{v} expresses the shift of the average valence-band energy $E_{\mathrm{v,av}}$ per relative change of the volume V. A similar relation applies for the shift of the conduction-band energy E_{c} under the action of hydrostatic pressure, with a_{v} and $E_{\mathrm{v,av}}$ in (4.3) being replaced by the deformation potential of the conduction band a_{c} and E_{c}, respectively. Consequently the change of the gap energy $E_{\mathrm{c}} - E_{\mathrm{v,av}}$ is also given by such a relation with a deformation potential a, which is equal to $a = a_{\mathrm{c}} - a_{\mathrm{v}}$. It must be noted that also another sign convention is widely used for a_{v}, yielding $a = a_{\mathrm{c}} + a_{\mathrm{v}}$. Furthermore, different conventions of the quantities b and d are used. They yield $d' = \sqrt{3}/2 \times d$ for trigonal distortions and refer to a term proportional $(J_x^2\varepsilon_{xx} + \text{c.p.})$ instead of $((J_x^2 - 1/3J^2)\varepsilon_{xx} + \text{c.p.})$ for tetragonal distortions.

Using the deformation potential defined by (4.3), the influence of the hydrostatic strain component on the offsets of the average valence-band and the conduction band is given by

$$\begin{aligned} \Delta E_{\mathrm{v,av}} &= a_{\mathrm{v}}\frac{\Delta V}{V} \quad \text{and} \\ \Delta E_{\mathrm{c}} &= a_{\mathrm{c}}\frac{\Delta V}{V}, \end{aligned} \tag{4.4}$$

respectively, with the fractional volume change $\Delta V/V = (\varepsilon_{xx} + \varepsilon_{yy} + \varepsilon_{zz})$. Eventually the spin-orbit splitting of the valence band is considered. In semiconductors with zincblende or diamond structure the edge of the topmost valence band is

$$E_{\mathrm{v}} = E_{\mathrm{v,av}} + \frac{\Delta_0}{3}, \tag{4.5}$$

Δ_0 being the spin-orbit parameter.

Splittings of the valence band in addition to those originating from the spin-orbit interaction arise from shear components of the strain. They depend on the strain direction and are proportional to the strain in the linear regime, which is expected to be a good approximation for pseudomorphic heterostructures. Taking the average valence-band energy $E_{v,av}$ as reference, the shifts of the heavy hole, the light hole, and the split-off band for uniaxial strain along the [001] direction are given by [7, 8]

$$\begin{aligned} E_{v,hh} &= \frac{\Delta_0}{3} - \frac{1}{2}\delta E_{001}, \\ E_{v,lh} &= -\frac{\Delta_0}{6} + \frac{1}{4}\delta E_{001} + \frac{1}{2}\sqrt{\Delta_0^2 + \Delta_0 \times \delta E_{001} + \frac{9}{4}\delta E_{001}^2}, \\ E_{v,so} &= -\frac{\Delta_0}{6} + \frac{1}{4}\delta E_{001} - \frac{1}{2}\sqrt{\Delta_0^2 + \Delta_0 \times \delta E_{001} + \frac{9}{4}\delta E_{001}^2}. \end{aligned} \tag{4.6}$$

In (4.6) the abbreviation

$$\delta E_{001} = 2b(\varepsilon_{zz} - \varepsilon_{xx}) \tag{4.7}$$

is used, with the shear deformation potential b for biaxial strain which induces a tetragonal distortion of the cubic unit cell. Equation (4.6) also holds for uniaxial strain along [111], if δE_{001} is replaced by

$$\delta E_{111} = 2\sqrt{3}d\varepsilon_{xy}, \tag{4.8}$$

where $\varepsilon_{xy} = 1/3(\varepsilon_{\perp} - \varepsilon_{\parallel})$, and d is the deformation potential for this strain direction.

The effect of uniaxial strain along the [001] direction, or similarly by biaxial strain along [110] and [$\bar{1}$10], is shown in Fig. 4.2. The unstrained valence band of a zincblende semiconductor at the Brillouin zone center $k = 0$ is split by the spin-orbit interaction Δ_0 into a fourfold degenerate heavy-hole (*hh*) and light-hole (*lh*) band with total angular momentum $J = 3/2$ ($M_J = \pm 3/2, \pm 1/2$), and a doubly degenerate split-off (*so*) band with $J = 1/2$ ($M_J = \pm 1/2$). The strain reduces the symmetry from T_d to D_{2d} and lifts the degeneracy of the $J = 3/2$ band, yielding a $J = 3/2$, $M_J = \pm 3/2$ *hh* band and a $J = 3/2$, $M_J = \pm 1/2$ *lh* band [9]. In addition, the hydrostatic component of the stress shifts the bandgap energy. In common zincblende materials the bandgap energy increases for compressive strain due to the nature of the atomic bonding. It is generally believed that most of the change occurs in the upward moving conduction band. Since the share of a_c and a_v is difficult to isolate experimentally it is usually based on theoretical predictions.

We illustrate strain effects in pseudomorphic structures for results obtained from the model-solid theory [10] outlined in Sect. 4.2.5. Calculated deformation potentials are given in Table 4.1 for some semiconductors. Data computed according to (4.3) refer to the direct gap at the Γ point of the Brillouin zone (index *dir*) or to the indirect gap (*indir*). The gap energies E_g are taken from low-temperature experiments, yielding with (4.5) the conduction-band values $E_c = E_v + E_g$.

To illustrate the effect of strain on the valence and conduction bands, we consider data for a thin pseudomorphically strained ZnS layer on a ZnSe substrate [10]. Both solids have zincblende structure with lattice parameters of 5.40 Å and 5.65 Å, respectively. ZnS has a smaller unstrained lattice parameter, and is tensely

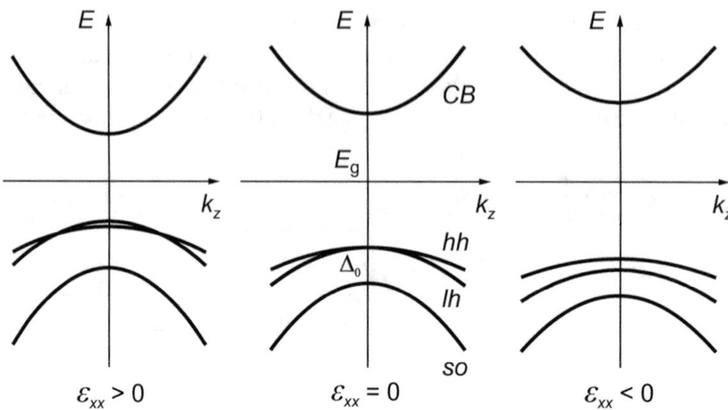

Fig. 4.2 Effect of strain on the valence bands and the lowest conduction band (*CB*) of a zincblende-type semiconductor. *hh*, *lh*, and *so* denote heavy-hole, light-hole, and split-off hole valence bands, Δ_0 is the spin-orbit splitting. $\varepsilon_{xx} = \varepsilon_{yy}$ is the in-plane strain

Table 4.1 Average valence-band energies $E_{\mathrm{v,av}}$ and deformation potentials of the valence band a_{v}, the conduction band a_{c}, and the gap energy a calculated from the model-solid theory. Δ_0 and E_{g} denote measured spin-orbit splitting and energy gap at 0 K, respectively. All values are given in eV. Data from [10]

Solid	$E_{\mathrm{v,av}}$	a_{v}	$a_{\mathrm{c}}^{\mathrm{dir}}$	a^{dir}	$E_{\mathrm{g}}^{\mathrm{dir}}$	$E_{\mathrm{c}}^{\mathrm{dir}}$	$a_{\mathrm{c}}^{\mathrm{indir}}$	a^{indir}	$E_{\mathrm{g}}^{\mathrm{indir}}$	$E_{\mathrm{c}}^{\mathrm{indir}}$	Δ_0
Si	−7.03	2.46	1.98	−0.48	3.37	−3.65	4.18	1.72	1.17	−5.85	0.04
Ge	−6.35	1.24	−8.24	−9.48	0.89	−5.36	−1.54	−2.78	0.74	−5.51	0.30
GaP	−7.40	1.70	−7.14	−8.83	2.90	−4.47	3.36	1.56	2.35	−5.02	0.08
AlAs	−7.49	2.47	−5.64	−8.11	3.13	−4.27	4.09	1.62	2.23	−5.17	0.28
GaAs	−6.92	1.16	−7.17	−8.33	1.52	−5.29					0.34
InAs	−6.67	1.00	−5.08	−6.08	0.41	−6.13					0.38
ZnS	−9.15	2.31	−4.09	−6.40	3.84	−5.29					0.07
ZnSe	−8.37	1.65	−4.17	−5.82	2.83	−5.40					0.43

strained parallel to the interface (x, y) and compressively strained perpendicular (z). According to (2.19a)–(2.19c) the respective strains are $\varepsilon_{xx} = \varepsilon_{yy} = 0.046$ and $\varepsilon_{zz} = -0.058$, leading to a fractional volume increase $\Delta V/V = 0.035$. The change of the volume affects the energy of valence band and conduction band. The band energies of the strained ZnS layer follow from (4.4), yielding $E_{\mathrm{v,av}}^{\mathrm{ZnS}} = -9.07$ eV and $E_{\mathrm{c}}^{\mathrm{ZnS}} = -5.43$ eV. The deformation potential b of ZnS is -1.25 eV, yielding with (4.7) $\delta E_{001} = 0.26$ eV. From (4.6) we finally obtain the energy shifts of the heavy hole, the light hole, and the split-off hole of the strained ZnS layer with respect to $E_{\mathrm{v,av}}^{\mathrm{ZnS}}$ being -0.11 eV, $+0.26$ eV, and -0.16 eV, respectively. The resulting alignment with $\Delta E_{\mathrm{v}} = -0.50$ eV and $\Delta E_{\mathrm{c}} = 0.03$ eV is depicted in Fig. 4.3.

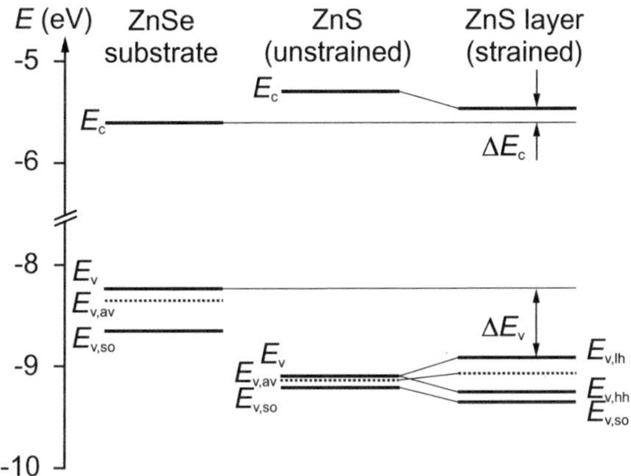

Fig. 4.3 Band alignment of a biaxially strained pseudomorphic ZnS layer on a (001)-oriented ZnSe substrate, calculated according to the model-solid theory using experimental values for gap energies and spin-orbit splittings. After [10]

Deformation potentials are experimentally obtained from optical spectroscopy. A comparison of the reflectivity and a two-photon absorption spectrum of a ZnSe bulk crystal is shown in Fig. 4.4a. The sharp nonlinear resonances of the $1S$ exciton allow for a direct measurement of splittings induced by stresses applied along various crystallographic low-index directions. The deformation potentials a, b, and d are contained in the Hamiltonian of the $1S$ orthoexciton and are evaluated by assigning the measured energies to the eigenenergies [11]. Results for uniaxial stress along [001] are given in Fig. 4.4b.

Application of uniaxial stress to a biaxially strained 5.3 μm thick ZnSe layer on GaAs substrate is shown in Fig. 4.4c. In the lowest spectrum the splitting of light-hole and heavy-hole $1S$ exciton due to pure biaxial strain of the epilayer is seen. Four resonances of the $1S$ exciton appear when an additional stress is applied along [110]. They are linearly polarized either parallel (π) or perpendicular (σ) with respect to the axis of external stress. The stress-induced strain lowers the symmetry to C_{2v} and creates dipole-allowed mixtures of paraexcitons with orthoexcitons, giving rise to 2×2 resonances [12].

Experimentally determined deformation potentials for some semiconductors are given in Table 4.2.

Wurtzite Crystals

We now consider semiconductors with *wurtzite structure* like the Column III nitrides or ZnO. The unstrained valence-band structure of wurtzite crystals shown in Fig. 4.1b. Due to a weak spin-orbit coupling the dispersions of the *hh*-, *lh*-, and *ch*-valence-bands are not strongly affected by strain, in contrast to effects in zincblende materials. Under biaxial strain in the basal plane the C_{6v} symmetry of the unit cell is preserved, but the crystal-field splitting changes. For compressive biaxial strain the energy which separates the crystal-hole band from the heavy-hole and light-hole bands is increased, for tensile strain it is decreased. Uniaxial strain in the basal plane reduces the symmetry to C_{2v}. Under compressive uniaxial strain along the Γ–K direction in

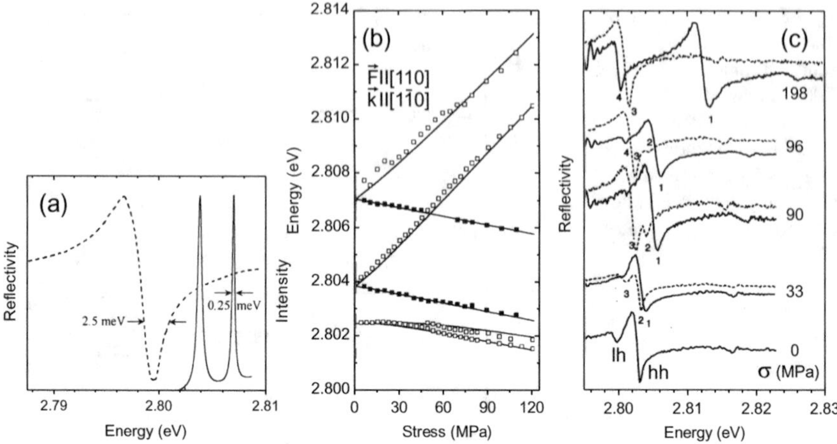

Fig. 4.4 **a** Two-photon excitation spectrum (*solid lines*) and reflectivity (*dashed*) of a ZnSe bulk crystal, recorded at 5 K and 2 K, respectively. **b** Shift and splitting of two-photon resonances for uniaxial stress applied along [001], incident light along [1$\bar{1}$0], and polarizations parallel to [001] (*filled symbols*) and [110] (*open symbols*). Reproduced with permission from [11], © 1995 APS. (**c**) Reflectance spectra of a biaxially strained epitaxial ZnSe/GaAs layer under an additional external stress σ applied along [110]. *Solid* and *dotted* curves refer to σ and π polarizations, respectively. Reproduced with permission from [12], © 1996 APS

Table 4.2 Deformation potentials a, b, and d, experimentally determined for the lowest bandgap E_g from optical spectroscopy (exp). Theoretical values (theo) are from the model-solid theory [10]. Solids denoted in bold have an indirect lowest bandgap. All values are given in eV

Solid	a^{exp}	a^{theo}	b^{exp}	b^{theo}	d^{exp}	d^{theo}
Si	+1.5 ± 0.3[a]	+1.72	−2.10 ± 0.10[a]	−2.35	−4.85 ± 0.15[a]	−5.32
Ge	−2.0 ± 0.5[b]	−2.78	−2.86 ± 0.15[b]	−2.55	−5.28 ± 0.50[b]	−5.50
GaP	−9.9 ± 0.3[c]	+1.56	−1.5 ± 0.2[c]		−4.6 ± 0.2[c]	
AlAs	−8.2[d]	+1.62	−2.3[d]		−3.4[d]	
GaAs	−8.5[d,e]	−8.33	−2.0[d,e]	−1.90	−4.8[d,e]	−4.23
InAs	−6.1[d]	−6.08	−1.8[d]	−1.55	−3.6[d]	−3.10
ZnS	−4.56[f]	−6.40	−0.75[f]	−1.25		
ZnSe	−4.7 ± 0.2[g]	−5.82		−1.20	−6.37 ± 0.07[g]	

[a]Reference [13], [b]Reference [8], [c]Reference [14], [d]Reference [3], [e]Reference [15], [f]Reference [9], [g]Reference [11] (factor $\sqrt{3}/2$ in d taken into account)

the first Brillouin zone ('y' direction) the lh band in this direction and the hh band in the perpendicular lateral x direction move to higher energy. The same effect has a tensile strain along the x direction, and a reverse effect has a tensile strain along the y direction.

4.1.3 Spontaneous Polarization and Piezoelectricity

In absence of external electric fields and stress, crystals of low symmetry may show a *spontaneous electric polarization*. No such effect occurs for zincblende crystals, but spontaneous polarization is observed with wurtzite crystals and structures of lower symmetry [6]. It originates from a mutual shift of positive and negative charges in the unit cell, and in wurtzite structure the main part depends on the deviation of the lattice parameter u listed in Table 2.3 from the ideal value 3/8 $c = 0.375\ c$: a larger u leads to an increase of the spontaneous polarization [16]. The polarization produces charges located at interfaces and surfaces, and the polarization vector $\mathbf{P}_{\mathrm{SP}}$ points from negative to positive charges. Experimental values of the spontaneous polarization are difficult to measure and scarce; values computed using density-functional theory are given in Table 4.4.

If stress is applied on a crystal which has no inversion centre in the unit cell, an electric polarization is produced; this effect is referred to as *piezoelectricity*. The piezoelectric polarization $\mathbf{P}_{\mathrm{PE}}$ is in first order proportional to the strain, but second-order piezoelectricity may be relevant for zincblende semiconductors [17]. For crystals showing spontaneous polarization, the total macroscopic polarization $\mathbf{P}$ is given by the sum of $\mathbf{P}_{\mathrm{SP}}$ and $\mathbf{P}_{\mathrm{PE}}$.

The piezoelectric polarization vector $\mathbf{P}_{\mathrm{PE}}$ is obtained from the product of the piezoelectric tensor and the strain ε; *in the linear regime* the components of $\mathbf{P}_{\mathrm{PE}}$ are given by

$$P_{\mathrm{PE},i} = \sum_{jk} e_{ijk}\varepsilon_{jk}; \tag{4.9}$$

here e_{ijk} are components (*piezoelectric stress constants*) of the third-rank piezoelectric tensor and ε_{jk} are components of the second-rank strain tensor defined in (2.4)[1]; the indices i, j, k stand for the x, y, and z axes of the Cartesian coordinate system. The piezoelectric tensor is commonly written in the matrix representation of the Voigt notation, where the indices j and k are replaced by a single index, yielding for (4.9) the notation

[1]The strain components denoted e_{jk} introduced in (2.3) and applied in (2.4) should not be confused with the piezoelectric tensor components e_{ijk} introduced in (4.9).

$$\mathbf{P}_{\mathrm{PE}} = \begin{pmatrix} P_{\mathrm{PE,x}} \\ P_{\mathrm{PE,y}} \\ P_{\mathrm{PE,z}} \end{pmatrix} = \begin{pmatrix} e_{11}\ e_{12}\ e_{13}\ e_{14}\ e_{15}\ e_{16} \\ e_{21}\ e_{22}\ e_{23}\ e_{24}\ e_{25}\ e_{26} \\ e_{31}\ e_{32}\ e_{33}\ e_{34}\ e_{35}\ e_{36} \end{pmatrix} \begin{pmatrix} \varepsilon_{xx} \\ \varepsilon_{yy} \\ \varepsilon_{zz} \\ \varepsilon_{yz} + \varepsilon_{zy} \\ \varepsilon_{zx} + \varepsilon_{xz} \\ \varepsilon_{xy} + \varepsilon_{yx} \end{pmatrix}. \tag{4.10}$$

The piezoelectric tensor has 18 independent components only for the low-symmetry triclinic crystals; for crystals with higher symmetry many components are zero. *All* $e_{ij} = 0$ in (4.10) for crystals with inversion symmetry, such as diamond Si and Ge, or rock-salt MgO. For zincblende structure the piezoelectric tensor has only a single independent constant and reads

$$\mathbf{e}_{\mathrm{zincblende}} = \begin{pmatrix} 0\,0\,0\ e_{14}\ 0\ \ 0 \\ 0\,0\,0\ \ 0\ \ e_{14}\ 0 \\ 0\,0\,0\ \ 0\ \ 0\ \ e_{14} \end{pmatrix}; \tag{4.11}$$

for wurtzite crystals three independent piezoelectric constants exist, yielding

$$\mathbf{e}_{\mathrm{wurtzite}} = \begin{pmatrix} 0 & 0 & 0 & 0 & e_{15} & 0 \\ 0 & 0 & 0 & e_{15} & 0 & 0 \\ e_{31} & e_{31} & e_{33} & 0 & 0 & 0 \end{pmatrix}. \tag{4.12}$$

In *zincblende crystals* only off-diagonal components of the strain tensor produce piezoelectricity, as we note when inserting (4.11) into (4.10). Strain along the cubic axes as, e.g., occurring in biaxially strained quantum wells grown on (001) oriented substrates, have consequently no effect—in contrast to strain in quantum wells on (111) substrates. Piezoelectricity is usually weak in zincblende material, even for growth of heterostructures along [111]; in such strained layers the x component of $\mathbf{P}_{\mathrm{PE}}$ is given by

$$P_{\mathrm{PE},x} = e_{14}\,\left(\varepsilon_{yz} + \varepsilon_{zy}\right) = 2e_{14}\varepsilon_{yz}. \tag{4.13}$$

The y and z components of $\mathbf{P}_{\mathrm{PE}}$ result from a cyclic change of x, y, and z. The three off-diagonal strain components are equal, and consequently also those of $\mathbf{P}_{\mathrm{PE}}$ which points in [111] direction; the absolute value of the polarization is hence $P_{\mathrm{PE}} = 2\sqrt{3}\ e_{14}\varepsilon_{xy}$. The off-diagonal strain components are related to the lattice mismatch f of [111] oriented pseudomorphic layers with $\varepsilon_{||} = f$ (see (2.20a)) by

$$\varepsilon_{xy} = -\frac{C_{11} + 2C_{12}}{C_{11} + 2C_{12} + 4C_{44}} f. \tag{4.14}$$

The common zincblende III–V semiconductors have negative components e_{14} and the II–VI semiconductors positive components (Table 4.3). A [111] compressively strained quantum well in a III–V semiconductor then produces a polarization vector pointing from the cation A face to the anion B face.

Table 4.3 Linear e_{14} and quadratic B coefficients of the piezoelectric tensor for zincblende semiconductors given in C/m^2. Linear coefficients with index *calc* and quadratic coefficients are calculated values from [19] with 0.05 C/m^2 precision unless explicitly indicated, *exp* denotes experimental values

Semiconductor	$e_{14}^{\rm exp}$	$e_{14}^{\rm calc}$	B_{114}	B_{124}	B_{156}
AlAs		−0.048	−1.5	−2.6	−1.2
AlSb	0.157, 0.068	−0.084	−0.7	−2.2	−0.7
GaP	−0.10, −0.15	−0.131	−0.7	−3.6	−0.9
GaAs	−0.16	−0.238	−0.4	−3.8	−0.7
GaSb	−0.12, −0.126	−0.247	0.2 ± 0.1	−3.2 ± 0.1	0.0
InP	−0.083	0.003	−1.1	−3.8	−0.5
InAs	−0.045, −0.12	−0.115	−0.6 ± 0.1	−4.1	0.2
InSb	−0.071, −0.097	−0.159	0.1	−3.5	0.6
ZnSe	0.049				
ZnTe	0.028				
CdTe	0.034				

The electric field produced by the polarization is given by [18, 19]

$$E_{\rm PE} = -\frac{P_{\rm PE}}{\varepsilon_{\rm r}\,\varepsilon_0} = -\frac{2\sqrt{3}}{\varepsilon_{\rm r}\,\varepsilon_0} e_{14}\,\varepsilon_{xy}, \tag{4.15}$$

where $\varepsilon_{\rm r}$ is the permittivity of the low-frequency dielectric constant.

Second-order piezoelectricity may be important for zincblende structures [17]. Considering both linear and quadratic terms in strain we extend (4.9) to

$$P_{{\rm PE},i} = \sum_{jk} e_{ijk}\varepsilon_{jk} + \frac{1}{2}\sum_{jkmn} B_{ijkmn}\varepsilon_{jk}\varepsilon_{mn}, \tag{4.16}$$

with the indices denoting the Cartesian axes. B_{ijkmn} is a fifth-rank tensor with – for zincblende symmetry – 24 nonzero components comprising three independent ones. In second order the polarization involves both diagonal and off-diagonal components according to [19],

$$\mathbf{P} = e_{14}\begin{pmatrix} 2\varepsilon_{yz} \\ 2\varepsilon_{xz} \\ 2\varepsilon_{xy} \end{pmatrix} + B_{114}\begin{pmatrix} 2\varepsilon_{xx}\varepsilon_{yz} \\ 2\varepsilon_{yy}\varepsilon_{xz} \\ 2\varepsilon_{zz}\varepsilon_{xy} \end{pmatrix} + B_{124}\begin{pmatrix} 2\varepsilon_{yz}(\varepsilon_{yy}+\varepsilon_{zz}) \\ 2\varepsilon_{xz}(\varepsilon_{zz}+\varepsilon_{xx}) \\ 2\varepsilon_{xy}(\varepsilon_{xx}+\varepsilon_{yy}) \end{pmatrix} + B_{156}\begin{pmatrix} 4\varepsilon_{xz}\varepsilon_{xy} \\ 4\varepsilon_{yz}\varepsilon_{xy} \\ 4\varepsilon_{yz}\varepsilon_{xz} \end{pmatrix}. \tag{4.17}$$

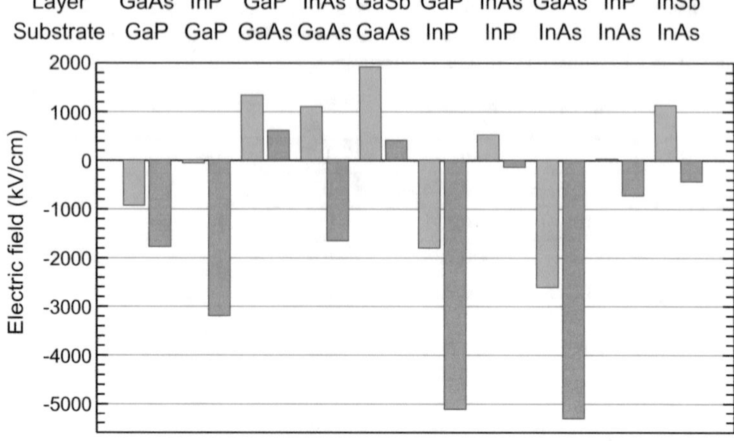

Fig. 4.5 Calculated electric fields in coherently strained [111] oriented quantum wells of zincblende A/B (A on B substrate) heterostructures produced by linear (*green*) and linear + quadratic (*red*) piezoelectricity. Data from [19]

For growth along [111] now also the diagonal strain components appear:

$$\varepsilon_{xx} = \varepsilon_{yy} = \varepsilon_{zz} = \frac{4C_{44}}{C_{11} + 2C_{12} + 4C_{44}} f, \tag{4.18}$$

and (4.16) yields the second-order polarization (in Voigt notation)

$$P_{\mathrm{PE}} = 2\sqrt{3} e_{14} \varepsilon_{yz} + 2\sqrt{3} \left[(B_{114}\varepsilon_{xx} + 2B_{124}\varepsilon_{yy})\varepsilon_{yz} + 2B_{156}\varepsilon_{xz}\varepsilon_{xy} \right], \tag{4.19}$$

with the linear first and the quadratic second summand. First-principles calculations indicate that the piezoelectric effect of second order in strain yields important contributions in the studied III–V semiconductors [19], see Fig. 4.5. Data of piezoelectric constants for some zincblende III–V and II–VI semiconductors are given in Table 4.3.

Wurtzite heterostructures of the column III nitrides or ZnO show strong piezoelectricity. We consider polarizations along the c axis, i.e., along the common [0001] growth direction. Biaxial strain in the basal plane is expressed by $\varepsilon_{xx} = \varepsilon_{yy} = \varepsilon_{||} = (a_{\mathrm{L}} - a_{\mathrm{L},0})/a_{\mathrm{L},0}$, and strain along the c axis correspondingly by $\varepsilon_{zz} = \varepsilon_{\perp} = (c_{\mathrm{L}} - c_{\mathrm{L},0})/c_{\mathrm{L},0}$. The indices L and L,0 denote the actual (strained) and natural lattice parameters of the layer, respectively. Using the relation between $\varepsilon_{||}$ and $\varepsilon_{\perp}$ (2.19c) we obtain for the z component of the piezoelectric polarization of a biaxially strained wurtzite crystal

Table 4.4 Coefficients e_{ij} of the piezoelectric tensor and spontaneous polarization P_{SP} of wurtzite semiconductors. All values are given in C/m^2

Semiconductor	e_{15}	e_{31}	e_{33}	P_{SP}
GaN	$-0.22\ldots-0.33$[a]	-0.35[b]	1.27[b]	-0.029[c]
AlN	-0.48[a]	-0.50[b]	1.79[b]	-0.081[c]
InN		-0.57[b]	0.97[b]	-0.032[c]
ZnO	$-0.35\ldots-0.59$[d]	$-0.35\cdots-0.62$[d]	$0.96\ldots1.56$[d]	-0.057[c]

[a]Reference [21], [b]Reference [3], [c]Reference [16], [d]Reference [22]

$$\mathbf{P}_{\mathrm{PE},z} = 2\varepsilon_{\|}\left(e_{31} - e_{33}\frac{C_{13}}{C_{33}}\right). \tag{4.20}$$

Literature data of piezoelectric tensor components and the spontaneous polarization for some wurtzite semiconductors are given in Table 4.4. According to the sign convention a positive c axis points from the metal cation to the adjacent anion. The semiconductors listed in Table 4.4 show a negative spontaneous polarization. The sign of the strain-induced piezoelectricity depends on the sign of strain. For $\mathrm{Al}_x\mathrm{Ga}_{1-x}\mathrm{N}$ the bracket in (4.13) yields negative values, and the piezoelectric polarization is positive for compressive strain ($\varepsilon_{\|} < 0$). The total polarization $P_z = P_{\mathrm{SP},z} + P_{\mathrm{PE},z}$ may then have either sign. For a review on electromechanical phenomena in semiconductor nanostructures see [20].

Piezoelectricity has a substantial effect on the electronic properties of devices made from GaN-based heterostructures. At the interface between two wurtzite semiconductors 1 and 2 the total polarization changes, giving rise to a sheet charge σ determined by the difference in total polarization,

$$\sigma = (P_{\mathrm{SP},z1} + P_{\mathrm{PE},z1}) - (P_{\mathrm{SP},z2} + P_{\mathrm{PE},z2}).$$

High-electron-mobility transistors based on strained AlGaN/GaN heterostructures achieved very high sheet carrier concentrations in a two-dimensional electron gas formed at the interface, enabling devices with excellent performance. The high values could be assigned to an additive effect of spontaneous and piezoelectric polarization in structures with tensely strained AlGaN barriers [23].

4.1.4 Temperature Dependence of the Bandgap

Energies of electronic bands are generally calculated for a temperature $T = 0$ K. To describe the temperature dependence of the important fundamental bandgap E_g a number of approaches was developed. The usually observed decrease of E_g for increased temperature originates from a change of both, the electron-phonon interaction and the interatomic bond distance. Instead of an explicit derivation from such interactions empirical formula are widely used to express the thermal behavior of E_g. Most popular is the empirical *Varshni formula* [24]

$$E_g(T) = E_g(T = 0) - \frac{\alpha T^2}{T + \beta}, \tag{4.21}$$

where the three parameters $E_g(T = 0)$, α, and β are fitted to experimental data. The dependence describes both, direct and indirect bandgaps. $E_g(T = 0)$ is the bandgap energy at 0 K. α is claimed to be related to the Debye temperature but may in certain cases be negative. Moreover, at very low temperatures a rather temperature-independent behavior of E_g was found instead of the quadratic dependence predicted from (4.21). Typical values for α and β are in the range (0.4–0.6) meV/K and (200–600) K, respectively. The thermal shift of the bandgap energy according to (4.21) is illustrated in Fig. 4.6.

A number of better motivated models was developed later, based on the occupation of phonon states and assuming an average phonon energy [25–27]. They particularly improve the description of low-temperature data but require at least one additional parameter. The physically well-founded model reported in [27, 29] assumes four parameters to account for the Bose-Einstein occupation of phonon modes and for phonon dispersion; it comprises the three-parameter approach of Varshni for the limit of very large phonon dispersion. The analytical four-parameter approximation reads

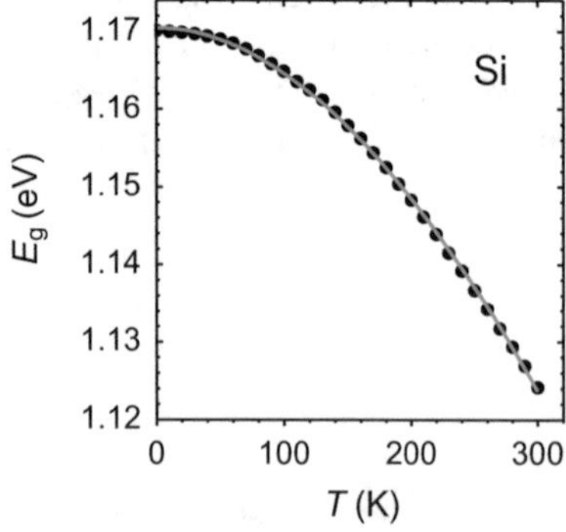

Fig. 4.6 Dependence of the indirect bandgap energy of Si on temperature. The *gray curve* is a fit to Varshni's formula (4.21), data are from [28]

$$E_g(T) = E_g(T=0) - \alpha\Theta \left\{ \frac{1-3\Delta^2}{\exp(\Theta/T)-1} + \frac{3\Delta^2}{2}\left(\sqrt[6]{1+\frac{\pi^2}{3(1+\Delta^2)}\left(\frac{2T}{\Theta}\right)^2 + \frac{3\Delta^2-1}{4}\left(\frac{2T}{\Theta}\right)^3 + \frac{8}{3}\left(\frac{2T}{\Theta}\right)^4 + \left(\frac{2T}{\Theta}\right)^6} - 1 \right) \right\} \quad (4.22)$$

Here Θ is the effective (average) phonon temperature, and Δ is a dispersion coefficient; α is the high-temperature limiting magnitude of the slope of the $E_g(T)$ dependence. The two dispersion-related parameters Θ and Δ are defined by introducing moments $\langle \varepsilon^m \rangle = \int \varepsilon^m \cdot w(\varepsilon)\, d\varepsilon$, $m = 1, 2$, of the total phonon energy spectrum with a normalized weighting function $w(\varepsilon) \geq 0$, $\int w(\varepsilon)\, d\varepsilon = 1$, yielding

$$\Theta = \langle \varepsilon \rangle / k, \quad (4.23)$$

$$\Delta = \sqrt{\langle \varepsilon^2 \rangle - \langle \varepsilon \rangle^2} / \langle \varepsilon \rangle, \quad (4.24)$$

$$\alpha = -\left(dE_g/dT\right)\big|_{T\to\infty}. \quad (4.25)$$

The dimensionless dispersion parameter Δ ranges from 0 (small dispersion) to ~3/4, while a reproduction of values resulting from Varshi's formula (4.21) require unrealistic large values of ~1.25 [29]. Table 4.5 lists parameter data for both Varshni's empirical model and the dispersion-related approach.

4.1.5 *Bandgap of Alloys*

The bandgap energy of a miscible random alloy of two (or more) semiconductors may continuously be varied by changing the composition. Unlike the lattice constant discussed in Sect. 2.1.10 the bandgap of the alloy is usually not obtained by a linear interpolation. Instead, it can normally be described by a quadratic dependence using a *bowing parameter b* which is mostly positive [32]. The bandgap energy $E_{g\ \mathrm{alloy}}$ of an alloy A_xB_{1-x} from two materials A and B with the same crystal structure and bandgaps $E_{g\ A}$ and $E_{g\ B}$, respectively, is expressed by

$$E_{g\ \mathrm{alloy}} = xE_{g\ A} + (1-x)E_{g\ B} - bx(1-x), \quad (4.26a)$$

x being the molar fraction of A in the alloy (cf. Sect. 2.1.10). The bandgap energy of a (pseudobinary) ternary alloy $A_xB_{1-x}C$ from binaries AC and BC is given by the same relation putting $E_{g\ A}$ and $E_{g\ B}$ to $E_{g\ AC}$ and $E_{g\ BC}$, respectively. For quaternary compounds of the type $A_xB_yC_{1-x-y}D$ (i.e., mixing of A, B, C atoms on the cation

Table 4.5 Parameterized temperature dependence of the bandgap energy E_g for various semiconductors according to the empirical Varshi formula and the dispersion-related model. Varshi parameters of III–V are from [3, 29], II–VI parameters are from [31]; dispersion-related parameters are from [29]

Semiconductor (lowest band)	$E_g(T=0)$ (eV)	Varshi's parameters		Dispersion-related parameters		
		α (10^{-4} eV/K)	β(K)	α (10^{-4} eV/K)	Θ (K)	Δ
Si (X)	1.170	4.73	636	3.23	446	0.51
Ge (L)	0.743	4.774	235	4.13	253	0.49
AlN (Γ_A)	6.25	17.99	1462	9.1	770	0.34
GaN(Γ_A)	3.510	9.09	830	6.14	586	0.40
InN(Γ_A)	0.78	2.45	624	2.3	590	0.35
AlP (X)	2.52	3.18	588			
GaP(X)	2.35	5.771	372	4.77	355	0.60
InP (Γ)	1.4236	3.63	162	3.96	274	0.48
AlAs (X)	2.24	7.0	530	3.90	256	0.48
GaAs (Γ)	1.519	5.405	204	4.77	252	0.43
InAs (Γ)	0.417	2.76	93	2.82	147	0.68
AlSb (X)	1.696	3.9	140	3.45	205	0.76
GaSb (Γ)	0.812	4.17	140	3.87	205	0.44
InSb (Γ)	0.235	3.2	170	2.54	155	0.36
ZnS (Γ)	3.91	10	600	5.49	285	0.37
CdS (Γ_A)	2.579	4.7	230	4.10	166	0.47
ZnSe (Γ)	2.820	5.58	187	5.00	218	0.36
CdSe (Γ_A)	1.84	17	1150	4.08	187	0.20
ZnTe (Γ)	2.39	5.49	159	4.68	170	0.37
CdTe (Γ)	1.606	3.10	108	3.08	104	0.69

sublattice) the bandgaps are described by the weighted sum of the related ternary alloys ABD, ACD, and BCD, yielding [33]

$$\begin{aligned} E_{\mathrm{g\ alloy}} &= xE_{\mathrm{g}\ AD} + yE_{\mathrm{g}\ BD} + (1-x-y)E_{\mathrm{g}\ CD} \\ &\quad - b_{AB}xy - b_{AC}x(1-x-y) - b_{BC}y(1-x-y). \end{aligned} \tag{4.26b}$$

Here b_{AB}, b_{AC} and b_{BC} are the three bowing parameters for the ternary alloys $A_xB_{1-x}D$, $A_xC_{1-x}D$, and $B_xC_{1-x}D$, respectively. The bandgap energy for quaternary alloys of the type $A_xB_{1-x}C_yD_{1-y}$ (i.e., mixing on the cation *and* anion

Table 4.6 Bowing parameters b of the direct bandgap of alloyed GaAs-based zincblende and GaN-based wurtzite semiconductors. Data from [3]

Semiconductor	b for alloy with GaAs (eV)	Semiconductor	b for alloy with GaN (eV)
AlAs	$-0.127 + 1.310 \times x_{\mathrm{Al}}$	AlN	1.0
InAs	0.477	InN	3.0
GaP	0.8		

sublattice) is calculated from the ternary parameters $E_{\mathrm{g}\ ABC}$, $E_{\mathrm{g}\ ABD}$, $E_{\mathrm{g}\ ACD}$, and $E_{\mathrm{g}\ ABD}$,

$$E_{\mathrm{g\ alloy}} = \frac{x(1-x)[yE_{\mathrm{g}\ ABC}(x) + (1-y)E_{\mathrm{g}\ ABD}(x)] + y(1-y)[xE_{\mathrm{g}\ ACD}(y) + (1-x)E_{\mathrm{g}\ BCD}(y)]}{x(1-x) + y(1-y)},$$
$$E_{\mathrm{g}\ ABC}(x) = xE_{\mathrm{g}\ AC} + (1-x)E_{\mathrm{g}\ BC} - b_{ABC}x(1-x), \tag{4.26c}$$
$$E_{\mathrm{g}\ ABD}(x),\ E_{\mathrm{g}\ ACD}(y),\quad \text{and}\quad E_{\mathrm{g}\ BCD}(y)\ \text{accordingly.}$$

We note that (4.26a)–(4.26c) are similar to (2.2a)–(2.2c) for $b = 0$. Bowing parameters for some ternary (pseudo-binary) alloys are given in Table 4.6. The value given for $Al_xGa_{1-x}As$ indicates that a *constant* bowing parameter does not always yield an appropriate description.

Bandgap energies for a number of important semiconductors are given in Fig. 4.7. The lines representing the bandgap of alloys in Fig. 4.7 sometimes have a kink. Such features originate from a transition of a direct to an indirect semiconductor due to a crossing of the lowest Γ conduction band and an X or L conduction band (for Si and Ge crossing of indirect X and L bands).

The bandgap energy of alloys composed of more than two semiconductors can be illustrated using diagrams with curves of constant energy versus composition parameters. A material of particular importance for optoelectronic devices is the quaternary alloy $Ga_xIn_{1-x}As_yP_{1-y}$. The bandgap of such quaternaries may be chosen independently from the lattice parameter by a proper selection of the two independent composition parameters x and y. For lattice matching conditions on InP substrates one composition parameter is independent and may be used to choose a bandgap energy, while the other parameter is given by

$$x = 0.1896y/(0.4176 - 0.0125y) \approx 0.47y \quad (0 \le y \le 1).$$

The diagram Fig. 4.8 shows the variation of the direct bandgap of $Ga_xIn_{1-x}As_yP_{1-y}$ in the full range of compositions x and y, along with lattice matching conditions for GaAs and InP substrates.

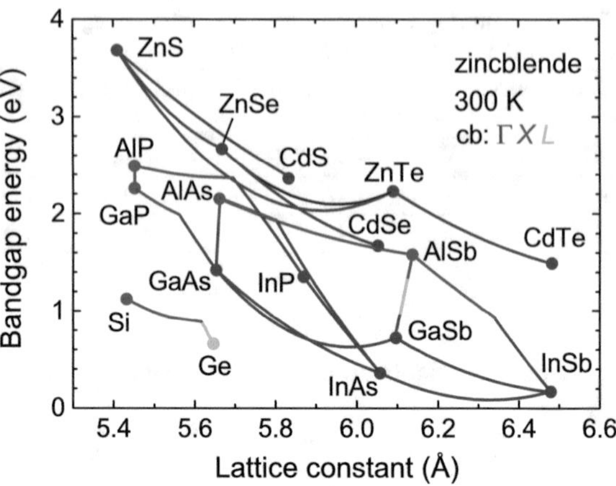

Fig. 4.7 Bandgap energy as a function of lattice constant for pure (*dots*) and alloyed zincblende and diamond semiconductors at room temperature. *Blue, red, and green* drawing denote direct Γ, indirect X, and indirect L bandgap, respectively

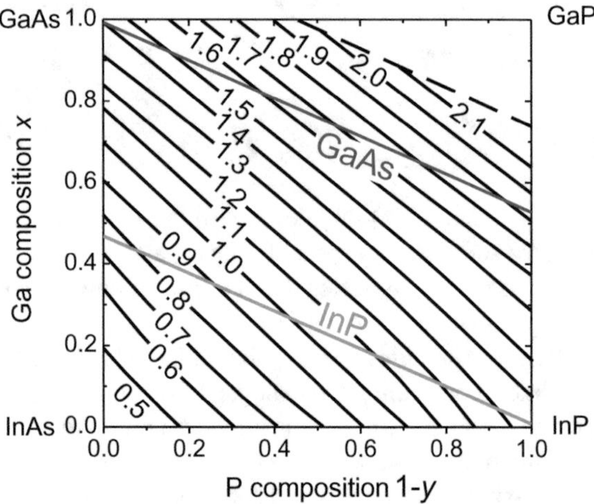

Fig. 4.8 Bandgap energy of the quaternary alloy $Ga_xIn_{1-x}As_yP_{1-y}$ in eV at 300 K as a function of compositions x and y. *Red* and *green lines* mark compositions which are lattice matched to GaAs and to InP. From [38]

Equations (4.26a)–(4.26c) show that the bandgap energy of an alloy usually deviates from a linear concentration-weighted interpolation by a quadratic term described by the bowing parameter b. It should be noted that the virtual-crystal approximation (Sect. 2.1.11) which describes a lattice parameter variation *without* bowing also yields such bowing for the bandgap energy [32, 34–36]. The reason is that the eigenvalues produced by band-structure methods are mostly nonlinear in the potential matrix elements. It was, however, pointed out that a given set of band energies can be fitted by widely different potential parameters in semiempirical models, permitting almost any value of b within the simple VCA approach [37].

The origin of optical bowing is widely associated with disorder in an alloy. The effect of alloy disorder alone may however be insufficient to describe bowing. In a first-principles approach evidence was given for three contributions to b in tetrahedrally coordinated compound semiconductors: A modification of the band structure due to the change of the lattice constant, a relaxation of the anion-cation bond length in the alloy, and a contribution of chemical electronegativity due to charge exchange in the alloy [37]. Respective calculations provided a good description of bowing parameters experimentally obtained from zinc chalcogenes. Bowing of nitride alloys is reported in [30, 39–41].

4.2 Band Offsets

Epitaxy allows to produce a pseudomorphic, atomically sharp transition from one solid to another. The intimate contact of two solids with different electronic properties forms a heterojunction. At the interface the electronic bands of the solids align on a scale of atomic nearest-neighbor distances. In addition, transfer of charges from one solid to the other and charge accumulation at the interface may lead to an electrostatic bending of the bands on a larger scale. The band alignment is of basic technological importance, because it controls the transport and confinement of charge carriers and hence the properties and performance of (opto-)electronic devices. Calculation and measurement of band offsets (also termed *band discontinuities*, *band lineups* or *band alignments*) are difficult, and a thorough understanding of the physics of semiconductor band-alignment is still missing. In the following we focus on the technologically important contact between two semiconductors. The junction between a semiconductor and a metal is treated in Sect. 10.3.

Various types of band alignments may occur at semiconductor junctions, usually labeled type I and type II. In a *type I* alignment the bandgap of one semiconductor lies completely within the bandgap of the other. If such straddled type I offset is applied to a *BAB* double heterostructure, electrons and holes can both be confined in material A with the smaller bandgap, see Fig. 4.9a. This feature is often employed in low-dimensional structures like quantum wells to tailor electronic properties, cf. Sect. 4.3. If the offset in the valence band and the conduction band has *the same sign* for an electron transfer from material A to B the band alignment is referred to as *type II*. In a *BAB* double heterostructure with such a staggered band lineup *either* electrons *or* holes are confined in material A. In Fig. 4.9b a lower conduction band edge in material A leading to electron confinement is assumed. When the bandgaps of the two semiconductors do not overlap at all a *misaligned* (or *broken gap*) configuration occurs. Such alignment appears if in Fig. 4.9 $E_{\mathrm{v}\ A}$ lies above $E_{\mathrm{c}\ B}$ or $E_{\mathrm{c}\ A}$ lies below $E_{\mathrm{v}\ B}$.

A case analogous to the broken gap configuration occurs if a junction is formed by a semiconductor and a zero-gap semiconductor like, e.g., HgTe. This kind of alignment is occasionally referred to as *type III*.

The type of band alignment forming in a semiconductor heterostructure depends on the position of the respective band edges. The prediction of this alignment is not trivial, because there exists no natural common reference energy. Such reference should be a property of a bulk crystal. Much theoretical and experimental work was devoted to predict or measure offsets within the required precision of about 0.1 eV

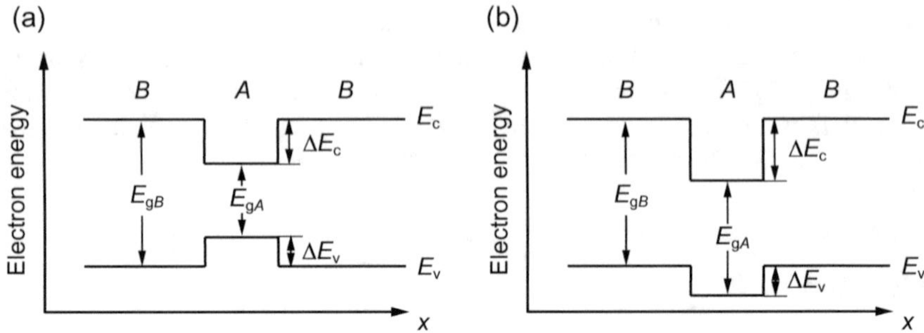

Fig. 4.9 Alignment of band edges in **a** type I and **b** type II double heterostructures built from a small-bandgap semiconductor A with a small extension along the spatial coordinate x and a wide-bandgap semiconductor B

or better. A comprehensive review as of 1991 was given in [42]. In the following we will consider some rules and more recent theoretical and experimental results.

4.2.1 Electron-Affinity Rule

If two semiconductors A and B are combined their Fermi levels E_F tend to align by transferring electrons from the solid with higher Fermi energy to the other. In the ideal case the vacuum level E_{vac} is a common reference energy [43, 44]. In the framework of this so-called *electron-affinity rule* the alignment of the conduction bands follows from their electron affinities $e\chi = E_{vac} - E_c$, E_c being the band edge of the (lowest) conduction band and e the elementary charge, see Fig. 4.10. The discontinuity of the conduction band at the interface is then

$$\Delta E_c = e(\chi_A - \chi_B), \tag{4.27}$$

and the corresponding offset of the (uppermost) valence band is given by

$$\Delta E_v = e(\chi_A + E_{g\,A}) - e(\chi_B + E_{g\,B}), \tag{4.28}$$

E_g being the bandgap.

Theoretical calculations suggest that only a single atomic layer away from the interface the electronic structure in a heterostructure becomes nearly bulk-like. The offset can hence be well assumed as being abrupt as illustrated in Fig. 4.10b. This does not apply for the long-range band bending (μm scale) originating from the electron

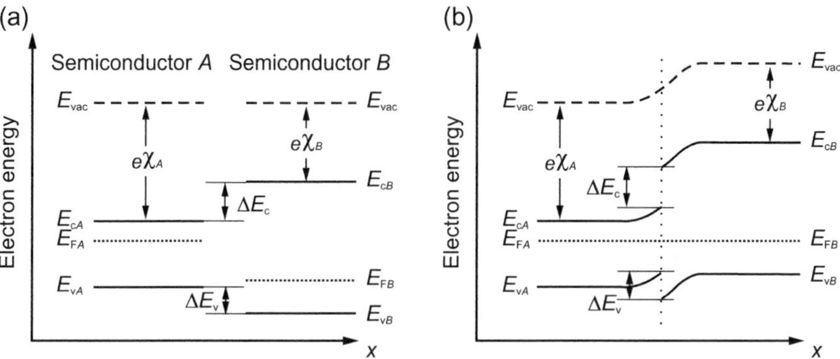

Fig. 4.10 Band alignment at a heterointerface between two semiconductors in the ideal case in absence of dipoles and interface states. **a** Before and **b** after formation of a heterojunction in thermal equilibrium

transfer for aligning E_F. Therefore double-heterostructures with semiconductors of small dimensions (nm scale) embedded in other semiconductors may be represented by flat bands as depicted in Fig. 4.9. The electron-affinity rule outlined above provides in many cases a good estimate of the band alignment of the heterointerface in organic semiconductors; this topic is discussed in Sect. 5.3.1.

A drawback of the classical electron-affinity rule is that the reference energy E_{vac} is not a bulk property. The electron affinity $e\chi$ is determined by experiments involving the surface. Therefore the structure of the surface and related charges may strongly affect the potential. Consequently the vacuum level is not a reliable reference.

4.2.2 Common-Anion Rule

Heterostructures are often fabricated from compound semiconductors like ZnS and ZnSe. In such cases a heuristic rule called *common-anion rule* has been applied to estimate whether the offset essentially occurs in the valence band or in the conduction band. The rule arises from the evidence that the valence-band states are essentially derived from p-states of the anions in sp-bonded AB compounds. The energy of the valence-band edge on an absolute scale is therefore expected to be basically determined by the valence electrons of the anions. Consequently the valence-band offset should be governed by the difference in anion electronegativity of the two semiconductors. A heterojunction with a common cation like Zn in a ZnS/ZnSe contact is expected to induce merely a small offset in the conduction band, leaving the majority of the bandgap difference for an offset in the valence band. Early models of band alignments largely complied with the common-anion rule [45, 46]. The models did not include d-orbitals of cations. Deviations from the common-anion rule were basically ascribed to contributions of these orbitals to the valence band [47]. It must, however, be noted that the rule fails in many cases. A basic shortcoming is the fact that the rule does not pay attention to an interface dipole formed from contributions of both, anions and cations [48].

4.2.3 Model of Deep Impurity Levels

Experiments indicate the existence of some "natural" reference potential which adopts the role of the vacuum level in the classical approach. An indication for such a reference is the transitivity rule for the valence-band offset found for some combinations of semiconductors A, B, and C, i.e., $\Delta E_{\mathrm{v}}^{AC} = \Delta E_{\mathrm{v}}^{AB} + \Delta E_{\mathrm{v}}^{BC}$ [49, 50]. Moreover, deep level impurities were found to have similar energy differences in different semiconductors. The observation was used for an empirical description of heterostructure band-offsets.

Transition metals like Fe form localized impurity states in semiconductors. They often possess several charge states separated by a fraction of the energy gap of the host crystal. A change of the impurity's charge state implies the transfer of a charge carrier from or to the host. This allows to determine the impurity level with respect to the band edges of the host material. Comparing such levels for a series of impurity ground states in various semiconductors yields an apparent similarity of both ordering and relative energy separations, cf. Fig. 4.11. The levels of transition-metal impurities are not pinned to either band edge like those of shallow impurities used as donors or acceptors (Sect. 10.1). This finding leads to the approach to take these levels within the bulk crystals as a reference for the alignment of the band edges in a heterojunction [51, 52]. The offset of the valence band at the interface ΔE_{v} is then given by the difference in the energy level positions of a given impurity in the two semiconductors forming the heterojunction. The constant separation of deep cationic impurity levels from the vacuum level was attributed to their antibonding character [52].

The energy levels of the transition-metal impurities depicted in Fig. 4.11 refer to a charge transition of an acceptor from a singly negative charge to the neutral state $(-/0)$. The measured transition energies were vertically shifted, so as to minimize the overall deviations, yielding the relative positions of the valence-band edges [51]. Positions of the conduction-band edges were experimentally obtained from low-temperature energy gaps. It should be noted that the method is restricted to heterojunctions formed by pairs of isovalent compound semiconductors, e.g., among III–V or among II–VI compounds, to ensure an electrically neutral interface without a dipole moment.

A comparable universal alignment of deep impurity levels as described above was also reported for hydrogen [53]. By computing the position of the Fermi energy where the stable charge state of interstitial hydrogen changes from the H^+ donor state to the H^- acceptor state, predictions of band alignments for a wide range of host zincblende and wurtzite compound semiconductors were given.

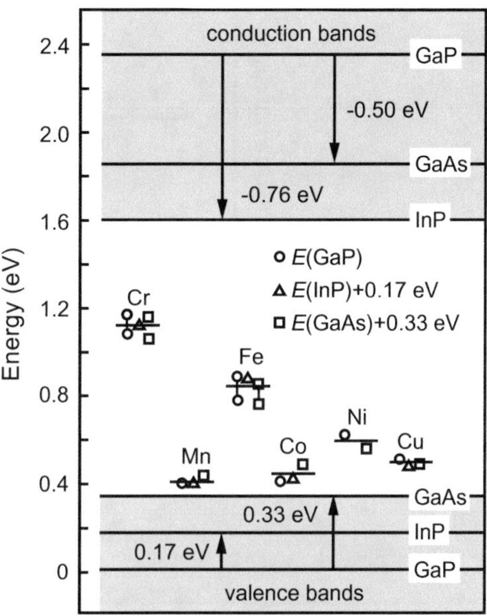

Fig. 4.11 Average energy levels of transition metal impurities forming deep acceptor states in GaP, InP, and GaAs, drawn with respect to the GaP valence-band edge. After [51]

4.2.4 *Interface-Dipol Theory*

The dominant role of charge accumulation at the interface rather than the effect of bulk properties was emphasized in the *interface-dipol theory* for the prediction of heterojunction band-offsets [54]. The approach was also applied to the theory of Schottky barrier heights at a metal-semiconductor interface [55] considered in Sect. 10.3.2. The initial point of the approach is that the discontinuity at a semiconductor interface induces electronic states in the bandgap of at least one semiconductor. The formation of such states is illustrated in Fig. 4.12 for the artificial case of a band discontinuity created in a homogeneous semiconductor by an external step potential. States lying near the conduction-band edge at side A of the interface have exponentially decaying tails into side B. At side B they lie in the gap of the semiconductor. Any state in the gap has a mixture of valence- and conduction-band character. Occupying such state leads locally to an excess charge, according to its degree of conduction character. Filling a state which lies near the top of the gap gives a large excess charge of almost one electron due to a large conduction character. Leaving that state empty gives an only slight charge deficit. Conversely filling a state near the bottom of the gap at side A results in a slight excess charge in proportion to its little conduction character, while leaving it empty leads to a charge deficit of almost one electron. Changing the band lineup hence induces a net dipole. The external potential assumed to create the potential step in Fig. 4.12 leads to an electron deficit at side B and an electron excess at side A of the interface. The resulting dipole tends to reduce the offset, i.e.,

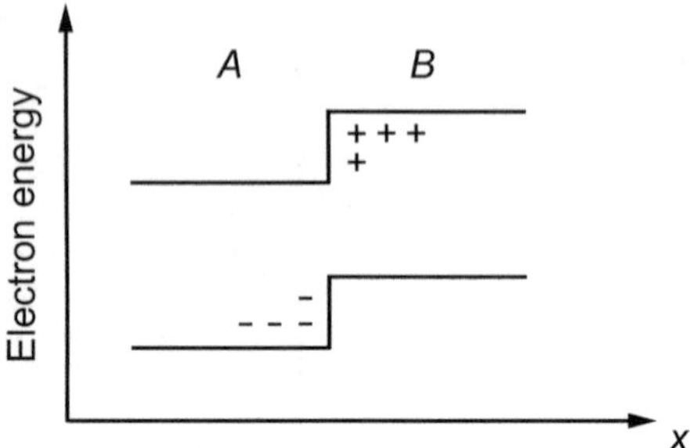

Fig. 4.12 Scheme of band alignment at an interface artificially induced in a homogeneous semiconductor. (+) and (−) represent net charges of unoccupied states with electron deficit and occupied states with electron excess, respectively. After [54]

Table 4.7 Differences of midgap energies $E_{B,A} - E_{B,B}$ according to the interface-dipol theory and experimental valence-band discontinuities for some semiconductor heterojunctions. Data from [54]

Heterojunction		$E_{B,A} - E_{B,B}$ (eV)	$\Delta E_{\mathrm{v,exp}}$ (eV)
A	*B*		
Si	Ge	0.36 − 0.18 = 0.18	0.20
GaAs	Ge	0.70 − 0.18 = 0.52	0.53
GaAs	InAs	0.70 − 0.50 = 0.20	0.17

the potential step is screened. By the action of the induced local charges the step is reduced by a factor of ε, the bulk dielectric constant.

The heterojunction of two different semiconductors is described analogous to the case considered above. Local states in the gap of one or both semiconductors lead to a dipole which screens the potential step and drives the lineup towards a value which minimizes the dipole. The zero-dipole lineup condition is not obvious. An effective midgap energy E_B for each semiconductor is introduced at which the gap states, on the average, cross over from valence to conduction character [54]. States at E_B are nonbonding on the average, the respective position is calculated from the band structure [55]. For a heterojunction E_B plays a role in analogy to the Fermi energy in metals: E_B is aligned for the respective semiconductors. Results of the interface-dipol theory are given in Table 4.7 for some heterojunctions, calculated values are claimed to be typically accurate to ~0.1 eV. Extensions of the interface-dipol theory distinguish between the long-range tails of the gap states considered above and polarization of the bonds which form the interface. More recently the approach of branch-point energies was also applied to zincblende-type nitrides [56, 57].

4.2.5 *Model-Solid Theory*

The more recently widely applied model-solid theory emulates the classical electron affinity rule illustrated in Fig. 4.10 by constructing a local reference level and avoiding dipoles. Within this approach the charge density in a semiconductor is composed by a superposition of neutral atoms [58]. The potential outside each such sphere goes exponentially to zero. This zero is taken as reference level. The construction leads to a well-defined electrostatic potential with respect to the vacuum level *in each atom*. By superposition the average electrostatic potential in a model solid composed of such atoms is hence specified on an absolute energy scale. The electron configuration of an atom in the solid is determined from a tight-binding calculation. This leads for, e.g., one Si atom in a silicon bulk crystal to $1.46s$ and $2.54p$ electrons, meaning that a part of the two s electrons of a Si atom are excited into the p band [58]. The result of the calculation is the position of the valence band on some absolute energy scale, allowing to relate it to the respective value of another semiconductor. For semiconductors with zincblende or diamond structure the value $E_{\mathrm{v,av}}$ represents an average of the heavy hole, light hole and split-off hole valence-bands. Spin-orbit effects are added a posteriori. Once $E_{\mathrm{v,av}}$ values are computed separately for a pair of semiconductors, their band discontinuities can be predicted for an unstrained heterojunction with a perfect interface, i.e., an abrupt change in the type of material without displacements of atoms from their ideal positions. The result of a calculation of the average valence-band energy $E_{\mathrm{v,av}}$ on an absolute scale in the framework of the model-solid theory is given in Fig. 4.13 for some unstrained semiconductors. Data are included in Table 4.1 along with the effect of strain. Data of $E_{\mathrm{v,av}}$ for *alloys* of III–V semiconductors are reported in [59].

4.2.6 *Offsets of Some Isovalent Heterostructures*

For many isovalent combinations of semiconductors with equal valence of the atoms on the two sides of a common junction the contribution of the interface dipole is not very large. In such cases the valence-band energy may be considered a bulk property on either side, and the simple difference of electron affinities (4.28) is a reasonable approximation to evaluate the valence-band offset. Data for a number of technologically important III–V compound semiconductors are summarized in Fig. 4.14. The values do not include effects of strain. Furthermore, the valence-band offset is taken to be independent of temperature, basically due to a lack of reliable data. Valence-band offsets at a junction of two semiconductors of Fig. 4.14 are given by the energy difference of their plotted band positions. The same applies for the offset in the direct conduction band. The direct bandgap energy is represented by the vertical lines and is given for low temperature (0 K). Note that some of the binaries have a smaller *indirect* bandgap and that the three nitrides usually crystallize in the wurtzite structure.

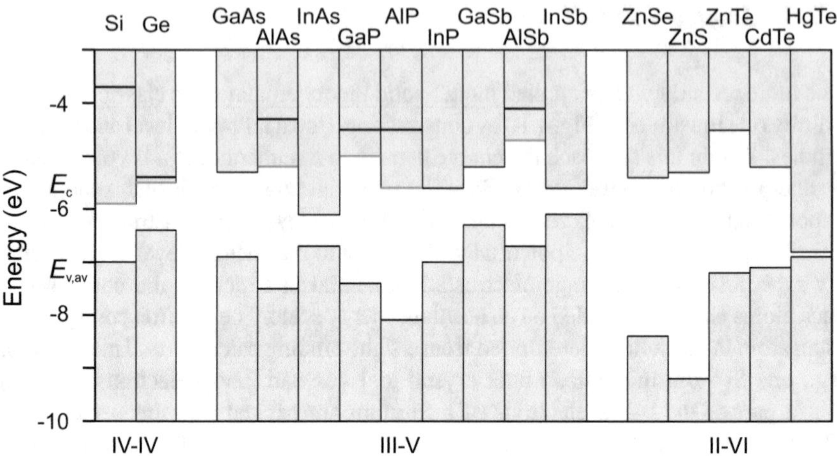

Fig. 4.13 Average valence-band energy $E_{v,av}$ of unstrained semiconductors on an absolute scale resulting from the model-solid theory. Experimental values of bandgap energies are used to depict the conduction-band energy E_c, *black* and *gray lines* at the *bottom* of the conduction bands denote direct and indirect conduction-band edges, respectively. Data taken from [10]

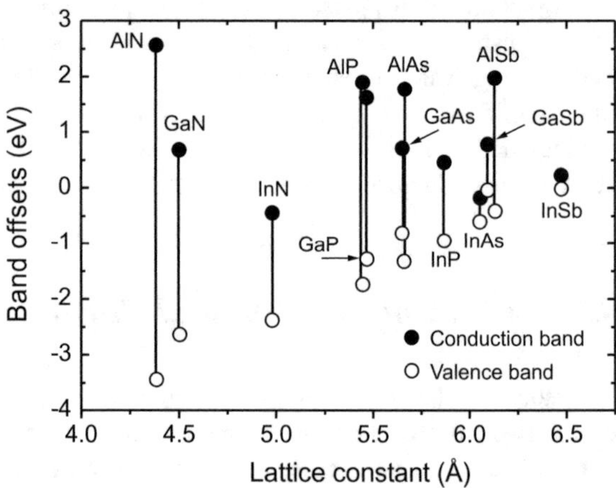

Fig. 4.14 Valence-band offsets (*open circles*) of binary semiconductors with zincblende structure, plotted with respect to the valence-band maximum of InSb. *Filled circles* indicate corresponding offsets of the lowest direct conduction band. From [3]

4.2.7 Band Offset of Heterovalent Interfaces

Models considered above in Sects. 4.2.1–4.2.5 considered abrupt interfaces and derived band offsets without detailed knowledge of the atomic interface structure. Theoretical work showed that the valence-band offset may actually depend on the microscopic arrangement of atoms at the interface [60]. A particularly strong effect was experimentally found for heterovalent interfaces, where—in contrast to isovalent systems—the atoms at the two sides of the interface have different chemical valence. We will consider the well-studied ZnSe/GaAs interface in more detail. The

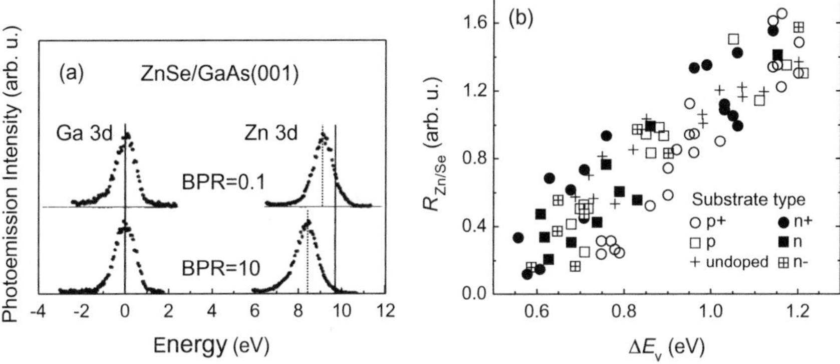

Fig. 4.15 **a** X-ray photoemission spectra of ZnSe/GaAs heterojunctions grown with different Zn/Se beam-pressure ratios (BPR). The energy origin was set to the center of the Ga *3d* core-level emission. **b** Experimental valence-band offsets for ZnSe/GaAs(001) interfaces with different type of substrate doping versus Zn/Se ratio obtained from integrated XPS emission intensities for 0.3 nm thick ZnSe layers. Reproduced with permission from [66], © 1994 APS

same findings were observed in the similar system ZnSe/AlAs [61] and many other heterovalent interfaces. Reviews with an emphasis on theoretical aspects are given in [62, 63].

ZnSe and GaAs have both zincblende structure and are well lattice-matched with a misfit below 0.5%. Due to the heterovalent nature of the II–VI and the III–V material an abrupt interface builds a strong dipole moment. Calculations indicate that such an abrupt transition at a heterovalent interface is energetically unstable and structures consisting of one or two intermixed layers are more favorable [60, 64, 65].

Experimental evidence for the impact of growth conditions on the band offset of the ZnSe/GaAs interface are given in Fig. 4.15. Using molecular beam epitaxy (Sect. 11.3) ZnSe layers were epitaxially grown on (001)-oriented GaAs substrates [66, 67]. Non-stoichiometric growth conditions were applied by controlling the composition via the Zn/Se beam-pressure ratio (BPR). The valence-band offset at the interface was measured using X-ray photoemission (XPS) related to the Ga *3d* and the Zn *3d* core levels, see Fig. 4.15a. The observed Zn *3d* core-level separation with respect to the ZnSe bulk value (difference between dotted and solid lines in Fig. 4.15a) gives directly the valence-band offset across the heterojunction. The corresponding Zn/Se ratio at the interface $R_{Zn/Se}$ was determined from the ratio of the integrated *3d* core-level emission intensities related to those of Zn and Se (Se not shown) for thin ZnSe layers. The measurements given in Fig. 4.15b clearly evidence a monotonous increase of the valence-band offset ΔE_v with increasing Zn/Se ratio from 0.58 eV (Se-rich) to 1.2 eV (Zn-rich).

The apparent dependence of the band offset from the interface stoichiometry can be understood in terms of differently mixed layers formed at the interface for varied growth conditions [65, 66]. Let us consider the sp^3 bonds of a binary zincblende semiconductor like GaAs. Each atom has 4 hybrid orbitals directed to the surrounding four nearest neighbors (Fig. 2.4a). Each of these bonds comprises 2 electrons. The *primitive* unit cell of GaAs contains one cation (Ga) and one anion (As) with a total number of 8 valence electrons: 3 from the Column III element Ga and 5 from the Column V element As. Each atom can be considered to donate one quarter of its valence electrons to its four bond orbitals. The number of valence electrons in one orbital referring to one Ga atom is then $3/4$, and the corresponding number per orbital of an As atom is $5/4$. One Ga-As bond thus contains $3/4 + 5/4 = 2$ electrons. The same applies for the II–VI semiconductor ZnSe: One Zn-Se bond contains $2/4 + 6/4 = 2$ electrons. At the heterovalent interface between GaAs and ZnSe we find Ga-Se or Zn-As bonds. A Ga-Se bond contains $3/4 + 6/4 = 2\frac{1}{4}$ electrons, i.e., it has an excess of a quarter electron. Such a bond acts like a donor. Likewise a Zn-As bond has a quarter electron deficiency and acts like an acceptor. This unbalanced charge generates the dependence of the valence band offset from the stoichiometry at the interface.

An abrupt interface contains either solely Ga-Se bonds or Zn-As bonds. In both cases a strongly localized, two-dimensional charge is created at the interface. Such a charge is connected to a high interface energy. Consequently the abrupt interface is thermodynamically unstable against intermixing [65]. Interface layers with a mixture of atoms from both semiconductors contain both kind of bonds; they accumulate less charge and are more stable. Figure 4.16 illustrates two examples of atomic configurations at intermixed heterovalent ZnSe/GaAs interfaces.

The total charge of an intermixed interface is reduced by charge transfer from donor-like to acceptor-like bonds. An interface with an equal number of uneven bonds is compensated. There may, however, remain a strong dipol moment at the interface: If the intermixed interface is built by a *single layer* of 50% Ga and 50% Zn cations (Fig. 4.16a) the acceptor-like Zn-As bonds lie towards the ZnSe and the donor-like Ga-Se bonds towards the GaAs. Note that the dipole moment is reversed, if such single-layer interface is formed on the anion sublattice (50% As + 50% Se, not shown in Fig. 4.16): The acceptor-like Zn-As bonds then lie towards the GaAs and the donor-like Ga-Se bonds towards the ZnSe. The valence-band offset is increased in the first case and decreased in the latter. Calculated values are +1.75 eV and +0.72 eV, respectively, when going from ZnSe to GaAs [65]. Formation of these mixings is favored in more Zn-rich growth conditions and Se-rich conditions, respectively, and describes correctly the experimentally observed tendency shown in Fig. 4.15. The average of the two offsets (1.25 eV) is close to the experimental value of 1.22 eV [68]. This indicates the possibility to compensate the dipole moment, namely by interface layers comprising more than a single layer.

An interface consisting of an intermixed cation layer and an intermixed anion layer is illustrated in Fig. 4.16b. The Ga atoms and the Se atoms in the interface double-layer create a quarter charge excess each, while the two As-Zn bonds have a quarter charge deficiency each. The dipole moment in this intermixed double layer is hence fully compensated. It should be noted that the dipole moment is also compensated *in average* if small domains of single-layer interfaces of both polarities occur in a ratio of 1:1 [65].

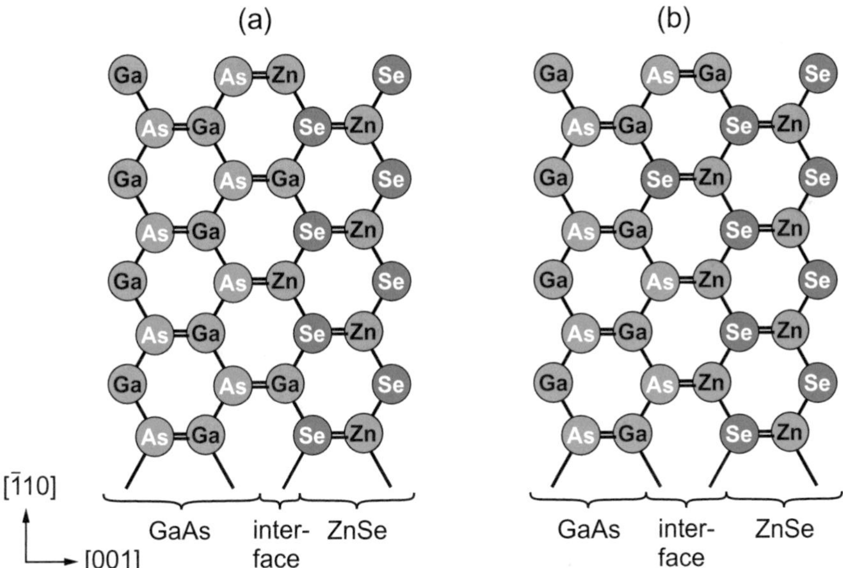

Fig. 4.16 Schematic arrangement of atoms at the heterovalent ZnSe/GaAs interface, viewed along the [110] direction. Each atom is bond by two bonds in the figure plane and by two bonds, which are directed out of and into the figure plane—the *horizontal double lines*. **a** Interface with a single intermixed cation layer comprising 50% Zn and 50% Ga. **b** Interface with an intermixed double layer containing 75% As and 25% Se anions, and 25% Ga and 75% Zn cations

The ZnSe/GaAs interface is well lattice-matched. Elastic relaxation as discussed in Sect. 2.2 does hence not play a significant role, and the effect was not considered here. Atomic relaxation does, however, play a crucial role in more mismatched heterostructures like GaN/SiC(001) [69].

Isovalent heterojunctions have, in contrast to the heterovalent interfaces discussed above, band offsets which are almost independent of the local atomic arrangement. This commonly accepted conclusion was initially established for common-ion systems, and later generalized [70]. It should be noted that intrinsic defects like antisites may limit the validity of this general statement and must be taken into account in case of low formation energies. The same applies for the formation of heterovalent interlayers at the interface.

4.2.8 Band Offsets of Alloys

The observation of a constant energy with respect to the vacuum level of deep transition-atom impurities according to the model of deep impurity levels (Sect. 4.2.3) was also used to measure the change of valence-band energy E_{v} in alloys. Experi-

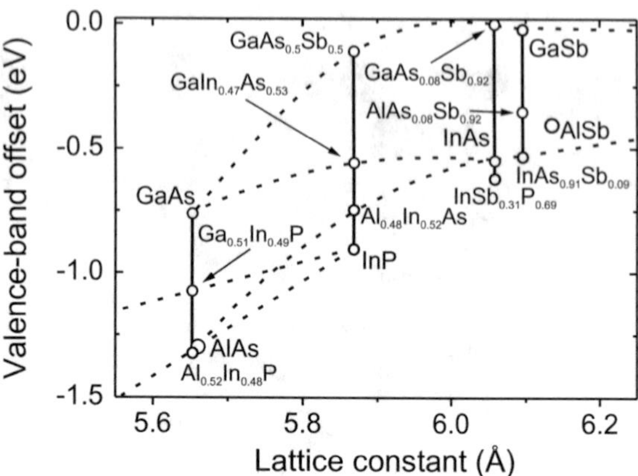

Fig. 4.17 Valence-band offset of alloy semiconductors, plotted with respect to the valence-band maximum of InSb. *Vertical lines* mark common substrate materials, points signify offsets for binaries and lattice-matched ternary alloys. From [3]

ments reveal mostly a linear dependence of the valence-band maximum in the alloy with respect to the deep levels for varied composition x. Studied materials are, e.g., $Ga_{1-x}Al_xAs$:Fe [51, 71], $Ga_{1-x}Al_xAs$:Cu [72], $GaAs_{1-x}P_x$:Cu [73], $In_{1-x}Ga_xP$:Mn [74]. The slope of the linear dependence was found to be largely independent from the impurity. Such linear variation may also be inferred from the interface-dipol theory (Sect. 4.2.4), where the effective midgap energy E_B of an alloy was estimated as a linear interpolation from the pure semiconductors [48]. Valence-band offsets of some zincblende semiconductor-alloys are summarized in Fig. 4.17 for composition parameters matching the lattice constants of the common substrate materials GaAs, InP, InAs, and GaSb.

4.3 Electronic States in Low-Dimensional Structures

The unique properties of low-dimensional structures originate essentially from the modification of the electronic density-of-states (DOS) produced by the confinement of charge carriers. To track such modification for the reduction of dimensionality from a three-dimensional (3D) bulk crystal to a 0D quantum dot we first recall the origin of 3D DOS and then consider the effect of potentials confining the mobility of charge carriers gradually to two, one, and eventually zero dimensions. Electronic properties of the solid are described in the framework of the effective-mass approximation by applying effective carrier masses and the relative permittivity as characteristic parameters.

4.3.1 Dimensionality of the Electronic Density-of-States

We first describe the energy of a single quasi-free electron confined in a bulk crystal by a simple square potential W given by the dimensions of the crystal L_x, L_y, and L_z. The periodic potential of the atoms which leads to the band structure is neglected. It may be treated in a second step as a perturbation of W. If we assume W being 0 inside the crystal and a constant $W_0 > 0$ outside we obtain energy and eigenstates of the electron inside the crystal by solving the Schrödinger equation

$$-\frac{\hbar^2}{2}\left(\frac{1}{m_x^*}\frac{\partial^2}{\partial x^2}+\frac{1}{m_y^*}\frac{\partial^2}{\partial y^2}+\frac{1}{m_z^*}\frac{\partial^2}{\partial z^2}\right)\psi(\mathbf{r})=E\psi(\mathbf{r}) \quad \text{(three dimensions)}. \tag{4.29}$$

The quantities m^* are the electron's effective masses along the three spatial directions x, y, z. Using periodic boundary conditions like $\psi(x \pm L_x, y, z) = \psi(x, y, z)$ we obtain the solutions of (4.29) given by plane waves $\psi_\mathbf{k}(\mathbf{r})$ with eigenenergies $E_\mathbf{k}$:

$$\psi_\mathbf{k}(\mathbf{r}) = \frac{1}{\sqrt{V}} e^{i(k_x x + k_y y + k_z z)}, \tag{4.30}$$

$$E_\mathbf{k} = \frac{\hbar^2 k_x^2}{2m_x^*} + \frac{\hbar^2 k_y^2}{2m_y^*} + \frac{\hbar^2 k_z^2}{2m_z^*}. \tag{4.31}$$

In (4.30) $V = L_x \times L_y \times L_z$ is the volume of the bulk crystal. If we apply the boundary conditions to (4.30) we yield allowed values for $\mathbf{k}$,

$$k_x = \frac{2\pi}{L_x}n_x, \qquad k_y = \frac{2\pi}{L_y}n_y, \qquad k_z = \frac{2\pi}{L_z}n_z, \quad n_x, n_y, n_z = 0, \pm 1, \pm 2, \ldots. \tag{4.32}$$

Each electron state is hence described by *discrete* values of $\mathbf{k}$ as illustrated in the scheme of the reciprocal space depicted in Fig. 4.18a. Each state marked by a dot in the figure is occupied by 2 electrons with opposite spin. The spacing between allowed adjacent values along k_x is $2\pi/L_x$. Since the crystal dimensions L_x, L_y, L_z are macroscopic quantities, a finite region of k-space contains a very high number of dense lying allowed states. $\mathbf{k}$ and likewise $E_\mathbf{k}$ are therefore quasi-continuous quantities. The number of allowed k-values per unit volume of $\mathbf{k}$ space, i.e., the density of states in $\mathbf{k}$ space, is given by the constant quantity $V/(2\pi)^3$.

The electronic density of states $g(E)$—expressed in units of $\mathrm{m}^{-3} \times \mathrm{J}^{-1}$ or $\mathrm{cm}^{-3} \times \mathrm{eV}^{-1}$—is obtained from the number of electron states dN per unit volume V and per energy interval dE,

$$g(E) = 2 \times \frac{1}{V}\frac{dN}{dE}. \tag{4.33}$$

The factor 2 in (4.33) accounts for the spin degeneracy, allowing for a two-fold occupancy of each state. We obtain dN from the volume in $\mathbf{k}$ space between two planes of constant energy at E and $E + dE$ multiplied by the constant density of

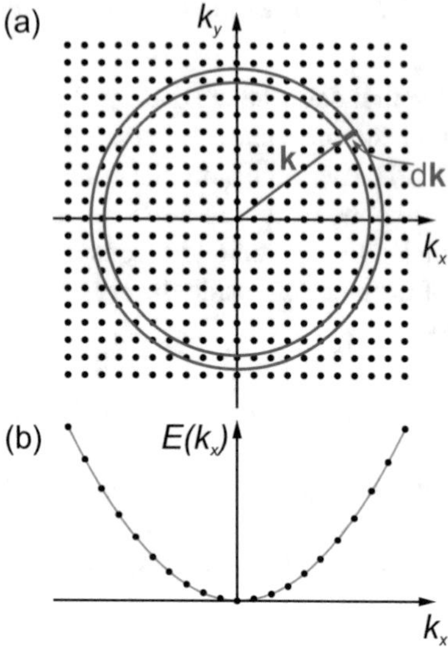

Fig. 4.18 **a** Cross section of **k** space in the k_x–k_y plane. **b** Values of $E(k_x)$ which fulfill the boundary condition for a quasi-free electron in a solid with finite dimension L_x

states in **k** space. Such a volume is illustrated in Fig. 4.18a by the spherical shell of thickness $d\mathbf{k}$. We note that the volume increases for a given dE (or correspondingly $d\mathbf{k}$) with the same power as the area of a sphere in **k** space increases as k augments. For a quasi-free electron we obtain

$$dN = \frac{V}{(2\pi)^3}\int_E^{E+dE} d^3k = \frac{V}{(2\pi)^3} \times 4\pi k^2 dk. \tag{4.34}$$

To keep expressions simple we assume an isotropic medium with equal electron masses m^* and equal crystal dimensions L in all three spatial directions. We then may put $E = (\hbar^2/2)(k_x^2/m_x^* + k_y^2/m_y^* + k_z^2/m_z^*) = (\hbar^2/2m^*)k^2$. From this we obtain $k = (2m^*/\hbar^2)^{1/2}\sqrt{E}$ and $kdk = (m^*/\hbar^2)dE$, yielding for (4.34)

$$dN = \frac{V}{4\pi^2}\left(\frac{2m^*}{\hbar^2}\right)^{3/2}\sqrt{E}dE \quad \text{(isotropic medium)}. \tag{4.35}$$

Inserting (4.35) into the definition (4.33) we obtain the square-root dependence of the electronic DOS for bulk crystals

$$g(E) = \frac{1}{2\pi^2}\left(\frac{2m^*}{\hbar^2}\right)^{3/2}\sqrt{E} \quad \text{(three dimensions)}. \tag{4.36}$$

We now consider the *two-dimensional case* by assuming an additional contribution to the potential $W(z)$ which confines the mobility of the electron to the xy plane. Within this two-dimensional plane it still moves quasi-free. The electron states are now described by the Schrödinger equation

$$-\left(\frac{\hbar^2}{2}\left(\frac{1}{m_x^*}\frac{\partial^2}{\partial x^2}+\frac{1}{m_y^*}\frac{\partial^2}{\partial y^2}+\frac{1}{m_z^*}\frac{\partial^2}{\partial z^2}\right)+eW(z)\right)\psi(\mathbf{r})$$
$$= E\psi(\mathbf{r}) \quad \text{(two dimensions)}. \tag{4.37}$$

Equation (4.37) can be separated into two equations describing movements either within the xy plane or perpendicular by using the approach for the solution

$$\psi_{\mathbf{k}}(\mathbf{r}) = \frac{1}{\sqrt{V}} e^{i(k_x x + k_y y)} \varphi_n(z). \tag{4.38}$$

The plane-wave term in $\psi_{\mathbf{k}}(\mathbf{r})$ is analogous to the three-dimensional case described by (4.30) and yields correspondingly the eigenvalues

$$E_{xy} = \frac{\hbar^2 k_x^2}{2m_x^*} + \frac{\hbar^2 k_y^2}{2m_y^*}. \tag{4.39}$$

For the z direction we obtain

$$-\left(\frac{\hbar^2}{2}\frac{1}{m_z^*}\frac{\partial^2}{\partial z^2} + eW(z)\right)\varphi_n(z) = E_n\varphi_n(z). \tag{4.40}$$

The eigenvalues E_n depend on the characteristics of the potential $W(z)$. If we assume a square potential with infinite barriers separated by a spacing L_z the eigenvalues follow from the condition for allowed waves $L_z = n\lambda_{z,n}/2$, $n = 1, 2, 3, \ldots$. Putting $k_{z,n} = 2\pi/\lambda_{z,n}$ the eigenvalues are given by

$$E_{z,n} = \frac{\hbar^2}{2m_z^*}\left(\frac{n\pi}{L_z}\right)^2, \quad n = 1, 2, 3, \ldots. \tag{4.41}$$

We note that the energy of the ground state with the quantum number $n = 1$ is increased by the *quantization energy* $\Delta E = E_{z,1}$. The eigenvalues of (4.37) are given by the sum of (4.39) and (4.41), $E = E_{xy} + E_{z,n}$. The band scheme $E(\mathbf{k})$ along k_x and k_y therefore consists of a series of parabola, each labeled by a particular value of n. The parabola are also referred to as *subbands*.

The two-dimensional electronic DOS follows from the equidistant states in $\mathbf{k}$ space similar to the three-dimensional case. For an area $L_x \times L_y$ the two-dimensional density of states in $\mathbf{k}$ space is given by $L_x L_y/(2\pi)^2$. The volume of a spherical shell

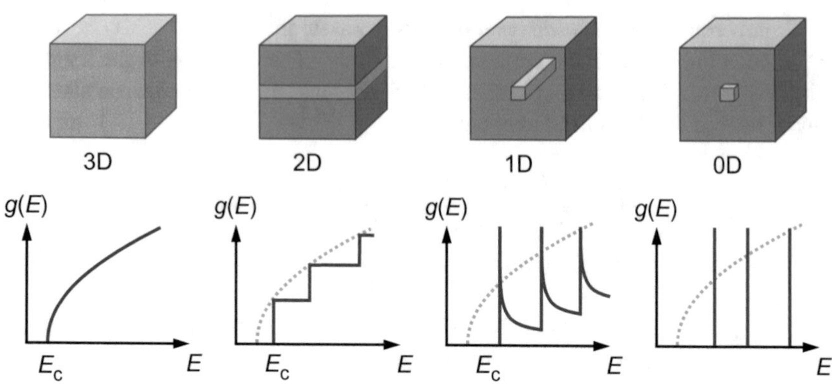

Fig. 4.19 Electronic density of states $g(E)$ in isotropic semiconductors (*red*) with different dimensionalities: 3D bulk semiconductor, 2D quantum well, 1D quantum wire, and 0D quantum dot. The environment drawn in *blue* provides potential barriers for the charge carriers. E_C denotes the conduction-band edge in the semiconductor

in **k** space is replaced in 2D by the area of a circular ring bounded by $E(\mathbf{k}) = \text{const}$ and $E + dE = E(\mathbf{k} + d\mathbf{k}) = \text{const}$. We assume again an isotropic medium and use the isotropic energy dispersion $E(k) = (\hbar^2/2m^*_{xy})k^2$. In analogy to (4.34) we obtain

$$dN = \frac{L^2}{(2\pi)^2}\int_E^{E+dE} d^2k = \frac{L^2}{(2\pi)^2}2\pi k dk \quad \text{(isotropic medium)}. \tag{4.42}$$

The general expression for the two-dimensional DOS $g(E)$ is obtained by inserting this into the definition (4.33), yielding for the isotropic DOS of the nth subband

$$g_n = \frac{m^*_{xy}}{\pi\hbar^2} = \text{const} \quad \text{(two dimensions)}. \tag{4.43}$$

The two-dimensional DOS in (4.43) is expressed in units of $\text{cm}^{-2} \times \text{eV}^{-1}$. The total electronic DOS follows from the sum of all subband contributions, which all have the same magnitude given by (4.43): $g(E) = \sum g_n$. This results in the staircase-like function of the total DOS for a two-dimensional (2D) semiconductor illustrated in Fig. 4.19 and expressed by

$$g(E) = \sum g_n = \frac{m^*_{xy}}{\pi\hbar^2}\sum_n \Theta(E - E_n) \quad \text{(two dimensions)}, \tag{4.44}$$

$\Theta(E - E_n)$ being the unit step-function. The subbands are consecutively occupied as the energy is increased.

The *one-dimensional case* follows from the two-dimensional case by an additional confinement $W(y)$. This leads to a quasi-free mobility only along the x axis. The energies along the y and z axes are quantized, and the subbands have two corresponding indices l and n. Analogous to (4.39) and (4.41) we may write

$$E = E_{l,n} = \frac{\hbar^2 k_x^2}{2m_x^*} + l^2 \frac{\hbar^2}{2m_y^*}\left(\frac{\pi}{L_y}\right)^2 + n^2 \frac{\hbar^2}{2m_z^*}\left(\frac{\pi}{L_z}\right)^2 \quad \text{(one dimension)}. \quad (4.45)$$

The one-dimensional electronic DOS is obtained from a one-dimensional "volume" element in **k** space simply given by dk, yielding for the subband l, n

$$g_{l,n}(E) = \sqrt{\frac{m_x^*}{2\pi^2\hbar^2}} \frac{1}{\sqrt{E - E_{l,n}}} \quad \text{(one dimension)}. \quad (4.46)$$

The one-dimensional DOS in (4.46) is expressed in units of $\text{cm}^{-1} \times \text{eV}^{-1}$. The total DOS is again given by the sum of all subband contributions, $g(E) = \sum_{l,n} g_{l,n}(E)$. The resulting function of the one-dimensional (1D) semiconductor is illustrated in Fig. 4.19. Note that the peaks are not necessarily equidistant since L_y and L_z are independent.

Adding a further confining potential $W(x)$ to the one-dimensional semiconductor leads to the *zero-dimensional case*. The mobility is now restricted in all three spatial dimensions. Accordingly the energy is quantized in all directions, and from (4.45) follows

$$E = E_{j,l,n} = \frac{\hbar^2\pi^2}{2}\left(\frac{j^2}{m_x^* L_x^2} + \frac{l^2}{m_y^* L_y^2} + \frac{n^2}{m_z^* L_z^2}\right) \quad \text{(zero dimensions)}. \quad (4.47)$$

The zero-dimensional electronic DOS is a sum of δ functions given by $g(E) = \sum 2\delta(E - E_{j,l,n})$. The function is shown in Fig. 4.19. Like in the two-dimensional case the peaks are not necessarily equidistant.

4.3.2 *Characteristic Scale for Size Quantization*

We considered above the modification of the electronic density-of-states for solids of reduced dimensionality. What is the characteristic scale required for the size quantization to become observable in experiment? Besides the size of the solid it is related to the effective mass of the considered charge carriers and to the temperature. For a *quasi-free charge carrier* with effective mass m^* size quantization gets distinguishable if the motion is confined to a length scale in the range of or below the de Broglie wavelength $\lambda = h/p = h/\sqrt{2m^*E}$. Assuming a room temperature thermal energy $E = (3/2)k_B T = 26$ meV and an effective mass of one tenth of the free electron mass, a typical length is in the 10 nm range.

Optical properties of semiconductors are often dominated by *excitons*. They consist of an electron and a hole in close proximity which attract each other via Coulomb interaction and form hydrogen-like bound states, cf. Sect. 5.1.3 [2]. The relevant length scale of these two-particle states is the exciton Bohr-radius. The exciton Bohr-radius is given by

$$a_X = \frac{h^2 \varepsilon\varepsilon_0}{\pi \mu e_0^2}, \tag{4.48}$$

where ε, ε_0, μ, and e_0 designate the relative permittivity of the solid and that of vacuum, the reduced mass of the exciton and the electron charge, respectively. The reduced mass of the exciton is defined by $1/\mu = 1/m_e^* + 1/m_h^*$. The hole mass m_h^* is often much heavier than the electron mass m_e^*, leading to a reduced mass close to m_e^*. The value of the exciton Bohr-radius is related to the binding energy (also termed Rydberg constant) of the exciton

$$E_X = \frac{\mu e_0^4}{8h^2(\varepsilon\varepsilon_0)^2}. \tag{4.49}$$

The product $a_X \times E_X$ is constant for three-dimensional excitons. The relation remains a good estimate also for two-dimensional excitons [75]. Values for some semiconductors are given in Table 4.8. A typical length to observe size quantization for excitons is also in the 10 nm range. It must be noted that exciton binding-energy and Bohr radius are significantly modified by a spatial localization [75].

Size-quantization effects were observed at surfaces and in thin layers of both, metals and semiconductors [81]. A review on early work was given in, e.g., [80]. In the following we will focus on semiconductor heterostructures. Clear quantum-size effects are particularly observed in GaAs-based heterostructures. Their electronic properties can be described almost purely quantum mechanically using the effective-mass approximation, with constituent materials represented by a few band parameters.

Table 4.8 Exciton Bohr radius a_X and binding energy E_X of excitons in some direct semiconductors with zincblende (ZB) or wurtzite (W) structure. A, B, and C in wurtzite material refer to the three valence bands

Semiconductor		a_X (Å)	E_X (meV)
GaAs[a]	ZB	115	4.7
InAs[d]	ZB	494	1.0
InP[b]	ZB	113	5.1
ZnTe[c]	ZB	11.5	13
ZnSe[c]	ZB	10.7	19.9
ZnS[c]	ZB	10.2	29
ZnO[d]	W		60 (A)
			57 (B)
GaN[e]	W		21 (A)
			21 (B)
			23 (C)

[a]Reference [76], [b]Reference [77], [c]Reference [1], [d]Reference [78], [e]Reference [79]

4.3.3 *Quantum Wells*

A quantum well is made from a thin semiconductor layer with a smaller bandgap energy clad by semiconductors with a larger bandgap forming barriers. Usually the same barrier material is used in such double heterostructure leading to a symmetrical square potential as illustrated in Fig. 4.9. The confinement is given by the band offsets. Since the potential is no longer infinite as assumed to obtain the eigenvalues (4.41) the wave functions of a confined charge carrier now penetrate into the barriers. For finite barrier energy $W(z) = W_0$ the eigenvalues are obtained from a transcendental equation

$$\tan\left(\sqrt{\frac{m_\mathrm{w} E_n L_z^2}{2\hbar^2}}\right) = \sqrt{\frac{m_\mathrm{w}}{m_\mathrm{b}} \frac{W_0 - E_n}{E_n}} \tag{4.50a}$$

for even wave functions, i.e. even values of quantum numbers n, and

$$\cot\left(\sqrt{\frac{m_\mathrm{w} E_n L_z^2}{2\hbar^2}}\right) = \sqrt{\frac{m_\mathrm{w}}{m_\mathrm{b}} \frac{W_0 - E_n}{E_n}} \tag{4.50b}$$

for odd wave functions, i.e., odd n [82]. m_w and m_b are the effective masses of the charge carrier in the well and the barriers, respectively, and L_z is the well width. Numerically obtained solutions are given in Fig. 4.20 [83]. Energies in the figure are scaled in units of the ground-state energy E_1 of a well with infinite barriers, cf. (4.41). The gray line signifies the top of the well at $E = W_0$. Discrete bound states are found for $E < W_0$, while continuum states exist for $E \geq W_0$. We note that the number of bound states of the confined particle decreases as W_0 decreases. Furthermore, the

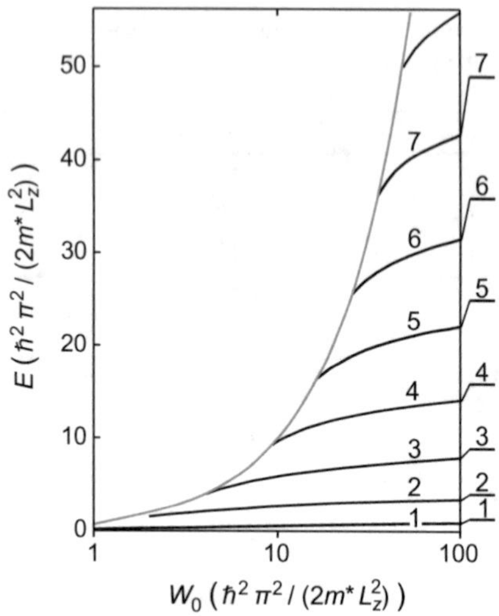

Fig. 4.20 Calculated bound-state energies of a particle in a symmetrical rectangular potential well of finite depth W_0 indicated by the *gray line*. *Bars* at the *right-hand side* mark energy levels with quantum numbers n for infinitely high barriers. After [84]

level spacing and consequently the energy of the levels decrease. The topmost bound level approaches the top of the well as W_0 is gradually reduced. It should be noted that at least one bound level exists in any quantum well.

A clear experimental observation of the quantum-size effect in quantum wells requires well-defined sharp interfaces. A vivid demonstration of subband formation was accomplished by imaging the local DOS in InAs/GaSb quantum wells [85]. Growth conditions ensured the formation of smooth InSb-like interfaces between InAs and GaSb. The two semiconductors form a misaligned staggered band alignment with the conduction-band edge of the InAs quantum well lying below the valence-band edge of the cladding GaSb. This broken-gap configuration provides a large confinement potential for electrons in the well. Using the tip of a low-temperature scanning tunneling microscope the local DOS was probed across the InAs well from the differential conductance dI/dV, cf. Fig. 4.21.

The differential conduction given in Fig. 4.21a clearly shows a standing-wave pattern originating from electron subbands in the InAs quantum well. The number of maxima increases with energy, i.e. sample bias voltage. The tunneling tip locally probes the probability amplitude of electrons across the well. The experimental result agrees remarkably well with the calculated subband wave-functions shown in Fig. 4.21b. The two-dimensional DOS of the quantum well given in Fig. 4.21c is experimentally deduced from integrating over the local DOS. The contributions of the subbands lead to apparent steps in the DOS being characteristic for a 2D heterostructure, cf. Fig. 4.19.

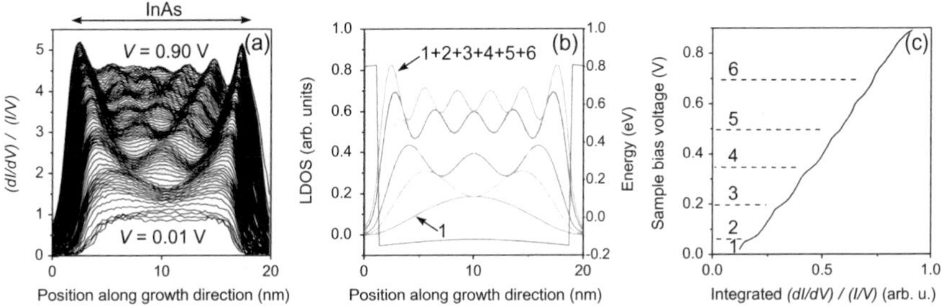

Fig. 4.21 **a** Experimental scanning tunneling $(dI/dV)/(I/V)$ spectra locally probed across a 17 nm wide InAs/GaSb quantum well for bias varied from 0.01 to 0.9 V in steps of 0.01 V. **b** Calculated local density-of-states given as the sum of the squared subband wave-functions. **c** Experimental density-of-states obtained from integration of each curve in **a** over the quantum well. Reproduced with permission from [85], © 2007 APS

Features of both confined electrons and holes are observed in optical spectra. Transitions measured from nearly unstrained GaAs quantum wells with $Al_{0.2}Ga_{0.8}As$ barriers and from strained GaAs wells with $GaAs_{0.5}P_{0.5}$ barriers on GaAs substrates are shown in Fig. 4.22. The structures consist of multiple wells, i.e. superlattices with barriers sufficiently thick to prevent electronic coupling of the well states. They were grown using molecular-beam epitaxy [86] and metalorganic vapor-phase epitaxy [87], respectively. The curves show the intensity of the ground-state emission under excitation at the displayed varied photon energy. The spectra exhibit series of peaks labeled according to the participating levels of the confined electrons and holes. Most intense peaks refer to allowed transitions between electron and hole states of two-dimensional excitons with an equal quantum number n. Such exciton states have largest electron-hole overlap if strain effects are not dominating. The strongest line labeled $E_{1\text{h}}$ originates from the radiation of an exciton with the electron in the $n_\text{e} = 1$ state and the $m_J = 3/2$ hole denoted heavy hole in the $n_\text{hh} = 1$ state. The corresponding transition of an electron and an $m_J = 1/2$ light hole in the $n_\text{lh} = 1$ state is labeled $E_{1\text{l}}$. The energy difference $E_{1\text{l}} - E_{1\text{h}}$ reflects the splitting of the $m_J = 3/2$ and $1/2$ valence bands due to the biaxial shear strain in the GaAs well arising from the different lattice constants of barriers and well (Sect. 4.1.2). Note that excited states gradually broaden, as indicated in Fig. 4.22 for the heavy-hole series labeled $E_{n\text{h}}$. The broadening is described by $\Delta E_n = \Delta E_1 \times n^2$ [86], reflecting a non-constant well width probed by a gradually increasing exciton diameter. Besides the two series of allowed transitions with $n_\text{electron} = n_\text{hole}$ for heavy and light hole there are also forbidden peaks associated with transitions where $n_\text{electron} \neq n_\text{hole}$: Lines $E_{\text{f}1}$ and $E_{\text{f}2}$ mark transitions $E_{1\text{e}} - E_{3\text{h}}$ and $E_{2\text{e}} - E_{4\text{h}}$, respectively [86, 88].

The transitions of the more strongly strained GaAs well clad by $GaAs_{0.5}P_{0.5}$ barriers shown in Fig. 4.22b exhibit a much larger splitting $E_{1\text{l}} - E_{1\text{h}}$ (44 meV) than observed for $Al_{0.2}Ga_{0.8}As$ barriers. Hydrostatic and shear-strain components of $\varepsilon_\text{h} = 0.008$ and $\varepsilon_\text{s} = 0.014$, respectively, were determined from the shift of the ground-state exciton with respect to a GaAs bulk exciton and from the light-hole—

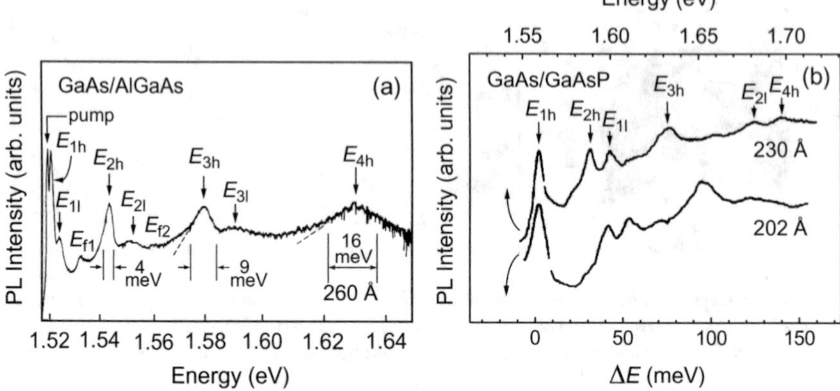

Fig. 4.22 Photoluminescence excitation spectra of GaAs quantum wells clad by **a** $Al_{0.2}Ga_{0.8}As$ and **b** $GaAs_{0.5}P_{0.5}$ barriers, recorded at low temperatures and low excitation density. The well width is indicated at the spectra. Labeled energies refer to optical transitions of bound electron states to bound hole states, see text. Absolute energy scale on top of panel (**b**) refers to upper spectrum, the lower spectrum is shifted in energy to align the E_{1h} transition. **a** After [86], **b** after [87]

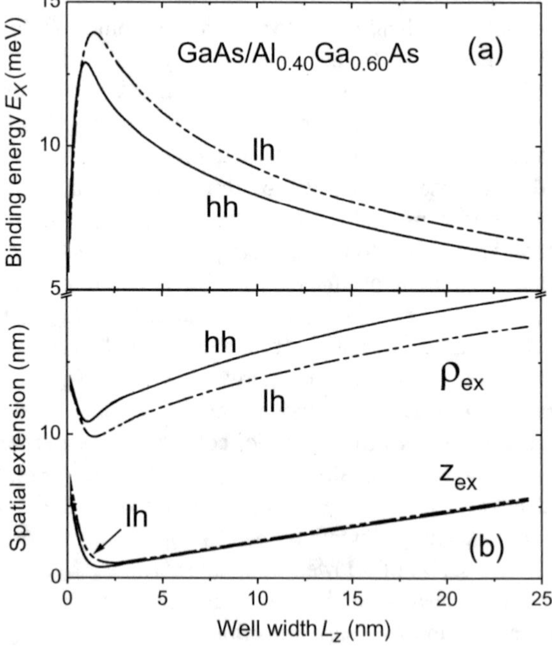

Fig. 4.23 **a** Calculated binding energy and **b** both lateral (ρ_{ex}) and perpendicular (z_{ex}) spatial extension of an exciton confined in a $GaAs/Al_{0.4}Ga_{0.6}As$ quantum well. *Solid* and *dashed lines* refer to heavy and light-hole exciton, respectively. Reproduced with permission from [75], © 1988 APS

heavy-hole splitting [87]. The quantum-size effect is expressed in Fig. 4.22b by a blue shift of the excited-state transitions in the thinner quantum well (lower spectrum).

The two-dimensional confinement changes spatial extension and binding energy of an exciton both within the quantum-well plane and perpendicular [75]. The effect is illustrated in Fig. 4.23 for an exciton confined in a GaAs/$Al_{0.4}Ga_{0.6}As$ quantum well. We note a strong increase of the binding energy E_X (4.48) in Fig. 4.23a with a maximum for very thin quantum wells near 1 nm. The smaller binding energy of the heavy hole originates from a larger in-plane mass of the light hole [75]. The (degenerate) bulk values of the binding energies of light-hole and heavy-hole excitons are approached for the limits of zero and infinite well width L_z. The binding energy of the excited $2s$ and $2p_\pm$ states of the lh and hh exciton vary similar to that of the $1s$ ground state shown in Fig. 4.23 [89]. The lateral and perpendicular extensions of the exciton given in Fig. 4.23b show a simultaneous squeezing inverse to the binding energy, according to the rule $a_X \times E_X = \text{const}$.

Heterostructure interfaces are usually not abrupt from one single atomic layer to the next layer in growth direction over a macroscopically large lateral scale. The resulting roughness of the interface is probed by excitons confined in a quantum well. Lateral fluctuations of the well thickness (and composition) are averaged within the spatial extent of the exciton which is given by twice the Bohr radius a_X. Such an effect leads to the gradual broadening of the exciton series shown in Fig. 4.22. If the fluctuations occur on a scale larger than the exciton diameter discrete transitions may be found. A model for interface roughness due to growth steps with a height of a single monolayer (ML) at both barriers of a quantum well is illustrated in Fig. 4.24. Since a step position at the lower interface is not expected to be reproduced by the upper interface three thicknesses arise from such fluctuations, namely L_z, $L_z + 1$ ML, and $L_z - 1$ ML.

Interface disorder of the kind illustrated in Fig. 4.24 may be observed particularly in narrow quantum wells. A clear observation in photoluminescence spectra of GaAs/$Al_xGa_{1-x}As$ quantum wells is shown in Fig. 4.25 [90]. In the (001)-oriented zincblende material one monolayer corresponds to half a lattice constant: 1 ML $= a/2$, a being the lattice constant of both materials, the quasi-unstrained GaAs well and the $Al_{0.3}Ga_{0.7}As$ barriers.

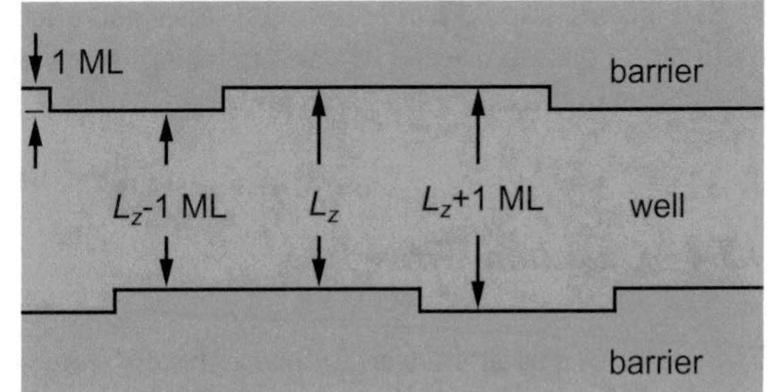

Fig. 4.24 Cross section of a quantum well/barrier double-heterostructure. Interface roughness by growth steps of single monolayer height is depicted at both barriers of the quantum well

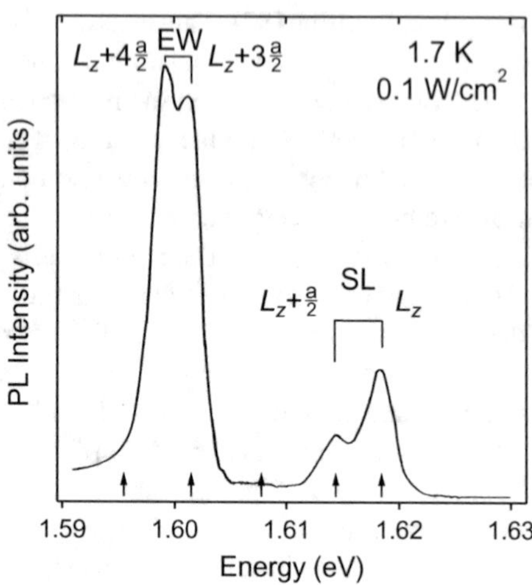

Fig. 4.25 Photoluminescence spectrum of a GaAs/$Al_{0.3}Ga_{0.7}As$ heterostructure comprising a superlattice with wells of 50 Å (SL) width and also wells with a width enlarged by additional three monolayers to 58.6 Å (EW). The emission was excited non-resonantly with a low excitation density at low temperature. *Arrows at bottom* indicate calculated transitions energies referring to $L_z + n \times a/2$. After [90]

The sample studied in Fig. 4.25 consists of a GaAs/$Al_{0.3}Ga_{0.7}As$ superlattice (SL) with $L_z = 50$ Å thick wells and 50 Å thick barriers. After growth of each 14 wells (1400 Å) one enlarged well (EW) with an additional thickness of three monolayers (50 Å + 8.6 Å) was introduced. The photoluminescence shows emissions from heavy-hole excitons of the superlattice and the enlarged wells, both split by an amount corresponding to about one-third of their mutual separation. (Corresponding transitions of the light-hole excitons were found in excitation spectra.) This agrees with an expected variation originating from a thickness difference of one monolayer ($a/2$). The position of the individual peaks matches calculated transitions energies of quantum wells with thicknesses $L_z + n \times a/2$, $n = 1$ to 4, and $a = 5.73$ Å, if the transition referring to L_z of the SL well is taken as a reference. The poor agreement of the leftmost peak is attributed to an increased disorder in the extended well. We note that the peaks corresponding to $L_z + 2a/2$ (middle arrow) is missing. Also no peak is found referring to $L_z - a/2$ (above 1.62 eV, no arrow drawn). These transitions would arise from thinner parts of the enlarged well and the superlattice, respectively. They are not observed due to a negligible thermal occupation at the low measurement temperature of 1.7 K.

Size-quantization effects of excited states and interface roughness are frequently observed in quantum wells of various materials systems, albeit usually not as pronounced as in the examples given above.

4.3.4 *Quantum Wires*

Fabrication of a quantum well follows naturally from the two-dimensional epitaxy of a double heterostructure. The potential of such a quantum well is basically defined by the well thickness and the (homogeneous) chemical composition of well and barrier

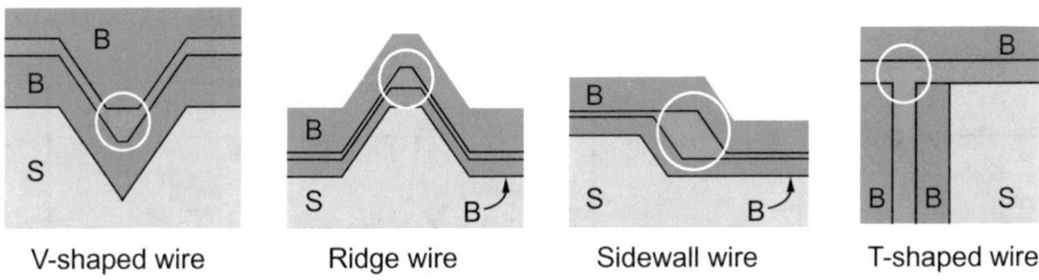

Fig. 4.26 Cross-section of typical types of epitaxial quantum wires (*encircled*). *B* and *S* signify barrier and substrate materials, respectively

materials, i.e., by the band discontinuities. A further reduction of dimensionality towards a one-dimensional quantum wire or a zero-dimensional quantum dot requires some patterning to define an additional lateral confinement. The small dimensions needed to obtain respective quantum size-effects (Sect. 4.3.2) can usually not be accomplished straightforward by patterning a quantum well structure using, e.g., lithography techniques [91]. The interface-to-volume ratio of 1D and 0D structures increases as compared to 2D quantum wells, and the electronic properties of such structures are largely governed by interface effects. A variety of techniques was developed instead to realize 1D and 0D structures with high optical quality, e.g., by employing growth on corrugated substrates [92], or whisker growth discussed in Sect. 12.2. Most of these techniques lead to complicate confinement potentials, and often an additional quantum well is coupled to the quantum wires or quantum dots. A few examples will be considered in this and the following sections.

Much progress in the formation of epitaxial 1D quantum wires was achieved using V-shaped wires or T-shaped wires. A schematic of some typical wire geometries is given in Fig. 4.26.

The V-shaped, ridge, and sidewall wires depicted in Fig. 4.26 are fabricated by employing the dependence of the growth rate on crystallographic orientation (Sect. 12.1). Using patterned substrates with various facet orientations thereby a locally enhanced thickness of a wire material clad by barrier materials can be grown. The T-shaped wire depicted in Fig. 4.26 is formed from an overgrowth of the cleaved edge of a quantum-well structure. The fabrication process of low-dimensional structures is described in more detail in Sect. 7.3.

Early demonstrations of quantum wires in the late 1980s suffered from thickness fluctuations on a length scale of the exciton Bohr radius, leading to 0D behavior at low temperatures. Clear 1D behavior was proved later using improved structures and also by employing micro-photoluminescence on single wires. Photoluminescence (PL) and corresponding excitation (PLE) spectra of V quantum wires are shown in Fig. 4.27 [93]. The given wire thickness refers to the thickest part in the center of the crescent-shaped GaAs, which has parabola-like interfaces to the cladding AlGaAs barriers. The wire emission is labeled QWR in Fig. 4.27a, the dominating emission originates from the quantum wells formed at the side walls of the V groove.

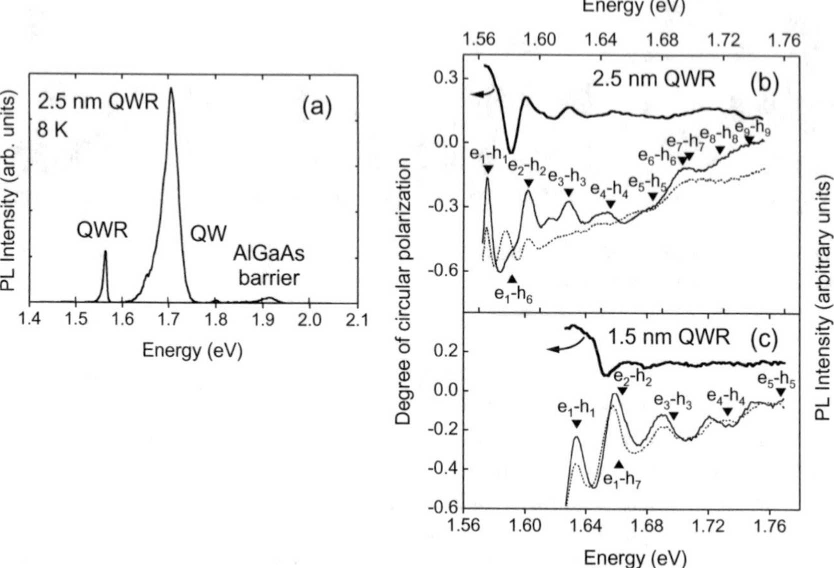

Fig. 4.27 **a** Photoluminescence of a V-shaped GaAs/$Al_{0.3}Ga_{0.7}As$ quantum wire, non-resonantly excited at 2.4 eV with a low excitation density of 25 W/cm^2. **b** PLE spectra for 2.5 nm and **c** 1.5 nm thick wires polarized parallel (*thin solid line*) or perpendicular (*dotted line*) the [110]-oriented wires. *Thick solid curves* represent the degree of circular polarization. *Arrows* mark calculated positions of excitonic e_n–h_m interband transitions. Reproduced with permission from [93], © 1997 APS

The PLE spectra given in Fig. 4.27b, c show a number of features also observed with quantum wells, cf. Fig. 4.22. Basically optical transitions with an equal quantum number n of electron and hole states are found. The e_n–h_n transitions shift to the blue as the wire thickness decreases, and their mutual energy spacings increase. The assignment of the peaks to corresponding exciton states is not trivial due to the complicate crescent-shaped potential of the wire. The transition energies marked in Fig. 4.27b, c were calculated using a 16-band **kp** model and a potential shape extracted from transmission-electron micrographs. We note a pronounced oscillator strength of the usually forbidden e_1–h_6 transition in the thicker well. Moreover, no distinction is drawn between contributions of light holes and heavy holes. Calculations reveal a strong mixing of light- and heavy-hole character in the valence-band states, particularly involved in the mentioned transition. The wave function of the ~70% light-hole part of the h_6 state has its maximum in the wire center similar to the e_1 state, yielding a large overlap [94] (for separate lh and hh assignments cf. [95, 96]). The valence-band mixing also gives rise to the polarization behavior observed in the samples. It should be noted that such mixing and polarization anisotropy may also originate from other sources like structural inhomogeneity, and care was taken to ensure the 1D origin.

For crescent-shaped V groove quantum-wires a simple model with infinite barriers and hyperbolic boundaries was considered to obtain approximative energy separations of the subbands in the wires [97]. The assumed potential leads to analytical solutions given by

$$E_{n,m} \cong \frac{\hbar^2 \pi^2 n^2}{2m^* t_0^2} + \frac{\hbar^2 \pi \sqrt{\alpha} n}{m^* t_0 \overline{\rho}} (m - 1/2), \quad m, n = 1, 2, 3, \ldots, \tag{4.51}$$

where m^* is the effective charge-carrier mass, t_0 is the crescent thickness at its center, and $\overline{\rho} = \sqrt{\rho_{\text{low}} \times \rho_{\text{up}}}$ is the geometric mean radius of the lower and upper radii of the crescent. The upper curvature ρ_{up} is given by $\rho_{\text{up}} = \rho_{\text{low}} + \alpha t_0$, with α describing the linear increase of the curvature as the wire thickness increases. The second summand in (4.51) describes the energy separation of the 1D subbands due to the lateral confinement.

Size-quantization effects of 1D wires are sometimes not well distinguished from those of a 2D well or a 0D dot. The lateral confinement potential of commonly used wire geometries (Fig. 4.26) is typically only on the order of 30–40 meV, giving rise to only 10 meV subband energy-separation [92]. The potential is hence much smaller than that of a quantum well. Furthermore, size fluctuations may give rise to an additional confinement along the wire axis on the same order of magnitude.

The potential depth of the lateral confinement in T-shaped GaAs/$Al_xGa_{1-x}As$ quantum wires was evaluated from sample series with varied well widths and Al compositions. Energies of exciton PL peaks related to the T quantum wire (QWR), the stem well QW1 (vertical quantum well in Fig. 4.26), and the arm well QW2 (horizontal) are shown in Fig. 4.28a for varied thickness of QW2 [98]. The exciton energy increases for thinner QW2 also in the quantum wire. Note that the slope of the QWR energy is smaller than that of the QW2 energy. The energy of the wire hence approaches that of QW1 for thinner QW2 and that of QW2 for thicker QW2. The reason is a convergence of QWR states into QW1 for $t_{\text{QW2}} \ll t_{\text{QW1}}$ and vice versa into QW2 for $t_{\text{QW2}} \gg t_{\text{QW1}}$. The spacings between the QWR PL peak energy and those of the quantum wells QW1 and QW2, i.e., the energy differences between the 1D exciton and the 2D exciton states of the neighboring wells, give directly the effective lateral confinement in the wire for given quantum-well thicknesses. This value includes possible changes of the exciton binding-energy. The lateral confinement gets maximum for identical thicknesses of QW1 and QW2. For such balanced T quantum wires Fig. 4.28b shows the dependence of the effective lateral confinement on the exciton energy in QW2.

The calculated dependence of lateral confinement energy on the recombination energy of well-excitons given in Fig. 4.28b shows a sublinear increase for finite barrier height. This feature arises from a penetration of the wave function into the wells. The dependence approaches linearity for increasing barriers. At infinite barriers both, QWR energy and QW2 energy scale according to t^{-2} leading to the linear relation. The strong experimental *super*linear increase found for increased barriers (x_{Al} from

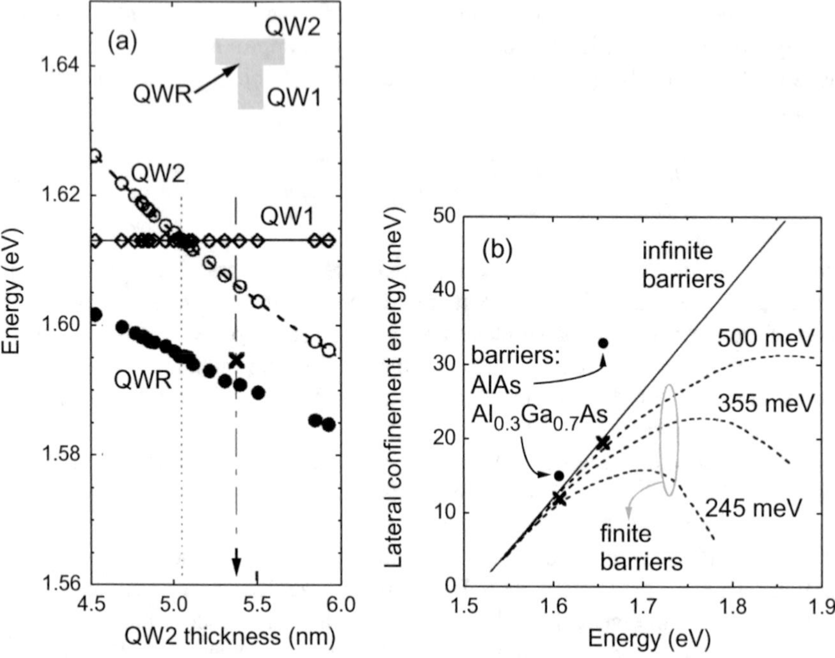

Fig. 4.28 **a** PL peak energies of excitons confined in a T quantum wire (QWR) and the neighboring stem quantum well (QW1) and arm well (QW2) of a GaAs/$Al_{0.3}Ga_{0.7}As$ heterostructure. *Dash-dotted* and *dotted vertical lines* indicate identical QW1 and QW2 well thicknesses and QW2 thickness at identical QW1 and QW2 exciton energy, respectively. **b** Calculated lateral confinement energy of electrons in balanced T quantum wires as a function of QW2 exciton energy. *Solid* and *dashed lines* refer to energies for infinite and finite barriers as indicated. *Crosses* and *filled circles*, respectively, mark calculated and experimental confinement energies extracted from the measurement of GaAs/Al_x $Ga_{1-x}As$ structures. Reproduced with permission from [98], © 1996 APS

0.30 to 1.00) is due to the substantial enhancement of Coulomb interaction in 1D excitons [98].

The excitonic absorption of a 1D quantum wire is expected to deviate significantly from that of a 2D quantum well or a 3D bulk crystal. For direct allowed transitions above band edge the intensity ratio of the unbound (continuum) exciton to the free electron-hole pair is found to be smaller than unity, in contrast to the 2D and 3D cases [99]. The feature may be understood by considering the eigenenergies of states bound in a bare Coulomb potential for d dimensions ($d = 1, 2, 3, \ldots$). The analytical solutions of the Schrödinger equation yield in each dimension d energies E_n^d of eigenstates with s symmetry in form of a Rydberg series [99]

$$E_n^d = -\mathrm{Ry}^*\left(n + \frac{d-1}{2}\right)^{-2}, \quad n = 0, 1, 2, \ldots, \tag{4.52}$$

Ry* being the effective Rydberg energy. For the three-dimensional case ($d = 3$) we recognize the well-known Rydberg series of hydrogen. For 1D we note a singularity for the lowest state $n = 0$, corresponding to infinite binding energy, in contrast to the 3D and 2D case. This suggests the attractive force between electron and hole being stronger in 1D than in 2D or 3D. In a descriptive idea a particle may move *around* the origin of a Coulomb potential in 2D or 3D, while it moves *through* the origin in 1D. In fact, the $1/r$ singularity of the Coulomb potential is removed upon integration in 2D and 3D, but it remains as a logarithmic singularity in 1D.

The 1D nature of a quantum wire leads also to a characteristic dynamics of the radiative decay of exciton population after pulse excitation [100]. The population decay-time was found to vary proportional to the square root of sample temperature, in contrast to the linear proportionality observed in quantum wells. The PL hence decays slower at low temperatures and faster at high temperatures in wires compared to wells.

4.3.5 Quantum Dots

A quantum dot represents the ultimate limit in charge-carrier confinement, leading to fully quantized electron and hole states like the discrete states in an atom. The most successful approach for the fabrication of dislocation-free semiconductor quantum-dots is the self-organized (also referred to as self-assembled) technique of Stranski–Krastanow growth (Sect. 7.3.1). This kind of growth mode may be induced by epitaxy of a highly strained layer, which initially grows two-dimensionally and subsequently transforms to three-dimensional islands due to *elastic* strain relaxation. Some part of the layer material does not redistribute but remains as a two-dimensional layer, due to a low surface free energy compared to the covered material. Since this layer wets the surface of the material underneath it is referred to as wetting layer. For practical use such islands are covered by a cap layer, which builds an upper barrier in addition to the lower barrier provided by the covered material underneath, cf. Fig. 4.29. It must be noted that a minimum size of a quantum dot exists to allow for confining a charge carrier, in contrast to structures of higher dimensionality. For a dot with spherical shape the minimum diameter $D_{\min}$ required to confine at least one bound state of a particle is given by [101]

$$D_{\min} = \frac{\pi \hbar}{\sqrt{2m^* W_0}}, \tag{4.53}$$

where W_0 is the confining potential and m^* the effective mass (assumed to be identical in dot and barrier). For a rough estimate of the minimum size to confine a single electron in a spherical InAs/GaAs dot we use an unstrained conduction band offset of $\sim$0.9 eV for W_0 and an effective electron mass in InAs of 0.03 m_0, yielding $D_{\min} \cong 6$ Å.

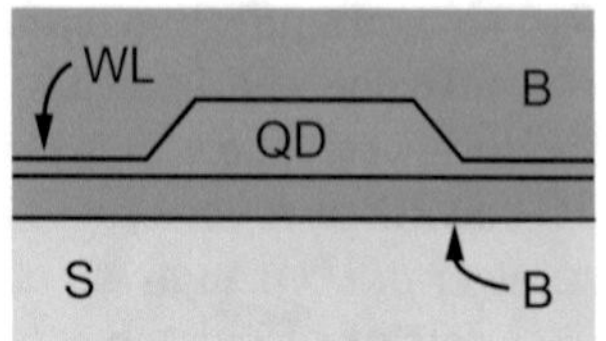

Fig. 4.29 Cross-section view of a quantum dot grown self-organized in the Stranski–Krastanow mode. QD, WL, B, and S signify quantum dot, wetting layer, barriers and substrate, respectively

Quantum dots grown in the Stranski–Krastanow mode often have a shape of a truncated pyramid. Due to the persistence of the wetting layer in the formation process such dots are coupled to a quantum well, similar to the epitaxial quantum wires depicted in Fig. 4.26. Self-organized Stranski–Krastanow growth may be induced by a strong mismatch of the heterostructure. InAs dots on GaAs with 7% mismatch represents the most studied model system. The type I band alignment to GaAs leads to a confinement of both, electrons and holes. The finite barrier height provided by the GaAs matrix allows for only few electronic states to be confined within the dot. Luminescence spectra of single quantum dots detected with high spatial resolution show radiative recombination of correlated electrons and holes, cf. Fig. 4.30. The narrow half-width of such confined-exciton recombination reflects the zero-dimensional DOS. Various lines from a single QD originate from states differently filled with few particles like, e.g., the negative trion (X^-) formed by two electrons and one hole, or the biexciton (XX) formed by two excitons.

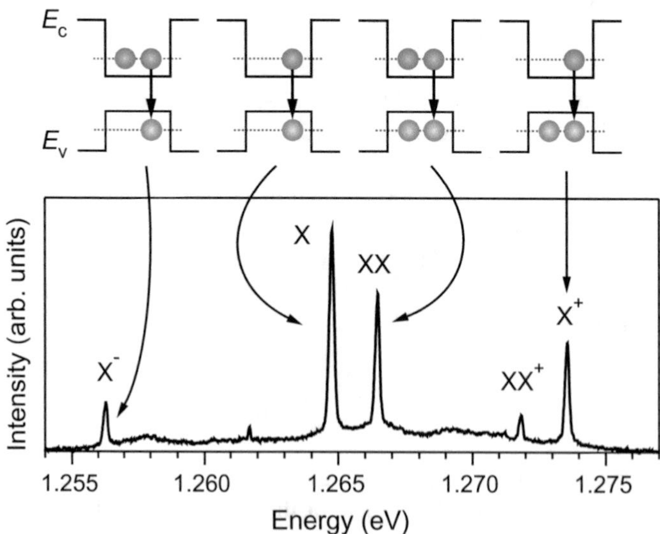

Fig. 4.30 Luminescence of a single quantum dot showing sharp recombination transitions of neutral and charged confined excitons, as depicted above the spectrum. E_c and E_v denote conduction and valence band edge, respectively. Spectrum reproduced with permission from [102], © 2005 APS

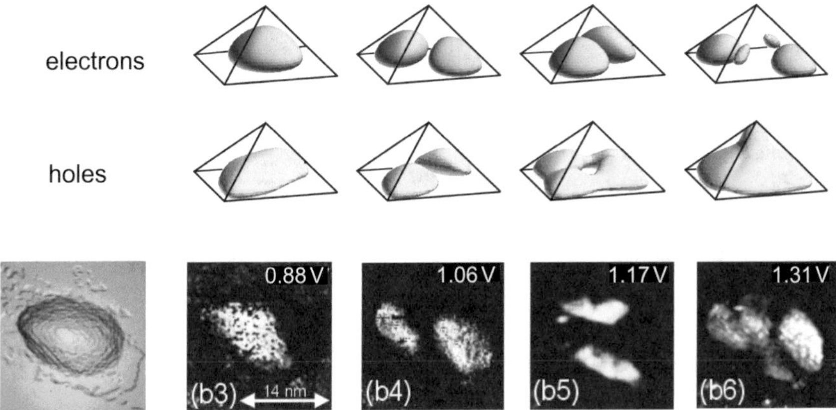

Fig. 4.31 *Top*: Calculated probability densities of electron and hole wave-functions confined in a *pyramid-shaped* InAs quantum dot in a GaAs matrix. Reproduced with permission from [106], © 1999 APS. *Bottom*: Low-temperature STM images of an uncovered InAs quantum dot on GaAs. *Left image*: Constant-current image showing the dot shape. *Right four images*: Single-electron densities of different excitation states sampled at different bias voltage. Reproduced with permission from [104], © 2003 APS

The wave functions of the confined states can be calculated by considering realistic size, shape, and composition of the dot, in addition to material properties like dielectric constants, strain tensors and piezoelectric tensors [103]. Results for one electron and one hole confined in an InAs dot with pyramidal shape of 11.3 nm base length and {101} side facets in a GaAs matrix are shown in Fig. 4.31. The iso-surfaces encase 65% probability and resemble atomic s-like ground states (left column), and p- and d-like excited states. Corresponding wave functions of single-electron states in uncovered InAs/GaAs dots were experimentally imaged using a low-temperature scanning-tunneling microscope. The images were obtained from spatially resolved differential voltage-current curves dI/dV taken at different sample voltage [104]. In the same way states of holes confined in InAs quantum dots embedded in GaAs matrix were imaged from cleaved samples [105].

The discrete energies of confined charge carriers sensitively depend on size, shape, and composition of the dot. Since these quantities show some variation among individual dots within an ensemble comprising many dots the eigenenergies vary from dot to dot. Optical spectra detected as a response of the entire ensemble are consequently inhomogeneously broadened due to a superposition of numerous transitions of different energies. The discrete nature of the transitions, i.e. the *homogeneous* line width, becomes apparent when only a single dot of an ensemble is spatially selected by using, e.g., micro-photoluminescence (micro PL) or opaque masks with a small aperture. The luminescence spectra shown in Fig. 4.32 illustrate the effect of a gradually increased spatial selection of the detected or excited area in a planar field of a quantum-dot ensemble. The broad luminescence spectrum in Fig. 4.32a measured using a conventional PL setup (macro PL) originates from a large number ($\sim 10^6$)

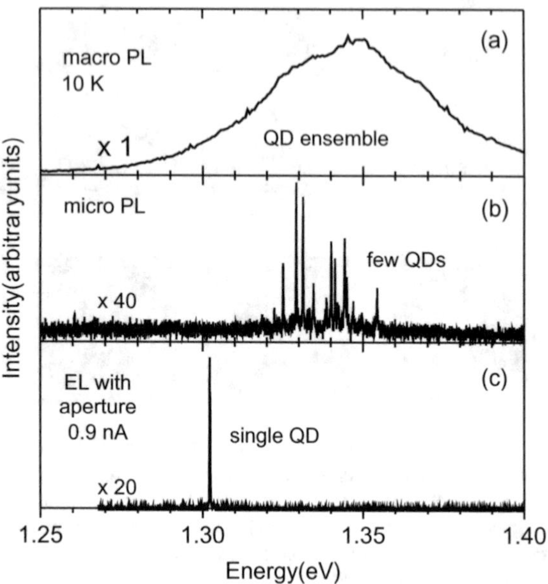

Fig. 4.32 a, b Luminescence of an InAs/GaAs quantum dot ensemble with $\sim 10^8$ cm^{-2} areal density, detected at 10 K using a conventional setup (**a**) and a setup with a small focus (**b**). **c** 10 K electroluminescence of a single dot of the same kind as in **a**, **b**, integrated into a diode structure with a current-confining aperture of sub-micron diameter. Spectra (**b**) and (**c**) adapted from [107]

of dots. The full width at half maximum (FWHM) of an ensemble PL is typically several 10 meV broad. The micro-PL setup used in Fig. 4.32b probes the photo-excited quantum-dot ensemble solely within a small focus, thereby detecting only $\rho_{\text{QD}} \times A_{\text{focus}}$ dots simultaneously, ρ_{QD} being the areal dot density. Further selection eventually leads to the detection of a single dot of the ensemble, cf. Figs. 4.32c and 4.30.

The homogeneous line width of a single-dot emission is extraordinary small. In absence of inhomogeneous contributions (e.g., by spectral diffusion) it is given by the lifetime τ of the excited charge carrier according to the $\Delta E \times \Delta\tau$ uncertainty relation. It is typically on the order of μeV and usually below the experimental detection limit.

The areal density of a quantum-dot ensemble grown using the Stranski–Krastanow mode is typically in the 1–10 $\times$ 10^{10} cm^{-2} range, cf. Sect. 7.3.1. The number of states to be occupied by charge carriers in a single dot layer (or few dot layers) is much smaller than in bulk or a quantum well. Occupation of quantum dots with more than one exciton or charge carrier hence occurs already at reasonably low optical or electrical excitation. Since the ground state may be occupied with only two charge carriers (of opposite spin) excited states are easily populated. Figure 4.33 shows the strong occupation of excited levels in a quantum-dot ensemble at increased excitation density. While at low excitation density only the inhomogeneously broadened emission from the ground state labeled I0 is observed, gradually additional emissions from excited states (I1 to I4) appear on the high-energy side for more intense excitation. The number of excited levels is high in this particular sample, because an $\text{Al}_{0.3}\text{Ga}_{0.7}\text{As}$ barrier was placed underneath the dot layer, separated by a 1 nm

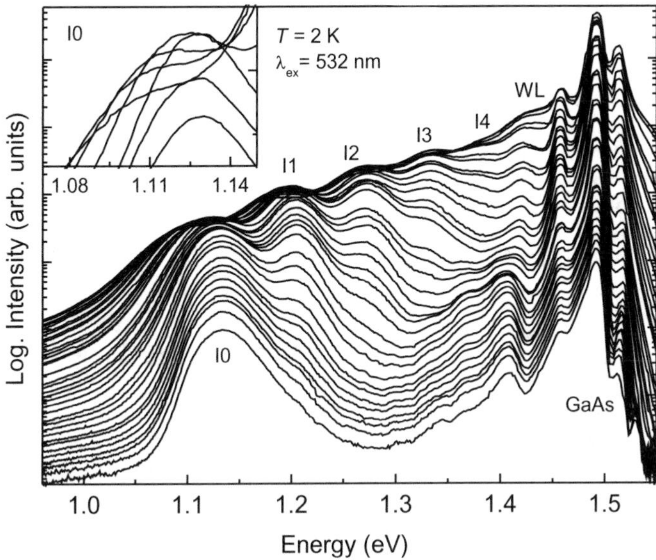

Fig. 4.33 Photoluminescence spectra of an $In_{0.6}Ga_{0.4}As/GaAs$ quantum-dot ensemble, excited into the GaAs matrix with an excitation density gradually increasing from 0.2 W/cm^{-2} (*bottom*) to 5 kW/cm^{-2}. *I0* marks ground-state emission, peaks *I2* to *I4* originate from excited states. *WL* and *GaAs* signify emission from the wetting layer and the *GaAs* matrix, respectively. *Inset*: Emission *I0* on a linear scale. Reproduced with permission from [108], © 2004 Elsevier

thick GaAs spacer, so as to increase the confinement potential. Taking the degeneracy $2(n + 1)$ of a lateral harmonic potential as a rough estimate for the exciton occupation in a quantum dot, the maximum overall occupation number of the confined excitonic states is about 30 in the studied case. The approach of an harmonic potential is supported by scanning tunneling spectroscopy [109, 110].

A close look to the energies of the quantum-dot emission shows that the increasing occupation of levels with charge carriers is accompanied by a small energy decrease. The ground-state transition, e.g., exhibits a red shift by a total of 35 meV (inset Fig. 4.33). Such behavior cannot be described in a single-particle picture, because the Coulomb interaction between the confined particles alters the overall energy in the system. The effect is referred to as *renormalization* of the band gap. The origin was traced back in bulk semiconductors to the exchange-correlation energy in dense charge-carrier ensembles, being independent on the characteristics of the electronic band structure, i.e., on the material [111]. It is also pronounced in quantum-well structures [112]. In quantum wires [113] and quantum dots [114] such renormalization may be smaller in case of a dominating strong confinement.

An ensemble of self-organized quantum dots usually exhibits a single distribution of dot sizes and compositions, leading to a single inhomogeneously broadened ensemble emission. Under certain growth conditions also distributions with several well separated emission maxima are formed. An insight into the interplay between

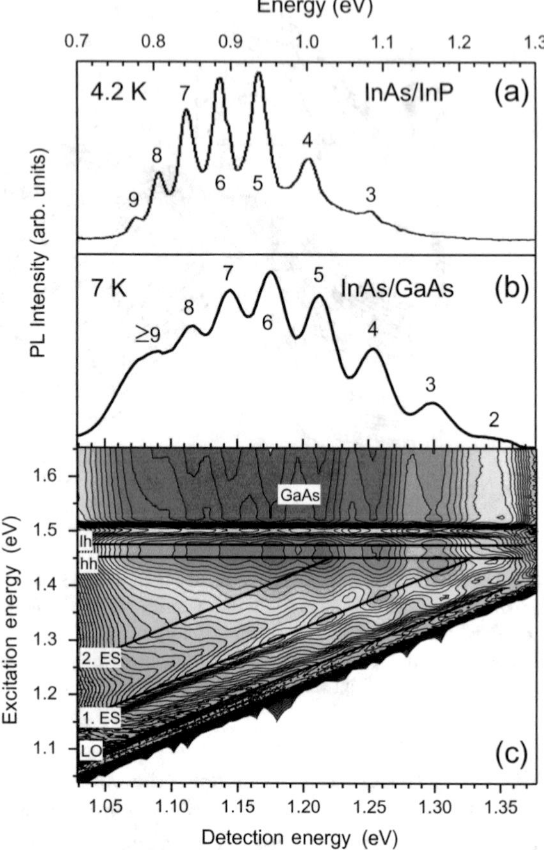

Fig. 4.34 Ensemble emission of **a** InAs/InP and **b** InAs/GaAs quantum dots with a multimodal size distribution. Numbers at the peaks refer to a common dot height within a subensemble in units of monolayers. **c** Combined PL and PL-excitation contour plot of the ensemble measured in **b** on a logarithmic color scale, abscissa also applies for (**b**). *Horizontal lines* mark light-hole (*lh*) and heavy-hole (*hh*) resonances of the wetting layer, *inclined lines* indicate resonances of the first and second excited state of the dots and an *LO* phonon transition. Data in **a** adapted from [118], data in **b** and **c** reproduced with permission from [119], © 2007 Elsevier

confined-particle interaction and confinement potential was obtained using dots with a multimodal size distribution, featuring a number of clearly resolved fairly narrow emission peaks of the ensemble [115]. The multiple peaks refer to subensembles (or families) of dots within the ensemble, which differ in height by integral numbers of InAs monolayers. Peak labels given in Fig. 4.34 indicate the height of the dots in the corresponding subensemble in units of monolayers. The assignment was proved by both structural characterization [116] and calculated exciton energies [117], revealing a truncated pyramidal shape of the dots. Spectra shown in Fig. 4.34a, b were excited with a low excitation density to ensure emission solely from the ground state.

PL excitation spectra of a multimodal dot ensemble are given in the contour plot Fig. 4.34c: A horizontal cut represents a photoluminescence spectrum at fixed excitation energy, a vertical cut represents an excitation spectrum for the selected detection energy. For large quantum dots with low emission energy an excitation spectrum shows two excited states ES1 and ES2 below the excitation of the dot emission via the wetting layer (labels hh and lh). We note that smaller dots with a

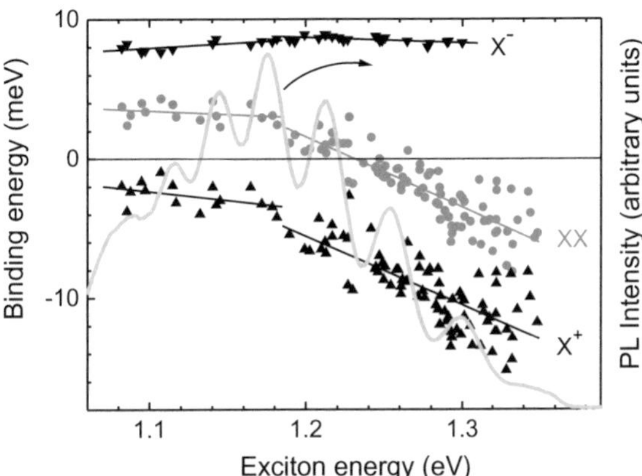

Fig. 4.35 Binding energy of few-particle complexes confined in InAs/GaAs quantum dots as a function of the neutral exciton emission energy. *0* refers to the neutral exciton recombination energy. X^-, X^+, and *XX* refer to negative trion, positive trion, and biexciton. *Right ordinate* corresponds to the *light-gray* emission spectrum. Reproduced with permission from [119], © 2007 Elsevier

height between 5 and 3 monolayers—and correspondingly a higher emission energy between 1.22 eV and 1.32 eV—have only a single excited state. Dots with even higher emission energy (>1.32 eV) have no excited state at all.

The well-defined size and shape of multimodal dots was employed to study the renormalization of few-particle transition energies acting when, e.g., one electron and hole recombine in the presence of additional charge carriers. Figure 4.35 shows relative binding energies of negative trions (consisting of 2 electrons and 1 hole), positive trions $(1e, 2h)$, and biexcitons $(2e, 2h)$ for many single-dot spectra recorded all over the inhomogeneously broadened ensemble peak [120]. The binding energy of these few-particle complexes is given with respect to the (neutral) exciton-recombination energy and the recombination energy of these complexes, i.e. by $E_{\mathrm{X}} - E_{\mathrm{complex}}$.

The three excitonic complexes show an obvious characteristic trend. The negatively charged exciton labeled X^- has always a positive binding energy being almost independent on the neutral exciton-recombination energy. Contrary the binding energies of the positively charged exciton X^+ and the biexciton XX clearly decrease for increasing exciton-recombination energy (decreasing dot size). Moreover, for the biexciton a transition from positive to negative binding energies is observed.

The non-zero binding energies originate from the Coulomb interaction C between the confined charge carriers. The binding energy of, e.g., X^- directly depends on the difference between the two direct Coulomb terms $C(e, h)$ and $C(e, e)$. The wave function of the hole is much stronger localized than that of the electron due to the larger effective mass of the holes, independent on the dot size as shown on top of Fig. 4.36. The absolute value of the Coulomb interaction between two electrons $|C(e, e)|$ is consequently smaller than that between two holes, and that between an electron and a hole is in between, i.e., $|C(e, e)| < |C(e, h)| < |C(h, h)|$, cf. Fig. 4.36a. All energies increase in smaller dots because the wave functions are slightly squeezed. A negative trion is formed by adding an electron to a dot which is filled with an exciton, thereby adding a repulsive interaction and additionally a

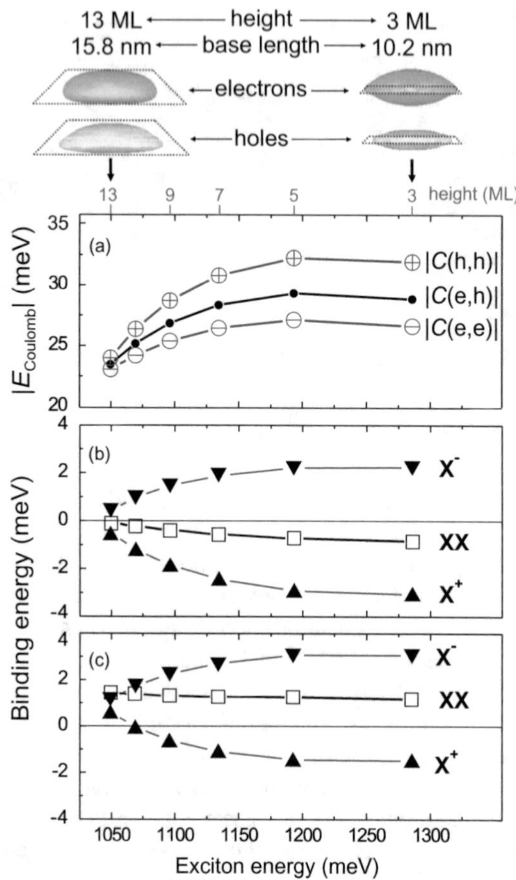

Fig. 4.36 *Top*: Electron and hole wave-function (65% probability surface) for confinement in a 15 monolayer high (*left*) and 3 monolayer high (*right*) InAs/GaAs quantum dot. **a** Absolute value of the Coulomb energy between two charge carriers confined in a quantum dot. **b, c** Calculated few-particle binding energies: **b** only ground state considered, **c** additionally 2 excited levels for both, electrons and holes included. Reproduced with permission from [121], © 2006 Elsevier

(larger) attractive interaction. The negative trion has therefore a positive binding energy $E_X - E_{X^-}$, while the positive trion has a negative binding energy.

The simple picture does not account for the trend of the biexciton XX. Including three instead of one level for both, electron and hole states into the calculation (i.e., considering also effects of *correlation*) changes the result from a nearly constant negative binding energy (Fig. 4.36b) to a nearly constant positive binding energy (Fig. 4.36c). Figure 4.34c demonstrates that large dots bind more confined levels than small dots. Consequently the degree of correlation decreases as the dot size decreases. We hence start with large dots (small exciton energy) in a case depicted in Fig. 4.36c, and end with small dots in a case similar to Fig. 4.36b. Thereby the XX binding energy changes from positive to negative values as found in the experiment Fig. 4.35. A more detailed consideration shows that the number of *hole* levels plays a major role in correlation [120].

4.4 Problems Chapter 4

4.1 The $Ga_{1-x}In_xP$ alloy is interesting due to its direct large bandgap energy compared to that of GaAs and InP.

(a) Calculate the bandgap energy of $In_xGa_{1-x}P$ at room temperature, assuming a layer which is lattice-matched on GaAs substrate. Check whether this alloy has a direct or an indirect conduction-band minimum. Compare the order of these bandgap energies to that expected from the virtual crystal approximation. Room-temperature bandgap parameters E^{Γ}, E^{X}, E^{L} (in eV) of GaP are 2.78, 2.27, 2.6, and parameters of InP are 1.34, 2.19, 1.93, respectively. The bowing parameters of the alloy are (in eV) $b^{\Gamma} = 0.65$, $b^{X} = 0.20$, $b^{L} = 1.03$.

(b) Determine the composition parameter x, where unstrained $Ga_{1-x}In_xP$ changes from a direct to an indirect semiconductor. Which bandgap energy exists at the crossing point?

4.2 Consider a lattice-matched $InP/In_xGa_{1-x}As/InP$ double heterostructure on (001) InP substrate. Find the offset in the conduction band at 300 K and at 77 K, if the valence-band offset of +0.36 eV from InP to $In_xGa_{1-x}As$ is assumed temperature-independent, and effects of thermally induced strain can be neglected. Band parameters E_g at 0 K (in eV), α (in meV/K), and β (in K) are for InAs 0.417, 0.276, and 93, for GaAs 1.519, 0.541, and 204, and for InP 1.424, 0.363, and 162, respectively, $b_{InGaAs} = 0.48$ eV.

4.3 A pseudomorphic hexagonal $Al_{0.2}Ga_{0.8}N$ layer grown on a relaxed thick GaN buffer layer produces a sheet charge-density at the interface induced by piezoelectric polarization. Apply in the following linearly weighted materials parameters.

(a) Calculate the sheet carrier-concentration per cm^2 of the AlGaN layer originating from the piezoelectric polarization.

(b) The piezoelectric polarization adds to the spontaneous polarization of a wurtzite semiconductors, the latter being -0.029 C/m^2 for GaN and -0.081 C/m^2 for AlN. Determine the lateral strain of the GaN layer required to yield a zero total polarization in the AlGaN layer. What is then the resulting total sheet carrier-density at the interface?

4.4 (a) Find the offset in the valence band and the conduction band for a transition from a GaAs (001) substrate to a thin pseudomorphic InP layer. Neglect the strain-induced splitting between heavy hole and light hole.

(b) What offsets occur in the inverse heterostructure of a thin pseudomorphic GaAs layer on InP(001) substrate, if the strain-induced splitting between heavy hole and light hole is neglected?

(c) Compare the energy splitting between the heavy-hole valence band and the light-hole valence band of GaAs in case (b) with the band offsets calculated for a neglected splitting. Which exciton has lowest energy? What are the actual band offsets obtained by including the energy splitting between heavy hole and light hole?

4.5 (a) The temperature dependence of the direct bandgap of GaAs is described by the parameters $\alpha = 0.54$ meV/K and $\beta = 204$ K. Which bandgap energy has GaAs at room temperature (300 K) and at the temperature of liquid nitrogen (77 K)?

(b) The effective bandgap changes in presence of a confining potential. It is then approximately given by the difference between the ground-state levels of the electron in the conduction band and the hole (with lowest energy) in the valence band. Calculate the effective bandgap of an unstrained GaAs quantum well of 9 nm thickness for infinite barriers at room temperature. Effective electron and hole masses are $0.067m_e$ and $0.082m_e$, respectively.

(c) The thickness of the quantum well in (b) varies by $\pm$ one monolayer due to interface roughness. To which variations in the effective bandgap translates this thickness fluctuation?

4.5 General Reading Chapter 4

E.T. Yu, J.O. McCaldin, T.C. McGill, Band offsets in semiconductor heterojunctions, in *Solid State Physics*, vol. 46, ed. by H. Ehrenreich, D. Turnbull (Academic Press, Boston, 1992), pp. 1–146

P.Y. Yu, M. Cardona, *Fundamentals of Semiconductors*, Physics and Materials Properties, 4th edn. (Springer, Berlin, 2010)

K.W. Böer, U.W. Pohl, *Semiconductor Physics* (Springer, Cham, 2018)

References

1. P.Y. Yu, M. Cardona, *Fundamentals of Semiconductors* (Springer, Berlin, 1996)
2. K.W. Böer, U.W. Pohl, *Semiconductor Physics* (Springer, Cham, 2018)
3. I. Vurgaftman, J.R. Meyer, L.R. Ram-Mohan, Band parameters for III-V compound semiconductors and their alloys. J. Appl. Phys. **89**, 5815 (2001)
4. M. Suzuki, T. Uenoyama, First-principles calculations of effective-mass parameters of AlN and GaN. Phys. Rev. B **52**, 8132 (1995)
5. M. Posternak, A. Baldereschi, A. Catellani, Ab initio study of the spontaneous polarization of pyroelectric BeO. Phys. Rev. Lett. **64**, 1777 (1990)

6. G.L. Bir, G.E. Pirkus, *Symmetry and Strain-Induced Effects in Semiconductors* (Wiley, New York, 1974)
7. F.H. Pollak, M. Cardona, Piezo-electroreflectance in Ge, GaAs, and Si. Phys. Rev. **172**, 816 (1968)
8. C.G. van de Walle, R.M. Martin, Theoretical calculation of heterojunction discontinuities in the Si/Ge system. Phys. Rev. B **34**, 5621 (1986)
9. K. Shahzad, D.J. Olego, C.G. van de Walle, Optical characterization and band offsets in ZnSe-ZnS$_x$ Se$_{1-x}$ strained-layer superlattices. Phys. Rev. B **38**, 1417 (1988)
10. C.G. van de Walle, Band lineups and deformation potentials in the model-solid theory. Phys. Rev. B **39**, 1871 (1989)
11. D. Fröhlich, W. Nieswand, U.W. Pohl, J. Wrzesinski, Two-photon spectroscopy of ZnSe under uniaxial stress. Phys. Rev. B **52**, 14652 (1995)
12. F. Kubacki, J. Gutowski, D. Hommel, M. Heuken, U.W. Pohl, Determination of deformation potentials in ZnSe/GaAs strained-layer heterostructures. Phys. Rev. B **54**, 2028 (1996)
13. L.D. Laude, F.H. Pollak, M. Cardona, Effects of uniaxial stress on the indirect exciton spectrum of silicon. Phys. Rev. B **3**, 2623 (1971)
14. H. Mathieu, P. Merle, E.L. Ameziane, B. Archilla, J. Camassel, Deformation potentials of the direct and indirect absorption edges of GaP. Phys. Rev. B **19**, 2209 (1984)
15. P. Pfeffer, I. Gorczyca, W. Zawadzki, Theory of free-electron optical absorption in *n*-GaAs. Solid State Commun. **51**, 179 (1984) (Table 1 with references therein)
16. F. Bernardini, V. Fiorentini, D. Vanderbilt, Spontaneous polarization and piezoelectric constants of III-V nitrides. Phys. Rev. B **56**, R10024 (1997)
17. G. Bester, X. Wu, D. Vanderbilt, A. Zunger, Importance of second-order piezoelectric effects in zinc-blende semiconductors. Phys. Rev. Lett. **96**, 187602 (2006)
18. D.L. Smith, Strain-generated electric fields in [111] growth axis strained-layer superlattices. Sol. State Commun. **57**, 919 (1986)
19. A. Beya-Wakata, P.-Y. Prodhomme, G. Bester, First- and second-order piezoelectricity in III-V semiconductors. Phys. Rev. B **84**, 195207 (2011)
20. L.C. Lew Yan Voon, M. Willatzen, Electromechanical phenomena in semiconductor nanostructures. J. Appl. Phys. **109**, 031101 (2011)
21. O. Ambacher, J. Smart, J.R. Shealy, N.G. Weimann, K. Chu, M. Murphy, W.J. Schaff, L.F. Eastman, R. Dimitrov, L. Wittmer, M. Stutzmann, W. Rieger, J. Hilsenbeck, Two-dimensional electron gases induced by spontaneous and piezoelectric polarization charges in N- and Ga-face AlGaN/GaN heterostructures. J. Appl. Phys. **85**, 3222 (1999)
22. U. Rössler (ed.), *Semiconductor II–VI and I–VII Compounds; Semimagnetic Compounds*. Landolt-Börnstein III/41B, revised and updated edition of Vols. III/17 and 22 (Springer, Heidelberg, 1999)
23. O. Ambacher, B. Foutz, J. Smart, J.R. Shealy, N.G. Weimann, K. Chu, M. Murphy, A.J. Sierakowski, W.J. Schaff, L.F. Eastman, R. Dimitrov, A. Mitchell, M. Stutzmann, Two-dimensional electron gases induced by spontaneous and piezoelectric polarization in undoped and doped AlGaN/GaN heterostructures. J. Appl. Phys. **87**, 334 (2000)
24. Y.P. Varshni, Temperature dependence of the energy gap in semiconductors. Physica **34**, 149 (1967)
25. L. Viña, S. Logothetidis, M. Cardona, Temperature dependence of the dielectric function of germanium. Phys. Rev. B **30**, 1979 (1984)
26. K.P. O'Donnell, X. Chen, Temperature dependence of semiconductor band gaps. Appl. Phys. Lett. **58**, 2924 (1991)
27. R. Pässler, Parameter sets due to fittings of the temperature dependencies of fundamental bandgaps in semiconductors. Phys. Status Solidi B **216**, 975 (1999)
28. W. Bludau, A. Onton, W. Heinke, Temperature dependence of the band gap of silicon. J. Appl. Phys. **45**, 1846 (1974)
29. R. Pässler, Dispersion-related description of temperature dependencies of band gaps in semiconductors. Phys. Rev. B **66**, 085201 (2002)

30. I. Vurgaftman, J.R. Meyer, Band parameters for nitrogen-containing semiconductors. J. Appl. Phys. **94**, 3675 (2003)
31. S. Adachi, *Properties of Group-IV, III-V and II-VI Semiconductors* (Wiley, Chichester, 2005)
32. J.A. Van Vechten, T.K. Bergstresser, Electronic structures of semiconductor alloys. Phys. Rev. B **1**, 3351 (1970)
33. C.K. Williams, T.H. Glisson, J.R. Hauser, M.A. Littlejohn, Energy bandgap and lattice constant contours of III-V quaternary alloys of the form $A_x B_y C_z D$ or $AB_x C_y D_z$. J. Electron. Mater. **7**, 639 (1978)
34. A. Baldereschi, E. Hess, K. Maschke, H. Neumann, K.-R. Schulze, K. Unger, Energy band structure of $Al_xGa_{1-x}As$. J. Phys. C Solid State Phys. **10**, 4709 (1977)
35. A.-B. Chen, A. Sher, Electronic structure of III-V semiconductors and alloys using simple orbitals. Phys. Rev. B **22**, 3886 (1980)
36. A.-B. Chen, A. Sher, Electronic structure of pseudobinary semiconductor alloys $Al_xGa_{1-x}As$, $GaP_x As_{1-x}$, and $Ga_x In_{1-x}P$. Phys. Rev. B **23**, 5360 (1981)
37. J.E. Bernard, A. Zunger, Electronic structure of ZnS, ZnSe, ZnTe, and their pseudobinary alloys. Phys. Rev. B **36**, 3199 (1987)
38. A.T. Gorelenok, A.G. Dzigasov, P.P. Moskvin, V.S. Sorokin, I.S. Tarasov, Dependence of the band gap on the composition of $In_{1-x}Ga_x As_{1-y}P_y$ solid solutions. Sov. Phys. Semicond. **15**, 1400 (1981)
39. M. Ferhat, F. Bechstedt, First-principles calculations of gap bowing in $In_xGa_{1-x}N$ and $In_xAl_{1-x}N$ alloys: relation to structural and thermodynamic. Phys. Rev. B **65**, 075213 (2002)
40. P.G. Moses, M. Miao, Q. Yan, C.G. Van de Walle, Hybrid functional investigations of band gaps and band alignments for AlN, GaN, InN, and InGaN. J. Chem. Phys. **134**, 084703 (2011)
41. L. Cláudio de Carvalho, A. Schleife, J. Furthmüller, F. Bechstedt, Distribution of cations in wurtzitic $In_xGa_{1-x}N$ and $In_xAl_{1-x}N$ alloys: consequences for energetics and quasiparticle electronic structures. Phys. Rev. B **85**, 115121 (2012)
42. E.T. Yu, J.O. McCaldin, T.C. McGill, Band offsets in semiconductor heterojunctions, in *Solid State Physics*, vol. 46 (Academic Press, New York, 1992), pp. 1–146
43. R.L. Anderson, Experiments on Ge-GaAs heterojunctions. Solid-State Electron. **5**, 341 (1962)
44. A.G. Milnes, D.L. Feucht, *Heterojunctions and Metal-Semiconductor Junctions* (Academic Press, New York, 1972)
45. W.R. Frensley, H. Kroemer, Theory of the energy-band lineup at an abrupt semiconductor heterojunction. Phys. Rev. B **16**, 2642 (1977)
46. W.A. Harrison, Elementary theory of heterojunctions. J. Vac. Sci. Technol. **14**, 1016 (1977)
47. S.-H. Wei, A. Zunger, Role of d orbitals in valence-band offsets of common-anion semiconductors. Phys. Rev. Lett. **59**, 144 (1987)
48. J. Tersoff, Band lineups at II-VI heterojunctions: failure of the common-anion rule. Phys. Rev. Lett. **56**, 2755 (1986)
49. A.D. Katnani, G. Margaritondo, Empirical rule to predict heterojunction band discontinuities. J. Appl. Phys. **54**, 2522 (1983)
50. A.D. Katnani, R.S. Bauer, Commutativity and transitivity of GaAs-AlAs-Ge(100) band offsets. Phys. Rev. B **33**, 1106 (1986)
51. J.M. Langer, H. Heinrich, Deep-level impurities: a possible guide to prediction of band-edge discontinuities in semiconductor heterojunctions. Phys. Rev. Lett. **55**, 1414 (1985)
52. A. Zunger, Electronic structure of 3D transition-atom impurities in semiconductors, in *Solid State Physics*, vol. 39 (Academic Press, New York, 1986), pp. 275–464
53. C.G. van de Walle, J. Neugebauer, Universal alignment of hydrogen levels in semiconductors, insulators and solutions. Nature **423**, 626 (2003)
54. J. Tersoff, Theory of semiconductor heterojunctions: the role of quantum dipoles. Phys. Rev. B **30**, 4874 (1984)
55. J. Tersoff, Shottky barrier heights and the continuum of gap states. Phys. Rev. Lett. **52**, 465 (1984)
56. D. Mourad, Tight-binding branch-point energies and band offsets for cubic InN, GaN, AlN, and AlGaN alloys. J. Appl. Phys. **113**, 123705 (2013)

57. M. Landmann, E. Rauls, W.G. Schmidt, Understanding band alignments in semiconductor heterostructures: Composition dependence and type-I-type-II transition of natural band offsets in nonpolar zinc-blende $Al_xGa_{1-x}N/Al_yGa_{1-y}N$ composites. Phys. Rev. B **95**, 155310 (2017)
58. C.G. van de Walle, R.M. Martin, Theoretical study of band offsets at semiconductor interfaces. Phys. Rev. B **35**, 8154 (1987)
59. M.P.C.M. Krijn, Heterojunction band offsets and effective masses in III-V quaternary alloys. Semicond. Sci. Technol. **6**, 27 (1991)
60. R.G. Dandrea, S. Froyen, A. Zunger, Stability and band offsets of heterovalent superlattices: Si/GaP, Ge/GaAs, and Si/GaAs. Phys. Rev. B **42**, 3213 (1990)
61. S. Rubini, E. Milocco, L. Sorba, E. Pelucchi, A. Franciosi, A. Garulli, A. Parisini, Y. Zhuang, G. Bauer, Structural and electronic properties of ZnSe/AlAs heterostructures. Phys. Rev. B **63**, 155312 (2001)
62. M. Peressi, N. Binggeli, A. Baldereschi, Band engineering at interfaces: theory and numerical experiments. J. Phys. D **31**, 1273 (1998)
63. M. Peressi, A. Baldereschi, Ab initio studies of structural and electronic properties, in *Characterization of Semiconductor Heterostructures and Nanostructures*, 2nd edn., ed. by C. Lamberti, G. Agostini (Elsevier, Amsterdam, 2013), pp. 21–73
64. W.A. Harrison, E.A. Kraut, J.R. Waldrop, R.W. Grant, Polar heterojunction interfaces. Phys. Rev. B **18**, 4402 (1978)
65. A. Kley, J. Neugebauer, Atomic and electronic structure of the GaAs/ZnSe (001) interface. Phys. Rev. B **50**, 8616 (1994)
66. R. Nicolini, L. Vanzetti, G. Mula, G. Bratina, L. Sorba, A. Franciosi, M. Peressi, S. Baroni, R. Resta, A. Baldereschi, Local interface composition and band discontinuities in heterovalent heterostructures. Phys. Rev. Lett. **72**, 294 (1994)
67. A. Bonanni, L. Vanzetti, L. Sorba, A. Franciosi, M. Lomascolo, P. Prete, R. Cingolani, Optimization of interface parameters and bulk properties in ZnSe-GaAs heterostructures. Appl. Phys. Lett. **66**, 1092 (1995)
68. N. Kobayashi, Single quantum well photoluminescence in ZnSe/GaAs/AlGaAs grown by migration-enhanced epitaxy. Appl. Phys. Lett. **55**, 1235 (1989)
69. M. Städele, J.A. Majewski, P. Vogl, Stability and band offsets of polar GaN/SiC(001) and AlN/SiC(001) interfaces. Phys. Rev. B **56**, 6911 (1997)
70. F. Bernardini, M. Peressi, V. Fiorentini, Band offsets and stability of BeTe/ZnSe(100) heterojunctions. Phys. Rev. B **62**, R16302 (2000)
71. Z.-G. Wang, L.-Å. Ledebo, H.G. Grimmeis, Optical properties of iron doped Al_x $Ga_{1-x}As$ alloys. J. Appl. Phys. **56**, 2762 (1984)
72. Z.-G. Wang, L.-Å. Ledebo, H.G. Grimmeis, Nuovo Cimento **2D**, 1718 (1983)
73. L. Samuelson, S. Nilsson, Z.-G. Wang, H.G. Grimmeis, Direct evidence for random-alloy splitting of Cu levels in $GaAs_{1-x}P_x$. Phys. Rev. Lett. **53**, 1501 (1984)
74. A.A. Reeder, J.M. Chamberlain, Optical study of the deep manganese acceptor in $In_{1-x}Ga_x$ P: evidence for vacuum-level pinning. Solid State Commun. **54**, 705 (1985)
75. M. Grundmann, D. Bimberg, Anisotropy effects on excitonic properties in realistic quantum wells. Phys. Rev. B **38**, 13486 (1988)
76. S. Adachi, GaAs, AlAs and $Al_xGa_{1-x}As$: material parameters for use in research and device applications. J. Appl. Phys. **58**, R1 (1985)
77. S. Adachi, *Physical Properties of III-V Semiconductor Compounds* (Wiley, New York, 1992)
78. D.C. Reynolds, D.C. Look, B. Jogai, C.W. Litton, G. Cantwell, W.C. Harsch, Valence-band ordering in ZnO. Phys. Rev. B **60**, 2340 (1999)
79. W. Shan, B.D. Little, A.J. Fischer, J.J. Song, B. Goldenberg, W.G. Perry, M.D. Bremser, R.F. Davis, Binding energy for the intrinsic excitons in wurtzite GaN. Phys. Rev. B **54**, 16369 (1996)
80. G. Dorda, Surface quantization in semiconductors, in *Festkörperprobleme*, ed. by H.J. Queisser. Advances in Solid State Physics, vol. 13 (Pergamon/Vieweg, Braunschweig, 1973), p. 215

81. U.W. Pohl, Low-dimensional semiconductors, in *Handbook of Materials Data*, 2nd edn., ed. by H. Warlimont, W. Martienssen (Springer, Cham, 2018), pp. 1077–1100
82. G. Bastard, J.A. Brum, Electronic states in semiconductor heterostructures. IEEE J. Quantum Electron. **QE-22**, 1625 (1986)
83. R. Dingle, W. Wiegmann, C.H. Henry, Quantum states of confined carriers in very thin $Al_x Ga_{1-x}As$-GaAs-$Al_x Ga_{1-x}As$ heterostructures. Phys. Rev. Lett. **33**, 827 (1974)
84. R. Dingle, Confined carrier quantum states in ultrathin semiconductor heterostructures, in *Festkörperprobleme*, ed. by H.J. Queisser. Advances in Solid State Physics, vol. 15 (Pergamon/Vieweg, Braunschweig, 1975), p. 21
85. K. Suzuki, K. Kanisawa, C. Janer, S. Perraud, K. Takashina, T. Fujisawa, Y. Hirayama, Spatial imaging of two-dimensional electronic states in semiconductor quantum wells. Phys. Rev. Lett. **98**, 136802 (2007)
86. R.C. Miller, D.A. Kleiman, W.A. Nordland Jr., A.C. Gossard, Luminescence studies of optically pumped quantum wells in GaAs-$Al_x Ga_{1-x}As$ multilayer structures. Phys. Rev. B **22**, 863 (1980)
87. P.L. Gourley, R.M. Biefeld, Quantum size effects in GaAs/$GaAs_x P_{1-x}$ strained-layer superlattices. Appl. Phys. Lett. **45**, 749 (1984)
88. R.C. Miller, D.A. Kleiman, O. Munteanu, W.T. Tsang, New transitions in the photoluminescence of GaAs quantum wells. Appl. Phys. Lett. **39**, 1 (1981)
89. R.L. Greene, K.K. Bajaj, D.E. Phelps, Energy levels of Wannier excitons in GaAs-$Ga_{1-x}Al_x As$ quantum-well structures. Phys. Rev. B **29**, 1807 (1984)
90. B. Deveaud, J.Y. Emery, A. Chomette, B. Lambert, M. Baudet, Observation of one-monolayer size fluctuations in a GaAs/GaAlAs superlattice. Appl. Phys. Lett. **45**, 1078 (1984)
91. A. Forchel, H. Leier, B.E. Maile, R. Germann, Fabrication and optical spectroscopy of ultra small III–V compound semiconductor structures, in *Festkörperprobleme*, ed. by U. Rössler. Advances in Solid State Physics, vol. 28 (Pergamon/Vieweg, Braunschweig, 1988), p. 99
92. X.-L. Wang, V. Voliotis, Epitaxial growth and optical properties of semiconductor quantum wires. J. Appl. Phys. **99**, 121301 (2006)
93. F. Vouilloz, D.Y. Oberli, M.-A. Dupertuis, A. Gustafsson, F. Reinhardt, E. Kapon, Polarization anisotropy and valence band mixing in semiconductor quantum wires. Phys. Rev. Lett. **78**, 1580 (1997)
94. F. Vouilloz, D.Y. Oberli, M.-A. Dupertuis, A. Gustafsson, F. Reinhardt, E. Kapon, Effect of lateral confinement on valence-band mixing and polarization anisotropy in quantum wires. Phys. Rev. B **57**, 12378 (1998)
95. Z.-Y. Deng, X. Chen, T. Ohji, T. Kobayashi, Subband structures and exciton and impurity states in V-shaped GaAs-$Ga_{1-x}Al_x$ As quantum wires. Phys. Rev. B **61**, 15905 (2000)
96. E. Martinet, M.-A. Dupertuis, F. Reinhardt, G. Biasiol, E. Kapon, O. Stier, M. Grundmann, D. Bimberg, Separation of strain and quantum-confinement effects in the optical spectra of quantum wires. Phys. Rev. B **61**, 4488 (2000)
97. E. Kapon, G. Biasiol, D.M. Hwang, E. Colas, M. Walther, Self-ordering mechanism of quantum wires grown on non-planar substrates. Solid-State Electron. **40**, 815 (1996)
98. T. Someya, H. Akiyama, H. Sakaki, Enhanced binding energy of one-dimensional excitons in quantum wires. Phys. Rev. Lett. **76**, 2965 (1996)
99. T. Ogawa, T. Takagahara, An exact treatment of excitonic effects. Phys. Rev. B **44**, 8138 (1991)
100. D. Gershoni, M. Katz, W. Wegscheider, L.N. Pfeiffer, R.A. Logan, K. West, Radiative lifetimes of excitons in quantum wires. Phys. Rev. B **50**, 8930 (1994)
101. D. Bimberg, M. Grundmann, N.N. Ledentsov, *Quantum Dot Heterostructures* (Wiley, Chichester, 1999)
102. S. Rodt, A. Schliwa, K. Pötschke, F. Guffarth, D. Bimberg, Correlation and few particle properties of self-organized InAs/GaAs quantum dots. Phys. Rev. B **71**, 155325 (2005)
103. A. Schliwa, M. Winkelnkemper, D. Bimberg, Impact of size, shape, and composition on piezoelectric effects and electronic properties of In(Ga)As/GaAs quantum dots. Phys. Rev. B **76**, 205324 (2007)

104. T. Maltezopoulos, A. Bolz, C. Meyer, C. Heyn, W. Hansen, M. Morgenstern, R. Wiesendanger, Wave-function mapping of InAs quantum dots by scanning tunneling spectroscopy. Phys. Rev. Lett. **91**, 196804 (2003)
105. A. Urbieta, B. Grandidier, J.P. Nys, D. Deresmes, D. Stiévenard, A. Lemaître, G. Patriarche, Y.M. Niquet, Scanning tunneling spectroscopy of cleaved InAs/GaAs quantum dots at low temperatures. Phys. Rev. B **77**, 155313 (2008)
106. O. Stier, M. Grundmann, D. Bimberg, Electronic and optical properties of strained quantum dots modeled by 8-band kp theory. Phys. Rev. B **59**, 5688 (1999)
107. M. Scholz, S. Büttner, O. Benson, A.I. Toropov, A.K. Bakarov, A.K. Kalagin, A. Lochmann, E. Stock, O. Schulz, F. Hopfer, V.A. Haisler, D. Bimberg, Non-classical light emission from a single electrically driven quantum dot. Opt. Express **15**, 9107 (2007)
108. F. Guffarth, S. Rodt, A. Schliwa, K. Pötschke, D. Bimberg, Many-particle effects in self-organized quantum dots. Phys. E **25**, 261 (2004)
109. B. Fain, I. Robert-Philip, C. David, Z.Z. Wang, I. Sagnes, J.C. Girard, Discretization of electronic states in large InAsP/InP multilevel quantum dots probed by scanning tunneling spectroscopy. Phys Rev. Lett. **108**, 126808 (2012)
110. K. Teichmann, M. Wenderoth, H. Prüser, K. Pierz, H.W. Schumacher, R.G. Ulbrich, Harmonic oscillator wave functions of a self-assembled InAs quantum dot measured by scanning tunneling microscopy. Nano Lett. **13**, 3571 (2013)
111. P. Vashishta, R.K. Kalia, Universal behavior of exchange-correlation energy in electron-hole liquid. Phys. Rev. B **25**, 6492 (1982)
112. G. Tränkle, E. Lach, A. Forchel, F. Scholz, C. Ell, H. Haug, G. Weimann, G. Griffiths, H. Kroemer, S. Subbanna, General relation between band-gap renormalization and carrier density in two-dimensional electron-hole plasmas. Phys. Rev. B **36**, 6712 (1987)
113. R. Ambigapathy, I. Bar-Joseph, D.Y. Oberli, S. Haacke, M.J. Brasil, F. Reinhard, E. Kapon, B. Deveaud, Coulomb correlation and band gap renormalization at high carrier densities in quantum wires. Phys. Rev. Lett. **78**, 3579 (1997)
114. R. Heitz, F. Guffarth, I. Mukhametzhanov, M. Grundmann, A. Madhukar, D. Bimberg, Many-body effects on the optical spectra of InAs/GaAs quantum dots. Phys. Rev. B **62**, 16881 (2000)
115. U.W. Pohl, InAs/GaAs quantum dots with multimodal size distribution, in *Self-assembled Quantum Dots*, ed. by Z.M. Wang (Springer, New York, 2008), pp. 43–66, Chap. 3
116. U.W. Pohl, K. Pötschke, A. Schliwa, F. Guffarth, D. Bimberg, N.D. Zakharov, P. Werner, M.B. Lifshits, V.A. Shchukin, D.E. Jesson, Evolution of a multimodal distribution of self-organized InAs/GaAs quantum dots. Phys. Rev. B **72**, 245332 (2005)
117. R. Heitz, F. Guffarth, K. Pötschke, A. Schliwa, D. Bimberg, N.D. Zakharov, P. Werner, Shell-like formation of self-organized InAs/GaAs quantum dots. Phys. Rev. B **71**, 045325 (2005)
118. S. Raymond, S. Studenikin, S.-J. Cheng, M. Pioro-Ladrière, M. Ciorga, P.J. Poole, M.D. Robertson, Families of islands in InAs/InP self-assembled quantum dots: a census obtained from magneto-photoluminescence. Semicond. Sci. Technol. **18**, 385 (2003)
119. S. Rodt, R. Seguin, A. Schliwa, F. Guffarth, K. Pötschke, U.W. Pohl, D. Bimberg, Size-dependent binding energies and fine-structure splitting of excitonic complexes in single InAs/GaAs quantum dots. J. Lumin. **122–123**, 735 (2007)
120. S. Rodt, A. Schliwa, K. Pötschke, F. Guffarth, D. Bimberg, Correlation of structural and few-particle properties of self-organized InAs/GaAs quantum dots. Phys. Rev. B **71**, 155325 (2005)
121. U.W. Pohl, R. Seguin, S. Rodt, A. Schliwa, K. Pötschke, D. Bimberg, Control of structural and excitonic properties of self-organized InAs/GaAs quantum dots. Phys. E **35**, 285 (2006)

Chapter 5
Electronic Properties of Organic Semiconductors

Abstract The highest occupied und lowest unoccupied molecular orbitals (HOMO and LUMO) form bands in organic semiconductors corresponding to valence and conduction bands of their inorganic counterparts. Organic crystals have comparatively narrow bands and large bandgaps, and they feature strong anisotropies. The Gaussian density of states of HOMO and LUMO supplemented by defect states produces a substantial tailing of the bands into the bandgap. The strong structural and electronic relaxation occurring when a molecule is charged leads to strongly bound Frenkel excitons and to a pronounced polaron character of mobile carriers. The small polaron radius and the high density of defects lead to a prevailing, thermally activated hopping conductance with a relatively low carrier mobility. Band conductance with an increased mobility at low temperature may be found at reduced scattering rates. Molecules at the interfaces between two organic semiconductors have often weak interaction, yielding a band alignment largely described by matching the vacuum levels. Molecular interaction at interfaces to metals is generally strong, creating surface dipoles and an often observed pinning of the Fermi level.

5.1 Band Structure of Organic Crystals

Organic semiconductors comprise small-molecule crystals and polymers. They are employed today in a wide field of applications, e.g., in organic LEDs (OLEDs) and displays, radio-frequency tags, and solar cells; reviews are given in, e.g., [1–5]. The electronic properties of organic crystals show characteristic features usually not observed in inorganic solids, originating from the molecular structure.

U. W. Pohl, *Epitaxy of Semiconductors*, Graduate Texts in Physics,
https://doi.org/10.1007/978-3-030-43869-2_5

5.1.1 *Highest Occupied and Lowest Unoccupied Molecular Orbitals*

Carriers in Organic and Inorganic Crystals

Organic crystals are composed of molecules as building blocks instead of atoms (Sect. 3.1). As a consequence, the periodicity approach used to describe free carriers in the periodic potential of inorganic solids does not apply well for describing the electronic properties of organic solids: while carriers may move well along the backbone of a linear molecule, they experience a substantial energy barrier for the transfer to the adjacent molecule. This barrier originates from the weak intermolecular interaction and the resulting large distance between atoms of neighboring molecules. In addition the strong disorder in organic crystals leads to a Gaussian broadening of the electronic density of states at the band edge with trap states in the bandgap. Transport properties are therefore often rather described by a hopping conductance than by a band-like conductance.

Free carriers in organic crystals differ from those in inorganic solids. The reason is the strong structural relaxation of the lattice occurring when a charge is introduced to an organic molecule. This relaxation accompanies the moving carrier, affecting his effective mass. Such a *phonon-dressed* carrier in an organic crystal is conveniently described as a *small polaron* (Sect. 5.2.1). Lattice polarization by free carriers is also observed in inorganic semiconductors, but the effect is much smaller; their weak electron-phonon interaction is described by a *large* polaron with a small Fröhlich coupling constant α, leading to a merely slight increase of the effective mass.

Electronic Levels of Molecules

The origin of a band structure in an organic crystal can be illustrated by comparing energy levels of a free molecule with levels of the organic crystal. The simplified level scheme of an isolated single molecule is depicted in Fig. 5.1. The molecule has an even number of electrons if it is electrically neutral and not a radical. The *highest occupied molecular orbital* (HOMO) in the electronic ground state of the molecule is occupied by 2 electrons with opposite spin. It is hence a singlet state (S_0 in Fig. 5.1)

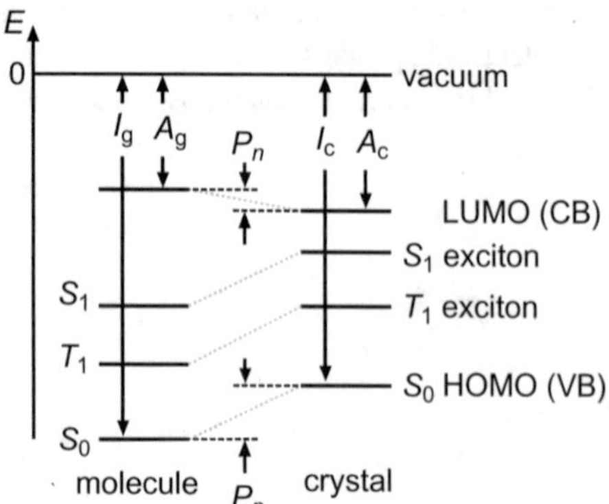

Fig. 5.1 Schematic of the energy levels of a single molecule (gas phase, *left*) and of a molecule crystal (*right*); I_g, A_g, I_c, and A_c denote respectively the ionization energies I and electron affinities A of the molecule in the gas phase and in the crystal; P_n and P_p signify the polarization energies of an electron and a hole in the molecule crystal

with total spin $S = 0$; the first excited singlet state is S_1. There exists also a triplet state labelled T_1 with one electron in the HOMO state and one electron in the excited state with *parallel* spin, yielding a total spin $S = 1$; the energy of T_1 is larger than that of S_0. Further excited states S_2 and T_2 may exist with an energy separation to S_0 below the ionization energy I_g, but the lifetime of an electron excited to such states is very small compared to the lifetime in the states S_1 and T_1. These higher excited states of the singlet and triplet manifolds are not shown in Fig. 5.1 for clarity.

The ionization energy I_g represents the energy required to remove the most weakly bound electron from the ground state of an isolated molecule and is readily accessible in experiment. On the other hand, an energy referred to as *electron affinity* A_g is released, if an additional electron is bound to an isolated molecule to form a negatively charged molecular ion.

If the electron is removed from a molecule in the environment of a *crystal lattice*, the required energy I_c is *smaller* than I_g. Organic molecules are strongly polarized when they are charged; removing an electron from the HOMO forms a positively charged radical with a mobile hole, and atoms in the molecule and in neighboring molecules rearrange. The related polarization energy P_p is large (typically above 1 eV) and represents the difference between the ionization energies: $P_p = I_g - I_c$. The polarization of the crystal is expressed by a dielectric constant of typically about 3; the reduced ionization energy can hence also be attributed to the effect of the polarizable medium when negative and positive charges are separated.

Due to relaxation upon ionization also the electron affinity A_g of an isolated molecule is altered in the crystal; it is increased by a polarization energy P_n to a value A_c. The electron added to the molecule of a crystal is mobile similar to the hole left by a removed electron. The energies of these two transport states are separated by an energy gap E_g, which in the case of $P_p = P_n \equiv P$ is given by [6]

$$E_g = I_c - A_c = I_g - A_g - 2P = 2I_c - I_g - A_g; \tag{5.1}$$

in (5.1) all quantities are assumed positive (see also Fig. 5.2). E_g is the bandgap energy for a single carrier and refers to carrier generation in either the HOMO state or the lowest unoccupied (LUMO) state. The energy gap E_g and its position with respect to the vacuum level depend on the spatial extent of the delocalized π electrons or holes in the molecules of the crystal; this dependence is illustrated for acenes (Sect. 3.1.2) in Fig. 5.2.

In an anthracene crystal the bandgap energy E_g is 4.1 eV; this energy is required to remove an electron from the HOMO level to a quasi-free state of the molecule, when the positive charge left in the HOMO is not correlated. The quasi-free state is the *lowest unoccupied molecular orbital*, the LUMO level; in analogy to inorganic semiconductors the LUMO in organic semiconductors is referred to as *conduction band*, and the HOMO is called *valence band*. The S_1 state indicated in Fig. 5.1 lies below the LUMO level; in this state the electron is still bound to this positive charge, forming an *exciton* (Sect. 5.1.3); the excitation energy of the S_1 exciton in anthracene is about 0.5 eV lower than E_g. This energy difference represents the

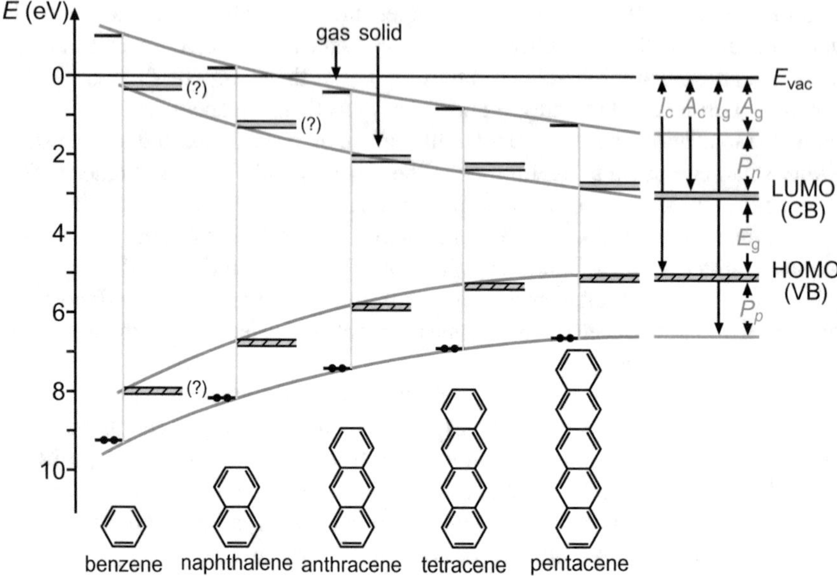

Fig. 5.2 Ionization energies of the highest occupied level and binding energies of the lowest unoccupied level for various oligoacenes in the gas phase (*horizontal bars left of gray vertical lines*, referring to I_g and A_g) and in a crystal (*right bars*, referring to I_c and A_c). Reproduced with permission from [6], © 1974 Springer Nature

binding energy of the exciton; its value is much larger than values found in inorganic semiconductors. A large binding energy corresponds to a strong spatial localization and a strong structural relaxation of the molecules, both being typical features of excitons in organic crystals.

5.1.2 *Bands and Bandgaps in Organic Crystals*

In organic semiconductors often band-like conduction similar to that of inorganic semiconductors is observed. The energy of carriers is then well characterized in terms of a band structure. The electronic band structure of an anthracene crystal calculated using the tight-binding approach is given in Fig. 5.3. The dispersions $E(\mathbf{k})$ refer to directions of the monoclinic Brillouin zone shown in the inset; $\mathbf{d}_1 = (\frac{1}{2}\frac{1}{2}0)$ and $\mathbf{d}_2 = (-\frac{1}{2}\frac{1}{2}0)$ are next-neighbor directions in the xy plane. Anthracene has two equivalent molecules with different orientations in the unit cell; both, the LUMO and HOMO levels are composed of symmetrical and asymmetric linear combinations of molecular orbitals of the two molecules, leading to a two-fold splitting. The degeneracy of k_x and k_y at the edge of the Brillouin zone in the xy plane originates from a crystal glide-plane symmetry; no such symmetry exists in triclinic crystals like

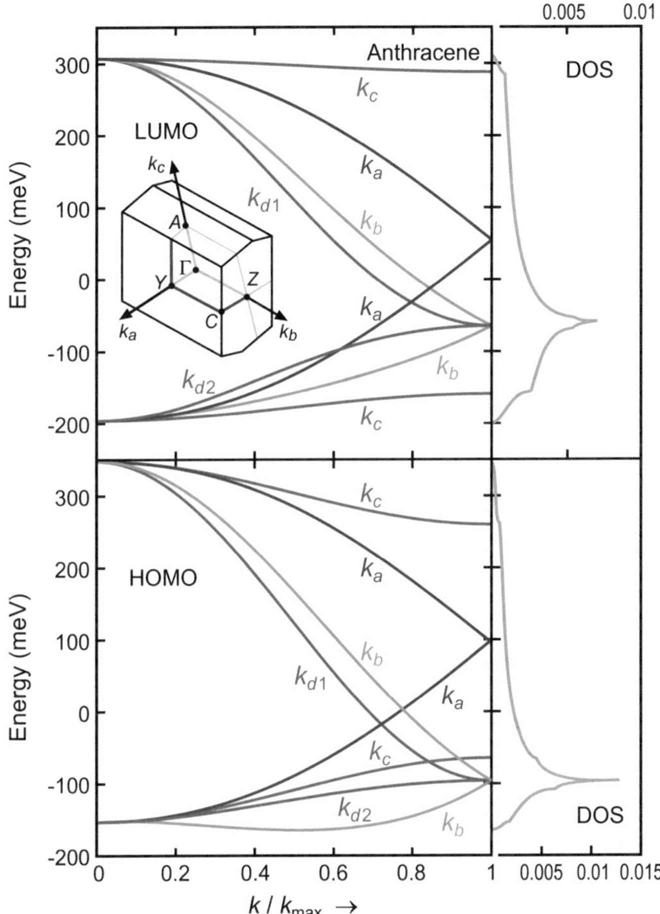

Fig. 5.3 Dispersion $E(k)$ and density of states (DOS) of the lowest unoccupied and the highest occupied molecular-orbital (LUMO and HOMO) bands for an anthracene crystal. k_x to k_z signify crystallographic directions of the monoclinic Brillouin zone depicted in the inset of the upper panel, k_{d1} and k_{d2} refer to nearest-neighbor directions. Reproduced with permission from [7] © 2003 AIP

tetracene and pentacene, where this degeneracy is lifted. Along the next-neighbor directions $\mathbf{d}_1$ and $\mathbf{d}_2$ there is a substantial overlap of orbitals of adjacent molecules; consequently the dispersions along $\mathbf{k}_{d1}$ and $\mathbf{k}_{d2}$ are quite large. The dispersion along k_z is very small, and a large gap occurs between upper and lower band. These features originate in this direction from small interactions of molecules located in adjacent layers and is often found in organic crystals with herringbone packing [7]. The density of states of the LUMO and HOMO bands plotted at the right side of Fig. 5.3 reflects the inverse slope of the contributing dispersion curves.

Rubrene crystals also show band-like conduction; the orthorhombic phase considered in Fig. 5.4 comprises two differently aligned molecule pairs per unit cell in

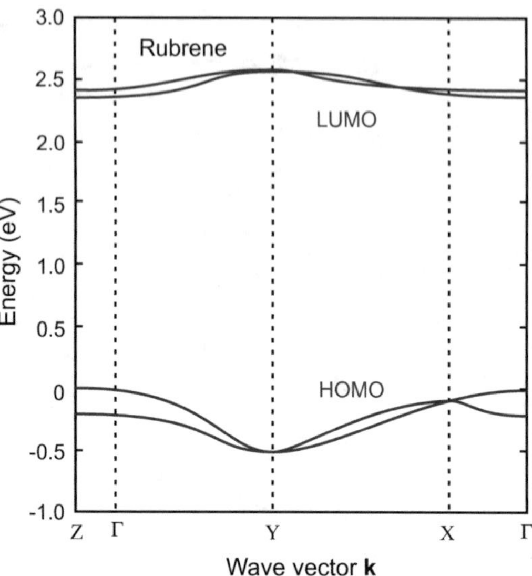

Fig. 5.4 Electronic dispersion of the LUMO and HOMO bands for a rubrene crystal, calculated using density functional theory and the G_0W_0 approximation; reproduced with permission from [9], © 2013 APS

a herringbone packing. The calculated direct bandgap of 2.34 eV agrees reasonably with an experimental optical bandgap of 2.2 eV. The HOMO bands of rubrene show a large dispersion along Γ-Y and only minor dispersion along Γ-Z, in accordance with angle-resolved photoelectron-spectroscopy data [8] and the experimental large anisotropy of carrier transport.

Width of Electronic Bands

The width of the conduction and valence bands in organic semiconductors is generally quite narrow due to the weak intermolecular van der Waals bonds. The intermolecular interaction is usually expressed in terms of a transfer integral outlined in Sect. 5.2.2. In the tight-binding approximation of a one-dimensional molecule chain, the total bandwidth equals four times the transfer integral between neighboring molecules. The bandwidth for any molecular packing can be expressed from the amplitude of the transfer integrals between the various interacting units [10]. Data of the lowest unoccupied and the highest occupied molecular-orbital (LUMO and HOMO) bands for some organic semiconductors are given in Table 5.1. The bandwidth is maximum at $T \to 0$. At increasing temperature the bandwidth is progressively reduced (Fig. 5.5) by an increased electron-phonon coupling and a consequently reduced transfer integral between adjacent molecules [11, 12].

The width of the bands illustrated in Fig. 5.5 is intimately connected with the temperature-dependent effective mass of electrons and holes. The minimum limit at low temperature is obtained from the curvature of the HOMO and LUMO bands resulting from band-structure calculations. Instead of effective masses generally the *mobility* is reported for organic semiconductors (Sect. 5.2.2). This quantity is readily

Table 5.1 Calculated total bandwidth W of excess electron and hole bands for crystalline organic semiconductors

Crystal	Bandwidth W (meV)	
	HOMO (meV)	LUMO
Naphthalene[a]	409	372
Anthracene[a]	509	508
Tetracene[a]	625	502
Pentacene[a]	738	728
Rubrene[b]	520	230
CuPc[c]	96	260

[a]Reference [7]
[b]Reference [9]
[c]Reference [13]

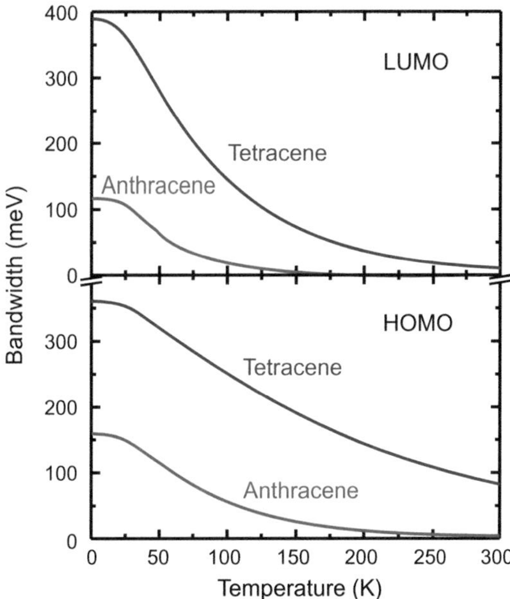

Fig. 5.5 Temperature-dependent narrowing of the LUMO and HOMO bands in anthracene and tetracene crystals, calculated from a semiempirical approach employing Buckingham atom-atom potentials. Reproduced with permission from [12] © 2004 APS

accessible in experiment and usually much larger for holes than for electrons. Furthermore, it is strongly anisotropic due to the low symmetry of molecules and crystal lattices, and hence the oriental dependence of the transfer integrals.

Transport Bandgap and Optical Bandgap

The polarization energy P in organic crystals in (5.1) has an electronic and a vibronic component. If an electron is added to a molecule, the neighboring molecules are polarized by the negative charge in their vicinity. The characteristic response time for this *electronic polarization* is on the order of the oscillation period in an optical transition.

This time is much shorter than that of a vibronic oscillation; the electronic polarization of the molecular neighborhood thus follows the movement of a quasi-free electron, thereby affecting its effective mass. The quasi-free electron in the LUMO and its surrounding polarization cloud combined form a negative polaron; correspondingly a positive charge in the HOMO builds a positive polaron. The polaron character of mobile carriers in organic crystals is usually not explicitly considered; in analogy to the quasiparticles of inorganic semiconductors the carriers are simply termed electrons and holes.

The *vibronic component* of the polarization energy is related to a relaxation of the crystal lattice in the environment of the charged molecule. The characteristic response time is given by the period of a phonon oscillation and therefore much longer than the electronic response. Vibronic relaxation provides a contribution on the order of 5–10% to the polarization energy. After ionization of a molecule in a crystal by a photon the ensuing bandgap energy is hence slightly decreased; the adiabatic bandgap resulting from ionization energy, electron affinity, and electronic polarization is somewhat larger than the *transport gap* $E_{\mathrm{g}}^{\mathrm{transport}}$ including the vibronic contribution.

When an electron is optically excited from the HOMO an onset of absorption is already observed well below the energy of the transport gap. The long-wavelength edge of absorption spectra or short-wavelength edge of luminescence spectra originates from $S_0 - S_1$ transitions of the singlet exciton, see Figs. 5.1 and 5.8. The energy of this edge is called *optical bandgap* $E_{\mathrm{g}}^{\mathrm{opt}}$. The charge-separation energy, $E_{\mathrm{g}}^{\mathrm{transport}} - E_{\mathrm{g}}^{\mathrm{opt}}$, is the *binding energy* E_{X} of the exciton. This quantity is far larger than in inorganic semiconductors [14]. Figure 5.6 shows the transport and optical bandgaps for many organic semiconductors; their difference is given by the separation of the straight line and the dotted line, illustrating an increase of the exciton binding energy for larger bandgap energies. This trend reflects the decrease of the permittivity ε for larger bandgap and the inverse dependence $E_{\mathrm{X}} \propto \varepsilon^{-1}$ [15], which differs from the dependence (4.49) observed for Wannier excitons.

The bandgap of organic semiconductors is generally large (typically well above 2 eV). The thermal activation of carriers from HOMO to LUMO states is therefore negligible: pure organic solids are often insulating. Since reliable and robust doping proved difficult for many organic compounds, charge carriers (both, electrons and holes) are generally injected from the contacts into organic semiconductors used for devices.

The energy of the transport gap $E_{\mathrm{g}}^{\mathrm{transport}}$ can be directly obtained from electron spectroscopy: ultraviolet photoelectron spectroscopy (UPS) measures the HOMO energy, and inverse photoelectron spectroscopy (IPES) determines the LUMO energy. Data for some organic semiconductors are given in Table 5.2. It should be noted that energy values reported for organic crystals show a large scatter due to differences in used samples, experimental conditions, or due to assumptions in calculations.

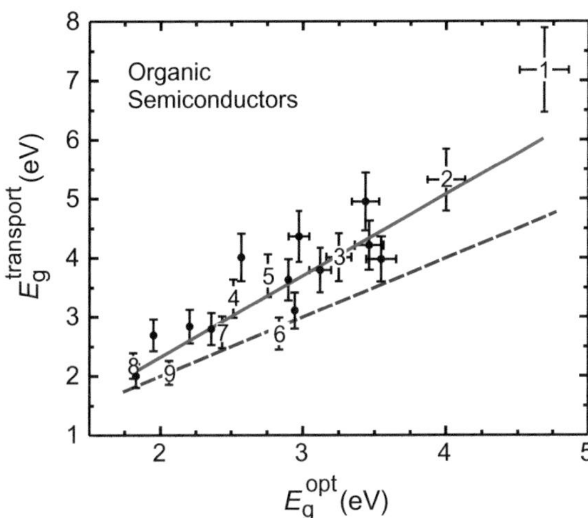

Fig. 5.6 Transport-bandgap energy $E_g^{transport}$ versus optical bandgap energy E_g^{opt} for 24 organic semiconductors. Numbers at data points refer to the compounds listed in Table 5.2. The solid line signifies a linear fit to the data, and the dashed straight line represents $E_g^{transport} = E_g^{opt}$. Reproduced with permission from [16], © 2009 Elsevier

Table 5.2 LUMO and HOMO energies for organic semiconductors measured by photoelectron spectroscopy (IPES and UPS, respectively), transport bandgap $E_g^{transport} = E^{LUMO} - E^{HOMO}$, and measured optical bandgap E_g^{opt}. The numbers in the first column refer to those in Fig. 5.6; data from [16]

No	Compound	E^{LUMO} (eV)	E^{HOMO} (eV)	$E_g^{transport}$ (eV)	E_g^{opt} (eV)
1	Benzene	−0.4	−7.58	7.2	4.68
2	Naphthalene	−1.1	−6.4	5.3	4.00
3	Anthracene	−1.7	−5.70	4.0	3.25
4	Tetracene	−1.8	−5.10	3.3	2.51
5	Alq3	−1.96	−5.65	3.69	2.75
6	Perylene	−2.5	−5.2	2.7	2.83
7	α-Hexathiophene	−2.57	−5.3	2.7	2.43
8	CuPc	2.65	−4.82	2.17	1.80
9	Pentacene	−2.8	−4.85	2.1	2.06

5.1.3 Excitons in Organic Crystals

Excitons are eigenstates of correlated electron-hole pairs with energies below the bandgap energy by an amount given by the binding energy of the exciton. In estimating the binding energy, the band structure of valence and conduction bands must be considered, entering into the effective mass and the dielectric function. An exciton can be modeled by an electron and a hole, circling each other much like the electron and proton in a hydrogen atom, except that they have almost the same mass; hence, in a semi-classical model, their center of rotation lies closer to the middle on their interconnecting axis as illustrated in Fig. 5.7. Depending on the reduced exciton mass and dielectric constant, we distinguish between Wannier excitons, which extend over many lattice constants, and Frenkel excitons, which have a radius comparable to the interatomic distance. Frenkel excitons become localized and resemble an atomic excited state [17].

The large *Wannier excitons* are common in inorganic semiconductors; they are well described by a hydrogen model with the proton and electron masses replaced by the effective masses of hole and electron, and a screened Coulomb potential. The screening of the Coulomb potential is appropriately described by the static dielectric constant $\varepsilon_{\mathrm{stat}}$ or, in case of a stronger lattice interaction including the optical dielectric constant $\varepsilon_{\mathrm{opt}}$ (also named ε_{∞}), by an *effective* dielectric constant $\varepsilon^{*} = (\varepsilon_{\mathrm{opt}}^{-1} + \varepsilon_{\mathrm{stat}}^{-1})^{-1}$, which provides less shielding. Wannier excitons are free to move through the lattice and may be bound to defects or may be confined in quantum structures as described in Sect. 4.3. The binding energy of free excitons is generally small and typically on the order of 10 meV, see Table 4.8.

Frenkel excitons [19] are well-known from ionic crystals with relatively small dielectric constants, large effective masses, and strong carrier-lattice coupling resulting in large structural relaxations. These excitons show relatively large binding energies, usually in excess of 0.5 eV; they cannot be described in a simple hydrogenic model. Frenkel excitons are observed particularly in the organic molecular crystals, where the binding forces within the molecule (covalent) are large compared to the binding forces between the molecules (van der Waals interaction, Sect. 3.1.1). Here localized excited states within a molecule are favored. The binding energy increases for larger bandgap energy from about 0.25 eV for 2 eV transport bandgap to 1.5 eV for $E_{\mathrm{g}}^{\mathrm{transport}} = 6\,\mathrm{eV}$, see Fig. 5.6.

In contrast to most inorganic semiconductors, the excited states in molecular semiconductors are highly localized, with the charge densities typically confined. The localization of electron and hole on the same molecule leads to a large exchange splitting, generating the distinct states of singlets S (spin 0) and triplets T (spin 1); The energy difference between the respective excited states S_1 and T_1 in organic

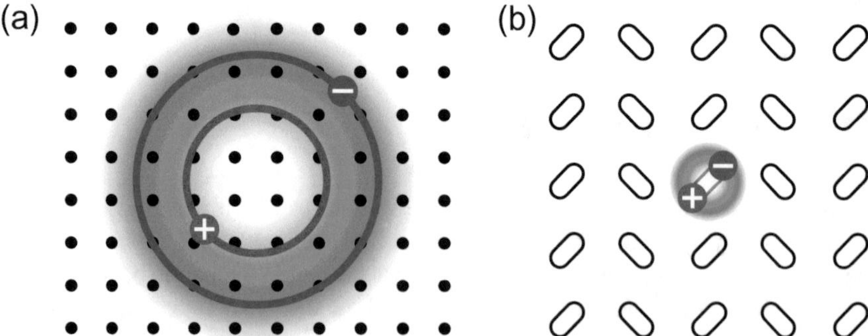

Fig. 5.7 Schematic model of an exciton. **a** Wannier exciton in an inorganic semiconductor with a large radius extending over many lattice constants. **b** Frenkel exciton with a small radius localized at a molecule in an organic crystal. Reproduced with permission from [18], (c) 2018 Springer

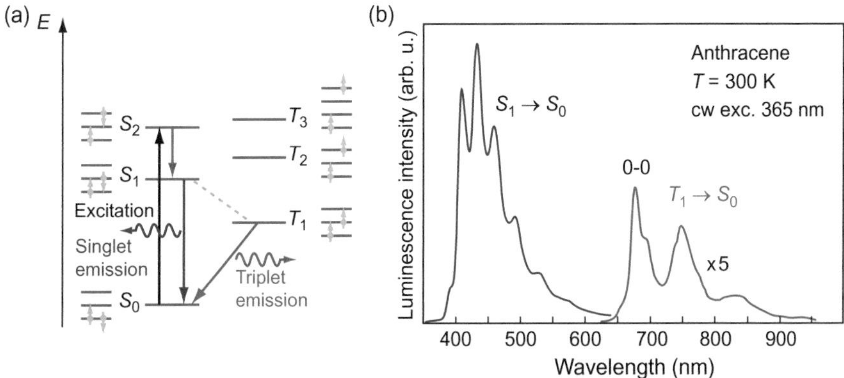

Fig. 5.8 **a** Singlet-triplet exciton schematics with ground state S_0, excited singlet states S_1 and S_2, and triplet states T_1 to T_3; occupied spin states are indicated beside the electronic levels. **b** Optical emission spectrum of singlet-singlet and singlet-triplet excitons in an anthracene crystal, measured at room temperature with excitation at 365 nm. Reproduced with permission from [25], © 2014, Springer Nature

molecules ranges from 50 meV to more than 1 eV [20, 21]. The ground state of neutral molecules is a singlet state—see Fig. 5.8a. In recombination, the singlet-singlet transition is allowed, yielding a strong luminescence called fluorescence in organic semiconductors. The triplet-singlet transition is spin-forbidden; consequently, the triplet state has a long lifetime, depending on possible triplet-singlet mixing, and a much weaker luminescence referred to as phosphorescence. Such singlet and triplet excitons are common in organic semiconductors [22]. The luminescence of singlet excitons is employed in organic LED (OLED) devices used to create displays for, e.g., mobile phones or TV screens, or for solid-state lighting [23, 24].

The luminescence of molecular singlet and triplet Frenkel excitons is shown in Fig. 5.8. Panel (a) gives the level scheme of organic molecules with an even number of π electrons (i.e., no ions or radicals); the photoluminescence spectrum for anthracene is shown in panel (b). The transition near 678 nm labelled 0-0 refers to the zero-phonon line of the $T_1 \rightarrow S_0$ emission; transitions at longer wavelength originate from additional emissions involving molecule and lattice vibrations. The spin-forbidden $T_1 \rightarrow S_0$ transition is much weaker than the allowed $S_1 \rightarrow S_0$ transition with an onset near 380 nm, and its lifetime of 8 ms is correspondingly much longer than the 6 ns of the allowed emission.

5.2 Transport in Organic Semiconductors

Organic solids are easily polarized, as evidenced by the large difference between transport gap and optical bandgap pointed out in Sect. 5.1.2. This property leads to a pronounced polaron character of mobile carriers with a low mobility, and to the large binding energy of excitons discussed above. In the following we consider the polaron behavior in more detail and consequences for the carrier transport in organic semiconductors.

5.2.1 *Polarons*

In the effective-mass picture of a free carrier its interaction with the induced motion of surrounding lattice atoms is usually neglected. This adiabatic approximation applies well for inorganic semiconductors, particularly for those with a high degree of covalent bonding; the coupling of the carrier to the lattice may then be treated as a small correction to the effective mass. The lattice polarization by the carrier becomes more significant in crystals with a strong ionic character and a large bandgap energy, such as in II–VI semiconductors and III-nitrides; it is especially pronounced in alkali halides.

The polarization of the lattice by a free carrier is commonly expressed in terms of an interaction via phonons. The moving carrier is considered a superposition of the moving charge and a phonon cloud, which accounts for the distortion of the surrounding lattice and travels with the carrier. This phonon-dressed carrier is described as a quasiparticle referred to as *polaron*.

Large Polarons

If the radius of the phonon cloud is larger than a lattice constant the carrier is termed a *large polaron*, in contrast to a small polaron with a strong lattice polarization occurring within a unit cell. The coupling via longitudinal optical phonons is found to be dominant and described by the dimensionless *Fröhlich coupling constant*

$$\alpha = \frac{1}{4\pi\varepsilon_0}\frac{e^2}{\hbar}\sqrt{\frac{m^*}{2\hbar\omega_{\mathrm{LO}}}}\left(\frac{1}{\varepsilon_{\mathrm{opt}}} - \frac{1}{\varepsilon_{\mathrm{stat}}}\right), \tag{5.2}$$

with the static and optical dielectric constants $\varepsilon_{\mathrm{stat}}$ and $\varepsilon_{\mathrm{opt}}$, and the energy of an effective phonon $\hbar\omega_{\mathrm{LO}}$; m^* denotes the effective mass of the carrier in the rigid lattice without polaron effect. In semiconductors without ionic character of the bonding (Si and Ge) the static and optical dielectric constants are equal, yielding $\alpha = 0$; in ionic compounds $\varepsilon_{\mathrm{stat}}$ and $\varepsilon_{\mathrm{opt}}$ differ substantially. Semiconductors have typically α values below unity, e.g., GaAs (0.068), GaP (0.201), GaN (0.48) or the II–VI compounds ZnSe (0.43), ZnS (0.63), ZnO (1.19); alkali halides have much stronger coupling: KI (2.50), KBr (3.05), KCl (3.44).

The effective mass of a polaron m_{pol} exceeds the effective mass m^* of a carrier moving through a rigid lattice. In the framework of Fröhlich coupling the mass of a large polaron is approximately given by

$$m_{\text{pol,large}} \cong \frac{m^*}{1 - \frac{\alpha}{6}} \text{ for } \alpha \leq 1. \tag{5.3}$$

With $\alpha \ll 1$ we note an only small deviation from m^*. For stronger coupling expressed by $\alpha \gg 1$ the approximation $m_{\text{pol, large}} \cong \left(\frac{2\alpha}{3\pi}\right)^4 m^*$ is used [26].

Small Polarons

When the ion displacement in the environment of the carrier by its Coulomb interaction is very large, the carrier gets trapped in a potential well created by this distortion. The charge is then typically confined to a volume of one unit cell or less, and the quasiparticle is called a *small polaron*. The strength of the electron-lattice interaction is a distinguishing parameter between the self-trapped small and the mobile large polaron: self-trapping is observed when the coupling constant is larger than 5. Small polarons are described by ab initio tight-binding calculations to account for the movement of all atoms in the environment.

While the large polaron moves much like a quasi-free carrier described by the Boltzmann equation with scattering events, the small polaron moves via hopping between neighboring ions. The effective mass of small polarons is approximately given by [27]

$$m_{\text{pol,small}} \cong \frac{m^* \alpha^2}{48}. \tag{5.4}$$

Examples for materials with small polarons are generally narrow-band semiconductors with large values of α (~10): the band conductivity is usually disturbed by phonon scattering if the bandwidth is narrow. This feature is found in organic crystals (see Sect. 5.1.2), and hopping mechanisms may prevail even if the material has crystalline structure. The bandwidth is maximum at $T \to 0$; at increasing temperature the bandwidth is progressively reduced by an increased electron-phonon coupling (Fig. 5.5), and the polaron mass consequently increases. The carriers get more localized over single molecules (or molecule chains), leaving only transport by a thermally activated hopping mechanism. The transport thus gradually changes from coherent band-like motion at low temperatures to phonon-assisted hopping transport at high temperature. This transition occurs in oligoacenes at about room temperature [7]. Small polarons occur also in inorganic solids, particularly in oxides such as NiO or $BaTiO_3$.

Following the generally accepted usage the mass of mobile carriers is in the following referred to as effective mass (denoted m^*), even if the polaron effect leads to self-trapping and hopping conductance with a low mobility.

5.2.2 Mobility of Carriers in Organic Semiconductors

Band Conductance in Organic Crystals

The mobility of mobile carriers is defined by

$$\mu = \frac{q}{m^*}\tau, \tag{5.5}$$

with the effective mass m^* and charge q of the carrier, and a mean time τ between scattering events. In organic crystals the mobility at room temperature is usually below 1 cm^2/(Vs), often orders of magnitude smaller, compared to values of 10^3 cm^2/(Vs) for inorganic semiconductors. A major reason is the poor structural perfection of organic crystals, which creates trap levels and scattering centers. Other reasons are a pronounced polaron character of carriers, and particularly the weak intermolecular contact in an organic crystal. The filled HOMO and the empty LUMO levels of each molecule are separated from those of neighboring molecules by a potential barrier. The barrier originates from the weak intermolecular van der Waals interaction, leading to a predominant localization of the HOMO and LUMO wave functions in each molecule. If the intermolecular barrier is low, bands similar to those in inorganic semiconductors are created; the stronger coupling is related to a larger overlap of wave functions of adjacent molecules. Higher barriers may still allow for conductivity by phonon-assisted hopping considered further below.

The intermolecular barriers are expressed in terms of intermolecular *transfer integrals*. These quantities describe the electronic coupling between molecular orbitals of adjacent molecules (HOMOs and LUMOs) and depend sensitively on the spacing and relative orientations of the molecules. For acenes typical values are in the range of some tens of meV [7]. In the regime of band conduction the bands are formed by linear combinations of the orbitals of each molecule, and the transfer integral V affects the effective mass of a mobile carrier moving in a given direction by

$$m^* = \frac{\hbar^2}{2|V|d^2}, \tag{5.6}$$

where d is the distance of adjacent molecules in the considered direction. In the regime of hopping conduction the transfer integral enters the charge-transfer rate k for hopping transitions between adjacent molecules [28],

$$k = \frac{|V|^2}{\hbar}\sqrt{\frac{\pi}{\lambda k T}}\exp\left(-\frac{\lambda}{4kT}\right). \tag{5.7}$$

In this *Marcus expression* λ is the is the reorganization energy describing the vibrational relaxation in the vicinity of the mobile charge carrier; this energy is proportional to the electron-phonon coupling strength γ and can be described by $\lambda = 2\gamma^2\hbar\omega$, with the characteristic phonon mode ω [29].

Band conductance is found particularly in small-molecule crystals. *Single-component semiconductors* such as the acenes are usually good insulators, but may become photoconductive with sufficient optical excitation and also conduct carriers injected from contacts. They show comparatively high mobilities for electrons and holes, typically in the 10^{-2} to 10 $cm^2/(Vs)$ range at 300 K; the mobility *falls* with increasing temperature as discussed below. *Two-component* (charge-transfer) *semiconductors* include highly conductive compounds and may even show semiconductor-metal transitions. The large conductivity originates from an incomplete electron exchange between donors and acceptors and consequently partially filled bands. The resistivity of these semiconductors lies between 10^2 and 10^6 Ω cm at room temperature with a transfer-energy gap in the 0.1–0.4 eV range [30]. The conductivity is usually highly anisotropic, with the electron transfer-integral in the stacking direction typically a factor of 10 larger than in the direction perpendicular to the stacks [31]. The charge-transfer crystals provide an opportunity for fine tuning of the semiconductive properties by replacing TTF-type donors and TCNQ-type acceptors (Fig. 3.5) with other similar molecules [32].

Organic molecules show a strong structural relaxation when a charge is introduced. The band structure calculated for a crystal composed of weakly bonded neutral molecules will hence usually not be preserved in the presence of carriers. As a rule of thumb, band conductance occurs despite lattice relaxation if the transfer integral between molecules is sufficiently large: a large transfer integral delocalizes the carrier wave-function over several molecules. A more quantitative estimate follows from the width W of the energy band for carrier transport. If the mean dwell time of a moving carrier on a molecule is substantially shorter than the characteristic time of a molecule vibration, no significant relaxation and consequential trapping is expected to occur and band conductance may prevail. Taking a characteristic time τ on the order of 10^{-15} s yields a band width

$$W \gg \frac{\hbar}{\tau}, \tag{5.8}$$

i.e., values exceeding several hundred meV; this condition is reasonably well fulfilled for crystals of acenes and comparable aromatic compounds (Table 5.1).

Another approach considers the scattering of a moving carrier. If the mean scattering time τ_{scatter} is in the range or smaller than $\hbar/W$, no wave vector $\mathbf{k}$ can be assigned to the carrier. The description in terms of conduction in a band with dispersion $E(\mathbf{k})$ hence requires $\tau \gg \hbar/W$. The bands in organic crystals are rather narrow due to a small amount of wave-function overlap of π electrons (Sect. 5.1.2). With typical bandwidths W in the range of some hundred meV we obtain $\tau_{\text{scatter}} > 10^{-15}$ s. Both molecular relaxation and narrow band widths impose limits for band-like conduction in organic semiconductors.

A clear indication for band conduction is provided by the temperature dependence of the mobility. In inorganic semiconductors scattering at acoustic phonons leads to a $T^{-3/2}$ dependence of the carrier mobility. A comparable decrease of mobility at higher temperature is observed for carriers in organic crystals with band conduction.

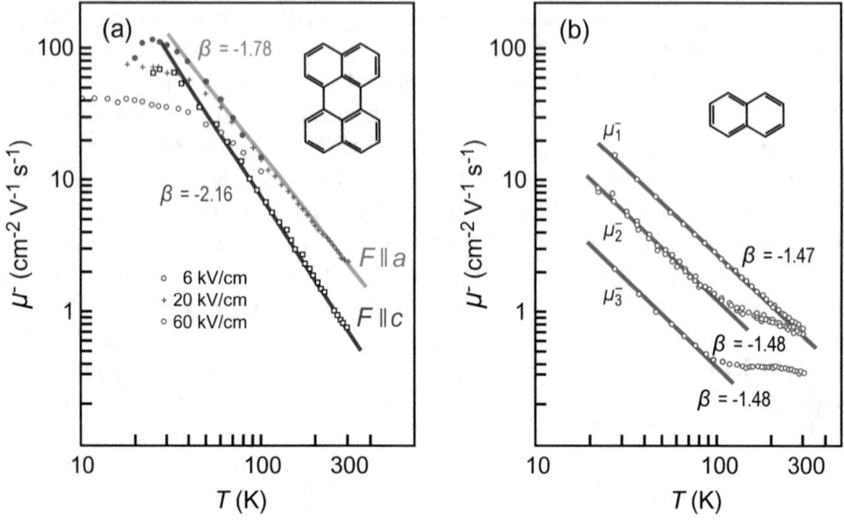

Fig. 5.9 Temperature dependence of the electron mobility **a** in a perylene crystal, and **b** in a naphthalene crystal. Reproduced with permission from [33, 34], © 1985 and 2001 Springer

The mobility data in Fig. 5.9 show typical features of band-like transport in organic crystals. The temperature dependence at low electric field is described by a power law

$$\mu(T) = \mu_0 \times (T/T_0)^\beta, \tag{5.9}$$

where μ_0 is the mobility at a reference temperature T_0 usually set to 300 K. The exponent β deviates somewhat from the ideal value of $-3/2$, see Table 5.3. The

Table 5.3 Mobility μ in units of cm^2 V^{-1} s^{-1} for organic crystals at 300 K and exponent β of the temperature dependence according to (5.9)

Crystal	Direction	Electrons		Holes	
		μ	β	μ	β
Naphthalene	*a*	0.62	−1.4	0.94	−2.8
	b	0.64	−0.55	1.84	−2.5
	c′	0.44	+0.04	0.32	−2.8
Anthracene	*a*	1.73	−1.45	1.13	−1.46
	b	1.05	−0.84	1.84	−1.26
	c′	0.39	+0.16	0.32	−1.43
Perylene	*a*	2.37	−1.78		
	b	5.53	−1.72		
	c′	0.78	−2.15		

low-field mobilities do not depend on the value of the electric field and show a pronounced anisotropy; the principal axes of the mobility tensor deviate slightly from the crystallograpic crystal axes. Typical mobilities are on the order of 10^{-2} to 10 $cm^2/(Vs)$. Data for a large number of organic semiconductors are tabulated in [35]; they decrease at higher temperature according to (5.9) with $0 > \beta > -3$.

Hopping Conductance in Organic Semiconductors

Conductance in organic semiconductors is strongly affected by disorder. Even in perfect small-molecule crystals the weak intermolecular van der Waals bonds give rise to a phonon-related *dynamical* disorder which affects the mobility of carriers. In addition, many organic compounds cannot be grown as single crystals and are prepared as paracrystalline thin films by evaporating or spin coating; for a specification of paracrystallinity see (3.7). In these semiconductors the *static* disorder dominates at most temperatures; their carrier mobility is by orders of magnitudes lower than that of crystalline compounds and *increases* for increasing temperature: the poor coupling between the molecules in the solid leads to a strong localization of the carriers on single molecules; transport occurs via a sequence of charge-transfer steps from one molecule to another, similar to the hopping between defect states in inorganic semiconductors. The larger mobility at higher temperature is due to the required thermal activation from the localized states. The temperature dependence of such a hopping conductance in organic semiconductors is hence inverse to the band conductance discussed above. Reviews on mobility and other physical data for organic semiconductors are given in [36, 37].

The transport properties are described by the formalism of hopping conductance developed for amorphous inorganic semiconductors: the charge carriers are assumed to hop in a time-independent disordered energy landscape. In heavily disordered or amorphous inorganic semiconductors defects or deviations from ideal topology lead to a continuous tailing of energy states into the bandgap. Deeper states occur less frequently because the centers which produce such states are less probable. When deep enough, each type of defect center will produce a localized level. Shallower centers are, however, close enough: their levels broaden into bands, and the resulting states are no longer localized. Overlapping levels and narrow bands all merge into the tailing states. Thus, when going from deep tail states to the band states a transition from localized to delocalized states is expected at a critical energy, referred to as the *mobility edge*. The smooth transition in the density of states of a highly disordered inorganic semiconductor is illustrated in Fig. 5.10a, with a mobility edge indicated for n-type conductivity. Carriers which are thermally activated above the mobility edge contribute to charge transport, while carriers at lower energy are localized in defect states. In Fig. 5.10b the density of states is illustrated for an organic semiconductor with band-like conductivity, indicating the mobility edge for electrons at the low-energy band edge of the narrow conduction band.

The basic difference between amorphous inorganic semiconductors and disordered organic semiconductors is the shape of the density of states (DOS) $g(E)$. In a disordered inorganic semiconductor the DOS is found to have a mobility edge and a tail of localized states with an *exponentially* decreasing distribution extending into the band gap. In contrast, the DOS in organic solids has a *Gaussian* shape:

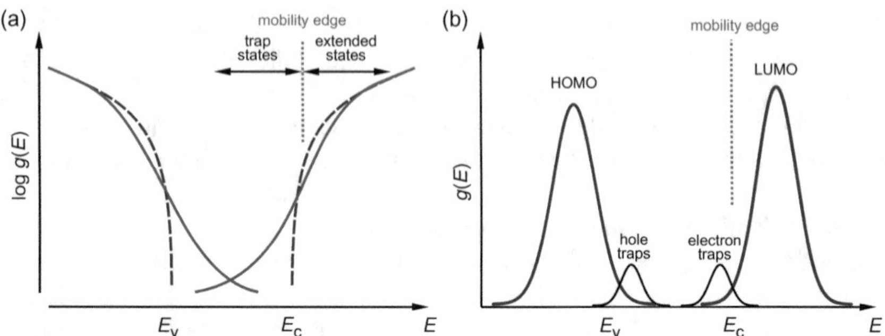

Fig. 5.10 Tailing of states into the bandgap of **a** a heavily disordered inorganic semiconductor (*red solid lines*) compared to the density of states $g(E)$ in the crystalline state (dashed), and **b** an organic semiconductor. Extended and localized states are separated by a mobility edge, indicated for electron conductance by the *dotted line*

$$g(E) = \frac{G_{\mathrm{tot}}}{\sqrt{2\pi}\sigma} \exp\left[\frac{(E - E_{\mathrm{center}})^2}{2\sigma^2}\right], \tag{5.10}$$

where G_{tot} is the total DOS, E_{center} is the center of the energy distribution, and σ is the variance of the distribution [38]. Hopping of carriers is determined by both, the energy difference ΔE and the spatial separation d of initial and final states; in addition, hopping is affected by an electric field F. The Gaussian DOS results in a thermally activated mobility [38]

$$\mu = \mu_0 \exp\left[-\left(\frac{2\sigma}{3kT}\right)^2 + C\left(\left(\frac{\sigma}{kT}\right)^2 - \Sigma^2\right)\sqrt{F}\right], \tag{5.11}$$

determined by the spread σ of the energy distribution in the conducting band, the structural disorder parameter Σ, the applied electric field F, and the parameter μ_0 representing the mobility of the hypothetic not-disordered semiconductor at high temperature. For a more detailed review of various transport models see [39], a review on experimental techniques for measuring transport properties is given in [40].

The mobility is widely measured using time-of-flight experiments. The experimental findings depend on the chemical purity, on the temperature range, and on the electric field; in many cases they can be described by the empirical dependence

$$\mu \cong \mu_0 \exp\left[-\frac{E_{\mathrm{A}}}{kT} + b\sqrt{F}\right], \tag{5.12}$$

with the zero-field mobility μ_0 and an activation energy E_A; the activation energy corresponds to the difference between the energy of the localized state and the mobility edge. The quantity b is the field-activation factor

$$b = B\left(\frac{1}{kT} - \frac{1}{kT_0}\right) \tag{5.13}$$

with empirical parameters B and T_0. Since polymers are stable only in a very limited temperature range, the temperature dependence can often be described in this interval by different relations, such as either (5.11) or (5.12).

The hole mobility of the poly(paraphenylene vinylene) (PPV) is shown in Fig. 5.11. Holes dominate the current in many polymers. The polymer PPV is a prominent compound due to its electroluminescence properties; PPV derivatives are soluble and can be spin coated for, e.g., fabrication of organic LEDs. The mobility is reasonably described by a $\log(\mu) \propto T^{-2}$ dependence, although also an Arrhenius dependence $\log(\mu) \propto T^{-1}$ dependence fits well [41]; in the limited temperature range a clear distinction is not possible. The activation energies E_A for PPV derivatives (5.12) range between 0.3 and 0.5 eV, and widths σ of the Gaussian DOS according to (5.11) are near 100 meV, with mean separations d of localization sites in the range 1.1–1.7 nm [42]. We note in Fig. 5.11 the comparatively low mobility of disordered organic semiconductors and the characteristic increase at higher temperature.

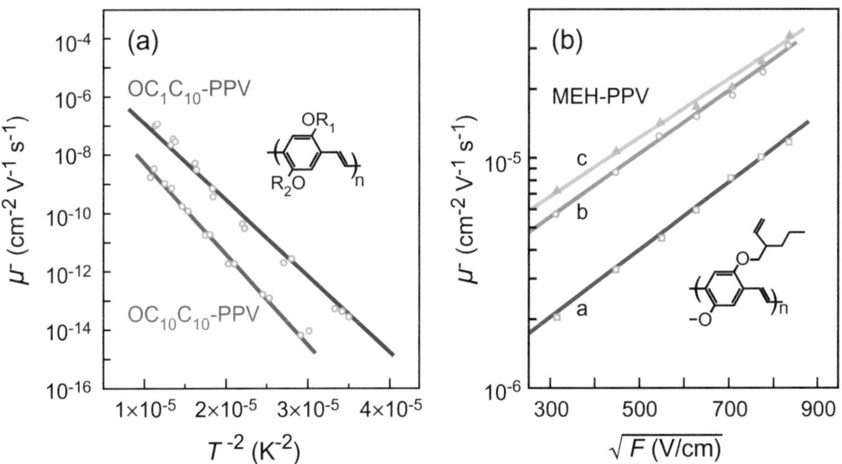

Fig. 5.11 Hole mobility in derivatives of PPV. **a** Mobility in derivatives with different side groups at zero electric field; OC_1C_{10}-PPV: $R_1 = CH_3$, $R_2 = C_{10}H_{21}$; $OC_{10}C_{10}$-PPV: $R_1 = R_2 = C_{10}H_{21}$; reproduced with permission from [42], © 2015 ACS. **b** Dependence of the hole mobility on the electric field F for MEH-PPV ($R_1 = C_8H_{16}$, $R_2 = CH_3$) prepared without (*curve a*) and with application of an electric field of 3 and 6 kV/cm for curves *b* and *c*; reproduced with permission from [43], © 2006 AIP

The enhancement of the mobility in an electric field F is shown in Fig. 5.11b, demonstrating the dependence on $F^{1/2}$. This behavior has also been observed from time-of-flight measurements in many molecularly doped polymers and amorphous glasses. An electrically induced polarization of MEH-PPV during the preparation of films significantly enhances the mobility.

The *density of states* in organic crystals is governed by disorder as noted above. Intrinsic defects (Sect. 3.2) and phonons contribute respectively to static and dynamical structural disorder, and impurities give rise to traps in the bandgap with a broad distribution of states. The density of trap states was experimentally derived from various measurements; results from the evaluation of the space-charge limited current in transistor structures are shown in Fig. 5.12. We observe a strong tailing from the valence-band edge (mobility edge at $E = 0$) into the bandgap. Particularly low trap densities have pentacene and rubrene single crystals (*sc bulk*) and field-effect transistors made with such crystals (*sc FET*); the increase of the trap density in FET structures is attributed to defects at the interface to the gate dielectric. The comparably low trap density corresponds to the large carrier mobility found in such semiconductors. The defect density in organic small-molecule transistors is comparable to that in hydrogenated amorphous silicon.

Polycrystalline thin films have significantly higher trap densities due to the high density of defects. Distortions in the (para-) crystalline regions of a polymer consist,

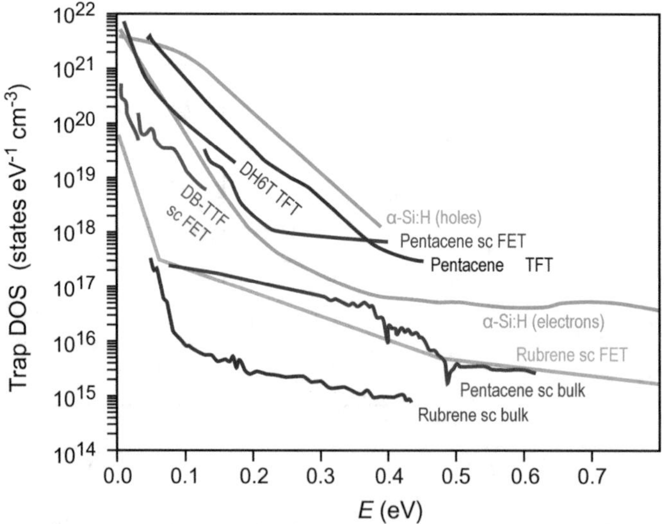

Fig. 5.12 Typical trap density in small-molecule semiconductors (*black, red, and blue curves*) and hydrogenated amorphous silicon (*green curves*). *Blue curves* refer to bulk single-crystals (*sc*), *red curves* to field-effect transistors (*FET*) made with single crystals; the *black curve* gives a result of a polycrystalline thin-film transistor (*TFT*). Energy values are relative to the valence-band edge E_v (for electrons in α-Si:H relative to the conduction-band edge E_c). Reproduced with permission from [44], © 2010 APS

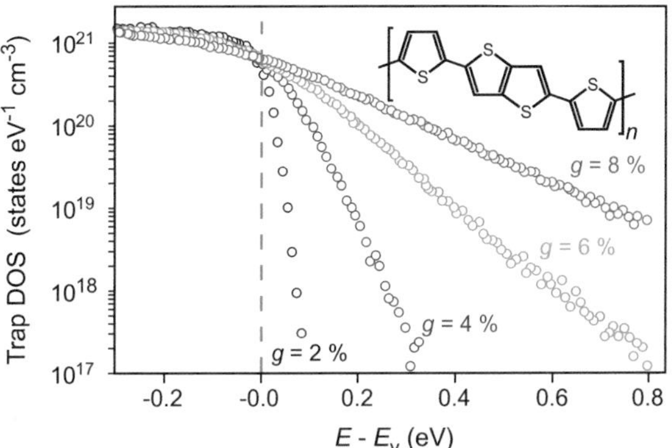

Fig. 5.13 Calculated density of states (DOS) of fused-ring polythiophene for different degrees of paracrystalline disorder expressed by the paracrystallinity parameter g. The valence-band maximum E_V refers to a perfectly crystalline region. Reproduced with permission from [45], © 2011 APS

e.g., of random variations in the spacings between the backbone lamella illustrated in Fig. 3.6. Such variations create band tailing into the bandgap comparable to the distortions shown in Fig. 3.8b for molecule shifts in a pentacene crystal. The magnitude of such variations was measured for aligned ribbons of fused-ring polythiophene delineated in the inset of Fig. 5.13. By analyzing data of grazing incidence X-ray diffraction, the effect on the density of states shown in Fig. 5.13 was calculated using density-functional theory [45]. In the calculations the distance between the chains (ideally a) was varied according to a Gaussian distribution with a standard deviation σ, yielding a paracrystallinity $g = \sigma / a$, see (3.5). The model predicts an exponential dependence of the distribution of trap-tailing states within the bandgap, similar to results of disordered inorganic solids depicted in Fig. 5.10. The characteristic energy E_0 in the DOS function $g(E) = g_0 \exp(-E/E_0)$ has values near 10 meV for well-ordered and 100 meV for strongly disordered polymers.

5.3 Interfaces in Organic Semiconductors

Any semiconductor device comprises heterojunctions between different semiconductor layers and to metal contacts. When two solids are brought into contact, their energy bands at the interface must align with respect to a common reference. This problem is discussed for interfaces between inorganic semiconductors in Sect. 4.2, and for a contact between an inorganic semiconductor and a metal in Sect. 10.3. The alignment leads in both cases to a flow of carriers across the interface, creating a space-charge region and a corresponding band bending. The basic principles of band alignment outlined for inorganic solids in Sect. 4.2 also apply for organic semiconductors. However, the molecular structure and disorder introduce additional conditions, which complicate the electronic structure of such interfaces. Even though a consistent model to describe organic interfaces is missing, some general rules exist and procedures to control the interface can be identified; for reviews see [46–48].

5.3.1 Organic-Organic Heterointerface

Vacuum-Level Alignment

Many organic semiconductors have no intramolecular total dipol moment and have intrinsically low concentrations of free carriers; moreover, the requirement of lattice matching found in inorganic heterointerfaces (Sect. 2.2.5) is largely relaxed for molecular crystals with weak van der Waals bonds (Sect. 3.2.1). A contact between two such solids often does not induce a significant redistribution of charges, i.e., no significant interface dipole is created. In the majority of these cases the classical electron-affinity rule introduced in Sect. 4.2.1 provides a good description of the band alignment at organic-organic semiconductor heterojunctions [49, 50]: the vacuum level E_{vac} of the individual semiconductors represents the common reference level across the interface, and the bands are flat without bending.

The electronic properties of organic-organic interfaces are widely measured using photoemission spectroscopy (PS): ultraviolet PS (UPS) yields the ionization energy I for the removal of an electron from the HOMO, and inverse PS (IPES) provides the electron affinity A required to add an electron to the LUMO, see Fig. 5.1. We discussed in Sect. 5.1.2 that organic molecules experience a strong relaxation upon charging; the quantities I and A do hence not yield the HOMO and LUMO levels of the neutral molecule, but the respective related energies of the negative and positive polarons.

There exist conditions where the simple rule of vacuum-level alignment fails. If the organic heterojunction is formed between a strong electron donor and a strong electron acceptor, electrons are transferred from the n-type to the p-type semiconductor and a space-charge region with accompanying band bending is created. Dipols at the interface or on individual molecules introduce potentials which affect the vacuum level across the interface; in addition, pinning of energy levels may occur. Such conditions are discussed below.

Molecular Charge Distribution

In a neutral molecule of an organic semiconductors the negative charge of the delocalized π electrons is balanced by the positive charge of the atomic nuclei and their electrons in core and σ orbitals. If we consider the planar pentacene molecule (Fig. 3.10a) the positive charge is located on the backbone built by the nuclei and their electrons, and the negative charges are symmetrically located above and below this plane (comparable to Fig. 3.1). This separation of charges creates two electric dipols, each pointing from the negative to the positive charge distribution. Since both dipols have the same magnitude but opposite orientation their dipol moments cancel, but a quadrupole moment remains. An organic crystal built from these molecules has an electrostatic surface potential, which depends on the orientation of the topmost molecules.

Effect of Molecular Orientation

In the *lying-down configuration* with the molecular plane parallel to the surface (Fig. 3.7) the potential energy for an electron is increased with respect to the vacuum

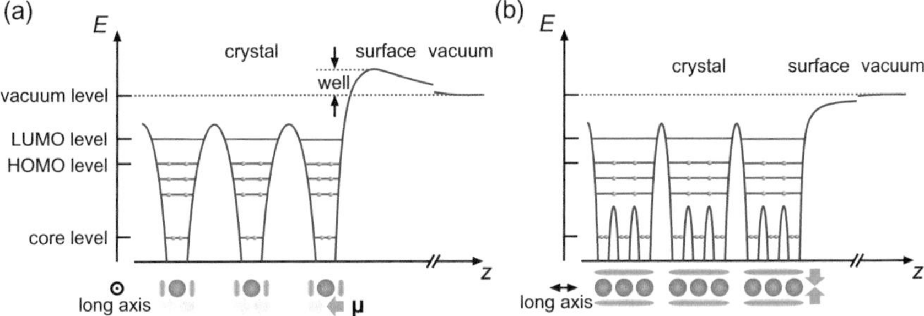

Fig. 5.14 Schematic of the potential energy and levels for electrons in an organic crystal made of rod-like molecules near the surface **a** for a laying-down configuration of the molecules parallel to the surface and **b** for a standing-up configuration. *Red dots* indicate atom cores, adjacent blue clouds signify π electrons, the *green arrow* labeled μ is the surface dipol in the laying-down configuration

level further away; the increase results from the terminating dipol layer, which leads to a potential well as illustrated in Fig. 5.14a. The values of the ionization energy I and electron affinity A rise by this potential well. In the *standing-up configuration* depicted in Fig. 3.8a no significant projection of the dipol with respect to the surface normal exists, and the values for I and A are smaller.

The orientation of the molecular dipol with respect to the interface plane of a heterojunction affects the electrostatic potential at the interface and consequently the alignment of the bands. This was clearly verified by UPS spectra obtained from perfluoropentacene (PFP) with different inclinations on surfaces. In PFP all 14 hydrogen atoms of pentacene are replaced by strongly electronegative fluorine atoms, which produce intramolecular dipoles in addition to the quadrupol introduced above; the dipol orientations at the ends of the PFP backbone points from the negative ends (comprising two F atoms) toward the molecule center. For PFP the *standing-up* configuration is expected to create a much stronger termination dipol than the laying-down configuration. PFP was prepared in a laying-down configuration of a single monolayer on a clean Au(111) surface, with tilted molecular long axes in a multilayer herringbone configuration, and in a standing-up configuration on SiO_2 [51]. UPS spectra yield respective ionization energies of 5.80, 6.00, and 6.65 eV, demonstrating the effect of the increasing projection of the terminating dipol on the surface normal. The example also shows that the ionization energy of an organic crystal depends on orientation and is no unique value as for an isolated molecule; this must be considered when using data for a tentative assignment based on vacuum-level alignment.

Top layers at organic-organic interfaces comprising planar molecules typically adopt the orientation of the underlying layer. A review on various heterointerfaces and the orientation dependence of their molecular arrangement is given in [52].

5.3.2 Organic-Metal Interface

The interaction between a clean metal surface and π electrons of organic semiconductors is strong (Sect. 3.2.1), resulting generally in a laying-down configuration of molecules for a single monolayer. If the molecules have a dipol moment p normal to the surface, the work function $e\phi_m$ of the metal is changed by an amount

$$e\Delta\phi = \frac{e}{\varepsilon_r \varepsilon_0} Np \tag{5.14}$$

where e, ε_r, and ε_0 are respectively the elementary charge and the permittivities of the solid and vacuum. The areal density N of the dipols gets maximum for a complete monolayer. According to (5.14) the resulting work function $e(\phi_m + \Delta\phi)$ of a metal can linearly be varied by the lateral density of a monolayer of dipolar molecules. The surface dipol can also be created by applying molecules which act as electron donors or acceptors. A donor molecule chemisorbed on a metal surface leads to a re-distributed electron concentration with a negative charge at the outermost surface, and consequently a dipol oriented towards the surface (acceptor vice versa). The introduction of suitable interlayers allows to control the effective work function of electrodes used as contacts in organic devices [53].

It must be noted that also unintentional surface contaminants like water, oxygen, or hydrocarbons affect the effective work function of metals. This easily leads to discrepancies between energies usually measured in ultrahigh vacuum (10^{-9} mbar) and those obtained with high-vacuum conditions (10^{-6} mbar) applied for device fabrication; in high vacuum the metal surface is inevitably covered by a monolayer thick chemically inert layer. The organic-metal interaction to an organic semiconductor deposited on top of such a metal is substantially weaker.

As a consequence, the energy levels at the contact generally align according to the dependence illustrated in Fig. 5.15. Barriers experienced by carriers at the contact between metal and organic semiconductor are depicted in panel (a); here an ideal

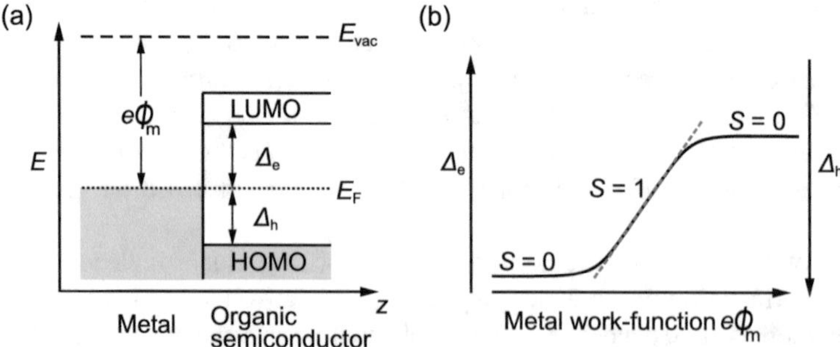

Fig. 5.15 a Work function $e\phi_m$, electron injection-barrier Δ_e, and hole injection-barrier Δ_h at an organic-metal contact without interface dipol. **b** Slope parameter S of the organic-metal contact for varied work function of the metal

junction according to the Schottky-Mott model (Sect. 10.3.1) without interface dipol is assumed. Electrons injected from the metal to the LUMO have to surmount the electron barrier Δ_e to reach unoccupied states, and holes injected from the metal to the HOMO experience the barrier Δ_h. When the work function of the metal $e\phi_m$ increases, Δ_e increases and Δ_h decreases. The slope parameter

$$S = \frac{\Delta_{e,h}}{e\Delta\phi_m} \tag{5.15}$$

for the ideal contact is unity. Figure 5.15b shows that such a linear dependence is observed for actual metal electrodes on organic semiconductors only in a limited range. High metal work-functions above this range are desirable to avoid the hole barrier Δ_h at p-type contacts, and low metal work-functions below this range can avoid the electron barrier Δ_e at n-type contacts. However, there is no change of the injection barriers observed when the metal-workfunction is varied; these ranges with strongly reduced S are referred to Fermi-level pinning.

Fermi-Level Pinning

Fermi-level pinning is well-known from inorganic semiconductors (Sect. 10.3.2). Gap-state models applied to explain the phenomenon are also used to describe organic-metal contacts, taking characteristic features of organic crystals into account [54–56]. The tailing of the density of states outlined in Sect. 5.2.2 provides localized states in the gap between HOMO and LUMO, which can accommodate carriers transferred across the interface when the contact is made.

If a metal with a low work function is applied to fabricate a low-barrier n-type contact, its Fermi energy will lie in a region where before contact empty states of the organic semiconductor with a lower energy exist; below the LUMO these are empty tail states. When the contact is made and tailing is pronounced, electrons will move from the metal across the barrier into localized states to reestablish equilibrium (Fig. 5.16). They thereby create an interface dipol, which raises the metal work-function according to (5.14). Similarly a metal with high work function applied to produce a p-type contact leads to a flow of holes across the interface to drain occupied gap states extending from the HOMO, thereby decreasing the work function. Tailing will hence generally reduce the slope parameter S; a nearly ideal dependence of injection barriers on the metal work-function can only occur in an energy range of negligible density of localized states in the bandgap. Since the degree of disorder and purity of organic semiconductors vary, usually no unique pinning energy is found.

5.4 Problems Chapter 5

5.1 For a neutral anthracene molecule in the gas phase respective ionization and electron affinity energies of 7.42 and 0.58 eV are measured; in an anthracene crystal a ionization energy of 5.77 eV is observed.

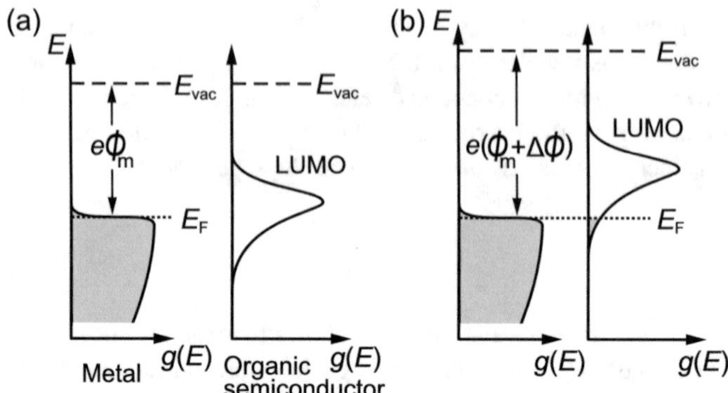

Fig. 5.16 Density of states $g(E)$ and respective occupation (*gray shading*) at an interface between metal and an organic semiconductor **a** before and **b** after contact

(a) Which electron affinity in the crystal results from these data, if the polarization energy upon adding an electron to a neutral molecule of the crystal equals that for removal of an electron?
(b) Which binding energy of the singlet exciton is inferred from these data, if the onset of the respective absorption is observed at 382 nm?

5.2 For the mobility of negative carriers in the band-like conduction of a tetracene crystal values of 0.85 and 25.5 cm^2/(Vs) are found at 300 K and 77 K, respectively. Which mobility results from this data at 0 °C (32 F)?

5.3 Ionization and electron-affinity energies of 5.8 and 3.1 eV are found in photoelectron spectroscopy for Alq_3 (Fig. 3.3g). We consider organic-organic interfaces between Alq_3 and BCP (4,4′-*N,N*′-dicarbazolyl-biphenyl), PTCDA (Fig. 3.3f), and α-NPD (*N,N*′-diphenyl-*N,N*′-bis(l-naphthyl)-l,l′ biphenyl-4,4″diamine).

(a) For a layer of BCP deposited on top of Alq_3 a ionization energy of 6.4 eV is measured and an energy difference of 3.1 eV between HOMO and LUMO levels is determined. How do the HOMO and LUMO levels align at this heterojunction if vacuum-level alignment applies?
(b) For the organic-organic interfaces PTCDA/Alq_3 and α-NPD/Alq_3 the vacuum-level of Alq_3 is lower than those of the organic layers in contact. How large is the step of the vacuum level at the PTCDA/Alq_3 interface (going from PTCDA toAlq_3) for a measured PTCDA ionization energy of 6.7 eV, if the LUMO level of PTCDA lies 2.2 eV above the HOMO level and 0.9 eV below the LUMO level of Alq_3? How do the HOMO levels align?
(c) Using the result of (b): How large is the expected step of the vacuum level for a PTCDA/α-NPD interface, if a dipol of −0.3 eV is determined for the α-NPD/Alq_3 interface (going from α-NPD toAlq_3) and a transitivity rule applies?

5.4 A layer of α-NPD is deposited either on Au with 5.2 eV work function or on Mg with 3.8 eV work function, and the position of the Fermi level in α-NPD is measured using photoelectron spectroscopy; a ionization energy of 5.4 eV is found for the organic layer.

(a) Which Fermi energies (with respect to the HOMO level) would be obtained in α-NPD for the two cases if vacuum-level alignment would apply?
(b) Measured values deviate from those obtained assuming vacuum-level alignment due to the formation of interface dipols, yielding 1.4 eV for Au and 2.0 eV for Mg (above the HOMO level). How large are the dipols at the interfaces to Au and to Mg? Which slope parameter results from these data?

5.5 General Reading Chapter 5

A. Köhler, H. Bässler, *Electronic Processes in Organic Semiconductors: An Introduction* (Wiley-VCH Verlag, Weinheim, 2015)
M. Schwoerer, H.C. Wolf, *Organic Molecular Solids* (Wiley-VCH Verlag, Weinheim, 2007)
W.R. Salaneck, K. Seki, A. Kahn, J.-J. Pireaux (eds.), *Conjugated polymer and molecular interfaces* (Marcel Dekker Inc., New York, 2002)

References

1. L.S. Hung, C.H. Chen, Recent progress of molecular organic electroluminescent materials and devices. Mater. Sci. Eng. **39**, 143 (2002)
2. M.C. Gather, A. Köhnen, K. Meerholz, White organic light-emitting diodes. Adv. Mater. **23**, 233 (2011)
3. A.C. Arias, J.D. MacKenzie, I. McCulloch, J. Rivnay, A. Salleo, Materials and applications for large area electronics: solution-based approaches. Chem. Rev. **110**, 3 (2010)
4. P. Peumans, A. Yakimov, S.R. Forrest, Small molecular weight organic thin-film photodetectors and solar cells. J. Appl. Phys. **93**, 3693 (2003)
5. A.W. Hains, Z. Liang, M.A. Woodhouse, B.A. Gregg, Molecular semiconductors in organic photovoltaic cells. Chem. Rev. **110**, 6689 (2010)
6. N. Karl, Organic Semiconductors, in *Festkörperprobleme/Advances in Solid State Physics*, vol. 14, ed. by H.J. Queisser (Vieweg, Braunschweig, 1974), pp. 261–290
7. Y.C. Cheng, R.J. Silbey, D.A. da Silva Filho, J.P. Calbert, J. Cornil, J.L. Brédas, Three-dimensional band structure and bandlike mobility in oligoacene single crystals: a theoretical investigation. J. Chem. Phys. **118**, 3764 (2003)
8. S.-I. Machida, Y. Nakayama, S. Duhm, Q. Xin, A. Funakoshi, N. Ogawa, S. Kera, N. Ueno, H. Ishii, Highest-occupied-molecular-orbital band dispersion of rubrene single crystals as observed by angle-resolved ultraviolet photoelectron spectroscopy. Phys. Rev. Lett. **104**, 156401 (2010)

9. S. Yanagisawa, Y. Morikawa, A. Schindlmayr, HOMO band dispersion of crystalline rubrene: effects of self-energy corrections within the GW approximation. Phys. Rev. B **88**, 115438 (2013)
10. J.L. Brédas, J.P. Calbert, D.A. da Silva Filho, J. Cornil, Organic semiconductors: a theoretical characterization of the basic parameters governing charge transport. Proc. Natl. Acad. Sci. USA **99**, 5804 (2002)
11. T. Holstein, Studies of polaron motion part II: the "small" polaron. Ann Phys. (N. Y.) **8**, 343 (1959)
12. K. Hannewald, V.M. Stojanović, J.M.T. Schellekens, P.A. Bobbert, G. Kresse, J. Hafner, The theory of polaron bandwidth narrowing in molecular crystals. Phys. Rev. B **69**, 075211 (2004)
13. Y. Yang, Y. Yang, F. Wua, Z. Wei, First-principles electronic structure of copper phthalocyanine (CuPc). Solid State Commun. **148**, 559 (2008)
14. I.G. Hill, A. Kahn, Z.G. Soos, R.A. Pascal Jr., Charge-separation energy in films of π-conjugated organic molecules. Chem. Phys. Lett. **327**, 181 (2000)
15. P.K. Nayak, Exciton binding energy in small organic conjugated molecule. Synth. Met. **174**, 42 (2013)
16. P.I. Djurovich, E.I. Mayo, S.R. Forrest, M.E. Thompson, Measurement of the lowest unoccupied molecular orbital energies of molecular organic semiconductors. Org. Electron. **10**, 515 (2009)
17. J. Singh, The dynamics of excitons, in *Solid State Physics*, vol. 38, ed. by H. Ehrenreich, D. Turnbull (Academic Press, Orlando, 1984), pp. 295–370
18. K.W. Böer, U.W. Pohl, Excitons, in *Semiconductor Physics* (Springer, Cham, 2018)
19. J.I. Frenkel, On the transformation of light into heat in solids II. Phys. Rev. **37**, 1276 (1931)
20. G. Klein, R. Voltz, M. Schott, Singlet exciton fission in anthracene and tetracene at 77 degrees K. Chem. Phys. Lett. **19**, 391 (1973)
21. H. Uoyama, K. Goushi, K. Shizu, H. Nomura, C. Adachi, Highly efficient organic light-emitting diodes from delayed fluorescence. Nature **492**, 234 (2012)
22. M. Pope, C.E. Swenberg, *Electronic Processes in Organic Crystals* (Oxford University Press, Oxford, UK, 1982)
23. J. Shinar (ed.), *Organic Light-Emitting Devices: A Survey* (Springer, New York, 2004)
24. K.T. Kamtekar, A.P. Monkman, M.R. Bryce, Recent advances in white organic light-emitting materials and devices (WOLEDs). Adv. Mater. **22**, 572 (2010)
25. S. Reineke, M.A. Baldo, Room temperature triplet state spectroscopy of organic semiconductors. Sci. Rep. **4**, 3797 (2014)
26. R.P. Feynman, Slow electrons in a polar crystal. Phys. Rev. **97**, 660 (1955)
27. J. Appel, Polarons, in *Solid State Physics*, vol. 21, ed. by F. Seitz, D. Turnbull, H. Ehrenreich (Academic Press, New York, 1968), pp. 193–391
28. R.A. Marcus, Electron transfer reactions in chemistry. Theory and experiment. Rev. Mod. Phys. **65**, 599 (1993)
29. A. Troisi, Charge transport in high mobility molecular semiconductors: classical models and new theories. Chem. Soc. Rev. **40**, 2347 (2011)
30. C.L. Braun, Organic semiconductors, in *Handbook on Semiconductors, Vol. 3, Materials Properties and Preparation*, ed. by T.S. Moss, S.P. Keller (North Holland Publ, Amsterdam, 1980), pp. 857–873
31. H.J. Keller (ed.), *Chemistry and Physics in One-Dimensional Metals* (Plenum Press, New York, 1977)
32. A.N. Bloch, T.F. Carruthers, T.O. Poehler, D.O. Cowan, The organic metallic state: some physical aspects and chemical trends, in *Chemistry and Physics in One-Dimensional Metals*, ed. by H.J. Keller (Plenum Press, New York, 1977), pp. 47–85
33. W. Warta, R. Stehle, N. Karl, Ultrapure, high mobility organic photoconductors. Appl. Phys. A **36**, 163 (1985)
34. N. Karl, Charge-carrier mobility in organic crystals, in *Organic Electronic Materials: Conjugated Polymers and Low Molecular Weight Organic Solids*, ed. by R. Farchioni, G. Grosso (Springer, Berlin, 2001), pp. 283–326

35. L.B. Schein, Temperature independent drift mobility along the molecular direction of As_2S_3. Phys. Rev. B **15**, 1024 (1977)
36. N. Karl, *Organic Semiconductors. Landoldt-Börnstein* (Springer, Heidelberg, 1984)
37. H. Bässler, A. Köhler, Charge transport in organic semiconductors. Top. Curr. Chem. **312**, 1 (2012)
38. H. Bässler, Charge transport in disordered organic photoconductors. Phys. Stat. Solidi B **175**, 15 (1993)
39. R. Noriega, A. Salleo, Charge transport theories in organic semiconductors, in *Organic Electronics II*, ed. by H. Klauk, (Wiley-VCH, Weinheim, 2012), pp. 67–104
40. V. Coropceanu, J. Cornil, D.A. da Silva Filho, Y. Olivier, R. Silbey, J.L. Bredas, Charge transport in organic semiconductors. Chem. Rev. **107**, 926 (2007)
41. P.W.M. Blom, M.J.M. de Jong, M.G. van Munster, Electric-field and temperature dependence of the hole mobility in poly(p-phenylene vinylene). Phys. Rev. B **55**, R656 (1997)
42. P.W.M. Blom, M.C.J.M. Vissenberg, Charge transport in poly(p-phenylene vinylene) light-emitting diodes. Mater. Sci. Eng. **27**, 53 (2000)
43. Q. Shi, Y. Hou, J. Lu, H. Jin, Yunbai Li, Yan Li, X. Sun, J. Liu, Enhancement of carrier mobility in MEH-PPV film prepared under presence of electric field. Chem. Phys. Lett. **425**, 353 (2006)
44. W.L. Kalb, S. Haas, C. Krellner, T. Mathis, B. Batlogg, Trap density of states in small-molecule organic semiconductors: a quantitative comparison of thin-film transistors with single crystals. Phys. Rev. B **81**, 155315 (2010)
45. J. Rivnay, R. Noriega, J.E. Northrup, R.J. Kline, M.F. Toney, A. Salleo, Structural origin of gap states in semicrystalline polymers and the implications for charge transport. Phys Rev B **83**, 121306 (2011)
46. W.R. Salaneck, K. Seki, A. Kahn, J.-J. Pireaux (eds.), *Conjugated Polymer and Molecular Interfaces* (Marcel Dekker Inc., New York, 2002)
47. N. Koch, Electronic structure of interfaces with conjugated organic materials. Phys. Stat. Solidi RRL **6**, 277 (2012)
48. K. Akaike, Advanced understanding on electronic structure of molecular semiconductors and their interfaces. Jpn. J. Appl. Phys. **57**, 03EA03 (2018)
49. I.G. Hill, D. Milliron, J. Schwartz, A. Kahn, Organic semiconductor interfaces: electronic structure and transport properties. Appl. Surf. Sci. **166**, 354 (2000)
50. H. Vázquez, W. Gao, F. Flores, A. Kahn, Energy level alignment at organic heterojunctions: role of the charge neutrality level. Phys. Rev. B **71**, 041306 (2005)
51. G. Heimel, I. Salzmann, S. Duhm, N. Koch, Design of organic semiconductors from molecular electrostatics. Chem. Mater. **23**, 359 (2011)
52. W. Chen, D.-C. Qi, H. Huang, X. Gao, A.T.S. Wee, Organic-organic heterojunction interfaces: effect of molecular orientation. Adv. Funct. Mater. **21**, 410 (2011)
53. N. Koch, S. Duhm, J.P. Rabe, A. Vollmer, R.L. Johnson, Optimized hole injection with strong electron acceptors at organic-metal interfaces. Phys. Rev. Lett. **95**, 237601 (2005)
54. H. Vázquez, F. Flores, R. Oszwaldowski, J. Ortega, R. Pérez, A. Kahn, Barrier formation at metal-organic interfaces: dipole formation and the charge neutrality level. Appl. Surf. Sci. **234**, 107 (2004)
55. J. Hwang, A. Wan, A. Kahn, Energetics of metal-organic interfaces: new experiments and assessment of the field. Mater. Sci. Eng. R **64**, 1 (2009)
56. J.-P. Yang, F. Bussolotti, S. Kera, N. Ueno, Origin and role of gap states in organic semiconductor studied by UPS: as the nature of organic molecular crystals. J. Phys. D: Appl. Phys. **50**, 423002 (2017)

Chapter 6
Thermodynamics of Epitaxial Layer-Growth

Abstract Growth requires some deviation from thermodynamic equilibrium. This chapter outlines the driving force for equilibrium-near growth of a crystal in terms of macroscopic quantities. We consider a thermodynamic description for the transition of a gaseous or liquid phase to the solid phase. The initial stage of layer growth requires a nucleation process. We discuss the energy of a surface and illustrate the nucleation of a layer and the occurrence of different growth modes.

Growth of a crystalline solid represents a transition from one phase, e.g., the vapor phase, to a crystalline phase. Such phase transition is phenomenologically described by *thermodynamics* using macroscopic quantities. Often the understanding of growth phenomena requires also consideration of the *kinetics*. A respective description considers transition states on an atomistic scale. Epitaxy may be performed close to thermodynamic equilibrium using, e.g., liquid phase epitaxy (Sect. 11.1). Growth is then well described in terms of thermodynamic properties of the system. Epitaxy far away from equilibrium by applying, e.g., processes occurring during molecular beam epitaxy (Sect. 11.3), may often more appropriately be described in terms of kinetics. In this chapter basics of growth processes employed in epitaxial methods are discussed in terms of thermodynamics.

6.1 Phase Equilibria

Thermodynamics studies the effect of changes in, e.g., temperature on a system at a macroscopic scale by analyzing the collective motion of its particles. The term particles comprises both, atoms and molecules, and the term system denominates a large ensemble of particles which is marked-off from environment in some defined manner. The system may either contain a single kind of particles (single-component system) or a mixture of different kind of particles (multi-component system). If the system is completely *homogeneous* (regardless of being a single- or multi-component system) it is called to consist of a single phase. Growth occurs in *heterogeneous* systems which are characterized by interfaces separating different phases within the system.

U. W. Pohl, *Epitaxy of Semiconductors*, Graduate Texts in Physics,
https://doi.org/10.1007/978-3-030-43869-2_6

6.1.1 Thermodynamic Equilibrium

If a system is left undisturbed by outside influences the interacting particles will share energy among themselves and reach a state, where the global statistics are unchanging in time. Parameters describing the system then have ceased to change with time. This allows single parameters like temperature or pressure to be attributed to the *whole* system.

Growth requires a force to drive particles from one phase of a system across the interface towards a solid phase. The crystal grower controls parameters in a way that a more volatile phase of the system is thermodynamically less stable than the crystalline phase, i.e., he adjusts some deviation from equilibrium. There are two kinds of state variables which describe the system:

- *Intensive parameters* like temperature T, pressure P, mole fraction x_i of component i, or chemical potential μ_i are independent on the size of the system.
- *Extensive parameters* like internal energy U, entropy S, volume V, or amount of substance n_i do depend on system size.

The amount of substance n_i designates the number of moles of component i in the system and is given by the mass m_i of the substance of component i divided by its mole mass M_i. The corresponding mole fraction x_i (also called *molar fraction*) denotes the number of moles of component i as a proportion of the total number of moles of all components in the system, i.e.,

$$x_i = \frac{n_i}{\sum_{j=1}^{N_c} n_j} = \frac{n_i}{n}, \tag{6.1}$$

where N_c is the number of components in the system and n is the total number of moles in the system.

All thermodynamic properties of a system may be derived from its internal energy U. For a system composed of N_c components this state function is given by

$$U = TS - PV + \sum_{i=1}^{N_c} \mu_i n_i. \tag{6.2}$$

S, V, and n_i are the proper variables for the internal energy, and μ is the chemical potential introduced in (6.4) below. For crystal growers the direct access to the intensive parameters temperature and pressure is more convenient than control of the extensive parameters entropy and volume. Therefore the state function *Gibbs energy* G with T, P, and n_i as proper variables is introduced,

$$G = U + PV - TS. \tag{6.3}$$

Gibbs energy G—also referred to as Gibbs function or Gibbs free energy—is an extensive function like the internal energy U. The chemical potential of component i in the system is given by the partial derivative of Gibbs energy,

$$\mu_i = \left(\frac{\partial G}{\partial n_i}\right)_{T,P,n_{j\neq i}} = \mu_i(T, P, x_1, \ldots, x_{N_c}). \tag{6.4}$$

Inserting (6.2) into (6.3), Gibbs energy can also be expressed by the N_c amounts of substance n_i and their respective chemical potentials,

$$G = \sum_{i=1}^{N_c} \mu_i n_i. \tag{6.5}$$

The total chemical potential μ then reads

$$\mu = \frac{G}{n} = \sum_{i=1}^{N_c} \mu_i x_i. \tag{6.6}$$

Thermodynamic equilibrium is characterized by the *minimum* of Gibbs energy G. Extensive functions of a system are similarly given as the sum of the corresponding quantities for each phase. For a system containing N_p phases, G is hence the sum of those for each phase σ_i,

$$\begin{aligned} G &= \sum_{j=1}^{N_p} G(\sigma_j) = G(\sigma_1) + G(\sigma_2) + \cdots + G(\sigma_{N_p}) \\ &= \sum_{j=1}^{N_p} \sum_{i=1}^{N_c} \mu_i(\sigma_j) n_i(\sigma_j) \\ &= \sum_{i=1}^{N_c} \mu_i(\sigma_1) n_i(\sigma_1) + \sum_{i=1}^{N_c} \mu_i(\sigma_2) n_i(\sigma_2) + \cdots + \sum_{i=1}^{N_c} \mu_i(\sigma_{N_p}) n_i(\sigma_{N_p}). \end{aligned} \tag{6.7}$$

The σ_j and $n_i(\sigma_j)$ identify the phases and the amount of substance of component i in phase σ_j, respectively. If G is minimum at equilibrium, then condition

$$\left(\frac{\partial G}{\partial n_i(\sigma_j)}\right)_{T,P,n(\sigma_j)_{j\neq i}} = 0 \tag{6.8}$$

must apply for all N_c components and all N_p phases of the system. Under given (constant) conditions for T, P, and n_i, Gibbs energy G hence assumes a minimum value with respect to any variation of these state parameters.

Let us assume a variation of one parameter, namely a transition of component 1 from phase α to phase β. The change is expressed by a variation of $n_1(\alpha)$ with the condition $n_1(\alpha) + n_1(\beta) = \text{constant}$. Applying condition (6.8) we obtain

$$\begin{aligned}\frac{\partial G}{\partial n_1(\alpha)} &= \frac{\partial(G(\alpha) + G(\beta) + G(\gamma) + \cdots)}{\partial n_1(\alpha)}\\ &= \frac{\partial G(\alpha)}{\partial n_1(\alpha)} + \frac{\partial G(\beta)}{\partial n_1(\alpha)} = \frac{\partial G(\alpha)}{\partial n_1(\alpha)} - \frac{\partial G(\beta)}{\partial n_1(\beta)} = 0.\end{aligned}$$

In the second fraction all apart from the first two summands are zero for all phases except for α and β due to the assumed transition, and the negative sign of the second summand originates from $\partial n_1(\beta) = -\partial n_1(\alpha)$. Inserting (6.4) applied to component 1 yields the equilibrium condition

$$\mu_1(\alpha) - \mu_1(\beta) = 0.$$

This condition applies for pairs of *all* phases and *all* components. The general conditions for equilibrium hence reads: The chemical potentials μ_i of all N_c components are equal among each other in each of the N_p phases,

$$\mu_i(\sigma_1) = \mu_i(\sigma_2) = \cdots = \mu_i(N_p), \quad i = 1, \ldots, N_c \quad \text{(equilibrium)}. \tag{6.9}$$

Equation (6.9) applies simultaneously for all N_c components of the system, yielding a set of N_c relations in case of an equilibrium between two phases. In equilibrium of a multi-component system much more phases than two may coexist. For N_p coexisting phases a set of $(N_p - 1) \times N_c$ relations must be fulfilled at equilibrium. The *composition* of the N_p phases is generally *not* equal at equilibrium to fulfill these conditions, i.e., usually $x_1(\sigma_1) \neq x_1(\sigma_2) \neq x_1(\sigma_3) \neq \cdots$.

Figure 6.1 illustrates the equilibrium between phase α and a solid phase β of a three-component system. Different species of particles are depicted by different symbols. Note different compositions indicated in phases α and β. The interface between α and β is the surface of the solid. Its structure is thermodynamically less favorable than that of the solid or of the more volatile phase α. The surface hence

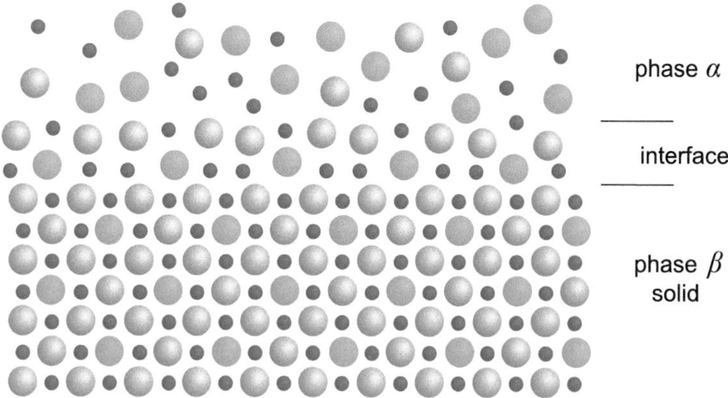

Fig. 6.1 Schematic of a three-component system with two phases α and β in equilibrium

tends to be thin. In practice, it extends over a few atomic layers. In a thermodynamic description the surface region is usually not regarded as a separate phase and assumed infinitely thin.

6.1.2 Gibbs Phase Rule

The number of phases which can coexist at equilibrium is limited. According to (6.4) a set of the variables T, P, and compositions x_i of the N_c components specifies a phase of a heterogeneous system. The chemical potential in this phase is fixed when $N_c - 1$ mole fractions are quoted besides T and P: The last of N_c mole fractions is given by the condition $\sum_i x_i = 1$ in each phase. We hence have $N_p \times (N_c - 1)$ independent mole fractions for a system with N_p phases. Therefore a total of $N_p \times (N_c - 1) + 2$ intensive variables are sufficient to fix all of the intensive variables of the system. Many of them are dependent, though. Considerations leading to (6.9) show that N_c relations according to $\mu_i(\alpha) = \mu_i(\beta)$ have to be fulfilled at equilibrium between two coexisting phases. For three coexisting phases $2 \times N_c$ relations, and for N_p phases a number of $(N_p - 1) \times N_c$ relations have to be fulfilled. There remains thus a number N_f of *independent* intensive variables given by $(N_p \times (N_c - 1) + 2) - ((N_p - 1) \times N_c)$, or

$$N_f = N_c - N_p + 2. \tag{6.10}$$

This number of independent intrinsic state parameters may be used to control growth near thermodynamic equilibrium without changing the number of phases N_p. N_f is called the *number of degrees of freedom* for a given thermodynamic condition and specifies how many control variables can be altered while maintaining this condition. This *Gibbs phase rule* was stated by Josiah Willard Gibbs in the 1870s. In literature the numbers of degrees of freedom, of components, and phases N_f, N_c, and N_p, are also labeled as F, C, and π, respectively.

N_f may not be negative. Consequently a maximum of $N_p = 3$ phases may coexist in equilibrium of a single-component system ($N_c = 1$). Since in this case $N_f = 0$ the variables T and P are fixed in a triple point. In a two-phase equilibrium of a single-component system we obtain $N_f = 1$. We hence can choose one parameter independently (within some limits), say T, while the other, P, adjusts accordingly to accomplish equilibrium. In a single-phase range of a single-component system $N_f = 2$. T and P may then be varied independently. In a two-component system a maximum of 4 phases may coexist, and in a two-phase equilibrium still two parameters may be controlled independently.

6.1.3 Gibbs Energy of a Single-Component System

A single-component system has a maximum of three phases in equilibrium (Sect. 6.1.2), and a state is fixed by fixing temperature, pressure, and volume of the system. The equilibrium conditions between the thermodynamically distinct phases are clearly represented in a phase diagram. A simple illustration is the pressure-temperature diagram given in Fig. 6.2. The diagram is a projection of P–V–T space on the P–T plane for a fixed volume and shows the phase boundaries between the three equilibrium phases of solid, liquid, and gas. On a boundary two phases coexist, and in the triple point three phases coexist. Dashed curves near the triple point indicate metastable states (cf. Sect. 6.1.5). The solid-liquid phase boundary of most substances has a positive slope. This is due to the solid phase having a higher density than the liquid, so that increasing the pressure increases the melting point. Prominent exceptions are, e.g., water and silicon.

To study the behavior of the system at transitions from one phase to another we consider a variation of one of the variables P or T while keeping the other fixed [1]. We first discuss the temperature dependence of Gibbs energy $G(T)_P$, i.e., $G(T)$ at constant pressure P, along path 1 drawn in Fig. 6.2. In a second step the pressure dependence $G(P)_T$ along path 2 will be discussed.

Considering a system consisting of one component yields a descriptive dependence of Gibbs energy G from temperature and pressure. There is only a single chemical potential,

$$\mu = \frac{\partial G}{\partial n} = \frac{G}{n} \equiv g \quad \text{(single-component system)}.$$

The relation shows that the chemical potential is identical to the Gibbs energy per mole g. Both μ and G depend solely on T and P.

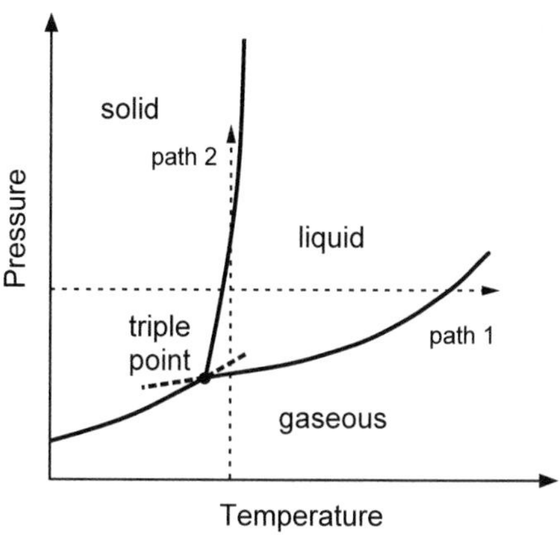

Fig. 6.2 *P–T* phase diagram for a single-component system. *Dashed curves* near the *triple point* indicate metastable states. The dependence of Gibbs energy $G(P, T)$ for variations along *paths 1* and *2* is discussed in the text

We first consider $G(T)_P$. The temperature dependence of G is evaluated from the respective dependence of the three summands in (6.3)

$$G = U + PV - TS.$$

The internal energy $U(T)$ starts at $T = 0$ K with some constant value U_0 which merely represents an offset of all parameters. In the low-temperature range the increase of $U(T)$ is essentially determined by the heat capacity of the (solid) system at constant volume, C_V^{S}. The slope increases due to an increase of $C_V^{\mathrm{S}}(T)$. We may neglect here the small volume change of the solid at constant pressure and the consequential small contribution of $U(V(T))$. At the melting temperature T_{SL} (index SL for solid → liquid) and also at the boiling temperature T_{LG} (liquid → gaseous) the internal energy $U(T)$ increases by material-dependent steps ΔU_{SL} and ΔU_{LG}, respectively: The bordering phases differ in internal energy. The thermal trend of U in the liquid and gaseous phases is likewise described using the heat capacities $C_V^{\mathrm{L}}(T)$ and $C_V^{\mathrm{G}}(T)$, respectively. The entire temperature dependence of the first summand of $G(T)$ in (6.3) is then given by

$$\begin{aligned} U(T) = & U_0 + \int_0^{T_{\mathrm{SL}}} C_V^{\mathrm{S}}(T)\, dT + \Delta U_{\mathrm{SL}} + \int_{T_{\mathrm{SL}}}^{T_{\mathrm{LG}}} C_V^{\mathrm{L}}(T)\, dT \\ & + \Delta U_{\mathrm{LG}} + \int_{T_{\mathrm{LG}}}^{T} C_V^{\mathrm{G}}(T)\, dT. \end{aligned} \tag{6.11}$$

For $T < T_{\mathrm{LG}}$ thereinafter summands are accordingly omitted. The trend of $U(T)$ is illustrated in Fig. 6.3.

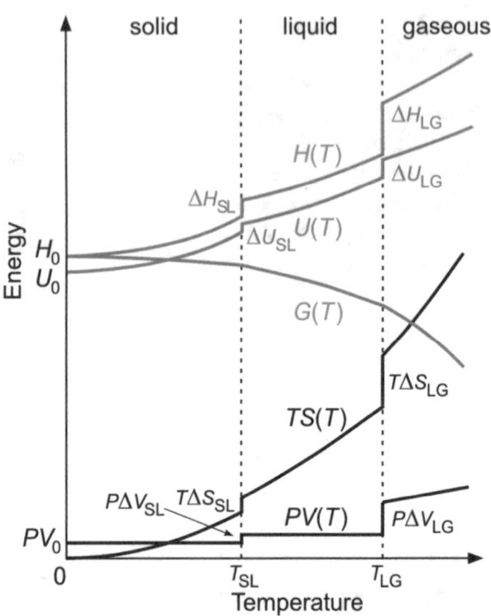

Fig. 6.3 Temperature dependence of Gibbs energy G, internal energy U, enthalpy H, and the energies TS and PV at constant pressure for a single-component system. After [1]

The trend of the second summand of G in (6.3) is given by $V(T)$ because P is assumed to be constant. Since the thermal expansion coefficients in the solid and liquid phases are very small (typ. of order 10^{-6}) we may assume a virtually constant volume. Melting and evaporation lead to a small step ΔV_{SL} and a large step ΔV_{LG}. The dependence in the gas phase is approximated by the ideal gas law $V(T) \cong nRT/P$, R being the universal molar gas constant.

Adding the first two summands of G in (6.3) yields the *enthalpy* $H(T) = U(T) + PV(T)$. The thermal dependence is similar to that of $U(T)$ with slightly larger values due to positive values of P and V, cf. Fig. 6.3. The step at the melting point $\Delta H_{SL} = \Delta U_{SL} + P\Delta V_{SL}$ is the heat of fusion absorbed during melting at T_{SL}. It is vice versa identical to the heat released during solidification. According to convention a positive sign of ΔH indicates an endothermic reaction, where the system receives heat supplied from the surroundings. Similar considerations apply for evaporation and condensation expressed by ΔH_{LG}. $H(T)$ may be expressed by a similar relation as (6.11) when U and C_V are replaced by H and C_P, respectively.

The third summand of G in (6.3) increases with T according to the *entropy* $S = \int C_P/T = \int C_P d\ln T$. S_0 is assumed to be zero. During the phase transitions at T_{SL} and T_{LG} the entropy S experiences a steplike increase $\Delta H/T$. It should be noted that these steps have the same height as those in the function $H(T)$, since $\Delta H = T\Delta S$. S may also be expressed by a similar relation as (6.11) when U, C_V, and dT are replaced by S, C_P, and $d\ln T$, respectively.

We now put the three summands together to obtain Gibbs energy $G(T)$ according to (6.3). As shown in Fig. 6.3 $G(T)$ is a monotonously decreasing function with a gradually increasing negative slope. $G(T)$ is continuous at the phase transitions but experiences a kink. Since $\mu = G/n$ we also know the trend of the chemical potential $\mu(T)$. We will consider this in more detail in Sect. 6.1.4.

In a second step we address the pressure dependence $G(P)_T$ evaluated from the respective dependence of the three summands in (6.3) at constant temperature, cf. Fig. 6.4.

Besides steps at phase transitions, $U(P)$ may be considered nearly constant in all phases at constant temperature. The second summand $PV(P)$ is approximately constant at low pressures due to the ideal gas law $V \cong nRT/P$. In the liquid and solid phases $V(P)$ is nearly constant. As P increases $PV(P)$ hence rises approximately linearly, with a larger slope in the liquid phase. The enthalpy $H(P) = U(P) + PV(P)$ reproduces these slopes. The steps ΔH_{LG} and ΔH_{SL} again represent heats of condensation and solidification, respectively.

The trend of the third summand of G in (6.3) $TS(P)$ is given by $S(P)$ at constant temperature. To obtain the trend at low pressure (gas phase) we apply the general Maxwell relation $(\partial S/\partial P)_T = -(\partial V/\partial T)_P$ and infer from the ideal gas law $(\partial V/\partial T)_P \cong nR/P$. Equalizing and integration yields $S(P) \cong -nR \ln P + S_0$, where S_0 is an integration constant. In the gas phase the trend decreases monotonously with a steep slope at small pressure due to the term $-\ln P$. In the liquid and solid

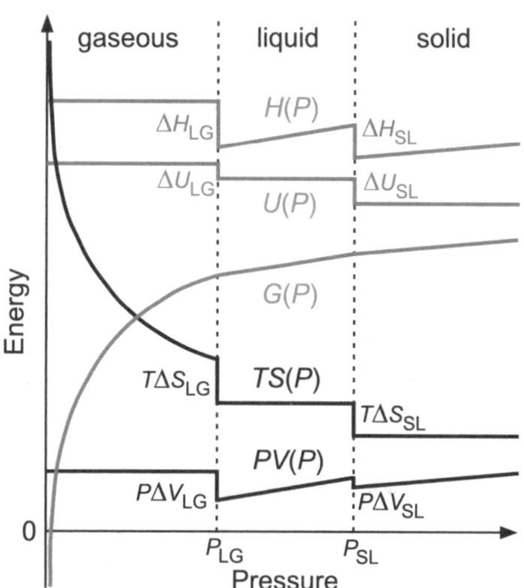

Fig. 6.4 Pressure dependence of Gibbs energy G, internal energy U, enthalpy H, and the energies TS and PV at constant temperature for a single-component system. After [1]

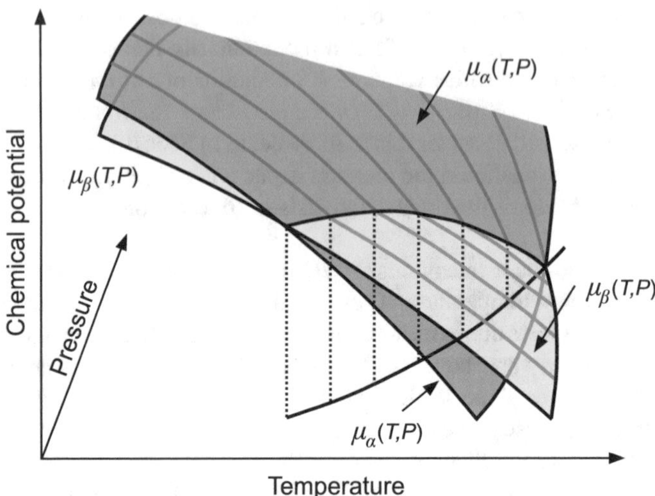

Fig. 6.5 Chemical potentials $\mu_\alpha(T, P)$ and $\mu_\beta(T, P)$ of two phases α and β near their intersection line, where both phases coexist. *Dotted vertical lines* indicate projection of the intersection line to the P–T plane. After [1]

phases $(\partial V/\partial T)_P$ is quite small due to the small thermal expansion coefficient. $S(P)$ is considered nearly constant in these phases.

The sum adds up to the pressure dependence $G(P)_T$ illustrated in Fig. 6.4. $G(P)$ is a monotonously increasing function. At phase transitions $G(P)$ is continuous but experiences a kink. The same applies for the chemical potential $\mu(P) = G(P)/n$.

6.1.4 Phases Boundaries in a Single-Component System

We consider the chemical potential $\mu(T, P)$ at the boundary between two phases α and β of a single-component system. The characteristics in the P–T phase diagram is obtained from $\mu(T)_P = G(T)_P/n$ and $\mu(P)_T = G(P)_T/n$. We know from Sect. 6.1.3 that $\mu(P, T)$ is a curved plane which in each phase monotonously decreases along T and increases along P. The slopes of $\mu_\alpha(T, P)$ and $\mu_\beta(T, P)$ differ in two different phases. The respective planes hence intersect. Along the intersection line the chemical potentials must be equal, $\mu_\alpha(T, P) = \mu_\beta(T, P)$. Adjacent to the line of intersection the respective lower plane refers to the stable phase. The stable phases change order upon crossing the section boundary: The respective other phase gets stable. Figure 6.5 illustrates the situation.

The projection of the intersection line of the $\mu_\alpha(T, P)$ and $\mu_\beta(T, P)$ planes on the P–T plane shown in Fig. 6.5 yields the boundary between phases α and β depicted in Fig. 6.2. The characteristics of this boundary is described by the Clapeyron relation or, if a transition between a condensed (liquid or solid) to a gas phase is involved, by the Clausius-Clapeyron relation, named after B.P. Émile Clapeyron and Rudolf Clausius. The general Clapeyron equation follows from a variation $d\mu(T, P)$ of a state on the coexistence line that must fulfill the identity $d\mu_\alpha = d\mu_\beta$, or, by inserting G/n for μ,

$$\frac{1}{n}\left(\frac{\partial G_\alpha}{\partial T}\right)_P dT + \frac{1}{n}\left(\frac{\partial G_\alpha}{\partial P}\right)_T dP = \frac{1}{n}\left(\frac{\partial G_\beta}{\partial T}\right)_P dT + \frac{1}{n}\left(\frac{\partial G_\beta}{\partial P}\right)_T dP.$$

In the relation we may replace $(\partial G/\partial T)$ by $-S$ and $(\partial G/\partial P)$ by V. This yields a condition for the proportion of the two variables for a variation along the coexistence line,

$$dP = \frac{S_\beta - S_\alpha}{V_\beta - V_\alpha}\, dT.$$

The relation gives the slope dP/dT of the coexistence curve. ΔS and ΔV are respectively the change of entropy and volume during a phase transition across the boundary. Since $\Delta S = \Delta H/T$ we may write

$$\frac{dP}{dT} = \frac{\Delta S}{\Delta V} = \frac{\Delta H}{T\Delta V}. \tag{6.12}$$

Here ΔH is the latent heat exchanged with the surroundings during the phase transition without change of temperature T. The *Clapeyron equation* (6.12) explains basic characteristics of the P–T phase diagram Fig. 6.2. A transition from the gaseous phase to the liquid or the solid phase is accompanied by a large volume change ΔV. According to (6.12) the slopes of these two phase boundaries are therefore small. The heat of evaporation ΔH is a positive quantity just as the volume increase ΔV. The slope dP/dT is therefore also positive. Furthermore, the slope increases as P increases due to a decrease of ΔV at higher pressure. Due to a larger heat of sublimation of a solid compared to the heat of evaporation of a liquid the slope of the solid-gas boundary is steeper than that of the liquid-gas boundary.

To obtain an idea of the relation $P(T)$ for a transition from a condensed phase to the gas phase, we roughly approximate the volume change by the volume of the gas formed, $\Delta V \cong V_{\text{gas}}$, and put $PV_{\text{gas}} \cong nRT$. Inserting the two approximations into (6.12) and taking ΔH as independent of T, we obtain $P(T)$ by integration, yielding

$$\ln\left(\frac{P}{P_0}\right) \cong \frac{-\Delta H}{nR}\left(\frac{1}{T} - \frac{1}{T_0}\right), \quad \text{or} \quad P \cong P_0 \exp\left(\frac{-\Delta H}{nR}\left(\frac{1}{T} - \frac{1}{T_0}\right)\right). \tag{6.13}$$

$\Delta H/n$ is the molar heat of evaporation or sublimation. We see that for the transition from a condensed to a gas phase the pressure approximately obeys an Arrhenius-like dependence over a limited temperature range. Vapor pressures are often tabled in terms of 2 parameters a and b, obtained from (6.13) by using $\ln(x) = \ln(10) \times \log(x)$ and putting $b = -\Delta H/(nR\ln(10))$. The relation then reads $\log(P) = a - b/T$, with the parameter a depending on the pressure unit; see Table 11.2.

The liquid-solid boundary is generally quite steep in the P–T phase diagram due to a very small volume change ΔV during melting. Since the liquid phase may be more dense or less dense than the solid phase ΔV may have either sign. The slope dP/dT of this boundary may hence be positive or negative.

6.1.5 *Driving Force for Crystallization*

To induce a transition from a stable phase to another the parameters temperature and pressure have to be controlled such that the chemical potential in the target phase is smaller than that in the initial phase. Particles from the initial phase will then cross the phase boundary toward the target phase to allow the system for approaching (the new) equilibrium conditions.

For a temperature variation at constant pressure the difference $\Delta T = T - T_e$ between the controlled temperature T and the equilibrium temperature T_e at the given pressure is a measure for the deviation from equilibrium. ΔT is a negative quantity and termed *supercooling*. In case of a liquid-solid equilibrium T_e is the melting temperature.

The phase transition may as well be induced by varying the pressure by an amount $\Delta P = P - P_e$ at constant temperature. If the initial phase is the gaseous phase ΔP and P_e are called *supersaturation* and saturation vapor-pressure, respectively. The achievement of equilibrium in the target phase via either of the two paths is illustrated in Fig. 6.6.

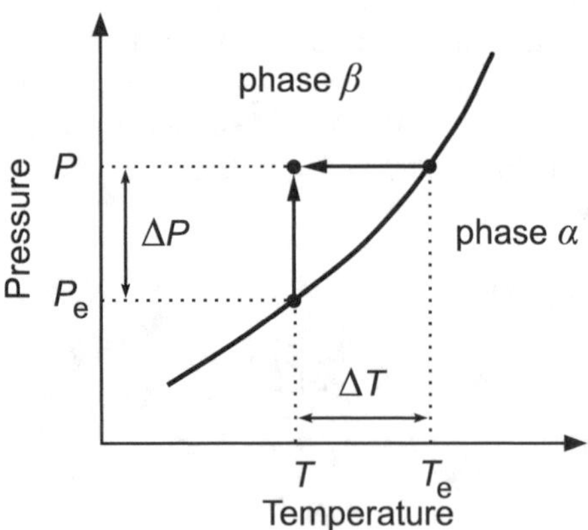

Fig. 6.6 Crossing the phase boundary from a less stable phase α to a stable phase β induced by supercooling ΔT at constant pressure or by supersaturation ΔP at constant temperature

We consider the phase transition induced by supercooling $\Delta T = T - T_e$ at constant pressure in more detail. At fixed pressure phase α is above the equilibrium temperature T_e more stabile than phase β, i.e. $\mu_\alpha(T) < \mu_\beta(T)$. Below T_e, however, $\mu_\beta(T) < \mu_\alpha(T)$ due to different slopes of the chemical potentials and their intersection at T_e, see Fig. 6.5. Phases α and β may exchange particles via the common phase boundary. Since $\mu_\beta(T) < \mu_\alpha(T)$ below T_e, the system can lower the total enthalpy $G = G_\alpha + G_\beta = n_\alpha \mu_\alpha + n_\beta \mu_\beta$ by transferring particles from phase α to phase β: The amount of substance n_β in the stable phase β increases on expense of the amount of substance n_α in the less stable phase α. The driving force for this phase transition is the difference in the chemical potential

$$\Delta\mu \equiv \mu_\beta - \mu_\alpha. \tag{6.14}$$

$\Delta\mu$ is called *growth affinity*, or *driving force for crystallization* if the stable phase β is a solid phase. $\Delta\mu$ is a negative quantity, though it must be mentioned that $\Delta\mu$ is often chosen positive in literature. The relation between the driving force $\Delta\mu$ and the supercooling ΔT is illustrated in Fig. 6.7. We see that an increase of supercooling raises $\Delta\mu$ and thereby enhances the rate of substance crossing the phase boundary.

It should be noted that the respective instable phases may sometimes be experimentally observed to some extent. Such metastable state of a system is indicated by dashed lines in Fig. 6.7. Well-known examples are heating of a liquid above the boiling point (bumping) or supersaturated vapor in Wilson's cloud chamber. The effect is particularly pronounced for undercooled metal melts which may still be liquid some tens of degrees °C below the temperature of solidification. The phenomenon should not be confused with the size effect of melting point depression in nanoscale materials that originates from a high surface to volume ratio.

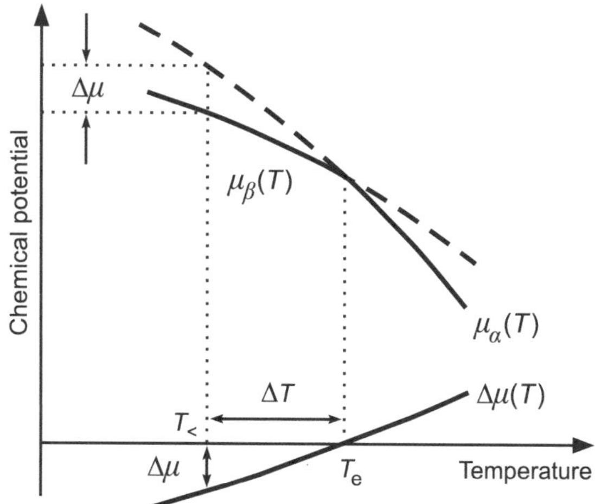

Fig. 6.7 Temperature dependence of the chemical potential μ of phases α and β. The growth affinity $\Delta\mu$ designates the difference of the chemical potentials μ_α and μ_β induced by a supercooling ΔT to a temperature $T_<$ below the equilibrium temperature T_e

The driving force (6.14) may likewise be induced by a supersaturation $\Delta P = P - P_e$ at constant temperature. For solidification from the liquid phase P must be increased if the liquid-solid boundary has a positive slope dP/dT and decreased otherwise. The driving force increases as ΔP is raised, similar to the dependence in supercooling discussed above.

To obtain an explicit relation for the temperature dependence of the driving force $\Delta\mu(\Delta T)$ we expand the function into a Taylor series at equilibrium temperature $T = T_e$,

$$\Delta\mu(\Delta T) = \mu_\beta - \mu_\alpha \cong \left(\frac{\partial\mu_\beta}{\partial T} - \frac{\partial\mu_\alpha}{\partial T}\right)\Delta T + \left(\frac{\partial^2\mu_\beta}{\partial T^2} - \frac{\partial^2\mu_\alpha}{\partial T^2}\right)\frac{(\Delta T)^2}{2} + \cdots .$$

Replacing $n(\partial\mu/\partial T)_P = (\partial G/\partial T)_P$ by $-S = -H/T$ we obtain for the expressions in the second summand

$$\left(\frac{\partial^2\mu}{\partial T^2}\right)_P = -\frac{1}{n}\left(\frac{\partial S}{\partial T}\right)_P = -\frac{1}{nT}\left(\frac{\partial H}{\partial T}\right)_P = -\frac{C_P}{nT}.$$

Insertion into $\Delta\mu$ yields

$$\Delta\mu(\Delta T) = \frac{1}{n}\left(-\frac{\Delta H_{\text{trans}}}{T_e}\Delta T - \frac{\Delta C_P}{2T_e}(\Delta T)^2 + \cdots\right). \tag{6.15}$$

In this relation ΔH_{trans} and ΔC_P denote the latent heat exchanged with the surroundings during the phase transition and the difference of the specific heat at both sides of the phase boundary, respectively. We note that the driving force $\Delta\mu$ is not a linear function of the supersaturation ΔT, particularly for significant deviation from the equilibrium temperature T_e and a large difference of the specific heat in phases α and β.

In a similar way we obtain an explicit relation for the pressure dependence of the driving force $\Delta\mu(P)$ by expanding the function into a Taylor series at equilibrium pressure $P = P_e$,

$$\Delta\mu(P) = \mu_\beta - \mu_\alpha \cong \left(\frac{\partial\mu_\beta}{\partial P} - \frac{\partial\mu_\alpha}{\partial P}\right)\Delta P + \left(\frac{\partial^2\mu_\beta}{\partial P^2} - \frac{\partial^2\mu_\alpha}{\partial P^2}\right)\frac{(\Delta P)^2}{2} + \cdots .$$

This approach yields a reasonable description of the transition between condensed phases like, e.g., liquid $\rightarrow$ solid. We replace $n(\partial\mu/\partial P)_T = V$ and $(-1/V)(\partial V/\partial P)_T = \kappa$ (isothermal expansion coefficient), obtaining

$$\Delta\mu(\Delta P) = \frac{1}{n}\left(\Delta V_e \Delta P - (V_{e,\beta}\kappa_{e,\beta} - V_{e,\alpha}\kappa_{e,\alpha})(\Delta P)^2/2 + \cdots\right). \tag{6.16}$$

$\Delta V_e/n$ designates the difference of the molar volumes at both sides of the phase boundary at equilibrium pressure P_e. This quantity controls the driving force induced by a supersaturation ΔP as a first approximation.

The driving force $\Delta\mu(P)$ of a pressure-induced transition between a gas phase α and a solid phase β may be obtained by integrating over the pressure. We neglect the much smaller solid molar volume and assume the validity of the ideal gas law,

$$\begin{aligned}\Delta\mu(\Delta P) &= \int_{P_e}^{P} \frac{\partial(\Delta\mu)}{\partial P}\, dP = \int_{P_e}^{P} \frac{(V_\beta - V_\alpha)}{n}\, dP \cong \int_{P_e}^{P} -\frac{V_\alpha}{n}\, dP \\ &\cong -\int_{P_e}^{P} \frac{RT}{P}\, dP = -RT\ln(P/P_e).\end{aligned}$$

In this case the driving force for crystallization is approximately proportional to the logarithm of the supersaturation ratio P/P_e of the gas phase α.

6.1.6 Two-Component System

Practical real systems usually comprise more than one component. The treatment of a multi-component system is quite complex. For simplicity we focus on binary systems, which consist of only two components. Such a system still allows for describing basic phenomena occurring in multi-component systems. Prominent examples are solidification from a solution, e.g., in liquid-phase epitaxy, or the solubility of impurities governing doping of semiconductors.

A *component* in a multi-component system is a chemically distinct constituent whose concentration may be varied *independently* in the various phases. The number of components N_c specifies how many substances are (at least) needed to describe the composition of the system in all phases. Thus, pure water (H_2O) forms a single-component system, though some dissociation into hydronium cations (H_3O^+) and hydroxyl anions (OH^-) occurs; the numbers of anions and cations are contrained to be equal due to charge neutrality and may hence not vary independently. If the substances of a system do not react the number of components N_c is simply given by the number of substances.

A system containing two components A and B with substance amounts n_A and n_B is usually characterized by its composition x. This quantity designates the mole fraction of one of the two components, say $x \equiv x_A$. Often x refers to the solute in a solution, i.e. to the minority component A in the substance formed by adding A to B. The respective remaining composition x_B follows from the condition $x_A + x_B = 1$, yielding $x_B = 1 - x$. $x = 0$ and $x = 1$ then designate the pure components B and A, respectively.

Equilibrium of a binary system is described by a minimum of Gibbs energy $G = H - TS$ which depends on three independent variables, namely temperature T,

Pressure P, and composition x. In the following we will consider the *molar* Gibbs energy g. According to (6.1) and (6.5) g is given by

$$\begin{aligned} g &= G/n = (\mu_A n_A + \mu_B n_B)/(n_A + n_B) = \mu_A x_A + \mu_B x_B \\ &= \mu_A x + \mu_B(1-x) = (\mu_A - \mu_B)x + \mu_B. \end{aligned}$$

We assume a solution of A and B with composition x formed at fixed temperature and pressure in a single phase. If the two components are just put together *without* intermixing or solving, the molar Gibbs energy is a linear function $\overline{g}$ of the composition x with the starting and ending vertex given by the chemical potentials $\mu_{0,B}$ and $\mu_{0,A}$ of the *pure* components B and A, respectively (Fig. 6.8),

$$\overline{g}(x) = (\mu_{0,A} - \mu_{0,B})x + \mu_{0,B}. \tag{6.17}$$

If now the two components are mixed their particles interact. This is taken into account by an additive term, the molar Gibbs free energy of mixing Δg_{M}. In general Δg_{M} contains contributions of both, a change of molar enthalpy Δh_{M} and a change of molar entropy Δs_{M} on mixing,

$$\Delta g_{\mathrm{M}} = \Delta h_{\mathrm{M}} - T\Delta s_{\mathrm{M}}. \tag{6.18}$$

A particularly simple case is given by an *ideal solution*, where the interactions between unlike and like particles in the solution are the same. If the interaction energies in the solution are identical to those in the pure components, mixing will not change the overall enthalpy. No heat is then exchanged with the surroundings on mixing, $\Delta h_{\mathrm{M}} = 0$. Due to molecular forces being independent on composition, mixing is not accompanied by a change in molar volume, $\Delta v_{\mathrm{M}} = 0$. There is, however, a change in molar entropy Δs_{M} by mixing. The molar Gibbs energy of an ideal solution is hence given by

$$g(x) = \overline{g} + \Delta g_{\mathrm{M}} = \overline{g} - T\Delta s_{\mathrm{M}} \quad \text{(ideal solution).} \tag{6.19}$$

Ideal solutions are completely miscible. For a statistically random distribution of A and B particles the change in molar entropy is

$$\Delta s_{\mathrm{M}} = -R(x_A \ln x_A + x_B \ln x_B) = -R\big(x \ln x + (1-x)\ln(1-x)\big) \quad \text{(ideal solution),}$$

where R is the molar gas constant. The mixing-induced change of molar entropy Δs_{M} is a positive quantity, because the ln function yields negative values for x ranging from 0 to 1. The term $-T\Delta s_{\mathrm{M}}$ in $g(x)$ hence effects a deviation from the linear dependence $\overline{g}$ towards lower values. This decrease of molar Gibbs energy gets stronger for increased temperature, as illustrated for phase α of a two-component system in Fig. 6.8a (gray lines at T_1 and $T_2 > T_1$).

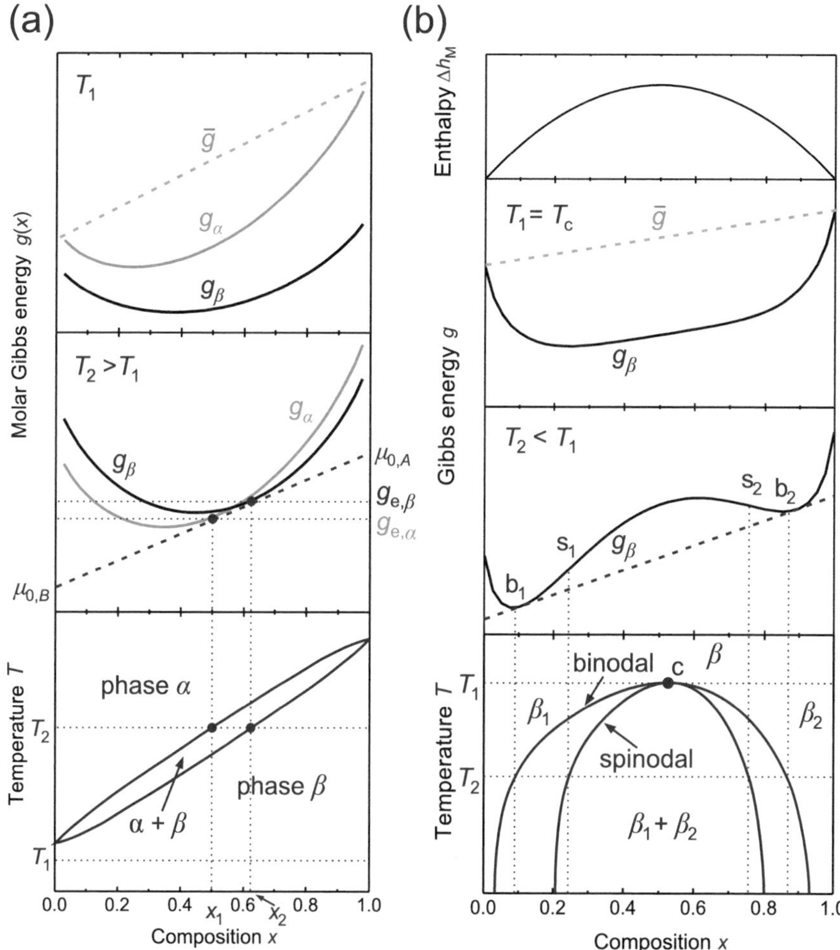

Fig. 6.8 **a** Composition-dependent molar Gibbs energy of an ideal solution with phases α and β at a low temperature T_1 (*top*) where only phase β is stable, and a higher Temperature T_2 where phases α and β coexist in equilibrium. *Bottom*: T–x phase diagram with phases α and β and a range of coexistence. **b** Real solution with contribution of an endothermal enthalpy on mixing (*top*), Gibbs energy of phase β at a critical temperature T_c and a lower temperature T_2. The phase diagram at the *bottom* shows two stable phases β_1 and β_2, an area of metastable compositions between the binodal and spinodal curves, and a range of spontaneous decomposition bounded by the spinodal

Figure 6.8a makes clear how the composition of a two-component system changes at a transition from a phase α (e.g., the liquid phase) to phase β (solid). At a low temperature T_1 the system is in a solid state β: The molar Gibbs energy of the solid $g_\beta(x)$ is smaller than $g_\alpha(x)$ for any composition x. As the temperature is increased both $g_\alpha(x)$ and $g_\beta(x)$ get smaller, cf. $g(T)$ depicted in Fig. 6.3. Since the temperature dependence in the liquid phase is stronger than in the solid phase $g_\alpha(x)$ decreases faster and will be tangent to $g_\beta(x)$ at a melting temperature $T_{\text{melt},B}$. (The melting temperature $T_{\text{melt},B}$ of component B lies above T_1 at the lower left corner of the lens-shaped area between phases α and β in Fig. 6.8a (bottom).) At $T_{\text{melt},B}$ the pure component with the lower melting point starts melting, say component B. Above $T_{\text{melt},B}$ the solid and liquid states will coexist within some concentration range Δx and temperature range ΔT up to a maximum temperature $T_{\text{melt},A}$, the melting point of pure component A. Above $T_{\text{melt},A}$ $g_\alpha(x) < g_\beta(x)$ for any composition x, i.e., the system is entirely in the liquid state α.

At an intermediate temperature T_2 the liquid (α) and solid (β) phases coexist between the compositions x_1 and x_2, see Fig. 6.8a. In this range $g_\beta(x) < g_\alpha(x)$ for smaller x, and $g_\alpha(x) < g_\beta(x)$ for larger x, see $g(x)$ at T_2. For a given composition x between x_1 and x_2 the system attains lowest molar Gibbs energy with a mole fraction x_1 in the liquid phase α and a mole fraction x_2 in the solid phase β: g takes an equilibrium value $g_\alpha(x_1) = g_{\text{e},\alpha}$ at x_1 in phase α, and in phase β accordingly $g_\beta(x_2) = g_{\text{e},\beta}$ at x_2. The molar Gibbs energy of the two-phase state lies on the common tangent $(\partial g_\alpha/\partial x)_{x1} = (\partial g_\beta/\partial x)_{x2}$ at the given composition x.

A different composition in two phases α and β of a binary system in equilibrium represents a general feature of multicomponent systems. For an average composition x a mole fraction x_1 is in phase α and a mole fraction x_2 in phase β. x_1 and x_2 are fixed for a given temperature for any value of x between x_1 and x_2. The fraction of substance in phase α is given by $x_\alpha = (x_2 - x)/(x_2 - x_1)$, and the fraction of substance in phase β is given by $x_\beta = (x - x_1)/(x_2 - x_1)$. In the example discussed above and in Fig. 6.8a $x_1 = 0.50$ and $x_2 = 0.62$ at T_2. This means a fraction $x_\alpha = 17\%$ of $A + B$ is in the liquid phase α and has a composition of 50% of component A in $A + B$, while $x_\beta = 83\%$ is in the solid phase β which has a composition of 62% of A in $A + B$. If α and β are respectively the liquid and solid phases, then the line between regions α and $\alpha + \beta$ is called liquidus curve. It is built from all points x_1 as the temperature is varied. The line separating regions $\alpha + \beta$ and β is then called solidus curve.

The quotient $k_0 \equiv x_\alpha/x_\beta$ is the equilibrium distribution-coefficient. It controls the solubility of diluted components and is used to describe any mixture, e.g. doping or alloying of crystals. Its value exceeds 1 if the coexistence boundary increases as the concentration of the considered component is increased.

In *real solutions* particle interactions A–A, A–B, and B–B are all different. There is hence an enthalpy change $\Delta h_{\text{M}} \neq 0$ upon mixing which may have either sign. A negative value signifies an exothermal reaction, where heat is generated during mixing due to an attractive interaction among the components. A positive value of Δh_{M} occurs for an endothermal reaction, where heat from the environment is consumed by the system for mixing. Such behavior indicates a diminishment of binding energy in the mixed system and a consequential tendency towards a separation into immiscible components.

Most real solutions are miscible over the entire or a certain range of composition x. If the components of the solution are randomly distributed (in crystalline solid solutions with a random distribution of atoms on lattice sites), then the mixture is termed *regular solution*. Such real solutions have the same mixture-induced change of entropy Δs_M as an ideal solution. Due to a finite Δh_M the Gibbs energy of a regular solution differs from that of an ideal solution,

$$g(x) = \overline{g} + \Delta g_M = \overline{g} + \Delta h_M - T\Delta s_M \quad \text{(regular, real solution).} \tag{6.20}$$

The enthalpy change Δh_M is proportional to the concentrations x and $(1 - x)$ of the components of the system A and B, respectively. It may be described by a parameter Ω characterizing the interaction among the components,

$$\Delta h_M(x) = \Omega x(1 - x) = \Omega\left(x - x^2\right). \tag{6.21}$$

Expression (6.21) represents a quadratic correction to the ideal Gibbs energy (6.19) for regular solutions with a maximum (or—for a negative Ω value—a minimum) at $x = 0.5$. The function is drawn for positive Δh_M in the top panel of Fig. 6.8b.

A negative enthalpy change Δh_M upon mixing—indicative for miscibility—leads to a decrease of Gibbs energy $\overline{g}$ in addition to that induced by the entropy change. There is then a quantitative change of Gibbs energy $g(x)$, but largely only a minor qualitative change in the shape of the function compared to the ideal case shown in Fig. 6.8a.

A positive enthalpy change $\Delta h_M > 0$ counteracts the entropy change. A large enthalpy change upon mixing leads to a contribution comparable to the term $-T\Delta s_M$ in the range of medium composition and at sufficiently low temperature. At low temperature the system is considered to be in the solid phase β. Below a critical temperature T_c, Gibbs energy $g_\beta(x)$ then has a negative curvature $(\partial^2 g_\beta/\partial x^2 < 0)$ in this range, leading to *two* minima b_1 and b_2.

The function $g_\beta(x)$ is drawn in Fig. 6.8b for a corresponding low temperature T_2. The minima b_1 and b_2 correspond to two chemical potentials $\mu_i = \frac{\partial g_\beta}{\partial x}|_{x_i}$ at the compositions x_i. In equilibrium $\mu_1 = \mu_2$ applies, and the chemical potentials are given by the common tangent drawn in Fig. 6.8b. Analogous to the case of the ideal system with two phases α and β discussed above and in Fig. 6.8a, the lowest Gibbs energy of the solid phase $g_\beta(x)$ is attained if the system decomposes into two solid phases β_1 and β_2 with compositions x_1 and x_2 at T_2, respectively. Such behavior is referred to as *spinodal decomposition*. The minima b_1 and b_2 represent points for the given temperature of a curve termed *binodal*.

The minima b_1 and b_2 are separated by a local maximum in the range of medium composition and two inflection points s_1, s_2. The maximum represents an instable composition of the solid solution. Such a maximum emerges as the temperature is decreased below a critical value T_c. The homogeneous solid phase β then spontaneously separates into two immiscible phases with compositions given by the binodal. The miscibility gap gets wider for lower temperature as illustrated at the bottom of Fig. 6.8b. Cooling of the solid therefore leads to a continuous change in the compositions of the two phases β_1 and β_2 along the binodal—provided the process is sufficiently slow to allow for establishing a thermal equilibrium. If the kinetics in the solid phase is not fast enough, the compositions of the solid phase β may be preserved in a metastable state below T_c. This feature may be employed in epitaxy to fabricate alloy layers, which cannot be grown using equilibrium-near techniques.

Regions of negative curvature of $g_\beta(x)$ lie within the inflection points of the curve $(\partial^2 g_\beta/\partial x^2 = 0)$. The points are called the spinodes, and their locus as a function of temperature defines the *spinodal* curve. For compositions bounded by the spinodal, a homogeneous solution is unstable against infinitesimal fluctuations in composition, and there is no thermodynamic barrier to the decomposition into phases β_1 and β_2. For regular solutions the spinodal is given by

$$x(1-x) = RT/2\Omega.$$

Epitaxy of multicomponent systems is often performed at low temperatures, and phase separation leading to compositional fluctuations on a small scale may become an issue. The solid may still be deposited in a metastable state when the kinetics is sufficiently slow.

The interaction parameter Ω in (6.21) was assumed independent on composition. Since bond length of alloy constituents usually differ significantly (Fig. 2.8a), a solid solution comprises inherent strain. This leads to a composition dependent $\Omega(x)$: it requires more energy to incorporate a large atom A into a small crystal B in the alloy A_xB_{1-x} at small x than vice versa at large x, yielding $\Omega(x=0) > \Omega(x=1)$. Calculated values of Ω for zincblende semiconductor alloys are given in Table 6.1. Data for zincblende and wurtzite nitrides are reported in [2, 9] and [3], respectively.

It must be noted that spinodal decomposition is significantly affected by elastic strain, which in epitaxial heterostructures is generated by lattice misfit. Experiments demonstrated that strained semiconductor alloys can be grown within the miscibility gap well below the critical temperature T_c [5, 6]. Lowering of T_c and narrowing of the immiscibility range bounded by the spinodal is described by an additive term in the bulk free energy that is proportional to the misfit f (2.20a)–(2.20c) of the in-plane lattice parameter parallel to the layer-substrate interface [7–10]. The calculations show that the suppression of phase separation is related to the nonlinearity of the interaction parameter $\Omega(x)$.

Table 6.1 Calculated interaction parameter Ω of cubic semiconductor alloys at T = 800 K; data are from [4]

Alloy	$\Omega(x=0)$	$\Omega(x=0.5)$	$\Omega(x=1)$
$Al_xGa_{1-x}As$	0.30	0.30	0.30
$In_xGa_{1-x}As$	3.56	2.35	2.68
$GaAs_xP_{1-x}$	1.07	0.86	0.53
$In_xGa_{1-x}P$	4.60	3.07	2.92
$GaAs_xSb_{1-x}$	3.78	3.96	4.51
$Zn_xCd_{1-x}Te$	2.24	2.29	2.87
$Hg_xCd_{1-x}Te$	0.45	0.38	0.31
$Zn_xHg_{1-x}Te$	2.13	1.88	2.15

6.2 Crystalline Growth

Growth of a crystalline layer proceeds by the attachment of particles (atoms, molecules) to the surface. The driving force is provided by the chemical potential treated in the previous section. The growth process is basically described in terms of a structural model and kinetic processes, which an impinging particle experiences until being incorporated into the crystal [11, 12]. In the *initial* stage of growth the formation of the new phase requires, however, some nucleation: small clusters of the particles forming nuclei of the solid. Once stable nuclei are formed the crystal grows according to conditions controlled by properties of the applied phase transition and adjusted growth parameters. In the following we first consider the initial step in terms of the classical theory of nucleation.

6.2.1 Homogeneous Three-Dimensional Nucleation

We discussed in Sect. 6.1 how a driving force may be created by varying temperature or pressure to make a solid phase more stable than a liquid or gas phase. If the system was initially in a homogeneous phase, e.g., entirely liquid, the solid phase is *not* spontaneously formed when the equilibrium boundary is crossed. Instead, the initial phase remains at first in a metastable state of undercooling (of a liquid below the freezing point) or supersaturation (of a vapor below the condensation point). These states are indicated by dashed lines in Figs. 6.2 and 6.7. The reason for the persistence of the no longer most stable phase is the need to create an interface at the boundaries of the new phase. The formation of such an interface consumes some energy, based on the *surface energy* of each phase. The stable new phase is only formed at sufficient undercooling or supersaturation. Any disturbance like, e.g., a slight agitation or charge, will reduce to some extent the required deviation from equilibrium. *Homogeneous nucleation* occurs spontaneously and randomly in a homogeneous initial phase. Nucleation is strongly facilitated in presence of preferential *nucleation sites*. Such a site is any inhomogeneity in the metastable phase, e.g., a substrate or suspended minute particles. In the case of such *heterogeneous nucleation*, some energy is released by the partial destruction of the pre-existing interface.

Nucleation is the onset of the phase transition in small regions called nuclei: regions of the new phase with an interface at the boundaries to the initial phase. If a potential nucleus is too small, the energy that would be released by forming its volume is not enough to create its surface, and nucleation does not proceed. The critical size of a nucleus required for growth in homogeneous nucleation is described by thermodynamics.

The creation of a nucleus is accompanied by a change of Gibbs energy ΔG_{N}. This quantity is composed of three contributions. First some amount of substance enters the new stable phase, liberating an energy ΔG_{V} proportional to the volume of the nucleus. This process is favorable for the system, and ΔG_{V} is hence a negative quantity. Secondly the interface between the new stable phase and the metastable surrounding phase must be created, yielding a positive cost ΔG_S. This contribution of surface free energy is proportional to the area of the interface.

The *surface free energy* γ may be regarded as reversible work to form a unit area of surface (or interface) at equilibrium with constant system volume and number of components. For a crystal γ is always a positive quantity, because otherwise the crystal phase would not be stable. In a solid the magnitude of the surface energy γ per atom is roughly given by half the heat of melting per atom. This rule of thumb originates from the imagination that melting breaks all bonds of an atom, while only half of the bonds of a surface atom are broken. The surface free energy γ is related to the surface stress tensor γ^S with the in-plane components γ_{ij}^S, $i, j = 1, 2$. The γ_{ij}^S describe the work per unit area for an elastic strain of the surface and may have either sign. A negative sign for compressive strain indicates that work is released when the surface area decreases. The surface stress γ_{ij}^S designates the difference between the stress field in the bulk of the crystal $\sigma_{ij}^{\text{bulk}}$ used in Hooke's law (2.10) and the modified stress field $\sigma_{ij}(z)$ in the region near the surface located at $z = 0$: $\gamma_{ij}^S = \int (\sigma_{ij}(z) - \sigma_{ij}^{\text{bulk}})\, dz$. The relation between surface stress γ_{ij}^S and the surface free energy γ is given by the *Shuttleworth relation*

$$\gamma_{ij}^S = \gamma \delta_{ij} + \left. \frac{\partial \gamma}{\partial \varepsilon_{ij}} \right|_{T,\mu_i},$$

δ_{ij} being the Kronecker symbol and ε_{ij} the in-plane components of the strain tensor [13]. In a liquid the second summand equals zero, because a liquid does not resist to strain: If the surface is increased (e.g. by tilting a half-filled vessel) material just flows from the bulk of the liquid to enlarge the surface. In a liquid surface energy γ and surface stress γ^S are equal.

A third term ΔG_E arises in the change of Gibbs energy ΔG_N upon creation of a nucleus if the nucleus is subjected to elastic stress. ΔG_E is a positive quantity which counteracts the favorable volume term ΔG_V and may even suppress nucleation if $\Delta G_E > |\Delta G_V|$. The total change of Gibbs energy is given by

$$\Delta G_N = \Delta G_V + \Delta G_S + \Delta G_E. \tag{6.22}$$

We consider an *unstrained* nucleus and neglect the last term ΔG_E to obtain a simple expression for the characteristics of homogeneous nucleation. Strain effects are considered in the next section. As a further simplification we assume an *isotropic* surface free energy γ expressed by a constant energy per unit surface area. In our approach we consider a nucleus with spherical shape, which is realized for a liquid nucleus in the bulk of a vapor phase. In this case we obtain $\Delta G_S = \gamma \Delta S = \gamma 4\pi r^2$ and $\Delta G_V = (\Delta g/v)(4/3)\pi r^3$. Here r is the radius of the nucleus and ΔS its surface;

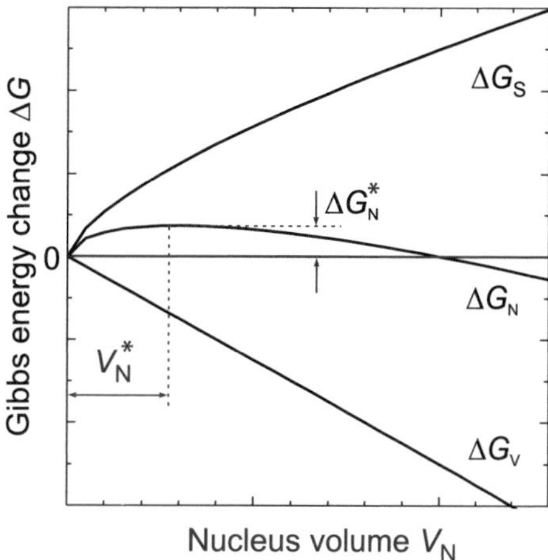

Fig. 6.9 Change of Gibbs energy for creating a spherical nucleus with volume V_N in homogeneous nucleation. V_N^* and ΔG_N^* are the critical nucleus size and the critical work of nucleus creation, respectively

v is the mole volume of the new phase and Δg is the change of Gibbs energy of the system by creating one mole of the new phase. Equation (6.21) then reads

$$\Delta G_N = \Delta G_S + \Delta G_V = 4\pi \left(\gamma r^2 + (\Delta g/v) r^3/3\right) \quad \text{(unstrained spherical nucleus).} \tag{6.23}$$

The molar change of Gibbs energy Δg is negative because the new phase is the stable phase for the set conditions. The sum $\Delta G_N = \Delta G_V + \Delta G_S$ is drawn in Fig. 6.9 as a function of nucleus volume V_N.

We see from Fig. 6.9 that Gibbs energy initially *increases* during formation of the stable phase due to the unfavorable creation of the boundary to the initial phase. The favorable volume term counterbalances the energy cost of the phase boundary as the nucleus grows. Since increase of the absolute value of the volume term ($\propto r^3$) preponderates that of the surface term ($\propto r^2$) Gibbs energy change ΔG_N passes a maximum at some critical radius r^* as the nucleus grows. The volume term eventually prevails over the energy cost of the phase boundary, and Gibbs energy change induced by the creation of the nucleus becomes negative.

The critical nucleus size denoted by the critical volume V_N^* follows from the maximum condition of ΔG_N in (6.23) and is attained at the critical radius

$$r^* = -\frac{2v\gamma}{\Delta g}. \tag{6.24}$$

v designates the molar volume in the new stable phase and Δg is the change of molar Gibbs energy for a transition to this phase. Inserting r^* into (6.23) yields the corresponding critical work required to create such nucleus,

$$\Delta G_N^* = (16/3)\pi \gamma^3 (v/\Delta g)^2. \tag{6.25}$$

The critical free energy of nucleation ΔG_N^* represents an activation energy for the formation of stable nuclei. The maximum condition of (6.23) $\partial(\Delta G_N)/\partial r = 0$ implies that the critical nucleus is in thermodynamic equilibrium with the metastable phase at $r = r^*$. The chemical potential of the nucleus is then identical to that of the surroundings. This equilibrium is labile: Gibbs energy degreases for both a decrease and an increase of nucleus size. Nuclei with a radius smaller than r^* are instable and tend to dissolve. Above r^* nuclei tend to grow since addition of particles to clusters larger than the critical radius releases, rather than costs, energy. Once ΔG_N gets negative the nucleus is stable. Growth is then no longer limited by nucleation, albeit it may still be limited by supply of particles, i.e. by a limited diffusion, or by reaction kinetics. Growth of the stable phase is now more favorable until thermodynamic equilibrium is restored.

If we apply (6.24) to supercooling expressed by (6.15) we obtain an approximate expression for the critical radius r^* of a nucleus depending on the deviation $\Delta T = T - T_e$ from the equilibrium temperature T_e, yielding

$$r^* \cong \frac{2\gamma v T_e}{\Delta h_{trans} \Delta T},$$

Δh_{trans} being the change of molar enthalpy by the phase transition, i.e., the molar heat of crystallization. In this relation only the linear term in (6.15) was considered, and we used $g = \mu$ for a single-component system. Implicitly we assumed that the properties of the system in the nucleus are identical to those in the macroscopic phase. We similarly may apply (6.24) to supersaturation conditions $\Delta P = P - P_e$ described by (6.16), yielding

$$r^* \cong \frac{2\gamma v}{\Delta v_e \Delta P}.$$

Here Δv_e is the difference of molar volume in the metastable and stable phases, respectively.

Homogeneous nucleation induces randomly a spontaneous phase transition in the initial phase. This may, e.g., occur in metalorganic vapor phase epitaxy if instable precursors react in the gas phase before being transported to the substrate. In epitaxy such a situation must generally be avoided to maintain control over growth. Epitaxial growth is therefore performed *below* the critical supersaturation or supercooling required for homogeneous nucleation.

6.2.2 Heterogeneous Three-Dimensional Nucleation

Nuclei form in heterogeneous nucleation at preferential sites. In epitaxy such sites are provided by the crystalline substrate. Due to the presence of a pre-existing solid phase the interface area to the ambient metastable phase is reduced. The energy barrier to nucleation ΔG_N^* is therefore substantially smaller. Since a part of the substrate is

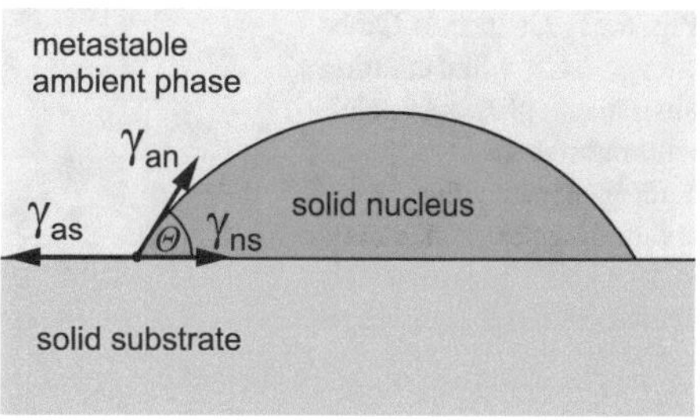

Fig. 6.10 Nucleus created on a substrate. The balance of surface energies γ leads to a wetting angle Θ

covered by the nucleus a new interface is created between nucleus and substrate. The total change of Gibbs energy ΔG_N in heterogeneous nucleation is described by the sum of volume, surface and stress terms given in (6.22) in the same way as in homogeneous nucleation.

We again at first neglect elastic stress term ΔG_E in (6.22) and consider its effect at the end of this section. We assume a nucleus with the shape of a spherical cap with radius r on a substrate as depicted in Fig. 6.10. Such a case is realized for a liquid nucleus on a structureless substrate. The balance of interface tensions at the line of contact between the three phases of metastable ambient (index a), nucleus (n), and substrate (s) is given by three quantities which represent the energies needed to create unit area of each of the three interfaces. From the figure we read *Young's relation* for the absolute values of tensions in balance $\gamma_{as} = \gamma_{ns} + \gamma_{an} \cos\Theta$, or

$$\cos\Theta = \frac{\gamma_{as} - \gamma_{ns}}{\gamma_{an}}. \tag{6.26}$$

The wetting angle Θ may vary between 0 and 180° depending on the degree of wetting corresponding to the affinity of nucleus and substrate materials. Θ hence determines the shape of the nucleus. The volume of a spherical cap is $(4/3)\pi r^3 \times f$, where the geometrical factor f depends on Θ according to

$$f = \left(2 - 3\cos\Theta + \cos^3\Theta\right)/4.$$

The shape factor f reduces the volume of the nucleus with respect to homogeneous nucleation and accordingly the volume term of Gibbs energy change ΔG_N, yielding

$$\Delta G_V^{hetero} = f \times \Delta G_V^{homo}.$$

The surface term ΔG_S is now composed of two parts related to the interfaces ambient-nucleus and nucleus-substrate. Including the corresponding geometrical factors we obtain the critical nucleus size r^*_{hetero} in the same way as in homogeneous nucleation,

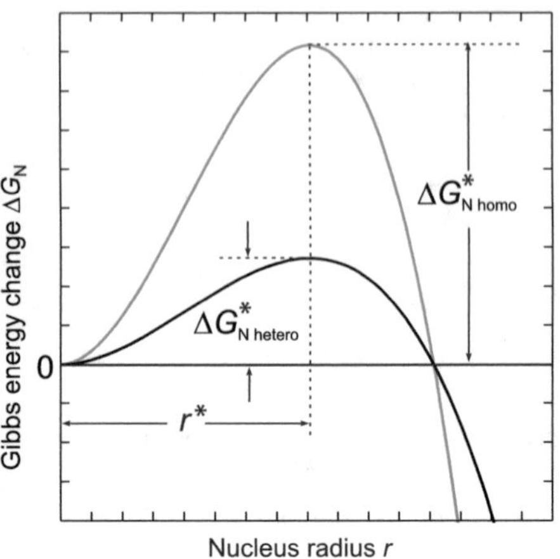

Fig. 6.11 Change of Gibbs energy ΔG_N when creating unstrained spherical nuclei with radius r in heterogeneous (*black curve*) and homogeneous nucleation (*gray*). r^* is the critical nucleus radius

leading to $r^*_{\text{hetero}} = r^*_{\text{homo}}$ given by (6.24). The critical free energy of nucleation $\Delta G^*_{\text{N hetero}}$ for (unstrained) heterogeneous nucleation is then

$$\Delta G^*_{\text{N hetero}} = \frac{16\pi}{3} \frac{\gamma^3_{\text{an}} v^2}{(\Delta g)^2} f. \tag{6.27}$$

Comparing (6.27) to (6.25) shows that $\Delta G^*_{\text{N hetero}}$ is reduced with respect to $\Delta G^*_{\text{N homo}}$ by the shape factor f which is controlled by the three surface tensions. Note that there is a smaller amount of substance in the nucleus expressed by f for given r^* in heterogeneous nucleation. The relation of Gibbs energy change in hetero- and homogeneous nucleation is illustrated in Fig. 6.11.

Epitaxy is preferred on substrates which are well wet by the epitaxial layer, so as to obtain a large separation to homogeneous critical supersaturation or undercooling. Furthermore, usually a two-dimensional atomically flat layer-by-layer growth is intended in epitaxy, favored by such a wetting.

Effect of Strain

The structureless surface of the substrate assumed above is certainly oversimplified. The surface structure of the substrate provides specific sites for nucleating atoms. On a plane substrate such sites are defined by the potential created by the topmost substrate atoms. Section 2.2 pointed out how layer atoms accommodate to the lateral spacing of the substrate atoms. Such an alignment also applies for nucleation at the initial stage of growth. The nucleus is consequently strained and accumulates a strain energy ΔG_E, which is—according to Hooke's law—proportional to the difference of the lateral equilibrium atom spacings of layer atoms and substrate atoms. The strain ε enters Gibbs energy ΔG_N of the nucleus in both the volume term ΔG_V and the surface term ΔG_S discussed in the previous section for the unstrained case. ΔG_V

is reduced by an increased chemical potential in the nucleus originating from the strained atomic bonds. The surface energy ΔG_S of the strained nucleus is increased in a complex way due to the anisotropic nature of the strain and the effect of a partial elastic strain relaxation at side faces (compare Fig. 7.35a). As a consequence, the energy ΔG_N^* to form a nucleus of critical size is *increased* in the presence of strain. A simple explicit expression for the effect of the strain ε on the critical free energy of nucleation ΔG_N^* was derived in [14] for the case of a Kossel crystal with first-neighbor interactions in absence of the Poisson effect,

$$\Delta G_{\mathrm{N\ hetero,strained}}^* = \Delta G_{\mathrm{N\ hetero,unstrained}}^* \left(1 - \frac{\varepsilon}{E_1}\right)^2 \left(1 - \frac{2\varepsilon}{\Delta\mu}\right)^{-2}. \tag{6.28}$$

The quantity E_1 is the first-neighbor interaction of the bond between two adjacent atoms, and $\Delta\mu$ is the supersaturation. We note a singularity in (6.28) for $\Delta\mu$ equals 2ε. At large strain the term ΔG_E in (6.23) which counteracts the favorable (negative) volume term ΔG_V in Gibbs energy ΔG_N will even suppress nucleation.

6.2.3 Growth Modes

The surface energies leading to Young's relation (6.26) affect the initial stage of layer deposition on a substrate of different material, i.e., in heteroepitaxy. Epitaxy usually aims at depositing a layer with a smooth growth surface. This corresponds to a wetting angle of 0 in Young's relation, or $\gamma_{as} = \gamma_{ns} + \gamma_{an}$. If this condition applies or γ_{as} exceeds the sum of the two other interface energies, we obtain complete wetting of the layer on the substrate surface. The condition implies that layer atoms are more strongly attracted to the substrate than to themselves. Growth may then proceed in an atomically flat layer-by-layer mode referred to as Frank-Van der Merve growth mode in heteroepitaxy. Figure 6.12 illustrates the initial stages of such two-dimensional layer growth for different thickness of deposited material. Nucleation in this case proceeds by the formation of *two-dimensional* islands or step advancement as outlined in Sects. 6.2.5 and 6.2.7.

A different surface morphology of the layer is observed if layer atoms are more strongly attracted to each other than to the substrate. This situation is expressed in Young's relation (6.26) by a wetting angle of π, or $\gamma_{ns} = \gamma_{as} + \gamma_{an}$. If this condition applies or γ_{ns} is even larger, then the layer does not wet the substrate surface. The surface energy of layer plus substrate is minimized if a maximum of substrate surface (with a low surface energy γ_{as}) is *not* covered by layer material (which has a large surface energy γ_{an}). This results in a three-dimensional growth of the layer referred to as Volmer-Weber growth mode in heteroepitaxy. Figure 6.12 shows that the layer material is deposited in form of islands, which for thicker deposits eventually coalesce. Here, nucleation proceeds by 3D nuclei as pointed out in the preceding sections.

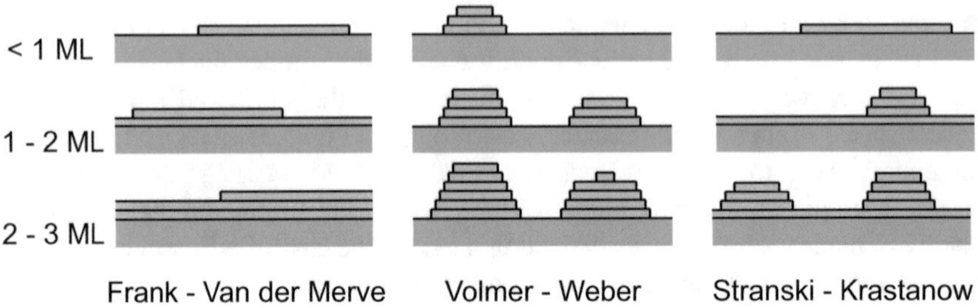

Fig. 6.12 Schematic of the three growth modes in heteroepitaxy, illustrated as a function of approximately equal coverage given in units of monolayers (ML)

Often an intermediate case is found referred to as Stranski-Krastanow growth mode (also termed layer-plus-island growth). Here the condition $\gamma_{as} \geq \gamma_{ns} + \gamma_{an}$ for Frank-Van der Merve growth applies solely for the first deposited monolayer (or the first few monolayers). After exceeding some critical coverage thickness the growth changes to the Volmer-Weber case where $\gamma_{ns} \geq \gamma_{as} + \gamma_{an}$ applies. Such change may be induced by the gradual accumulation of strain in the epitaxial layer (Sect. 2.2.6). The layer material then resumes growth in form of three-dimensional islands, leaving a two-dimensional wetting layer underneath. It should be noted that the critical thickness pointed out here lies usually below that required for the formation of misfit dislocations treated in Sects. 2.2.6 and 2.3. Stranski-Krastanow growth has gained much advertence in recent years, because it may be employed for growth of defect-free quantum dots. We therefore will consider this specific growth mode in more detail separately in Sect. 7.3.1.

The three growth modes depicted in Fig. 6.12 arise from a thermodynamic consideration of the interface energies. It should be noted that layer-by-layer and three-dimensional growth also occur in homoepitaxy, i.e., in growth on a substrate of the same material without a wetting issue. Island growth may here originate from a too high growth rate and a related too short adatom diffusion length; such conditions occur at very low growth temperature or high supersaturation. Furthermore, additional growth modes have been introduced to account for frequently observed surface morphologies. On terraced surfaces *step-flow growth* may be found under suitable conditions; depending on growth parameters also *step bunching* may occur (Sect. 7.2.4). In the presence of screw dislocations *spiral growth* (Sect. 6.2.7) or *screw-island growth* is observed. Eventually for materials with low surface mobility of adatoms *columnar growth* may be obtained. Similar to Volmer-Weber growth islands nucleate, but they do not merge to continuous layers when growth proceeds. A mosaic crystal composed of numerous slightly tilted and twisted columnar crystallites is formed instead.

6.2.4 Equilibrium Surfaces

The Stranski-Krastanow growth mode described above indicates that under certain conditions a rough or faceted surface is more stable than a flat surface. The reason is the pronounced dependence of the surface energy γ of crystalline solids on the orientation of the surface. In absence of kinetic effects and defects the macroscopic shape of a solid is determined by a minimum of Gibbs free energy G which includes a term of its surface G_{surf}. The equilibrium shape is therefore given by a minimum of G_{surf} for a given volume (or amount of substance),

$$G_{\text{surf}} = \oint_A \gamma(\mathbf{n})\, dA \rightarrow \text{minimum}, \tag{6.29}$$

where $\mathbf{n}$ is the surface normal of the respective part dA of the surface considered. In liquids $\gamma(\mathbf{n}) = \gamma$ is isotropic, and the equilibrium shape (in absence of gravity) is consequently a sphere which has minimum surface area. This is illustrated in Fig. 6.13 for a small lead particle annealed slightly below and above the melting point. A spherical shape is observed above the melting point (Fig. 6.13b), whereas the equilibrium shape of solid lead Pb in Fig. 6.13a shows sharp facets of the fcc structure.

In crystalline solids $\gamma(\mathbf{n})$ has pronounced minima for specific faces denoted by their Miller indices (hkl). The equilibrium shape is therefore not given by a minimum surface area but by a minimum of G_{surf}. Such a minimum is attained by a polyhedron built from these particular faces of minimum surface energy. While γ of liquids may be readily obtained from capillary experiments, determination of $\gamma(\mathbf{n})$ from crystal surfaces is difficult. We employ a simple approach to obtain a qualitative estimate on the orientation dependence of γ.

We consider a Kossel crystal which assumes atoms as cubes that are bond by their six faces. Only next-neighbor bonding and only a single species of atoms are presumed. The surface energy is evaluated in the framework of the terrace-step-kink model (Sect. 7.2.1). According to this model a surface may be classified into the three

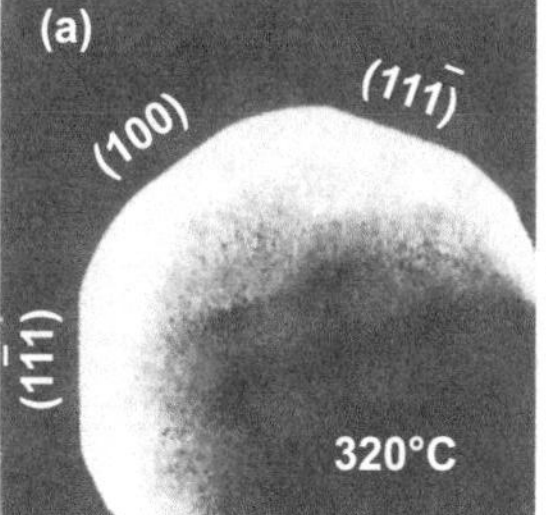

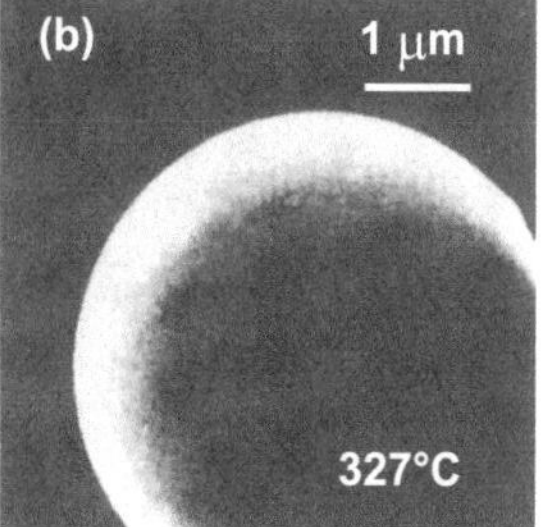

Fig. 6.13 In-situ imaged scanning electron micrographs of a small lead particle slightly below (**a**) and above (**b**) the melting point. Indices in (**a**) signify equilibrium facets. Reproduced with permission from [15], © 1989 Elsevier

categories *singular* (i.e., perfectly oriented low index plane), *vicinal* (i.e., slightly inclined with respect to a singular surface), and *rough*. Let us first consider a singular (001) surface of a Kossel crystal. The surface energy γ is given by the sum of the energies of all bonds broken to create this surface, i.e.

$$\gamma = \frac{E_B}{2}\nu. \tag{6.30}$$

ν is the number of bonds on the surface per unit area, E_B is the bond energy and the factor $\frac{1}{2}$ takes into account that 1 bond connects 2 atoms (i.e., actually *two* surfaces are created when a bulk crystal is cleaved into two parts). E_B is related to the sublimation energy ΔH_S: Releasing an atom from the bulk requires breaking 6 next-neighbor bonds, leading to the relation

$$\Delta H_S = \frac{6}{2}E_B, \quad \text{or} \quad E_B = \frac{1}{3}\Delta H_S. \tag{6.31}$$

The number ν of bonds per unit area of a (001) surface with $n_1 \times n_2$ atoms along the two orthogonal lateral directions is just $\nu = n_1 \times n_2/(n_1 \times n_2 \times a^2) = 1/a^2$.

To obtain an expression for the orientation dependence of γ we evaluate a vicinal surface which is inclined by a small angle Θ with respect to the (001) plane. Figure 6.14 shows such a vicinal surface of a Kossel crystal. The surface consists of flat terraces separated by steps. We assume the steps are regularly spaced (i.e., absence of step-step interactions) and of monoatomic height.

The number ν of bonds per unit area of the vicinal surface is larger than that of the singular surface and given by

$$\nu = \frac{1}{a^2}(\cos\Theta + \sin\Theta).$$

Inserting this expression into (6.30) and using (6.31) yields

$$\gamma = \frac{\Delta H_S}{6a^2}(\cos\Theta + \sin\Theta). \tag{6.32}$$

Since the angle dependence in the brackets is always greater than unity, the surface energy of a vicinal surface always exceeds that of the corresponding singular surface.

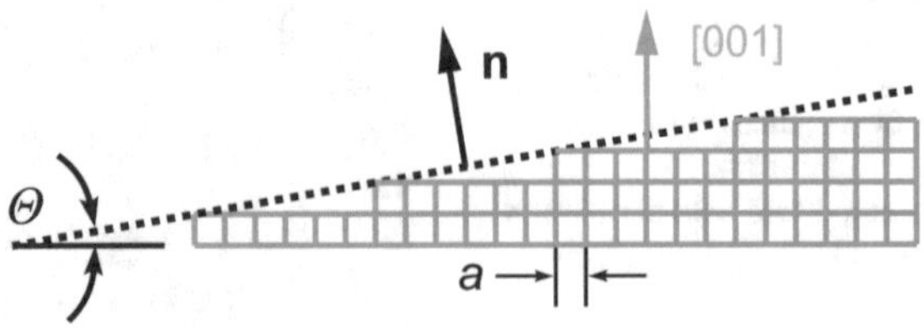

Fig. 6.14 **a** Cross section of a Kossel crystal (*gray*) with a vicinal (016) surface (*black, dotted*), inclined by a small angle Θ with respect to the singular (001) plane

The excess energy originates from the ledge atoms located at the steps. These atoms have an additional missing bond with respect to atoms located on a terrace site.

Relation (6.32) leads to a geometrical construction to represent the surface energy of a solid: the *Wulff plot*. Equation (6.32) describes a circle (in three dimensions a sphere) with diameter $2r = \Delta H_S/(6a^2)$. In polar coordinates a circle which passes through the origin is given by

$$\gamma = 2r\cos(\Theta - \varphi) = 2r(\cos\Theta\cos\varphi + \sin\Theta\sin\varphi),$$

see Fig. 6.15a. If the origin of the circle lies on a line inclined by $\varphi = \pi/4$ with respect to a coordinate axis as illustrated in the figure we obtain an expression like (6.32). Due to the symmetry of the considered cubic crystal the entire three-dimensional representation of the surface energy $\gamma(\Theta)$ is actually described by eight interpenetrating spheres, the origins of which are lying on the four $\langle 111\rangle$ axes. Figure 6.15 is a two-dimensional cross section of $\gamma(\Theta)$ for an azimuth angle of 0. We note that the surface energy γ has pronounced minima (cusps) on the low-index $\langle 100\rangle$ axes.

The equilibrium shape of the Kossel crystal with next-neighbor bonds is found from the surface energy given in Fig. 6.15a by drawing lines from the origin to all possible crystallographic directions through the γ plot. At a point of intersection with the surface of the plot, a plane is constructed perpendicular to the line. Such *Wulff plane* has an equal γ as the crystal at the intersection point for the orientation given by the line. The equilibrium shape of the crystal is the inner envelope of all such possible Wulff planes. The construction is called *Wulff plot*. Figure 6.15b shows the Wulff plot of the Kossel crystal with next-neighbor bonds. Since the surface energy of the $\{100\}$ faces is much lower than that of all other crystallographic faces, the polyhedron given by the inner envelope of all planes is a cube. This is no longer the case if also second-next neighbor interactions of the atoms are taken into account. Atoms in the considered cubic Kossel crystal have 12 second-next neighbor atoms located along the $\langle 110\rangle$ directions. Their bonds are weaker than those of next-neighbor atoms. Consequently their contribution to the surface energy is also weaker. Figure 6.16

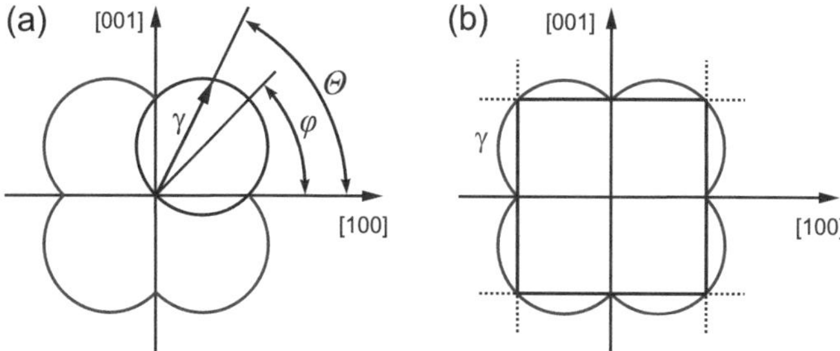

Fig. 6.15 **a** Construction of the surface energy $\gamma(\Theta)$ of a Kossel crystal (*blue curves*) by *circles* passing through the origin. **b** Wulff plot of the equilibrium crystal shape (*black lines*), constructed from the surface energy given in (**a**)

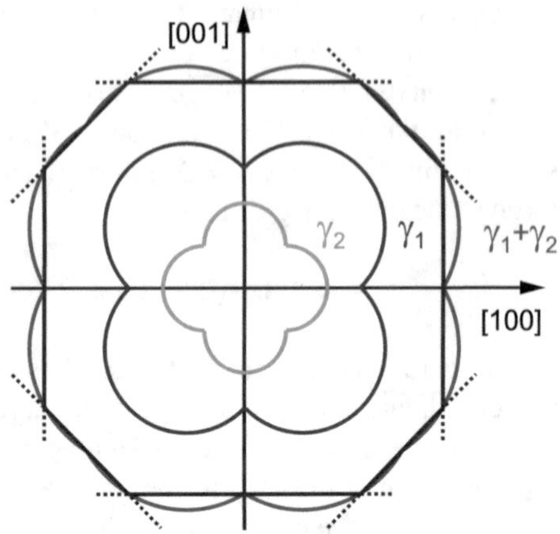

Fig. 6.16 Surface energy $\gamma = \gamma_1 + \gamma_2$ of a cubic Kossel crystal with first (γ_1) and second (γ_2) nearest neighbor interactions taken into account. The *black polygon* represents a cross section of the equilibrium shape of the crystal

shows a two-dimensional representation of the first and second-nearest neighbor contributions to γ. The contribution γ_2 of the weaker second-nearest neighbor bonds to the surface energy γ is constructed in the same way as that of nearest neighbor bonds γ_1, leading to a smaller $\gamma(\Theta)$ plot with cusps along the $\langle 110\rangle$ bond directions. Consequently the sum $\gamma = \gamma_1 + \gamma_2$ has also (local) minima along these directions, and the Wulff plot results in a polyhedron with more faces than obtained in the case with only next-neighbor bonds.

The equilibrium crystal shape is generally composed of more crystallographic inequivalent facets if the distance of bonding forces increases. Surfaces of small surface energy γ lead to larger facets. On the other hand the facets referring to surfaces with too large surface energy γ lie far away from the origin in a Wulff plot and can hence not contribute to the equilibrium shape. The condition to appear in the equilibrium crystal shape of a cubic crystal in addition to the $\{001\}$ facets reads for any surface hkl

$$\gamma_{hkl} < \frac{h+k+l}{\sqrt{h^2+k^2+l^2}} \times \gamma_{001}.$$

According to this relation e.g. $\{111\}$ facets appear on a cubic crystal if $\gamma_{111} < \sqrt{3}\gamma_{001}$ applies.

A general expression of the minimum surface-energy condition (6.29) may be given in terms of *Wulff's theorem*

$$\frac{\gamma_1}{r_1} = \frac{\gamma_2}{r_2} = \frac{\gamma_3}{r_3} = \cdots = \text{const.} \tag{6.33}$$

The expression is a concise summarization of Wulff's construction illustrated above. The indices in (6.33) represent the (hkl) sets of the relevant surfaces, and r is the absolute value of the vectors r_{hkl} parallel to the corresponding surface normals $\mathbf{n}_{hkl}$ with $r_{hkl} \propto \gamma_{hkl}$.

Table 6.2 Calculated surface energies γ of low-index surfaces, data are given in J/m^2. Type of reconstruction noted at the listed data indicate the corresponding lowest-energy surface selected from various calculated configurations, (1×1) relaxed denotes an unreconstructed cleavage surface

Solid	(100)		(110)		(111)	
Si[a]	1.41	$c\,(4\times4)$	1.70	(1×1) relaxed	1.36	7×7
Ge[a]	1.00	$c\,(4\times4)$	1.17	(1×1) relaxed	1.01	$c(2\times8)$
GaAs[b]	~0.96	$\beta\,2(2\times4)$	0.83	(1×1) relaxed	0.87	(2×2) Ga vacancy
InAs[c]	0.75	$\beta\,2(2\times4)$	0.66	(1×1) relaxed	0.67	(2×2) In vacancy
InP[d]	~0.99	$\beta\,2(2\times4)$	0.88	(1×1) relaxed	0.99	(2×2) In vacancy

References [16–19]

Surface energies $\gamma(\mathbf{n})$ of solids are difficult to measure. During the 1990s first-principles calculations based on slab configurations have reached a maturity to yield sufficient numerical precision. It must be noted that $\gamma(\mathbf{n})$ for a given orientation depends significantly on the structure of the surface. The equilibrium structure may in turn depend on the chemical potential of the ambient. Data given in Table 6.2 represent the lowest values determined for different reconstructions. In the case of GaAs the reconstruction for the broadest range of the anion potential is given. Calculations of data given for the (100) face assumed ambient conditions $\mu_{\mathrm{As}} = \mu_{\mathrm{As}}^{\mathrm{bulk}} - 0.3$ eV. Details of surface reconstructions are considered in Chap. 7.

The equilibrium crystal shape follows from the surface energies $\gamma(\mathbf{n})$. Applying Wulff's construction and using calculated surface energies for As-rich conditions we obtain the equilibrium shape of a GaAs crystal shown in Fig. 6.17a [17]. The polyhedron is composed of low-index planes reflecting the symmetry of the zincblende lattice. It should be noted that the (111) face and the $(\bar{1}\bar{1}\bar{1})$ face show a different dependence on the ambient chemical potential and hence vary differently in size as the chemical potential is changed. The cross section of the equilibrium crystal shape shown in Fig. 6.17b demonstrates the dependence on the chemical potential μ_{As}. A high value of μ_{As} corresponds to an As-rich (i.e., Ga-poor) ambient. The corresponding shape (black polyhedron) is smaller than that expected for a Ga-rich environment (gray polyhedron): As-terminated surfaces have surfaces energies about 20% smaller than those in a Ga-rich ambient [17]. For equilibrium crystal shapes of wurtzite crystals see [20, 21].

Growth occurs at some deviation from thermodynamic equilibrium as pointed out in Sect. 6.1.5. For a single-component system we obtained homogeneous critical nuclei with radius r^* for a given driving force $\Delta\mu$. The respective relation (6.24) was derived assuming unstrained spherical nuclei and an *isotropic* surface energy γ.

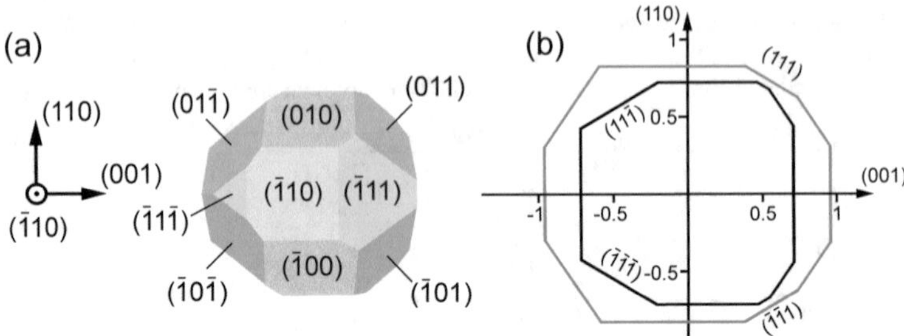

Fig. 6.17 **a** Calculated equilibrium crystal shape of GaAs in As-rich ambient ($\mu_{\text{As}} = \mu_{\text{As(bulk)}}$). The polyhedron is constructed from the different surface energies of (100), (110), (111), and ($\bar{1}\bar{1}\bar{1}$) facets. **b** Cross sections calculated for different chemical potentials: *black lines* $\mu_{\text{As}} = \mu_{\text{As(bulk)}}$, *gray lines* $\mu_{\text{As}} = \mu_{\text{As(bulk)}} - 0.3$ eV. Energies are given in J/m^2. After [17]

We now take different surface energies γ_i of inequivalent facets into account, modifying (6.24) to

$$|\Delta\mu| = \frac{2\nu\gamma_i}{r_i}, \tag{6.34}$$

where i denotes (hkl) sets of the facets. At equilibrium the chemical potentials of all facets must be equal, leading to constant ratios γ_i/r_i, and consequently to Wulff's theorem (6.33). Equation (6.34) says that the critical size of stable nuclei decreases with increasing deviation $|\Delta\mu|$ from equilibrium.

The shape of a *growing* crystal deviates from the equilibrium crystal shape. From (6.34) we note that the differences to the equilibrium get smaller for large dimensions r_i. The driving force to maintain the equilibrium crystal shape therefore gets negligible for macroscopic crystals. Their shape is instead determined by kinetic conditions. Surfaces which grow most slowly eventually build the facets of a crystal, as pointed out below. The consequence of a slow growth rate of a facet is illustrated in Fig. 6.18. We assume slowest growth of {111} facets, and, for simplicity, an equal growth rate of a facet along a given crystallographic direction and the inverse direction, i.e., for $\{hkl\}$ and $\{\bar{h}\bar{k}\bar{l}\}$ facets. Starting from an initial stage 1 with almost spherical crystal shape made from the facets indicated in Fig. 6.18, we note that successively all faster growing facets disappear. At stage 4 only the slowly growing {111} facets remain. Differences in the growth rates of inequivalent facets can be utilized to fabricate faceted structures. This topic is discussed in Sect. 12.1.3.

Real crystals may deviate from the ideal growth form discussed above. Local differences in the supersaturation and transport to growing surfaces lead to differences in the growth rate of equivalent facets and consequently to different areas of these facets, i.e., to a parallel displacement of the facets with respect to the ideal form. Moreover, crystal defects may significantly accelerate the growth rate of facets. Particularly screw dislocations build a continuous source of surface steps which facilitate the attachment of particles impinging on the surface and the subsequent incorporation into the crystal. Kinetic processes in the growth at steps are considered in Sect. 7.2.4.

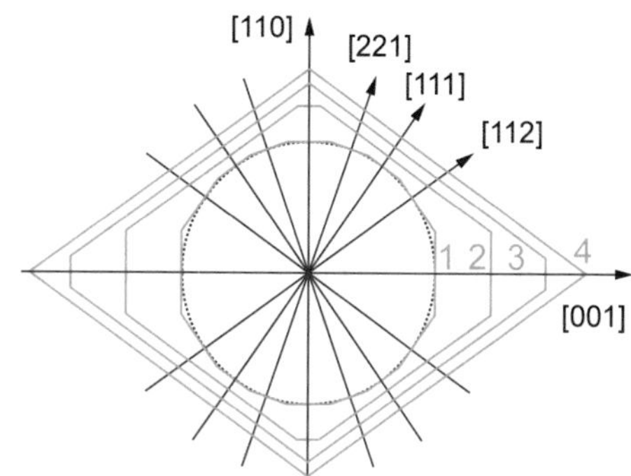

Fig. 6.18 Facets of a growing crystal at successive stages. Different growth rates are assumed along the indicated directions, with a minimum along ⟨111⟩

6.2.5 *Two-Dimensional Nucleation*

Flat surfaces have generally slow growth rates, because they hardly offer favorable sites like steps for the attachment of atoms from the ambient phase. In absence of terrace steps and defects like screw dislocations, growth of atomically flat surfaces in the layer-by-layer mode proceeds by clusters of adsorbed atoms with a sufficient size. They form *two-dimensional* (2D) nuclei which provide preferential sites for the attachment of further atoms at their perimeter. Which mode of nucleation, 2D or 3D, is thermodynamically preferred depends on the difference in interface energies and strain, cf. Sect. 6.2.3. 2D nuclei must exceed a critical size for stable growth in analogy to three-dimensional nuclei treated in the previous section.

We consider the critical size of a two-dimensional nucleus for a one-component system. For simplicity we assume a circular shape of the 2D nucleus with radius ρ and monoatomic height as depicted in Fig. 6.19.

The formation of a 2D cluster comprising an amount of substance n changes the Gibbs energy due to the transition from the ambient phase to the more stable solid phase by $\Delta G_n^{2D} = n\Delta\mu$. Since in the assumed one-component system the 2D cluster grows on a surface made of atoms of the same kind, no change of the interface energy of the system occurs. There is, however, a contribution ΔG_γ^{2D} originating from the ledge atoms of the cluster. They have fewer neighbors and thus more unsaturated bonds, which add a positive destabilizing boundary free energy to the total free energy of the 2D cluster. The contribution ΔG_γ^{2D} is the 1D analog of the surface free energy, which is generally measured as the surface tension. ΔG_γ^{2D} can hence be considered as a line tension and is proportional to the length of the 2D cluster perimeter $2\pi\rho$, ρ being the radius of the nucleus. A specific free step energy γ_p assumed to be isotropic

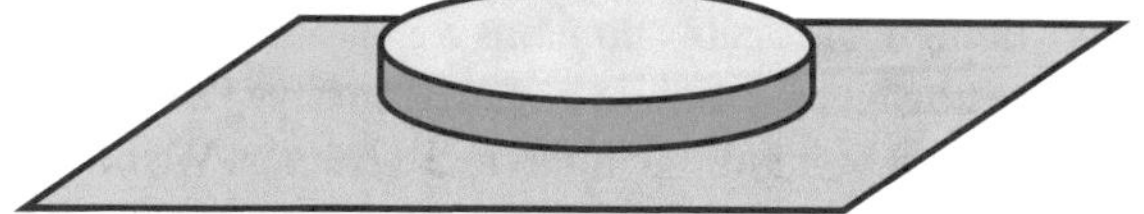

Fig. 6.19 Two-dimensional nucleus of circular shape with monoatomic height on a flat surface

is taken as proportionality factor. If one mole of atoms in a circular 2D island covers an area a, the change of Gibbs energy by the formation of a 2D cluster is given by

$$\Delta G_{\mathrm{N}}^{\mathrm{2D}} = \Delta G_{n}^{\mathrm{2D}} + \Delta G_{\gamma}^{\mathrm{2D}} = \frac{\pi \rho^2}{a} \Delta\mu + 2\pi\rho\gamma_{\mathrm{p}}. \tag{6.35}$$

The critical 2D nucleus size is obtained similar to the 3D case from the maximum condition of $\Delta G_{\mathrm{N}}^{\mathrm{2D}}$, yielding the critical radius of a two-dimensional nucleus

$$\rho^* = \frac{a\gamma_{\mathrm{p}}}{|\Delta\mu|}, \tag{6.36}$$

and the activation energy

$$\Delta G_{\mathrm{N}}^{\mathrm{2D}\,*} = \pi \frac{a\gamma_{\mathrm{p}}^2}{|\Delta\mu|}. \tag{6.37}$$

a designates the specific area covered by one mole of atoms, and γ_{p} the excess energy per length of atoms located at the perimeter of the in 2D nucleus with respect to the energy of atoms located on a flat surface. The prefactor π in (6.37) applies for a circular shape of the 2D nucleus and is to be replaced by another form factor for a different shape (e.g., $(1+\alpha)^2/\alpha$ for rectangular nuclei, α being the ratio of the side lengths). Relations (6.36), (6.37) are analogous to (6.24) and (6.25) obtained for three-dimensional nuclei. We note that the driving force $\Delta\mu$ enters the 2D activation linearly in the denominator, while a quadratic dependence is found in the 3D case (6.25). For a given small deviation from equilibrium $\Delta\mu$ the formation energy of a 2D nucleus is hence smaller than in homogeneous 3D nucleation.

The shape of 2D nuclei and subsequently growing islands is not necessarily given by the equilibrium crystal shape, due to the required deviation from equilibrium. Furthermore, the shape is affected by the structure of the substrate surface and kinetic effects considered in more detail in Chap. 7.

Evidence for such effects are given in Fig. 6.20. The in situ recorded scanning tunneling micrographs show two Si islands growing on a Si(111) substrate [22]. During growth the islands clearly show a triangular shape (Fig. 6.20a). When the growth is interrupted and the temperature is raised, the lateral form gradually turns into a hexagon-like shape (Fig. 6.20b). Here, edges perpendicular to $[11\bar{2}]$ are somewhat shorter than those perpendicular to $[\bar{1}\bar{1}2]$, indicating a lower energy of the latter. When growth is resumed, the islands quickly adopt the initial triangular shape (Fig. 6.20c).

To provide some visual evidence for critical 2D nucleus formation, we consider the assembly of Co-Si clusters at 400 °C on a Si(111)-(7 × 7) surface. High speed scanning tunneling microscopy represents one such cluster as a single bright protrusion, though it probably contains 3 Si atoms and 6 Co atoms [23]. STM images show that the clusters are mobile on the Si surface [24]. Figure 6.21 shows a series of STM scans over 8 × 8 nm^2 captured in 5 s frames. We note a steady change of the imaged configuration, until the cluster labeled Y occupies the vacancy visible in Fig. 6.21c.

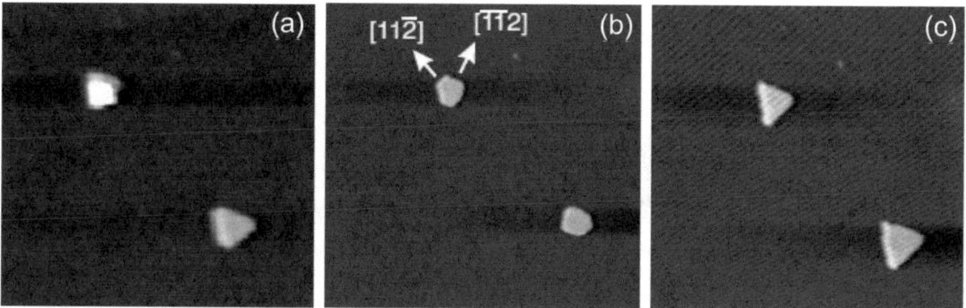

Fig. 6.20 Scanning tunneling images of two-dimensional Si islands on Si(111) substrate, recorded during molecular beam epitaxy at 725 K (**a**, **c**) and after 18 min annealing at 775 K (**b**). The image size is 55×55 nm^2. Reproduced with permission from [22], © 2004 IOP

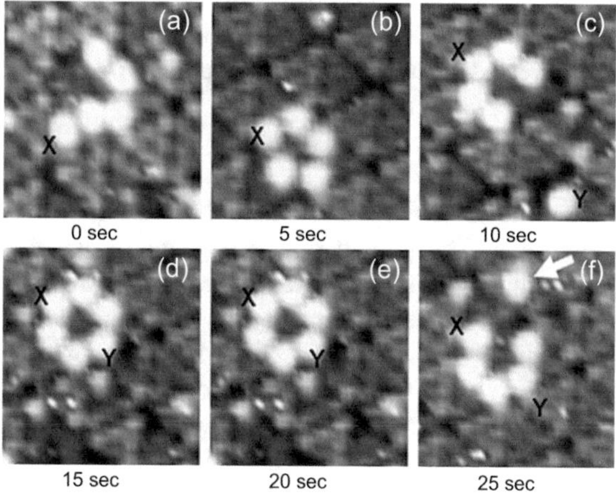

Fig. 6.21 In situ recorded formation of a critical nucleus of Co-Si clusters (*white protrusions*) on a Si(111)-(7×7) surface. X and Y label two particular clusters. At (**d**) cluster Y has moved to complete a ring-like structure consisting of 6 clusters. Reproduced with permission from [24], © 2007 RSC Publishing

The ring composed of 6 clusters is fairly stable. Eventually a cluster detaches (arrow in Fig. 6.21f) and the ring-structure decomposes. Studies reported in [24] show that the $i = 6$ configuration represents a critical nucleus for the studied system. Once one of the 6 clusters moves to occupy the vacant site in the center and a seventh cluster attaches to the ring, a stable nucleus with $i = 7$ clusters is formed.

The critical free energy of nucleation $\Delta G_{\mathrm{N}}^{\mathrm{2D}*}$ (6.37) represents an activation energy for the formation of 2D nuclei. The rate of formation per unit area of such nuclei is given by the Arrhenius dependence

$$j^{\mathrm{2D}} = j_0^{\mathrm{2D}} \exp\left(\frac{\Delta G_{\mathrm{N}}^{\mathrm{2D}*}}{kT}\right). \tag{6.38}$$

The prefactor j_0^{2D} follows from kinetic considerations of the nucleation process on an atomic scale.

6.2.6 Island Growth and Coalescence

Using j^{2D} (given in units of $\mathrm{m}^{-2} \times \mathrm{s}^{-1}$) we obtain the total nucleation rate J^{2D} on a surface of a given area $L \times L$ by the product $J^{2D} = j^{2D} \times L^2$. Two-dimensional growth of a layer proceeds by a lateral growth of 2D nuclei, which eventually coalesce and form a complete coverage of the surface by a new monolayer. The 2D nuclei grow by the attachment of atoms at the step formed at their perimeter. The time T required to grow a monolayer thick complete coverage may therefore be expressed in terms of the velocity of step advancement v_{step} by the relation $T \cong L/v_{\mathrm{step}}$. The number of nuclei N created during the time interval T of one monolayer formation is then given by

$$N \cong j^{2D} \times L^2 \times T \cong j^{2D} \times L^3/v_{\mathrm{step}}.$$

From this relation we may derive an estimate for the velocity R of the growth front along the surface normal. The growth rate R depends on whether only a single nucleus grows and eventually completes a new monolayer on the considered area L^2, or multinuclear (or even multilayer) growth occurs. In the first case a complete monolayer originates from only a single nucleus, i.e., the number of new forming additional nuclei is $N < 1$. This condition corresponds to $L < \sqrt[3]{v_{\mathrm{step}}/j^{2D}}$, i.e., it is favored for a small area. The two-dimensional growth of the epilayer proceeds by a formation of a nucleus and its lateral growth until the completion of a monolayer, before the subsequent nucleus is created. The growth rate R is then given by the total nucleation rate J^{2D} times the thickness d of the 2D nucleus, i.e.,

$$R = j^{2D} L^2 d \quad (N < 1,\ \text{single nucleus}). \tag{6.39}$$

Generally the condition $N < 1$ does not apply, because the area of an epilayer is large (compared to the diffusion length of an atom on the surface) and supersaturation is not thus small to allow only for the creation of just a single nucleus. The more general case $N > 1$ for the number of nuclei N created during the time interval T of monolayer growth corresponds to $L > \sqrt[3]{v_{\mathrm{step}}/j^{2D}}$. More than just one nucleus is formed, and all nuclei contribute to the growing layer. In such a multinuclear layer-by-layer growth nuclei have a mean time t to grow to 2D islands with a mean radius l before meeting another growing 2D island originating from another nucleus. This time interval is of the order $t \cong l/v_{\mathrm{step}} \cong 1/J^{2D} \cong (j^{2D} \times \pi l^2)^{-1}$. The mean radius may therefore be written $l \cong \sqrt[3]{v_{\mathrm{step}}/(\pi j^{2D})}$. The growth rate R is in this case given by

$$R \cong j^{2D} \pi l^2 d \cong d \sqrt[3]{\pi v_{\mathrm{step}}^2 j^{2D}} \quad (N > 1,\ \text{many nuclei}), \tag{6.40}$$

where d is the thickness of the 2D nuclei. If growth is initiated on an flat surface without steps, the first nuclei appear simultaneously. When the isolated nuclei (which subsequently become islands) grow, their total perimeter increases. The total perimeter represents also the total step length. Layer growth proceeds by the overall advancement of this step via the attachment of atoms, which arrive at the surface from the ambient and diffuse laterally. Once the growing islands come into contact, they coalesce and their steps annihilate. The total step length therefore decreases at the onset of island coalescence. If no additional nucleation occurs on top of large and still growing 2D islands, the total step length reaches a minimum (0 without any nucleation) when the layer is completed. The total step length oscillates between a minimum and a maximum. Since the incorporation of atoms into the solid occurs at surface steps, also the growth rate oscillates. The oscillation period corresponds to the growth duration of one monolayer, i.e., $T \cong (\pi v_{\text{step}}^2 j^{2D})^{-1/3}$. Usually after deposition of a few monolayers there occurs gradually additional nucleation on large and still growing 2D islands, giving rise to growth on various heights. As a result, the oscillations are gradually damped and the growth rate eventually approaches a constant value. Such a behavior is often observed in the intensity of the specular reflectivity of reflection high energy electron diffraction (RHEED) applied during growth in molecular-beam epitaxy (Sect. 8.3): A high diffraction intensity reflected from a smooth completed layer changes periodically with a low intensity reflected from a rough layer composed of many 2D islands.

Some of the growth features mentioned above are illustrated in the series of scanning tunneling micrographs Fig. 6.22, which were recorded in situ during homoepitaxial growth of Si/Si(111) [25]. The two-dimensional islands have a triangular shape, similar to those shown in Fig. 6.20a, c and those studied in more detail in Sect. 7.2.8. In Fig. 6.22 epitaxy does not proceed in a layer-by-layer mode at the given growth conditions. Instead, additional nucleation occurs already before islands start to coalesce (arrow in the middle of the frames). We still observe lateral growth of the islands at their perimeter. Eventually the islands coalescence and the total perimeter decreases (Fig. 6.22d). The left arrow in Fig. 6.22c marks an island, which is rotated by 180° with respect to the other islands. This orientation corresponds to a stacking fault along the [111] growth direction. The chosen growth temperature is significantly

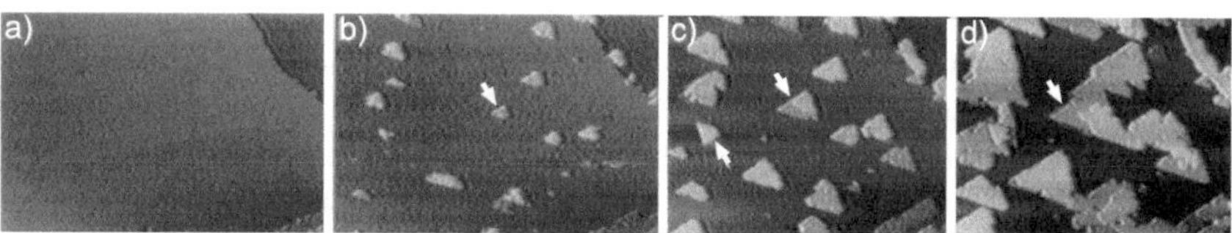

Fig. 6.22 Growth of Si on Si(111) during chemical vapor deposition at 485 °C. The *arrows* in frames (**b**) to (**d**) mark an island which grows by an additional layer each image. The *left arrow* in (**c**) designates an island, which is rotated by 180° and represents the onset of a stacking fault. The image size is 200 × 180 nm^2. Reproduced with permission from [25], © 1993 Springer

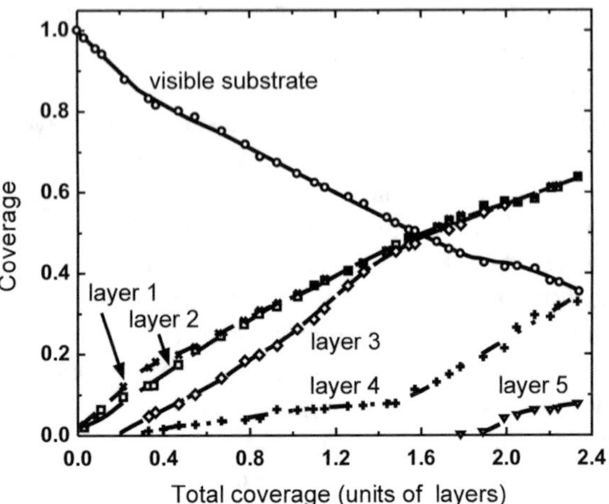

Fig. 6.23 Buildup of the surface coverage in each layer during epitaxial growth. Distribution of coverages in the first five layers as counted from the Si/Si(111) growth sequence shown in Fig. 6.22. Reproduced with permission from [25], © 1993 Springer

lower than that usually applied in growth of silicon. This leads to structural defects, which are not annealed during further growth. In the case shown in Figs. 6.22c, d an antiphase domain boundary is formed when islands coalesce, cf. the schematics Figs. 2.26 and 2.33.

A measure for layer-by-layer growth is obtained if the coverage buildup of each layer is plotted as a function of the total coverage or of growth time. The curves in Fig. 6.23 correspond to the simultaneous growth of multiple monolayers visible in the series of Fig. 6.22. We note a set of curves with a slow progression: Even after deposition of 2.4 layers (which actually are bilayers in the stacking *AaBbCc* of the diamond structure of Si) 40% of the substrate are still not covered. This is a further indicative for a low surface mobility of adatoms due to the low growth temperature applied for in situ characterization.

6.2.7 *Growth without Nucleation*

The nucleation work pointed out in Sects. 6.2.2 and 6.2.5 (3D and 2D nucleation) impose a significant supersaturation required for nucleation and consequential layer growth. This is in remarkable disagreement with frequent experimental findings of growth occurring already at negligible supersaturation. The reason for such apparent discrepancy is found in the fact that growth occurs at *steps* on the surface, and surfaces without any steps are hardly realized. There are two major sources of persisting monoatomic surface steps: A vicinal orientation of the surface and screw dislocations.

Wafer crystals cut with respect to a specific crystallographic orientation always have a slight misorientation. The result of such offcut is the formation of a vicinal surface composed of terraces and steps as illustrated in Fig. 6.14. For a typical offcut of $\Theta = 0.1$ degree the terrace width is $l = \tan^{-1}\Theta \times a \cong 2$ μm for an assumed

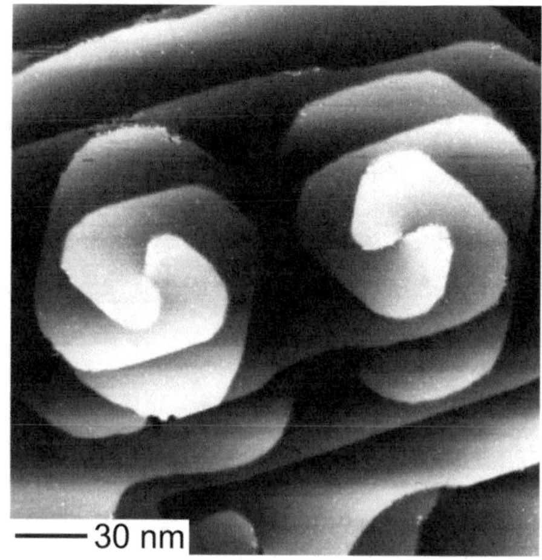

Fig. 6.24 Screw dislocations on the Ga-face of a (0001) oriented GaN/sapphire layer with total Burgers vector $c[0001]$. The image size is $230 \times 230\ \text{nm}^2$. Reproduced with permission from [27], © 1998 AVS

typical thickness $a \cong 0.3$ nm of an epitaxial monolayer (for, e.g., zincblende material $a_{001} = \frac{1}{2}a_0$) and equally spaced steps. Such terrace width is usually well below the mean diffusion length of adatoms, allowing for atoms which arrived from the ambient at the surface to be attached at the terrace step. Consequently there is no need for nucleation to grow on a vicinal surface. The resulting growth proceeds by the advancement of more or less parallel steps and is referred to as *step-flow growth*. The effect is often designedly applied to facilitate layer growth in epitaxy. For this purpose wafers with a specified offcut in a desired direction to choose the terrace width and the kind of vicinal surface are used. Step-flow growth is discussed in more detail in terms of growth kinetics in Sect. 7.2.3.

A second source of steps persisting on a surface during growth is provided by screw dislocations (Sect. 2.3) [26]. It was found that the distance between two neighboring arms of the spiral is directly proportional to the size of a 2D nucleus which depends on the supersaturation. The slope of the growth spiral around the core of the screw dislocation is then direct proportional to the supersaturation. An example of screw dislocations in the metalorganic vapor-phase epitaxy of GaN/Al_2O_3 is given in Fig. 6.24 [27]. The scanning tunneling micrograph shows two adjacent screw dislocations with Burgers vectors $c[0001]$ pointing in the opposite direction to that of the screw dislocation on an N-face of GaN shown in Fig. 2.32b. Their spirals combine, yielding a doubled Burgers vector.

We should note that even perfect crystal surfaces are generally not atomically flat and may intrinsically provide sites for a preferential attachment of atoms. Atomistic details are considered in Sect. 7.2.

6.2.8 Ripening Process After Growth Interruption

The interruption of the growth process corresponds to a change of the chemical potential of the surface with respect to that of the ambient. Since growth occurs at a depart

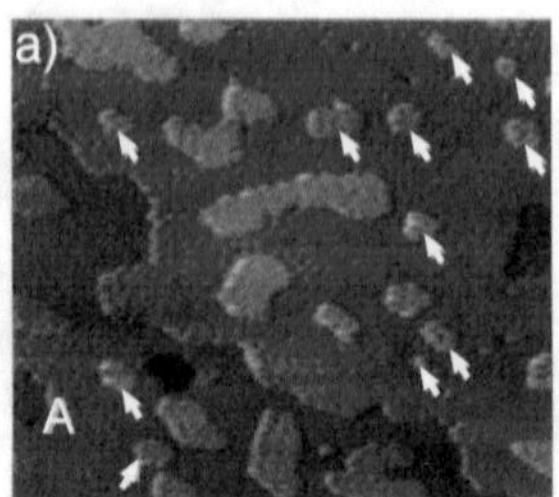

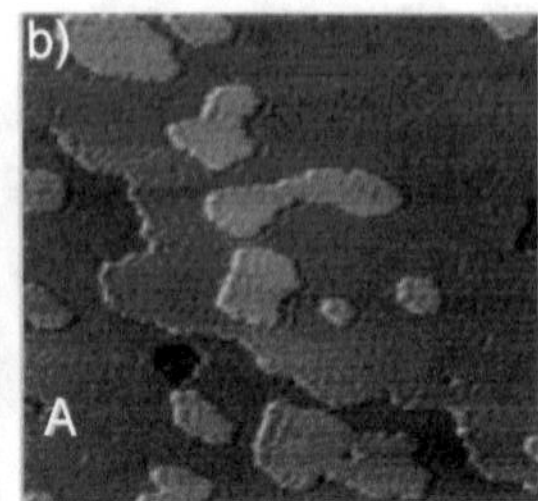

Fig. 6.25 Ostwald ripening of 2D Si islands on Si(111). (**a**) Image scanned during growth at 520 °C, (**b**) image scanned after a 23 min interruption of the deposition at the same temperature. From [28]

from thermal equilibrium, a sudden interruption of this process leaves the surface in a non-equilibrium state. An equilibration is therefore initiated comprising diffusion and evaporation–condensation processes. Changes of the surface morphology occurring during equilibration are often described by *Ostwald ripening*. The term denotes the evolution of an inhomogeneous structure (also in solid or liquid solutions) over time first described by Wilhelm Ostwald in 1896. Generally speaking, large particles (with a lower surface to volume ratio) grow in size, drawing material from smaller particles, which shrink. In the process, many initially formed small particles (or crystals, islands) slowly disappear, except for a few that grow larger, at the expense of the small particles. As a result, the broad size distribution of particles gradually gets more narrow as the mean particle size increases.

An example of Ostwald ripening initiated by an interruption of the homoepitaxial chemical vapor deposition of Si is shown in Fig. 6.25 [28]. All small islands with an unfavorable ratio of perimeter over area marked by an arrow in Fig. 6.25a dissolve. Their material feeds the larger islands remaining 23 min later, cf. Fig. 6.25b. In addition, holes in the main terrace tend to close (see, e.g., the hole near label A).

6.3 Problems Chapter 6

6.1 For solid silicon, the molar entropy is 18.8 J/(mol K) and the molar volume is 12.1 cm^3/mol at 1 bar and 298 K. Estimate the change in the chemical potential $\mu(P, T)$ for

(a) a decrease in temperature by 5 K at $P = 1$ bar.
(b) an increase in pressure by 5 bar at $T = 298$ K.

6.2 Differences in the bonding of InN compared to GaN or AlN lead to a large miscibility gap of solid solutions. We consider an unstrained $In_xGa_{1-x}N$ alloy described by a regular solution with an interaction parameter of 32 kJ/mol.

(a) Calculate the molar Gibbs free energy of mixing for an indium composition $x = 0.2$ at 800 °C and 1200 °C.
(b) At which temperature is Gibbs free energy of mixing zero at the given composition?

(c) Determine the indium compositions for the locus of the spinodal at 800 °C.

6.3 Construct the cross section of a Wulff plot for GaAs in a {110} plane, using the surface energies given in Table 6.2. Consider only the listed facets, assuming for simplicity equal surface energies for $\{hkl\}$ faces of different polarity. Verify that the considered facets fulfill the condition $\gamma_{hkl} < \frac{h+k+l}{\sqrt{h^2+k^2+l^2}} \times \gamma_{011}$.

6.4 Consider a rectangular two-dimensional silicon island on a flat Si(001) surface, bound by two S_A and two S_B single-layer steps with step energies of 0.02 eV/a and 0.1 eV/a, respectively, a being the 1×1 surface lattice constant of 3.8 Å. Assume that the shape of the island reflects the ratio of the edge energies, and the length of an S_b side is 1.9 nm.

(a) Calculate the free step energy of the island boundary, neglecting the small contribution of the four edges. What would be the free step energy, if the island had the same area, but S_B steps also had the step energy of S_A steps?
(b) In a very simplified description the island of (a) is considered to have a circular shape of equal area bounded by a step with 0.02 eV/3.8 Å step energy. Calculate the number of atoms in the island, if the specific area of a layer is 3×10^4 m^2/mol. Determine the activation energy (in kJ/mol) for the formation of a stable two-dimensional nucleus, if the island exceeds the critical radius by a factor of 10.

6.5 The conditions for the epitaxy of (001)-oriented GaAs are set to a growth rate of 2 μm/h.

(a) Express the growth rate in units of monolayers per second.
(b) Estimate the approximate step velocity for layer-by-layer growth from multiple nuclei, if the nucleation rate is 1×10^{13} m^{-2} $\times$ s^{-1}. What is the mean radius of the 2D islands just before coalescence?

6.4 General Reading Chapter 6

T. Nishinaga (Ed.), *Handbook of Crystal Growth, vol. I Fundamentals: Part A Thermodynamics and Kinetics, Part B Transport and Stability* (Elsevier, Amsterdam 2014)
I.V. Markov, *Crystal Growth for Beginners,Fundamentals of Nucleation, Crystal Growth, and Epitaxy*, 3rd edn., (World Scientific, Singapore, 2017)
A.A. Chernov (ed.), *Modern Crystallography, vol. III, Crystal Growth*. Springer Series Sol. State Sci., vol. 36 (Springer, Berlin, 1985)

References

1. K.-Th. Wilke, J. Bohm, *Kristallzüchtung* (Harry Deutsch, Thun/Frankfurt a. M., 1999). (Crystal Growing, German)
2. I. Ho, G.B. Stringfellow, Solid phase immiscibility in GaInN. Appl. Phys. Lett. **69**, 2701 (1996)
3. T. Takayama, Theoretical analysis of unstable two-phase region and microscopic structure in wurtzite and zinc-blende InGaN using modified valence force field model. J. App. Phys. **88**, 1104 (2000)

4. S.-H. Wei, L.G. Ferreira, A. Zunger, First-principles calculation of temperature-composition phase diagrams of semiconductor alloys. Phys. Rev. B **41**, 8240 (1990)
5. M. Quillec, C. Daguet, J.L. Benchimol, H. Launois, $In_xGa_{1-x}As_yP_{1-y}$ alloy stabilization by the InP substrate inside an unstable region in liquid phase epitaxy. Appl. Phys. Lett. **40**, 325 (1982)
6. M.J. Jou, Y.T. Cherng, H.R. Jen, G.B. Stringfellow, Organometallic vapor phase epitaxial growth of a new semiconductor alloy: $GaP_{1-x}Sb_x$. Appl. Phys. Lett. **52**, 549 (1988)
7. W.A. Jesser, D. Kuhlmann-Wilsdorf, On the theory of interfacial energy and strain of epitaxial overgrowths in parallel alignment on single crystal substrates. Phys. Status Solidi **19**, 95 (1967)
8. D.M. Wood, A. Zunger, Epitaxial effects on coherent phase diagrams of alloys. Phys. Rev. B **40**, 4062 (1989)
9. SYu. Karpov, N.I. Podolskaya, I.A. Zhmakin, A.I. Zhmakin, Statistical model of ternary group-III nitrides. Phys. Rev. B **70**, 235203 (2004)
10. SYu. Karpov, Suppression of phase separation in InGaN due to elastic strain. MRS Internet J. Nitride Semicond. Res. **3**, 16 (1998)
11. W. Kossel, Zur Theorie des Kristallwachstums. Nachr. Akad. Wiss. Gött. Math.-Wiss. Kl. **135–143**, (1927). (in German)
12. I.N. Stranski, Zur Theorie des Kristallwachstums. Z. Phys. Chem. **136**, 259 (1928). (in German)
13. R. Shuttleworth, The surface tension of solids. Proc. Phys. Soc. A **63**, 444 (1950)
14. I.V. Markov, *Crystal Growth for Beginners* (World Scientific, Singapore, 1995)
15. J.J. Métois, J.C. Heyraud, SEM studies of equilibrium forms: roughening transition and surface melting of indium and lead crystals. Ultramicroscopy **31**, 73 (1989)
16. A.A. Stekolnikov, J. Furthmüller, F. Bechstedt, Absolute surface energies of group-IV semiconductors: dependence on orientation and reconstruction. Phys. Rev. B **65**, 115318 (2002)
17. N. Moll, A. Kley, E. Pehlke, M. Scheffler, GaAs equilibrium crystal shape from first principles. Phys. Rev. B **54**, 8844 (1996)
18. N. Moll, M. Scheffler, E. Pehlke, Influence of stress on the equilibrium shape of strained quantum dots. Phys. Rev. B **58**, 4566 (1998)
19. Q.K.K. Liu, N. Moll, M. Scheffler, E. Pehlke, Equilibrium shapes and energies of coherent strained InP islands. Phys. Rev. B **60**, 17008 (1999)
20. H. Li, L. Geelhaar, H. Riechert, C. Draxl, Computing equilibrium shapes of wurtzite crystals: the example of GaN. Phys. Rev. Lett. **115**, 085503 (2015)
21. B.N. Bryant, A. Hirai, E.C. Young, S. Nakamura, J.S. Speck, Quasi-equilibrium crystal shapes and kinetic Wulff plots for gallium nitride grown by hydride vapor phase epitaxy. J. Crystal Growth **369**, 14 (2013)
22. B. Voigtländer, M. Kawamura, N. Paul, V. Cherepanov, Formation of Si/Ge nanostructures at surfaces by self-organization. J. Phys. Condens. Matter **16**, S1535 (2004)
23. M.A.K. Zilani, Y.Y. Sun, H. Xu, L. Liu, Y.P. Feng, X.-S. Wang, A.T.S. We, Reactive Co magic cluster formation on Si(111)-7 $\times$ 7. Phys. Rev. B **72**, 193402 (2005)
24. W.J. Ong, E.S. Tok, Configuration dependent critical nuclei in the self assembly of magic clusters. Phys. Chem. Chem. Phys. **9**, 991 (2007)
25. U. Köhler, L. Andersohn, B. Dahlheimer, Time-resolved observation of CVD-growth of silicon on Si(111) with STM. Appl. Phys. A **57**, 491 (1993)
26. W.K. Burton, N. Cabrera, F.C. Frank, The growth of crystals and the equilibrium structure of their surface. Philos. Trans. R. Soc. Lond. A **243**, 299 (1951)
27. A.R. Smith, V. Ramachandran, R.M. Feenstra, D.W. Grewe, M.-S. Shin, M. Skowronski, J. Neugebauer, J.E. Northrup, Wurtzite GaN surface structures studied by scanning tunneling microscopy and reflection high energy electron diffraction. J. Vac. Sci. Technol. A **16**, 1641 (1998)
28. U. Köhler, Kristallwachstum unter dem Rastertunnelmikroskop. Phys. Bl. **51**, 843 (1995). (in German)

Chapter 7
Atomistic Aspects of Epitaxial Layer-Growth

Abstract Crystal growth far from thermodynamic equilibrium is affected by kinetic barriers. This chapter describes nucleation and growth in terms of atomistic processes, which are characterized by such energy barriers. We consider the ideal and the real structure of a crystal surface, and discuss kinetic steps occurring during nucleation and growth. At the end of the chapter phenomena of self-organization employed for epitaxial growth of nanostructures are presented.

The description of growth treated in the previous chapter was based on macroscopic quantities: Thermodynamics describes collective motions of particles in terms of a macroscopic growth affinity. It accounts well for the direction a system tends to reach and describes conditions particularly near thermal equilibrium. Crystal growth is actually governed by a competition between kinetic and thermodynamic processes. Growth conditions in epitaxy are sometimes far from equilibrium and processes may largely be determined by kinetics. Particularly on a short time scale a kinetic description of growth on an atomic level may be more appropriate. The kinetics of epitaxial growth can usually be described by only a few categories of atomistic rate processes. Such processes strongly depend on the specific location of an atom on or near the surface. We therefore first consider the surface structure of a solid.

7.1 Surface Structure

The equilibrium conditions for surface atoms are different from those of bulk atoms due to the absence of neighboring atoms on one side. Therefore the atomistic structure of the surface does usually not coincide with that of the bulk. This is particularly true for semiconductors, which have pronounced directional bonds. By contrast metals have a chemical bond which is basically not directed, and in many cases the surface lattice corresponds to the bulk lattice. Starting with a simple model we will consider some examples of semiconductor-surface reconstructions to provide a basis for the description of growth on the atomic scale.

U. W. Pohl, *Epitaxy of Semiconductors*, Graduate Texts in Physics,
https://doi.org/10.1007/978-3-030-43869-2_7

7.1.1 The Kink Site of a Kossel Crystal

The Kossel crystal is a simple model which assumes the atoms as cubes (more generally the building blocks which may be atoms, ions, or molecules). In the most simple approach only nearest-neighbor interaction is assumed, i.e., only the bond at the face between adjacent cubes. Let us consider a surface with a monoatomic step as shown in Fig. 7.1. Atoms at different positions on the surface have different numbers of neighboring atoms. Consequently these atoms are differently bound to the surface. An atom embedded in the uppermost layer of a flat surface—at position 1 in Fig. 7.1—has 4 lateral bonds and 1 bond to the layer below, leaving just a single unsaturated bond. The atom embedded into the step (site 2) has only 3 lateral bonds and 1 vertical bond, leaving *two* bonds unsaturated. This atom is hence less tightly bound as that at site 1 and may more easily be detached from the surface.

A site of particular importance is position 3, the so-called *kink site*. We note that 3 bonds are unsaturated and 3 bonds are attached to neighboring atoms of the solid. This means that exactly one half of the bonds is unsaturated. The three saturated bonds attach the atom to a half row of atoms (the row with the atom embedded at site 2), to a half crystal plane (the plane with the atom embedded at site 1), and to a half bulk crystal (the crystal underneath the atom), respectively. The kink position is therefore also referred to as half-crystal position. Crystal growth basically proceeds via the incorporation of atoms at this site. The work $\varphi_{\frac{1}{2}}$ required to detach an atom from a kink position is just half the work required to detach an atom located in the bulk.

We note that an attachment or detachment of an atom at a kink site leaves the number of unsaturated bonds at the surface unchanged: the initial configuration is reproduced. Since the number of unsaturated bonds is associated with the surface energy no change occurs for an occupation of this site. This means that the work required to add or remove a kink atom is equal to the chemical potential of the crystal. This implies that the chemical potential of the atom at a kink site is equal to

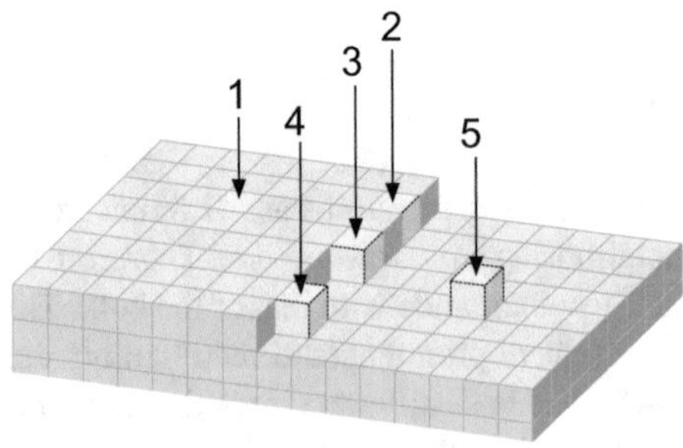

Fig. 7.1 Important sites of an atom on a crystal surface. Numbers signify unsaturated nearest-neighbor bonds of atoms located at the marked sites: *1*—atom embedded in the uppermost surface layer, *2*—atom embedded into a step edge, *3*—atom at a kink site, *4*—atom attached to a step, *5*—atom adsorbed on the surface (adatom)

that of the crystal. Transferring an atom from the ambient to a kink site of the crystal surface (or vice versa) is therefore equal to the difference of the chemical potentials of the ambient and the crystal.

In thermodynamic equilibrium of a (large) crystal with an ambient the probabilities of attachment and detachment of a kink atom are equal. Sites which provide more bonds (like, e.g., a hole in a surface at site 1 in Fig. 7.1) have a larger probability for attachment with respect to detachment. In equilibrium such sites are hence probably occupied. On the other hand sites with less bonds as a kink position like sites 4 and 5 have a smaller probability to be occupied in thermal equilibrium. As a consequence a crystal in thermal equilibrium with an ambient is bounded by flat faces: the equilibrium surfaces discussed in Sect. 6.2.4. It should be noted that perfect atomically flat surfaces do not exist at finite temperatures: The action of entropy always gives rise to some roughening.

7.1.2 Surfaces of a Kossel Crystal

Crystal faces of different crystallographic orientation differ in their surface structure. We first consider the low-index faces of a Kossel crystal. A schematic of this model crystal with {100}, {110}, and {111} faces is given in Fig. 7.2.

The figure illustrates that the depicted low-index planes differ significantly with respect to their roughness on an atomic scale: {100} faces are perfectly flat and {111} faces are rough. Generally crystal surfaces can be classified into three groups, namely flat (F), stepped (S), and kinked (K). Low-index faces of all groups represent singular faces. The classification follows from a consideration of rows of atoms with a most dense arrangement. The direction of a densely packed row of atoms is referred to as nearest-neighbor periodic-bond-chain. Faces of any lattice which are parallel to at least *two* nearest-neighbor periodic-bond-chains are called F (flat) faces. Usually these faces have the highest surface density of atoms. In a Kossel crystal close-packed

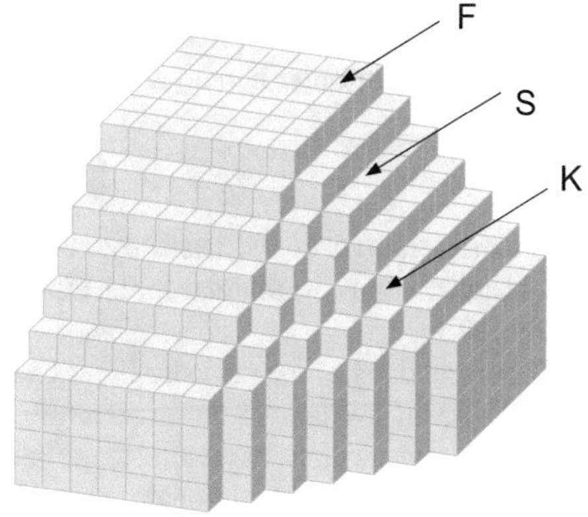

Fig. 7.2 Kossel crystal with flat (*F*), stepped (*S*), and kinked (*K*) surfaces

atoms are aligned along a ⟨100⟩ direction. F faces are therefore {100} faces. Each atom on such a surface offers one unsaturated nearest-neighbor bond. Faces which comprise *one* nearest-neighbor periodic-bond-chain are called S (stepped) faces. S faces of a Kossel crystal with lowest indices are {110} faces like those depicted in Fig. 7.2. Each atom on such an S face offers two unsaturated nearest-neighbor bonds. Faces which comprise *no* nearest-neighbor periodic-bond-chain are called K (kinked) faces. K faces of a Kossel crystal with lowest indices are {111} faces. Each atom of a K face offers three unsaturated nearest-neighbor bonds.

In Sect. 6.2 we pointed out the importance of a step and, in particular, of a kink for growth of a crystal layer. Obviously, an F face does not offer such sites, and growth requires the formation of stable 2D nuclei. By contrast, the atomically rough K face offers a large number of kink sites. Growth on such a face does not require the presence of 2D nuclei or structural defects. The growth rate is simply expected to increase with supersaturation, without any barrier to be surmounted. For a given supersaturation the growth rate of K faces will therefore be highest. The S face offers kink sites once an atom is attached. The density of kink sites is lower than on F faces, the growth rate will therefore be smaller. F faces will have the smallest growth rate due to the need of nucleation. According to our discussion of the facets of a growing crystal illustrated in Fig. 6.18 we note that K faces are expected to disappear first, followed by S faces, leaving eventually the F faces. A growing Kossel crystal will thus eventually be bounded by {100} faces.

The model discussed so far may be extended by taking higher-order interactions into account. If we include also second and third-nearest neighbor interactions, the detachment energy $\varphi_{\frac{1}{2}}$ from a kink position is given by the sum of the interaction energies E_i and their respective coordination numbers Z_i,

$$\varphi_{\frac{1}{2}} = \frac{1}{2}(Z_1 E_1 + Z_2 E_2 + Z_3 E_3). \tag{7.1}$$

The coordination number is the number of ith nearest neighbors. A bulk atom in a cubic Kossel crystal has $Z_1 = 6$ nearest neighbors along ⟨100⟩, $Z_2 = 12$ second-nearest neighbors along ⟨110⟩, and 8 third-nearest neighbors along ⟨111⟩. At a kink site half of these neighbors exist, respectively. Our consideration of the half-crystal site and (7.1) may be applied to extend relation (6.31) between the bond energy (including higher order bonds) and the enthalpy of evaporation,

$$\Delta H_S \cong \varphi_{\frac{1}{2}}. \tag{7.2}$$

The considerations above basically also apply for other lattices. The numbers of neighboring atoms of ith order for an atom on a kink site of various lattices are listed in Table 7.1.

The directions of the periodic-bond-chains in the cubic diamond structure differ from those in the simple cubic structure: They are oriented along the six ⟨110⟩ directions and are thus not parallel to the next neighbor bonds. The related zincblende lattice has six periodic-bond-chains along ⟨110⟩ for each of the two ion types. F faces

Table 7.1 Numbers of nearest neighbors Z_i for an atom on a kink site for some lattices

Lattice	Z_1	Z_2	Z_3
Simple cubic	3	6	4
Diamond	2	6	6
Hexagonal close-packed	6	3	1

of the diamond and zincblende structures are {111} faces, while {100} faces and also {110} faces are stepped. K faces are, e.g., {311} faces. Also the diamond structure and the hexagonal close-packed (hcp) structure are related: Both are derived from stacking net planes along a [0001] stacking axis (cf. Sect. 2.1). F faces of the hcp structure are hence the (0001) and $(000\bar{1})$ faces which correspond to the {111} faces of the diamond structure.

7.1.3 Relaxation and Reconstruction

The atomic bonds in the uppermost layer of the surface of a solid differ significantly from those in the bulk due to the absence of neighboring atoms on one side. The altered equilibrium conditions for surface atoms lead to shifted atomic positions with respect to the bulk lattice. Two different effects in the rearrangement are to be distinguished: relaxation and reconstruction. *Relaxation* signifies a pure shift of atoms *normal* to the surface due to the missing attractive forces above the surface. Usually a compression occurs in the topmost few layers as illustrated in Fig. 7.3a. Relaxation leaves the lateral periodicity of the atom positions unchanged with respect to the bulk, i.e., the 2D surface unit cell corresponds to the unit cell of a truncated bulk crystal.

Usually surface atoms tend to restore some of the bonds broken to create the surface. Such rearrangement is accompanied by *lateral* shifts of atoms, i.e., a shift

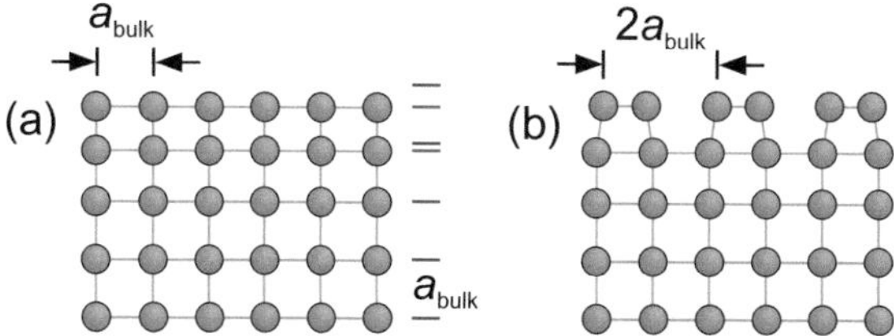

Fig. 7.3 **a** Relaxation and **b** reconstruction of atoms at the surface of a crystalline solid. *Blue* and *red horizontal bars* at the side of schematic (**a**) mark vertical positions of the atomic net planes corresponding to the bulk and the relaxed surface, respectively

parallel to the surface. This kind of reordering is termed *reconstruction*. The 2D surface unit cell of a reconstructed surface does *not* correspond to the lateral part of the bulk 3D unit cell. In most cases the dimension of the surface unit cell differs from the bulk unit cell as depicted in Fig. 7.3b. Reconstruction includes also surface unit cells of unchanged lateral dimensions with a modified lateral order compared to the bulk.

The modified order of surface-near atoms is related to a change of the surface energy. Surface energy may be regarded as the energy difference between an ideal surface without any relaxation and reconstruction, and the actual surface. Vice versa, it is the energy required to rearrange all surface-near atoms of a relaxed and reconstructed surface to the order of the bulk lattice to form an ideal surface.

The ordering of atoms near the surface depends on the kind of bonds and is different for crystalline semiconductors and metals. Semiconductors usually have strongly directed covalent bonds, particularly those with a tetrahedral coordination. Atoms at the surface tend to saturate at least some of the unsaturated bonds to minimize the density of unfavorable dangling bonds. Semiconductors show a large variety of often quite complex reconstructions, some are illustrated below in Sect. 7.1.6. Metals, by contrast, have a delocalized electron gas yielding basically no directed bonds. As a consequence many metal surfaces do not reconstruct. It must be noted that the type of reconstruction does not only depend on the considered solid and (*hkl*) surface plane, but also to a large extent on preparation conditions.

A simple picture on the formation of a reconstructed surface is illustrated in Fig. 7.4. The figure assumes a tetrahedrally coordinated crystal bonded by sp^3 hybrides. The ideal surface (Fig. 7.4a) may be considered to be created from a cut of the bulk crystal, leaving half-filled dangling bond orbitals at the surface. Each dangling bond is occupied by one electron. Such ideal surface is not stable, because a dangling bond represents an unfavorable energy state of the surface. The number of dangling bonds can be significantly reduced by the formation of surface dimers (Fig. 7.4b): Two unsaturated bond orbitals form one filled bridge-bond orbital parallel to the surface, respectively. The surface energy may be further minimized by a charge redistribution among the remaining dangling bonds. The formation of a more-than-half filled s-like and a less filled p-like dangling bond eventually leads to asymmetric dimers (Fig. 7.4c). The electron-charge transfer from the “down” atom to the “up” atom of the dimer changes the covalent bond of the symmetric dimer to a partially ionic bond [1]. Such a relaxation of the dimer bond is referred to as *buckling*. The vertical shifts of the atoms are accompanied by lateral displacements towards one another.

7.1.4 Electron-Counting Model

For tetrahedrally coordinated compound semiconductors such as GaAs a simple *electron-counting model* has been developed to explain a wide variety of surface reconstructions [2]. This model accounts for the occupation of surface states and

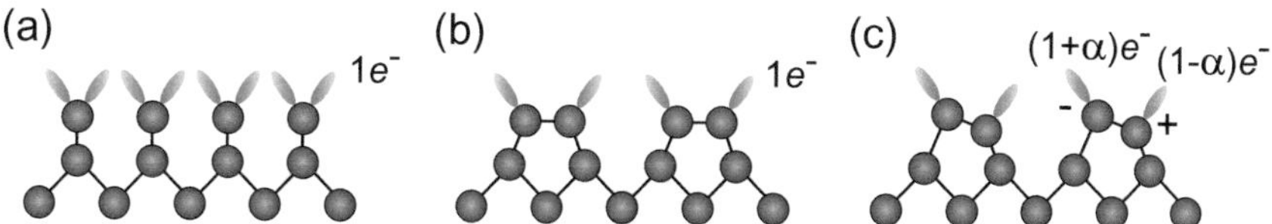

Fig. 7.4 Schematic on the gradual formation of a surface reconstruction from an ideal surface (**a**) via dimerization with symmetric dimers (**b**) to relaxed, buckled dimers (**c**)

assumes no net surface charge for reordered surfaces with minimum energy. Bonding and nonbonding surface states *below* the Fermi energy must be filled, while the antibonding and nonbonding states *above* the Fermi energy must be empty. For the given number of available electrons in the surface layer all dangling bonds on the electronegative element will then be full, and all those on the electropositive element of the compound semiconductor will be empty. This rule is generally applicable for any surface of a crystalline solid.

Formation and occupation of bond orbitals at the surface is illustrated for GaAs in Fig. 7.5 [3]. In the bulk crystal, bonds oriented towards their nearest neighbors are constructed from sp^3 hybrids with energies $\varepsilon_\mathrm{h} = (3\varepsilon_\mathrm{s} + \varepsilon_\mathrm{p})/4$ on each atom. The bond energy ε_b within each bond results from the linear combination of the two hybrids ε_h (Ga) and ε_h (As) with lowest energy. Eventually the occupied valence band with maximum E_v is formed from linear combinations of these orbitals. Correspondingly, the empty conduction band with minimum E_c is built from antibonding linear combinations. At the surface, some hybrid orbitals cannot form bonds, and partially filled dangling bonds remain. Details depend on the considered surface. On (111) surfaces the surface atoms are bonded to three atoms of the layer underneath, yielding three hybrid orbitals used for bonds and one dangling hybrid directed out of the solid. For the electropositive Ga atom the energy of this hybrid lies near the conduction-band edge, leading to an empty state (Fig. 7.5). The hybrid of the electronegative As atoms on a $(\bar{1}\bar{1}\bar{1})$ surface (also denoted $(111)B$ surface) lies below the valence-band edge and is consequently expected to be occupied. On the (001) surface *two* of the four sp^3 hybrids are used to form bonds to the crystal underneath. Linear combinations of the two remaining dangling-bond hybrids with lowest and highest energy yield a pure p state and an sp hybrid [3]. Both, Ga and As p states lie above the conduction band according to Fig. 7.5, leaving the p dangling bond unoccupied. The *sp*-hybrid of As lies below the valence-band edge. In a relaxed state this applies also for the *sp*-hybrid of the Ga [3].

Any structure obeying the electron-counting model exactly fills the electrons available in the surface layer into all dangling-bond states in the valence band, leaving all dangling-bond states in the conduction band empty. The surface will then be semiconducting, while partially filled dangling bonds may lead to a metallic surface. The electron-counting rule also assures that no charge accumulates at the surface. The model can successfully explain principal reconstructions on many surfaces. It can, however, not decide among alternatives which fit the model. Moreover, also

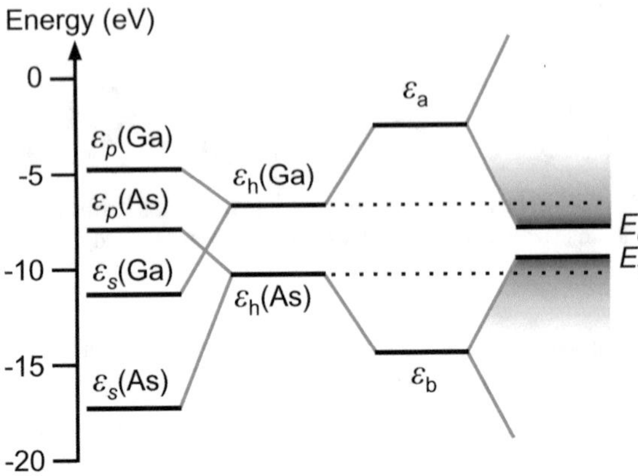

Fig. 7.5 Energy levels ε_h of sp^3 hybride dangling-bond orbitals of GaAs. The hybride energies are derived from the energies ε_s and ε_p of the atomic s and p orbitals of Ga and As, respectively. E_c and E_v denote conduction and valence bands of the GaAs bulk crystal. Data from [3]

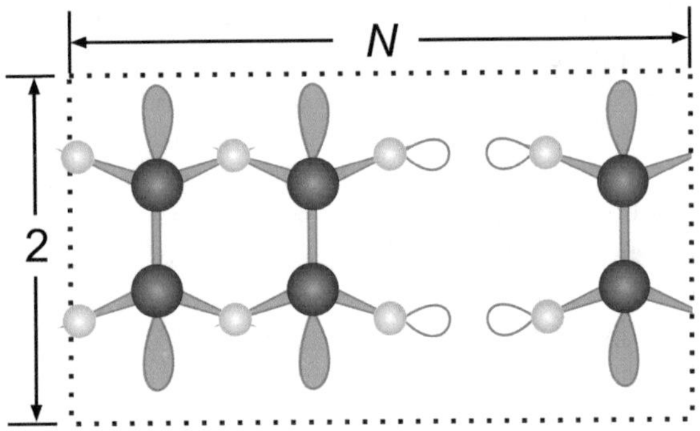

Fig. 7.6 $(2 \times N)$ surface unit-cell with a missing dimer on a (001) face of a compound semiconductor with zincblende structure. Electronegative and electropositive elements are *dark* and *bright*, respectively. Bonds and dangling bonds are shaded if filled, empty bonds are open. After [2]

reconstructions may exist which are connected to a surface charge and do not obey the electron-counting rule.

We illustrate the application of the electron-counting model for the (As-rich) GaAs(001) surface [2]. Since numerous structures comply with the model, we need some reasonable assumption on the nature of the reconstruction as a starting point, obtained, e.g., from scanning tunneling micrographs. Such images show a surface unit cell in one specific direction being twice as long as the lateral part of the bulk unit cell, originating from the formation of surface dimers as depicted in Fig. 7.4. In the other lateral direction the surface unit-cell is a factor of N longer than that of the bulk (Fig. 7.9). This periodicity arises from periodically missing dimers. The reconstruction is termed $(2 \times N)$ and leaves D dimers in one unit cell with $D \leq N$, cf. Fig. 7.6.

The relation between D and N follows from a balance between the number of electrons required to fill the bonds and that of available electrons. Each As dimer bond requires 2 electrons, and another 2 for the filled dangling bond at each of the two As atoms of the dimer, yielding a total of 6 electrons per dimer, or $6D$ electrons per unit cell. Furthermore, a total of $8D$ electrons per unit cell is required to bond all dimer As atoms to the Ga layer underneath. The dangling bonds on the electropositive

Ga atoms in the second layer are empty. On the other hand, the number of available electrons results from the number of valence electrons V_n of the electronegative element and V_p of the electropositive element. The number of electrons in one unit cell available from the topmost three layers are therefore:

$2V_nD$ from the topmost layer comprising the As dimers, and
$\frac{1}{2}2V_pN$ from the second layer comprising Ga atoms, with the factor $\frac{1}{2}$ because half of the total electrons are involved in bonding to the bulk crystal underneath.

To balance the numbers of required and available electrons in a $(2 \times N)$ surface-unit cell, we thus yield the condition

$$6D + 8D = 2V_nD + V_pN. \tag{7.3}$$

In the case of the studied GaAs surface $V_n = 5$ and $V_p = 3$ applies. Inserting these numbers we eventually obtain

$$4D = 3N \quad \text{(for GaAs)}.$$

The smallest periodicity fulfilling this condition and $D \le N$ is $N = 4$ and $D = 3$. This yields a (2×4) reconstruction of 3 dimers and one missing dimer in a unit cell (Fig. 7.6). If we consider a (Se-rich) (001) surface of zincblende ZnSe we have $V_n = 6$ and $V_p = 2$. The energies of the hybrid orbitals of the electropositive Zn and that of the electronegative Se lie in the conduction and valence bands of the solid, respectively, similar to the GaAs case [3]. The electron counting model (7.3) yields in this case $N = D$. We therefore expect a (2×1) reconstruction with no missing dimers as a favorable candidate. The Zn-rich ZnSe(001) surface actually forms a $c(2 \times 2)$ reconstruction. Such a reconstruction with Zn dimers is also consistent with the model, since a $c(2 \times 2)$ periodicity is formed from a complete layer of Zn dimers if each dimer row is displaced by one spacing in the $2\times$ direction with respect to the previous dimer row [2].

The surface reconstruction of a semiconductor may significantly be modified in the presence of metal adsorbates. Metals then act as an electron reservoir which donates or accepts the right number of electrons, when the surface assumes a specific reconstruction to fulfill the electron-counting model. The model was therefore more recently extended to account also for metal-induced reconstructions [4].

A couple of prominent surface reconstructions assumed by GaAs and Si surfaces are discussed in Sect. 7.1.6. Prior to that we point out the notation used to designate surface unit-cells of reconstructions.

7.1.5 Denotation of Surface Reconstructions

The surface of a solid represents a three-dimensional structure. The symmetry properties may be described by two-dimensional operations, though. We therefore con-

Table 7.2 The 4 crystal systems and the 5 Bravais lattices in two-dimensional space. **a** and **b** are lattice vectors spanning the unit cell, α is the angle between the vectors

System	Unit cell	Bravais lattices	Symmetry axes
Cubic	$\mathbf{a} = \mathbf{b}$, $\alpha = 90°$	Cubic	1 fourfold axes of rotation or inversion parallel to $\mathbf{a} \times \mathbf{b}$
Rectangular	$\mathbf{a} \neq \mathbf{b}$, $\alpha = 90°$	Primitive rectangular, centered rectangular	1 twofold axes of rotation or inversion parallel to $\mathbf{a} \times \mathbf{b}$
Hexagonal	$\mathbf{a} = \mathbf{b}$, $\alpha = 120°$	Hexagonal	1 sixfold axes of rotation or inversion parallel to $\mathbf{a} \times \mathbf{b}$
Oblique	$\mathbf{a} \neq \mathbf{b}$, $\alpha \neq 90°$	Oblique	1 axes of inversion parallel to $\mathbf{a} \times \mathbf{b}$

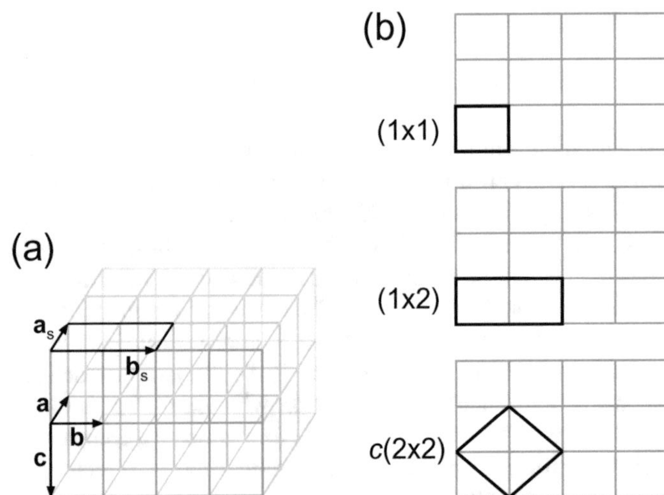

Fig. 7.7 **a** Unit cells of surface and bulk lattices, defined by their corresponding vectors. **b** Some unit cells of the surface lattice (*black*) on the bulk lattice (*gray*)

sider a two-dimensional lattice, which constitutes a 2D translational periodicity, and a basis representing the atomic structure of the recurring surface unit-cell. There are only 5 Bravais lattices in two dimensions, which constitute 4 crystal systems of differently shaped surface-unit cells. Table 7.2 gives the symmetry properties of the 5 two-dimensional Bravais lattices.

The 2D lattice describing the surface periodicity is related to the 3D lattice of the bulk crystal underneath. The relation is illustrated in Fig. 7.7.

The surface lattice vectors $\mathbf{a}_s$, $\mathbf{b}_s$ may be expressed in terms of the two vectors **a**, **b** which span the surface unit cell of a truncated bulk lattice,

$$\begin{pmatrix} \mathbf{a}_s \\ \mathbf{b}_s \end{pmatrix} = \mathbf{M} \begin{pmatrix} \mathbf{a} \\ \mathbf{b} \end{pmatrix} = \begin{pmatrix} m_{11} & m_{12} \\ m_{21} & m_{22} \end{pmatrix} \begin{pmatrix} \mathbf{a} \\ \mathbf{b} \end{pmatrix}. \tag{7.4}$$

The (2×2) matrix $\mathbf{M}$ provides an unambiguous relation between the surface and the bulk lattices. Usually the vectors $\mathbf{a}_s$ and $\mathbf{b}_s$ are multiples of the vectors $\mathbf{a}$, $\mathbf{b}$ (by convention $a_s < b_s$ if $a_s \neq b_s$). This leads to more convenient shorthand terms (*Wood's notation*), which comprise these multipliers and, if necessary, an angle by which the surface lattice is rotated with respect to the bulk lattice [5]. Let us consider, e.g., the $\{hkl\}$ surface of a solid X with a surface lattice, which fulfills $\mathbf{a}_s = 2\mathbf{a}$, $\mathbf{b}_s = \mathbf{b}$. Such a structure may be formed by surface dimers as depicted in Fig. 7.4. The corresponding surface is then described by the shorthand term $X\{hkl\}$ (2×1). Primitive and centered cells are indicated by adding p or c, respectively, and a rotated cell is denoted by adding R and the angle of rotation in units of degrees. Usually the p for a primitive cell is suppressed. The relation between the matrix representation and corresponding shorthand terms is given in Table 7.3. It should be noted that the simplified notation is not always unambiguous. A centered cubic unit cell $c(2 \times 2)$ may as well be described by the unit cell $(\sqrt{2} \times \sqrt{2})R45°$.

7.1.6 Reconstructions of the GaAs(001) Surface

The electron-counting rule pointed out in Sect. 7.1.4 selects favored structures for a specific surface. The actual reconstruction is further determined by experiments and calculations. Different reconstructions with different surface stoichiometries become thermodynamically stable, when the chemical potential of the ambient is varied. Moreover, different reconstructions may coexist on a given surface due to varied surface preparations and kinetic barriers for a reordering of surface atoms.

Table 7.3 Relation between matrix representation and corresponding shorthand terms for some surface lattices

$\mathbf{M}$	Shorthand term
$\begin{pmatrix} 1 & 0 \\ 0 & 1 \end{pmatrix}$	(1×1)
$\begin{pmatrix} 1 & 0 \\ 0 & 2 \end{pmatrix}$	(1×2)
$\begin{pmatrix} 2 & 0 \\ 0 & 2 \end{pmatrix}$	(2×2)
$\begin{pmatrix} 1 & 1 \\ -1 & 1 \end{pmatrix}$	$c(2 \times 2)$ or $(\sqrt{2} \times \sqrt{2})R45°$

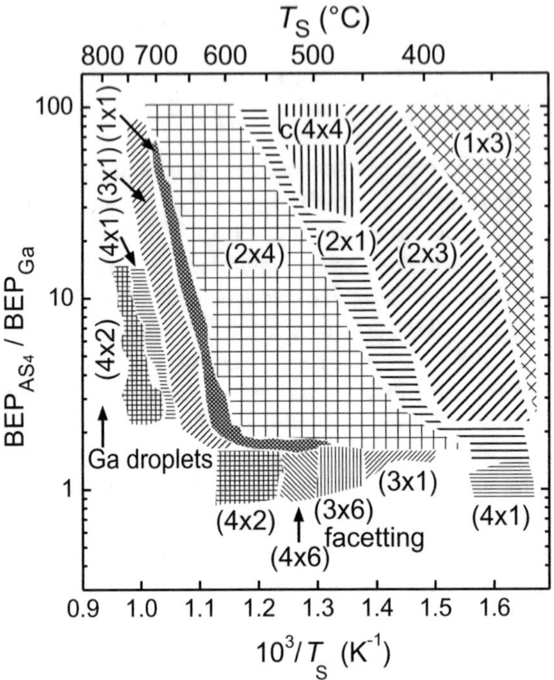

Fig. 7.8 Existence range of surface reconstructions forming during the MBE of GaAs for various temperatures T_S and beam equivalent pressure (BEP) ratios of the As_4 and Ga fluxes. Reproduced with permission from [6], © 1990 Elsevier

As an example we consider the technologically important (001) surface of GaAs, already treated above for illustrating the electron-counting rule. Experimentally, many surface reconstructions were found depending on ambient conditions. Figure 7.8 shows experimental results for a growing GaAs(001) surface composed as a phase diagram [6]. Here, a vicinal surface tilted by 2° toward the $(1\bar{1}1)$ As plane was used to facilitate growth. This tilt leads to As-terminated steps. The structure was analyzed by reflection high-energy electron diffraction (RHEED) during molecular-beam epitaxy at a fixed Ga flux corresponding to a growth rate of 0.7 monolayer/sec GaAs. In the experiment the chemical potential was varied by the supplied flux of As_4 molecules, expressed in the diagram in terms of the beam equivalent pressure (BEP) ratio with respect to that of the Ga flux. In addition the temperature was varied as an independent basic growth parameter. We note that the conditions do not correspond to equilibrium.

The diagram Fig. 7.8 shows 14 different reconstructions which may be distinguished as As-rich surfaces ((2 × 4) or (2 × 1)), as a transition range ((3 × 1) or (1 × 1)), as Ga-rich surfaces ((4 × 1), (4 × 2), (4 × 6), or (3 × 6)), as absorption structures of the type $c(4 \times 4)$, (2 × 3), (1 × 3), and surfaces where degradation occurs [6]. The transitions between all regions were found to be reversible and could be described by an Arrhenius law. Within a transition range often a coexistence of adjoining patterns was observed.

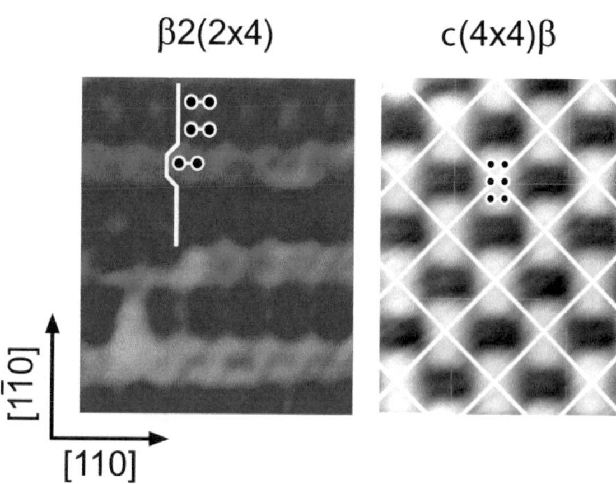

Fig. 7.9 Filled-state STM image of **a** the $\beta 2(2 \times 4)$ and **b** the $c(4 \times 4)\beta$ reconstructions of the GaAs(001) surface. In **a** two As-As top dimers and an As-As trench dimer are marked. In **b** three As-As top dimers and the $c(4 \times 4)$ lattice mesh are marked. Reproduced with permission from [7] and [8], © 1999 and 2004 APS, respectively

The periodicity of the surface unit cell is usually readily accessible in MBE from electron diffraction experiments. The arrangement of atoms within the unit cell often poses, however, a tricky problem. A variety of methods such as scanning tunneling microscopy, low-energy electron diffraction, and angle-resolved photoelectron spectroscopy has to be applied to obtain an unequivocal structural model. Two prominent reconstructions of the GaAs(001) surface imaged using high-resolution STM are shown in Fig. 7.9. Though many details can be identified on such images, the arrangement of the atoms at the surface may hardly be extracted.

Calculations of surface energies for various orientations and reconstructions help substantially to find and to understand the actual atomic structure of the surface unit cell. The stable equilibrium surface-reconstruction is that with the lowest surface free energy γA. The surface of a compound semiconductor like GaAs may be nonstoichiometric with respect to the number of atoms of different species, and the surface free energy depends on the chemical potential μ_i for each species i. The surface may exchange atoms with the ambient which acts as a reservoir, and μ_i is the free energy per particle of species i in the reservoir. In the experiment μ_i can be varied within the limits given by the bulk chemical potentials of the condensed phases of the species, $\mu_i < \mu_{i(\text{bulk})}$, because otherwise the elemental phase of species i will form on the surface. In the case of GaAs this will be bulk As and bulk Ga. Furthermore, in equilibrium the sum of the chemical potentials of the species Ga and As must be equal the bulk energy per GaAs pair [9],

$$\mu_{\text{Ga}} + \mu_{\text{As}} = \mu_{\text{GaAs(bulk)}} = \mu_{\text{Ga(bulk)}} + \mu_{\text{As(bulk)}} + \Delta H_{\text{f}}^{\text{GaAs}},$$

$\Delta H_{\text{f}}^{\text{GaAs}}$ being the heat of formation of the GaAs bulk crystal. Using this relation the surface energy may be expressed as a function of the As chemical potential μ_{As}. If μ_{As} is varied, different surface stoichiometries and related reconstructions get most stable. A large variety of atomic configurations exists for surface unit cells of each of the reconstruction periodicities shown in Fig. 7.8. Surface structures with low

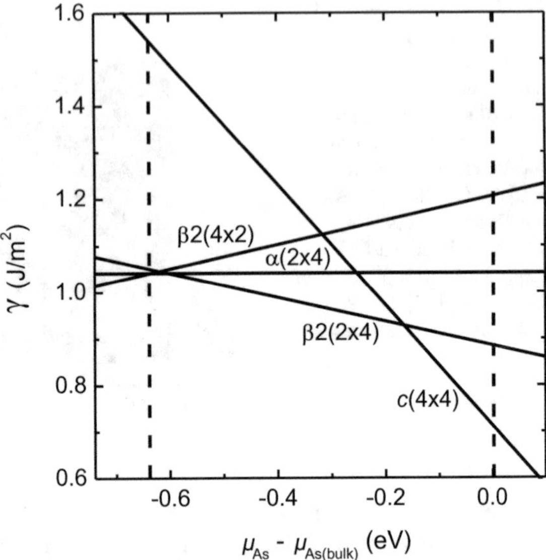

Fig. 7.10 Surface energy of different low-energy GaAs(001) reconstructions, depending on the chemical potential of the arsenic. *Dashed lines* mark limits of the chemical potential difference $\mu_{\mathrm{As}} - \mu_{\mathrm{As(bulk)}}$. Data from [9]

surface energies within some range of μ_{As} are given in Fig. 7.10. These structures all fulfill the electron counting rule (Sect. 7.1.4) and are semiconducting (filled anion and empty cation dangling bonds).

The surface energies shown in Fig. 7.10 are calculated from the total energy E_{tot} according to [9]

$$\gamma A = E_{\mathrm{tot}} - \mu_{\mathrm{GaAs}} N_{\mathrm{Ga}} - \mu_{\mathrm{As}} (N_{\mathrm{As}} - N_{\mathrm{Ga}}).$$

The stoichiometry of the surface $N_{\mathrm{As}} - N_{\mathrm{Ga}}$ determines the slope in a linear dependence on the chemical potential μ_{As}. The reconstructions noted in Fig. 7.10 are those with a low surface energy within the indicated allowed limits of μ_{As}. Particular phases of GaAs(001) reconstructions with given periodicity of the unit cell are indicated using a leading label like, e.g., α or $\beta 2$ [10, 11]. Starting from the Ga-rich $\beta 2(4 \times 2)$ reconstruction we note a progressive trend to a negative slope in the sequence towards the most As-rich $c(4 \times 4)$ surface for high $\mu_{\mathrm{As}} - \mu_{\mathrm{As(bulk)}}$ values. The $\alpha(2 \times 4)$ surface is stoichiometric, i.e., $N_{\mathrm{As}} = N_{\mathrm{Ga}}$, and does not depend on the chemical potential of As. The low-energy As-rich $\beta 2(2 \times 4)$ and $c(4 \times 4)$ surfaces are those generally used in the molecular-beam epitaxy and the metalorganic vapor-phase epitaxy of (001)-oriented GaAs, respectively.

The atomic structure of the four most favorable reconstructions indicated above are shown in Fig. 7.11. Both, Ga-Ga and As-As dimer bonds occur in the topmost layer of the displayed unit cells. They both have filled bonding and empty antibonding states.

The Ga-rich $\beta 2(4 \times 2)$ structure consists of two Ga-Ga dimers per unit cell oriented along [110] in the top layer and two missing As atoms in the second layer.

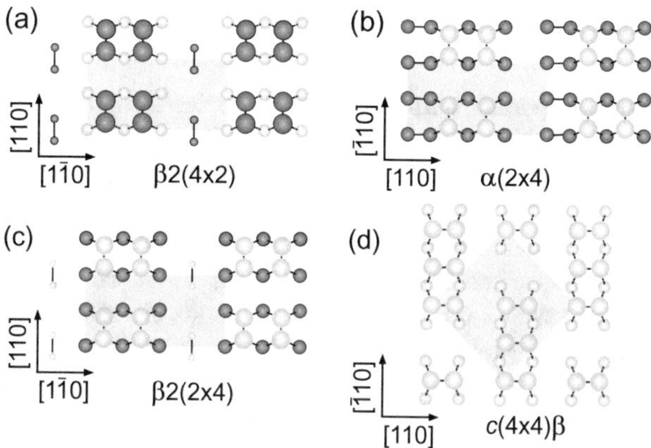

Fig. 7.11 Models of the four GaAs(001) surface reconstructions considered in Fig. 7.10. The size of the atoms indicates the vertical position, *bright* and *dark atoms* represent As and Ga, respectively. *Shaded areas* indicate the surface unit-cells

An electron counting within the unit cell [12] yields a surface stoichiometry of $N_{\text{As}} - N_{\text{Ga}} = -\frac{1}{4}$ per (1×1) unit cell.

The stoichiometric $\alpha(2 \times 4)$ surface consists of two As-As top-layer dimers oriented along $[\bar{1}10]$, adsorbed on a complete Ga monolayer underneath. Two As-As dimers are missing in the row of top-layer dimers, and two Ga-Ga bonds are formed within the Ga monolayer where one As-As dimer is missing. It should be noted that another stoichiometric structure with a single As-As top-layer dimer termed $\alpha 2(2 \times 4)$ was found with an even lower energy than that depicted here [11, 13, 14].

The more As-rich $\beta 2(2 \times 4)$ structure is obtained from the $\alpha(2 \times 4)$ structure by removing two Ga atoms in the missing As-As dimer region, resulting in a stoichiometry of $N_{\text{As}} - N_{\text{Ga}} = \frac{1}{4}$ per (1×1) unit cell. The structure represents the counterpart of $\beta 2(4 \times 2)$, with Ga atoms exchanged for As atoms and vice versa.

The most As-rich $c(4 \times 4)\beta$ surface represents a double-layer structure consisting of As-As dimers oriented along the [110] direction and adsorbed on a full As monolayer, yielding a stoichiometry of $N_{\text{As}} - N_{\text{Ga}} = \frac{5}{4}$ per (1×1) unit cell. The commonly accepted model comprises a block of three dimers in a row along $[\bar{1}10]$, being interrupted by a dimer vacancy. As-Ga heterodimers are assumed to appear in the $c(4 \times 4)$ reconstruction as the chemical potential $\mu_{\text{As}} - \mu_{\text{As(bulk)}}$ is lowered towards less As-rich conditions and the $\beta 2(4 \times 2)$ structure is approached [15]. The $c(4 \times 4)$ reconstruction with three As-Ga top-layer dimers is referred to as $c(4 \times 4)\alpha$.

The reconstructions of the GaAs(001) surface considered above as some typical examples give an impression about the wealth of surface structures. More detailed information on surface reconstructions is found in, e.g., [16, 17].

7.1.7 The Silicon (111)(7 × 7) Reconstruction

Silicon is the most important semiconductor for device technology. Chips for integrated circuits are fabricated from (001)-oriented Si wafers. Atoms in the uppermost layer of the unreconstructed {100} surface of the diamond structure have two dangling bonds each. The surface energy is lowered by formation of asymmetric buckled dimers as illustrated in Fig. 7.4c, leading to a Si(001)(2 × 1) reconstruction. Usually Si(001) wafers are chemically treated using a dilute HF-etch to remove surface defects. During such a procedure dangling bonds are saturated by hydrogen. The H-terminated surface relaxes but does not reconstruct.

Some applications such as electric high-power applications employ the (111) surface of silicon. (111) is the primary cleavage plane of Si. When Si is cleaved in vacuum along this plane with a wedge kept parallel to the $\langle 11\bar{2}\rangle$ direction, a Si(111)(2 × 1) reconstruction is created. This reconstruction differs from that of the (100) surface by forming long pi-bonded chains in the first and second surface layers. Annealing above 300 °C converts the structure irreversibly into a (7 × 7) reconstruction, which represents an equilibrium phase. Much effort was spent over more than two decades to unravel the complex structure of this surface.

The Si(111)(7 × 7) surface unit-cell is 49 times larger than the unit cell of the ideal surface. An unreconstructed cell of this size would have 98 dangling bonds. The formation of the actual (7 × 7) structure comprises an extensive rearrangement of atoms and the addition of adatoms on top. Scanning tunneling micrographs show a pattern of 12 protruding adatoms in a unit cell and vacancies at the deep corner holes, see Fig. 7.12. Each adatom saturates three dangling bonds from the layer underneath and leaves one dangling bond. STM images recorded with a suitable bias show that the rhombic unit cell is composed of two triangular subunits.

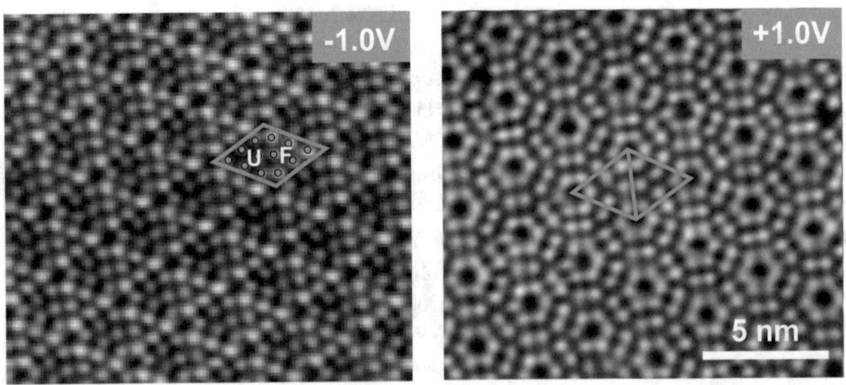

Fig. 7.12 Scanning tunneling images of the Si(111)(7 × 7) surface for negative (*left, filled states*) and positive (*right, empty states*) sample bias. The *orange rhombs* mark a unit cell comprising 12 protruding adatoms, *U* and *F* denote unfaulted and faulted half-unit cell, respectively. Courtesy of M. Dähne, TU Berlin

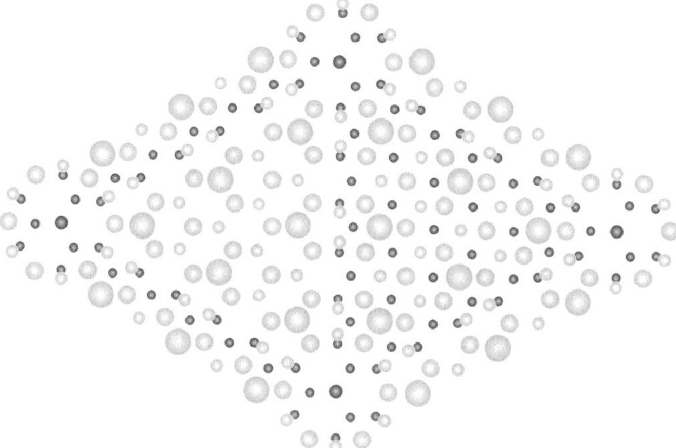

Fig. 7.13 Structure of the Si(111)(7 × 7) surface unit-cell according to the dimer-adatom-stacking fault model. Atoms in layers of decreasing height are represented by decreasing size, *dark atoms* represent unreconstructed atoms of the uppermost bulk layer underneath the surface layers

The structure of the Si(111)(7 × 7) surface was eventually explained by the generally accepted DAS model derived from a detailed analysis of transmission electron diffraction [18]. The name DAS refers to the basic structural elements: 15 dimers (D), 12 adatoms (A), 1 stacking fault (S). The model has also been confirmed by computation, yielding an energy gain of 60 meV per (1 × 1) unit cell with respect to the (2 × 1) reconstruction [19]. A top view of the unit cell according to the DAS model is given in Fig. 7.13. The uppermost atoms are indicated by largest size, and we recognize the 12 adatoms visible in the STM image Fig. 7.12. We also note the corner holes in the top layer. The model shows an apparent difference in the left and right triangles of the unit cell. In the right half we see the uppermost atoms of the bulk crystal underneath the surface layers drawn with a dark shading. This yields an ABC stacking order with respect to the first two surface layers, thereby continuing the cubic diamond structure of bulk Si. The uppermost bulk atoms are not visible in the left triangle, because they are hidden by the surface layers due to an $ABAB$ stacking order. The left part of the unit cell has consequently a hexagonal order, i.e., the sequence is faulted with respect to the structure of bulk Si. The faulted and non-faulted triangular subunits are each surrounded by 3 vacancies at the corners and 9 dimers, three of them located at the separating domain boundary, which is oriented along the vertical short diagonal in Fig. 7.13.

The DAS model leaves only 19 unsaturated dangling bonds per unit cell: 12 for the adatoms, 6 from three-fold coordinated rest atoms lying in the layer below the adatoms, and 1 from the atom below the vacancy at the corner. A number of similar DAS reconstructions in a $(2n + 1) \times (2n + 1)$ pattern have also been observed in non-equilibrium conditions, including 5×5 and 9×9 reconstructions [20]. The preference for the (7 × 7) reconstruction is attributed to an optimal balance of charge transfer and stress.

7.2 Kinetic Process Steps in Layer Growth

The non-equilibrium process of growth is governed by a competition between kinetics and thermodynamics. As the size of heterostructures approaches the nanometer-scale regime, atomic-level control is becoming crucial. Kinetic growth processes may to a large extend be described by the terrace-step-kink model already used in Sect. 6.2.4 to characterize surface energies. We will first employ this simplified model to consider growth kinetics, and include the effect of the actually more complex surface structure as described above in a second step.

7.2.1 Kinetics in the Terrace-Step-Kink Model

The *terrace-step-kink* (TSK) *model* of a surface [21] (also termed *terrace-ledge-kink* (TLK) *model*) is based on the idea that the energy of an atom's position on a crystal surface is determined by its bonding to neighboring atoms. Processes on an atomistic scale hence involve the counting of broken and formed bonds. In the framework of the TSK model the kinetics of growth is described by rates of transition steps, which atoms undergo on the surface. The complex process of epitaxy is largely determined by only a few categories of such processes. Some processes of particular importance are illustrated for the surface of a Kossel crystal in Fig. 7.14.

Growth proceeds by a number of consecutive steps as indicated in Fig. 7.14. Atoms arrive from the ambient and are adsorbed on the surface. They may then diffuse over the surface until they cease to diffuse by one of several processes. Such processes are re-evaporation (or re-solution in case of a liquid ambient), nucleation of (2D or 3D) islands, attachment to existing islands or to defects like a step on the surface. In the framework of the considered model the adatom diffusion on the surface is described by rates of hopping from one site to an adjacent site. The vibrational

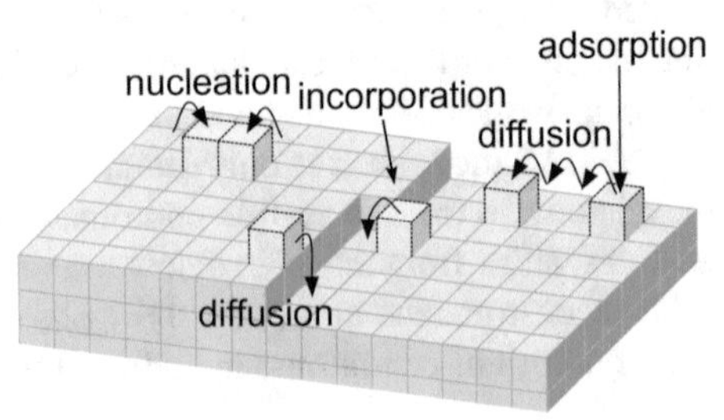

Fig. 7.14 Schematic of atomistic rate processes in epitaxial growth. The step-down diffusion differs from the indicated lateral diffusion

motion of an adatom on the surface is regarded as an attempt at such a hop. Since many attempts are required to produce a single hop, a factor is introduced representing the probability per attempt to hop, yielding an effective hopping rate $k(T)$. The rate may also be expressed in terms of a mean residence time $\tau(T) = k^{-1}(T)$. Each thermally activated kinetic process of epitaxy is governed by characteristic parameters entering an Arrhenius dependence with an activation energy E,

$$\tau^{-1}(T) = \nu_0 e^{-\frac{E}{k_B T}}. \tag{7.5}$$

E is the barrier which has to be surmounted in the process. The prefactor ν_0 represents an attempt-rate constant for the given process.

Rate equations referring to a few basic processes are used in numeric Monte Carlo simulations to model the dynamics of growth and the evolution of the growth surface. Only the rate-limiting steps are included in the calculations. Faster processes are accounted for in average by using *effective* kinetic parameters. Parameters to control the supersaturation $\Delta\mu$ are usually the experimental variables T, the arrival rate of atoms R, and the material parameters of the kinetic processes describing diffusion, re-evaporation, and nucleation. The approach does not require a detailed knowledge of the atomic interactions and permits simulations including large time scales. Values for ν_0 and E (on the order of 10^{12} s^{-1} and 1 eV, respectively) for each process are estimated from, e.g., molecular dynamics, or they are taken as parameters to fit experimental results.

7.2.2 Atomistic Processes in Nucleation and Growth

The atomistic processes indicated in Fig. 7.14 are now described in more detail to account for nucleation and growth analogous to the thermodynamic approach discussed in Sect. 6.2. In our simplified approach [22] we neglect surface reconstructions, interdiffusion, and chemical reactions between substrate and deposit material. The assumed processes are indicated in Fig. 7.15.

Atoms arriving from the ambient on the surface at an externally controlled flux F (atoms per unit area and per unit time) diffuse on the surface. They either meet other atoms to nucleate, or they are captured at a growing island (also denoted cluster) after a mean time τ_c, or they re-evaporate after a mean adsorption residence time τ_a. A nucleus will either loose adatoms after a mean time τ_n, or grows to a critical size comprising i atoms. The critical nucleus also may dissolve or grow to a stable island comprising $x > i$ atoms. Such islands will capture further adatoms but—under growth conditions—hardly loose atoms.

We trace an atom which arrives at the surface. At a high temperature the adatom will only stay on the surface for a short residence time τ_a. This time is determined by the adsorption energy E_a and may be written

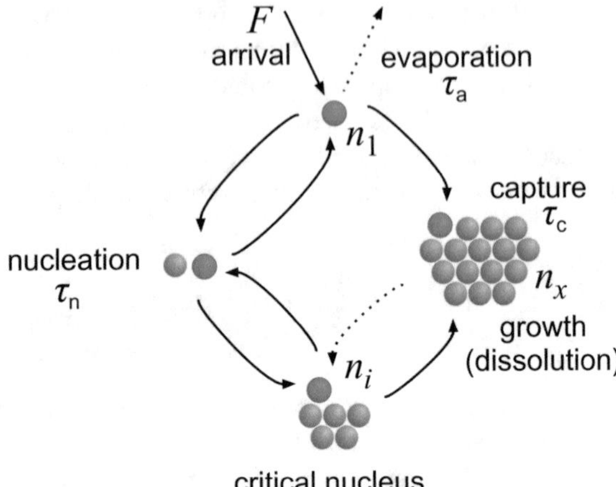

Fig. 7.15 Processes considered for the kinetic description of growth

$$\tau_a^{-1} = \nu_{a0} \exp\big(-E_a/(k_B T)\big); \tag{7.6}$$

the prefactor ν_{a0} corresponds to an atomic surface vibration-frequency and represents an attempt rate for desorption. For some materials relation (7.6) is complemented by a factor, which accounts for a dependence of the residence time on the material coverage on the substrate. During τ_a the adatom migrates in a random-walk process consisting of a series of jumps to respective adjacent substrate sites. The mean-square displacement of the adatom during a period of duration t depends on the hopping rate (or, mean hopping frequency) ν_d and is given by

$$\langle \lambda^2 \rangle = \nu_d a^2 t, \tag{7.7}$$

a being the mean jump distance. Usually a is the distance of two neighboring lattice sites on the surface. The number of hops during the considered period is $\nu_d t$. Hopping to a neighboring site requires surmounting a potential barrier, cf. Fig. 7.23. This diffusion barrier E_d is surmounted more easily at increased temperature due to an Arrhenius dependence of the hopping rate,

$$\nu_d = \nu_{d0} \exp\big(-E_d/(k_B T)\big); \tag{7.8}$$

The prefactor ν_{d0} is an attempt rate for a diffusion hop. The ratio of the mean-square displacement (7.7) with respect to the duration of the period is the time-independent *diffusion coefficient* (or *diffusivity*) D,

$$D = \frac{\langle \lambda^2 \rangle}{\eta t} = \frac{\nu_d a^2}{\eta}, \tag{7.9}$$

where η is the number of neighboring sites reachable by a single jump: For one-dimensional diffusion $\eta = 2$, while for 2D surface diffusion $\eta = 4$ on a square lattice and 6 on a hexagonal lattice. Using Einstein's relation of the diffusion length, $\lambda = \sqrt{D\tau}$, and (7.9), the displacement (root-mean-square value) from the arrival site to the site of eventual evaporation or incorporation reads

$$\lambda = \sqrt{D\tau} = \lambda_0 \exp\big((E_a - E_d)/(2k_B T)\big), \qquad (7.10)$$

τ being the mean time of surface diffusion (basically τ_a at high T, τ_c at lower T). The pre-exponential factor λ_0 is a merged *effective* elementary jump distance, e.g., for $\tau = \tau_a$ given by $\lambda_0 = \sqrt{\nu_{d0}/\nu_{a0}} \times a/\sqrt{\eta}$. Values of ν_{d0} are typically somewhat less than those of the corresponding parameter for adsorption ν_{a0}, but of the same order. The barrier to be surmounted for desorption, E_a, is however usually much larger than the diffusion barrier E_d, and also several times exceeding thermal energies $k_B T$ at typical growth temperatures. The adatom will therefore migrate over a quite long distance $\lambda \gg a$ before evaporation. For small values of $1/T$ (in the high temperature range) the surface diffusion length λ increases exponentially with $1/T$ (or, decreases with T) as indicated by (7.10) and the straight line in the Arrhenius plot Fig. 7.16. The residence time in this desorption regime is short, and adatoms are likely to evaporate before being incorporated. At large values $1/T$ (low T) the slope in the Arrhenius dependence of $\lambda(T)$ changes sign. The residence time is large according to (7.6); at low T adatoms are incorporated after diffusion and the competition by desorption gets negligible. Best epitaxial growth is often achieved for large diffusion length. This is obtained just below the onset of significant re-evaporation.

During surface diffusion the adatom encounters other atoms on the surface. The probability of such a meeting depends on the areal density n_1 of single migrating atoms, and the areal density of clusters containing more than one atom. In our approach we assume the clusters to be stationary, i.e., single adatoms are the only species which are mobile.

The clusters on the surface are divided according to their size into subcritical clusters with atom numbers $j \leq i$, and stable clusters, $j > i$, the density of which is summed as

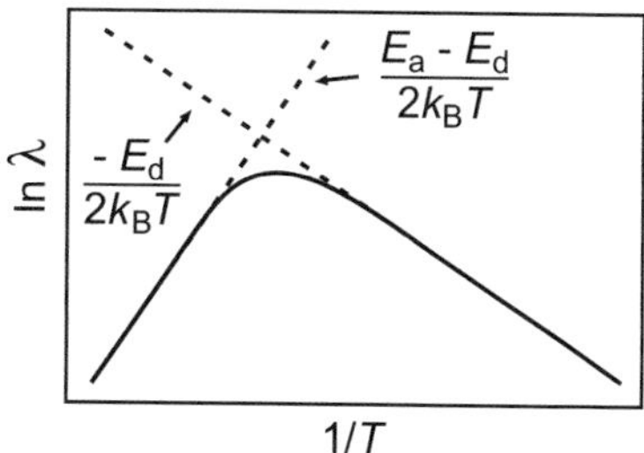

Fig. 7.16 Temperature dependence of the surface diffusion length λ

$$n_x = \sum_{j=i+1}^{\infty} n_j.$$

After an initial nucleation of stable clusters the subcritical clusters nucleate and dissolve in an approximate detailed balance. They are therefore not considered further in the evolution due to the local equilibrium condition $dn_j/dt = 0$ for $2 \le j \le i$. In the following we focus on the density of single atoms n_1 and that of stable clusters n_x.

The density n_1 of the mobile single atoms follows from a balance of processes described by an appropriate rate equation. As we read from Fig. 7.15, atoms arriving on the surface at a rate F increase n_1, while nucleation (τ_n), capture (τ_c), and evaporation (τ_a) decrease n_1. Furthermore, when larger stable islands cover a significant fraction Z of the surface, arrival on top of an island is possible, thereby decreasing the part of atoms on the uncovered surface. The evolution of the single-atom density n_1 is therefore described by

$$dn_1/dt = +F(1-Z) - n_1/\tau_n - n_1/\tau_c - n_1/\tau_a = F(1-Z) - n_1/\tau, \quad (7.11)$$

with $\tau^{-1} = \tau_n^{-1} + \tau_c^{-1} + \tau_a^{-1}$. The sum may be extended when additional loss processes for diffusing adatoms are to be taken into account. On the other hand, also less processes may be considered.

In the limit of *high temperature* the residence time τ_a gets very short, and nucleation and capture processes become negligible. Since also $Z \cong 0$ in this case, we obtain $dn_1/dt = F - n_1/\tau_a$. This yields for long times $t \gg \tau_a$ a stationary state ($dn_1/dt = 0$) with the solution

$$n_1 = F\tau_a.$$

The constant density in the high temperature limit reflects the balance of arrival and evaporation without nucleation or growth.

Below the high temperature limit loss of adatoms by attachment to growing stable islands is relevant. Once stable islands exist, the term $-n_1/\tau_c$ in (7.11) is generally much larger than the term $-n_1/\tau_n$. Neglecting the nucleation term $-n_1/\tau_n$ for $t \gg \tau_a$ the solution gets

$$n_1 = F(1-Z)(1/\tau_a - 1/\tau_c)^{-1}.$$

The mean time τ_c of the adatom capture at a stable island depends on their density n_x and the diffusional flow,

$$\tau_c^{-1} = \sigma_x D n_x. \quad (7.12)$$

The capture numbers σ_k of the stable clusters summed to a mean number σ_x express the local decrease of $n_1(\mathbf{r}, t)$ near a k-sized cluster due to capture [23]. They are typically of order 5–10 and may regarded for a first appreciation as slowly varying quantities.

The density n_x of stable clusters increases by a nucleation rate J involving capture at critical clusters with i atoms. The creation of new stable clusters is proportional to the density n_i of the critical clusters with their capture number σ_i, and the density of single adatoms n_1 and their diffusivity D, yielding $J = \sigma_i D n_1 n_i$. On the other hand, as growth of a layer proceeds the density will reduce due to an impingement of stable clusters on each other. Such coalescence is accounted for by including a negative term proportional to the temporal change of coverage Z, leading to the rate equation [22]

$$dn_x/dt = +\sigma_i D n_1 n_i - 2n_x dZ/dt. \tag{7.13}$$

To solve the coupled equations (7.11) and (7.13) a relation between Z and n_x is needed. If the clusters grow two-dimensionally, we may put [22]

$$dZ/dt = \Omega^{2/3}\big((i+1)n_1/\tau_i + n_1/\tau_c + FZ\big), \tag{7.14}$$

where Ω is the atomic volume of the deposit and the stable-cluster nucleation rate $\tau_i^{-1} = \sigma_i D n_i$. The solution of the rate equations leads to the cluster densities schematically shown in Fig. 7.17.

Growth on a flat surface starts at coverage $Z = 0$. At a *high temperature* T_1 we note an initial rise of the single-atom density $n_1 \sim Ft$ lasting for $t < \tau_a$, followed by a constant value described by the high-temperature limit. The density of stable clusters n_x starts at a negligible value at $t = \tau_a$, and increases for $t > \tau_a$ as given by the first term in (7.13), i.e., before coalescence. Both, n_1 and n_x decrease as coalescence sets in. The condensation coefficient $\alpha(t)$, which denotes the fraction of atomic dose impinging on the surface and being incorporated into the deposit, is initially very small at this high temperature, and so also the total deposit $Ft\alpha$.

At a *low temperature* T_2 no re-evaporation occurs, i.e., $\alpha = 1$. In this temperature range the single-atom density n_1 plotted in Fig. 7.17 increases linearly until capture

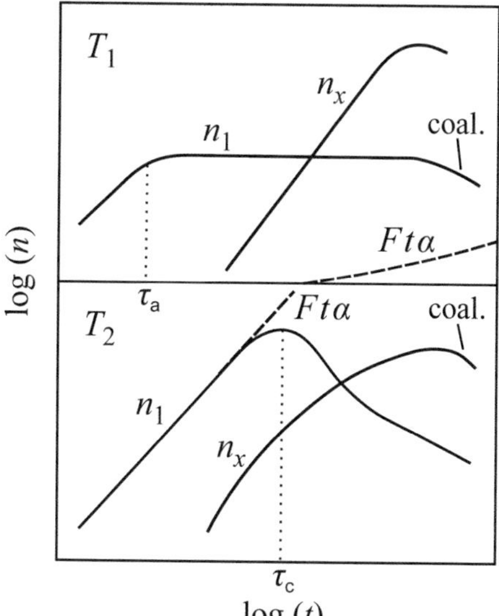

Fig. 7.17 Density of single atoms n_1, stable clusters n_x, and the total number $Ft\alpha$ of atoms condensed on the surface as a function of time t for a high temperature T_1 and a low temperature T_2. τ_a and τ_c indicate re-evaporation and capture times, respectively, the label *coal.* marks a decrease of n_1 and n_x due to coalescence. After [22]

Table 7.4 Parameters p and E of (7.15) for the maximum density n_x of stable two-dimensional clusters in various regimes of condensation. i is the number of atoms in the critical cluster. From [22]

Condensation regime	p	E
Extremely incomplete	i	$E_i + (i+1)E_a - E_d$
Initially incomplete	$i/2$	$\frac{1}{2}(E_i + iE_a)$
Complete	$i/(i+2)$	$(E_i + iE_d)/(i+2)$

by previously nucleated clusters sets in, causing n_1 to pass a maximum at the mean capture time τ_c and to decrease subsequently. The stable-cluster density n_x increases after a nucleation period and eventually decreases due to coalescence.

The maximum density of stable clusters has the general form

$$n_x \sim (F/\nu)^p \exp\big(-E/(k_B T)\big), \tag{7.15}$$

where the materials parameters p and E lead to different regimes of condensation [22]. Values for the regimes of extremely incomplete condensation, initially incomplete in an intermediate range, and complete condensation are given in Table 7.4. The three regimes refer to the conditions $\sigma_x D\tau_a n_x \ll Z$, $Z < \sigma_x D\tau_a n_x < 1$, and $\sigma_x D\tau_a n_x \gg 1$, respectively, and define the meaning of a *high* and a *low* temperature for growth of the considered material.

We note from Table 7.4 that the condensation regimes are essentially determined by the relation between the materials parameters E_i, which signifies the energy difference between i atoms in the adsorbed state and in stable clusters, the adsorption energy E_a, and the critical cluster size i. At *extremely incomplete* condensation growth is slow due to strong re-evaporation and proceeds essentially by direct impingement. In the intermediate, *initially incomplete* regime stable clusters grow by diffusive capture in the initial state. Direct impingement becomes relevant at large coverage Z, leading to more complete condensation. This regime will often occur in practice. At *complete* condensation growth is fast, since re-evaporation is negligible.

7.2.3 Adatoms on a Terraced Surface

Growth of a flat surface proceeds basically by the attachment of adatoms at steps. The steps may originate either from 2D nucleation as discussed above, or from screw dislocations, or from the terrace structure of a vicinal surface inclined with respect to a singular face by a small tilt angle. The rate of layer growth r, which represents the velocity of a parallel displacement of the singular faces, is hence determined by the advancement of steps.

To obtain an expression for the velocity of step advance we consider growth conditions with an only minor contribution of nucleation. Such conditions may be found in step-flow growth of a vicinal surface (Sects. 6.2.7, 7.2.4). This surface consists of terraces of width l which are separated by steps of height a (Fig. 6.14). Let us assume for simplicity that the steps are equidistant and straight. If the steps remain straight all the time, we obtain a one-dimensional problem. The steps are assumed perfect sinks for diffusing adatoms arriving from either side and have monoatomic height. Furthermore, the terrace width l is assumed smaller than the mean displacement λ of an adatom after arrival on the surface. If now a constant flux F deposits a low density of adatoms n on the surface (in the previous Sect. 7.2.2 labeled n_1), the adatoms will diffuse with some finite probability to an adjacent step where they are incorporated, and we may neglect nucleation on the terrace. For the given conditions the concentration of adatoms on the surface $n(x, t)$ is described by

$$\dot{n}(x,t) = Dn''(x,t) - \frac{1}{\tau_a}n(x,t) + F, \qquad (7.16)$$

where x is the lateral direction perpendicular to the steps, dot and double primes designate temporal derivative $\partial/\partial t$, and spatial derivative $\partial^2/\partial x^2$, respectively, D is the surface diffusion coefficient (7.9), and $1/\tau_a$ denotes the evaporation probability (7.6). The change of the adatom density n expressed by (7.16) is composed of the three terms surface diffusion, loss by evaporation, and deposition by the external flux F.

To solve (7.16) we need boundary conditions, which are to be fulfilled by $n(x, t)$. We consider steady-state conditions, where F is constant and n does not change in time, i.e., $\dot{n}(x, t) = 0$. Steps are considered as perfect sinks. A fast incorporation of adatom at steps is fulfilled by the condition that n equals a constant concentration n_{eq} at a step to be reached at thermal equilibrium. Choosing the origin of x at the center of a terrace, the condition reads

$$n(\pm l/2) = n_{eq}. \qquad (7.17)$$

The steady-state solution of (7.16) is then given by

$$n(x) = \frac{\cosh(\kappa x)}{\cosh(\kappa l/2)}(n_{eq} - F\tau_a) + F\tau_a. \qquad (7.18)$$

In (7.18) κ is the inverse of the mean displacement, $\kappa = 1/\lambda = 1/\sqrt{D\tau_a}$. If the residence time τ_a is long, re-evaporation described by the second term in (7.16) becomes negligible. κ is then small. If also $\kappa l/2$ is small, solution (7.18) can be approximated by a Taylor expansion, yielding

$$n(x) \cong n_{eq} + \left(F\tau_a\kappa^2/8\right)\left(l^2 - 4x^2\right) = n_{eq} + (F/8D)\left(l^2 - 4x^2\right) \quad (\kappa l/2 \text{ small}). \qquad (7.19)$$

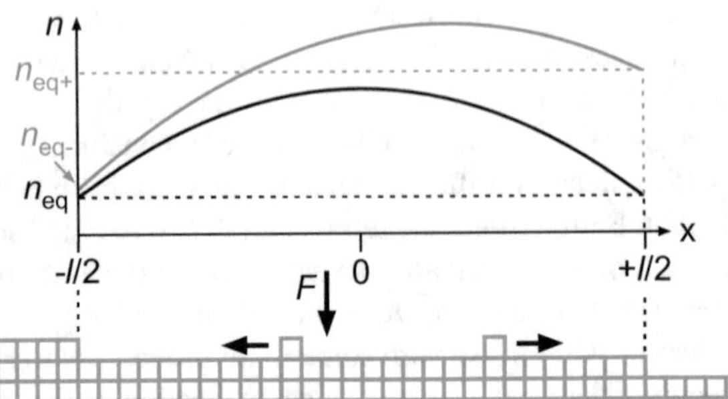

Fig. 7.18 Density profile $n(x)$ of adatoms on a terraced surface under steady-state conditions provided by a constant flux F and steps acting as sinks (*black curve*). *Gray curve*: $n(x)$ resulting from an uneven capture probability of up-steps and down-steps

The steady-state solution (7.19) for negligible re-evaporation is shown in Fig. 7.18 (black curve). It is a parabola with a maximum centered in the middle of a terrace at $x = 0$ with values $n(\pm l/2) = n_{eq}$.

The assumptions made to obtain the simple geometry expressed by (7.16) and the approximate solution (7.19) are quite extensive and may in practice often not be fulfilled. The requirement of a perfect sink may, e.g., be fulfilled by the left step in Fig. 7.18 but less well by the right step, because the latter requires a less probable down-step diffusion for incorporation. An unequal capture probability can be expressed by *two* equilibrium adatom-concentrations at the steps n_{eq+} and n_{eq-}, leading to a similar solution as (7.19) with a parabola maximum being displaced from $x = 0$ by $\Delta x_{max} = D/(Fl)(n_{eq+} - n_{eq-})$. The respective profile $n(x)$ is drawn in Fig. 7.18 in gray; values for n_{eq+} and n_{eq-} are assumed larger than n_{eq}.

The mean adatom density n in steady-state conditions is obtained from (7.19) by integration over $n(x)$, yielding

$$n = \frac{1}{l}\int_{-l/2}^{+l/2} n(x)\,dx = n_{eq} + Fl^2/(12D). \tag{7.20}$$

In case of unequal capture probability at the steps $n_{eq} = (n_{eq+} + n_{eq-})/2$ in the right-hand side of (7.20).

7.2.4 Growth by Step Advance

We may treat the capture of adatoms at steps similar to the capture at stable clusters of density n_x. If we assume the steps acting as strong sinks, i.e., $n_{eq} = 0$, such a treatment is accomplished by adding a term $-n_1/\tau_s$ to the right side of (7.11). τ_s is the mean time of adatom capture at a step. The mean single-adatom density n_1 given by (7.20) then corresponds to $F\tau_s$. Capture at steps now compete with capture at stable clusters and nucleation. Capture at a cluster occurs for an adatom arrival within a root-mean-square distance $\lambda_c = (D\tau_c)^{1/2}$. Near a step (ideal sink) nucleation is depressed. This zone on either side of a step is called *denuded zone*. Its width is given by λ_c. This consideration allows to express a condition for the transition from nucleation and 2D island growth at lower temperatures to step flow growth at higher temperatures. For dominating capture at clusters, n_1 is expressed by $F\tau_c$ with τ_c given by (7.12). Step flow becomes more dominant than nucleation on a terrace if $\tau_s < \tau_c$. Inserting n_1 from (7.20) and (7.12), respectively, we see that this condition is met if

$$l^2/(12D) < (\sigma_x D n_x)^{-1}. \tag{7.21}$$

This inequality may be expressed in terms of the denuded-zone width λ_c with τ_c from (7.12), yielding

$$\frac{l^2}{12}\frac{\sigma_x n_x}{1} = \frac{l^2}{\lambda_c^2}\frac{1}{12} < 1. \tag{7.22}$$

According to (7.22) the transition from nucleation to step flow occurs if the width of the denuded zone λ_c (which increases as the temperature increases) becomes larger than a fraction of about one third of the terrace length l.

The surface morphology of homoepitaxial GaAs layers grown at different temperatures is shown in Fig. 7.19. The sample grown in step-flow mode at a high temperature exhibits terraces with steps of atomic height, originating from an off-orientation of the substrate with respect to the [001] growth direction. The morphology of the sample grown at decreased temperature shown in Fig. 7.19b exhibits 2D island growth on terraces. The terrace edges are irregular, the large terrace width indicates a small off-orientation of the substrate. Note that no islands nucleate close to a step.

The transition between island growth to step-flow growth can be studied experimentally using diffraction techniques, see Sect. 8.3. For GaAs(001) surfaces the diffusion length of Ga adatoms in MBE growth was determined from the transition temperature obtained from the onset of RHEED intensity oscillations; the transition was in fact found to depend on the surface-diffusion length with respect to the terrace width [24]. This was also observed for AlN(0001) [25].

A simulation of the molecular-beam epitaxy of GaAs on GaAs(001) in the step-flow growth mode for appropriate growth conditions is given in Fig. 7.20. The growing (001) surface of the GaAs zincblende structure is polar: surfaces with terminations of Ga cations and As anions alternate periodically. For a slow As_2 incorporation rate assumed in the simulation, the As coverage (even numbered gray curves) lags behind

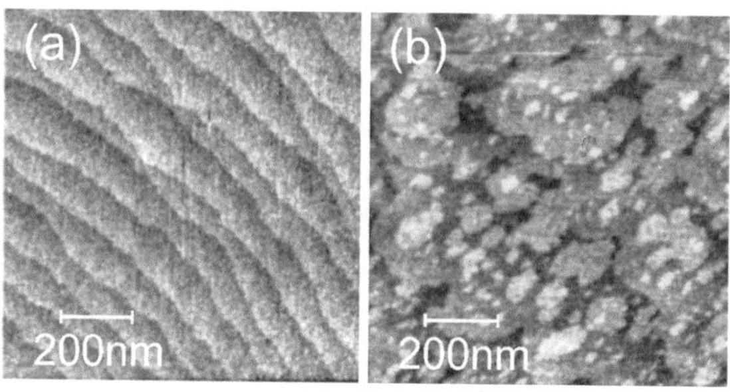

Fig. 7.19 Atomic-force micrographs of 0.5 μm thick homoepitaxial GaAs layers grown at **a** 570 °C, and **b** 530 °C

the directly preceding Ga layer coverage (odd numbered black curves) until the near completion of a layer [26]. We note that the S-shaped curves of subsequent (bi)layers overlap: Nucleation of the next layer starts before the layer underneath is completed. Of the first monolayer's worth of Ga material delivered (at $\tau_{\mathrm{ML}} \cong 0.9/\mathrm{sec}$) about 95 % goes into the first Ga layer and 5 % into the next for the considered growth conditions. At fast As_2 incorporation, coverages of As layers *lead* the coverage of the preceeding Ga layers [26]; for a comprehensive treatment see [27].

The advancement of a step results from the capture of adatoms from the upper and lower terrace. Let us consider the velocity of lateral advance r_{step} of the step at the position $x = -l/2$. We again assume steady-state conditions, i.e., adatoms diffuse much faster than steps move. Furthermore, we again consider terraces of equal width and an equal probability for capture from either terrace. The current density j of adatoms arriving at the step follows from the gradient of the adatom density $n(x)$ at the step position $x = -l/2$,

$$j(-l/2) = -D\frac{d}{dx}n(-l/2). \tag{7.23}$$

Inserting the symmetric adatom density $n(x)$ from (7.18) we obtain

$$j(-l/2) = D\kappa(F\tau_{\mathrm{a}} - n_{\mathrm{eq}})\tanh(-\kappa l/2). \tag{7.24}$$

The step velocity r_{step} results from the (equal) contributions from the upper and the lower terrace separated by the step,

$$r_{\mathrm{step}} = a^2\left(j_{\mathrm{upper}}(-l/2) + j_{\mathrm{lower}}(-l/2)\right) = 2a^2 j(-l/2), \tag{7.25}$$

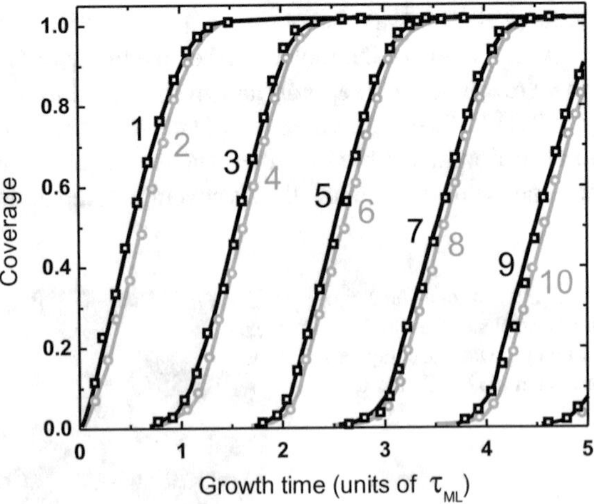

Fig. 7.20 Simulated coverage in the molecular-beam epitaxy of GaAs/GaAs(001) for slow As_2 incorporation. *Odd* and *even numbered curves* refer to Ga and As layers, respectively. Reproduced with permission from [26], © 1986 APS

a^2 being a unit area of the terrace. The velocity r_{step} depends on the terrace width, because larger terraces collect more adatoms from the ambient, cf. Fig. 7.21. For small $\kappa l/2$ (long residence time τ_a) (7.24) simplifies to

$$j(-l/2) = D(F\tau_a - n_{\text{eq}})\kappa^2 l/2. \tag{7.26}$$

The linear relation $j \propto l$ is reflected in the constant slope of r_{step} for small terrace width l in Fig. 7.21.

We now consider the growth rate r of a flat crystal surface. Growth proceeds essentially by the attachment of adatoms at steps. The steps may either originate from 2D nucleation, or screw dislocations, or from the terrace structure of a vicinal surface inclined with respect to a singular face by a small tilt angle Θ (Fig. 6.14). The advancement of steps therefore determines the rate of layer growth r, which represents the velocity of a parallel displacement of the singular faces. The relation between the lateral advancement of steps and growth rate r is illustrated for a vicinal surface in Fig. 7.22. The figure depicts a cross section of the terraced surface at a time t_1 (black lines) and a later time t_2 (gray lines). During $t_2 - t_1$ the steps are laterally displaced via adatom attachment by some amount s, yielding the velocity $r_{\text{step}} = s/(t_2 - t_1)$.

The growth rate r of the singular layer (i.e., the mean vertical displacement of the horizontal terraces in Fig. 7.22a) follows from r_{step} via the tilt angle Θ of the vicinal layer. For steps of height d and terraces of width l the tilt angle is given by $\tan\Theta = d/l$. Since also $\tan\Theta = r/r_{\text{step}}$, we obtain

$$r = r_{\text{step}} \tan\Theta = (d/l) r_{\text{step}}. \tag{7.27}$$

The growth rate r_{vicinal} of the vicinal layer (i.e., the velocity of a parallel displacement of the layer indicated by dotted lines in Fig. 7.22) may be read from the enlarged scheme Fig. 7.22b. Since Θ is also the angle between r and r_{vicinal}, the relation is given by

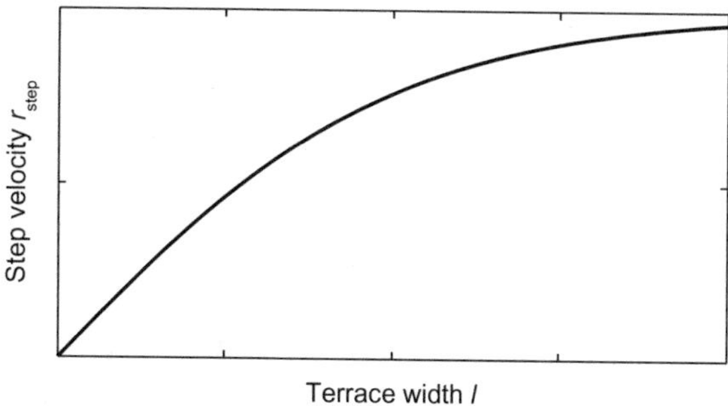

Fig. 7.21 Steady-state velocity of lateral step advance r_{step} depending on the terrace width l for equidistant and straight steps

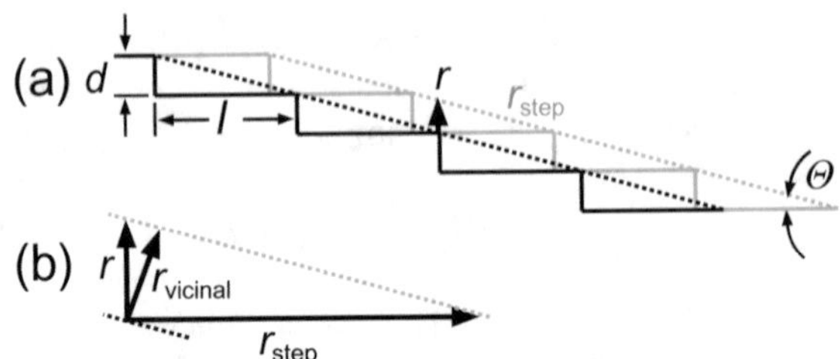

Fig. 7.22 **a** Relation between the lateral growth rate r_{step} of steps on a vicinal surface and the growth rate r of the related singular layer. *Black* and *gray lines* represent cross sections of the surface at two times, the *dotted lines* indicate the vicinal surface. **b** Magnified scheme showing the relation between the growth rates of steps, singular surface and vicinal surface

$$r_{\text{vicinal}} = r \cos \Theta = r_{\text{step}} \sin \Theta. \tag{7.28}$$

7.2.5 *The Ehrlich-Schwoebel Barrier*

Diffusion of an adatom on a flat or terraced surface is of particular importance in the kinetics of growth. Surface diffusion by a random-walk process on a *flat* surface was described by (7.10). Until now we assumed that steps or stable 2D clusters capture adatoms from their top and from their bottom with the same probability. In fact, diffusion across a flat surface generally differs from diffusion across a step, as first pointed out by Ehrlich [28], Schwoebel [29], and co-workers. Adatom capture from the upper and lower terrace at a step hence also differ. Such unequal capture probability causes the asymmetry in the adatom density discussed in Sect. 7.2.3. We consider the diffusion barrier provided by a step and consequences for two-dimensional growth.

An intuitive picture of the so-called Ehrlich-Schwoebel barrier of diffusion across a step is given in Fig. 7.23. As an adatom moving on the upper terrace approaches the step it has to overcome a pronounced maximum of the potential to step down. The reason is a fewer number of nearest neighbors in the transitions state of the hop over the ledge compared to that in flat-surface diffusion. We note from Fig. 7.23 that the Ehrlich-Schwoebel barrier E_{ES} is actually the *difference* in the activation energy for hopping across the step and hopping across a flat surface. Once the adatom has surmounted the barrier E_{ES} it enters a position at the step edge with an increased binding energy E_{s}. The increase originates from the larger number of nearest neighbors at the lower edge of the step.

Step down of an adatom may also occur by another process with a potentially lower barrier: The adatom located at the position on top drawn in Fig. 7.23 pushes the edge atom underneath away to the right and takes its place. This *exchange process* may provide an efficient parallel channel to the step-down process discussed above.

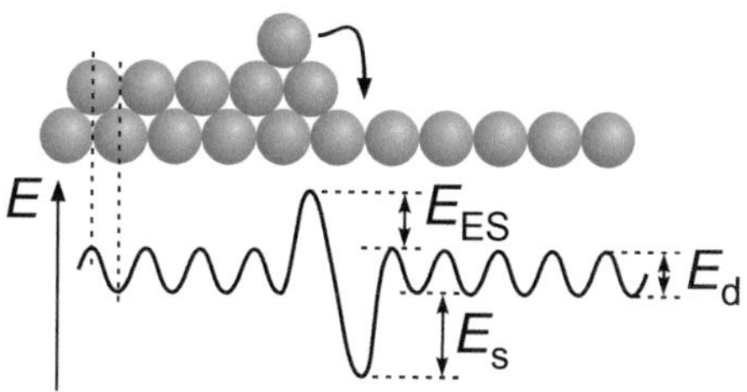

Fig. 7.23 Cross section of a monoatomic surface step and schematic of the potential associated with the diffusion of an adatom across the surface. Energies E_d, E_s, and E_{ES} refer to the activation of surface diffusion, binding to a step, and the Ehrlich-Schwoebel barrier, respectively

Both processes have always higher barriers than diffusion on a flat surface. In the following we focus on the usual Ehrlich-Schwoebel barrier depicted in Fig. 7.23.

We note from Fig. 7.23 a distinct difference for step-down and step-up diffusion. In the first case the adatom approaches a *descending* step and experiences a repulsive step-edge barrier of height E_{ES}, which tends to reflect the adatom. By contrast, an adatom approaching an *ascending* step (i.e., migrating from the right to left towards the step in Fig. 7.23) experiences a trapping potential E_s accompanied by a subsequent barrier E_{ES}. Such potential landscape tends to capture the adatom at the step.

Incorporation of diffusing adatoms into the crystal was mentioned in Sect. 7.1.1 to occur preferentially at kink positions. For actual incorporation an adatom caught at a step by the binding potential illustrated in Fig. 7.23 needs to diffuse along the step until meeting a kink position. We consider this one-dimensional diffusion analogous to the surface diffusion on a terrace. Let us assume a single adatom diffusing along a step towards a kink formed by an additional row of atoms (i.e., a kink with *positive* sign, cf. Fig. 7.1). To *pass* the kink site and continue diffusion along the step the adatom has to hop around the kink. Obviously the kink provides a barrier in addition to the activation barrier for the one-dimensional diffusion along a straight step, and also in addition to the Ehrlich-Schwoebel barrier discussed above. Qualitatively, the potential looks like that for two-dimensional diffusion *across* the step depicted in Fig. 7.23. Due to this analogy the barrier is termed the *Kink Ehrlich-Schwoebel barrier*, and the constraint of the 1D diffusion by this barrier is referred to as the *Kink Ehrlich-Schwoebel effect*.

The barriers for diffusion *across* a step and *along* a step both affect the morphology of the surface. The first effect may particularly affect the distribution of steps, while the latter may control the structure of a step. Both effects are addressed in the following.

7.2.6 *Effect of the Ehrlich-Schwoebel Barrier on Surface Steps*

In the discussion on growth by step advance Sect. 7.2.4 we assumed a terraced surface with equidistant and straight steps. Qualitative arguments show that both conditions may be oversimplified. It has in fact been observed in experiments as well as in computer simulations that the step-flow mode is only metastable. Let us consider step-flow growth of an ideal vicinal surface with equidistant steps, and *absence* of an Ehrlich-Schwoebel barrier (Fig. 7.24a). Adatoms from an upper and a lower terrace will then be captured at a step with equal probability, and due to a constant width of all terraces all steps advance with the same velocity r_{step}. We recall from (7.25) and (7.26) that the velocity is proportional to the widths of upper and lower terrace, $r_{step} \propto (l_{upper} + l_{lower})$. If now one step lags behind its regular position for any reason (e.g., some fluctuation) the area of the lower terrace in front increases and that of the upper terrace behind decreases, see Fig. 7.24b. The lower, wider terrace in front of the step will collect more adatoms, and distributes one half of this surplus to the considered step and the other half to the step ahead. The upper, smaller terrace accordingly collects less adatoms. The lack is equal to the surplus at the lower terrace. Less adatoms are hence supplied from the upper terrace to the considered step and also to that behind. As a consequence, the decreased velocity of the considered step remains unchanged. In addition, the velocity of the step ahead is slightly accelerated, and the velocity of the step behind is slightly delayed. The equidistant arrangement of steps is therefore not stable in absence of an Ehrlich-Schwoebel barrier. Eventually bunches of steps appear on the surface, cf. Fig. 7.24c. Such a process is termed *step bunching*.

In *presence* of an Ehrlich-Schwoebel barrier adatoms are basically captured from the lower terrace at a step. A delayed step will then be supplied with more adatoms from the lower (wider) terrace ahead and its velocity will increase. On the other hand,

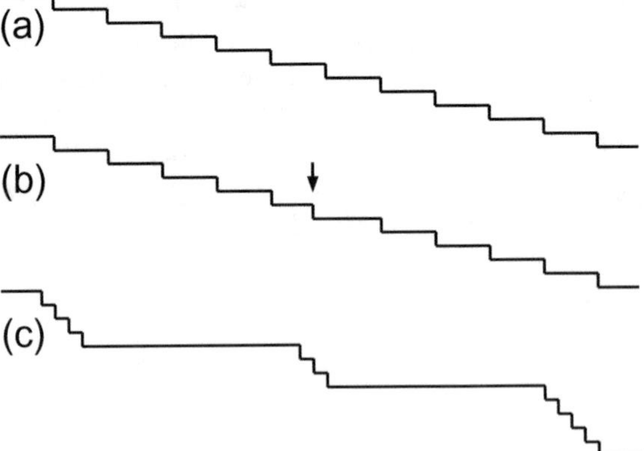

Fig. 7.24 **a** Surface cross-section during stable step-flow growth in presence of Ehrlich-Schwoebel barriers. **b** Step delayed behind its regular position (*arrow*) and **c** eventually forming bunches of steps in absence of Ehrlich-Schwoebel barriers

a faster step is supplied less and its speed decreases. The Ehrlich-Schwoebel barrier therefore stabilizes an equidistant train of steps during step-flow growth.

Conditions are reversed if no flux is supplied to the surface. During evaporation of adatoms the Ehrlich-Schwoebel barrier destabilizes a regularly stepped surface and leads to the formation of step bunches. Steps move in reverse direction due to detachment of atoms and subsequent evaporation. If now a step is delayed, the lower terrace gets more narrow than the upper terrace. The flux of evaporation is then reduced, leading to a further decrease of step velocity.

Similar to the transition between island and step-flow growth (Sect. 7.2.4), a transition between step-flow and step-bunching growth depending on sample offcut and growth conditions is found for many semiconductors, e.g., for Si(001) [30], GaAs [31], InP [32], GaN [33], and AlN [25, 34]. The experimental and theoretical studies indicate that the transition also depends on the balance of surface diffusivity and terrace width, where the diffusion can be controlled by supersaturation (or anion/cation ratio) and temperature [31, 34].

Surface steps were by now considered straight and assumed to move during step-flow growth on the hole like a rigid entity. Experiments and theoretical analyses show that such a condition is not stable in practice. A step will generally be perfectly straight only at $T = 0$ K. At finite temperature the action of entropy produces a roughening addressed in the next section. The overall shape of a step is basically preserved by such change which occurs on a small scale.

On a larger scale the shape of a straight step was found to be unstable during step-flow growth, if the adatom attachment from the upper and lower terraces differ. G.S. Bales and A. Zangwill pointed out that surface diffusion in presence of an Ehrlich-Schwoebel barrier gives rise to a morphological instability of straight steps, leading to a distinct wavy shape [35]. The qualitative reason for the *Bales-Zangwill instability* is analogous to that considered above to explain the debunching effect of the Ehrlich-Schwoebel barrier on a sequence of steps during step-flow growth.

We recall from (7.25) that the step velocity r_{step} is proportional to the sum of the current density of adatoms j_{upper} from the upper terrace to the step and j_{lower} from the lower terrace, $r_{\text{step}} = a^2(j_{\text{upper}} + j_{\text{lower}})$. Furthermore, we note that the current density $\mathbf{j}$ on either side of the step is proportional to the gradient of the adatom density n near the step, $\mathbf{j} = -D\partial n/\partial \mathbf{r}$. Let us now assume a step deviates at one section from the perfect straight shape and has, say, a warpage in the direction along r_{step} (like point A in Fig. 7.25). The lines of isoconcentration of the adatom density n on the lower terrace are then more dense in front of the warpage than on the straight parts of the step. The adatom density is even less dense in front of notches (like point B in Fig. 7.25). A high density of isoconcentration lines in front of point A corresponds to an increased current density j_{lower} towards the step at this point. In *absence* of an Ehrlich-Schwoebel effect this surplus of adatoms from the lower terrace is exactly balanced by a lack of adatoms from the upper terrace: Behind the warpage on the upper terrace the density of isoconcentration lines of n is decreased, and so j_{upper}. Without an Ehrlich-Schwoebel effect all parts of the distorted step will travel with the same velocity r_{step}. The *presence* of an Ehrlich-Schwoebel barrier will, however, reduce the compensating contributions from the upper terrace. As a result,

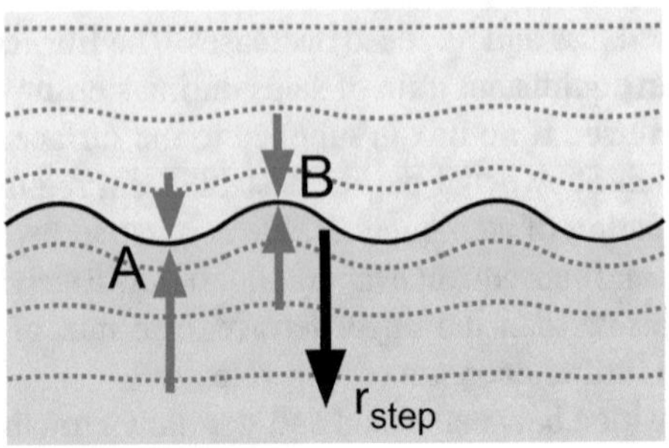

Fig. 7.25 Monoatomic wavy step (*solid black line*) viewed from top. *Light gray* and *gray shaded area* indicate upper and lower terrace, respectively. The *dotted lines* are isoconcentration lines of adatoms, *gray arrows* represent adatom fluxes towards the step being proportional to the gradient of adatom concentration. *A* and *B* denote respectively points of enhanced and diminished adatom inflow at the step compared to a straight step

the dominating contributions from j_{lower} lead to an increased growth rate at a convex section of the step (point A) and a decreased growth rate a concave section (point B). This positive feedback amplifies perturbations of a straight step. Eventually, the line tension at the step that increases with the step length limits the feedback, and a wavy shape as indicated in Fig. 7.25 develops.

The Bales-Zangwill instability is discussed quantitatively and in more detail in [35, 36]. We confine ourselves to the qualitative arguments given above. A wavy step structure can be recognized in Fig. 7.19; the effect is particularly strong at decreased temperature (Fig. 7.19b), where also low barriers act more effective due to a lowered adatom mobility. It should be noted that a meander pattern of a step may also be induced by a Kink-Ehrlich-Schwoebel effect [37, 38].

7.2.7 Roughening of Surface Steps

A surface step will generally be perfectly straight only at $T = 0$ K. As the temperature is increased, a finite contribution of entropy will decrease Gibbs free energy and kinks separated by straight parts will appear. This roughening provides kink sites for the incorporation of adatoms into the layer during growth. We consider the structure of a straight step of monoatomic height on the (001) surface of a cubic Kossel crystal oriented along the [100] direction [21]. Figure 7.26 shows a top view on a straight step at position (a), where one atom was moved from an embedded step site to a site attached to the step. To evaluate the energy cost of this operation in the framework of the terrace-step-kink model we count the number of broken and formed next-neighbor bonds. Three bonds with an energy E_{B} each were broken, and one bond for the attachment at the step was formed, yielding a net amount of $2E_{\text{B}}$. Counting

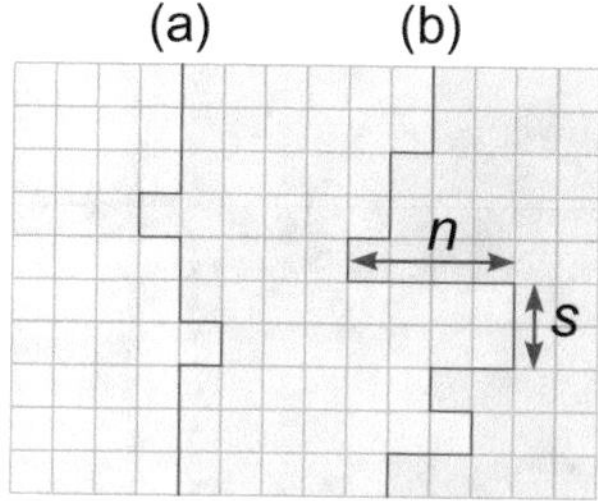

Fig. 7.26 Schematic of monoatomic steps on a Kossel crystal viewed from top, a *lighter gray shading* indicates higher lying terraces. **a** Kinks generated by moving one embedded ledge atom of a straight step to a position attached to the step. **b** Step with kinks of several atom units

of the kinks created yields (from bottom to top in Fig. 7.26 at position (a)) 1^+, 1^-, 1^-, 1^+, i.e., two positive and two negative kinks. We thus obtain an energy of $E_B/2$ necessary to form a kink.

The probability of having a kink or a straight part of the step depends on the energy per kink. Denoting the probabilities for (monoatomic) positive kinks, negative kinks, and straight parts p_+, p_-, and p_0, respectively, we obtain [21]

$$\frac{(1/2)(p_+ + p_-)}{p_0} = \exp\left(-\frac{E_B/2}{kT}\right), \quad p_+ + p_- + p_0 = 1. \tag{7.29}$$

The mean distance between kinks is given by $y_0 = a/(p_+ + p_-)$, a being the unit spacing. Inserting (7.29) yields

$$y_0 = \frac{a}{2}\exp\left(\left(\frac{E_B/2}{k_B T}\right) + 2\right) \approx \frac{a}{2}\exp\left(\frac{E_B/2}{k_B T}\right). \tag{7.30}$$

The mean distance y_0 between kinks decreases as the temperature is raised. Note that at common growth temperatures usually $E_B \gg k_B T$ applies. Considerations above assumed kinks of a single atom unit and steps without any mutual interaction arising from long-range strain fields.

We illustrate the effect of step roughening for the interesting vicinal surface of Si(001). The Si(001) surface reconstructs to a (2 × 1) surface unit-cell by forming rows of dimerized atoms. Dimer rows on terraces that are separated by a monoatomic step (or by an odd number of such steps) are perpendicular to each other due to the structure of the diamond lattice (Sect. 2.1.4). If the singular Si(001) surface is tilted by a small angle (below ∼1°) toward a [011] direction, adjacent monoatomic steps are hence not equivalent. It should be mentioned that larger miscuts (2–10°) and heating to typical growth temperatures above 600 °C leads to double-layer steps, which are not considered here. There exist two kind of monoatomic steps: those labeled S_A steps (single A steps) have dimers on the lower terrace directed parallel to the step, while the dimerization direction on the lower terrace at S_B steps runs perpendicular

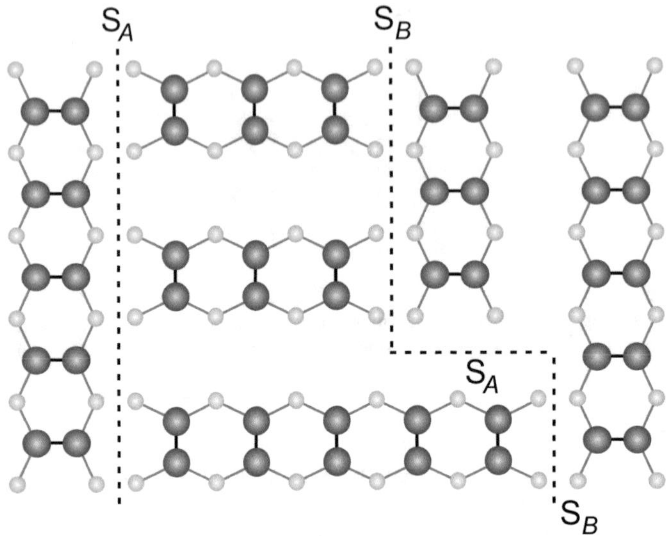

Fig. 7.27 Top view on monoatomic steps of a vicinal Si(001) surface. The height decreases from *left* to *right*. On each terrace *dark gray dimers* and *gray atoms* lie on the topmost layer and the layer underneath, respectively

to the steps, cf. Fig. 7.27. Calculations show that the formation energy per length of S_A steps is much lower ($\sim$0.01 $\pm$ 0.01 eV/atom) than that of S_B steps ($\sim$0.15 $\pm$ 0.03 eV/atom) [39]. This finding was basically confirmed by scanning-tunneling microscopy, yielding an upper-bound estimate of $\sim$0.028 $\pm$ 0.002 eV/atom for S_A steps and $\sim$0.09 $\pm$ 0.01 eV/atom for S_B steps [40].

Due to the symmetry of the Si(001) surface each kink must have a length of multiples of $2a$, and the same holds for the distance between kinks. We note from Fig. 7.27 that kinks in one type of step are made of segments of the other type of step. The two types of steps have different energies associated with the formation of kinks and consequentially also different morphology: S_A steps are smooth and S_B steps are rough, see Fig. 7.28.

The equilibrium distribution of steps and kinks was analyzed from images like that of Fig. 7.28 [40]. In the analysis a kink is any inside corner followed by an outside corner or vice versa. Separations s and lengths n of kinks are defined as shown in Fig. 7.26. Resulting distributions of kink separations s and kink lengths n of an S_B step are displayed in Fig. 7.29. The probability $p(s)$ of finding two adjacent kinks separated by s atoms follows from the probability p_k that a kink exists times the probability that *no kink* (i.e., $(1 - p_k)$) is nearby in a range $\pm s$ atoms. The exact relation is given by

$$p(s) = p_k (1 - p_k)^{s/2-1}. \tag{7.31}$$

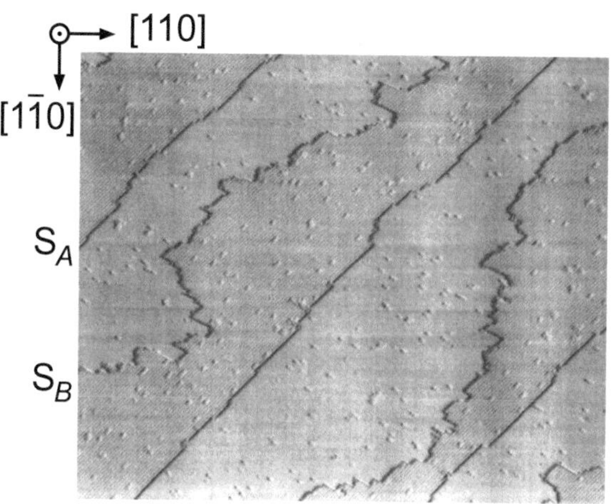

Fig. 7.28 Scanning-tunneling micrograph of a Si(001) surface miscut 0.3° towards [100]. The surface was annealed at 600 °C for 5 min prior to quenching to room temperature. Steps are descending from the *upper left* to the *lower right corner*. Reproduced with permission from [40], © 1990 APS

This function is drawn in Fig. 7.29a. In the plot p_k equals the number of kinks of S_B steps counted from STM images divided by the total number of possible kink sites. We observe that (7.31) describes the measured probability distribution $p(s)$ very well, indicating that kinks may be considered independent in this experiment. The validity of this assumption is also reflected in Fig. 7.29b, showing an exponential dependence of the number N of kinks of length n atoms from the length, i.e., $N \sim \exp(-E(n)/(kT))$, where $E(n)$ is the energy of a kink of length na. $E(n)$ was shown to be related to the formation energy per length of a step E_S according to $E(n) = nE_S + \text{const}$ [40]. E_S corresponds to an S_A step for kinks in an S_B step (and vice versa) due to the symmetry of the diamond lattice as illustrated in Fig. 7.27. The $E(n)$ offset, i.e., the const $= 0.08 \pm 0.02$ eV, may be considered as an additional energy due to the corner of a kink. The lower formation energy per length E_S of S_A steps leads to a small kink energy $E(n)$ in S_B steps, and consequently according to (7.30) to a smaller mean distance y_0 between kinks as compared to S_A steps.

7.2.8 Growth of a Si(111)(7 × 7) Surface

Surface kinetics considered so far assumed adatoms arriving on the surface to migrate and eventually nucleate islands or being incorporated on a regular lattice site at a step. The nucleation theory outlined above in Sect. 7.2.2 is often successfully applied to describe epitaxy of metals. Metal surfaces often do not reconstruct, and surface diffusion and capture to form a stable nucleus may be the only relevant processes which determine the growth morphology. We know, however, from Sect. 7.1 that the arrangement of surface-near atoms may strongly deviate from that inside the bulk solid. Such reconstruction is particularly prominent on semiconductor surfaces. Compared to an unreconstructed surface an additional energy barrier must hence be

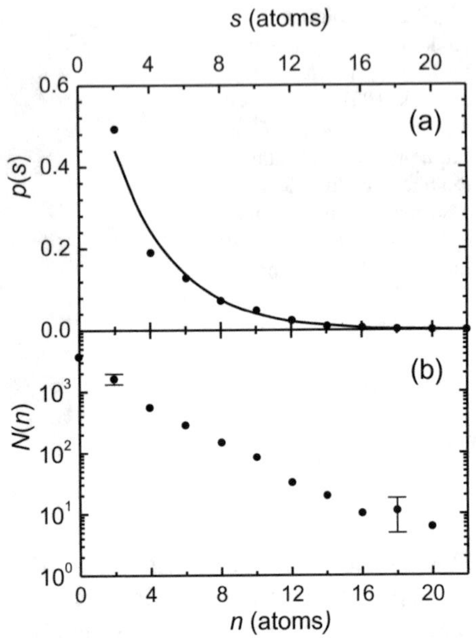

Fig. 7.29 **a** Probability of two adjacent kinks separated by s atoms depending on their separation for an S_B step on Si(001) prepared at 600 °C. The *solid line* is function (7.31) with p_K taken from the measured kink density. **b** Number of kinks of length n as a function of kink length n. Reproduced with permission from [40], © 1990 APS

surmounted to form a stable nucleus [41]. How are surface atoms rearranged from sites in reconstruction to regular lattice sites during growth? The complexity of surface reconstructions pointed out in Sects. 7.1.6 and 7.1.7 indicate that this question does not have a general answer. Only few studies provide a microscopic insight into the complex growth kinetics of reconstructed semiconductors. We consider sequential growth steps for the two previously treated examples to illustrate the dynamics of surface reconstructions.

The (7×7) reconstruction of the Si(111) surface is a prominent surface structure. According to the DAS model the complex reconstruction comprises 15 dimers, 12 adatoms, and 1 stacking fault, cf. Fig. 7.13. During growth the atoms of the surface reconstruction have to be rearranged to the bulk structure. We first consider an experimental observation of the epitaxial growth of a two-dimensional island. Figure 7.30 shows a sequence of STM images recorded at 575 K during MBE growth on a (7×7) reconstructed Si(111) surface [42]. Details of the reconstruction are not resolved in the images. We note a pronounced macro kink at the right edge of the island that gradually moves downwards, thereby eventually completing a laterally enlarged triangular island. Atomically resolved images proved that the fast growing stripe has the width of a (7×7) unit cell (27 Å).

The stability of the island during the sequence shown in Fig. 7.30 indicates a fast growth of the additional row. This is confirmed by the growth dynamics evaluated from such in situ STM images given in Fig. 7.31 [42]. The lower curve shows the experimentally observed size of a single island expressed in units of half the rhombic (7×7) unit cell. We note pronounced plateaus indicating a stable configuration of

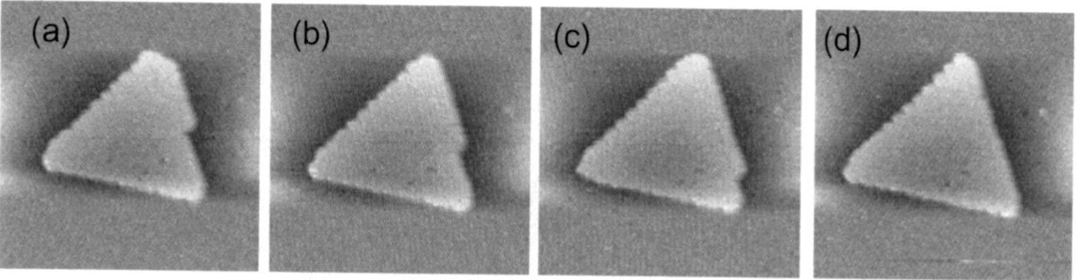

Fig. 7.30 Consecutive stages of the homoepitaxial growth of a two-dimensional island on a (7×7) reconstructed Si(111) surface, recorded at 575 K using scanning tunneling microscopy. The image size is 500×500 Å^2. Reproduced with permission from [42], © 1998 APS

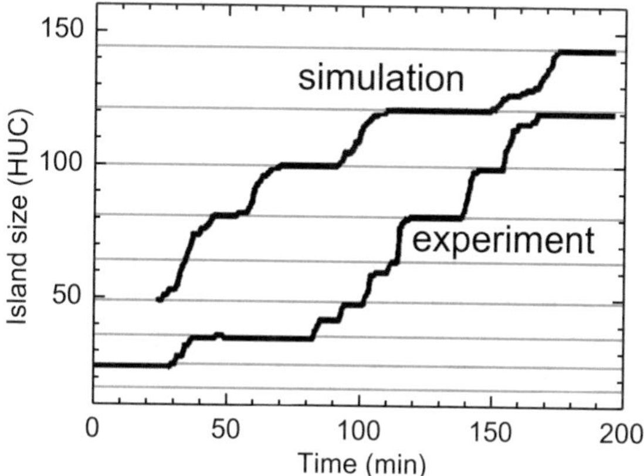

Fig. 7.31 Evolution of a single island on a Si(111) (7×7) surface. *Lower* and *upper curve* refer to experimental data and a kinetic Monte Carlo simulation, respectively. The island size is given in numbers of half-unit-cells (HUC), *gray horizontal lines* mark sizes of n^2 HUC. Reproduced with permission from [42], © 1998 APS

the island, and a rapid increase of the island size between the plateaus. These periods of fast growth correspond to the addition of a row like that shown in the sequence of Fig. 7.30. The stable islands occur at triangular "closed shell" configurations, i.e., at sizes of n^2 half-unit-cells. At these numbers isosceles triangles appear, cf. Fig. 7.32.

The observed island growth is related to the structure of the surface reconstruction. The rhombic (7×7) unit cell of the Si(111) (7×7) surface is composed of two triangular half-unit-cells (HUC): One unfaulted half (U-HUC) with a near-surface layer sequence identical to the bulk underneath, and a faulted half (F-HUC) with a stacking fault (the right and left triangles in Fig. 7.13, respectively). During growth the reconstructed atoms have to rearrange toward the bulk structure. In the unfaulted HUC these are the atoms in the uppermost layer. In the faulted HUC the stacking fault must be removed in addition. Atom rearrangement in the faulted half is therefore associated with a higher energy barrier than in the unfaulted half of the (7×7) unit cell [43].

A model of the growth sequence is illustrated in Fig. 7.32. A triangular two-dimensional island on Si(111) (7×7) is surrounded by *faulted* half-unit-cells denoted F in Fig. 7.32 (the initial nucleation is assumed to be favored on an unfaulted unit cell). Growth of the island hence requires the overgrowth of a faulted HUC. Surmounting the unfavorable high energy barrier causes the delay observed in the

Fig. 7.32 Arrangement of faulted (F) and unfaulted (U) half-unit-cells of a (7 × 7) reconstructed Si(111) surface around a triangular two-dimensional Si island (*brown*)

dynamics Fig. 7.31. Once an F-HUC nucleates the adjacent U-HUC can be overgrown more easily. Overgrowth of the next F-HUC is facilitated by the macro kink depicted in Fig. 7.32: The new F-HUC has a shorter edge length than the F-HUC which nucleated before (1 side instead of 2). Neighboring faulted and unfaulted half-unit-cells aside the island are overgrown in a quick succession until an enlarged island comprising n^2 half-unit-cells in total is completed.

The experimentally observed growth behavior is well reproduced by a kinetic Monte Carlo simulation [42]. The simplified model assumes a honeycomb lattice consisting of alternating F and U sites, and material to be transported towards the island in HUC units. The attachment barrier E is assumed to depend on both, the type of the underlying HUC (F or U), and the number n_{edge} of nearest-neighbor HUC already attached to the island, $E = E_{\text{U/F}} - n_{\text{edge}} \times E_{\text{edge}}$. Rates for hopping and attachment are modeled in terms of Arrhenius expressions with energies chosen to yield the best agreement with experimental data. The result shown in Fig. 7.31 features a similar dynamics as the island evolution observed in the experiment.

Nucleation of the half-unit-cells was addressed in spatially more refined models [44, 45]. Nucleation of a 2D island is considered to proceed in three steps, starting with a small stable cluster in a HUC of the Si(111) (7 × 7) surface. The cluster raises the adatom binding energy in adjacent HUCs, leading to a preferential formation of a second cluster in a neighboring HUC of the same unit cell. Eventually an additional adatom is attached to the cluster pair, and the (7 × 7) reconstruction is locally removed [45]. The species cluster, cluster pair, and 2D island are well reproduced in the STM image Fig. 7.33.

7.2.9 *Growth of a GaAs(001) β2(2 × 4) Surface*

A very detailed scenario of kinetic processes occurring during nucleation and growth of reconstructed surfaces was obtained by combining experimental STM results with kinetic Monte Carlo simulations [46, 47]. Advanced modeling emloyed rates obtained from density-functional theory (DFT) calculations [47]. Monte Carlo simulations can bridge the time scales from individual kinetic steps to macroscopic

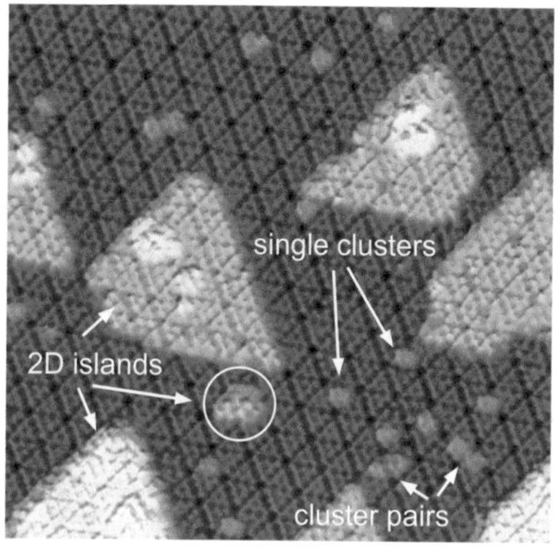

Fig. 7.33 STM image of a (7 × 7) reconstructed Si(111) surface after deposition of 0.2 bilayers of Si showing clusters, cluster pairs, and 2D islands. The smallest island (*circle*) has the size of one unit cell. The image size is 425 × 425 Å^2. Reproduced with permission from [45], © 2007 APS

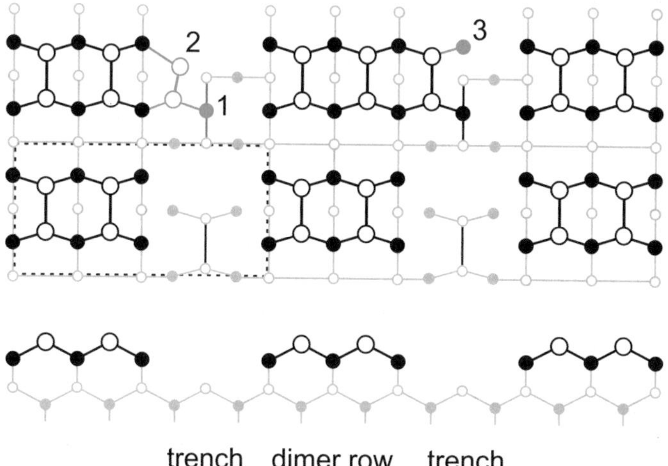

Fig. 7.34 Side and plan views of a $\beta2(2 \times 4)$ reconstructed GaAs(001). The unit cell is indicated by *dashed lines*, *filled* and *open symbols* represent Ga and As atoms, respectively. *Numbers* label initial growth steps

growth, while DFT calculations yield kinetic parameters from first principles. The studies focus on homoepitaxial growth of GaAs on a As-rich GaAs(001)-$\beta2(2 \times 4)$ surface from beams of atomic Ga and As_2 molecules. The two species behave quite different on the surface. Ga atoms adsorb with a sticking coefficient of unity on the As-rich surface [48], leading to a growth rate which is controlled by the Ga flux. On the other hand, the chemically rather stable As_2 binds only weakly to an As-rich GaAs surface [48, 49]. It adsorbs only after deposition of Ga [50] and diffusion to adjacent sites [47]. Diffusion and incorporation of Ga has thus a strong impact on the adsorption and incorporation of arsenic.

The $\beta 2(2 \times 4)$ reconstructed GaAs(001) surface consists of rows with topmost As-As dimers, separated by trenches, cf. Fig. 7.34. Diffusion of Ga adatoms is anisotropic with lowest activation energy for hopping processes along the trenches [51]. From calculations it was found that the terminating As_2 dimers of the $\beta 2(2 \times 4)$ reconstruction, particularly those in the trenches, act as traps for diffusing Ga adatoms: Ga atoms are immobilized for about 10^{-8} s, much longer than the time scale for hopping of order 10^{-12}–10^{-9} s. At such Ga adsorption site a gas-phase As_2 molecule can readily adsorb by forming one bond to the adsorbed Ga atom and two bonds to Ga atoms with unsaturated bonds that are located at the side walls of the $\beta 2(2 \times 4)$ trenches. The new As-As dimer complex becomes more stable by attaching another diffusing Ga adatom in the trench. Growth on the reconstructed surface hence preferentially nucleates at the side walls of the trench and largely proceeds by partially filling the trenches, followed by island nucleation on surface regions where a filling of the trenches has occurred [52]. The first steps of the mentioned process are depicted in Fig. 7.34. The surface unit-cell marked in the figure contains an As-As dimer in the trench. In one of the favored Ga diffusion channels the bond of these dimers is broken by Ga as depicted at position 1 in the top panel of Fig. 7.34 [51]. The new bonds relax the position of the two As atoms to their bulk position. The Ga atom at position 1 favors a subsequent As_2 adsorption at the trench side-wall as indicated at position 2. A dangling bond of As may then be saturated by a further Ga atom in the trench at position 3.

The outlined nucleation steps describe the main route occurring at standard conditions of molecular beam epitaxy of GaAs(001) at 800 K. The prevailing processes change when the growth temperature is changed [47].

The examples of homoepitaxial growth on the $\beta 2(2 \times 4)$ reconstructed GaAs(001) surface and the (7×7) reconstructed Si(111) surface indicated that the classical nucleation theory may sometimes be too simplified for describing the complex processes of semiconductor growth. Still the simple theory provides useful guidelines to interpret growth phenomena and to optimize growth conditions.

7.3 Self-organized Nanostructures

The reduction of the dimensionality of a solid leads to a modification of the electronic density-of-states. The effect of size quantization requires characteristic length scales in the range of typically 10 nm defined by the de-Broglie wavelength as outlined in Sect. 4.3.2. Fabrication of such small structures with high material quality may hardly be accomplished by lateral patterning of quantum wells. Etching or implantation techniques inevitably introduce defects, which deteriorate the electronic properties of a nanostructure. Therefore a number of techniques were developed particularly in the 1990s employing self-organization phenomena during epitaxial growth. The approaches are based on an anisotropy of surface migration of supplied atoms, originating from a non-uniform driving force like, e.g., strain, that tends to minimize the total energy of the system. Thereby structurally or compositionally non-uniform

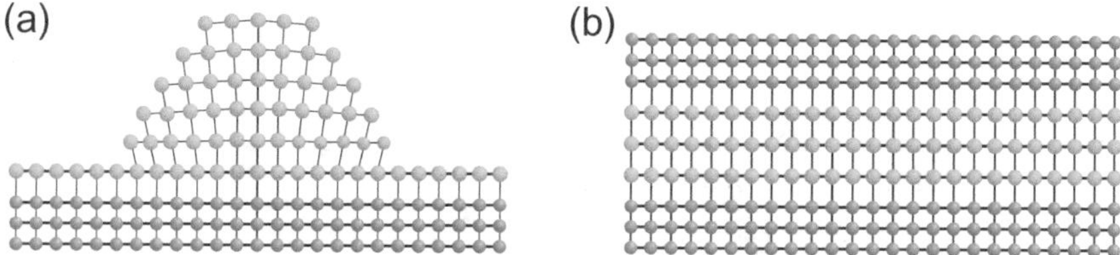

Fig. 7.35 Scheme illustrating elastic strain relaxation in **a** a quantum dot and **b** a quantum well. *Yellow* and *blue circles* represent atoms of size-quantized materials and barrier materials, respectively

crystals with dimensions in the nanometer range may be coherently formed without structural defects. Some of these techniques have attracted much attention for the fabrication of quantum wires or quantum dots, and are outlined in the following.

7.3.1 Stranski-Krastanow Island Growth

Stranski-Krastanow growth is one of the three fundamental growth modes of heteroepitaxy introduced in Sect. 6.2.3. The characteristic feature is a transition of an initially two-dimensional layer-by-layer growth to three-dimensional growth. The size of the three-dimensional islands formed by such a transition lies for many semiconductors in the range required for quantum dots (QDs) as expressed by (4.53). The transition of 2D to 3D layer growth may be induced by strain upon depositing a 2D layer on a substrate with a different lateral lattice constant. Since a coherent layer adopts the lateral atomic spacing of the substrate, strain accumulates with increasing thickness. Above a critical layer thickness the strain is relaxed plastically by introduction of misfit dislocations (Sect. 2.2.6). Below this critical thickness under suitable conditions a considerable part of the strain may be relaxed *elastically*, i.e. without introduction of dislocations, by the formation of three-dimensional surface structures. Such reorganization of a flat surface into a structure with tilted facets implies an enlargement of the surface area. The driving force of the Stranski-Krastanow growth mode is therefore a minimization of the total strain and surface energy [53]. The elastic strain relaxation in an uncovered 3D island is illustrated in Fig. 7.35a. We note that the strain is inhomogeneous, giving rise to shear deformations in addition to hydrostatic strain components. This feature is in contrast to a strained quantum well depicted in Fig. 7.35b. The quantum well exhibits a constant, homogeneous biaxial strain in the entire layer.

Elastic strain relaxation generally leads to an energy decrease of the structure. The contribution of the surface energy upon faceting of a flat surface may, by contrast, be positive or negative. The sign follows from a simple advisement. Consider, e.g., a (001) surface with area A and a surface energy γ_{001} as illustrated in Fig. 7.36. This surface will gain energy upon a faceting with {110} facets, if the surface energy γ_{011}

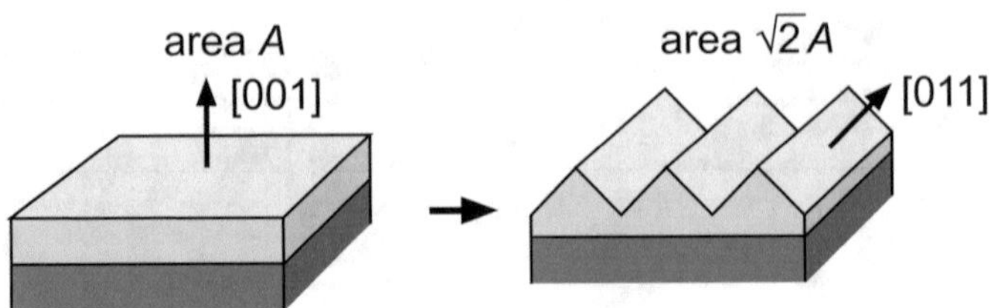

Fig. 7.36 Faceting of a (001) surface with area A to a surface of total area $\sqrt{2}A$ with {011} facets

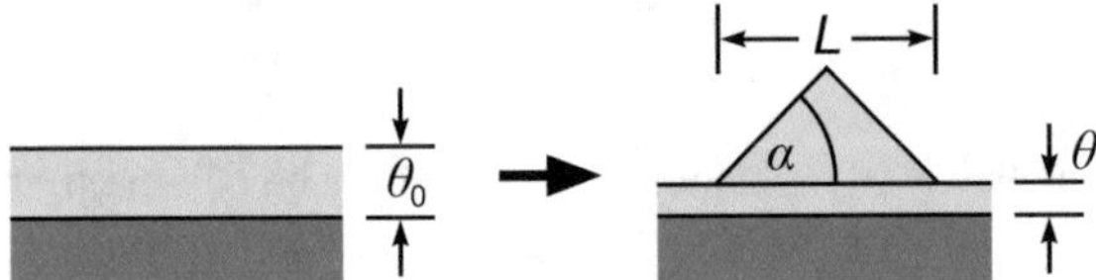

Fig. 7.37 Reorganization of a uniformly strained layer of θ_0 monolayers thickness to an island with pyramidal shape on a remaining wetting-layer thickness of θ monolayers. L and α are the base length of the island and the tilt angle of the island facets, respectively

of these facets fulfills the condition $\sqrt{2}\gamma_{011} < \gamma_{001}$. The prefactor $\sqrt{2}$ is the increase of surface area of a faceted {011} surface with respect to the flat (001) surface.

The total energy gain of three-dimensional islands with respect to a two-dimensional deposition is given by strain and surface energy-contributions of both, the reorganized part of the material and the part remaining in the wetting layer after the Stranski-Krastanow transition. The wetting layer is a fraction of the deposited strained material of at least one monolayer thickness. This part remains as a two-dimensional layer on the substrate and does not migrate to the faceted part due to a lower surface energy of the deposit as discussed in Sect. 6.2.3.

We consider the total energy gain per unit volume obtained, if a two-dimensional deposit of nominal thickness θ_0 (in units of monolayers) reorganizes to a single faceted three-dimensional island on a wetting layer (WL) of thickness θ, cf. Fig. 7.37. The contributions sensitively depend on the shape of the island. For simplicity we assume an island with the shape of a pyramid with a quadratic base of length L as often found in experiment. The gain can be expressed as [54]

$$\frac{E_{\text{total}}}{V} = \varepsilon_{\text{island}}^{\text{elast}} - \varepsilon_{\text{layer}}^{\text{elast}} + \frac{A\gamma_{\text{facet}} - L^2\gamma_{\text{WL}}(\theta_0)}{V} + \left(\frac{1}{\rho} - L^2\right)\frac{\gamma_{\text{WL}}(\theta) - \gamma_{\text{WL}}(\theta_0)}{V}, \tag{7.32}$$

with the elastic energy densities $\varepsilon_{\text{island}}^{\text{elast}}$ and $\varepsilon_{\text{layer}}^{\text{elast}}$ of the island and the uniformly strained layer, respectively. The third term describes the change in surface energy due to the island: γ_{facet} is the surface energy of the island facets, A is their area, and L the island base length. The volume of the pyramidal shaped island V follows from mass conservation and is given by

$$V = \frac{1}{\rho}(\theta_0 - \theta)d_{\mathrm{ML}} = \frac{1}{6}L^3 \tan\alpha,$$

where ρ, d_{ML}, and α are the areal density of the islands (in units cm^{-2}), the thickness of one monolayer, and the tilt angle of the facets. The fourth term in (7.32) accounts for the change of wetting-layer energy as the thickness decreases from θ_0 to θ. The relation $\gamma_{\mathrm{WL}}(\theta)$ computed for thin reconstructed InAs layers on GaAs substrate is shown in Fig. 7.38a [54]; for similar findings with Ge/Si(001) see [55]. The prefactor $(1/\rho - L^2)$ effectuates that this term solely applies for the free surface *besides* the island (i.e., the area *not* covered by the island). The sum of all four contributions is given in Fig. 7.38b. Values refer to pyramidal InAs islands with relaxed, but unreconstructed $\{110\}$ facets and a $\beta2(2 \times 4)$ reconstructed (001) surface of the wetting layer, grown on (001) oriented GaAs.

The calculated energy gain per unit volume upon island formation given in Fig. 7.38b shows that major contributions originate from the elastic energy relief, i.e., from the first two terms in (7.32). This negative part scales with the island volume. The surface energy described by the third term in (7.32) is an unfavorable positive energy with a decreasing contribution for larger islands due to a $\propto V^{2/3}$ dependence. The wetting-layer contribution represented by the fourth term in (7.32) is also positive. Besides island volume it depends on island density and the coverage. The importance of facet edges was controversial in literature [53, 56]; the calculation presented in Fig. 7.38b yields an only minor contribution.

The total energy density given by the sum of all contributions has an energy minimum for a particular island size as indicated by an arrow in the figure. Such a minimum is important for obtaining a narrow size distribution for an ensemble of islands. It should be noted that a theory considering only the two contributions of elastic relaxation (negative, $\propto V$) and island surface-energy (positive, $\propto V^{2/3}$) does not yield a finite equilibrium size. For sufficiently high coverages the volume term always prevails and favors steady island growth as previously discussed in nucleation and ripening (Sects. 6.2.1, 6.2.8).

Stranski-Krastanow growth induced by strain represents a quite universal behavior. Self-organized formation of islands is found for both, compressively and tensely-strained layers in various materials systems and crystal structures. Table 7.5 and Fig. 7.39 give some examples. The mismatch is defined by (2.20a), yielding a negative sign for compressively strained islands. Usually substrate material is also employed for covering the islands after formation. The material is then generally termed matrix. Often the island material is alloyed with matrix material to reduce the strain, yielding a parameter for controlling the transition energy of confined excitons.

Self-organized Stranski-Krastanow growth was particularly studied for fabricating InAs and $\mathrm{In}_{1-x}\mathrm{Ga}_x\mathrm{As}$ quantum dots in a GaAs matrix [61]. The InAs lattice constant exceeds that of GaAs by $\sim$7 %, and the critical thickness for elastic relaxation of such a highly strained 2D InAs layer on a GaAs(001) substrate to 3D islands is only about 1.5 InAs layers. Once this critical layer thickness is exceeded, islands form with a high areal density of typically $10^{10} - 10^{11}\ \mathrm{cm}^{-2}$.

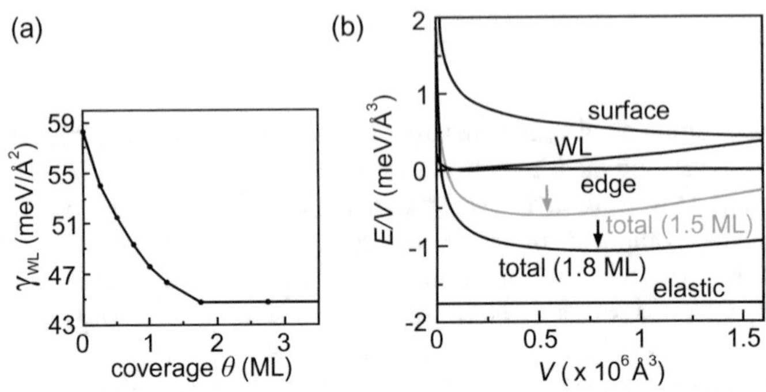

Fig. 7.38 **a** Calculated formation energy of a strained InAs wetting layer on (001)-oriented GaAs as a function of coverage θ. **b** Total energy gain by island formation with various contributions for $\rho = 10^{10}$ cm^{-2} areal density and $\theta_0 = 1.8$ monolayers InAs layer coverage on GaAs (*black lines*). The *gray line* refers to the total energy gain for $\theta_0 = 1.5$ monolayers. *Arrows* mark the minima of the total energy curves, *WL* denotes the contribution of the wetting layer. Reproduced with permission from [54], © 1999 APS

Table 7.5 Some semiconductor materials used for strain-induced, self-organized Stranski-Krastanow formation of islands

Island/matrix	Ge/Si	InAs/GaAs	GaN/AlN	PbSe/PbTe
Structure	Diamond	Zincblende	Wurtzite	Sodium chloride
Orientation	(001)	(001)	(0001)	(111)
Mismatch (%)	−3.6	−7	−2.5	+5.5

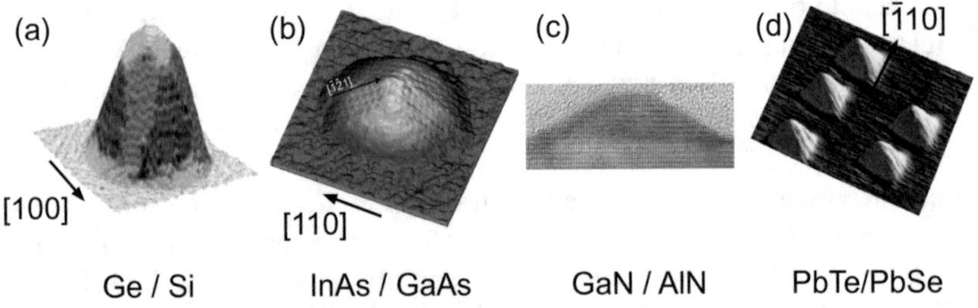

Fig. 7.39 Free standing self-organized islands formed by Stranski-Krastanow growth in various strained heteroepitaxial materials: **a** Ge/Si(001) [57], **b** InAs/GaAs(001) [58], **c** GaN/AlN(0001) [59], **d** PbTe/PbSe(111) [60]. The AFM images are vertically not to scale with respect to the lateral scale

The Stranski-Krastanow transition from 2D growth of a strained layer to 3D growth of islands occurs quite abrupt, albeit it may be delayed by kinetic barriers constituted by suitable growth conditions. During molecular-beam epitaxy the 2D–3D transition can be monitored in situ using reflection high-energy electron diffraction: The streaky reflection pattern indicating reflection from a flat 2D surface changes to a spotty pattern created by three-dimensional surface structures [62], see Fig. 8.7. Atomic force micrographs taken for different coverages of InAs yield virtually no islands below the critical coverage θ_c, and a sharp increase of the density of islands ρ above θ_c following a relation $\rho \sim (\theta - \theta_c)^\alpha$ [63]. The dependence of island density on InAs coverage θ is shown in Fig. 7.40.

The shape of islands created by Stranski-Krastanow growth depends on the material system and also on the growth conditions. For uncovered islands in several materials systems the shape is found to undergo a transition upon increase of volume, which increases for thicker coverage. Generally a transition with a continuous introduction of steeper facets at the island edge is expected [64], particularly due to the strain concentrating at the base perimeter. Experimental results for the two most studied materials InAs/GaAs(001) and Ge/Si(001) are given in Fig. 7.41. Small and large islands, referred to as pyramids with shallow facets and domes with steep facets, respectively, were found. The gray scale in the figure indicates the local slope of the facets.

Figure 7.41 indicates that islands with distinctly different sizes may coexist. A bimodal size distribution of island ensembles with well separated maxima of the mean sizes is often observed. The finding indicates some departure from thermodynamic equilibrium due to the presence of kinetic barriers in the formation process. Island formation using Stranski-Krastanow growth is usually performed at quite low growth temperatures.

For quantum-dot applications in devices Stranski-Krastanow islands are covered by a capping layer. Usually the same material as that underneath the islands is employed. The morphology of the islands is generally strongly modified during the capping procedure, unless special deposition conditions are found to preserve

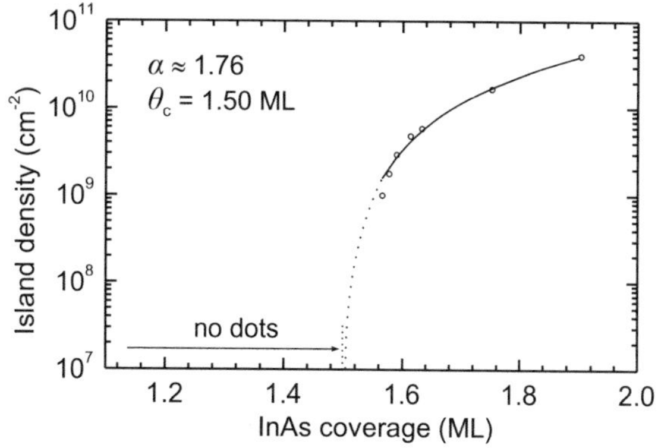

Fig. 7.40 Density of self-organized islands formed by Stranski-Krastanow growth during molecular-beam epitaxy at 530 °C with a low growth rate of 0.01 ML/s. Reproduced with permission from [63], © 1994 APS

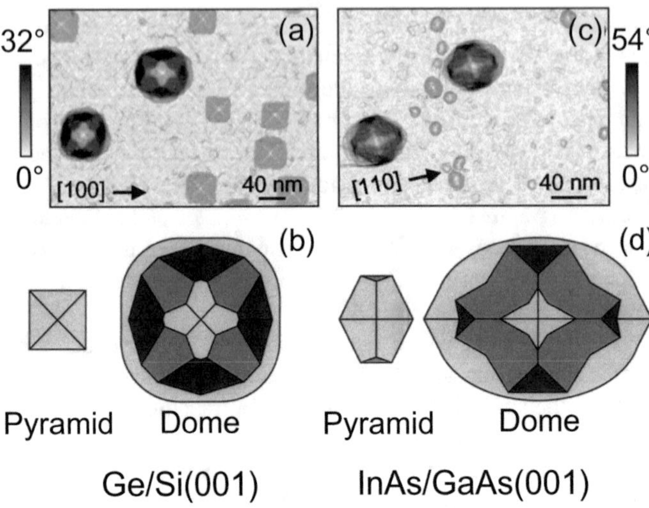

Fig. 7.41 Shape of uncapped Ge/Si (**a**, **b**) and InAs/GaAs islands (**c**, **d**). Steeper facets are marked by a *darker gray tone*: *light gray*, *gray* and *dark gray* denote {105}, {113}, and {15 3 23} facets for Ge/Si and {137}, {101}, and {111} facets for InAs/GaAs, respectively. Reproduced with permission from [68], © 2004 APS

the initial shape. While strain favors the formation of steep facets in the case of free standing islands, deposition of a mismatched material on top of the islands reverses this trend: Generally the islands tend to become flat during cap-layer deposition. Often quantum dots with a shape of truncated pyramids are formed. A microscopic picture of the capping dynamics of InAs/GaAs/(001) islands was obtained from scanning tunneling microscopy [65]. The process is illustrated in Fig. 7.42. Initially the deposited Ga atoms tend to migrate away from the islands' apex and accumulate at the base due to a better lattice match of the GaAs cap-layer material. Indium from the apes starts to alloy with Ga near the base, thereby releasing strain and increasing entropy. Eventually GaAs covers the remaining material of the island, yielding a surface morphology depending on the deposition rate of the cap layer.

The effect of cap-layer deposition on the island shape is illustrated in Fig. 7.43. The STM images show (110) cross sections of InAs islands in GaAs matrix material grown using MBE under the same conditions as applied for the free standing island shown in Fig. 7.39b. The GaAs cap layer was deposited with a rate of 0.15 ML/s at the same low temperature of 510 °C as the InAs islands after applying 10 s growth interruption. The shape changed from a pyramid with shallow facets to a truncated pyramid with a flat (001) top and steeper side facets [66]. Such a shape was also found for structures grown using MOVPE [67] and is typical for buried InAs/GaAs islands.

Often a growth interruption is applied after deposition of the island material and prior to deposition of the cap layer. The interruption intends to equilibrate the ensemble of islands for obtaining a narrow distribution of sizes. Usually the mean

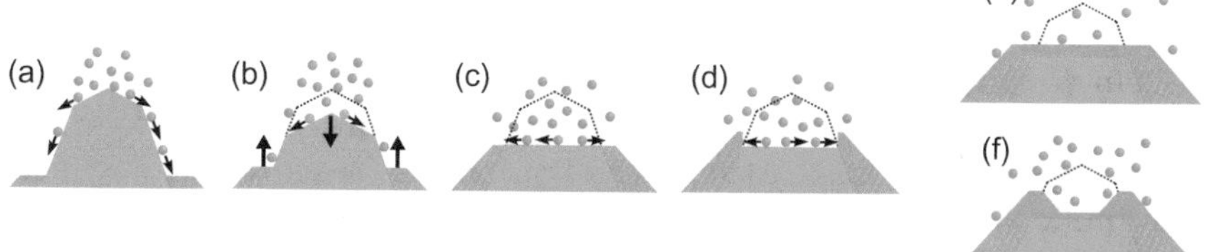

Fig. 7.42 Schematic of the overgrowth process of InAs/GaAs islands. *Orange* and *blue* mark In and Ga Column-III elements, respectively. *Shaded regions* indicate intermixed InGaAs materials. Cap-layer morphology obtained for a low capping rate is illustrated in (**e**), for a high capping rate in (**f**). Reproduced with permission from [65], © 2006 APS

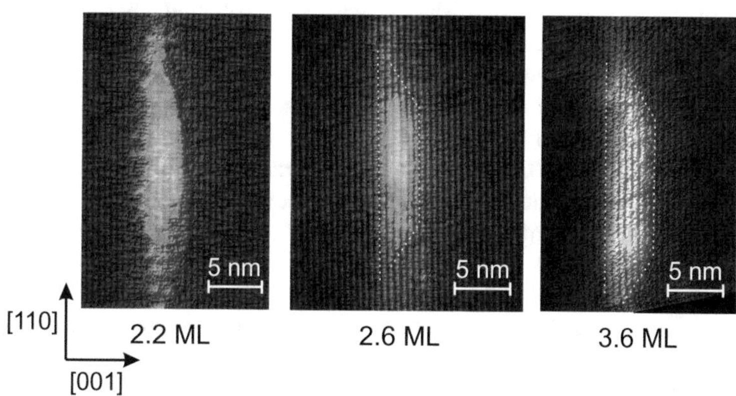

Fig. 7.43 Cross section STM images of small InAs/GaAs(001) islands formed from various InAs coverages and buried by GaAs. The *dotted white lines* indicate the truncated pyramid shape. Courtesy of M. Dähne, TU Berlin

size of the resulting quantum dots increases as the duration of the growth interruption increases. Such a ripening is a further indicative of kinetic limitations in the usually applied growth procedure.

7.3.2 Thermodynamics Versus Kinetics in Island Formation

Stranski-Krastanow growth in a regime with a major influence of kinetics sometimes leads to experimental findings which appear inconsistent. Size and density of islands for a given deposition thickness are important parameters which are readily obtained experimentally. At low deposition temperature the diffusion length of adatoms is short and the nucleation rate is high. Consequently small islands form with a high areal density. In this *kinetic regime* the islands grow larger and the density accordingly decreases for a given deposition thickness as the temperature is increased. Such a trend is usually observed in island formation using Stranski-Krastanow growth. On the other hand, in the *thermodynamic regime* at high temperature the size is expected to decrease in favor of a larger areal density as the deposition temperature is increased [69, 70]. Numerical results of corresponding equilibrium calculations are given in the inset of Fig. 7.44a. The maximum of the distribution function shifts to smaller island sizes and broadens as the temperature is increased. The same result is obtained from kinetic Monte Carlo (KMC) simulations shown in the main panel of Fig. 7.44a [71]. In the KMC simulation a long duration for equilibration of the island ensemble (35 s) was assumed.

The temporal evolution of the island sizes is displayed in Fig. 7.44b. The KMC simulation was performed on a lattice with 250 × 250 atoms and assumed an initial coverage of 4 %, randomly deposited at a flux of 1 ML/s. Adatoms diffuse by nearest neighbor hopping and may cross island edges by surmounting a Schwoebel barrier. Strain near an island is accounted for by including a position-dependent energy correction in the Arrhenius expression of the hopping rate. In the initial stage we find small islands at low temperature, whereas larger (and fewer) islands are formed at higher temperatures: The mean diffusion length increases at higher temperature, and nucleation of new islands is suppressed. Nucleation is hence the dominating process for a short duration of the evolution, a clear indication for a kinetically controlled growth. Right after deposition the island ensemble begins to equilibrate. At low temperature a slow increase of the mean island size is found. Such ripening proceeds much faster at higher temperature. Eventually the size distribution attains an equilibrium value. We note from Fig. 7.44b that the average island size now *decreases* as the temperature is increased, in agreement with equilibrium results shown in Fig. 7.44a. The study thus evidences a crossover from an initial kinetically controlled regime to thermodynamic equilibrium after an equilibration period.

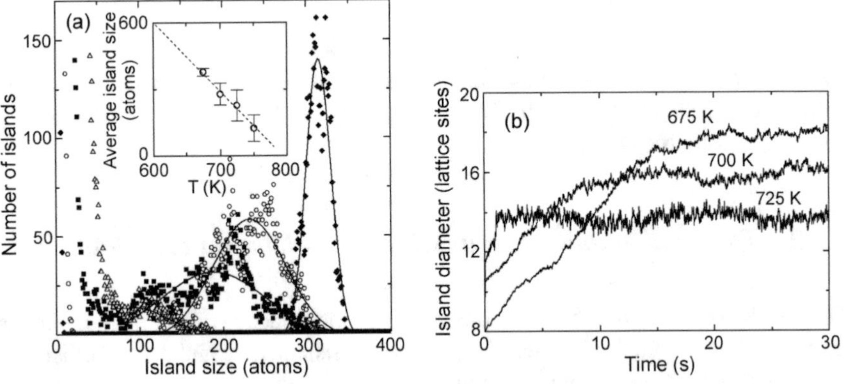

Fig. 7.44 **a** Equilibrium distribution of island sizes for $T = 675$ K (*diamonds*), 700 K (*circles*), 725 K (*squares*), and 750 K *triangles*) obtained from Monte Carlo simulations on a 250 × 250 grid. *Solid lines* are numerical fits. *Inset*: Results from thermodynamic theory. **b** Simulated evolution of the mean island size for various temperatures. Reproduced with permission from [71], © 2001 APS

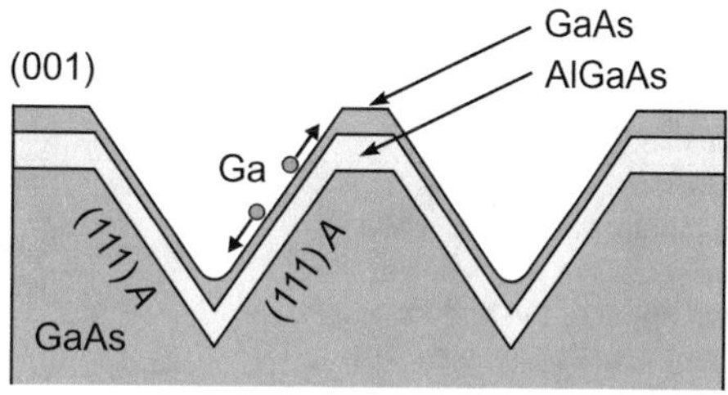

Fig. 7.45 Schematic of GaAs wire formation on a V-shaped patterned substrate with an AlGaAs barrier layer

7.3.3 Wire Growth on Non-planar Surfaces

One-dimensional structures in the nanometer-scale may be fabricated by employing a variety of growth-related phenomena [72]. Two different methods are pointed out here. For wire growth employing the vapor-liquid-solid mechanism see Sect. 12.2.

The realization of quantum wires is a challenging task. A 1D wire has a larger interface/volume ratio than a 2D quantum well and consequently makes high demands for structural quality. Interface fluctuations on a length scale of the exciton Bohr radius easily lead to localization referred to as zero-dimensional regime. Progress and true 1D properties of confined carriers was particularly achieved using V-shaped wires fabricated using patterned substrates as depicted in Fig. 4.26. The techniques utilizes the dependence of the Ga surface diffusion-length and GaAs growth rate on the crystallographic orientation of GaAs surfaces.

The diffusion length λ_{Ga} of Ga adatoms is—in a certain temperature range—approximately described by a dependence $\lambda_{\mathrm{Ga}} \propto \exp(-E_{\mathrm{eff}}/(k_{\mathrm{B}}T))$, with an effective activation energy E_{eff} depending on the orientation of the surface. From molecular beam epitaxy of GaAl/AlGaAs superlattices on various GaAs facets grown under Column-III limited flux conditions the following order of diffusion lengths was concluded [73]:

$$\begin{aligned}\lambda_{\mathrm{Ga}}(001) &\approx \lambda_{\mathrm{Ga}}(\bar{1}\bar{1}\bar{3})\mathrm{B}\\ &< \left\{\lambda_{\mathrm{Ga}}(\bar{1}\bar{1}\bar{1})\mathrm{B}, \lambda_{\mathrm{Ga}}(\bar{3}\bar{3}\bar{1})\mathrm{B}, \lambda_{\mathrm{Ga}}(013), \lambda_{\mathrm{Ga}}(113)A\right\}\\ &< \lambda_{\mathrm{Ga}}(159) \approx \lambda_{\mathrm{Ga}}(114)\mathrm{A} \approx \lambda_{\mathrm{Ga}}(111)\mathrm{A}\\ &< \lambda_{\mathrm{Ga}}(110).\end{aligned}$$

The Ga diffusion length increases in the order of GaAs surfaces related to (001), $(\bar{1}\bar{1}\bar{1})$B, (111)A, and (110) orientations. $\lambda_{\mathrm{Ga}}(001)$ was about 0.5 µm for the investigated growth at 620 °C.

On a non-planar GaAs substrate Ga adatoms migrate towards a surface with the minimum λ_{Ga} and are incorporated there. The growth rate of facets with a larger diffusion length is therefore decreased. This effect was employed to fabricate V-shaped quantum wires [75] and also ridge quantum-wires [76, 77]. Figure 7.45 illustrates the effect of wire formation. Growth is performed on a GaAs substrate with V-shaped grooves fabricated using lithography and anisotropic wet etching. Usually grooves

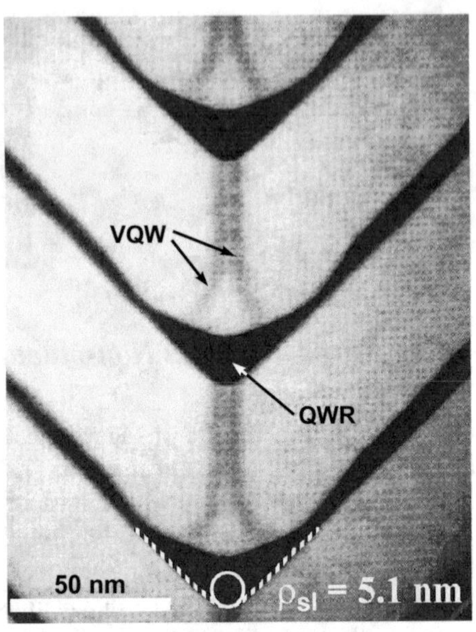

Fig. 7.46 Cross-section transmission-electron micrograph of a vertically stacked GaAs/$Al_{0.42}Ga_{0.58}As$ quantum wires. The *white circle* marks the radius of curvature of the bottom wire interface. Reproduced with permission from [74], © 1995 AIP

on a GaAs(001) substrate oriented along the $[\bar{1}10]$ direction and composed of two {111}A side walls are used. A lower AlGaAs barrier layer is then grown on the patterned substrate. The adatom diffusion-length during AlGaAs growth is quite short and does not show a pronounced facet formation; the V-groove bottom therefore remains quite sharp. In the subsequent GaAs growth Ga adatoms impinging on the {111}A side walls tend to migrate with a long diffusion length to facets with a short diffusion length. Thereby the growth rate is enhanced at the bottom of the V-groove, and a (001) facet and accompanying small {311} facets are generated. Adatom migration towards the center at the V-groove bottom is supported by an additional capillary effect [78]. Eventually the GaAs layer is capped by an upper AlGaAs barrier, leaving buried regions of an enhanced thickness which act as quantum wires.

During growth of the upper AlGaAs layer the diffusion length of Ga adatoms is again quite short. This leads to a sharpening of the V-groove bottom, and eventually to a shape similar to that existing before the deposition of the GaAs layer. The growth sequence may then be repeated, so as to create a vertical stack of quantum wires as shown in Fig. 7.46. The dark regions in the AlGaAs layers labeled VQW (vertical QW) represent Ga-rich parts with a lower bandgap.

Self-organized growth was also applied on non-patterned substrates to fabricate quantum wires. We consider the use of step bunches which may appear on surfaces as pointed out in Sect. 7.2.6. Quantum wires forming at step bunches were studied for virtually unstrained GaAs/AlGaAs [80], but as well for strained wire/barrier materials such as InGaAs/GaAs [79], InGaAs/InP, and SiGe/Si [81] on various low-index surfaces. The strain-induced interaction among steps in superlattice wire-structures was shown to favor the ordering of step bunching [82], leading to wires with regular

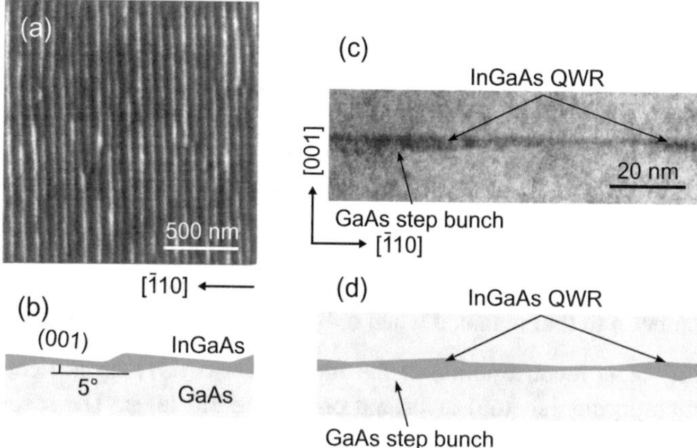

Fig. 7.47 Formation of quantum wires from 3 nm thick InGaAs deposition on GaAs (001) misoriented 5° towards [$\bar{1}$10]. **a** Atomic force micrograph of an $In_{0.1}Ga_{0.9}As$ layer grown on the stepped GaAs surface; adopted from [72]. **b** Magnified schematic of the cross section of image (**a**). **c** Cross-section transmission-electron micrograph and **d** schematic of an $In_{0.15}Ga_{0.85}As$ layer covered by a thick GaAs barrier layer. Reproduced with permission from [79], © 1997 Elsevier

lateral spacing [81]. The formation of wire material at steps of the barrier material may be attributed to the process of step-flow growth (Sect. 7.2.4) and the lower surface energy of the wire material. Furthermore, elastic relaxation of strained wire material is enhanced at step edges.

Formation of InGaAs quantum wires on a vicinal GaAs (001) surface misoriented by 5° towards the [$\bar{1}$10] direction is illustrated in Fig. 7.47. The vicinal GaAs surface forms step bunches of ~6.1 nm height and 70 nm spacing running along [110]. The terraces correspond to singular (001) faces. Metalorganic vapor-phase epitaxy of 3 nm thick InGaAs on this GaAs surface basically reproduces the stepped surface morphology [79]. InGaAs quantum wires form after growth of an upper GaAs barrier layer. The cross-sectional TEM image Fig. 7.47c shows ~6 nm thick quantum wires with a lateral width of ~25 nm. The thickness of the persisting quantum well on the (001) facets is ~1.3 nm, yielding a growth enhancement of a factor of 4 at the step edges.

7.4 Problems Chapter 7

7.1 Consider a Kossel crystal with the shape of a cube with edge length $n \times a$, consisting of atom cubes with edge length a.

(a) Determine the number of atoms N_i which differ in energy. Assume only nearest-neighbor bonds and distinguish between atoms in the bulk, on the faces (but not at corners or edges), at the edges (not at a corner), and at corners. Calculate N_i for n equals 10 and 1000, and check that the total number equals n^3.

(b) The number of bonds for the different types of atoms noted in (a) differs. With 1 bond of energy E_b shared by 2 neighbored atoms, a bulk atom has an energy $E = 6 \times E_b \times \frac{1}{2} = 3E_b$. Calculate the respective energies for the different types of atoms. The contribution of atoms at the surface (= faces, corners, and edges) to the total energy E_{tot} of all atoms decreases as the size of the crystal increases. Find the fraction of their energy contribution to E_{tot} for $n = 10^1$ to 10^4 in steps by a factor of 10. Compare the result for a large number n to that obtained when only bulk and face atoms are considered.

7.2 The $\beta2(2 \times 4)$ reconstruction of the As-rich GaAs(001) surface contains two As dimers in the 1st (top) layer and one in the 3rd layer. The reconstruction complies with the electron-counting model similar to the missing-dimer (2×4) surface discussed in the text (cf. Fig. 7.11).

(a) Apply electron counting for the $\beta2(2 \times 4)$ reconstruction by listing the required and the available electrons in the topmost 4 layers, assuming dangling Ga bonds empty and dangling As bonds filled. How many electrons are involved?

(b) Repeat (a) for the related Ga-rich $\beta2(4 \times 2)$ reconstruction. In this surface the As atoms are exchanged for Ga atoms and vice versa, yielding two Ga dimers in the first and one in the third layer (cf. Fig. 7.11). Does the $\beta2(4 \times 2)$ reconstruction comply with the electron-counting rule?

7.3 The surface diffusion of Ga is the rate-determining process in the epitaxy of GaAs under excess arsenic pressure. An experiment on the temperature-dependent surface-diffusion length λ yields the following four data points for an Arrhenius plot of λ/cm versus 1000/T/K: 5.3×10^{-6}, 1.00; 7.5×10^{-6}, 1.05; 7.0×10^{-6}, 1.20; 5.5×10^{-6}, 1.25. Assume that the diffusion length of the first two data points in the high-temperature range is solely governed by the diffusion barrier, and the latter data points also reflect a pure exponential behavior which is additionally affected by the adsorption energy.

(a) Calculate the energy of the diffusion barrier and the adsorption energy.

(b) Find two different temperatures where the diffusion length equals 4.0×10^{-6} cm.

7.4 We consider adatoms on a Si surface.

(a) How long is the residence time of an adatom at 1000 °C? Assume 2.0 eV adsorption energy and a typical atomic vibration frequency of 10^{13} Hz. Find the surface diffusivity, if the mean diffusion length of 4×10^{-5} cm is limited by the residence time. What is the diffusion length at 800 °C, if the diffusion barrier is half the adsorption energy?

(b) Step-flow growth occurs, if the surface diffusion-length exceeds well the terrace width. Find the height of a (not reconstructed) single Si(001) monolayer and the angle for a small miscut along [100] required to produce equidistant terraces with 200 nm width and 1 monolayer high steps. Find the rate for the advancement of the step for 1 μm/h growth rate on the Si surface.

7.5 An InAs island of pyramidal shape with 10 nm base length, bound by unreconstructed (110) side facets, is formed on a 1.40 monolayer thick wetting layer on GaAs substrate. The lattice parameter of InAs is approximately 6.1 Å. Neglect the strain in the following assessments.

(a) Estimate the number of In and As atoms in the island.
(b) Which thickness has the initially deposited two-dimensional InAs layer before the formation of the islands, if 4×10^{10} cm^{-2} islands are generated?
(c) What were the thickness of the initially deposited two-dimensional InAs layer, if islands of the same areal density as in (b) were generated, but with a shape of a spherical cap? The diameter and contact angle of these islands are assumed to be 10 nm and 45°, respectively (compare to Fig. 6.10). How many In and As atoms contains one of these islands?

7.5 General Reading Chapter 7

A. Pimpinellei, J. Villain, *Physics of Crystal Growth* (Cambridge University Press, Cambridge, 1998)
W.K. Burton, N. Cabrera, F.C. Frank, The growth of crystals and the equilibrium structure of their surfaces. Philos. Trans. R. Soc. Lond. A **243**, 299–358 (1951)
K. Oura, V.G. Lifshits, A.A. Saranin, A.V. Zotov, M. Katayama, *Surface Science—An Introduction*, 1st edn. (Springer, Berlin, 2003)
H. Ibach, *Physics of Surfaces and Interfaces* (Springer, Berlin, 2006)
F. Bechstedt, *Principles of Surface Physics* (Springer, Berlin 2003)

References

1. D.J. Chadi, Atomic and electronic structures of reconstructed Si(100) surfaces. Phys. Rev. Lett. **43**, 43 (1979)
2. M.D. Pashley, Electron counting model and its application to island structures on molecular-beam epitaxy grown GaAs(001) and ZnSe(001). Phys. Rev. B **40**, 10481 (1989)
3. W.A. Harrison, Theory of polar semiconductor surfaces. J. Vac. Sci. Technol. **16**, 1492 (1979)
4. S. Yuan, L. Zhang, H. Chen, E. Wang, Z. Zhang, Generic principle for the prediction of metal-induced reconstructions of compound semiconductor surfaces. Phys. Rev. B **78**, 075305 (2008)
5. E.A. Wood, Vocabulary of surface crystallography. J. Appl. Phys. **35**, 1306 (1964)
6. L. Däweritz, R. Hey, Reconstruction and defect structure of vicinal GaAs(001) and and $Al_xGa_{1-x}As(001)$ surfaces during MBE growth. Surf. Sci **236**, 15 (1990)
7. V. P. LaBella, H. Yang. D. W. Bullok, P. M. Thibado, P. Kratzer, M. Scheffler, Atomic structure of the GaAs(001)-(2 × 4) surface resolved using scanning tunneling microscopy and first-principles theory, Phys. Rev. Lett. **83**, 29892 (1999)
8. A. Ohtake, P. Kocán, J. Nakamura, A. Natori, N. Koguchi, Kinetics in surface reconstructions on GaAs(001). Phys. Rev. Lett. **92**, 236105 (2004)

9. N. Moll, A. Kley, E. Pehlke, M. Scheffler, GaAs equilibrium crystal shape from first principles. Phys. Rev. B **54**, 8844 (1996)
10. J.E. Northrup, S. Froyen, Structure of GaAs(001) surfaces: The role of electrostatic interactions. Phys. Rev. B **50**, 2015 (1994)
11. W.G. Schmidt, S. Mirbt, F. Bechstedt, Surface phase diagram of (2 × 4) and (4 × 2) reconstructions of GaAs(001). Phys. Rev. B **62**, 8087 (2000)
12. N. Chetty, R.M. Martin, Determination of integrals at surfaces using the bulk crystal symmetry. Phys. Rev. B **44**, 5568 (1991)
13. A. Ohtake, Surface reconstructions on GaAs(001). Surf. Sci. Rep. **63**, 295 (2008)
14. A.V. Bakulin, S.E. Kulkova, S.V. Eremeev, O.E. Tereshchenko, Ab-initio study of new Ga-rich GaAs(001) surface (4 × 4) reconstruction. Surf. Sci. **615**, 97 (2013)
15. E. Penev, P. Kratzer, M. Scheffler, Atomic structure of the GaAs(001) surface: first-principles evidence for diversity of heterodimer motifs. Phys. Rev. Lett. **93**, 146102 (2004)
16. F. Bechstedt, R. Enderlein, *Semiconductor Surfaces and Interfaces* (Akademie Verlag, Berlin, 1988)
17. K. Oura, V.G. Lifshits, A.A. Saranin, A.V. Zotov, M. Katayama, *Surface Science: An Introduction* (Springer, Berlin, 2003)
18. K. Takayanagi, Y. Tanishiro, S. Takahashi, M. Takahashi, Structure analysis of Si(111)-7 × 7 reconstructed surface by transmission electron diffraction. Surf. Sci. **164**, 367 (1985)
19. K.D. Brommer, M. Needels, B.E. Larson, J.D. Joannopoulos, Ab initio theory of the Si(111)-(7 × 7) surface reconstruction: a challenge for massively parallel computation. Phys. Rev. Lett. **68**, 1355 (1992)
20. R.S. Becker, J.A. Golovchenko, G.S. Higashi, G.S. Swartzentruber, New reconstructions on Silicon (111) surfaces. Phys. Rev. Lett. **57**, 1020 (1986)
21. W. K. Burton, N. Cabrera, F. C. Frank, The growth of crystals and the equilibrium structure of their surfaces, Phil. Trans. R. Soc. Lond. A **243**, 299 (1951)
22. J.A. Venables, G.D.T. Spiller, M. Hanbrücken, Nucleation and growth of thin films. Rep. Prog. Phys. **47**, 399 (1984)
23. J.A. Venables, Rate equation approaches to thin film nucleation kinetics. Phil. Mag. **27**, 693 (1973)
24. T. Shitara, T. Nishinaga, Surface diffusion length of gallium during MBE growth on the various misoriented GaAs(001) substrates. Jpn. J. Appl. Phys. **28**, 1212 (1989)
25. I. Bryan, Z. Bryan, S. Mita, A. Rice, J. Tweedie, R. Collazo, Z. Sitar, Surface kinetics in AlN growth: A universal model for the control of surface morphology in III-nitrides. J. Crystal Growth **438**, 81 (2016)
26. S.V. Ghaisas, A. Madhukar, Role of surface molecular reactions in the growth mechanism and the nature of nonequilibrium surfaces: a Monte Carlo study of molecular-beam epitaxy. Phys. Rev. Lett. **56**, 1066 (1986)
27. A. Madhukar, S.V. Ghaisas, The nature of molecular beam epitaxial growth examined via computer simulations, CRC Crit. Rev. Solid State Mater. Sci. **14**, 1 (1988)
28. G. Ehrlich, F.G. Hudda, Atomic view of surface self-diffusion: tungsten on tungsten. J. Chem. Phys. **44**, 1039 (1966)
29. R.L. Schwoebel, E.J. Shipsey, Step Motion on Crystal Surfaces. J. Appl. Phys. **37**, 3682 (1966)
30. J. Myslivecek, C. Schelling, G. Springholz, F. Schaffler, B. Voigtländer, P. Smilauer, On the origin of the kinetic growth instability of homoepitaxy on Si(001). Mater. Sci. Eng. B **89**, 410 (2002)
31. A.L.-S. Chua, E. Pelucchi, A. Rudra, B. Dwir, E. Kapon, A. Zangwill, D.D. Vvedensky, Theory and experiment of step bunching on misoriented GaAs(001) during metalorganic vapor-phase epitaxy. Appl. Phys. Lett. **92**, 013117 (2008)
32. A. Gocalinska, M. Manganaro, E. Pelucchi, D.D. Vvedensky, Surface organization of homoepitaxial InP films grown by metalorganic vapor-phase epitaxy. Phys. Rev. B **86**, 165307 (2012)
33. X.Q. Shen, H. Okumura, Surface step morphologies of GaN films grown on vicinal sapphire (0001) substrates by RF-MBE. J. Crystal Growth **300**, 75 (2007)

34. K. Bellmann, U.W. Pohl, C. Kuhn, T. Wernicke, M. Kneissl, Controlling the morphology transition between step-flow growth and step-bunching growth. J. Crystal Growth **478**, 187 (2017)
35. G. S. Bales, A. Zangwill, Morphological instability of a terrace edge during step-flow growth, Phys. Rev. B **41**,, 5500 (1990). Erratum referring to one equation cf. G. S. Bales, A. Zangwill, Phys. Rev. B **48**, 2024 (1993)
36. A. Pimpinelli, J. Villain, *Physics of Crystal Growth* (Cambridge University Press, Cambridge, UK, 1998)
37. O. Pierre-Louis, M.R. D'Orsogna, T.L. Einstein, Edge diffusion during growth: the Kink-Ehrlich-Schwoebel effect and resulting instabilities. Phys. Rev. Lett. **82**, 3661 (1999)
38. M. V. Ramana Murty, B. H. Cooper, Instability in molecular beam epitaxy due to fast edge diffusion and corner diffusion barriers, Phys. Rev. Lett. **83**, 352 (1999)
39. D.J. Chadi, Stabilities of single-layer and bilayer steps on Si(001) surfaces. Phys. Rev. Lett. **59**, 1691 (1987)
40. B.S. Schwartzentruber, Y.-W. Mo, R. Kariotis, M.G. Lagally, M.B. Webb, Direct determination of Step and kink energies on vicinal Si(001). Phys. Rev. Lett. **65**, 1913 (1990)
41. R.G.S. Pala, S. Liu, Critical Epinucleation on reconstructured surfaces and first-principle calculation of homonucleation on Si(100). Phys. Rev. Lett. **95**, 136106 (2005)
42. B. Voigtländer, M. Kästner, P. Smilauer, Magic islands in Si/Si(111) homoepitaxy. Phys. Rev. Lett. **81**, 858 (1998)
43. W. Shimada, H. Tochihara, Step-structure dependent step-flow: models for the homoepitaxial growth at the atomic steps on Si(111)7 $\times$ 7. Surf. Sci. **311**, 107 (1994)
44. H. Tochihara, W. Shimada, The initial process of molecular beam epitaxial growth of Si on Si(111)7 $\times$ 7: a model for the destruction of the 7 $\times$ 7 reconstruction. Surf. Sci. **296**, 186 (1993)
45. S. Filimonov, V. Cherepanov, Y. Hervieu, B. Voigtländer, Multistage nucleation of two-dimensional Si islands on Si(111)-7 $\times$ 7 homoepitaxy during MBE growth. Phys. Rev. B **76**, 035428 (2007)
46. M. Itoh, G.R. Bell, B.A. Joyce, D.D. Vvedensky, Transformation kinetics of homoepitaxial islands on GaAs(001). Surf. Sci. **464**, 200 (2000)
47. P. Kratzer, M. Scheffler, Reaction-limited island nucleation in molecular beam epitaxy of compound semiconductors. Phys. Rev. Lett. **88**, 036102 (2002)
48. C.T. Foxon, B.A. Joyce, Interaction kinetics of As_2 and Ga on 100 GaAs surfaces. Surf. Sci. **64**, 293 (1977)
49. P. Kratzer, C.G. Morgan, M. Scheffler, Density-functional theory studies on microscopic processes of GaAs growth. Prog. Surf. Sci. **59**, 135 (1998)
50. E.S. Tok, J.H. Neave, J. Zhang, B.A. Joyce, T.S. Jones, Arsenic incorporation kinetics in GaAs(001) homoepitaxy revisited. Surf. Sci. **374**, 397 (1997)
51. A. Kley, P. Ruggerone, M. Scheffler, Novel diffusion mechanism on the GaAs(001) surface: The role of adatom-dimer interaction. Phys. Rev. Lett. **79**, 5278 (1997)
52. P. Kratzer, C.G. Morgan, M. Scheffler, Model for the nucleation in GaAs homoepitaxy derived from first principles. Phys. Rev. B **59**, 15246 (1999)
53. V.A. Shchukin, N.N. Ledentsov, P.S. Kop'ev, D. Bimberg, Spontaneous ordering of arrays of coherent strained islands. Phys. Rev. Lett. **75**, 2968 (1995)
54. L.G. Wang, P. Kratzer, M. Scheffler, N. Moll, Formation and stability of self-assembled coherent islands in highly mismatched heteroepitaxy. Phys. Rev. Lett. **82**, 4042 (1999)
55. G.-H. Lu, M. Cuma, F. Liu, First-principles study of strain stabilization of Ge(105) facet on Si(001). Phys. Rev. B **72**, 125415 (2005)
56. N. Moll, M. Scheffler, E. Pehlke, Influence of surface stress on the equilibrium shape of strained quantum dots. Phys. Rev. B **58**, 4566 (1998)
57. A. Rastelli, M. Kummer, H. Von Känel, Reversible shape evolution of Ge islands on Si(001). Phys. Rev. Lett. **87**, 256101 (2001)
58. J. Márquez, L. Geelhaar, K. Jacobi, Atomically resolved structure of InAs quantum dots. Appl. Phys. Lett. **78**, 2309 (2001)

59. T. Xu, L. Zhou, Y. Wang, A.S. Özcan, K.F. Ludwig, GaN quantum dot superlattices grown by molecular beam epitaxy at high temperature. J. Appl. Phys. **102**, 073517 (2007)
60. M. Pinczolits, G. Springholz, G. Bauer, Direct formation of self-assembled quantum dots under tensile strain by heteroepitaxy of PbSe on PbTe (111). Appl. Phys. Lett. **73**, 250 (1998)
61. S. Ruvimov, P. Werner, K. Scheerschmidt, U. Gösele, J. Heydenreich, U. Richter, N.N. Ledentsov, M. Grundmann, D. Bimberg, V.M. Ustinov, AYu. Egorov, P.S. Kop'ev, ZhI Alferov, Phys. Rev. B **51**, 14766 (1995)
62. Y. Nabetani, T. Ishikawa, S. Noda, A. Sasaki, Initial growth stage and optical properties of a three-dimensional InAs structure on GaAs. J. Appl. Phys. **76**, 347 (1994)
63. D. Leonard, K. Pond, P.M. Petroff, Critical layer thickness for self-assembled InAs islands on GaAs. Phys. Rev. B **50**, 11687 (1994)
64. I. Daruka, J. Tersoff, A.-L. Barabási, Shape transition in growth of strained islands. Phys. Rev. Lett. **82**, 2753 (1999)
65. G. Costantini, A. Rastelli, C. Manzano, P. Acosta-Diaz, R. Songmuang, G. Katsaros, O.G. Schmidt, K. Kern, Interplay between thermodynamics and kinetics of InAs/GaAs(001) quantum dots. Phys. Rev. Lett. **96**, 226106 (2006)
66. H. Eisele, A. Lenz, R. Heitz, R. Timm. M. Dähne, Y. Temko, T. Suzuki, K. Jacobi, Change of InAs/GaAs quantum dot shape and composition during capping, J. Appl. Phys. **104**, 124301 (2008)
67. U.W. Pohl, K. Pötschke, A. Schliwa, F. Guffarth, D. Bimberg, N.D. Zakharov, P. Werner, M.B. Lifshits, V.A. Shchukin, D.E. Jesson, Evolution of a multimodal distribution of self-organized InAs/GaAs quantum dots. Phys. Rev. B **72**, 245332 (2005)
68. G. Costantini, A. Rastelli, C. Manzano, R. Songmuang, O.G. Schmidt, K. Kern, H. von Känel, Universal shapes of self-organized semiconductor quantum dots: Striking similarities between InAs/GaAs(001) and Ge/Si(001). Appl. Phys. Lett. **85**, 5674 (2004)
69. V.A. Shchukin N.N. Ledentsov, D. Bimberg, Entropy effects in self-organized formation of nanostructures, in *NATO Advanced Workshop on Atomistic Aspects of Epitaxial Growth*, ed. by M. Kortla et al. (Kluwer, Dordrecht, 2002)
70. V.A. Shchukin, N.N. Ledentsov, D. Bimberg, *Epitaxy of Nanostructures* (Springer, Berlin, 2004)
71. M. Meixner, E. Schöll, V.A. Shchukin, D. Bimberg, Self-assembled quantum dots: Crossover from kinetically controlled to thermodynamically limited growth. Phys. Rev. Lett. **87**, 236101 (2001)
72. X.-L. Wang, V. Voliotis, Epitaxial growth and optical properties of semiconductor quantum wires. J. Appl. Phys. **99**, 121301 (2006)
73. T. Takebe, M. Fujii, Y. Yamamoto, K. Fujita, T. Watanabe, Orientation-dependent Ga surface diffusion in molecular beam epitaxy of GaAs on GaAs patterned substrates. J. Appl. Phys. **81**, 7273 (1997)
74. A. Gustafsson, F. Reinhardt, G. Biasiol, E. Kapon, Low-pressure organometallic chemical vapor deposition of quantum wires on V-grooved substrates. Appl. Phys. Lett. **67**, 3673 (1995)
75. R. Bhat, E. Kapon, D.M. Hwang, M.A. Koza, C.P. Yun, Patterned quantum well heterostructures grown by OMCVD on non-planar substrates: Applications to extremely narrow SQW lasers. J. Crystal Growth **93**, 850 (1988)
76. S. Koshiba, N. Noge, H. Akiyama, T. Inoshita, Y. Nakamura, A. Shimizu, Y. Nagamune, M. Tsuchiya, H. Kano, H. Sasaki, Formation of GaAs ridge quantum wire structures by molecular beam epitaxy on patterned substrates. Appl. Phys. Lett. **64**, 363 (1994)
77. T. Sato, I. Tamai, H. Hasegawa, Growth kinetics and modeling of selective molecular beam epitaxial growth of GaAs ridge quantum wires on pre-patterned nonplanar substrates. J. Vac. Sci. Technol. B **22**, 2266 (2004)
78. G. Biasiol, A. Gustafsson, K. Leifer, E. Kapon, Mechanisms of self-ordering in nonplanar epitaxy of semiconductor nanostructures. Phys. Rev. B. **65**, 205306 (2002)
79. S. Hara, J. Motohisa, T. Fukui, Formation and characterization of InGaAs strained quantum wires on GaAs multiatomic steps grown by metalorganic vapor phase epitaxy. J. Crystal Growth **170**, 579 (1997)

80. S. Hara, J. Motohisa, T. Fukui, H. Hasegawa, Quantum well wire fabrication method using self-organized multiatomic steps on vicinal (001) GaAs surfaces by metalorganic vapor phase epitaxy. Jpn. J. Appl. Phys. **34**, 4401 (1995)
81. K. Brunner, Si/Ge nanostructures. Rep. Progr. Phys. **65**, 27 (2002)
82. L. Bai, J. Tersoff, F. Liu, Self-organized quantum-wire lattice via step flow growth of a short-period superlattice. Phys. Rev. Lett. **92**, 225503 (2004)

Chapter 8
In Situ Growth Analysis

Abstract Sensors to analyse layer structures already during epitaxial growth provide valuable information for developing device structures and for ensuring the reproducibility of run-to-run conditions. Most analytical online tools are applicable to all major growth techniques. Today a variety of probes is routinely integrated into growth systems for monitoring in situ sample temperature, growth rate, layer thickness, composition, strain, and other parameters of the growth process. Sensors measure either the ambient in the vicinity of the growing sample or the sample surface. *Ambient analysis* comprises mass spectrometry and optical probes; they provide information about the mass transport, the kind and density of species, their temperature, and potential mutual reactions. *Surface probes* include diffraction techniques and various optical tools. Surface sensitivity for diffraction is achieved by applying grazing-incidence angles, and the diffracted electron and X-ray beams disclose the surface morphology and reconstructions. Optical probes are widely applied in gaseous growth ambient. The selectivity for the surface may strongly be enhanced by taking advantages of symmetry-related surface properties, and several optical probes can resolve the growth of single monolayers. The chapter describes prominent techniques for in situ analysis of epitaxy. After discussing ambient analysis using mass spectrometry and optical spectroscopy, surface probes are considered. Structural analysis by reflection high-energy electron diffraction and by X-ray diffraction is outlined, and optical probes by pyrometry and deflectometry yielding data on temperature and strain are presented. The text then focuses on reflectometry, ellipsometry, and reflectance-difference spectroscopy (reflectance-anisotropy spectroscopy); these techniques provide both chemical and structural information.

8.1 Surface and Ambient Probing

Semiconductor devices consist of many epitaxial layers with different composition. The demand to control the perfection of growth in situ, i.e., already during epitaxy, led to the development of various analytical tools. Some of these sensors are today routinely integrated into commercial growth systems and allow for online monitoring

U. W. Pohl, *Epitaxy of Semiconductors*, Graduate Texts in Physics,
https://doi.org/10.1007/978-3-030-43869-2_8

of sample temperature, composition, growth rate, strain, and other parameters of the growth process.

There are mainly two different kinds of probes for in situ monitoring: electrons and photons. Using these probes either the ambient near the sample or the sample surface is monitored during growth. To achieve sufficient sensitivity for the small interaction volume of the probe with the surface, the penetration depth into the sample should be small. This condition is well fulfilled for electrons with a low energy as depicted in Fig. 8.1a; the given mean free path λ is largely independent of the material; a least squares analysis of data in Fig. 8.1a shows that the measurements for energies below 15 eV are described by a power law near E^{-2} and those above 75 eV by $E^{1/2}$, the solid line is a fit to $\lambda_i = a_i\, E^{-2} + b_i\, E^{1/2}$. The universal behavior makes electron diffraction an ideal tool for obtaining structural information on the surface. This is routinely applied in vacuum-based deposition techniques (Sect. 8.3.1). In the environment of vapor-phase deposition electron probes cannot be used due to absorption in the gas phase, but photons in the visible spectral range can penetrate the gas phase. A drawback of light is its comparably large penetration depth as illustrated in Fig. 8.1b, yielding a dominant contribution from the bulk when simple optical techniques like reflectometry are applied. Surface sensivity can be preserved when polarization effects or specific properties of the surface are utilized. Such optical probes are discussed in Sect. 8.4.

In addition to electrons or photons also other in situ probes are applied. A prominent technique is mass spectrometry, mainly used for analysing vacuum conditions. Direct information on surface morphology is obtained from in situ scanning probe microscopy, implemented in a specially designed growth setup.

In situ probing techniques were generally developed for ex situ analyses, and often the application in a growth apparatus requires some modification to provide access

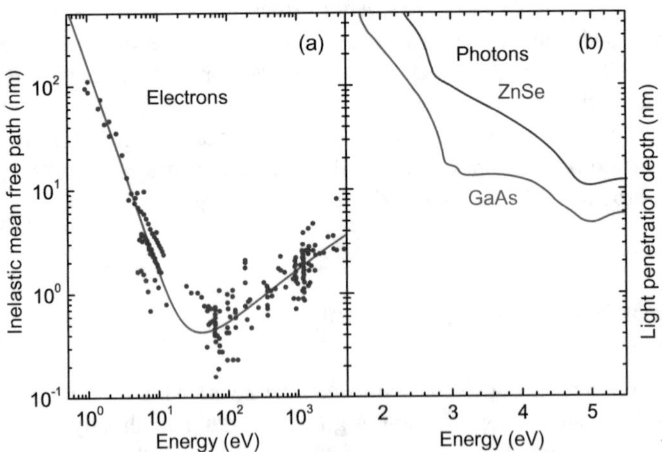

Fig. 8.1 **a** Inelastic mean free path of electrons in solids. **b** Penetration depth of light in GaAs and ZnSe. Reproduced with permission from [1, 2], © 1979 Wiley and 1996 Springer, respectively

to the sample surface or its environment. Still most of the widely applied probes have negligible impact on the growth process; a survey of in situ probes is given in Table 8.1.

The instrumentation of a growth apparatus with in situ sensors depends on the growth method and the material to be grown. Virtually any chamber for molecular beam epitaxy is equipped with RHEED and QMS, often also with pyrometry and reflectometry. Vapor-phase methods like MOVPE usually employ reflectometry and pyrometry. Pyrometers in commercial MOVPE reactors are today integrated into a closed-loop temperature control, and reflectometry is often applied for online correction of the growth rate. In the epitaxy of nitrides additionally deflectometry is routinely applied. In situ sensors employed in production of devices also monitor the compliance with specifications within certain tolerance levels, thereby increasing yield and reducing down times; often a fingerprint record of the sensor signal is sufficient to keep constant run-to-run growth conditions. On the other hand, growth

Table 8.1 Techniques for in situ analyses of epitaxial growth

Probes	Acronym	Information
Probes using electrons		
Reflection high energy electron diffraction	RHEED	Surface geometry
Probes using photons		
Deflectometry		Strain, composition
Pyrometry		Surface temperature
Reflectometry		Growth rate, layer thickness, optical constants, composition
Spectroscopic ellipsometry	SE	Optical constants, strain, layer thickness, composition
Reflectance difference/anisotropy spectroscopy	RDS, RAS	Surface and interface anisotropy, reconstructions
Raman scattering, coherent anti-Stokes Raman scattering	RS, CARS	Species, strain
Grazing incidence X-ray scattering/reflection	GIXS, GIXR	Surface structure, layer thickness
Surface photo absorption, *p*-polarized reflectance spectroscopy	SPA, PRS	Surface coverage
Laser light scattering	LLS	Macroscopic surface irregularities
Second harmonic generation	SHG	Surface symmetry
Laser-induced fluorescence (LIF), LIF spectroscopy	LIF, LIFS	Ambient species, partial pressure
Other probes		
Quadrupole mass spectrometry	QMS	Ambient species, partial pressure
Scanning tunneling microscopy	STM	Surface structure
Atomic force microscopy	AFM	Surface structure

systems used in research are often equipped with additional and specialized sensors, and the signals are carefully analyzed to achieve an insight into the growth process.

8.2 In Situ Ambient Analysis

Sensors measure either the (surface-near) sample ambient or the sample surface. In the following we discuss some widely applied methods for ambient monitoring. Analysis of the ambient in the vicinity of the growing sample provides information about the mass transport, i.e., about the species moving toward the surface and those desorbing into the ambient. Besides the identification of the species and their density also their temperature, hydrodynamics, and potential reactions may be disclosed, thereby offering valuable insights into the growth process.

8.2.1 Quadrupole Mass Spectrometry

Mass spectrometry is routinely employed in vacuum systems for detecting leakage and contaminations. The application in ambient monitoring requires a high resolution of masses to distinguish fragments of species, detection up to high masses, and differential pumping to provide the high vacuum for the detector; generally *quadrupole mass spectrometry* (QMS) is employed ambient analysis by QMS was applied in both molecular beam epitaxy (MBE) and (metalorganic) vapor phase epitaxy (VPE, MOVPE); studies focusing on desorption are also referred to as *desorption mass spectrometry*.

Mass spectrometry in MBE benefits from the UHV conditions in the growth chamber. By applying a line-of-sight configuration the mass spectrometer is sensitive only to particles emitted from the sample surface; this is achieved by appropriate apertures with cryo-pumping walls, such that the line of sight is established only between a small patch on the sample surface and the ionization volume in the QMS [3, 4]. The flux of species incident on the surface can be identified and monitored as well; for this purpose an inert surface is used, or the experimental surface is made inert. The number of particles within the ionization volume of line-of-sight QMS is small due to the geometry with two apertures defining the line of flight; this requires the use of a secondary electron multiplier within the mass spectrometer.

An example for the application of QMS to the MBE of GaN is given in Fig. 8.2. The UHV conditions of MBE leads to GaN growth in a thermodynamically metastable regime; deposition must hence overcompensate desorption. Figure 8.2a shows the temperature dependence of Ga desorbing from a GaN (0001) surface in vacuum [5]. Preceding calibration was performed by measuring the ^{69}Ga partial pressure for various Ga fluxes evaporated onto sapphire, which was kept at very high temperature to ensure that all impinging Ga desorbs; this yields a linear relation of 2.53×10^{-10} mbar/monolayer s^{-1} [6], used to scale Ga desorption in units of 1 nm/min; the measured Ga flux provides directly the GaN decomposition rate due to congruent

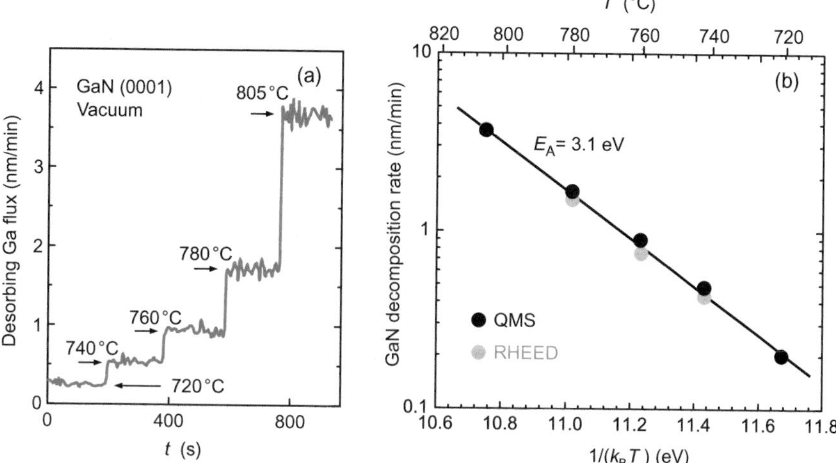

Fig. 8.2 **a** Ga flux desorbing from heated GaN in vacuum at various temperatures. **b** Decomposition rate of GaN as derived from the data in (**a**) and additional RHEED measurements as a function of T^{-1}; reproduced with permission from [5], © 2008 AIP

decomposition of GaN in vacuum. The Arrhenius plot Fig. 8.2b of the measured flux data for the GaN decomposition yields an activation energy of (3.1 ± 0.1) eV, also obtained from the simultaneously measured RHEED oscillations originating from a layer-by-layer decomposition in this experiment.

The same QMS technique was applied to confirm experimentally [6] the theoretically predicted existence of a Ga double layer on a Ga-polar (0001) face forming under Ga-rich conditions [7]. This metallic surface proved to be important for the epitaxy of smooth GaN layers [8].

Mass spectrometry in MOVPE is more demanding due to the high ambient pressure (typically 20–100 mbar from the carrier gas) and the large number of different species. Analysis of the gas composition behind the reactor yields results from a gas composition, which is not representative for the hot zone at the sample due to reactions during cooling. Therefor a gas probe is ingested in vicinity of the sample surface by a capillary [9, 10] or, in an early experiment, via a small orifice in the sample susceptor [11]; the latter arrangement yields a short sample path length, so that short-lived reactive species like free radicals are not removed by wall reactions and gas-phase recombination reactions. This provides the most realistic pattern of the growth ambient.

In the 1990s mass spectrometry and gas-phase spectroscopy were important means to study the pyrolysis reactions of MOVPE sources in the ambient, taking the frequently coupled pyrolysis pathways for both the cation and anion precursor molecules in the MOVPE of compound semiconductors into account [12, 13]. Since implementation of mass spectrometry in a MOVPE reactor requires a major modification, most of these studies were performed in an ersatz reactor, i.e., a thin heated flow tube, which quite roughly represents the hot zone of the reactor and neglected transport phenomena of a MOVPE reactor; they still helped finding the standard precursors routinely used today.

Mass spectrometry of tertiarybutylphosphine (TBP, $(C_4H_9)PH_2$) in D_2 flow-tube ambient is shown in Fig. 8.3a; the standard H_2 carrier gas was replaced by D_2 in order to label the products and to distinguish reactions among reactants and those with the ambient [14]. We observe the decrease of TBP as the temperature or the partial pressure increases, accompanied by the production of C_4H_8, C_4H_{10}, H_2, and PH_3 from the pyrolysis. At higher TBP partial-pressure the ratio of isobutene C_4H_8 and isobutane C_4H_{10} is reversed; this indicates a prevailing transition from a unimolecular pyrolysis by bond cleavage $(C_4H_9)PH_2 \rightarrow (C_4H_9)\bullet + \bullet PH_2$ and $(C_4H_9)\bullet \rightarrow (C_4H_8) + H\bullet$ at low partial TBP pressure, to a bimolecular reaction $(C_4H_9)\bullet + (C_4H_9)PH_2 \rightarrow HC_4H_9 + (C_4H_9)PH$ at high pressure, where a tertiarybutyl radical $(C_4H_9)\bullet$ removes a H from TBP [13, 14].

Pyrolysis of the nitrogen sources tertiarybutylhydrazine (TBHy, $(C_4H_9)NNH_2$) and dimethylhydrazine (DMHy, $(CH_3)_2NNH_2$) in a flow tube is shown in Fig. 8.3b; the given mass peak intensities of the undecomposed molecules ($TBHy^+$, $DMHy^+$) decrease at increasing temperature under production of products indicating a dominating N–N bond cleavage for both molecules; some detected masses such as

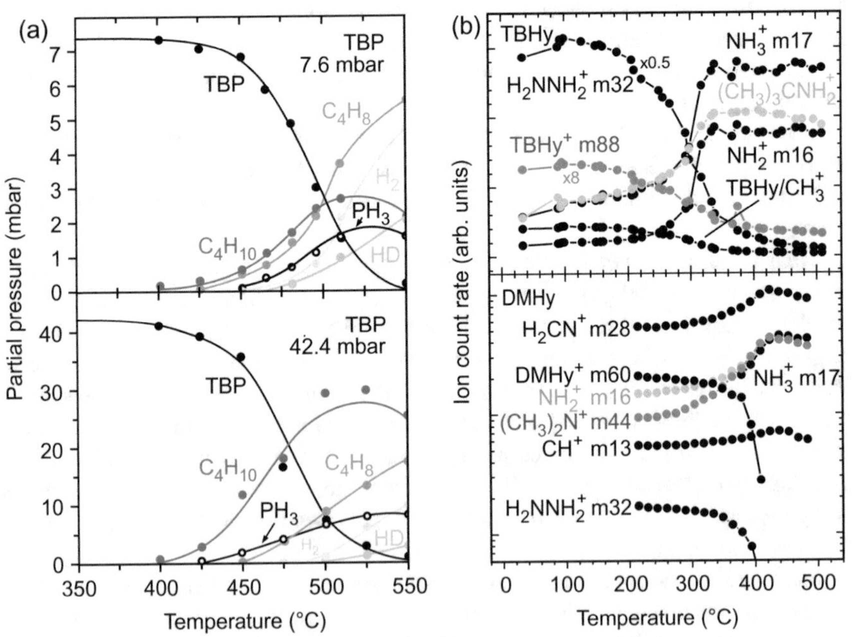

Fig. 8.3 **a** Temperature-dependent partial pressure of TBP and pyrolysis products measured using mass spectrometry. **b** Peak intensities in the temperature-dependent mass spectra of TBHy and DMHy. Reproduced with permission from [14, 15], © 1989 Springer and 1999 Elsevier, respectively

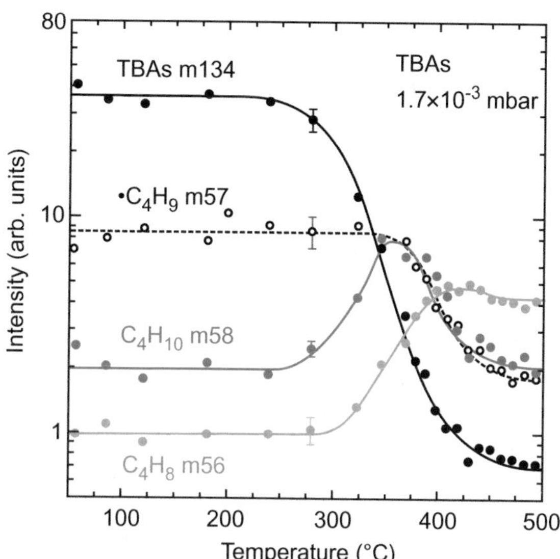

Fig. 8.4 Decomposition of TBAs measured using ion-trap mass spectrometry. Reproduced with permission from [10], © 2018 Springer Nature

$H_2NNH_2^+$ or TBHy/CH_3^+, which follow the abundance of the mol peak, originate from fragmentation in the mass spectrometer [15].

In a more recent study of tertiarybutylarsine (TBAs, $(C_4H_9)AsH_2$) pyrolysis a sensitive ion-trap mass spectrometer was fed by a nozzle in an MOVPE reactor to probe the gas phase in an *epitaxial* environment and largely avoid fragmentation by ionization [10]. The results given in Fig. 8.4. provide evidence for the $\bullet C_4H_9$ production by both thermal TBAs decomposition and some fragmentation. The production of C_4H_{10} preceding that of C_4H_8 similar to the finding with TBP indicates the existence of two pathways of TBAs decomposition: a free radical process generating C_4H_{10} and a β-H elimination process producing C_4H_8 [11, 16, 17].

8.2.2 *Optical Ambient Probing*

Optical characterization of the environment near the growing sample is of particular importance in gaseous ambient due to the strong interaction of particles in the gas phase and related reactions. Such interactions do not occur in molecular beams of vacuum-based epitaxy such as MBE and related techniques, where the growth species are given by the flux emitted from the source cells; they can be well measured using mass spectrometry.

Optical access to the reaction zone is generally provided by windows or fibre ports, purged by carrier gas or inert gas to prevent deposition.

An important aspect of gas-phase growth is hydrodynamics in the reactor. In the early days of reactor development in the 1980s *laser doppler anemometry* (LDA) was used to measure the velocity of gaseous particles. In these studies the actual molecules were replaced by micrometer-sized particles (TiO_2), and the Doppler shift of laser radiation scattered from the focus by Mie scattering was measured. An introduction to LDA is given in [18], application to MOVPE reactors is reviewed in [2]; for a visualization of the particle flow see Fig. 11.16a. Today a realistic insight into transport phenomena in the reaction zone (including major chemical reactions) is achieved by advanced modelling, and LDA measurements in MOVPE reactors were recently no longer applied.

Optical spectroscopy of molecules and atoms in the gas phase may identify species and roughly respective concentrations by observing electronic or vibronic transitions. Electronic transitions mostly occur in the ultraviolet and visible spectral range and can be probed by absorption or fluorescence spectroscopy. Vibrational excitations are either measured using Raman or infrared spectroscopy. The analysis is usually limited by the sensitivity for the detection of species with low concentration. Most optical in situ studies of the growth ambient were made in the 1980s and 1990s, when MOVPE was established as growth technique for mass production of (opto-) electronic devices. We briefly point out some features of these investigations.

Absorption spectroscopy requires a comparably simple setup. Electronic transitions may be observed using a single pass through the growth ambient, while the smaller absorption constants of vibrational transitions usually require a multi-pass configuration. Ultraviolet absorption spectroscopy was most popular for measuring gas-phase concentrations, with an emphasis on alkyl and hydride precursors used in the MOVPE of arsenides [19, 20], phosphides [21], and nitrides [22]. Applying Beer's law

$$I/I_0 = \exp(-\sigma c l), \tag{8.1}$$

with the absorption coefficient σ, the concentration c, and the absorption path-length l, the partial pressure of the species can be measured. In a multicomponent mixture the spectrally broad absorption bands must carefully be analyzed to obtain pure component partial pressures. Spectral selection is more convenient in the IR range, and the photon energies are too low to cause photolysis. Monitoring the IR absorption of In, Ga, Al akyls, and AsH_3 near $2 \dots 3\ \mu m$ in an ersatz reactor, effects of gas dynamics and chemistry were monitored online [23]. Measurement of species concentrations yielded precursor decomposition and reaction products, and transients measured velocities and diffusion effects. Besides conventional absorption spectroscopy also Fourier transform IR spectroscopy was applied, achieving reasonably low detection limits [24].

Most metalorganic precursors comprise C–H bonds; the stretching oscillation of this bond produces an absorption peak near 3.4 μm. Using this absorption also simple single-channel (i.e., non-dispersive) spectroscopy was applied for monitoring gas concentrations [25], albeit also other frequencies were used, e.g., for monitoring bis-cyclopentadienyl magnesium (Cp_2Mg), the standard *p*-type dopant source for GaN [26].

Raman spectroscopy benefits from a high spatial resolution, and signals from vibrational modes are specific for each species; in addition, the temperature of molecules is directly obtained from the occupation of rotational states. A drawback is the low sensitivity of the technique to species with low partial pressure such as decomposition products. The decomposition of trimethylindium (TMIn, $In(CH_3)_3$) with products of monomethylindium and indium, measured using Raman spectroscopy, is shown in Fig. 8.5. When approaching the heated susceptor (decreasing distance *d*) we observe a decrease of the TMIn signal, accompanied by increasing signals of In and monomethylindium; the latter decreases at smaller distances (spectra not shown in the figure).

Laser-induced fluorescence (LIF or LIFS, LIF spectroscopy) is a further optical technique for ambient measurement. A tunable laser selectively excites electronic transitions of molecules or atoms in the gas-phase, and the resulting fluorescence is analyzed. The method provides a high spatial, spectral, and (for pulsed excitation) temporal resolution. It is very sensitive for the detection of small molecules or atoms with a high quantum yield, while large molecules produce substantially smaller yield due to nonradiative relaxation or photo dissociation. Laser-induced fluorescence was widely used for monitoring processes in chemical vapor deposition (CVD) and etching [28]; there are, however, only few reports on applications in epitaxy, e.g. [29, 30].

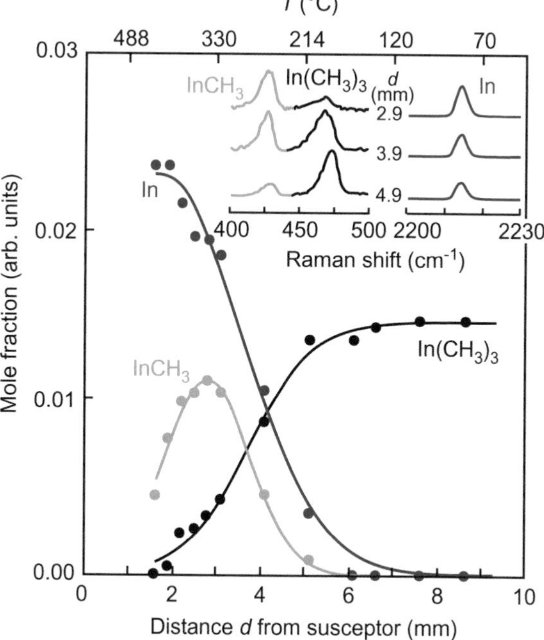

Fig. 8.5 Decomposition of trimethylindium measured using Raman spectroscopy. Reproduced with permission from [27], © 2002 RSC Publishing

8.3 Surface-Sensitive Diffraction Techniques

The structure of the surface during epitaxial growth can be best characterized by applying diffraction techniques. Waves of electrons and X-rays yield minimum perturbation of growth conditions and provide informations on the arrangement of atoms. Conditions for the diffraction of X-rays were discussed in Sect. 2.5; they also apply for electron diffraction, which is more widely used in vacuum-based growth. While the description in Sect. 2.5 focused on bulk diffraction, in situ monitoring of growth requires probes sensitive only to the structure of the surface. This sensivity is obtained for electrons by either using low energies (cf. Fig. 8.1a) or by applying grazing angles of incidence; the latter technique also allows for using X-rays, which otherwise have a large penetration depth in solids.

The diffraction of electrons with low energies yields precise informations on the surface structure. Low energy electron diffraction (LEED) is a powerful standard technique in surface science [31]. A serious drawback of the somewhat bulky LEED setup is, however, the mounting in front of the sample; thereby the path for molecular beams in MBE is blocked. LEED is therefor not applied in MBE and related vacuum-based epitaxy, and we focus on the deployed methods based on grazing incidence.

8.3.1 Reflection High-Energy Electron Diffraction (RHEED)

The diffraction of electrons with a very shallow (grazing) incidence angle is a standard method in any MBE apparatus, and is also well established in sample preparation for surface studies [32, 33]. The technique is compatible with vacuum, non-intrusive, and simple. Moreover, in MBE the RHEED assembly does not disturb the arrangement of the Knudsen cells opposite to the wafer, see Fig. 11.21. Electrons with an energy of 5–50 keV are directed to the surface at a low angle (<5°), and the diffracted beams leave the sample also near grazing incidence; the diffraction pattern is observed on a phosphor screen.

For a quick assessment of the surface status usually the position and shape of the diffraction spots is considered. A rough surface with small protruding crystallites yields sharp diffraction spots, which are also observed in bulk diffraction. Most surfaces produce, however, diffraction patterns consisting of elongated spots or rods.

The pattern also depends on the electron energy. At low energy of about 5 keV, higher incident angles can be used, still maintaining surface sensitivity. Step edges and scattering between terraces are then less detrimental, and good conditions to determine the surface morphology are given; such scattering is well described by a kinematic (single scattering) approach. When the electron energy is increased, the diffraction pattern is scaled down by a factor proportional to the square root of the energy (cf. 8.2). Scattering then occurs mainly in forward direction and requires a dynamical analysis.

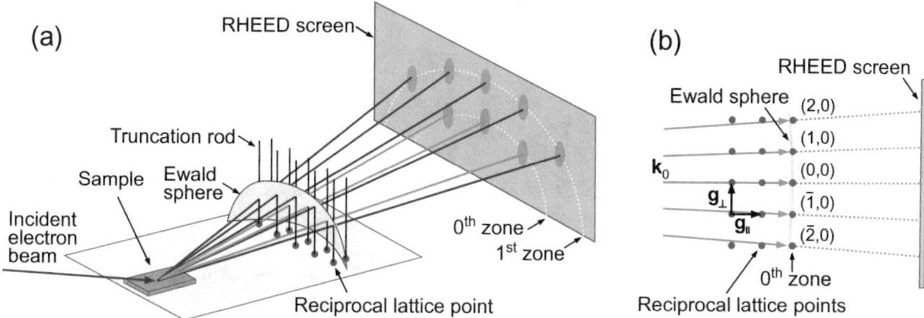

Fig. 8.6 Schematic of RHEED system with diffraction spots of a cubic lattice. **a** *Red dots* indicate the two-dimensional reciprocal lattice of the sample surface-lattice, the intersection points of the vertical *black truncation rods* with the Ewald sphere define the directions of diffracted beams. *Yellow dotted semicircles* on the screen indicate Laue circles. **b** Top view, not to scale

The elongated shape of the spots originates from diffraction conditions with a periodicity mainly in only two dimensions. Electrons impinging on the sample surface are scattered only from the topmost atoms due to the low incident angle. The Ewald construction discussed for a 3D lattice in Sect. 2.5.4 is illustrated for elastic 2D diffraction at a cubic lattice in Fig. 8.6. The *reciprocal* 2D lattice is a quadratic array represented by the red dots in the figure; their spacing is $2\pi/a$, where a is the atom spacing in real space. Since there is no periodicity in the vertical direction, constructive interference may occur along continuous vertical rods, referred to as crystal truncation rods. Wherever such a rod intersects the Ewald sphere, the condition for constructive interference of a scattered electron beam is fulfilled. These crossing points in **k** space hence determine the directions of constructive interference in real space, observed as spots on the phosphor screen. The set of spots on the screen lies on semicircles referred to as Laue circles, which are numbered starting with zero. The electron gun producing the incident electron beam and the fluorescent screen are usually in a distance of 10–30 cm from the sample.

The Ewald sphere is quite large due to the high energy of the electrons. The absolute value of the electron wave-vector is

$$|\mathbf{k}_0| = \frac{1}{\hbar}\sqrt{2m_0E + E^2/c^2}, \tag{8.2}$$

where the relativistic correction E^2/c^2 from Dirac's equation is ~1% for $E = 20$ keV. The corresponding radius of the Ewald sphere is then 400 times larger than the unit cell in reciprocal space of most cubic semiconductors, and the cut through the reciprocal lattice is almost planar.

RHEED provides much information about the surface. From the diffraction pattern the surface unit-cell and thereby the extent of surface reconstructions can be determined; the epitaxial relation between substrate and layer is observed, and also details of the reconstruction can be extracted from an analysis of reflection intensities [34]. Moreover, from the intensity of the RHEED pattern the roughness of

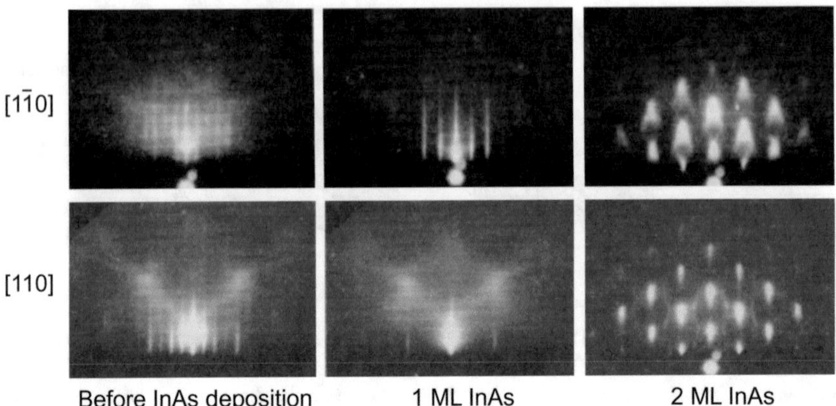

Fig. 8.7 RHEED patterns taken during InAs/GaAs molecular-beam epitaxy at 480 °C; the electron beam is incident either to $[1\bar{1}0]$ (*upper row*) or to [110] (*lower row*). Reproduced with permission from [35], © 1994 AIP

the surface can be assessed; this feature is routinely employed to assure a suitable starting surface for MBE, since the occurrence of periodicity indicates the removal of surface oxides from the wafer. It was also used to monitor the 2–3D transition in Stranski-Krastanow growth, applied to fabricate self-organized quantum dots discussed in Sect. 7.3; respective RHEED patterns observed during deposition of highly strained InAs on GaAs are shown in Fig. 8.7. We observe the streaky pattern of the pristine GaAs surface, that remains after the deposition of 1 monolayer (ML) InAs on top. The spotty RHEED pattern after deposition of 2 ML indicates the formation of three-dimensional islands; the oblique streaks with 55° angle in the pattern taken for an electron beam incident to $[1\ \bar{1}\ 0]$ originate from the {113}A facets of the InAs islands, cf. Fig. 7.41.

The dependence of the RHEED intensity on surface roughness can be used to monitor the growth of successive monolayers, if the epitaxy is performed in the layer-by-layer island growth-mode outlined in Sect. 6.2.6 (cf. also Fig. 7.20). The growth then leads to characteristic intensity oscillations of the RHEED spots during the growth process: a surface with many nucleating islands scatters electrons out of the regular beam direction, while a completed smooth monolayer yields an intense diffracted spot; a single oscillation hence corresponds to the completion of one monolayer [36]. The oscillations look similar to those observed in X-ray diffraction shown in Fig. 8.8. In the more often applied step-flow growth at increased temperature such oscillations do not appear.

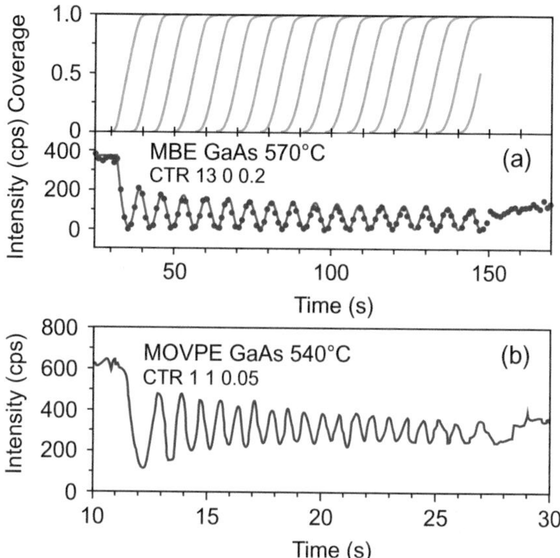

Fig. 8.8 X-ray intensity oscillations measured during epitaxy of GaAs(001). **a** MBE growth with β(2 × 4) reconstructed surface, intensity of the 13 0 0.2 crystal truncation rod. The *red solid line* is a model fit using the layer coverage shown on top (*green lines*). **b** MOVPE growth at ~50 mbar total reactor pressure, intensity of the 1 1 0.05 crystal truncation rod. Reproduced with permission from [43] and [44], © 2003 and 1993 Elsevier, respectively

8.3.2 *Grazing Incidence X-Ray Diffraction (GIXD) and Reflection*

X-rays interact with matter much weaker than electrons. Their diffraction in solids can hence also be applied in the gaseous atmosphere of vapor-phase epitaxy. By using grazing incidence angles, the penetration depth of X-rays inside the solid is reduced by about three orders of magnitude from typically (1–10) μm to (1–10) nm; thereby the required surface sensitivity is achieved. Usually synchrotron-based X-ray sources are used to obtain sufficient brightness for real-time measurements. The technique was employed for in situ monitoring of both MOVPE [37, 38] and MBE [39]. In either case the growth chamber must be equipped with beryllium windows for both the entrance of the X-ray beam and the exit of the scattered X-rays. When the incident beam impinges on the sample surface, it produces a beam diffracted at atomic planes perpendicular to the sample surface and also a specular reflection; these beams are utilized to either measure grazing incidence X-ray diffraction (GIXD, also referred to as GIXS with S for scattering) or grazing incidence X-ray reflection (GIXR).

In situ GIXR measurements basically yield growth rates and layer thicknesses; it can also be applied for measuring the critical thickness for surface roughening and related relaxation of InGaN/GaN layers [40]; such measurements demonstrated that the strain relaxation of $Ga_{1-x}In_xN$/GaN layers depends on the density of pre-existing threading dislocations in the GaN layer beneath, yielding a critical thickness of GaInN varying inversely with dislocation density [41].

In situ GIXD measurements provide informations similar to those obtained using RHEED described above. Data analysis of grazing incidence X-ray diffraction is facilitated by the weak interaction with matter, reducing dynamical effects and multiple scattering events; under most conditions a kinematic treatment is sufficient for a quantitative analysis. GIXD was employed to study surface reconstructions of homoepitaxial GaAs(001) developing under various conditions in MOVPE [42] and MBE [43]. In the layer-by-layer growth mode the oscillations of the crystal truncation-rods resolve the depositions of individual Ga-As bilayers as shown in Fig. 8.8. Growth temperatures were kept below the transition to the usually applied step-flow growth at higher temperature.

In situ GIXD studies of the growth-mode transitions between layer-by-layer and step-flow growth of GaAs(001) show significant differences between MOVPE and MBE. Analysis of the *diffuse* scattering in layer-by-layer growth of MOVPE yields island spacings being larger than for MBE [37], indicating a higher surface mobility of the active species. This is also reflected in a relatively low temperature of the crossover to step-flow growth. Despite a small offcut of 0.25° transition temperatures between 500 and 550 °C were derived from the disappearance of the X-ray intensity oscillations [45], significantly lower than the temperatures well above 550 °C observed in MBE for even large offcuts (i.e., smaller terrace width), cf. Fig. 7.19a.

8.4 Optical in Situ Surface Probes

Optical probes are applicable to virtually all major growth techniques; they are widely applied in techniques performed in gaseous ambient, and results can be compared to growth in vacuum-based systems—similar to the X-ray studies described above. There is a large variety of optical methods available for monitoring growth parameters and surface processes. Optical probes comprise monochromatic techniques such as reflectance or laser-light scattering, which mainly yield structural information, and spectral analysis such as spectroscopic ellipsometry or reflectance-difference (reflectance-anisotropy) spectroscopy, which also yield chemical information on the growing material. Most optical probes work essentially in the visible range; in gas ambient UV radiation is absorbed by molecular electronic transitions and IR radiation by molecular vibrations.

The surface sensitivity of optical probes is usually low due to absorption coefficients on the order of roughly 10^6 cm^{-1}, yielding a penetration depth of 10 nm or somewhat above—see Fig. 8.1. The surface response contributes consequently only to a small fraction to the total optical signal. Employing symmetry-related surface properties may enhance the surface selectivity, and even enable optical in situ control of single-monolayer growth.

In the following quite different kinds of optical probes are discussed, all providing valuable information about the current growth process.

8.4.1 Pyrometry

The temperature of the surface is of particular importance for the growth process, since it affects growth rate, interface abruptness, alloy composition, doping level, and other properties. The generally used thermocouples measure the susceptor temperature, which usually deviates by a substantial offset from the temperature of the sample surface due to insufficient thermal contact and thermal gradients. The offset cannot be corrected reliably, since it may change during an epitaxy run, e.g., due to wafer bowing.

Pyrometry measures the thermal emission from the sample, and relates the emitted power to the thermodynamic temperature using Planck's law of black-body radiation. The total black-body radiance is given by the Stefan-Boltzmann law $I(T) = \varepsilon\sigma T^4$, where σ is the Stefan-Boltzmann constant and $\varepsilon(T)$ the emissivity of the object with $\varepsilon_{\text{black body}} = 1$. This total radiance is measured by a bolometer. More often the measurement is performed in a small wavelength range (spectral pyrometer) instead of the wavelength-integrated Stefan-Boltzmann dependence. Using Wien's approximation of black-body radiation, the spectral radiance per unit area $I(\lambda,T)$ at a given wavelength λ is given by

$$I(\lambda, T) = \varepsilon\, c_1\, \lambda^{-5} \exp(-c_2/(\lambda\, T)) \tag{8.3}$$

the constants are $c_1 = 2\,h\,c^2 = 1.19{\cdot}10^8$ W μm^4 m^{-2} sr^{-1}, $c_2 = h\,c/k_B = 1.44{\cdot}10^4$ μm K, and $\varepsilon = \varepsilon(\lambda,T)$. Most spectral pyrometers use a single wavelength λ_0. The measured signal s of a thermal emission at λ_0 for temperature T is then

$$s = \varepsilon C \exp(-c_2/(\lambda_0 T)), \tag{8.4}$$

where the constant C includes all geometry and sensitivity factors. C may be determined by a calibration using a blackbody furnace with temperature T_{cal} known from a thermocouple, yielding the calibrated signal s_{cal}. Inverting (8.4) we obtain the temperature T from a measured signal s,

$$\frac{1}{T} = \frac{1}{T_{\text{cal}}} - \frac{\lambda_0}{c_2} \ln\left(\frac{s}{s_{\text{cal}}}\right). \tag{8.5}$$

However, the emissivity during epitaxy is subject to constant change: the reflectivity of a layer structure changes periodically due to Fabry-Pérot interferences from internal reflections, and emissivity and reflectivity of a surface are interconnected by Kirchhoff's law; for opaque materials (i.e., without transmission) it reads $\varepsilon = \alpha = 1 - R$, with the absorptivity α and the reflectivity R. Equation (8.5) may be complemented for describing a reflectance-correcting pyrometer [46],

$$\frac{1}{T} = \frac{1}{T_{\text{cal}}} - \frac{\lambda_0}{c_2} \ln\left(\frac{s}{s_{\text{cal}}} \frac{1 - R_{\text{cal}}}{1 - R}\right). \tag{8.6}$$

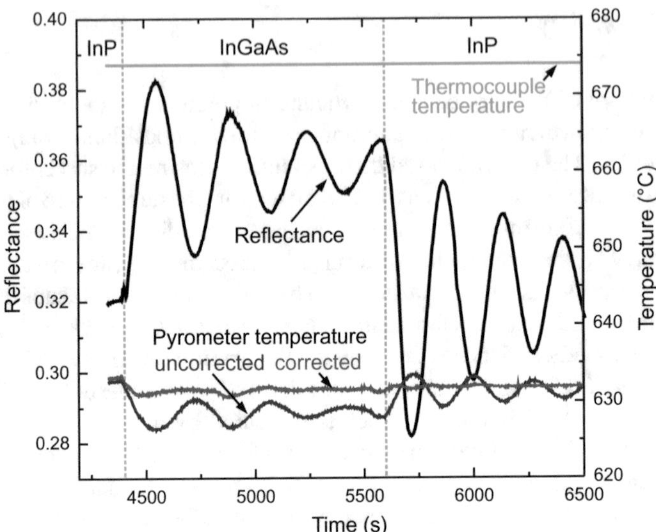

Fig. 8.9 Measurement of the wafer temperature during growth of a InP/InGaAs/InP sequence. The pyrometer temperature was corrected using the reflectance measured at 950 nm. Courtesy of K. Haberland, LayTec Berlin

By simultaneously measuring both the spectral radiance and the reflectivity at the same wavelength, the pyrometer signal can be corrected. A quasi-simultaneous measurement of spectral radiance and reflectance can be achieved by pulsing the probing light. Figure 8.9 shows the apparent temperature variations measured during growth of an InGaAs/InP structure using an uncorrected pyrometer, occurring despite a fixed process temperature controlled by a thermocouple [47]; we observe the simultaneously oscillating signals (with inverted phase) of the uncorrected pyrometer and the reflectance, along with the pyrometer temperature corrected for the varying reflectance; note the substantial deviation from the nominal process temperature.

8.4.2 Deflectometry

The curvature of a wafer induced by the stress of a mismatched layer can be precisely measured by evaluating the deflection of a laser beam. Usually two parallel beams are directed at the surface, and the change in relative spacing between the beams reflected off the curved surface are measured by a position-sensitive detector, see Fig. 8.10. Also three beams in a nonlinear arrangement are applied to measure non-spherical wafer bow, or multiple beams in an array to obtain a two-dimensional profile of the curvature. In vapor-phase reactors the reflected beams are usually directed to a small optical viewport (often only 3 mm in diameter) with an attached CCD detector. In the following we consider a two-beam setup.

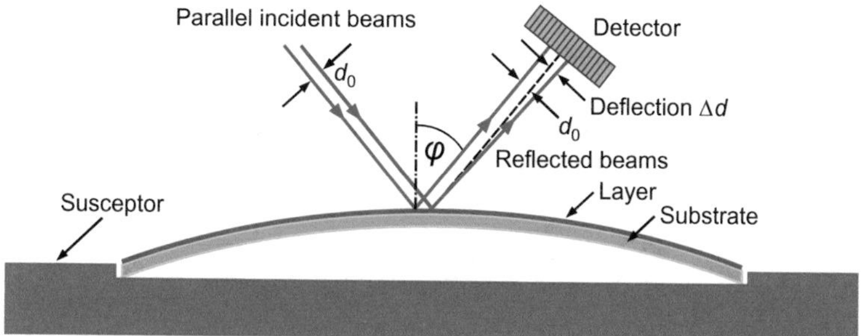

Fig. 8.10 Schematic of a wafer-bowing measurement. The *dashed line* indicates the right reflected beam for a planar wafer, the *deflection* Δd is related to the wafer bowing

If the layer is compressively strained by the substrate—either due to lattice mismatch or thermal mismatch—the wafer is bend in a convex curvature as illustrated in Fig. 8.10. A convex wafer bowing refers to a negative curvature κ, expressed by a negative curvature radius $R_c = 1/\kappa$; a tensely strained layer produces correspondingly a concave bowing with a positive curvature. The curvature κ of the wafer can be evaluated from the deflection Δd of the beam according to [48]

$$\kappa = \frac{1}{R_c} = \frac{1}{2L}\frac{\Delta d}{d_0}\cos\varphi, \quad (8.7)$$

where L is the distance between the surface and the detector.

Bowing is an issue for strongly mismatched substrate-layer combinations and particularly for thick layers. There are various origins of bowing [49, 50]. In the epitaxy of III nitrides four main contributions of total wafer bowing were found: the initial bowing of the substrate $1/R_{c,\text{initial}}$, the bowing caused by a vertical temperature gradient $1/R_{c,\Delta T\text{vertical}}$, the lattice mismatch-induced change in bowing during growth at constant temperature $1/R_{c,\text{growth}}$, and bowing contributions due to different linear thermal expansion coefficients of substrate and layer $1/R_{c,\alpha}$ [51]; apart from the initial bowing all contributions vary during the growth process. Wafer bowing leads to a spatially varying distance of the wafer back-side to the heated substrate and hence to a laterally and also vertically inhomogeneous temperature. A slight thermally induced wafer bowing is even observed for the pure substrate without a layer on top due to the (small) vertical temperature gradient; it is approximately given by

$$\frac{1}{R_{c,\Delta T_\text{vertical}}}(T) = \alpha_S \frac{T_\text{backside} - T_\text{surface}}{t_S}; \quad (8.8)$$

α_S and t_S are the linear expansion coefficient and thickness of the substrate. The temperature difference between substrate backside and surface depends on thermal conductivity; for the poorly conducting sapphire a curvature radius of 45 km was measured, corresponding to an estimated difference of 2 °C [51].

Deposition of a layer on top of the substrate creates additional strain and consequential bowing. For infinitesimally small deformations induced by the layer the resulting curvature can be directly converted into stress using Stoney's equation [52]; for larger deformations the equation was extended [53, 54]. The curvature for a given layer thickness t_L can be calculated from

$$\frac{1}{R_c} = \frac{1}{R_{c,\text{initial}}} + \frac{1}{R_{c,\Delta T_{\text{vertical}}}} + \frac{1}{R_{c,\text{growth}}}, \tag{8.9}$$

where the last term is given by

$$\frac{1}{R_{c,\text{growth}}}(t_L) = \frac{6t_L}{t_S^2}\frac{M_L}{M_S}\varepsilon. \tag{8.10}$$

Here t_S is the thickness of the substrate, M_L and M_S are the biaxial layer and substrate elastic moduli, and ε is the is the biaxial strain using the sign convention defined by (2.20a). Equation (8.10) relates the curvature $1/R_{c,\text{growth}}$ directly to the stress in the layer $\sigma_L = M_L \times \varepsilon$.

Curvature measurements were reported particularly for the highly strained epitaxial layers of nitrides. For basal-plane growth of hexagonal crystals the elastic moduli are given by their elastic constants, $M = C_{11}+C_{12}-2(C_{13}^2/C_{33})$ [55]. The evolution of curvature during metalorganic vapor-phase epitaxy of InGaN and AlGaN layers on sapphire (0001) is shown in Fig. 8.11a. We observe the steady increase of curvature

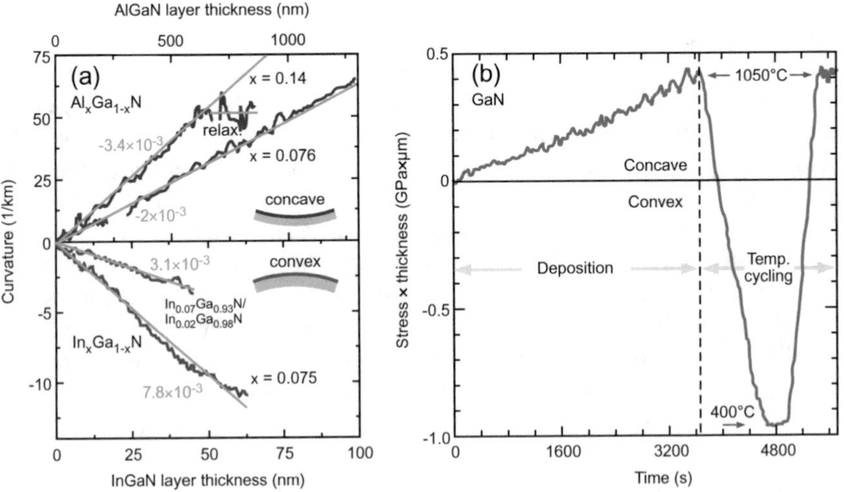

Fig. 8.11 **a** Evolution of the wafer curvature during MOVPE of AlGaN (*top*) and InGaN (*bottom*) single layers on sapphire substrate as a function of layer thickness. Evaluated strain values are indicated by the *green straight lines*. **b** Temporal evolution of stress × thickness during MOVPE and subsequent thermal cycling of a 2.2 thick μm GaN/sapphire layer from 1050 to 450 to 1050 °C. Reproduced with permission from [56, 57], © 2008 Elsevier and 1999 AIP, respectively

and related stress during growth at constant temperature; note the deviation from the linear increase for the most stressed $Al_{0.14}Ga_{0.86}N$ layer indicating relaxation. From the measured curvature and the materials parameters the ternary composition of the layers was calculated in good agreement with ex situ X-ray diffraction measurements [56]. The effect of thermal-expansion mismatch between the GaN layer and the sapphire (0001) substrate is illustrated in Fig. 8.11b. While growth at 1050 °C steadily increases the negative curvature and the deduced stress, cooling to 400 °C reverses the stress; the return of stress to the same value after restabilizing the temperature to 1050 °C indicates that no relaxation of the stress occured during cooling [57].

8.4.3 Reflectance and Ellipsometry

The technique of reflectometry and ellipsometry can both measure the state of polarization of light reflected from the sample surface and allow to extract growth rates and optical constants during deposition. Ellipsometry is performed at an oblique angle near the Brewster angle and hence requires a respectively designed growth chamber with two strain-free windows. The same applies for reflectance of light polarized parallel to the plane of incidence (*p* polarization). However, reflectometry can also operate at (near) normal incidence. This technique is very popular, although it is not as surface-sensitive as ellipsometry or reflectometry near Brewster angle. It requires only a single window for optical access and is robust with respect to the incidence angle in near-normal direction and to light polarization; therefor no strain-free window is required, and also optical fibers which do not preserve polarization can be used. Normal-incidence reflectance is hence easily measured in setups for layer growth.

Epitaxial structures generally consist of more than one single layer grown on a substrate. The optical response therefor contains contributions from all layers and interfaces buried by the uppermost layer in addition to that of the substrate. By replacing the structure beneath the topmost layer by a virtual substrate or a virtual interface, which represent the total effect of the entire subsurface structure, still growth rate and optical constants of the top layer can be extracted during sample growth.

The ***virtual substrate model*** and the related ***virtual interface model*** are concepts based on the idea that any multilayer film is mathematically identical to a single layer on an *effective* substrate as illustrated in Fig. 8.12. The models do not require the precise positions of the buried interfaces; instead, the exact dielectric function $\tilde{\varepsilon} = \varepsilon_1 + i\varepsilon_2$ of the structure is replaced by a *pseudodielectric function* $\langle\tilde{\varepsilon}\rangle = \langle\varepsilon_1\rangle + i\langle\varepsilon_2\rangle$. There are excellent reviews discussing the models in detail; the initially introduced virtual substrate model [58] is reviewed in [59, 60], and the closely related virtual interface model [61] is discussed in [62]. We hence only briefly point out the algorithm, which can be applied to both ellipsometry and reflectance.

Normal incidence reflection; we focus on reflection with normal incidence and follow arguments of [62]. In this most simple case only *s* polarization appears, and

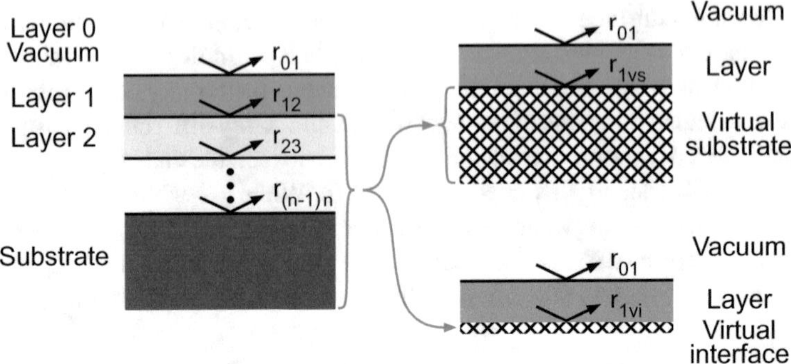

Fig. 8.12 Model of the virtual substrate or the virtual interface. The multilayer stack on the actual substrate (*left*) is reduced to a single layer on a virtual substrate or a virtual interface (*right*). *Arrows* represent incident and reflected beams

no propagation angle needs to be calculated. For a single reflection the measured real reflectance R is related to the complex reflectance $\tilde{r}$ given by the backward propagating complex electric field amplitude divided by the forward amplitude,

$$R = |\tilde{r}|^2. \tag{8.11}$$

For an infinite thick layer the reflectance (to vacuum) is given by

$$\tilde{r}_\infty = \frac{1-\tilde{n}}{1+\tilde{n}}, \tag{8.12}$$

where $\tilde{n} = n - i\kappa = \sqrt{\tilde{\varepsilon}}$ is the complex refractive index with the real refractive index n and the extinction coefficient κ. The reflectance of a layer with finite thickness is affected by the refractive indices of the layer(s) beneath. The behavior of each layer within the stack indicated in Fig. 8.12 may be described by the Jones-matrix formalism using 2×2 matrices [63]; the interface of a layer i to the layer $i-1$ above it is described by the matrix

$$I_i = \frac{1}{2\tilde{n}_{i-1}} \begin{bmatrix} \tilde{n}_{i-1} + \tilde{n}_i & \tilde{n}_{i-1} - \tilde{n}_i \\ \tilde{n}_{i-1} - \tilde{n}_i & \tilde{n}_{i-1} + \tilde{n}_i \end{bmatrix}, \tag{8.13}$$

and a matrix describing the phase shift of the electric field in the layer

$$L_i = \begin{bmatrix} \exp(i\beta) & 0 \\ 0 & \exp(-i\beta) \end{bmatrix}, \beta = 2\pi\tilde{n}_i d_i/\lambda; \tag{8.14}$$

here d_i is the thickness of the i-th layer and β is the layer phase-thickness at wavelength λ. Assuming that no light comes from the inside of the substrate

$\left(\tilde{E}_{\text{backward}} = 0\right)$, the electric-field of the n-layer stack is described by [62]

$$\begin{aligned}\begin{bmatrix}\tilde{E}_{\text{forward}}\\ \tilde{E}_{\text{backward}}\end{bmatrix}_{\text{surface}} &= I_1 L_1 I_2 L_2 \ldots I_n L_n I_{\text{substrate}} \begin{bmatrix}\tilde{E}_{\text{forward}}\\ 0\end{bmatrix}_{\text{substrate}} \\ &= I_1 L_1 I_2 L_2 \ldots I_n L_n \begin{bmatrix}\tilde{E}'_{\text{forward}}\\ \tilde{E}'_{\text{backward}}\end{bmatrix}_{\text{substrate}} \\ &= I_1 L_1 \begin{bmatrix}\tilde{E}''_{\text{forward}}\\ \tilde{E}''_{\text{backward}}\end{bmatrix}_{\text{virtual substrate}}\end{aligned}. \quad (8.15)$$

In the last step of (8.15) the double-primed field amplitudes may be considered to result from a single interface of layer 1 to a virtual substrate. Each new layer may again be considered as a single layer on a (changed) virtual substrate with the previously deposited layer lumped in.

The reflectance $\tilde{r}_i = \tilde{E}''_{\text{backward}}/\tilde{E}''_{\text{forward}}$ inside the layer may formally be described by a refractive index of the virtual substrate $\tilde{n}_{\text{vs}}$ yielding $\tilde{r}_i = (\tilde{n}-\tilde{n}_{\text{vs}})/(\tilde{n}+\tilde{n}_{\text{vs}})$, with the layer refractive-index $\tilde{n}$ on top of the virtual substrate; $\tilde{n}_{\text{vs}}$ has, however, only a physical meaning in the case of a single layer on top of the actual substrate.

The real reflectance R of a layer with real refractive index n and extinction coefficient κ on top of a virtual substrate, growing with a growth rate g, can now be written

$$R(t) = \frac{R_\infty - 2\sqrt{R_\infty R_i}\exp(-\gamma t)\cos(\delta t - \phi - \varphi) + R_i \exp(-2\gamma t)}{1 - 2\sqrt{R_\infty R_i}\exp(-\gamma t)\cos(\delta t - \phi + \varphi) + R_\infty R_i \exp(-2\gamma t)}, \quad (8.16)$$

with

$$R_\infty = |\tilde{r}_\infty|^2 = \frac{(1-n)^2 + \kappa^2}{(1+n)^2 + \kappa^2}, \quad \phi = \tan^{-1}\left(\frac{2\kappa}{n^2 + \kappa^2 - 1}\right) \quad (8.17)$$

$$R_i = |\tilde{r}_i|^2 = \frac{(n - n_{\text{vs}})^2 + (\kappa - \kappa_{\text{vs}})^2}{(n + n_{\text{vs}})^2 + (\kappa + \kappa_{\text{vs}})^2}, \quad \varphi = \tan^{-1}\left(\frac{2(n\kappa_{\text{vs}} - n_{\text{vs}}\kappa)}{n^2 - n_{\text{vs}}^2 + \kappa^2 - \kappa_{\text{vs}}^2}\right) \quad (8.18)$$

$$\gamma = 4\pi\kappa g/\lambda, \quad \delta = 4\pi n g/\lambda. \quad (8.19)$$

The reflectance R in (8.16) depends on the parameters n, κ, g, R_i, and ϕ, where the reflectance R_i and the phase shift ϕ refer to the total optical response of the layer stack beneath the topmost layer of interest; in practice they are fitting parameters to obtain the optical constants n and κ, and the growth rate g. $R(t)$ may continuously be recorded during growth; for a given wavelength λ at least five reflectance measurements at different times are required to obtain n, κ, and g [62].

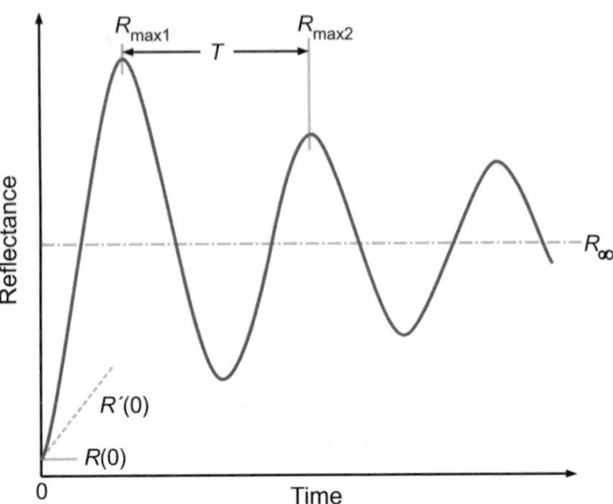

Fig. 8.13 Reflectance transient $R(t)$ with quantities to obtain approximately the optical constants and the growth rate of the topmost layer of a multiple-layer structure; after [63]

Equation (8.16) may be substantially simplified in many cases. If film and substrate have large refractive indices (about 4) and differ not too much $R_i \ll 1$ applies. In addition, the phase shift defined by φ becomes negligible for weak absorption. If these conditions hold (8.16) reads

$$R(t) \cong R_\infty - 2\sqrt{R_\infty R_i}(1 - R_\infty)\exp(-\gamma t)\cos(\delta t - \phi). \tag{8.20}$$

This is an oscillatory reflectance with a frequency δ and a decay constant γ, see Fig. 8.13. The parameter R_∞ is approximately given by the mean value of the reflectance $R(t)$, and estimates for α, R_i, and g are obtained from $R\ (t = 0)$ and the derivative $R'\ (t = 0)$, yielding [62, 64]

$$\begin{aligned}
&R_\infty \cong \langle R(t)\rangle, \quad \delta \cong 2\pi/T, \quad \gamma \cong \ln(R\text{max}_1/R\text{max}_2)/T,\\
&g \cong \frac{\lambda\delta}{4\pi}\left[\frac{1+R_\infty}{1-R_\infty} - \sqrt{\left(\frac{1+R_\infty}{1-R_\infty}\right)^2 - \frac{\gamma^2+\delta^2}{\delta^2}}\right],\\
&n \cong \delta\lambda/(4\pi g), \quad \kappa \cong \gamma\lambda/(4\pi g),\\
&\phi \cong \tan^{-1}\left(\frac{-R'(0)/\delta}{R_\infty - R(0)}\right),\\
&R_i \cong \frac{(R_\infty - R(0))^2 + (R'(0)/\delta)^2}{4R_\infty(1-R_\infty)^2}.
\end{aligned} \tag{8.21}$$

Estimates (8.21) may be used as starting points to fit the exact expression (8.16). For a fit of all five parameters at least a full-wave thick layer should be grown.

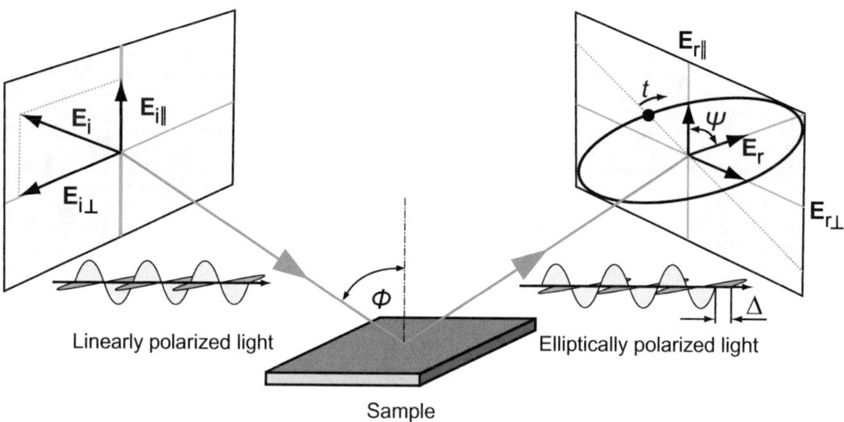

Fig. 8.14 Schematic of ellipsometry. Incident linearly polarized light is reflected from the sample surface with elliptical polarization

Spectroscopic ellipsometry operating near the Brewster angle is substantially more sensitive to the state of the surface than normal incidence reflectance described above—and it yields more information about the sample. It played an important role in the 1980s and 1990s, when a basic understanding of epitaxial growth was developed. Since the technique requires *two strain-free* windows, which furthermore have to be located at proper angels for incident and reflected beams, it is only applied in growth setups dedicated to basic growth studies.

The principle of an ellipsometric measurement is illustrated in Fig. 8.14, assuming a collimated monochromatic beam of light directed to the sample surface. The beam arrives an incidence angle ϕ, and the reflected light is analyzed for the ratio of the *polarization states* χ_{in} and χ_{out}—in contrast to spectral reflectometry, where the ratio of the *amplitudes* of incident and reflected beams I_{in} and I_{out} is analyzed. Spectroscopic ellipsometry is a well-developed traditional technique [63], and applications for real-time monitoring are reviewed in [60, 65, 66].

When a light beam is incident on a surface at an oblique angle, the reflectance R_p with polarization parallel to the plane of incidence is smaller than R_s, the reflectance with a perpendicular polarization; R_p is minimal at the (pseudo) Brewster angle. The corresponding complex reflection coefficients, $\tilde{r}_p$ and $\tilde{r}_s$, are therefore different; their amplitudes and phases experience different changes, and the reflected light has consequently an elliptical polarization as illustrated in Fig. 8.14. Ellipsometry is based on the complex reflectance ratio

$$\tilde{\rho} = \frac{\tilde{r}_p}{\tilde{r}_s} = |\tilde{\rho}| \exp(i\Delta) = \tan\psi \exp(i\Delta). \tag{8.22}$$

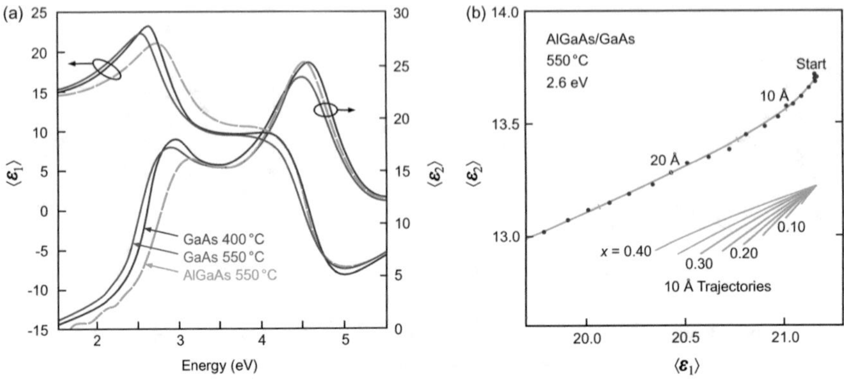

Fig. 8.15 **a** Pseudodielectric function of GaAs(001) at 400 °C (*blue*) and 550 °C (*red, solid lines*) and of $Al_{0.36}Ga_{0.64}As$ at 550 °C (*dashed green line*). **b** Trajectory of $\langle \varepsilon_2 \rangle$ versus $\langle \varepsilon_1 \rangle$ measured during growth of an $Al_{0.36}Ga_{0.64}As$/GaAs layer. Data points (*blue dots*) were obtained at 1.6 s intervals, 7 data points correspond to 10 Å layer thickness; the green solid line is calculated from a 3-phase model. Inset at bottom: Loci of $\langle \varepsilon \rangle$ for the deposition of a 10 Å thick $Al_xGa_{1-x}As$ layer on GaAs in increments of composition steps $\Delta x = 0.05$; reproduced with permission from [67], © 1990 AIP

The actually measured parameters are the absolute amplitude ratio tan ψ according to

$$|\tilde{\rho}| = \frac{|\tilde{r}_p|}{|\tilde{r}_s|} = \tan \psi, \tag{8.23}$$

and the difference Δ between the phase shifts, which p and s polarized light experience upon reflection; more precisely the measurement yields cos Δ.

The surface sensitivity of spectroscopic ellipsometry is illustrated in Fig. 8.15. Panel (a) shows the real and imaginary parts of the pseudodielectric function $\langle \tilde{\varepsilon} \rangle = \langle \varepsilon_1 \rangle + i \langle \varepsilon_2 \rangle$ of GaAs and of a thick $Al_{0.36}Ga_{0.64}As$/GaAs layer grown using MOVPE; the spectra were calculated from the measured ellipsometric angles ψ and Δ, based on the modelled complex reflectance ratio of a two-layer system comprising the surface of a bulk solid and the ambient above [67]. The calculated pseudodielectric function is identical to the bulk dielectric function $\tilde{\varepsilon} = \varepsilon_1 + i\, \varepsilon_2$, if the measured bulk is optically thick enough so that no significant response from buried layers contributes the spectra, and if no overlayers such as native oxides are present. The spectra of Fig. 8.15a are related to optical transitions to higher conduction bands of GaAs and show characteristic thermal changes. Selecting a fixed energy of 3.25 eV where the variation of $\langle \tilde{\varepsilon} \rangle$ with T is minimum, a detailed study of thermal desorption of the native oxide by monitoring $\langle \varepsilon_1 \rangle$ during sample heating under As stabilization was given in [67]. When GaAs is alloyed with Al, the spectra gradually change with increasing composition parameter x. Figure 8.15a indicates that the change is particularly pronounced at 2.6 eV. Plotting the trajectory of $\langle \varepsilon_2 \rangle$ versus $\langle \varepsilon_1 \rangle$ derived from ellipsometry at this energy during deposition of an $Al_{0.36}Ga_{0.64}As$/GaAs layer, both composition parameter and layer thickness can be obtained in real time, see Fig. 8.15b; the continuous line is calculated from a three-phase substrate-layer-ambient model, applying the empirical function $\varepsilon(x) = \varepsilon_{\mathrm{GaAs}} + (9.22x - 24.14x^2) + i(-16.94x - 0.33x^2)$. The inset shows the slopes for the initial 10 Å of layer growth that depend solely on the composition parameter x. A precision of 0.03 is achieved for $x > 0.20$, and the initially varied slope of the $x = 0.36$ trajectory reveals that the target composition in this deposition is only reached after growing 15 Å of AlGaAs.

8.4.4 Reflectance-Difference Spectroscopy

Reflectance-difference spectroscopy (RDS), also referred to as reflectance-anisotropy spectroscopy (RAS), measures the *difference* of normal-incidence reflectivity with polarizations oriented along two principal axes. In solids with optically isotropic bulk this technique is very surface sensitive, because light reflected in normal incidence from such a bulk can basically not depend on polarization, i.e., the light probes the anisotropy of the surface. This anisotropy originates essentially from surface reconstructions, which usually form dimer bonds along specific directions to minimize the total energy (Sects. 7.1.3 and 7.1.4). An anisotropic surface geometry leads to an anisotropic probability for optical transitions involving electronic surface states, and consequently to a response in the difference of respective RDS reflectivities.

The technique was introduced in the 1980s [68] and is widely applied for in situ monitoring of the epitaxial growth of cubic semiconductors, such as zincblende III–V and II–VI compound semiconductors; for reviews see [60, 69]. For zincblende semiconductors the two principal polarization axes x and y of a (001) surface are given by the $\left[\bar{1}10\right]$ and [110] directions. There are various configurations of RDS systems discussed in more detail in [60]; the schematic of a phase-modulated RDS setup is given in Fig. 8.16. The beam of a white light source is linearly polarized by a polarizer and directed onto the sample at near normal incidence. The reflected light is generally elliptically polarized; its state of polarization is analyzed by a photoelastic modulator with a high modulation frequency (typ. 50 kHz) and an analyzer, which converts the phase modulation generated by the modulator into an intensity modulation. This light is dispersed in a monochromator and detected.

The optical response of the sample can be described by introducing an anisotropic surface layer (overlayer) with a surface dielectric constant ε_{o} in addition to the bulk dielectric constant ε_{b}. The normal-incidence RDS response is then given by [70]

$$\frac{\Delta\tilde{r}}{\tilde{r}} = \frac{\tilde{r}_x - \tilde{r}_y}{(\tilde{r}_x + \tilde{r}_y)/2} = 2\pi i \frac{d_o}{\lambda/2} \frac{\tilde{\varepsilon}_{o,x} - \tilde{\varepsilon}_{o,y}}{\tilde{\varepsilon}_b - 1}, \tag{8.24}$$

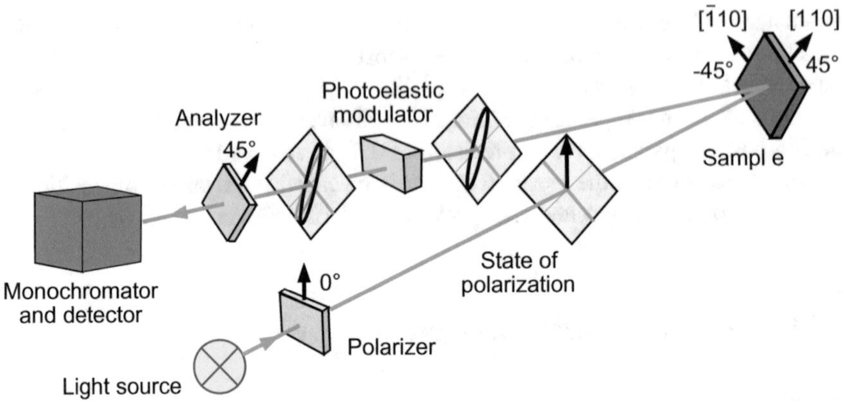

Fig. 8.16 Schematic of a setup for reflectance-difference spectroscopy on the (001) surface of a sample with cubic crystal structure

where i is the imaginary unit and $d_o \ll \lambda$ is the thickness of the anisotropic overlayer. The actually measured quantity is the real part of (8.24), i.e., for zincblende semiconductors $\mathrm{Re}\left[2\left(\tilde{r}_{[\bar{1}10]} - \tilde{r}_{[110]}\right)/\left(\tilde{r}_{[\bar{1}10]} + \tilde{r}_{[110]}\right)\right]$, which is often simply written $\Delta r/r$; the imaginary part is usually not measured. Typical RDS signals are on the order of 10^{-3}.

The interpretation of RDS spectra is complicated by the fact that the bulk also contributes to the RDS response. This occurs not only due to buried interfaces within the penetration depth of the light or due to imperfections such as an unbalanced density of α and β 60° dislocations and their related strain [71] or alloy ordering in the bulk [72]. Even in the ideal case transitions between surface-perturbed bulk states contribute strongly to RDS spectra near critical-point energies of the band structure as shown by calculated spectra given in Fig. 8.17. The spectra were obtained from DFT calculations in the local density approximation [73]; note that in this scheme the transition energies are underestimated due to neglected self-energy effects.

Contributions from bulk-related states to the RDS response appear in the spectra of Fig. 8.17 particularly near the E_1 and E_0'/E_2 critical points. Surface contributions are pronounced near 1.2 eV and near 3.4 eV, particularly for the c(4 × 4) reconstruction; these transitions are related to states of As dimers in this As-rich surface. Since layers in different depth beneath the surface provide different contributions to the spectra, also layer-by-layer decomposition of the RAS signal were computed to unravel the complex response [74].

Experimental RDS spectra of GaAs(001) surfaces for various reconstructions are given in Fig. 8.18. The spectra are typical fingerprints of the respective surfaces and do basically not depend on the growth method. Samples prepared using either MBE or MOVPE yield similar RDS response [75]; RHEED studies of MBE samples and comparative GIXD measurements on MOVPE samples reported in [76] confirmed the formation of similar well-ordered reconstructed surfaces in MOVPE despite the high ambient pressure.

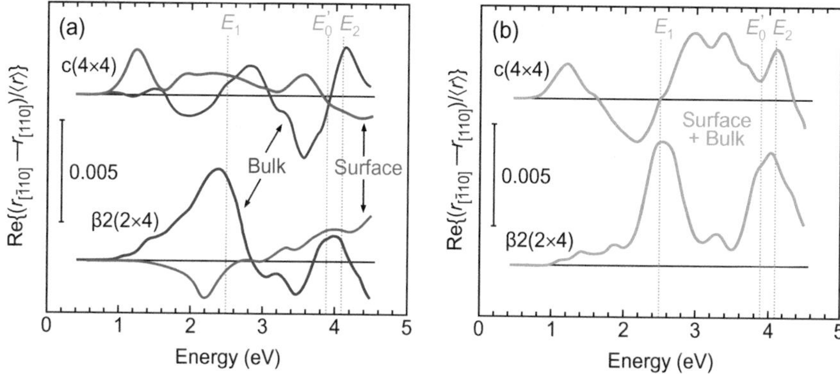

Fig. 8.17 RDS spectra of GaAs(001) with c(4 × 4) and β2(2 × 4) surface reconstructions, calculated using DFT-LDA. **a** Separate contributions of bulk and surface, **b** total RDS response. Dotted lines indicate transition energies of critical points. Reproduced with permission from [73], © 2001 Wiley

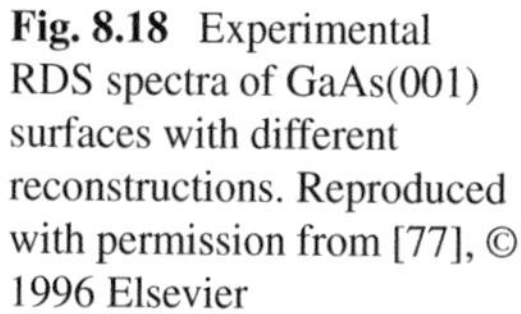
Fig. 8.18 Experimental RDS spectra of GaAs(001) surfaces with different reconstructions. Reproduced with permission from [77], © 1996 Elsevier

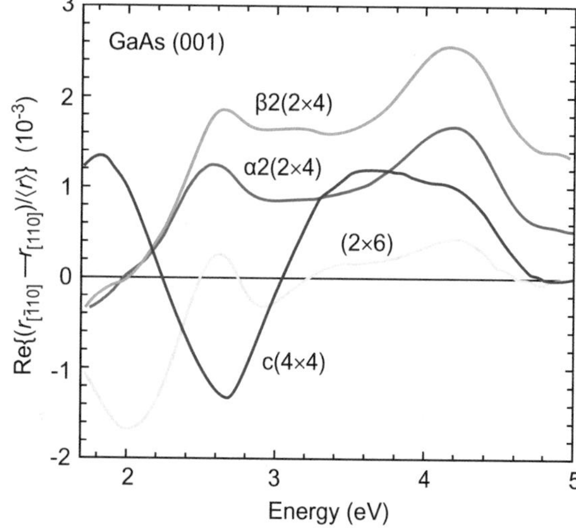

Today RDS is established as a valuable standard tool for monitoring metalorganic vapor-phase epitaxy, where RHEED is not applicable. Its excellent surface sensivity is widely employed to obtain information on the state of the surface, or simply to ensure the reproducibility of run-to-run conditions. Usually RDS transients at a selected fixed photon energy are recorded to trace the response during removal of the native oxide prior to deposition, to monitor the subsequent formation of a reconstructed surface [78], and to track the numerous subsequent process steps during epitaxy of heterostructures [79, 80]. Thereby layer thicknesses, compositions of alloyed layers and doping levels (typically those above mid 10^{17} cm^{-3}) can be determined already during growth of the semiconductor structures.

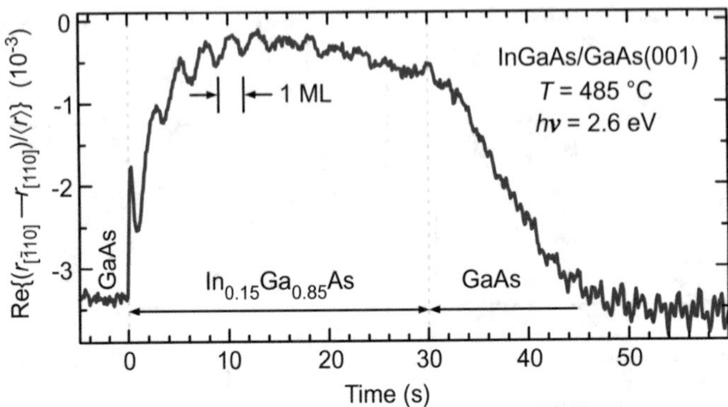

Fig. 8.19 Transient RDS response at 2.6 eV recorded during metalorganic vapor-phase epitaxy of $In_{0.15}Ga_{0.85}As$/GaAs(001) layers at a rate of 0.4 monolayers/second. Reproduced with permission from [82], © 2004 Elsevier

The high surface sensivity of RDS is proved by the obervation of monolayer oscillations during layer-by-layer island growth [42], analogous to the respective response of RHEED and GIXD discussed in Sects. 8.3.1 and 8.3.2. Oscillations of the RDS response during deposition of an InGaAs/GaAs layer are shown in Fig. 8.19. The Origin of RDS oscillations is less obvious than those of the morphology-related responses of RHEED and GIXD, which are due to long-range order; RDS studies for GaAs at two energies 2.6 and 1.9 eV, which according to calculations are respectively characteristic of dimers of As and Ga, indicate that oscillations are related to the periodic change of dimer orientations particularly near step edges [42]. More recent reports also emphasize the importance of the periodic modulation of orthorhombic surface strain associated to surface reconstruction [81].

Oscillations of RDS are generally well observed under low-temperatures conditions that are As-lean compared to standard growth conditions. But even at typical growth conditions the RDS response of nanometer-thick layers may provide valuable informations; without monolayer-oscillations the response recorded during quantum-well growth could be well related to the emission energy of this quantum structure [79].

8.5 Problems Chapter 8

8.1 A heated and not stabilized c-plane GaN surface decomposes in vacuum. Losses of 3.3 monolayers/minute (ML/min) at 760 °C sample temperature and 6.5 ML/min at 780 °C are measured by the desorbing Ga flux (≡ GaN flux) using calibrated mass spectrometry. Note: 1 ML corresponds to $c/2$ of GaN.

(a) Determine the activation energy E_d resulting from this measurement, if the thermally activated desorption follows an Arrhenius dependence.

(b) Calculate the Ga flux (expressed in ML/min), which is expected at 800 °C.
(c) How many Ga atoms per cm^2 of the surface desorb in one minute at 780 °C?

8.2 The electron gun of a RHEED setup is operated with 10 kV.

(a) Calculate the fraction of relativistic correction of the electron momentum with respect to the classical calculation. How large is this fraction at 50 kV?
(b) How many times larger is the Ewald sphere of the electrons at 10 kV compared to the unit cell of GaAs?

8.3 The calibration of a pyrometer yields a signal of 1.00 nA at the surface temperature of 577 °C with a real reflectance of 0.89 at a wavelength of 930 nm.

(a) Determine the pyrometer signal at 800 °C, if the reflectance did not change. To which temperature corresponds this pyrometer signal, if the reflectance changed to 0.85? Calculate the actual pyrometer signal for 800 °C and a reflectance of 0.85.
(b) Find the temperature of the surface, if a pyrometer signal of 1051.8 nA and a reflectance of 0.70 are measured.

8.4 A 2 μm thick (0001) oriented GaN layer is grown at 1050 °C on 0.45 mm thick basal plane sapphire. A (negligibly thin) properly recrystallized GaN nucleation layer on sapphire has provided an effective substrate for the GaN epitaxy with a relaxed lateral GaN lattice constant at 1050 °C but with elastic properties of sapphire; these are given by a thermal expansion coefficient of 8.2×10^{-6} K^{-1} and an in-plane biaxial elastic modulus of 607 GPa.

(a) Find the thermally induced lattice mismatch between the effective substrate and the GaN layer, when the temperature is lowered to 620 °C (a typical temperature for growing InGaN on top).
(b) Calculate the biaxial elastic modulus of the GaN layer. How large is the stress in the layer at 620 °C?
(c) Determine the curvature of the wafer resulting from the stressed layer. Calculate the corresponding curvature radius.

8.5 The real reflectance recorded during growth of a layer is described by the fit function $R \cong 0.4 + 0.2 \times \exp(-1.8 \times 10^{-3} s^{-1} t) \times \cos(2.0 \times 10^{-2} s^{-1} t)$.

(a) Find the reflectance measured at $t = 0$ and after 2.5 and 10 min according to the fit. Calculate the reflectance attained at an infinite thick layer.
(b) Determine the approximate the period of the oscillating reflectance. When is the amplitude of the oscillation reduced to 10% of its initial value?
(c) Calculate approximated values of the real reflective index and the extinction coefficient of the layer.
(d) Determine the growth rate for a recording wavelength of 830 nm.

8.6 General Reading Chapter 8

G. Bauer, W. Richter, *Optical Characterization of Epitaxial Semiconductor Layers* (Springer, Berlin, 1996)
M.A. Hermann, W. Richter, H. Sitter, *Epitaxy* (Springer, Berlin, 2004)

References

1. M.P. Seah, W.A. Dench, Quantitative electron spectroscopy of surfaces: a standard data base for electron inelastic mean free paths in solids. Surf. Interface Anal. **1**, 2 (1979)
2. W. Richter, D. Zahn, Analysis of epitaxial growth, in *Optical Characterization of Epitaxial Semiconductor Layers*, ed. by G. Bauer, W. Richter (Springer, Berlin, 1996), pp. 12–67
3. A.S.Y. Chan, M.P. Skegg, R.G. Jones, Line of sight techniques: providing an inventory of all species arriving at and departing from a surface. J. Vac. Sci. Technol., A **19**, 2007 (2001)
4. S.G. Hessey, R.G. Jones, Line-of-sight mass spectrometry: principles and practice. Surf. Interface Anal. **47**, 587 (2015)
5. S. Fernández-Garrido, G. Koblmüller, E. Calleja, J.S. Speck, In situ GaN decomposition analysis by quadrupole mass spectrometry and reflection high-energy electron diffraction. J. Appl. Phys. **104**, 033541 (2008)
6. G. Koblmüller, R. Averbeck, H. Riechert, P. Pongratz, Direct observation of different equilibrium Ga adlayer coverages and their desorption kinetics on GaN (0001) and (000$\bar{1}$) surfaces. Phys. Rev. B **69**, 035325 (2004)
7. J.E. Northrup, J. Neugebauer, R.M. Feenstra, A.R. Smith, Structure of GaN(0001): the laterally contracted Ga bilayer model. Phys. Rev. B **61**, 9932 (2000)
8. J. Neugebauer, T.K. Zywietz, M. Scheffler, J.E. Northrup, H. Chen, R.M. Feenstra, Adatom kinetics on and below the surface: the existence of a new diffusion channel. Phys. Rev. Lett. **90**, 056101 (2003)
9. M. Yoshida, H. Watanabe, F. Uesugi, Mass spectrometric study of $Ga(CH_3)_3$ and $Ga(C_2H_5)_3$ decomposition reaction in H_2 and N_2. J. Electrochem. Soc. **132**, 677 (1985)
10. L. Nattermann, O. Maßmeyer, E. Sterzer, V. Derpmann, H.Y. Chung, W. Stolz, K. Volz, An experimental approach for real time mass spectrometric CVD gas phase investigations. Sci. Rep. **8**, 319 (2018)
11. P.W. Lee, T.R. Omstead, D.R. McKenna, K.F. Jensen, In situ mass spectroscopy and thermogravimetric studies of GaAs MOCVD gas phase and surface reactions. J. Cryst. Growth **85**, 165 (1987)
12. G.B. Stringfellow, Alternate sources and growth chemistry for OMVPE and CBE processes. J. Cryst. Growth **105**, 260 (1990)
13. G.B. Stringfellow, *Organometallic Vapor-Phase Epitaxy*, 2nd edn. (Academic Press, San Diego, 1999)
14. S.H. Li, C.A. Larsen, N.I. Buchan, G.B. Stringfellow, Pyrolysis of tertiarybutylphosphine. J. Electron. Mater. **18**, 457 (1989)
15. U.W. Pohl, C. Möller, K. Knorr, W. Richter, J. Gottfriedsen, H. Schumann, A. Fielicke, K. Rademann, Tertiarybutylhydrazine: a new precursor for the MOVPE of III nitrides. Mater. Sci. Engin. B **59**, 20 (1999)
16. C.A. Larsen, S.H. Li, N.J. Buchan, G.B. Stringfellow, Mechanisms of GaAs growth using tertiarybutylarsine and trimethylgallium. J. Cryst. Growth **94**, 673 (1989)
17. A. Stegmüller, R.A. Tonner, Quantum chemical descriptor for CVD precursor design: predicting decomposition rates of TBP and TBAs isomers and derivatives. Chem. Vap. Depos. **21**, 161 (2015)

18. F. Durst, A. Melling, J.H. Whitelaw, *Principles and Practice of Laser-Doppler Anemometry*, 2nd edn. (Academic Press, London, 1981)
19. G.A. Hebner, K.P. Killeen, R.M. Biefeld, In situ measurement of the metalorganic and hydride partial pressures in a MOCVD reactor using ultraviolet absorption spectroscopy. J. Cryst. Growth **98**, 293 (1989)
20. H. Itoh, M. Watanabe, S. Mukai, H. Yajima, Ultraviolet absorption spectra of metalorganic molecules diluted in hydrogen gas. J. Cryst. Growth **93**, 165 (1988)
21. R.F. Karlicek, B. Hammarlund, J. Ginocchio, UV absorption spectroscopy for monitoring hydride vapor-phase epitaxy of InGaAsP alloys. J. Appl. Phys. **60**, 794 (1986)
22. M.C. Johnson, K. Poochinda, N.L. Ricker, J.W. Rogers Jr., T.P. Pearsall, In situ monitoring and control of multicomponent gas-phase streams for growth of GaN via MOCVD. J. Cryst. Growth **212**, 11 (2000)
23. J.E. Butler, N. Bottka, R.S. Sillmon, D.K. Gaskill, In situ, real-time diagnostics of OMVPE using IR-diode laser spectroscopy. J. Cryst. Growth **77**, 163 (1986)
24. S. Salim, C.A. Wang, R.D. Driver, K.F. Jensen, In situ concentration monitoring in a vertical OMVPE reactor by fiber-optics-based Fourier transform infrared spectroscopy. J. Cryst. Growth **169**, 443 (1996)
25. S.P. Watkins, T. Pinnington, J. Hu, P. Yeo, M. Kluth, N.J. Mason, R.J. Nicholas, P.J. Walker, Infrared single wavelength gas composition monitoring for metalorganic vapour-phase epitaxy. J. Cryst. Growth **221**, 166 (2000)
26. D. Hayashi, A. Teraoka, Y. Sakaguchi, M. Minami, H. Nishizato, Real-time measurement of Cp_2Mg vapor concentration using the non-dispersive infrared spectroscopy. J. Cryst. Growth **453**, 54 (2016)
27. C. Park, W.-S. Jung, Z. Huang, T.J. Anderson, In situ Raman spectroscopic studies of trimethylindium pyrolysis in an OMVPE reactor. J. Mater. Chem. **12**, 356 (2002)
28. I.P. Herman, *Optical Diagnostics for Thin Film Processing*, Chapter 7 Laser-induced fluorescence (Academic Press, New York 1996)
29. B. Zhou, X. Li, T.L. Tansley, K.S.A. Butcher, Growth mechanisms in excimer laser photolytic deposition of gallium nitride at 500 °C. J. Cryst. Growth **160**, 201 (1996)
30. V.M. Donnelly, R.F. Karlicek, Development of laser diagnostic probes for chemical vapor deposition of InP/InGaAsP epitaxial layers. J. Applied Physics **53**, 6399 (1982)
31. M.A. Van Hove, W.H. Weinberg, C.-M. Chan, *Low-Energy Electron Diffraction* (Springer, Berlin 1986)
32. A. Ichimiya, P.I. Cohen, *Reflection High Energy Electron Diffraction* (Cambridge University Press, Campridge, 2004)
33. W. Braun, *Applied RHEED, Reflection High-Energy Electron Diffraction During Crystal Growth* (Springer, Berlin, 1999)
34. Y. Ma, S. Lordi, J.A. Eades, Dynamical analysis of a RHEED pattern from the Si(111)-7 $\times$ 7 surface. Surf. Sci. **313**, 317 (1994)
35. Y. Nabetani, T. Ishikawa, S. Noda, A. Sasaki, Initial growth stage and optical properties of a three-dimensional InAs structure on GaAs. J. Appl. Phys. **76**, 347 (1994)
36. T. Sakamoto, H. Funabashi, K. Ohta, T. Nakagawa, N.J. Kawai, T. Kojima, Y. Bando, Well defined superlattice structures made by phase-locked epitaxy using RHEED intensity oscillations. Superlattices Microstruct. **1**, 347 (1985)
37. D.W. Kisker, G.B. Stephenson, P.H. Fuoss, S. Brennan, Characterization of vapor phase growth using X-ray techniques. J. Cryst. Growth **146**, 104 (1995)
38. G. Ju, M.J. Highland, A. Yanguas-Gil, C. Thompson, J.A. Eastman, H. Zhou, S.M. Brennan, G.B. Stephenson, P.H. Fuoss, An instrument for in situ coherent X-ray studies of metal-organic vapor phase epitaxy of III-nitrides. Rev. Sci. Instrum. **88**, 035113 (2017)
39. B. Jenichen, W. Braun, V.M. Kaganer, A.G. Shtukenberg, L. Däweritz, C.-G. Schulz, K.H. Ploog, A. Erko, Combined molecular beam epitaxy and diffractometer system for in situ X-ray studies of crystal growth. Rev. Sci. Instrum. **74**, 1267 (2003)
40. G. Ju, S. Fuchi, M. Tabuchi, H. Amano, Y. Takeda, Continuous in situ X-ray reflectivity investigation on epitaxial growth of InGaN by metalorganic vapor phase epitaxy. J. Cryst. Growth **407**, 68 (2014)

41. G. Ju, M. Tabuchi, Y. Takeda, H. Amano, Role of threading dislocations in strain relaxation during GaInN growth monitored by real-time X-ray reflectivity. Appl. Phys. Lett. **110**, 262105 (2017)
42. I. Kamiya, L. Mantese, D.E. Aspnes, D.W. Kisker, P.H. Fuoss, G.B. Stephenson, S. Brennan, Optical characterization of surfaces during epitaxial growth using RDS and GIXS. J. Cryst. Growth **163**, 67 (1996)
43. W. Braun, B. Jenichen, V.M. Kaganer, A.G. Shtukenberg, L. Däweritz, K.H. Ploog, Layer-by-layer growth of GaAs(001) studied by in situ synchrotron X-ray diffraction. Surf. Sci. **525**, 126 (2003)
44. D.W. Kisker, G.B. Stephenson, P.H. Fuoss, F.J. Lamelas, S. Brennan, Atomic scale characterization of organometallic vapor phase epitaxial growth using in-situ grazing incidence X-ray scattering. J. Cryst. Growth **124**, 1 (1992)
45. D.W. Kisker, G.B. Stephenson, J. Tersoff, P.H. Fuoss, S. Brennan, Atomic scale studies of epitaxial growth processes using X-ray techniques. J. Cryst. Growth **163**, 54 (1996)
46. W.G. Breiland, Reflectance-correcting pyrometry in thin film deposition applications, Technical Report, Sandia National Laboratories, Albuquerque, NM 87185, SAND2003-1868
47. K. Haberland, J.T. Mullins, T. Schenk, T. Trepk, L. Considine, A. Pakes, A. Taylor, J.-T. Zettler, First real-time true wafer temperature and growth rate measurements in a closed-coupled showerhead MOVPE reactor during growth of InGa(AsP). In International Conference on InP and Related Materials. Santa Barbara, CA, USA (2003)
48. J.A. Floro, E. Chason, S.R. Lee, Real time measurement of epilayer strain using a simplified wafer curvature technique. Mat. Res. Soc. Symp. Proc. **405**, 381 (1996)
49. S. Terao, M. Iwaya, R. Nakamura, S. Kamiyama, H. Amano, I. Akasaki, Fracture of $Al_xGa_{1-x}N$/GaN heterostructure—compositional and impurity dependence. Jpn. J. Appl. Phys. **40**, L195 (2001)
50. A. Krost, A. Dadgar, G. Strassburger, R. Clos, GaN-based epitaxy on silicon: stress measurements. Phys. Stat. Sol. A **200**, 26 (2003)
51. F. Brunner, V. Hoffmann, A. Knauer, E. Steimetz, T. Schenk, J.-T. Zettler, M. Weyers, Growth optimization during III-nitride multiwafer MOVPE using realtime curvature, reflectance and true temperature measurements. J. Cryst. Growth **298**, 202 (2007)
52. G. Stoney, The tension of metallic films deposited by electrolysis. Proc. R. Soc. London A **82**, 172 (1909)
53. L.B. Freund, J.A. Floro, E. Chason, Extensions of the Stoney formula for substrate curvature to configurations with thin substrates or large deformations. Appl. Phys. Lett. **74**, 1987 (1999)
54. C.A. Klein, How accurate are Stoney's equation and recent modifications. J. Appl. Phys. **88**, 5487 (2000)
55. C. Kisielowski, J. Krüger, S. Ruvimov, T. Suski, J.W. Ager III, E. Jones, Z. Liliental-Weber, M. Rubin, E.R. Weber, M.D. Bremser, R.F. Davis, Strain-related phenomena in GaN thin films. Phys. Rev. B **54**, 17745 (1996)
56. F. Brunner, A. Knauer, T. Schenk, M. Weyers, J.-T. Zettler, Quantitative analysis of in situ wafer bowing measurements for III-nitride growth on sapphire. J. Cryst. Growth **310**, 2432 (2008)
57. S. Hearne, E. Chason, J. Han, J.A. Floro, J. Figiel, J. Hunter, H. Amano, I.S.T. Tsong, Stress evolution during metalorganic chemical vapor deposition of GaN. Appl. Phys. Lett. **74**, 356 (1999)
58. D.E. Aspnes, Minimal-data approaches for determining outer-layer dielectric responses of films from kinetic reflectometric and ellipsometric measurements. J. Opt. Soc. Am. A **10**, 974 (1993)
59. D.E. Aspnes, Optical approaches to determine near-surface compositions during epitaxy. J. Vac. Sci. Technol., A **14**, 960 (1996)
60. J.-T. Zettler, Characterization of epitaxial semiconductor growth by reflectance anisotropy spectroscopy and ellipsometry. Prog. Cryst. Growth Charact. **35**, 27 (1997)
61. F.K. Urban III, M.F. Tabet, Virtual interface method for in situ ellipsometry of films grown on unknown substrates. J. Vac. Sci. Technol., A **11**, 976 (1993)

62. W.G. Breiland, K.P. Killeen, A virtual interface method for extracting growth rates and high temperature optical constants from thin semiconductor films using in situ normal incidence reflectance. J. Appl. Phys. **78**, 6726 (1995)
63. R.M.A. Azzam, N.M. Bashara, *Ellipsometty and Polarized Light* (North Holland, New York, 1987)
64. W.G. Breiland, H.Q. Hou, H.C. Chui, B.E. Hammons, In situ pre-growth calibration using reflectance as a control strategy for MOCVD fabrication of device structures. J. Cryst. Growth **174**, 564 (1997)
65. C. Pickering, In situ optical studies of epitaxial growth, in *Handbook of Crystal Growth 3, Part B: Thin Films and Epitaxy, Growth Mechanisms and Dynamics*, ed. by D. Hurle (Elsevier, Amsterdam, 1994), pp. 819–878
66. R.W. Collins, I. An, H. Nguyen, Y. Li, Y. Lu, Realtime spectroscopic ellipsometry studies of the nucleation, growth, and optical functions of thin films, Part I: Tetrahedrally bonded materials, in *Optical Characterization of Real Surfaces and Films*, ed. by K. Vedam (Academic Press, Orlando, 1994), pp. 49–125
67. D.E. Aspnes, W.E. Quinn, S. Gregory, Application of ellipsometry to crystal growth by organometallic molecular beam epitaxy. Appl. Phys. Lett. **56**, 2569 (1990)
68. D.E. Aspnes, Above-bandgap optical anisotropies in cubic semiconductors: a visible—near ultraviolet probe of surfaces. J. Vacuum Sci. Technol. B **3**, 1498 (1985)
69. W.G. Schmidt, Calculation of reflectance anisotropy for semiconductor surface exploration. Phys. Stat. Sol. B **242**, 2751 (2005)
70. K. Hingerl, D.E. Aspnes, I. Kamiya, L.T. Florez, Relationship among reflectance-difference spectroscopy, surface photoabsorption, and spectroellipsometry. Appl. Phys. Lett. **63**, 885 (1993)
71. L.F. Lastras-Martínez, A. Lastras-Martínez, Reflectance anisotropy of GaAs(100): Dislocation-induced piezo-optic effects. Phys. Rev. B **54**, 10726 (1996)
72. M. Zorn, P. Kurpas, A.I. Shkrebtii, B. Junno, A. Bhattacharya, K. Knorr, M. Weyers, L. Samuelson, J.T. Zettler, W. Richter, Correlation of InGaP(001) surface structure during growth and bulk ordering. Phys. Rev. B **60**, 8185 (1999)
73. W.G. Schmidt, F. Bechstedt, K. Fleischer, C. Cobet, N. Esser, W. Richter, J. Bernholc, G. Onida, GaAs(001): Surface structure and optical properties. Phys. Stat. Sol A **188**, 1401 (2001)
74. C. Hogan, R. Del Sole, Optical properties of the GaAs(001)-c(4×4) surface: direct analysis of the surface dielectric function. Phys. Stat. Sol B **242**, 3040 (2005)
75. I. Kamiya, D.E. Aspnes, H. Tanaka, L.T. Florez, J.P. Harbison, R. Bhat, Surface science at atmospheric pressure: Reconstructions on (001) GaAs in organometallic chemical vapor deposition. Phys. Rev. Lett. **68**, 627 (1992)
76. D.W. Kisker, G.B. Stephenson, I. Kamiya, P.H. Fuoss, D.E. Aspnes, L. Mantese, S. Brennan, Investigation of the relationship between reflectance difference spectroscopy and surface structure using grazing incidence X-ray scattering. Phys. Stat. Sol. A **152**, 9 (1995)
77. W. Richter, J.-T. Zettler, Real-time analysis of III-V-semiconductor epitaxial growth, Appl. Surf. Sci. **100/101**, 465 (1996)
78. C. Kaspari, M. Pristovsek, W. Richter, Deoxidation of (001) III–V semiconductors in metal-organic vapour phase epitaxy. J. Appl. Phys. **120**, 085701 (2016)
79. M. Zorn, M. Weyers, Application of reflectance anisotropy spectroscopy to laser diode growth in MOVPE. J. Cryst. Growth **276**, 29 (2005)
80. N.A. Kalyuzhnyy, V.V. Evstropov, V.M. Lantratov, S.A. Mintairov, M.A. Mintairov, A.S. Gudovskikh, A. Luque, V.M. Andreev, Characterization of the manufacturing processes to grow triple-junction solar cells, Int. J. Photoenergy **2014**, 836284 (2014)
81. J. Ortega-Gallegos, L.E. Guevara-Macías, A.D. Ariza-Flores, R. Castro-García, L.F. Lastras-Martínez, R.E. Balderas-Navarro, R.E. López-Estopier, A. Lastras-Martínez, On the origin of reflectance-anisotropy oscillations during GaAs (001) homoepitaxy. Appl. Surf. Sci. **439**, 963 (2018)
82. U.W. Pohl, K. Pötschke, I. Kaiander, J.-T. Zettler, D. Bimberg, Real-time control of quantum dot laser growth using reflectance anisotropy spectroscopy. J. Crystal Growth **272**, 143 (2004)

Chapter 9
Application of Surfactants

Abstract Surface-active species referred to as surfactants modify the surface energy in semiconductor epitaxy. Applying an appropriate combination of surfactant and semiconductor materials, the three-dimensional island growth occurring during heteroepitaxy can be suppressed, and the layer grows smoothly in a two-dimensional layer-by-layer mode. In this process the surfactant layer floats on the semiconductor layer during growth without significant incorporation. The effect is well established for many layer/substrate materials and particularly well studied for surfactant-mediated epitaxy of Ge layers on Si substrate: two-dimensional growth is maintained far beyond the critical thickness for the usually observed Stranski–Krastanow island formation with Sb or Bi acting as surfactant. The surfactant is found to substantially alter adatom diffusivity and nucleation rate. In addition, different processes for strain relaxation in the epitaxial layer are found. Similar findings are obtained with compound semiconductors. Both thermodynamic and prevailing kinetic effects account for the experimental observations. Generally accepted models consider a kinetically retarded approach to an equilibrium state of the layer comprising three-dimensional islands. Modelling of adatom diffusion processes on the bare surface without surfactant and on the surfactant-covered surface indicate the prominent role of the passivation of step edges at nucleating monolayer islands. The passivation is connected to a suppression of de-excitation processes with atoms below the surfactant layer reexchanging with surfactant atoms. Exchange pathways of adatoms and surfactant atoms with low activation energy were identified for specific surfaces and participating species using ab initio calculations.

9.1 The Surfactant Effect

9.1.1 *Concept of Surfactant-Mediated Growth*

Most device structures grown by epitaxy need sharp interfaces, a requirement usually met by growing with a smooth surface in a layer-by-layer growth mode (Frank–Van der Merve mode in heteroepitaxy). Such two-dimensional (2D) growth is often prevented in the epitaxy of a layer A on another material B (referred to here as

U. W. Pohl, *Epitaxy of Semiconductors*, Graduate Texts in Physics,
https://doi.org/10.1007/978-3-030-43869-2_9

substrate): if the surface energy of *B* is smaller than the sum of the surface energy of *A* and the interface energy of *A*/*B*, the layer does not wet the substate surface. This leads to 3D island growth in the Volmer–Weber mode pointed out in Sect. 6.2.3. Moreover, layer *A* is generally strained by structural or thermal mismatch to *B*. Thus, even if a wetting layer is formed, strain easily induces the 3D Stranski–Krastanow island growth as shown in Figs. 6.12 and 7.39.

These conditions leads to a fundamental obstacle for the growth of a *B*/*A*/*B* heterostructure such as a quantum well. One of the two solids *A* or *B* must have a lower surface energy. We disregard here the interface energy, since it refers to both *A*/*B* and *B*/*A* structures. If *A* with a lower surface energy grows on *B* in the Frank–Van der Merve or Stranski–Krastanow mode, then *B* tends to grow on *A* in the Volmer–Weber mode [1]. Hence, when an embedded layer is grown, either the growth mode of this layer or that of the capping layer tends to Volmer–Weber, i.e., a 3D growth with immediate islanding leading to a rough interface. Any growth of a quantum well or a superlattice must overcome this limitation. This is generally achieved by limiting the growth kinetics, using a decreased growth temperature or an increased growth rate. Since such means may also reduce the epitaxial quality, an alternative is desirable.

Early investigations indicated that the growth mode can be altered in the presence of impurities, see, e.g., [2]. Such an effect of contaminants was observed with many layers *A* on substrates *B* and discussed in an early review in terms of metastable layers [3].

Surfactant-mediated growth implies a controlled modification of the surface energy. A third material *C* is introduced which lowers the surface energy of both *A* and *B*: a thin film of a surfactant *C* is deposited on *B* before growing layer *A* on *B*. With an appropriate choice of material *C* it will float on top of *A* during growth without significant incorporation into layer *A*. Growth of *A* may now proceed in the layer-by-layer (Frank–Van der Merve) mode. This applies as well for the subsequent deposition of material *B* on layer *A*. By using a surfactant the formation of 3D islands can be avoided, and a 2D layer-by-layer growth can be maintained for layer thicknesses much beyond those obtained without the surfactant.

It must be noted that the *strain* introduced in the heteroepitaxy is still present, even if the 3D growth is suppressed. Similar to the growth without surfactant the strain is relieved by misfit dislocations for layers exceeding a critical thickness. The mechanism and the kind of misfit introduction may, however, differ significantly.

The term surfactant signifies the application of a *surface-active species*. The term is widely used in semiconductor growth, but its meaning differs substantially from that commonly used in chemistry. For semiconductors it describes the effect of an adsorbate layer on the growth of a layer deposited on a substrate, while in chemistry it usually denotes a material which improves emulsifying, dispersing, wetting, or other surface properties of liquids.

The surfactant effect is also found in the epitaxy of metals on metals [4, 5]. *B*/*A*/*B*-type layer structures with a monolayer (ML) of a high-surface-energy transition metal (Ni, Fe, Co) and a low-surface-energy noble metal (Au, Ag, Cu) are basic building blocks for a broad category of magnetic multilayers; these tend to grow with rough and intermixed interfaces, but smoothly in a layer-by-layer mode with the addition of a surfactant like Pb or Bi. Still the effect differs from that observed for semiconductors. In metal growth a small fraction of an adsorbate monolayer may change the surface kinetics, leading to modified nucleation rates and step-edge barriers [6]. In semiconductor growth typically an entire monolayer is required to produce the surfactant effect. This chapter focuses on the effect observed for semiconductors.

9.1.2 Evidence for the Surfactant Effect

The surfactant effect was demonstrated for many different semiconductor heterostructures. Most studies were made with group-IV layers on group-IV substrates, particularly with Ge on Si; in this system the investigations focused on (001) and (111) surfaces and employed various surfactants, mainly group-III elements (Ga, In) or group-V elements (As, Sb, Bi).

Ge and Si both crystallize in diamond structure, with the lattice constant of Ge (5.6576 Å) exceeding that of Si by 4.2%, see Sect. 2.1. Epitaxial growth of Ge/Si(001) proceeds in the Stranski–Krastanow mode (Figs. 7.39a and 9.1a) with a typically ~3 monolayers thick wetting layer [7] and a maximum thickness of 2D layer-by-layer growth of six monolayers [8]. In the 3D Stranski–Krastanow growth coalescence of the islands produces inevitably a highly defective Ge layer. Si grows on Ge and on Ge/Si in the 3D Volmer–Weber mode [9]. Growth of a Si/Ge/Si quantum well structures hence shows strong islanding and interdiffusion effects [10].

Growth of Ge on Si and Si on Ge/Si was found to be completely altered, if one monolayer of As is deposited on a Si(001) surface prior to the Ge deposition [1, 11]. With As as a surfactant, thick Ge layers can be grown in a clear layer-by-layer mode, see Figs. 9.1b and 9.2. Furthermore, Si deposited on a thin pseudomorphic Ge/Si(001) layer—and hence not undergoing significant strain—is also found to grow

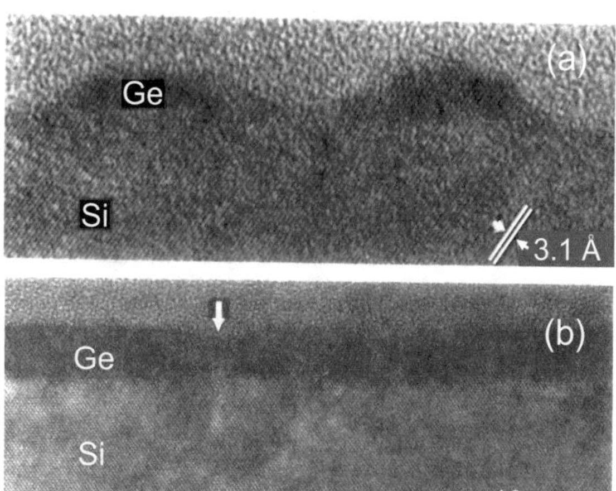

Fig. 9.1 **a** Cross sectional TEM micrograph of 8 ML Ge/Si(001) grown without surfactant; Ge forms pseudomorphic Stranski–Krastanow islands. **b** 50 ML Ge/Si(001) grown in a layer-by-layer mode with 1 ML As surfactant; the *arrow* marks a V-shaped defect. Reproduced with permission from [12], © 1990 APS

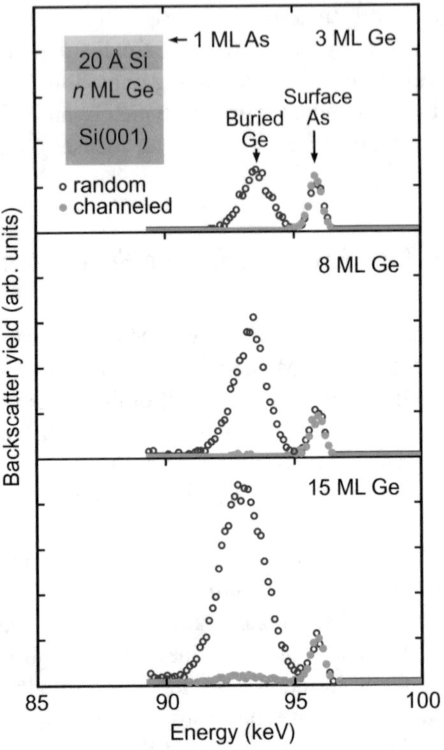

Fig. 9.2 He^+ ion backscattering spectra of Ge layers on Si(001) substrate, buried underneath a 20 Å thick Si cap layer; one monolayer of As was deposited before growth of the Ge layer and floates on the surface during Ge and Si depositions. *Blue circles* denote randomly scattered He^+ ions, *green dots* denote back-scattered channeled ions of 100 keV He^+ incident in the $[11\bar{1}]$ direction. Reproduced with permission from [11], © 1990 APS

as a smooth layer in the presence of As, while clear indications for island formation is observed if no As is layer is applied. In this study no As was incorporated in the Ge or Si layers within the resolution of 1%; instead, As was found to constantly segregate to the surface during growth [11].

The findings summarized above were first concluded from ion backscattering spectra given in Fig. 9.2, accompanied by photoemission and electron microscopy studies [11]. In these ion backscattering experiments He^+ ions accelerated to 100 keV were either impinged along the $[11\bar{1}]$ direction of the sample—leading to channeling—or some degrees off for random spectra; the energy of backscattered ions was analyzed with high resolution. The spectra for Ge layers embedded in Si show a peak at 96 keV due to backscattering from As at the surface, and a lower-energy peak for the buried Ge. The Ge peak increases for thicker Ge layers but remains at lower energy; the continuous increase of signal without saturation at thicknesses well above Stranski–Krastanow island formation indicates that no such islands formed, and the persisting lower energy of the peak indicates a continuous Si layer on top [11]. The strongly diminished channeling signal of the Ge peak results from the overlying Si cap and indicates that both Ge and Si are epitaxial. When islands are formed (typically after depositing 3 ML Ge), the ratio of channeled to random signals increases significantly [8]; this is not observed with As surfactant. Channeling and random spectra of the As peak are identical, because there is no shadowing cap layer on top. The conclusions drawn from the backscattering spectra were confirmed by electron micrographs and by core-level photoemission spectra, where the As-related lines did not diminish as the Ge layer thickness increases [11].

9.1.3 Surfactant-Mediated Ge/Si Epitaxy

The surfactant effect of Ge/Si on a (001) substrate described above is also found for [111] oriented heterostructures and for many other surfactant elements. These are elements of group I (H), III (Ga, In), IV (Sn, Pb), V (As, Sb, Bi), VI (Te), or noble metals with a low surface energy such as Au. A comprehensive review is given in [13].

The initial publication of surfactant-mediated Ge/Si(001) growth [1] was followed by many experimental studies with Ge/Si confirming the early results and providing indications for the origin of the pronounced effects. In a simple view it might be assumed that a surfactant passivates the surface by saturating dangling bonds, leading to an increased diffusion length of adatoms. Such enhanced adatom diffusivity was observed for group III and IV surfactants with a consequential 3D islanding in Ge/Si epitaxy, while group V and VI surfactants reduce the diffusivity with a suppression of islanding.

The diffusion length can be determined by studying the surface morphology: near the edge of a growing island or that of an advancing step no stable nuclei can be formed, because the adatom density is reduced due to incorporation at the edge (denuded zone, see Sect. 7.2.3). The mean distance between 2D islands or likewise the width of the depleted zone at step edges is hence a measure for the diffusion length, and for 2D island growth this distance R is related to the areal density of islands N by $R = 1/\sqrt{N}$.

Based on this context the diffusion length of adatoms in the homoepitaxy of Si/Si(111) and changes in the respective surfactant-mediated homoepitaxy was investigated by scanning tunneling microscopy (STM) [14]. Figure 9.3a shows the effect of the surfactants on the density of 2D islands and the related island distance for a submonolayer Si coverage; at increased substrate temperature the density decreases and the distance increases due to a higher adatom diffusivity in all studied cases. The dependence is well described by an Arrhenius law, yielding various barriers for adatom diffusion deviating from the Si/Si reference. A similar behavior is obtained from the width of the denuded zone depleted of islands at step edges shown in Fig. 9.3b.

These results lead to a factor g for the adatom diffusion-length in presence of a surfactant with respect to that of the pure Si/Si epitaxy at a given temperature. A factor $g \geq 1$ is found for Ga, In (group III), and Sn (group IV). The increased diffusion length supports lateral mass transport; such transport is required for the formation of large Stranski–Krastanow islands in the heteroepitaxy of Ge/Si. From the similar chemical nature of Si and Ge also a favored adatom diffusion of Ge on Si

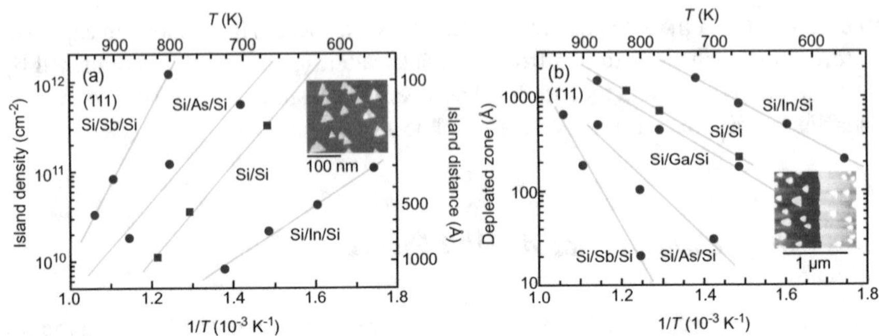

Fig. 9.3 **a** Density of 2D islands in the homoepitaxy of Si on Si(111) grown with and without surfactants as indicated at the curves. Inset: STM image of Si/Si grown at 500 °C. **b** Width of depleated zone at the step edge in Si on Si(111) epitaxy. Inset: STM image of Si/Ga/Si grown at 630 °C. Reproduced with permission from [14], © 1995 APS

was suggested, and in fact for these surfactants island formation is observed in Ge/Si epitaxy. On the other hand $g < 1$ is found for Sb and As (group V) surfactants, and the 3D island formation for Ge/Si is suppressed [14]. This led to the conclusion that the surfactant-mediated layer-by-layer growth originates from a kinetically reduced mass transport; in fact, in the surfactant-induced 2D growth of Ge/Sb/Si(111) a temperature increase leads to the reappearance of 3D growth [15, 16].

An alternative interpretation of the experimental findings was suggested by considering a passivation effect of a surfactant on the island edges [17]; such a passivation was actually confirmed for As-covered Si(111) [18]. If a surfactant passivates the island edges, they do not act as sinks for adatoms diffusing on the surface (Sect. 7.2.3). Consequently the width of depleted zones is reduced. We consider this model in more detail below in Sect. 9.2.3.

Strain relaxation is required also in the surfactant-mediated 2D heteroepitaxy. The relief of strain in the *pure* Ge/Si epitaxy with the formation of 3D Stranski–Krastanow islands differs for growth on (111) and (001) surfaces [19]. On both surfaces the 2D → 3D transition is preceded by the formation of surface structures with a decreased layer density and a related partial strain relaxation, and after island formation strain is elastically more efficiently relieved by a bending of the lattice in the islands; misfit dislocations are introduced from island edges when the thickness is increased and when islands coalesce (at ~100 ML Ge coverage with ~10^{12} cm^{-2} threading dislocations). If the elastic 3D relaxation mechanism is disabled by a surfactant-mediated 2D growth, misfit dislocations must accommodate the strain accumulated in the 2D layers.

Antimony is a preferred surfactant for the Ge/Si(111) epitaxy. Sb forms on Si(111) a self-limiting coverage of 1 ML with a $\left(\sqrt{3} \times \sqrt{3}\right)30°$ reconstruction comprising Sb trimers; each Sb is bonded with one electron to a Si top atom and with two electrons to the two neighboring Sb atoms, leaving the remaining two electrons in a lone-pair orbital [20]. The efficient segregation of Sb on the growing Ge layer was proved, yielding an upper limit for Sb incorporation into the Ge bulk below 2×10^{19} cm^{-3} by XPS [21] and below $2 \times 10^{18}/cm^3$ by SIMS [15].

Using antimony as surfactant for the Ge/Si(111) epitaxy, surface studies applying STM and LEED show successive stages for strain relief as the Ge coverage is increased. Up to a thickness of 8 ML Ge grows smooth and defect-free. Above 8 ML thickness, the analysis of LEED patterns recorded in situ during deposition shows the onset of a periodic dislocation network; near 8 ML 70% of the lattice mismatch of the Ge layer is immediately relaxed by dislocations with an average distance of ~15 nm [21]. After 10 ML additional Ge coverage the average distance attains its final value of 10.4 nm, leaving a completely strain-relieved Ge layer with bulk lattice constant. The satellite structure of the LEED spots indicates a relief by 60° dislocations.

In the 3D growth of pure Ge/Si, dislocations nucleate at the edge of the islands above a critical thickness of about 6 ML [8, 22]; they glide as Shockley partial dislocations (SPDs) and thread to the surface when the islands coalesce. The surfactant-mediated 2D epitaxy constrains the nucleation of strain-relieving dislocations; this extends the critical thickness for dislocation formation to about 8 ML. Since no sources for dislocation nucleation exist at the interface, they must nucleate at the surface.

A detailed TEM investigation shows a quite complex relaxation mechanism proceeding in two steps [20, 23]: In the first step an SPD nucleates at the surface and glides down to the interface, where it cross slips laterally along the interface; this creates stacking faults in the layer and at the interface. In the second step at thicker layers (with increased strain) another SPD nucleates at the surface at the same location where the first SPD intersects the surface; it glides on the same plane to the interface, eliminating the portion of the first SPD's stacking fault that threads through the epilayer. Such a restoring of the perfect lattice is very unlikely in presence of 3D islands due to the numerous low-energy nucleation sites at the island edges; if the dislocations are not at exactly the same plane, they leave a high density of stacking faults. The cross-slip process at the interface leads to an edge dislocation, which can climb to the surface by absorbing atoms of the Ge layer. As a result shown in Fig. 9.4, this relaxation mechanism creates a defect-free relaxed Ge(111) layer on Si(111) with a dense array of dislocations confined in the (111) plane of the interface.

The structure of the periodic defect array at the Ge/Si interface was also revealed in the TEM study [23]. The honeycomb pattern shown in Fig. 9.4b is composed of alternating faulted and unfaulted regions. The small faulted regions with a size below 5 nm can be described by dislocation lines oriented along the three $\langle 110 \rangle$ directions of the (111) interface plane indicated in Fig. 9.5. There are three dislocations of $\frac{a}{2}\left[1\bar{1}0\right]$ type with Burgers vectors in the interface plane (required to be glissile there), and each is equivalent to a sum of two partial dislocations D1 (90° edge dislocation) and D2 (30° dislocation), see Sect. 2.3.3. For the imaging condition of Fig. 9.4b with diffraction vector $\mathbf{g} = \left[02\bar{2}\right]$ all dislocations fulfilling $\mathbf{g} \cdot \mathbf{b} = 0$ are invisible, e.g., $\frac{a}{6}\left[\bar{2}11\right]$; visible are, e.g., the partial dislocations $\frac{a}{6}\left[11\bar{2}\right] + \frac{a}{6}\left[\bar{1}2\bar{1}\right] = \frac{a}{2}\left[01\bar{1}\right]$. In the network we hence observe two sets of dislocations 81 Å apart and one set of dislocations 40.5 Å apart. The dissociation of the perfect dislocation into two Shockley

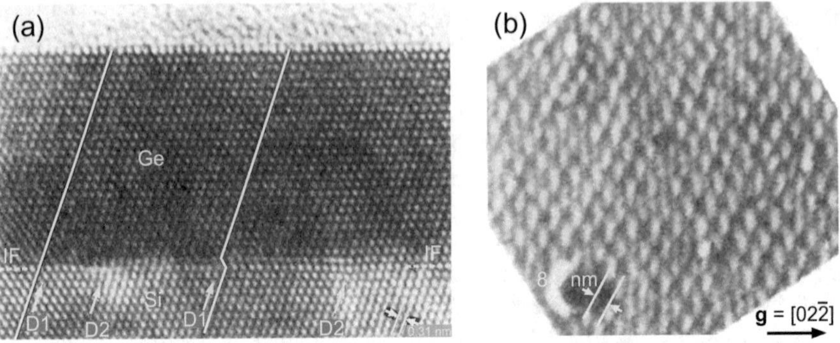

Fig. 9.4 **a** High-resolution TEM image along $[1\bar{1}0]$ of 50 ML Ge grown with Sb surfactant on Si(111). *D1* and *D2* denote alternating partial dislocations, the interface is marked by *IF*. **b** Plan-view weak-beam dark-field image showing a honeycomb array of small dislocations. Reproduced with permission from [23], © 1991 APS

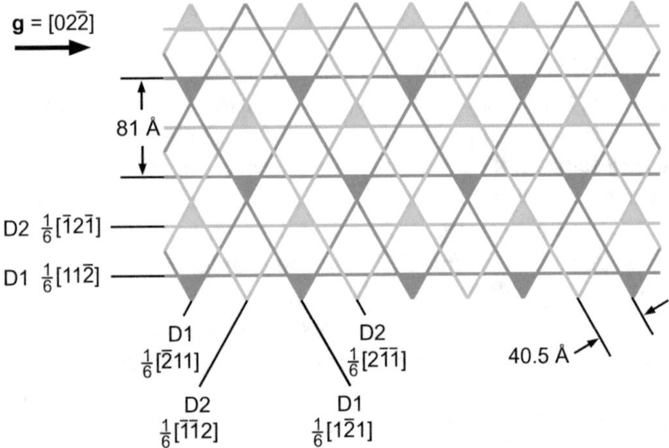

Fig. 9.5 Schematic of a strain-relieving dislocation array with a symmetry like that in Fig. 9.4b. *Blue and green lines* signify respective dislocation lines of 90° (D1) and 30° (D2) partial dislocations at the interface with Burgers vectors specified at the labels, *shaded areas* mark faulted regions delineated by the same type of dislocations. After [23]

partials creates a stacking-fault ribbon in between like that shown in Fig. 2.23. In Fig. 9.5 triangular faulted areas formed by the same type of bordering partial dislocations are indicated by shading. The spacing in the dislocation network corresponds to one plane less in the Ge layer every 25 planes of the Si substrate, yielding the relaxed Ge bulk lattice constant.

The relaxed Ge films can be grown further to arbitrary thickness. Sb sticks still at the surface, though; it may be desorbed at high temperature (≥720 °C), but such a treatment leads to the reappearance of Ge islands. This effect may arise from a strain-induced Ge–Si intermixing at the interface and subsequent glide of dislocations to the surface [21].

9.1.4 *Surfactant-Mediated Epitaxy of III–V Semiconductors*

The surfactant effect described above for the Ge/Si model system was also observed for compound semiconductors and particularly studied for III–V layers on III–V substrates. While the effect has clearly been demonstrated, modelling of the results is much more complex. The presence of anions and cations leads to a larger variety of surface reconstructions, which in addition to temperature also depend on partial pressures. Also the dynamics of growth is more complex even without surfactant. Still some basic correlations could be identified.

InAs on GaAs(001) has a lattice mismatch of about -7% and is well known to grow in the Stranski–Krastanow mode with a critical thickness of 1.5 ML for the 2D $\rightarrow$ 3D transition [24], see also Fig. 8.7. This property is widely employed for the fabrication of quantum dots, see Sect. 7.3.1. When the GaAs surface is covered with 1 ML Te prior to InAs deposition, the coherent 2D growth can be maintained up to a thickness of 6 ML [25]. The in situ measurement of the lateral lattice constant $a_{\mathrm{InAs}_\parallel}$ of InAs by RHEED given in Fig. 9.6 shows coherent 2D growth up to 1.5 ML without Te and up to 6 ML with Te acting as surfactant (the mismatch is given here according to (2.20b)). Accompanying XPS analysis proved the respective 3D and 2D growth mode and a floating of the Te layer on top of the InAs surface.

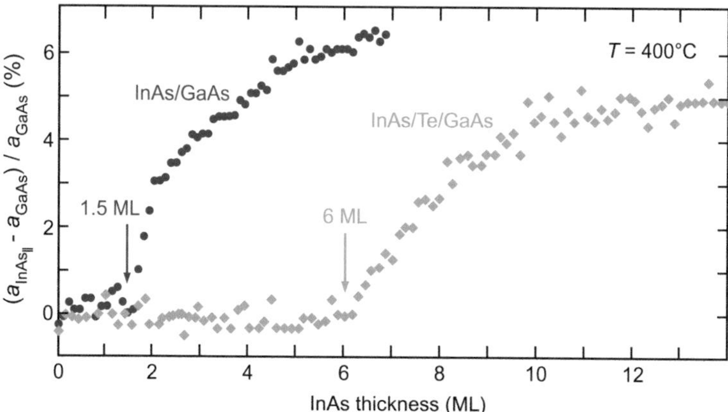

Fig. 9.6 Normalized difference of in-plane lattice parameters $a_\parallel$ of InAs and GaAs, measured in situ during growth of InAs on GaAs(001) without (*blue circles*) and with Te surfactant (*green diamonds*); the onset of plastic lattice relaxation is indicated by *arrows*. Reproduced with permission from [25], © 1992 APS

The effect of Te on the diffusion of adatoms was studied by monitoring the specular beam intensity of RHEED oscillations during homoepitaxial growth of GaAs: the oscillations disappear at increased temperature when the 2D island-nucleation growth mode changes to step-flow growth, see Sect. 8.3.1. This transition temperature is 580 °C for GaAs standard MBE on GaAs(001) 5° off toward (111)A; the offcut is used to obtain a terrassed surface [26]. For Te-assisted growth the transition occurs at 600 °C, proving the reduced diffusion length at a given temperature in presence of the Te surfactant.

A reduced diffusion length $\lambda = \sqrt{D\tau}$ may also be obtained by growth without surfactant; λ decreases if either the diffusion coefficient D or the average dwelling time τ are diminished. A surfactant reduces D by introducing energy barriers. A reduced diffusivity is also obtained at decreased temperature, as experimentally demonstrated by a persistent 2D growth mode of $In_xGa_{1-x}As$/GaAs(001) at low substrate temperature [27]; however, such procedure has generally a detrimental effect on the material quality. Also the residence time τ can be altered by growth conditions; an increase of the supersaturation leads to an increased growth rate, i.e., a reduced τ. The 2D growth mode of $In_xGa_{1-x}As$/GaAs(001) was in fact maintained up to twice the critical thickness t_c for epitaxy at increased growth rate [28]; here strain and t_c depend on the In composition x, and the extension of t_c disappears as x approaches unity. For III–V epitaxy also the V/III ratio applied during growth affects the adatom surface-diffusion and consequently the growth mode. Changing from anion-rich (V/III > 1) to cation-rich conditions, island formation of InAs/GaAs(001) is suppressed as well—an effect originating from a reduced diffusion length of the growth-limiting As species [29].

For InAs/GaAs(001) different types of surfactants were found, similar to the findings for Ge/Si; Te is shown above to decrease the diffusion length, while Pb *increases* the diffusion length, and the critical thickness for the 2D → 3D transition is correspondingly decreased. The different behavior was interpreted in terms of so-called reactive and non-reactive surfactants, respectively [26]. The surfactant effect may in fact differ from that of Ge/Si epitaxy: a large concentration of incorporated Te corresponding to ½ ML was found at the interface of Te-mediated InAs/GaAs(001) epitaxy [30]. Similar results were observed with thallium as surfactant [31]. Such a high local density of impurities is expected to affect the relaxation mechanism.

InGaAs on InP(001) can be grown lattice matched with 47% Ga composition on the cation sublattice; this applies as well for InAlAs with 48% Al and the quaternary $Al_xGa_{0.48-x}In_{0.52}As$. Such layers are interesting for optoelectronic devices operating at 1.5 μm telecom wavelength and for high-electron-mobility transistors. Surfactant-mediated growth was studied using Bi in MBE of InGaAs/InP [32]; at sufficiently high growth temperature (450 °C) Bi acts as a surfactant; it may change the growth mode from 3D to 2D and improve surface smoothness. This was employed for fabricating highly strained InGaAs/InP quantum wells [33] for lasers with improved performance [34]. At lower growth temperature Bi is rapidly incorporated into InGaAs by forming an InGaAsBi alloy [32, 35]. Such a behavior is in contrast to that of Sb [36]; Sb on InP(001) forms at elevated temperature (>300 °C) a relaxed InSb layer with 1 ML Sb coverage, while 1 ML Sb deposition at room temperature leads to an epitaxial Sb terminated surface [37], which could be employed for surfactant-mediated 2D growth of Ge/InP(001) up to 5 ML (instead of only 2 ML without Sb [38]).

GaN on AlN(0001) has an in-plane lattice mismatch $f = -2.3\%$. Epitaxy is often performed on basal-plane sapphire covered with thick relaxed buffer layers as pseudosubstrates. Growth at sufficiently high temperature and usual N-rich conditions proceeds in the Stranski–Krastanow mode with a GaN thickness of ~2 ML for the 2D $\rightarrow$ 3D transition. Similar to the findings for InAs/GaAs(001), 3D growth is suppressed at decreased temperature. Furthermore, 2D growth is maintained at low V/III ratio, i.e., under Ga-rich conditions [39].

Indium has a surfactant effect on the cation-polar basal plane of GaN and AlN [40, 41]. The change from Ga to N rich conditions in the epitaxy of GaN(0001) leads to a rough surface morphology; the smooth-to-rough transition shifts to much higher V/III ratios when In is added [42]. For AlN/GaN(0001) the AlN layer is tensely strained, and strain relief often occurs by cracks, see Fig. 9.7a. Supplying In as a surfactant during deposition of the AlN layer, step-flow growth evolves around the typical threading cores of screw dislocations—an effect attributed to an increased adatom diffusion length [43]. In the superlattice AlN/GaN structure shown in Fig. 9.7b an In incorporation below 2% was observed.

Indium is also an important component in InGaN quantum wells, where an In-rich composition is intended. Applying Sb as surfactant during growth of InGaN layers, an abrupt increase in the In content is observed above a threshold concentration of ~1% Sb [44]. This effect is accompanied by a change of surface morphology; above threshold an increased density of smaller island is found.

In the homoepitaxy of GaN(0001), Sb is expected to increase the effective diffusion length of N. Calculations based on density functional theory indicate that the presence of Sb leads to the formation of SbN intermediates, which are much more mobile than atomic N; since N diffusion is the rate-limiting step in the overall GaN growth process, such increased diffusivity favors a smooth surface morphology and an improved N incorporation during growth [45].

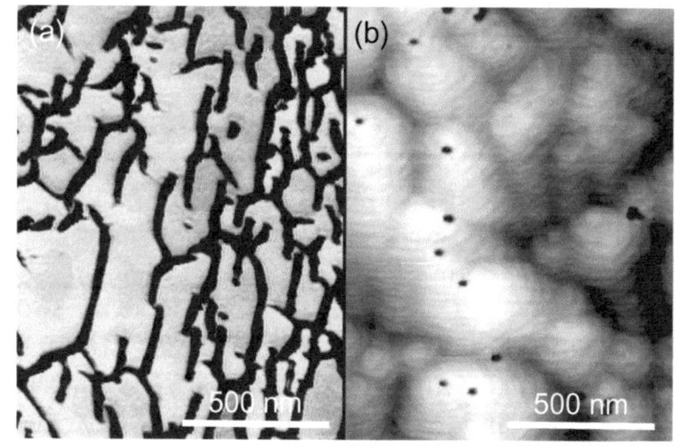

Fig. 9.7 Atomic force micrographs of the 30 nm thick cap layer of **a** AlN on 30 periods of (1.8 nm GaN wells/2 nm AlN barriers) and **b** Al(In)N on a similar superlattice with Al(In)N barriers. Reproduced with permission from [43], © 2006 AIP

9.2 Models of Surfactant-Assisted Epitaxy

There are some common observations in the applications of surfactants to various materials *A* and *B*. Introduction of the surfactant leads generally to an increased nucleation rate. This means that the density of 2D islands nucleating on a surface becomes larger and their size becomes smaller. This finding is connected to a decrease of the adatom diffusion length.

The surfactant layer removes the typical reconstruction of the pure surface. The surfactant-related reconstruction is usually passivated, i.e., dangling bonds are filled and the surface becomes chemically more inert. The new reconstruction is often more simple; e.g., the (7×7) reconstruction of Si(111) changes to a (1×1) reconstruction when As is applied as surfactant.

The combination of Ge and Si is a model system for the study of surfactants. This choice has several reasons. Compound semiconductors consist of (at least) two species with different chemical identity. This leads to complicated surface reconstructions, which depend on the cation/anion ratio and on temperature (see Fig. 7.8). Anions are typically not provided as monomers during growth; e.g., in the MBE of GaAs the As species are dimers As_2 or tetramers As_4. Growth hence requires a decomposition of these species, making the dynamics quite complex.

Studies such as those presented in Sect. 9.1 show that the surfactant changes both the thermodynamics and the kinetics of the involved processes. Several models were developed, emphasizing either a prevalence of the thermodynamic or the kinetic aspect. In the following we discuss approaches explaining the surfactant effect on semiconductors emphasizing these two aspects.

9.2.1 Thermodynamic Considerations

The basic assumption of a thermodynamic approach is that the equilibrium state of a layer in presence of the surfactant is a smooth *two-dimensional* surface. Such a state is formed if the equilibrium shape of a 3D crystal of material *A*, accommodated epitaxially on a misfit substrate *B*, has a vanishing aspect ratio of height *h* over lateral extension *l* [46]. This *Kern–Müller criterion* may be fulfilled by introducing a surfactant, which reduces surface stress and surface energy and consequently leads to a wetting of *A* on *B* that otherwise may not form.

We consider the change of free energy during the epitaxial deposition of a small island *A* on the substrate *B* at constant temperature and vapor pressure P_A; the process is illustrated in Fig. 9.8 [46]. In terms of simple cubic Kossel crystals *A* and *B* with a mismatch *f* according to (2.20a) there are two bulk and three surface terms contributing to the total change of free energy ΔF for the formation of the *A* island on substrate *B*. In the first step of Fig. 9.8a to b chemical work is gained by forming the solid *A* with all its faces from the gas phase of *A* species. This creates the negative bulk contribution ΔF_1 and the positive surface contribution ΔF_2 containing the

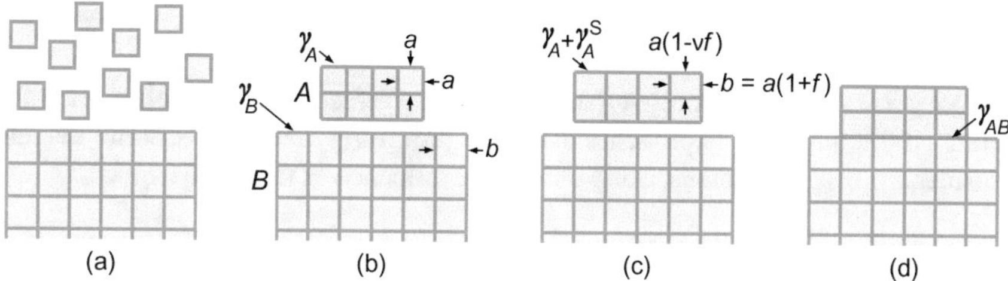

Fig. 9.8 Thermodynamical process for the formation of a small crystal A on a substrate B: **a** supersaturated gas phase of A species, **b** formation of the crystal and all crystal surfaces, **c** accommodation of A to B and corresponding deformation of crystal A surfaces, **d** adhesion of A to B

surface free energy γ_A. In the next step to Fig. 9.8c the (initially relaxed) crystal A is elastically strained for a coherent fit to substrate B. The deformation energy of the corresponding bulk contribution ΔF_3 contains the misfit f, Young's modulus, and Poisson's ratio of A. A surface contribution ΔF_4 accounts for the simultaneously occurring deformation of the surfaces and the resulting surface stress γ_A^S of the basal and side surfaces. In the last step to Fig. 9.8d the crystal A adheres to the substate B; the related contribution ΔF_5 contains the adhesion energy γ_{AB}.

The apect ratio h/l for a 3D crystal A on B in equilibrium is now considered for the case without elastic relaxation. The contributions ΔF_i mentioned above lead to a fraction for h/l with the numerator given by the so-called wetting factor [46]

$$2\gamma_A - \gamma_{AB} + 4f\left(\gamma_A^S - \gamma_A\right); \qquad (9.1)$$

this is the Kern–Müller criterion for 2D layer-by-layer growth, obtained from the Bauer criterion [47] (first two summands) by adding a strain term containing the misfit f. A decreasing ratio h/l, i.e., a 3D $\rightarrow$ 2D transition, is obtained when the adhesion energy γ_{AB} increases or if the surface energy γ_A decreases. The surface stress γ_A^S may drive the system away from or towards to the 3D $\rightarrow$ 2D transition.

By introducing a surfactant the wetting factor (9.1) is complemented by two additional summands, which respectively decrease the surface free energy γ_A and account for a differently stressed surface of A when some quantity of the surfactant is adsorbed on A; the adhesion energy γ_{AB} is not affected by the surfactant, which by definition floats on A [46]. The new summands depend both on the mismatch f and the quantity of surfactant Γ. For $f > 0$ (i.e., lattice constant $b > a$) the system can be driven from 3D to 2D growth by increasing the surfactant coverage Γ; for negative f this applies only for special parameter values within this model.

The equilibrium model discussed so far may well indicate whether a surfactant species drives the system into the right direction to suppress island formation. However, it does not account for the Stranski–Krastanow growth mode, where the layer A wets the surface B and still proceeds with 3D growth. Moreover, the effect of elastic strain relaxation of A islands is not included. This contribution is negative and proportional to the island volume; it is known to substantially reduce the cost of 3D island formation [48, 49].

Today it is generally accepted that the surfactant effect is not of purely thermodynamic origin, and most approaches assume a prevailing kinetic effect. The models consider atomistic processes, often with some influence of thermodynamics.

9.2.2 Kinetic Approach

A kinetic model assumes that the equilibrium state of a layer is a *three-dimensional* surface even in presence of the surfactant. The surfactant leads to layer-by-layer growth by kinetically prolonging the approach to equilibrium beyond the time scale of fabricating the structure. We first discuss a thermodynamically motivated atomistic model including kinetic contributions. A generally accepted kinetic description is outlined in Sect. 9.2.3.

The Markov Model

The commonly observed increase of nucleation rate and decreased adatom diffusion length for surfactant-mediated growth is described in a simple atomistic model of homoepitaxial nucleation, including both thermodynamic and kinetic aspects [50, 51]; accounting for strain would introduce an additional term into the work of nucleus formation ΔG containing the wetting factor (9.1). We consider the nucleation of a 2D cluster according to the process illustrated in Fig. 9.9 [52].

The process implies the isothermal and reversible evaporation of all surfactant atoms from the crystal surface, then producing a cluster with i crystal atoms on the surface, and finally condensing back the surfactant atoms. If we assume a cubic Kossel crystal and a surface cluster with a square shape and edge length l, the work for cluster formation on the surfactant-free surface (Fig. 9.9b) reads

$$\Delta G_0 = -i\,\Delta\mu + 4\,l\,\gamma_c; \tag{9.2}$$

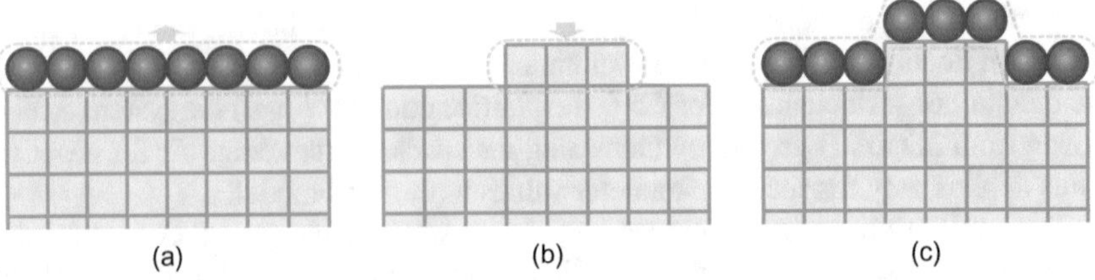

Fig. 9.9 Process for evaluating the Gibbs free energy change by the formation of a homoepitaxial small nucleus on a surface covered with a surfactant monolayer: **a** the surfactant atoms (*filled circles*) are evaporated, **b** the nucleus is formed, **c** the surfactant layer is condensed back

here $\Delta\mu$ is the supersaturation, i.e., the chemical potential difference between the ambient and the new phase, and γ_c is the edge energy per length of crystal atoms (index c) at the perimeter of the cluster. For a 2D cluster this is the energy of the dangling bonds of the perimeter atoms, and for the cluster on top of the Kossel crystal it is simply the number of dangling bond times half the bond energy between crystal atoms. When in the step of Fig. 9.9c the surfactant atoms are condensed back, the surfactant atoms beside the edge of the cluster saturate the dangling bonds, gaining an energy $-4ls\gamma_c$; the parameter s is introduced as a measure for the energy efficiency of this process. Additionally an edge energy per length $+4l\gamma_s$ must be paid for the surfactant atoms (index s) *on top* of the cluster at the edge. The total work for the formation of the cluster hence reads

$$\Delta G = \Delta G_0 - 4\,l\,s\,\gamma_c + 4\,l\,\gamma_s. \tag{9.3}$$

The parameter s for the surfactant efficiency is obtained from

$$s = 1 - \frac{\frac{1}{2}(E_{cc} + E_{ss}) - E_{sc}}{\frac{1}{2}E_{cc}}; \tag{9.4}$$

here E_{cc} is the bond energy between two crystal atoms, E_{ss} that between surfactant atoms, and E_{sc} that between a crystal atom and a surfactant atom; for the factor ½ see (6.30). In the fraction of (9.4) the numerator is the energy of a dangling bond saturated by a surfactant, and the denominator that of an unsaturated dangling bond. Without surfactant $E_{ss} = E_{sc} = 0$ and $s = 0$; on the other hand, for perfectly efficient saturation of dangling bonds $E_{cc} + E_{ss} = E_{sc} + E_{sc} = 2E_{sc}$ and $s = 1$. s must be positive and below unity to cause floating of the surfactant on the surface. s represents the energy parameter for the mixing of two species; for $s > 1$ alloying of surfactant and crystal atoms occurs [52].

The work ΔG for surfactant-mediated cluster formation (9.3) contains two counteracting terms added to the work on the pure surface ΔG_0. The first term $-4ls\gamma_c$ is the edge energy of the cluster with saturated lateral dangling bonds, and the second term $+4l\gamma_s$ is the edge energy of the surfactant atoms on top of the cluster. These terms produce a dependence of the Gibbs energy change for cluster formation on the surfactant efficiency s. This dependence is shown in Fig. 9.10 for parameters corresponding to cluster formation on the (111) surface of a metallic fcc crystal with constant supersaturation $\Delta\mu$ and a bond-strength ratio $\frac{E_{ss}}{E_{cc}} = 0.2$.

The model of small surface clusters consisting of individual atoms turns the smooth curve of the capillary theory in Fig. 6.9 to the curve with $s = 0.0$ composed of straight lines depicted in Fig. 9.10. ΔG is maximum at the critical nucleus, which for $s = 0$ is also considered in Sect. 7.2.2. This maximum is the critical work ΔG^* to create the critical nucleus, which contains $i = i^*$ atoms; without surfactant these are only two atoms for the assumed parameters of an fcc metal surface. As the efficiency of the surfactant s increases, both ΔG^* and the critical nucleus size i^* decrease. There is an exception for a very small efficiency $s = 0.05$; in such a case the edge energy of the surfactant cluster on top of the crystal cluster $4l\gamma_s$ is

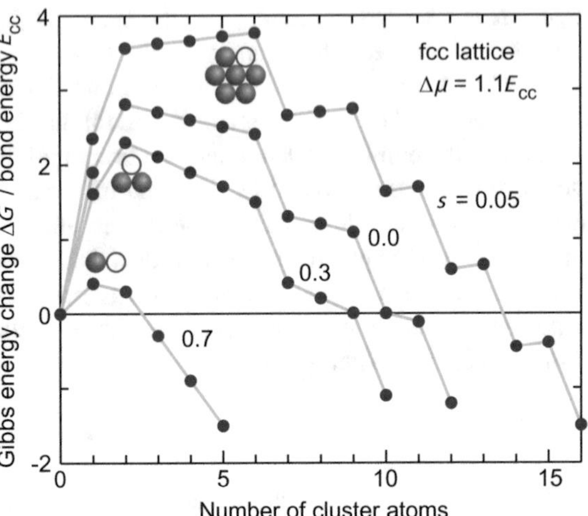

Fig. 9.10 Change of Gibbs free energy ΔG per energy of bond strength E_{cc} for the formation of a small cluster on the fcc(111) surface of equal crystal atoms. $\Delta G / E_{cc}$ depends on the efficiency s of the applied surfactant given at the curves. The insets depict the structures of the critical nuclei with *blue filled circles* denoting cluster atoms, the *red open circles* indicate atoms which convert the critical nuclei into smallest stable clusters. Reproduced with permission from [52], © 2010 Springer

larger than the edge energy of the crystal cluster with inefficiently saturated dangling bonds $4ls\gamma_c$. ΔG^* and critical nucleus size i^* consequently increase. It should be noted that all these changes induced by the surfactant occur at otherwise constant growth conditions.

The usually smaller critical cluster size and decreased work for critical nucleus formation leads also to a higher nucleation rate and an increased saturation nucleus-density N_s in the presence of a surfactant. For the density N_s of surfactant-mediated nucleation the relation

$$N_s = N_{s,0} \exp\left(-c\frac{E_s}{kT}\right) \tag{9.5}$$

holds, with $c = 1/(i^* + 2)$ if no significant step-edge barrier impedes the incorporation of adatoms into the nucleus (diffusion-limited case), and $c = 2/(i^* + 3)$ if such a barrier exists (kinetically limited case) [51]. The energy E_s contains all contributions introduced by the surfactant,

$$E_s = -4ls\gamma_c + 4l\gamma_s + E^*_{exc} - i^*\left[(E_{de-exc} - E_{exc}) - \left(E_{d,bare} - E_d\right)\right]; \tag{9.6}$$

here the first two terms are given by the thermodynamic aspects discussed above, while the two following terms are purely kinetic barriers. They refer to the processes of exchange at the edge of the nucleus $\left(E^*_{exc}\right)$, exchange (E_{exc}) and de-exchange (E_{de-exc}) far from the nucleus, and diffusion on the surfactant-free crystal surface $\left(E_{d,bare}\right)$ and on top of the surfactant monolayer (E_d). Since two terms are positive and two are negative, E_s may have either sign. $E_s < 0$ stimulates nucleation and promotes growth by 2D nucleation; in addition, the larger density of smaller nuclei favors layer-by-layer growth and hampers second-layer nucleation.

9.2.3 The Diffusion—De-exchange—Passivation (DDP) Model

Adatom Diffusion With De-Exchange Processes

The importance of adatom diffusion is emphazised in most models on surfactants. In early work the reduced adatom diffusion-length discussed for surfactant-mediated epitaxy of Ge/Si in Sect. 9.1.3 was considered the reason for the suppression of 3D growth. The argument was the following: epitaxy in presence of a surfactant layer requires a *site exchange* of an adatom and a surfactant atom to allow for crystal growth beneath the floating surfactant layer. The energy barrier for this exchange E_{exc} was assumed to be smaller than the barrier E_d for adatom diffusion on top of the surfactant layer. The diffusion length is then quite small and limited by the exchange process: underneath the surfactant layer the atom cannot diffuse. The short diffusion length leads to a high density of nucleating 2D islands. Consequently these islands coalesce before second-layer islands nucleate. Thereby growth proceeds layer-by-layer, and 3D growth is suppressed.

While this explanation appears reasonable it was found to be incomplete. It turned out that also the reverse process—the de-exchange—is important for describing experimental findings. The energy barrier $E_{de\text{-}exc}$ for a crystal atom on the topmost crystal layer to become an adatom on top of the surfactant layer is significantly larger than E_{exc} and E_d. Still a rough estimate shows that such a process can affect the growth process [17]. Analogous to (7.6) the time scale corresponding to a kinetic process with energy barrier E is given by $\nu^{-1}\exp(-E/(kT))$, where ν is the attempt rate, here typically on the order of 10^{13} s^{-1}. $E_{de\text{-}exc} = 1.6$ eV at a typical growth temperature $T = 600$ °C yields a characteristic time of 0.2 ms; this time is much shorter than the growth rate for a monolayer, being typically larger than one second. A de-exchange process therefore usually occurs before another crystal atom binds the buried atom permanently to the growing layer. The description of surfactant-mediated epitaxy including the de-exchange process is referred to as diffusion—de-exchange—passivation (DDP) model.

The combination of exchange and de-exchange processes makes the effective diffusion of adatoms more complex. The effective diffusion coefficient D_{eff} is related to the diffusion coefficient D_{bare} and the related energy barrier $E_{d,bare}$ for bare adatom diffusion without surfactant by [13]

$$D_{eff} = D_{bare}\frac{\exp\left((E_{d,bare} - E_d)/(kT)\right)}{1 + \exp((E_{de\text{-}exc} - E_{exc})/(kT))}. \tag{9.7}$$

The ratio of E_d and E_{exc} does hence not provide an information whether D_{eff} is smaller than D_{bare} or larger. If the surfactant passivates the surface, then $E_{d,bare} > E_d$ and $E_{de\text{-}exc} > E_{exc}$; which of the energy differences in (9.7) is larger cannot be decided a priori. First principles DFT calculations for surfactant-covered Si(111) yield $E_{de\text{-}exc} - E_{exc} = 0.8$ eV, while the difference of $E_{d,bare} - E_d$ is much smaller

[13]; this results in $D_{\text{bare}} \gg D_{\text{eff}}$, and hence a decreased diffusivity in presence of the surfactant. It should be noted that a less efficient passivation of the surface may reverse the ratios and the diffusion length could be increased; since a decreased diffusivity in presence of the surfactant may also occur for $E_{\text{exc}} > E_{\text{d}}$, a surfactant can simultaneously increase the diffusivity on top of a surfactant layer and decrease the *effective* diffusivity.

Passivation of Island Edges

Kinetic Monte-Carlo simulations in the DDP model indicate that the effective diffusion expressed by the coefficient D_{eff} is actually *not* the relevant process for inducing either 2D or 3D growth: two different surfactants leading to the same value of D_{eff} may induce different growth modes [13, 17]. These studies show that the crucial effect of a surfactant is the *passivation of step edges*. These edges occur at terraces of vicinal surfaces (Fig. 7.22) and also at the perimeter of nucleated 2D islands, see Sect. 7.2.

The passivation of a step edge can be accounted for by an increased energy barrier E_{det} for the detachment of an atom, which after an exchange process has been attached to a 2D island underneath the surfactant layer. The detachment requires the breaking of lateral bonds; in the simulation described below this process is allowed if only a single lateral bond has to be broken. The result of such a kinetic Monte-Carlo simulation for different detachment energies in homoepitaxy is shown in Fig. 9.11; for simplicity a simple cubic lattice is assumed. The simulations have equal parameters determining the adatom diffusion and differ only in the detachment energy. If the detachment process is suppressed by a respective high barrier (Fig. 9.11a) deposition of 0.15 monolayer leads to a high density of islands, and rough edges are formed at the islands. When the detachment is enabled the island density is substantially decreased and the island edges are faceted (Fig. 9.11b).

In Fig. 9.11a, b $E_{\text{d}} < E_{\text{exc}}$ was assumed. It must be noted that similar results may be obtained for a parameter set with $E_{\text{d}} > E_{\text{exc}}$ and appropriately chosen $E_{\text{de-exc}}$ and E_{det} [13]: island densities and shapes are not significantly affected. The results of homoepitaxy simulation are in agreement with experimental observations. Surfactants which suppress 3D growth increase the island density; also rough island edges and an increase of island density at the expense of their size at lower temperature are observed for Si homoepitaxy [14]. In the simulation the increased island density does not originate from a reduced diffusivity, but from the ability for adatoms to detach again from an island edge and subsequently nucleate on a flat part of the surface.

In heteroepitaxy strain is more efficiently elastically relaxed in 3D growth than in 2D growth, yielding a driving force for the formation of 3D islands. For 3D growth atoms must get detached from the island edge and migrate to the top of the island. If this detachment is suppressed by the presence of surfactant atoms the islands will grow wide and flat, i.e., in the layer-by-layer mode. Still strain affects growth; its effect is included in the simulation of Fig. 9.11c, d by assuming an altered bond strength in the strained islands. Consequently also the activation energy for the detachment of atoms from island edges is affected by strain; the other barriers are less affected and kept constant. For the detachment energy a dependence on the island

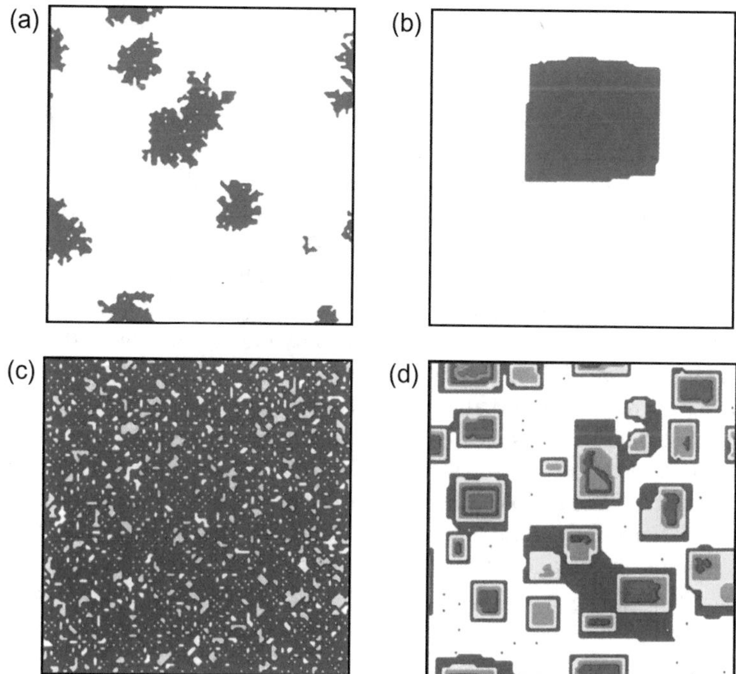

Fig. 9.11 Shape of 2D islands according to kinetic Monte-Carlo simulations of surfactant-mediated epitaxy with (**a**, **c**) and without (**b**, **d**) a passivation of the island edges; the initially flat substrate surface is *white*. **a** Homoepitaxy at 600 °C with activation energies $E_d = 0.5$eV, $E_{exc} = 0.8$eV, $E_{de-exc} = 1.6$eV, and edge passivation enforced by $E_{det} = 3.0$eV. **b** Same as **a** but with $E_{det} = 1.6$eV to enable atom detachment at edges. **c** Heteroepitaxy of 1 ML at 300 °C with same energies as in **a** but $f = 4\%$. **d** Same as **c** but with E_{det} as in **b**; *colors* indicate different surface heights; *blue*, *yellow*, and *green* indicate 1, 2, and 3 ML height. Reproduced with permission from [13], © 2000 Academic Press

size was assumed, according to the elastic strain energy of rectangular islands with lateral lengths l_1 and l_2 [53]:

$$E_{det} = \varepsilon_0 + \varepsilon_1 \left(\frac{\ln l_1}{l_1} + \frac{\ln l_2}{l_2} \right); \tag{9.8}$$

here $\varepsilon_0 = E_{de\text{-}exc}$ for surfactants passivating island edges and $\varepsilon_0 = 0$ without such passivation; ε_1 is set to 3.0 eV in the simulations of Fig. 9.11c, d.

The simulation without island-edge passivation Fig. 9.11d shows 3D growth with islands up to 7 ML height after deposition of 1 ML and still large uncovered areas of the substrate, while islanding is suppressed for passivated island edges (Fig. 9.11c); the latter surface is covered with 1 ML of the deposit, with some small holes (white) and some small 1 ML high islands (yellow) on top. Similar to homoepitaxy no significant changes occur when a parameter set with $E_d > E_{exc}$ is applied instead of $E_d < E_{exc}$.

The simulated surfactant-mediated 2D growth at low temperature changes quite abruptly to 3D growth at higher temperature with a transition around 400 °C for the given parameters; such a transition is experimentally observed for Sb-mediated Ge/Si(111) epitaxy [15, 16] and was studied in detail for Bi-mediated Ge/Si(111) epitaxy [54]. The size of islands nucleating during epitaxy usually follows a distribution function with a single peak at the mean island size [55]; in addition a denuded zone is found near step edges where hardly any island nucleates. Such distributions were measured for Si islands grown by Bi-mediated epitaxy on Si(111) and also for Ge islands, the latter, however, only at sufficiently high temperature (480 °C) [54]; at lower temperature (440 °C) a monotonously decreasing distribution function with a maximum at very small islands is observed, and correspondingly no pronounced denuded zone is found. These findings obtained from scanning tunneling microscopy of submonolayer depositions are well described by kinetic Monte-Carlo simulations based on the DDP model [54]. The temperature threshold for the Ge growth marks a transition from epitaxy limited by surface diffusion at high temperature to epitaxy limited by attachment kinetics at lower temperature, where the adatom distribution between step edges becomes uniform [56]. An analysis based on rate equations indicates that the transition originates from adatom exchange and de-exchange processes [57].

9.2.4 Exchange Pathways in Surfactant-Mediated Epitaxy

How does an adatom diffusing on a complete floating monolayer of surfactant atoms manage to exchange its site with a surfactant atom to incorporate below the surfactant? Such exchange pathways can hardly be observed experimentally and are generally studied using ab initio calculations, mostly by density functional theory in the local density approximation (DFT-LDA). The atomic configurations and related activation energies are specific for any surface, participating species, and pathways. Exchange paths are quite complex and consist of sections with higher-energy saddle points and local energy minima. We consider examples of the (111) and (001) surfaces of the Ge/Si model system.

For **Ge/Si(001)** epitaxy mediated by Sb or Bi surfactants, DFT-LDA calculations identified low-energy exchange processes where surfactant dimers are broken by adsorbed Ge atoms [58]. Most stable adsorption sites for a Ge dimer on the Si(001) surface covered with Sb dimers are illustrated in Fig. 9.12a, b, where the latter has a lower energy; the Ge dimer initially adsorbed above the trench (panel a) can move to the position above the Sb dimers below by breaking the bonds indicated by the crosses in Fig. 9.12a and rolling over to the position above the lower Sb dimer row. The Ge dimer there buckles (Fig. 9.12b) and—in a complex process—exchanges with two Sb atoms, yielding the configuration of Fig. 9.12c below an uplifted Sb dimer. There is a saddle-point energy of 0.78 eV between the sites of panels *a* and *b*,

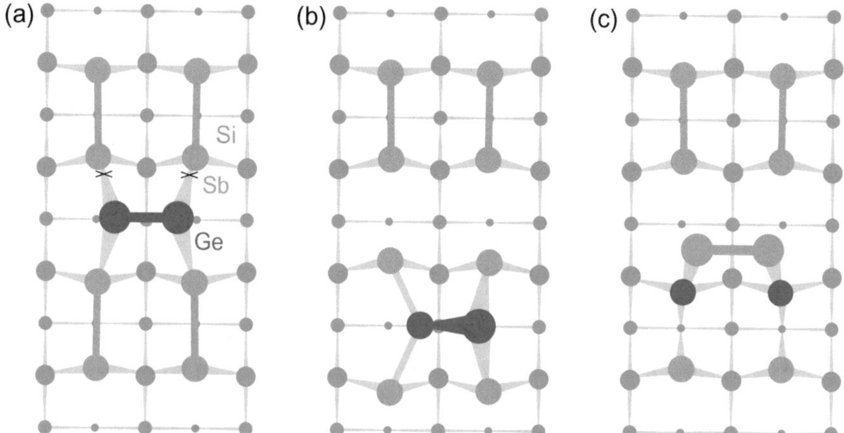

Fig. 9.12 Plan view on a Si(001) surface covered with Sb surfactant atoms (*green circles*) and a Ge dimer (*red circles*) adsorbed **a** above the horizontal trench between Sb dimers and **b** on a row of Sb dimers. **c** Final configuration with Ge atoms exchanged with surfactant atoms; atom sizes indicate the height above the surface. After [58]

and an exchange barrier of 1.05 eV with two local maxima between the sites of panels b and c; the total energy gain is 0.9 and 0.47 eV for $a \rightarrow b$ and $b \rightarrow c$, respectively; for the Bi surfactant a different pathway $a \rightarrow c$ is favorable with somewhat smaller energies [58].

Diffusion and exchange processes of surfactant-mediated **Ge/Si(111)** epitaxy were studied using ab initio molecular dynamic calculations [59] and DFT-LDA [13]. A low-energy diffusion path of a Ge adatom on Si(111) covered with 1 ML of As surfactant proceeds from a minimum position above site H_3 (position Ge_{ad} in Fig. 9.13) to a saddle point above a neighboring T_4 site and further to a position above

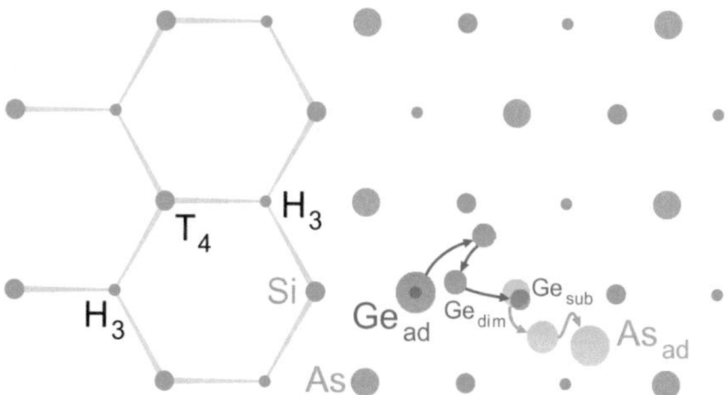

Fig. 9.13 Plan view on a Si(111) surface without (left part, *gray lines* indicate Si bonds) and with As surfactant atoms (*green circles*, right part); atom sizes indicate the height above the surface. The exchange process of a Ge adatom (*red*, Ge_{ad}) with an As surfactant atom is illustrated at the bottom right; it yields Ge on a substitutional site (Ge_{sub}) and As on an adatom site (As_{ad}). After [59]

the next H_3 site (path not shown in Fig. 9.13). On the saddle point the energy is 0.25 eV higher than at position Ge_{ad}; this difference represents the barrier for diffusion E_d. The path given in Fig. 9.13 illustrates an exchange process, where a Ge adatom gets incorporated under the As layer [59]: the Ge adatom approaches a saddle-point site and then moves downwards, pushing the As adatom located on a substitutional site (above Ge_{sub}) away. It then proceeds to the Ge_{dim} site where it forms a dimer bond to the moved As atom. Finally the Ge atom moves to a substitutional site Ge_{sub} and the As atom moves upwards to an adatom site As_{ad}.

The complex exchange path from Ge_{ad} to Ge_{sub} comprises local energy maxima and minima; the highest saddle point occurs in the last step and represents the exchange energy barrier $E_{exc} = 0.7$ eV [59]. Since E_{exc} is substantially larger than E_d the Ge adatom performs many diffusion steps before getting incorporated under the As layer. The large exchange barrier originates basically from strain, while the diffusion barrier is not significantly affected by strain; calculations of exchange barriers in surfactant-mediated homoepitaxy Si/As/Si(111) and Ge/As/Ge(111) yield small E_{exc} of 0.3 and 0.4 eV, respectively [59, 60].

The exchange processes for an entire Ge double layer (=1 ML) in Sb-mediated Ge/Si(111) epitaxy were studied in [13]. The modelled pathway implies a concerted exchange of Ge dimers on a floating Sb monolayer as illustrated in Fig. 9.14. In the initial configuration shown in panel *a* the Ge layer is only three-fold coordinated and the layer of the pentavalent Sb is fourfold coordinated. Arrows indicate the movement of atoms via the intermediate configuration *b* to the final configuration *c*, where all atoms are coordinated properly: Ge four-fold, and Sb three-fold. Consequently the energy of this state is lower than that of the initial and of the metastable intermediate

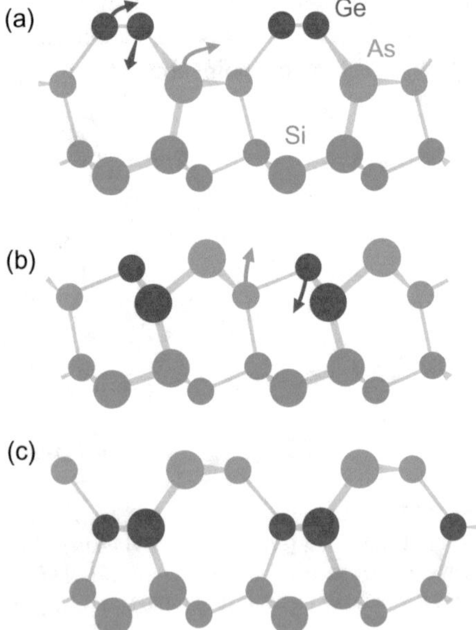

Fig. 9.14 Pathway for a concerted exchange that brings a Ge layer (*red circles*) on top of an Sb surfactant layer (*green*) on Si(111) beneath the surfactant, for a cross-sectional view along $[1\bar{1}0]$. After [13]

symmetry, where only half of the Ge atoms has exchanged site. The intermediate configuration is separated by saddle-point energies from the start and end configurations, and the activation energy for such a concerted exchange was found to be only 0.8 eV [13]. This model considers a final configuration, which is ready for the deposition of the next Ge layer; the growth process can hence proceed with similar steps. The model thus provides an idea of surfactant-mediated epitaxy, even if detailed studies of single exchange events may be physically more realistic.

9.3 Conclusion

The beneficial effect of a surfactant on smooth epitaxial growth is evident, and also a useful effect on doping (not discussed in this chapter) was reported [61–63]. Sill surfactants are hardly employed in device fabrication. A rare example is a Ge *p*-channel MOSFET fabricated on Si(111) substrate using Sb-mediated MBE; the device showed a quite high channel drift mobility of 430 cm^2/Vs [64]. One reason preventing the wide application of surfactants in device production is the finite incorporation of surfactant species into the layer structure. The concentration is usually in the range of 10^{-3} to 10^{-4}, but may be substantially lower; for Sb-mediated Ge/Si(111) epitaxy an incorporation below the SIMS detection limit of 2 $\times$ 10^{17} cm^{-3} was reported [65].

9.4 Problems Chapter 9

9.1 In the surfactant-mediated epitaxy of Ge on Si(111) a triangular network of dislocations confined to the interface is formed, see Figs. 9.4b and 9.5.

(a) Compute the lattice mismatch of Ge on Si and the remaining mismatch, if a coincident lattice with 25 planes in the Si substrate corresponds to one plane less of the Ge layer.

(b) The three dislocations of $\frac{a}{2}\left[1\bar{1}0\right]$ type with Burgers vectors in the interface plane dissociate in three sums of two partial dislocations D1 and D2 each, labelled in Fig. 9.5. Which of these dislocations are visible for the imaging conditions of Fig. 9.4b and which not? Which dislocations are consequently observed 81 Å apart, which are observed 40.5 Å apart?

9.2 The diffusion length of adatoms is usually decreased in presence of a surfactant; the change is described by the ratio of the diffusion coefficients for adatom diffusion on the bare surface (without surfactant) and on the surfactant layer.

(a) Consider a single diffusion step of a Ge adatom on an As covered Si(111) surface along the path $H_3 \rightarrow T_4 \rightarrow H_3$ of Fig. 9.13 with a barrier of 0.25 eV. Calculate the length a_{hop} of one diffusion hop and the diffusion constant

D_0 for a diffusion along this path at an attempt rate of $10^{13}\ s^{-1}$. How many diffusion hops occur in 10 ns at a temperature of 400 °C, and how long is the path for diffusion of this kind in this time?

(b) For a diffusion without de-exchange process the diffusion length of an adatom on a surfactant layer is limited by the (first) exchange process with a surfactant atom. Assuming Ge adatom diffusion only along the path and at the temperature given above, which diffusion length λ results for an exchange barrier of 0.7 eV? How many hops does the adatom perform on this path?

(c) If de-exchange processes occur, a dependence on the respective energy for de-exchange enters the effective diffusion coefficient. Given the quantities above and a diffusion barrier of 0.35 eV on the bare surface (without surfactant), compute the ratio of the effective diffusion coefficient with respect to the coefficient of bare diffusion for a de-exchange barrier of 1.1 eV.

(d) Assume the above given values for barriers for diffusion on the bare surface and on the surfactant, and for the exchange process; compute the energy barrier for de-excitation required to obtain an effective diffusion coefficient twice as large as the coefficient of bare diffusion.

9.3 (a) Determine the approximate activation energy for Si adatom diffusion on the bare Si(111) surface and on the Sb-covered surface from the Arrhenius plot Fig. 9.3b.

(b) Taking the width of the depleated zone plotted in Fig. 9.3b as the characteristic diffusion length and the activation energies determined above, calculate the diffusion length of Si on the bare Si(111) surface at 850 K and the factor g by which the diffusion length is changed when the Sb surfactant layer is introduced; calculate the temperatures for a diffusion length of 100 nm on the bare Si(111) surface and on the Sb-covered surface.

9.4 We consider a cubic Kossel model of an Sb-covered "Si(001)" surface and assume a Si–Si binding energy of 1.7 eV and corresponding Sb–Sb energy of 0.7 eV.

(a) Compute the Sb–Si binding energy for a surfactant efficiency expressed by a parameter of 0.85. How much is the energy of an unsaturated dangling Si bond lowered by surfactant saturation?

(b) Determine the total energy gain by surfactant passivation at the edge of a nucleus containing just a single crystal atom. Which gain results if the nucleus contains two atoms?

9.5 Application of a Te surfactant prolongates the onset of plastic relaxation of InAs on GaAs(001) as shown in Fig. 9.6. By which factor has the relaxed lattice parameter of InAs increased from 300 K to 400 °C? Which normalized lattice parameter $(a_{InAs_\parallel} - a_{GaAs})/a_{GaAs}$ has the completely relaxed InAs layer at 400 °C (in linear approximation)? Determine the actual normalized lateral lattice parameters of InAs/GaAs and of InAs/Te/GaAs for saturated relaxation from Fig. 9.6 and compute the respective lateral strains in InAs.

9.5 General Reading Chapter 9

I.V. Markov, Crystal Growth for Beginners, 3rd edn. (World Scientific, New Jersey, 2017)

References

1. M. Copel, M.C. Reuter, E. Kaxiras, R.M. Tromp, Surfactants in epitaxial growth. Phys. Rev. Lett. **63**, 632 (1989)
2. J.W. Matthews, E. Grünbaum, The need for contaminants in the epitaxial growth of gold on rocksalt. Appl. Phys. Lett. **5**, 106 (1964)
3. R. Kern, G. Le Lay, J.J. Metois, Basic mechanisms in the early stages of epitaxy, in *Current Topics in Materials Science*, vol. 3, ed. by E. Kaldis (North Holland Publ. Co., Amsterdam, 1979), pp. 130–419, Sect. 9
4. J. Camarero, T. Graf, J.J. de Miguel, R. Miranda, W. Kuch, M. Zharnikov, A. Dittschar, C.M. Schneider, J. Kirschner, Surfactant-mediated modification of the magnetic properties of Co/Cu(111) thin films and superlattices. Phys. Rev. Lett. **76**, 4428 (1996)
5. J.J. de Miguel, R. Miranda, Atomic aspects in the epitaxial growth of metallic superlattices and nanostructures. J. Phys.: Condens. Matter **14**, R1063 (2002)
6. Z. Zhang, M.G. Lagally, Atomic-scale mechanisms for surfactant-mediated layer-by-layer growth in homoepitaxy. Phys. Rev. Lett. **72**, 693 (1994)
7. A. Rastelli, M. Kummer, H. von Känel, Reversible shape evolution of Ge islands on Si(001). Phys. Rev. Lett. **87**, 256101 (2001)
8. J. Bevk, J.P. Mannaerts, L.C. Feldman, B.A. Davidson, A. Ourmazd, Ge-Si layered structures: artificial crystals and complex cell ordered superlattices. Appl. Phys. Lett. **49**, 286 (1986)
9. P.M.J. Marée, K. Nakagawa, F.M. Mulders, J.F. Van der Veen, K.L. Kavanagh, Thin epitaxial Ge-Si (111) films: study and control of morphology. Surf. Sci. **191**, 305 (1987)
10. E. Kasper, H. Jorke, Growth kinetics of molecular beam epitaxy, in *Chemistry and Physics of Solid, Surfaces*, vol. VII, ed. by R. Vanselow, R.F. Howe (Springer, Berlin, 1988), pp. 557–582
11. M. Copel, M.C. Reuter, M. Horn von Hoegen, R.M. Tromp, Influence of surfactants in Ge and Si epitaxy on Si(001). Phys. Rev. B **42**, 11682 (1990)
12. F.K. LeGoues, M. Copel, R.M. Tromp, Microstructure and strain relief of Ge films grown layer by layer on Si(001). Phys. Rev. B **42**, 11690 (1990)
13. D. Kandel, E. Kaxiras, The surfactant effect in semiconductor thin-film growth, in *Solid State Physics*, vol. 54, ed. by H. Ehrenreich F. Spaepen (Academic Press, San Diego, 2000), pp. 219–262
14. B. Voigtländer, A. Zinner, T. Weber, H.P. Bonzel, Modification of growth kinetics in surfactant-mediated epitaxy. Phys. Rev. B **51**, 7583 (1995)
15. B. Voigtländer, A. Zinner, Surfactant-mediated epitaxy of Ge on Si(111): the role of kinetics and characterization of the Ge layers. J. Vac. Sci. Technol. A **12**, 1932 (1994)
16. I. Portavoce, A. Berbezier, Ronda, Sb-surfactant-mediated growth of Si and Ge nanostructures. Phys. Rev. B **69**, 155416 (2004)
17. D. Kandel, E. Kaxiras, Surfactant mediated crystal growth of semiconductors. Phys. Rev. Lett. **75**, 2742 (1995)

18. A. Antons, R. Berger, K. Schroeder, B. Voigtländer, Structure of steps on As-passivated Si(111): ab initio calculations and scanning tunneling microscopy. Phys. Rev. B **73**, 125327 (2006)
19. S.A. Teys, Different growth mechanisms of Ge by Stranski-Krastanow on Si (111) and (001) surfaces: an STM study. Appl. Surf. Sci. **392**, 1017 (2017)
20. M. Horn-von Hoegen, F.K. LeGoues, M. Copel, M. C. Reuter, R. M. Tromp, Defect self-annihilation in surfactant-mediated epitaxial growth. Phys. Rev. Lett. **67**, 1130 (1991)
21. M. Horn-von Hoegen, Surfactants: perfect heteroepitaxy of Ge on Si(111). Appl. Phys. A **59**, 503 (1994)
22. M. Kammler, M. Horn-von Hoegen, Transition in growth mode by competing strain relaxation mechanisms: surfactant mediated epitaxy of SiGe alloys on Si. Appl. Phys. Lett. **85**, 3056 (2004)
23. F.K. LeGoues, M. Horn-von Hoegen, M. Copel, R.M. Tromp, Strain-relief mechanism in surfactant-grown epitaxial germanium films on Si(111). Phys. Rev. B **44**, 12894 (1991)
24. D. Leonard, K. Pond, P.M. Petroff, Critical layer thickness for self-assembled InAs islands on GaAs. Phys. Rev. B **50**, 11687 (1994)
25. N. Grandjean, J. Massies, V.H. Etgens, Delayed relaxation by surfactant action in highly strained III-V semiconductor epitaxial layers. Phys. Rev. Lett. **69**, 796 (1992)
26. J. Massies, N. Grandjean, Surfactant effect on the surface diffusion length in epitaxial growth. Phys. Rev. B **48**, 8502 (1993)
27. G.L. Price, Critical-thickness and growth-mode transitions in highly strained $In_xGa_{1-x}As$ films. Phys. Rev. Lett. **66**, 469 (1991)
28. N. Grandjean, J. Massies, Epitaxial growth of highly strained $In_xGa_{1-x}As$ on GaAs(001): the role of surface diffusion length. J. Crystal Growth **134**, 51 (1993)
29. E. Tournié, A. Trampert, K.H. Ploog, Interplay between surface stabilization, growth mode and strain relaxation during molecular-beam epitaxy of highly mismatched III-V semiconductor layers. Europhys. Lett. **25**, 663 (1994)
30. W.N. Rodrigues, V.H. Etgens, M. Sauvage-Simkin, G. Rossid, F. Sirotti, R. Pinchaux, F. Rochet, Heteroepitaxial growth of InAs on GaAs(100) mediated by Te at the interface. Solid State Commun. **95**, 873 (1995)
31. D.F. Storm, M.D. Lange, T.L. Cole, Surfactant effects of thallium in the epitaxial growth of indium arsenide on gallium arsenide(001). J. Appl. Phys. **85**, 6838 (1999)
32. G. Feng, K. Oe, M. Yoshimoto, Temperature dependence of Bi behavior in MBE growth of InGaAs/InP. J. Crys. Growth **301**, 121 (2007)
33. Y. Gu, Y.G. Zhang, X.Y. Chen, S.P. Xi, B. Du, Y.J. Ma, Effect of bismuth surfactant on InP-based highly strained InAs/InGaAs triangular quantum wells. Appl. Phys. Lett. **107**, 212104 (2015)
34. W.Y. Ji, Y. Gu, Y.G. Zhang, Y.J. Ma, X.Y. Chen, Q. Gong, B. Du, Y.H. Shi, InP-based pseudomorphic InAs/InGaAs triangular quantum well lasers with bismuth surfactant. Appl. Opt. **56**, H10 (2017)
35. Y. Zhong, P.B. Dongmo, J.P. Petropoulos, J.M.O. Zide, Effects of molecular beam epitaxy growth conditions on composition and optical properties of $In_xGa_{1-x}Bi_yAs_{1-y}$. Appl. Phys. Lett. **100**, 112110 (2012)
36. D. Rioux, H. Höchst, Sb/InP(100) interface: a precursor to surfactant-mediated Ge epitaxy. Phys. Rev. B **46**, 6857 (1992)
37. R.R. Wixom, N.A. Modine, G.B. Stringfellow, Theory of surfactant (Sb) induced reconstructions on InP(001). Phys. Rev. B **67**, 115309 (2003)
38. D. Rioux, H. Höchst, Enhanced structural and electronic properties of strained Ge(100) films grown by molecular-beam epitaxy with a Sb surfactant layer. J. Vac. Sci. Technol. A **10**, 759 (1992)
39. G. Mula, C. Adelmann, S. Moehl, J. Oullier, B. Daudin, Surfactant effect of gallium during molecular-beam epitaxy of GaN on AlN (0001). Phys. Rev. B **64**, 195406 (2001)
40. J.E. Northrup, C.G. Van de Walle, Indium versus hydrogen-terminated GaN(0001) surfaces: Surfactant effect of indium in a chemical vapor deposition environment. Appl. Phys. Lett. **84**, 4322 (2004)

41. Q. Zhuang, W. Lin, J. Kang, Effect of In-adlayer on AlN (0001) and (000-1) polar surfaces. J. Phys. Chem. C **113**, 10185 (2009)
42. H. Chen, R.M. Feenstra, J.E. Northrup, T. Zywietz, J. Neugebauer, D.W. Greve, Surface structures and growth kinetics of InGaN(0001) grown by molecular beam epitaxy. J. Vac. Sci. Technol. B **18**, 2284 (2000)
43. S. Nicolay, E. Feltin, J.-F. Carlin, M. Mosca, L. Nevou, M. Tchernycheva, F.H. Julien, M. Ilegems, N. Grandjean, Indium surfactant effect on AlN/GaN heterostructures grown by metal-organic vapor-phase epitaxy: applications to intersubband transitions. Appl. Phys. Lett. **88**, 151902 (2006)
44. J.L. Merrell, F. Liu, G.B. Stringfellow, Effect of surfactant Sb on In incorporation and thin film morphology of InGaN layers grown by organometallic vapor phase epitaxy. J. Cryst. Growth **375**, 90 (2013)
45. A.A. Gokhale, T.F. Kuech, M. Mavrikakis, A theoretical comparative study of the surfactant effect of Sb and Bi on GaN growth. J. Cryst. Growth **303**, 493 (2007)
46. R. Kern, P. Müller, Three-dimensional towards two-dimensional coherent epitaxy initiated by surfactants. J. Cryst. Growth **146**, 193 (1995)
47. E. Bauer, Phänomenologische Theorie der Kristallabscheidung an Oberflächen. Z. Krist. **110**, 372 (1958) (Theory of crystalline deposition on surfaces, in German)
48. L.G. Wang, P. Kratzer, N. Moll, M. Scheffler, Size, shape, and stability of InAs quantum dots on the GaAs(001) substrate. Phys. Rev. B **62**, 1897 (2000)
49. J.E. Prieto, I. Markov, Stranski-Krastanov mechanism of growth and the effect of misfit sign on quantum dots nucleation. Surf. Sci. **664**, 172 (2017)
50. I.V. Markov, Kinetics of nucleation in surfactant-mediated epitaxy. Phys. Rev. B **53**, 4148 (1996)
51. I.V. Markov, Influence of surface active species on kinetics of epitaxial nucleation and growth. Mater. Chem. Phys. **49**, 93 (1997)
52. I.V. Markov, Nucleation at surfaces, in *Handbook of Crystal Growth*, ed. by G. Dhanaraj, K. Byrappa, V. Prasad, M. Dudley (Springer, Berlin, 2010), pp. 17–52
53. J. Tersoff, R.M. Tromp, Shape transition in growth of strained islands: spontaneous formation of quantum wires. Phys. Rev. Lett. **70**, 2782 (1993)
54. V. Cherepanov, S. Filimonov, J. Myslivеček, B. Voigtländer, Scaling of submonolayer island sizes in surfactant-mediated epitaxy of semiconductors. Phys. Rev. B **70**, 085401 (2004)
55. M.C. Bartelt, J.W. Evans, Scaling analysis of diffusion-mediated island growth in surface adsorption processes. Phys. Rev. B **46**, 12675 (1992)
56. I.V. Markov, Nucleation and step-flow growth in surfactant mediated homoepitaxy with exchange/de-exchange kinetics. Surf. Sci. **429**, 102 (1999)
57. D. Wang, Z. Ding, X. Sun, Nucleation transition and nucleus density scaling in surfactant-mediated epitaxy. Phys. Rev. B **72**, 115419 (2005)
58. E.-Z. Liu, C.-Y. Wang, J.-T. Wang, Dimer-breaking-assisted exchange mechanism in surfactant-mediated epitaxial growth of Ge on Si(001): ab initio total energy calculations. Phys. Rev. B **76**, 193301 (2007)
59. K. Schroeder, A. Antons, R. Berger, S. Blügel, Surfactant mediated heteroepitaxy versus homoepitaxy: kinetics for group-IV adatoms on As-passivated Si(111) and Ge(111). Phys. Rev. Lett. **88**, 046101 (2002)
60. K. Schroeder, B. Engels, P. Richard, S. Blügel, Reexchange controlled diffusion in surfactant-mediated epitaxial growth: Si on As-terminated Si(111). Phys. Rev. Lett. **80**, 2873 (1998)
61. A.D. Howard, G.B. Stringfellow, Effects of low surfactant Sb coverage on Zn and C incorporation in GaP. J. Appl. Phys. **102**, 074920 (2007)
62. J. Zhu, F. Liu, G.B. Stringfellow, Enhanced cation-substituted p-type doping in GaP from dual surfactant effects. J. Cryst. Growth **312**, 174 (2010)
63. L.B. Karlina, A.S. Vlasov, B.Y. Ber, D.Y. Kazantsev, Diffusion of zinc in gallium arsenide with the participation isovalent impurities. J. Cryst. Growth **432**, 133 (2015)

64. D. Reinking, M. Kammler, N. Hoffmann, M. Horn-von Hoegen, K.R. Hofmann, Fabrication of high-mobility Ge *p*-channel MOSFETs on Si substrates. Electron. Lett. **35**, 503 (1999)
65. D. Reinking, M. Kammler, M. Horn-von Hoegen, K.R. Hofmann, High electron mobilities in surfactant-grown Ge on Si substrates. Jpn. J. Appl. Phys. **36**, L1082 (1997)

Chapter 10
Doping, Diffusion, and Contacts

Abstract The ability to control the conductivity is an essential feature of semiconductors. This chapter points out the basics for the control of the free carrier concentration and discusses the nature of limiting factors. The integrity of doping profiles and of interfaces in heterostructures depends on the stability of atoms against a change of lattice site. We briefly consider fundamentals of diffusion and discuss some basic mechanisms governing the diffusivity of atoms in a crystal. The chapter concludes with concepts for ohmic metal-semiconductor contacts.

Device applications of semiconductors require control of the free carrier concentration and application of ohmic contacts for carrier injection or extraction. The conductivity of a semiconductor can be varied over a wide range from semi-insulating to semi-metallic by the introduction of impurities. The substitutional replacement of a semiconductor atom by, e.g., a dopant atom with a chemical valence incremented by one introduces an additional electron and a positive charge at the ion core. For suitable dopant species the electron may thermally be released at room temperature due to a small binding energy, allowing for adjusting the conductivity via the concentration of impurities. The same applies for the creation of free holes by substitutional dopants with a lower chemical valence. The simple concept works particularly well for low doping levels and many semiconductors with a bandgap below about 2 eV. Section 10.1 points out the essentials for conductivity control, limits imposed by thermodynamics, and the nature of limiting factors.

Doping profiles are often created via diffusion by providing a concentration gradient of the doping species. Diffusion phenomena also control the abruptness of semiconductor interfaces. The underlying concepts are treated in Sect. 10.2.

Ohmic contacts are usually fabricated by evaporation of a contact metal on heavily doped semiconductor layers. Besides such non-epitaxial techniques also epitaxial contact structures may provide a solution in special cases. Basic concepts for ohmic metal-semiconductor contacts are discussed in Sect. 10.3.

U. W. Pohl, *Epitaxy of Semiconductors*, Graduate Texts in Physics,
https://doi.org/10.1007/978-3-030-43869-2_10

10.1 Doping of Semiconductors

The use of semiconductors in electronic and optoelectronic devices requires the reliable control of unipolar or bipolar conductivity. Typical concentrations of impurities employed for doping are in the 10^{15}–10^{20} cm^{-3} range, compared to about 5×10^{22} atoms/cm^3 of the host semiconductor. Doping of shallow impurities for both, donors and acceptors is usually well achieved for semiconductors with a sufficiently small bandgap energy such as Si. In contrast, wide-bandgap semiconductors with a gap energy above ~2 eV like, e.g., many II–VI compounds and group-III nitrides can typically be doped either n-type or p-type, but not both. It is exceedingly difficult to achieve n-type conductivity in ZnTe or p-type conductivity in ZnO. Fundamental problems arise from various origins. The solubility of dopants imposes limits for incorporation, a large ionization energy may hamper activation, native defects may compensate an intentional doping, dopants may change their character depending on the incorporation site or lattice relaxation, and eventually hydrogen may passivate dopants. Many of these processes are related to the bandgap energy, and are hence particularly pronounced in wide bandgap semiconductors. The effect is illustrated in Fig. 10.1 for the compensation of p-type doping by a native donor defect. The intentional p-type doping moves the Fermi level E_F to the valence-band edge E_v. The creation of a point defect made of host atoms like, e.g., an interstitial atom consumes some formation energy, but the transfer of the electron from its occupied donor level E_{defect} near the conduction band E_c to the Fermi level may recoup this energy. Since the energy gained by such a compensation is on the order of the bandgap energy, the tendency for native defect compensation increases as the bandgap increases.

Dopant incorporation during epitaxial growth may occur far from thermal equilibrium, potentially allowing to achieve a net doping which cannot be achieved under equilibrium-near conditions. In the following we treat basic concepts to obtain conductivity control.

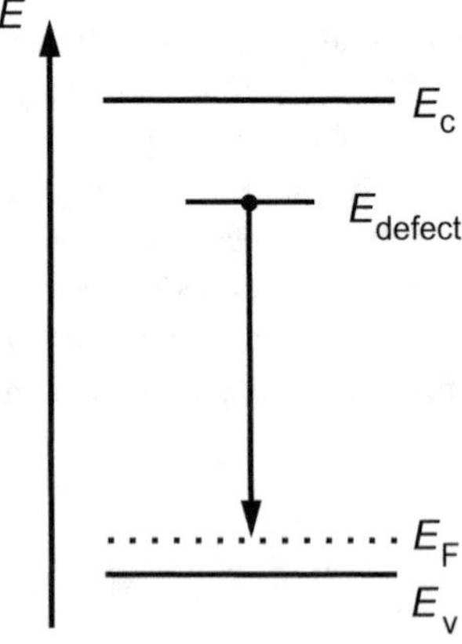

Fig. 10.1 Compensation of p-type doping by a native donor defect, which introduces an occupied level E_{defect} near the conduction band

10.1.1 Thermal Equilibrium Carrier-Densities

The density of electrons n in the conduction band at temperature T is given by the Fermi-Dirac distribution function f_n and depends on the density of energy levels in the conduction band $D_c(E)$ and the value of the Fermi energy E_F,

$$n = \int_{E_c}^{\infty} D_c(E) \frac{1}{e^{(E-E_F)/k_B T} + 1} dE. \tag{10.1}$$

The Fermi energy E_F of charge carriers is also referred to as chemical potential μ. A similar expression as (10.1) holds for the density of holes p in the valence band with density-of-states D_v and a distribution function $f_h = 1 - f_n$,

$$p = \int_{-\infty}^{E_v} D_v(E) \left(1 - \frac{1}{e^{(E-E_F)/k_B T} + 1}\right) dE. \tag{10.2}$$

Equations (10.1) and (10.2) are simplified, if the position of the chemical potential is not in the vicinity of one of the band edges, i.e., if conditions $E_c - E_F \gg k_B T$ or $E_F - E_v \gg k_B T$ hold, respectively. Such semiconductors are called nondegenerate. The Fermi distribution may then be replaced by the Boltzmann distribution by omitting the term "+1" in the denominator. Due to the exponential factors in the simplified integrals only energies within $k_B T$ at the band edges give significant contributions. The band edges are then well described by a quadratic approximation, yielding the density of states

$$D_{c,v}(E) = \frac{\sqrt{2} m_{e,h}^{*\ 3/2}}{\hbar^3 \pi^2} \sqrt{|E - E_{c,v}|} \tag{10.3}$$

with the effective masses of electrons m_e^* and holes m_h^*. The effective masses include potentially occurring degeneracy and anisotropy. For Si, e.g., the conduction band is 6-fold degenerate and comprises longitudinal and transversal mass components, yielding $m_e^* = 6^{2/3} \times (m_t^2 \times m_l)^{1/3}$. Similarly, the hole mass includes heavy-hole and light-hole contributions (the split-off hole is usually not occupied), yielding $m_h^* = (m_{hh}^{3/2} + m_{lh}^{3/2})^{2/3}$.

Using the density of states (10.3), (10.1) and (10.2) for the carrier densities read

$$n = 2\left(\frac{m_e^* k_B T}{2\pi \hbar^2}\right)^{3/2} e^{-(E_c - E_F)/k_B T} = D_c^{eff} e^{-(E_c - E_F)/k_B T}, \tag{10.4a}$$

$$p = 2\left(\frac{m_h^* k_B T}{2\pi \hbar^2}\right)^{3/2} e^{-(E_F - E_v)/k_B T} = D_v^{eff} e^{-(E_F - E_v)/k_B T}. \tag{10.4b}$$

The prefactors of the exponential functions in (10.4a), (10.4b) are the effective densities of states at the edges of conduction and valence bands D_c^{eff} and D_v^{eff}, respectively. The dependence on the until now unknown position of the chemical potential disappears from the *product* of the carrier densities,

$$n \times p = 4\left(\frac{k_B T}{2\pi \hbar^2}\right)^3 \left(m_e^* m_h^*\right)^{3/2} e^{-E_g/k_B T}, \tag{10.5}$$

where $E_g = E_c - E_v$ is the bandgap energy. This relation is called the *law of mass action* and applies for nondegenerate semiconductors. It means that at a given temperature the density of one carrier type is given by the density of the other. The position of the Fermi energy E_F within the bandgap follows from (10.4a), (10.4b). If we resolve (10.4a) for $E_c - E_F$ we obtain

$$E_c - E_F = k_B T \ln\left(\frac{D_c^{\text{eff}}}{n}\right). \tag{10.5a}$$

In devices the electron density n, and consequently E_F, is controlled by doping and basically given by the density of ionized donors N_D and acceptors N_A, i.e., $n \approx N_D - N_A$. But also in an undoped semiconductor the Fermi level is not fixed.

If in a pure crystal the contribution of impurity donors or acceptors to the carrier densities is negligible, the semiconductor is called *intrinsic*. In this case each electron in the conduction band originates from the valence band, $n = p \equiv n_i$. The value of the intrinsic carrier density n_i at any temperature is given by the square root of (10.5). We note that the intrinsic carrier concentration decreases exponentially as the bandgap increases. At $T = 300$ K we find values of, e.g., Ge ($E_g = 0.67$ eV, $n_i \approx 10^{13}$ cm^{-3}), Si (1.12 eV, $\sim 10^{10}$ cm^{-3}), GaAs (1.43 eV, $\sim 10^6$ cm^{-3}), and GaP (2.26 eV, $\sim 10^0$ cm^{-3}), illustrating the trend. In the intrinsic case the position of the chemical potential, E_F, follows from the charge neutrality condition $n = p$ and (10.1), (10.2), yielding

$$E_F = E_v + \frac{E_g}{2} + \frac{3}{4} k_B T \ln\left(\frac{m_h^*}{m_e^*}\right) \quad \text{(intrinsic case).} \tag{10.6}$$

At $T = 0$ the chemical potential lies in the middle of the energy gap. Since m_h^* is typically less than an order of magnitude larger than m_e^*, the last term in (10.6) is of order $k_B T$. This is usually much less than E_g, leading to a near midgap position of E_F at finite temperatures.

To control conductivity *impurities* are introduced in the crystal as sources of free carriers. The semiconductor is then called *extrinsic*. For efficient doping the energy level introduced by a dopant should lie in the bandgap near the band edge to allow for thermal activation at the intended operating temperature. Such impurities are referred to as *shallow impurities*. In a nondegenerate semiconductor the density of band states $D_v(E)$ or $D_c(E)$ is not considerably altered by doping.

Shallow donor impurities are well described in analogy to a hydrogen atom. A substitutional donor has a higher chemical valence than the replaced host atom. A single donor introduces a positive charge at the ion core and an additional electron. The binding energy of the electron is strongly reduced compared to the Rydberg energy $R_\infty = 13.6$ eV of hydrogen due to the electron motion in the medium of the semiconductor with a dielectric constant ε_r. Replacing in the hydrogen problem ε_0

by $\varepsilon = \varepsilon_r \varepsilon_0$, ε_r being the (relative) permittivity of the semiconductor, and the free electron mass m_0 by the effective mass m_e^*, we obtain the binding energy of the donor electron in the ground state

$$E_D^b = \frac{m_e^*}{m_0} \frac{1}{\varepsilon_r^2} \times R_\infty . \tag{10.7}$$

E_D^b is also referred to as donor Rydberg energy or donor ionization energy. The position of the donor energy in the bandgap is $E_D = E_g - E_D^b$. Taking typical values $m_e^* \approx m_0/10$ and $\varepsilon_r \approx 10$ we obtain binding energies of the order 10 meV. Donors which fulfill the approximation (10.7) are referred to as *effective-mass donors*.

The effective-mass and dielectric constant corrections lead to a donor Bohr radius

$$a_B^* = \frac{\varepsilon_r m_e^*}{m_0} \times a_B , \tag{10.8}$$

with the hydrogen Bohr radius $a_B = 0.53$ Å. Inserting typical values noted above yields an order of 50 Å. The electron orbit hence extends over many lattice constants. In such a case the description of a donor Coulomb potential screened by the effective dielectric constant ε_r is a good approximation. The usually small differences found for different shallow donors, termed chemical shift, are accounted for by a central-cell correction to the Coulomb potential.

An acceptor impurity has a lower chemical valence than a host atom. The missing electron is represented by a hole, which is bound to the excess negative charge of the acceptor core. An application of the hydrogen analogy to a shallow acceptor impurity must account for the more complicated structure of the valence band (Sect. 4.1). The dispersion of holes near the center of the Brillouin zone of semiconductors with zincblende or diamond structure is often described by the Luttinger parameters γ_1, γ_2, and γ_3 [1]. $1/\gamma_1$ is a multiplying factor (of order 10^{-1}) in the acceptor binding-energy E_A^b described analogous to (10.7) for the light and heavy holes. To obtain the acceptor ionization-energy an additional multiplication by a function $f(\gamma_1, \gamma_2, \gamma_3)$ is required, f covering values between 1 and 5. A more detailed treatment is given, e.g., in [2]. Rydberg energies of some dopants in semiconductors are given in Table 10.1.

We restrict ourselves to n-type doping with shallow donors to discuss the effect of doping. Electrons in the conduction band originate either from ionized donors or from the valence band, $n = N_D^+ + p$. N_D^+ denotes the density of positively charged donors, which lost their electron by thermal activation. The value follows from the total density of donors N_D and an electron occupation of donors depending on the activation energy E_D^b. Furthermore, the degeneracy g for the occupation of the donor ground-state must be included in the Fermi distribution-function, yielding

$$N_D^+ = N_D \left(1 - \frac{1}{\frac{1}{g} e^{(E_D - E_F)/k_B T} + 1} \right) \quad \text{(only donors)}. \tag{10.9}$$

Table 10.1 Binding energies of donors (E_D^b) and acceptors (E_A^b) in Si, GaAs (donor data from [3], C_Ga and acceptors from [4]), GaN (donor data from [5], acceptor data from [6], low-doping limits), and ZnSe (donor data from [7], acceptor data from [8], $\mathrm{Li_i}$ from [9])

Host	Bandgap E_g (eV)	Donor Rydberg E_D^b (meV)		Acceptor Rydberg E_A^b (meV)	
Si	1.12	$\mathrm{P_{Si}}$	45	$\mathrm{B_{Si}}$	45
		$\mathrm{As_{Si}}$	49	$\mathrm{Al_{Si}}$	5745
		$\mathrm{Sb_{Si}}$	39	$\mathrm{Ga_{Si}}$	65
GaAs	1.42	$\mathrm{S_{As}}$	5.9	$\mathrm{Be_{Ga}}$	28
		$\mathrm{Te_{As}}$	5.8	$\mathrm{Mg_{Ga}}$	29
		$\mathrm{Sn_{Ga}}$	5.8	$\mathrm{Zn_{Ga}}$	31
		$\mathrm{C_{Ga}}$	5.9	$\mathrm{C_{As}}$	27
GaN	3.42	$\mathrm{Si_{Ga}}$	∼22	$\mathrm{Mg_{Ga}}$	∼208
ZnSe	2.67	$\mathrm{Al_{Zn}}$	25.7	$\mathrm{N_{Se}}$	∼110
		$\mathrm{Ga_{Zn}}$	27.5	$\mathrm{P_{Se}}$	∼87
		$\mathrm{In_{Zn}}$	28.2	$\mathrm{As_{Se}}$	∼105
		$\mathrm{Li_i}$	∼26	$\mathrm{Li_{Zn}}$	∼114

For effective-mass donors $g = 2$ due to an occupation with an electron of either spin (up or down). We now consider the semiconductor at low temperature and neglect the small intrinsic contribution of p electrons, i.e., $n_\mathrm{i} \ll N_\mathrm{D}^+$, yielding

$$n = N_\mathrm{D}^+ \quad \text{(low temperature)}. \tag{10.10}$$

At low temperature most donors are not ionized, and the occupation of the conduction band is described by Boltzmann statistics used in (10.4a). We hence may express the chemical potential in (10.9) by

$$e^{E_\mathrm{F}/k_\mathrm{B}T} = \left(n/D_\mathrm{c}^\mathrm{eff}\right) e^{E_\mathrm{c}/k_\mathrm{B}T}. \tag{10.11}$$

This leads to a quadratic equation for the free carrier concentration

$$n^2 + \frac{1}{g} D_\mathrm{c}^\mathrm{eff} e^{-E_\mathrm{D}^\mathrm{b}/k_\mathrm{B}T} n - \frac{1}{g} D_\mathrm{c}^\mathrm{eff} e^{-E_\mathrm{D}^\mathrm{b}/k_\mathrm{B}T} N_\mathrm{D} = 0 \tag{10.12}$$

with the solution

$$n = 2N_\mathrm{D} \left(1 + \sqrt{1 + 4g \frac{N_\mathrm{D}}{D_\mathrm{c}^\mathrm{eff}} e^{E_\mathrm{D}^\mathrm{b}/k_\mathrm{B}T}} \right)^{-1}. \tag{10.13}$$

In the limit of very low temperatures the condition $4g(N_\mathrm{D}/D_\mathrm{c}^\mathrm{eff}) e^{E_\mathrm{D}^\mathrm{b}/k_\mathrm{B}T} \gg 1$ applies. Equation (10.13) then simplifies to

$$n \cong \sqrt{(1/g)N_\mathrm{D}D_\mathrm{c}^\mathrm{eff}}\, e^{-E_\mathrm{D}^\mathrm{b}/2k_\mathrm{B}T} \quad \text{(ionization regime)}. \tag{10.14}$$

In this low-temperature range sufficient donors still have their electron and may be ionized if the temperature is increased. The ionization energy can be derived from the slope of an Arrhenius plot of the carrier density versus reciprocal temperature, see Fig. 10.2. From (10.14) we obtain $E_\mathrm{D}^\mathrm{b} = -2k_\mathrm{B}d(\ln n)/d(1/T)$.

Once all donors are ionized, the carrier density saturates. n is given by the donor concentration and remains constant, independent on temperature,

$$n = N_\mathrm{D}^+ = N_\mathrm{D} \quad \text{(saturation regime)}. \tag{10.15}$$

As the temperature is further increased the carrier concentration again raises due to a thermal activation of electrons from the valence band. The intrinsic carrier concentration increases with a much steeper slope $-E_\mathrm{g}/2k_\mathrm{B}$ in the Arrhenius plot given in Fig. 10.2. In this high-temperature range we hence find

$$n \propto e^{-E_\mathrm{g}/(2k_\mathrm{B}T)} \quad \text{(intrinsic regime)}. \tag{10.16}$$

Pure doping of solely one kind of carriers does hardly occur. In practice dopant atoms of one kind are partially compensated by a smaller number of dopants of the other kind. Let us assume donors with a concentration N_D being partially compensated by residual acceptors with concentration N_A. Since acceptors provide low-energy states for donor electrons, even at lowest temperatures all acceptors are ionized, i.e. $N_\mathrm{A}^- = N_\mathrm{A}$. Consequently the number of donors which may release their electron to the conduction band is reduced by this number, yielding instead of (10.10) now $n = N_\mathrm{D}^+ - N_\mathrm{A}^-$. The most evident change in the free carrier concentration n is given by a decrease of the plateau in the saturation regime to the value $n = N_\mathrm{D} - N_\mathrm{A}$. Furthermore, for a given carrier concentration the mobility is reduced in a compen-

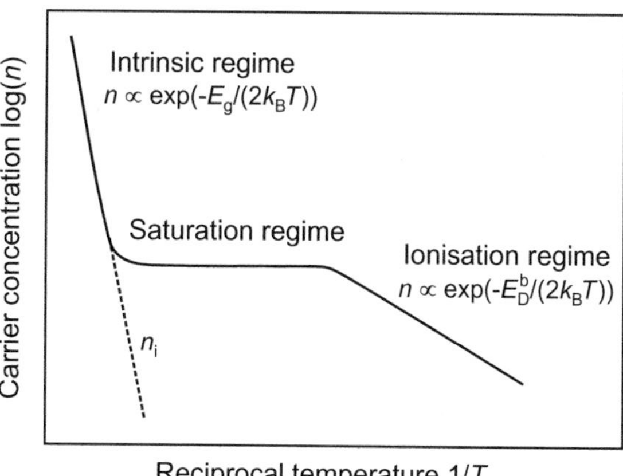

Fig. 10.2 Arrhenius plot of the carrier concentration n in the conduction band of an uncompensated n-type semiconductor. n_i denotes the contribution of intrinsic carriers

sated semiconductor due to additional scattering at ionized dopants. An additional effect is the appearance of a second ionization regime with a slope $-E_{\rm D}^{\rm b}/k_{\rm B}$ instead of $-E_{\rm D}^{\rm b}/2k_{\rm B}$ in the Arrhenius plot at the low-temperature end.

10.1.2 Solubility of Dopants

Doping of a semiconductor host with impurities requires a sufficiently high solubility for the introduction on the intended lattice site. The solubility of a dopant is inherently connected to the dissociation of the impurity atom into an ionized state and a carrier, and is therefore associated with the Fermi level [10]. If, e.g. a donor atom is introduced, energy is gained by incorporating the electron of the donor at the Fermi level. The energy gain decreases at higher doping concentration, because the Fermi level increases by doping. A comparable situation occurs for acceptors. Low dopant concentrations are therefore more readily achieved than high concentrations, and the problem gets increasingly severe as the bandgap energy increases. Incorporation of the dopant at other sites than intended may then get more favorable, and even the formation of phases composed of impurity and host atoms with a configuration deviating from that of the semiconductor may occur.

The creation of a defect is generally connected to the incorporation or removal of atoms from other parts of the crystal, its surface, or the environment. Since this finding and the treatment below apply for both, impurities and native defects, we will just denote them defects in the following. The creation—or annihilation—of a defect is considered by a coupling to subsystems or reservoirs, which donate or accept atoms and, in case of charged defects, also electrons. The equilibrium is described by respective chemical potentials.

We consider the introduction of an impurity by connecting the semiconductor to a reservoir of dopant atoms. The probability of incorporation is related to the formation energy of a respective defect in the crystal. In thermal equilibrium the concentration $[D_i]$ of the dopant (or any defect) in the semiconductor is given by

$$\left[D_i^q\right] = N \exp\left(-\frac{E_{\rm form}(D_i^q)}{k_{\rm B}T}\right). \tag{10.17}$$

$E_{\rm form}(D_i^q)$ is the formation energy of the defect D_i in the charge state q. N is the number of sites where the defect can form, e.g., the number of substitutional cation A sites of an AB compound semiconductor in case of dopant incorporation of this site. The defect-formation energy depends on the chemical potential of the species involved in the creation of the defect and the change of Gibbs free energy required to create the defect. An additional term accounts for the charge of the defect. The general expression reads

$$E_{\text{form}}\left(D_i^q\right) = \Delta G_{\text{form}}\left(D_i^q\right) + \sum_j n_j \mu_{j\,(\text{reservoir})} + q E_{\text{F}}, \tag{10.18}$$

where q is the charge state (e.g., -1) and E_{F} is the Fermi energy with respect to the valence-band edge E_{v}. The index j in the sum of (10.18) goes over all chemical species involved in the defect creation. The number n_j of species j is positive if an atom is removed from the semiconductor and negative if it is added. μ_j are the respective chemical potentials. The energy to substitute, e.g., a negatively charged Li acceptor for Zn in a ZnSe semiconductor reads

$$E_{\text{form}}\left(\text{Li}_{\text{Zn}}^-\right) = \Delta G_{\text{form}}\left(\text{Li}_{\text{Zn}}^-\right) + \mu_{\text{Zn}} - \mu_{\text{Li}} - E_{\text{F}}.$$

The change of Gibbs free energy $\Delta G_{\text{form}}(D_i^q)$ is given by

$$\Delta G_{\text{form}}\left(D_i^q\right) = \Delta E_0 - T\Delta S + P\Delta V, \tag{10.19}$$

ΔE_0 being the difference between the electronic ground-state energy of the system with and without the defect. Correspondingly ΔS and ΔV are the respective entropy and volume changes due to the creation of the defect. The contributions related to a volume change and the entropy change are usually expected to yield only minor corrections and are not considered. A positive sign of ΔG_{form} denotes an energy cost to create the defect.

The charge state of the defect will be that with lowest formation energy. Depending on E_{F} the defect may be in different charge states. The energy where the defect changes the charge state from q to q' is referred to as charge-transfer energy $E^{q/q'}$. Figure 10.3 illustrates how the formation energy depends on the charge state q and the Fermi energy E_{F}, which is varied between valence and conduction band. In the example given in Fig. 10.3 the defect is negatively charged if E_{F} lies above $E^{0/-}$, because in this range $E_{form}(D^-) < E_{\text{form}}(D^0)$. Below $E^{0/-}$ the defect is in its neutral charge state. The level in the bandgap $E^{0/-}$ is the value of the Fermi energy at which the two charge states $E_{\text{form}}(D^0)$ and $E_{\text{form}}(D^-)$ have the same energy.

The quantities noted in the equations above are accessible by first-principles calculations. The electrical interaction of the defect is treated analogous to the chemical interaction of atoms with their atomic reservoirs: The defect donates or accepts elec-

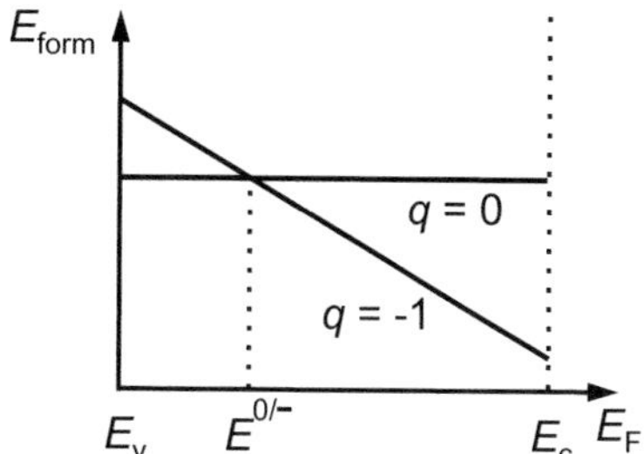

Fig. 10.3 Formation energy of a defect in a neutral and a negative charge state. The Fermi level E_{F} is varied between the energy of the valence band E_{v} and the conduction band E_{c}

trons from the electron reservoir of the semiconductor. A negative charge denotes that the defect accepts q electrons. The position of the Fermi energy is calculated self-consistently by applying (10.18) for all relevant defects and accounting for the condition of charge neutrality

$$\text{net charge} = 0 = p - n - \sum_i n_i^{\text{electron}} \left[D_i^q \right]. \tag{10.20}$$

p and n are the hole and electron densities, and n_i^{electron} is the number of excess electrons in the defect D_i^q. Details of the computational approach are given in [11] for the example of ZnSe doping. Within this framework performed in a supercell geometry the change of Gibbs free energy ΔG_{form} in (10.18) is the total energy of a supercell containing the defect minus the energy of a supercell of the pure bulk semiconductor. For an elementary semiconductor like Si the chemical potential, μ_{Si}, is fixed, while for an AB compound semiconductor this applies for the *sum* of the constituents $\mu_{AB} = \mu_A + \mu_B$, leaving an individual summand variable.

Bounds on chemical potentials arise from the phases that can be formed by the constituents. For the *host atoms* the upper bound is given by formation of the respective elements. In ZnSe, e.g., $\mu_{\text{Zn}}^{\max} = \mu_{\text{Zn(bulk)}}$: Further increase above this level will preferentially lead to the formation of Zn metal. The same applies for $\mu_{\text{Se}}^{\max}$. A lower bound is imposed by the heat of formation ΔH_{form} of the semiconductor and the fixed sum of the chemical potentials noted above. For our example we obtain $\mu_{\text{ZnSe}} = \mu_{\text{Zn(bulk)}} + \mu_{\text{Se(bulk)}} + \Delta H_{\text{form}}(\text{ZnSe})$, yielding $\mu_{\text{Zn}}^{\min} = \mu_{\text{Zn(bulk)}} + \Delta H_{\text{form}}(\text{ZnSe})$. Note that $\Delta H_{\text{form}} < 0$ for the stable semiconductor. For the *dopant* the various compounds which the impurity can form including host atoms must be considered. For the example of a p-type doping of ZnSe using Li the most stringent bound is found to be given by the formation of the LiSe_2 compound [12], yielding the constraint on the chemical potential $2\mu_{\text{Li}} + \mu_{\text{Se}} = \mu_{\text{Li}_2\text{Se}} = 2\mu_{\text{Li(bulk)}} + \mu_{\text{Se(bulk)}} + \Delta H_{\text{form}}(\text{Li}_2\text{Se})$.

To obtain the solubility and doping effect of a dopant we have to include all relevant configurations and charge states of the impurity as expressed by (10.18). To be specific we continue to analyze the Li acceptor in ZnSe [12]. Besides the substitutional site Li_{Zn}^- where Li acts as an acceptor we particularly find two interstitial sites Li_{i}^+ were it acts as a shallow donor. Calculations show that the tetrahedral site surrounded by Se is by 0.2 eV more favorable than the respective site with Zn atoms. The substitutional Se site may be excluded due to a very large formation energy.

Results of first-principles calculations for Li doping of ZnSe are displayed in Fig. 10.4 [12]. The contour plot Fig. 10.4a shows the total concentration of Li in ZnSe at an equilibrium temperature of 600 K, corresponding to the typical low growth temperature of ZnSe. Figure 10.4b gives the resulting Fermi energy.

In Fig. 10.4a the bounds of the chemical potentials of Zn and Li are indicated by straight gray lines. The formation of the Li_2Se compound leads to the bound for μ_{Li} with a slope of 2 due to the dependence of $\mu_{\text{Li}_2\text{Se}}$ from μ_{Li} noted above. Li_2Se islands were actually found to form during molecular beam epitaxy of heavily

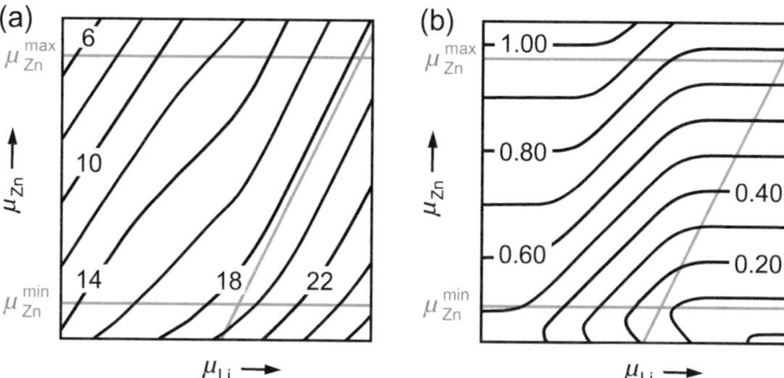

Fig. 10.4 **a** Total Li concentration $\log_{10}$[Li] in a ZnSe:Li semiconductor doped at 600 K. Values depend on the chemical potentials of Li and Zn, and are given in cm^{-3}. **b** Fermi level resulting from the Li doping shown in (**a**). Values are given in eV and refer to the top of the valence band. Reproduced with permission from [12], © 1993 APS

doped ZnSe:Li [13]. The calculated maximum of Li concentration of ZnSe:Li is slightly above 10^{18} cm^{-3} [12]. The coinciding slope of the contour lines near the Li_2Se-related bound is accidental and originates from the incorporation of *two* Li atoms (one substitutional, one interstitial) for the removal of *one* Zn atom. The highest Li concentration (1.7×10^{19} cm^{-3}) is obtained at the crossover point of a minimum accessible Zn chemical potential μ_{Zn}^{min} and a maximum accessible Li chemical potential before formation of Li_2Se. At this point the calculation yields fewer than 3% of the introduced Li atoms on an interstitial site and the most abundant native ZnSe defect of the Se antisite Se_{Zn}^{2+} (a donor) with two orders of magnitude concentration below that of Li. Native defect concentrations become sizeable only below the physically meaningful limit μ_{Zn}^{min} and lead to a bending of the contour lines. If the equilibrium temperature is lowered the concentrations of all defects reduce. Reduction factors of about 5, 10^1, and 10^2 for substitutional Li, interstitial Li, and the Se antisite, respectively, are found as the temperature decreases by 100 °C [12].

The Fermi energy resulting from the Li concentration of Fig. 10.4a is shown in Fig. 10.4b. We note that the Fermi level decreases at fixed μ_{Li} as the Zn chemical potential is lowered. This trend results from the increasingly favored incorporation of Li on a Zn site and is also reflected in the Li concentration of Fig. 10.4a. If μ_{Li} is increased at fixed μ_{Zn} we note from Fig. 10.4b that the decrease of the Fermi level tends to saturate despite the further increase of the total Li concentration shown in Fig. 10.4a. The reason is a steady increase by compensating interstitial Li donors that limits the hole concentration. The effect is minimized by decreasing the Zn chemical potential.

Results outlined above for the specific case of Li exemplify the general behavior of dopant impurities. For any dopant limits of the solubility are imposed by the compounds that can be formed by the participating atoms. The relevant bound of the dopant chemical potential is imposed by the phase with minimum heat of formation.

Saturation of the Fermi-level position occurs when compensating species are formed. Such species are extrinsic species of amphoteric dopants or intrinsic native defects and are addressed in the following sections.

In addition to isolated point defects treated above basically also *complexes* may affect the electrical properties. The concentration of a given complex (e.g., a pair of substitutional and an interstitial Li denoted (Li^-_{Zn}, Li^+_i)) gets significant if the binding energy exceeds the larger of the two formation energies of the individual defects out of which the complex is formed [12]. This makes it less likely that complexes have a significant influence, since complexes with a low binding energy tend to dissociate at growth temperatures.

10.1.3 Amphoteric Dopants

An amphoteric dopant is one that can act as either a donor or an acceptor. The word “amphoteric” originates from the Greek word amphoteroi (αμφοτεροι) and means “having two characters”. Amphoteric behavior can occur if the dopant occupies different lattice sites like lithium in ZnSe as illustrated above. Prominent examples are group-IV impurities (C, Si, Ge, Sn) in III–V semiconductors that form donors on group-III sites and acceptors on group-V sites. Obviously incorporation of an undesired site limits the intended doping effect. Such a compensation due to the same chemical species is referred to as *autocompensation*.

The discussion of the amphoteric behavior of Li in the preceding section indicates that the degree of compensation depends on the applied equilibrium conditions. Anion-poor conditions applied during doping of III–V semiconductors by group-IV impurities favor incorporation on an anion site, while anion-rich conditions favor incorporation on a cation site. This behavior is demonstrated in Fig. 10.5a for the incorporation of carbon in GaAs during metalorganic vapor-phase epitaxy [14]. C doping originates from the organic CH_3 ligands of the Ga source trimethylgallium $Ga(CH_3)_3$. A low V/III ratio in the gas phase of the applied arsenic source AsH_3 with respect to $Ga(CH_3)_3$ leads preferentially to C_{As} acceptors, while a high ratio favors C_{Ga} donors.

Ge is an impurity with a pronounced amphoteric character in III–V semiconductors. Both *n*- and *p*-type conductivity can hence be obtained depending on growth conditions. Besides the V/III ratio incorporation is also strongly affected by kinetics on the growth surface. GaAs surfaces stabilized by As favor incorporation on a Ga site yielding a Ge_{Ga} donor. This represents the usual case of molecular beam epitaxy on (001) GaAs surfaces at temperatures below 630 °C, i.e., below the congruent sublimation temperature of GaAs. Above this temperature Ga-rich surfaces occur leading to Ge_{As} acceptors. Even below 630 °C a gradual increase of the concentration of Ge on As is observed as the temperature is raised [15]. This behavior is shown in Fig. 10.5b. Temperature increase leads to a gradual reevaporation of As from the

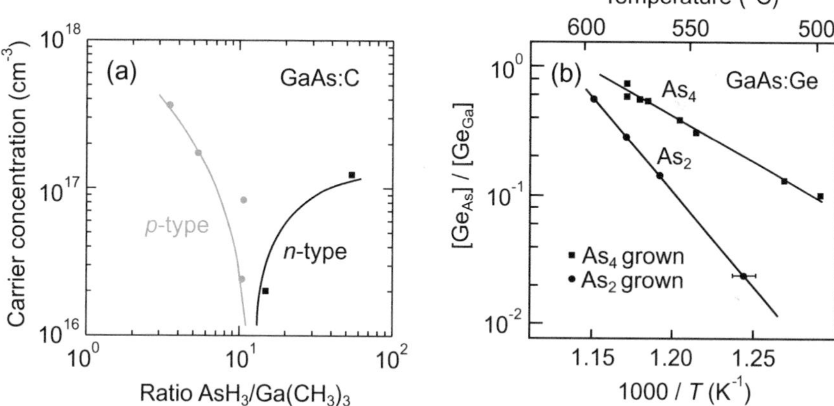

Fig. 10.5 **a** Doping of GaAs with C from organic ligands in metalorganic vapor-phase epitaxy for a varied ratio of arsenic and gallium sources in the gas phase. Data points are from [14]. **b** Autocompensation ratio of the doping of GaAs with Ge in molecular beam epitaxy at varied temperature for different arsenic species. Reproduced with permission from [15], © 1982 Springer

surface and hence increased occupation of Ge on As sites, i.e., to compensating acceptors. The more reactive As_2 species are expected to yield a higher As coverage of the surface and hence a smaller compensation ratio.

Amphoteric behavior may also occur for a defect on a given lattice site, i.e., without occupation of different inequivalent sites. If the defect has several levels in the bandgap like, e.g., Au in Ge, its character depends on the position of the Fermi level. We illustrate such a behavior below for a native defect pair.

10.1.4 Compensation by Native Defects

Native defects represent an equilibrium phenomenon and occur in any solid. The abundance of intrinsic charged point defects has a strong impact on the electronic properties of semiconductors. Such defects comprise vacancies, interstitial atoms, and, in case of compound semiconductors, antisites. In compound semiconductors the abundance of native defects is particularly sensitive to a deviation from stoichiometry. With about 5×10^{22} cm^{-3} atomic sites even a slight deviation from perfect stoichiometry as small as 10^{-4} leads to a defect concentration in the range 10^{18} cm^{-3}. Generally native defects which accommodate deviation from stoichiometry are those that compensate *majority* carriers. Since the energy gained by such a compensation increases as the bandgap energy increases, early work on wide-gap semiconductors focused particularly on self-compensation by native defects [16]. Certainly the effect contributes significantly to doping problems, albeit autocompensation and solubility may impose more rigorous limits.

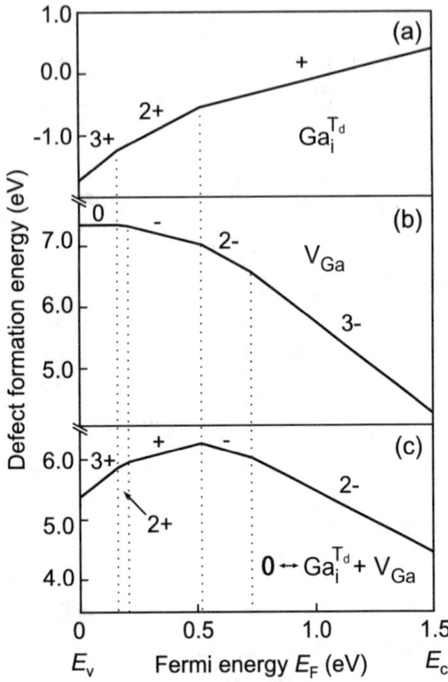

Fig. 10.6 Calculated formation energies of defects in GaAs. **a** Isolated interstitial Ga atom Ga_i at a tetrahedral site with 4 nearest As atoms, **b** isolated vacancy at the Ga site V_{Ga}, and **c** a vacancy-interstitial pair $Ga_i + V_{Ga}$. Defect energies in (**a**) and (**b**) are arbitrarily normalized. The origin of the Fermi-energy scale is set to the maximum of the valence band E_V. After [17]

We study the effect of a native defect for the vacancy-interstitial pair in GaAs created by a Ga atom which moved from a regular lattice site to an interstitial position. The pair formation is described by the reaction $0 \leftrightarrow Ga_i + V_{Ga}$. Calculated energies are given in Fig. 10.6 [17]. The ionization levels are in qualitative agreement with more recent calculations [18, 19]. Both isolated defects of the vacancy-interstitial pair have several charge-transfer energies in the band gap. For a Fermi level near the valence-band edge in p-type GaAs, the interstitial Ga_i defect with three-fold positive charge has lowest formation energy. The defect hence traps holes, thereby compensating the extrinsic doping. As the Fermi level is increased, less holes are trapped by Ga_i due to an increased formation energy and gradually more electrons are trapped by the vacancy V_{Ga} defect. As a result, the charge state of the vacancy-interstitial pair gets more negative. We note that the formation energy depicted in Fig. 10.6c decreases for *either* high doping. Both, p-type and n-type doping are hence compensated by the vacancy-interstitial pair.

The equilibrium concentration of intrinsic defects is given by (10.17) which was also used to describe extrinsic defects. The abundance of native defects hence depends on temperature, doping level, and the deviation from ideal stoichiometry. The dependence on the doping level, i.e., on the electron chemical potential, is illustrated in Fig. 10.6.

We now turn toward the dependence on the chemical potentials of the atoms. Limits for the stoichiometry are imposed by thermodynamics. We express these bounds by applying the approach discussed in Sect. 10.1.2 for impurity doping of

ZnSe. The chemical potentials of the constituents Ga and As may not exceed their respective bulk values, i.e., $\mu_{\mathrm{Ga}}^{\max} = \mu_{\mathrm{Ga(bulk)}}$, $\mu_{\mathrm{As}}^{\max} = \mu_{\mathrm{As(bulk)}}$, and their sum equals the chemical potential of bulk GaAs, $\mu_{\mathrm{Ga}} + \mu_{\mathrm{As}} = \mu_{\mathrm{GaAs}}$. The difference of the chemical potentials $\Delta\mu = \mu_{\mathrm{Ga}} - \mu_{\mathrm{As}} - (\mu_{\mathrm{Ga(bulk)}} - \mu_{\mathrm{As(bulk)}})$ is then limited by the heat of formation ΔH_{form}(GaAs) of bulk GaAs from elemental Ga and As, $|\Delta\mu| \le |\Delta H_{\mathrm{form}}|$, with $|\Delta H_{\mathrm{form}}| = |\mu_{\mathrm{GaAs}} - \mu_{\mathrm{Ga(bulk)}} - \mu_{\mathrm{As(bulk)}}|$. We recall that $\Delta H_{\mathrm{form}} < 0$, because GaAs is a stable compound. Ga-rich and As-rich bounds of $\Delta\mu$ are given by $\Delta\mu = +|\Delta H_{\mathrm{form}}|$ and $\Delta\mu = -|\Delta H_{\mathrm{form}}|$, respectively.

We consider some native defects in GaAs for a fixed doping level and temperature. Figure 10.7a shows the concentration of the Ga vacancy V_{Ga} and the Ga antisite $\mathrm{Ga_{As}}$ in n-type GaAs [20]. For high doping V_{Ga} is triply negatively charged as shown in Fig. 10.6. In the As-rich limit V_{Ga}^{3-} is the dominant defect with a concentration of about 1/3 of the effective doping level N_{d} [20]. Under these conditions the formation energy is so low that about one such defect is formed for every three electrons introduced by doping. We find a strong decrease of the V_{Ga}^{3-} abundance by more than 10 orders of magnitude as $\Delta\mu$ is increased toward Ga-rich conditions. Two reasons account for this finding. First, the formation energy of V_{Ga} increases. This effect is accompanied by a decrease of the electron chemical potential that further increases the formation energy of V_{Ga}^{3-} [20]. Second, As-rich conditions provide an effective sink for removed Ga atoms at the surface. This counteracts thermal equilibrium for Ga interstitial atoms and consequently, by the law of mass action, enhances the concentration of vacancies.

The decrease of V_{Ga}^{3-} in Ga-rich n-type GaAs is accompanied by an increase of the antisite $\mathrm{Ga_{As}}$, the formation energy of which linearly decreases as $\Delta\mu$ is raised. It becomes the dominat defect in the Ga-rich limit and acts also as a compensating electron trap under these conditions.

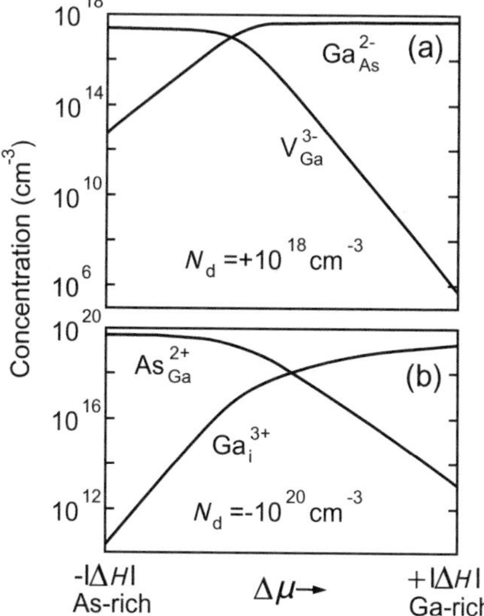

Fig. 10.7 **a** Calculated equilibrium concentration of the Ga vacancy and the Ga antisite as a function of the difference of the chemical potentials of Ga and As atoms under n-type conditions at 827 °C. **b** Concentration of the Ga interstitial and the As antisite for p-type conditions at 827 °C. Reproduced with permission from [20], © 1991 APS

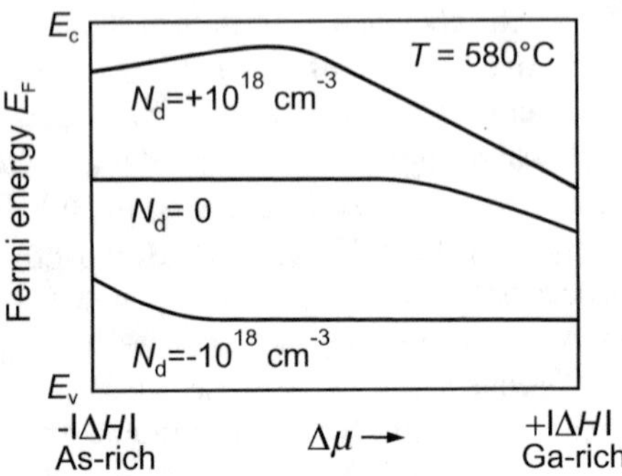

Fig. 10.8 Calculated position of the Fermi energy in GaAs as a function of the difference of the chemical potentials of Ga and As atoms for three fixed doping levels $N_d = N_D^+ - N_A^-$. Reproduced with permission from [20], © 1991 APS

In strongly p-type doped GaAs the dominant native defect under Ga-rich conditions is the interstitial Ga atom Ga_i at a tetrahedral site with 4 nearest As atoms. At this doping level it acts as a triple hole trap. Below 10^{17} cm^{-3} p-type doping it mainly appears in the singly charged state [20]. Decreasing $\Delta\mu$ toward As-rich conditions leads to a decrease of the Ga_i^{3+} abundance, accompanied by an increase of the As antisite As_{Ga}^{2+}, which also acts as a hole trap.

Charged native defects have an influence on the position of the Fermi level due to the charge neutrality condition (10.20). The effect on E_F becomes obvious by explicitly expressing the hole and electron densities in terms of Boltzmann statistics,

$$\begin{aligned} N_d &= N_D^+ - N_A^- \\ &= D_c^{eff} \exp\bigl(-(E_c - E_F)/k_B T\bigr) - D_v^{eff} \exp\bigl(-(E_F - E_v)/k_B T\bigr) \\ &\quad + \sum_i n_i^{electron} \bigl[D_i^q\bigr]. \end{aligned} \tag{10.21}$$

Here $[D_i^q]$ is the concentration of a native defect, and D_c^{eff} and D_v^{eff} are the effective conduction-band and valence-band densities of states defined in (10.4a)–(10.4b). Figure 10.8 shows the calculated dependence of the Fermi-level position on native defects created by a variation of the stoichiometry [20]. We note significant variations of E_F for fixed concentrations of excess free charge carriers. The strong decrease for Ga-rich conditions in n-type GaAs ($N_d = +10^{18}$ cm^{-3}) originates from charge compensation by the antisite defect Ga_{As}^{2-}, the slight increase on the As-rich side originates essentially from the vacancy V_{Ga}^{3-}.

10.1.5 Hydrogen Compensation and Passivation

Hydrogen is easily incorporated in semiconductors, particularly in MOVPE and related gas-source techniques. Due to its generally high reactivity and diffusivity it has a strong effect on electronic properties. Interstitial hydrogen is amphoteric and assumes the positive charge state H^+ in p-type and the negative state H^- in n-type semiconductors [21]. It thereby counteracts the prevailing conductivity: an acceptor dopant has a negative charge state A^- if it has created a free hole in the valence band; hydrogen then acts as a donor according $H^0 \rightarrow H^+ + e$. Vice versa it acts as an acceptor if donors create n-type material according to $H^0 + e \rightarrow H^-$. Hydrogen hence tends to *compensate* doping. The position of the Fermi level where hydrogen changes directly from the H^+ donor state to the H^- acceptor state depends on the band structure of a semiconductor on an absolute energy scale (cf. Sect. 4.2.3) and lies for most semiconductors and insulators in the bandgap. For many solids this energy was determined by first-principles calculations based on density-functional theory and ab initio pseudo potentials [21].

The positively charged, highly mobile H^+ in p-type material is attracted by the Coulomb field of the negatively charged ionized acceptor. This may easily lead to the formation of a neutral $(A\mathrm{H})^0$ complex, a process referred to as *passivation*. A similar process may occur in n-type material with ionized donors D^+ and H^-, yielding a neutral $(D\mathrm{H})^0$ complex. Passivation and compensation both reduce the density of free carriers, but the effects are actually different. In a Hall measurement the creation of neutral complexes leads to a reduced ionized-impurity scattering and hence an increased free-carrier mobility by passivation; in compensation, however, ionized hydrogen and dopants are located at different (random) positions and both contribute to ionized-impurity scattering—the mobility is hence *decreased*.

Passivation of shallow dopants by hydrogen is known for many technologically important semiconductors. Often the effect is stronger for p-type doping than for n-type doping. Prominent examples are p-GaN:Mg [22], p-GaAs:Zn [23], p-Si:B and n-Si:P [24], 6H p-SiC:Al, p-SiC:B and n-SiC:N [25].

10.1.6 DX Centers

A number of impurities introduce levels in the bandgap that are far away from the conduction-band and valence-band edges. Due to their large ionization energy they are able to trap free charge carriers, thereby increasing the resistivity of a semiconductor. Prominent examples are transition metals like Cr or Fe in GaAs or InP. According to the position of the charge-transfer level in the bandgap they are classified into donor-like or acceptor-like centers. A common feature is a strong localization of their wave-function compared to that of shallow impurities. Deep centers affect many properties of semiconductors, e.g., they compensate doping, reduce minority-carrier lifetime and diffusion length, and lead to a reduced carrier mobility.

There exists a large variety of different kind of centers introducing deep levels in semiconductors. We focus here on a specific kind of deep centers related to a structural instability of impurities. The motivation to emphasize these so-called DX centers is their amphoteric character, which converts an intentional shallow dopant to a deep-level impurity.

The label DX refers to a complex of a donor D and an unknown (at the time of discovery) intrinsic defect X. A comparable case for acceptors is referred to as AX

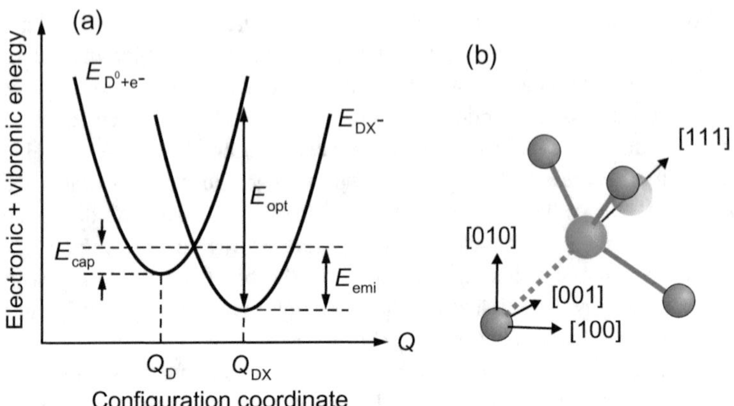

Fig. 10.9 **a** Configuration-coordinate diagram representing the total energy of a DX center and a substitutional donor. **b** Structural model of a DX center. The *dotted line* indicates the bond which breaks when the tetrahedrally coordinated impurity (*red*) is displaced along a ⟨111⟩ direction to a position depicted by the transparent atom

center. Experiments found a large energy difference between the optical and thermal ionization energies, indicating that the deep state is strongly coupled to the crystal lattice [26].

The generally accepted model of a DX center in a tetrahedrally coordinated compound semiconductor is illustrated in Fig. 10.9. The donor has two configurations: A substitutional site where it acts as a shallow donor, and a relaxed site where one of the four bonds with two electrons is broken and the impurity is displaced along one of the ⟨111⟩ directions [27]. The broken bond creates two dangling bonds which may be occupied by up to four electrons. The (neutral) donor thus can hold its own extra electron and accept an additional electron according to the reaction $D^0 + e^- \rightarrow DX^-$. Since the ionization of a shallow donor is described by $D^0 \rightarrow D^+ + e^-$, an electron transfer between the donors is given by the sum of the two reactions, yielding

$$2D^0 \rightarrow D^+ + DX^- . \tag{10.22}$$

The two donors do not need to be in close proximity to each other. Reaction (10.22) indicates that one half of all donor atoms may exist in the broken-bond configuration and compensates thereby the free electrons released from the other half in the substitutional shallow configuration.

Donor centers which yield an exothermic reaction according to (10.22) are also referred to as negative-U centers. U denotes the effective electron-electron correlation energy of the two centers. A negative correlation energy is obtained if the energy gained by the lattice distortion is larger than the energy cost for the electron repulsion.

The instability of specific dopants against formation of DX centers is assigned to a large Jahn-Teller distortion of the highly symmetric substitutional site. The strength of the effect depends on the bonding character and hence on both, the dopant *and* the host crystal. Ga, e.g., forms a stable substitutional donor on a Zn site in ZnSe, but is instable against formation of a DX center in ZnTe [28]. For a (meta-stable) Ga-related DX center in ZnSe the corresponding parabola (E_{DX^-}) in the configuration-coordinate diagram has a minimum at Q_{DX} with a *higher* energy than the minimum of the parabola ($E_{D^0+e^-}$) at Q_D.

Substitutional Si in GaAs forms a stable donor on a Ga site, while it builds a well-studied DX center in $Al_xGa_{1-x}As$ for compositions $x > 0.22$ (below this value the DX level lies in the conduction band) [26]. Electron capture into the DX center is thermally activated with a composition-dependent energy barrier E_{cap} (>0.2 eV) as depicted in Fig. 10.9a. The barrier E_{emi} for thermal emission of carriers from the DX center is larger (~0.4 eV). Carriers may as well be released by optical absorption with a large photon energy E_{opt} (>1 eV). Optically released carriers remain in the conduction band (for days) due to a very low recapture rate and generate a persistent photoconductivity. Other donors like, e.g., Te, Se, and Sn show similar DX characteristics. The model outlined in Fig. 10.9 accounts well for the experimental findings.

10.1.7 Fermi-Level Stabilization Model

The amphoteric character of intrinsic defects and the formation of related deep levels illustrated above leads to a phenomenological model to account for the widely differing doping ability of semiconductors. The *Fermi-level stabilization model*, also termed *amphoteric defect model* or *doping pinning rule*, points out trends for intrinsic limitations to account for, e.g., the difficulties for achieving n-type ZnTe or p-type ZnO. The model ties in with the empirical rule of naturally fixed energy levels of transition-metal impurities in different semiconductors (Sect. 4.2.3). Similarly, clear evidence for the localized nature of native defects is found. The Fermi-level stabilization model therefore assumes that doping limitations reflect the absolute position of the valence and conduction band edges with respect to a fixed reference energy like, e.g., the vacuum level. Doping restrictions are hence *not* assigned to the size of the bandgap per se or to properties of particular dopants.

The reference level for the natural alignment of the band edges with respect to the ability of doping is termed Fermi-level stabilization energy E_{FS}. Indication for such an internal reference was concluded from semiconductors, which were heavily damaged with gamma rays or electrons [29]. For a high density of damage, where material properties are controlled by native defects, the Fermi level was found to stabilize at a certain energy and becomes insensitive to further damage. This energy E_{FS} is located at an energy of about 4.9 eV below the vacuum level for the studied tetrahedrally bonded semiconductors [30]. The internal reference E_{FS} allows

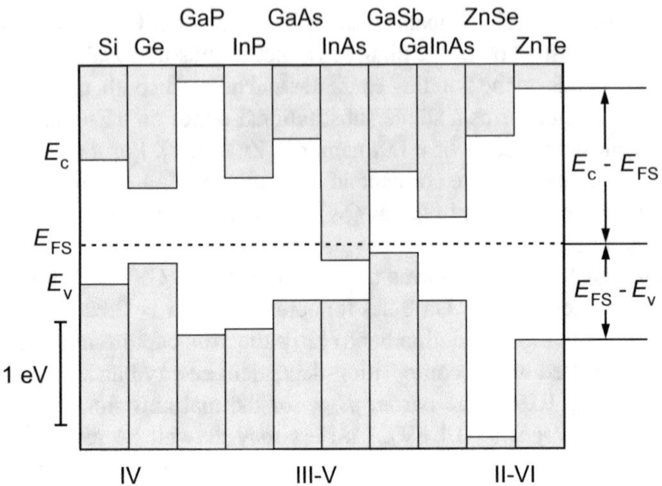

Fig. 10.10 Valence band maxima and conduction band minima of various semiconductors aligned with respect to their internal Fermi-level stabilization energy E_{FS}. Data from [31]

to arrange the bands of all semiconductors on a common energy scale. Figure 10.10 shows the band alignment of some tetrahedrally coordinated semiconductors.

The position of E_{FS} with respect to the band edges depicted in Fig. 10.10 affects the ability to achieve high carrier concentration of a given type. The formation energy of compensating native defects depends on the difference between the Fermi energy and the stabilization energy E_{FS}. If E_{FS} lies close to the valence band, the Fermi level should easily be moved into the valence band by p-type doping without significant formation of compensating defects. Consequently high p-type carrier concentration is expected. On the other hand, a large energy difference to the conduction band will lead to a high abundance of compensating defects if E_F is raised by n-type doping, resulting in solely poor n-type carrier concentration. The asymmetry of p and n-type doping increases as the bandgap increases.

Experimental data of maximum carrier concentrations achieved are given in Fig. 10.11. Electron and hole concentrations are normalized by the effective density of states D_c^{eff} and D_v^{eff}, respectively. Normalized values are used, because electron concentrations are given by $n = D_c^{eff} \times F_{1/2}[(E_F - E_c)/k_BT]$ and hole concentrations p accordingly with D_v^{eff} and $(E_v - E_F)$, where $F_{1/2}$ is the Fermi integral. Figure 10.11 clearly shows that the achieved normalized doping levels decrease as the energy separation between the stabilization energy E_{FS} and the band edges E_c or E_v increase. The data can be roughly described by the empirical relation [31]

$$cc_{max}/D^{eff} = a \times \exp\bigl(b|E_{FS} - E_{band\ edge}|\bigr).$$

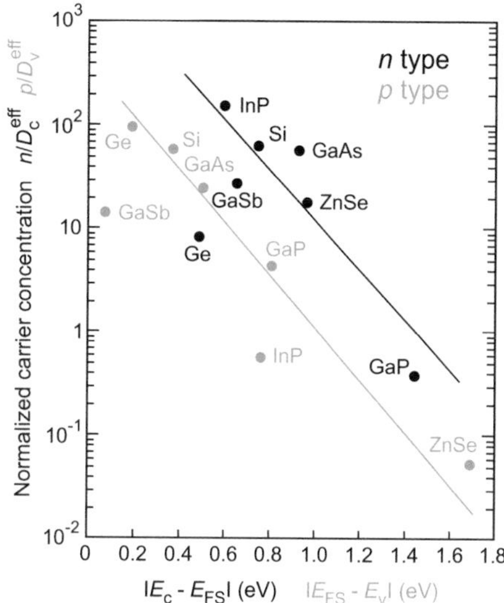

Fig. 10.11 Experimental maximum carrier concentrations for n-type (*black circles*) and p-type semiconductors (*gray circles*) as a function of the energy difference between E_{FS} and either the conduction-band edge E_c, or the valence-band edge E_v, respectively. The solid lines are fits. Data taken from [31]

Here, cc_{max} is the maximum carrier concentration n or p with the related effective density of states and the related band-edge energies E_c or E_v, and the parameters are $a_n = 2.7 \times 10^3$, $b_n = -5.5\ \text{eV}^{-1}$, $a_p = 4.0 \times 10^2$, $b_p = -6.1\ \text{eV}^{-1}$.

The maximum carrier concentrations achieved in experiments can be expressed in terms of pinning energies $E_{pin}^{(n)}$ and $E_{pin}^{(p)}$ of the Fermi energy in n-type and p-type semiconductors, respectively. In the approximation of single parabolic bands the maximum net free carrier concentration cc_{max} is given by [32]

$$cc_{max} = \frac{(2m^*_{(n/p)})^{3/2}}{2\pi^2} \int_0^\infty \frac{\sqrt{E}\, dE}{\exp((E - E_{pin}^{(n/p)})/k_B T) + 1}.$$

The equation is inverted to obtain the pinning energies $E_{pin}^{(n)}$ and $E_{pin}^{(p)}$ from the experimental cc_{max} data. The resulting pinning energies are given in Fig. 10.12 for various III–V and II–VI compound semiconductors. We note a fairly small scatter in the pinning energy, if the bands are aligned with respect to an absolute reference energy.

10.1.8 Delta Doping and Modulation Doping

Epitaxial growth allows for fabricating highly doped buried layers with a very small thickness in the range of atomic monolayers referred to as *delta doping* (δ doping) [33, 34]. The spacially inhomogeneous doping profile provides free carriers and an electrical potential which differs from that of the conventional homogeneous doping.

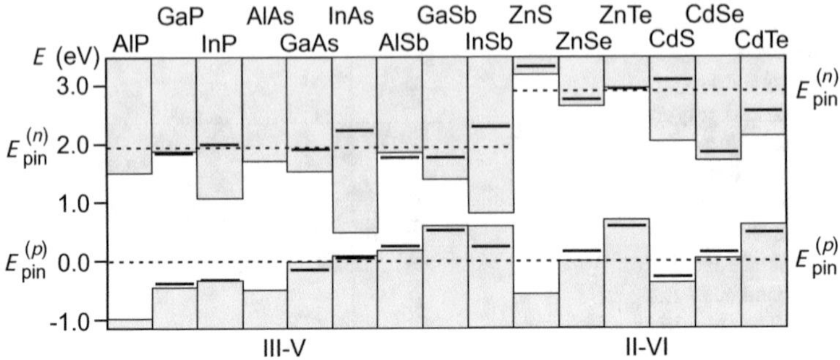

Fig. 10.12 Pinning energies (*black bars*) calculated from experimental maximum carrier concentrations. *Dotted lines* are averaged pinning energies. Bands are aligned with respect to calculated offsets, zero is set at the valence band edge of GaAs for III–V and that of ZnSe for II–VI compounds. Data are from [32]

Many characteristics of semiconductor structures related to potential fluctuations like, e.g., free carrier mobility of luminescence linewidth, perform better applying delta doping. The most prominent application is the enhancement of conductivity in a layer, where the carriers are not subjected to impurity scattering at the ionized dopant cores. The mobility of carriers in this layer is strongly increased, similar to the modulation-doping technique based on band bending at heterojunctions.

The effect of delta doping requires a confinement of the doping atoms within a layer thickness well below a relevant length scale, which usually is given by the de Broglie wavelength of free carriers. In practice doping profiles with a width of 3 nm and below can be considered like a delta function. Note that this width corresponds to only five lattice constants. To achieve such narrow profiles, impurity redistribution processes like diffusion or segregation must largely be suppressed. This usually implies deposition at lowered temperature to avoid thermally activated diffusion. The doping procedure basically proceeds by an initial interruption of the epitaxy of the undoped semiconductor, followed by the deposition of the highly doped layer. Eventually the growth of the undoped material is resumed. Growth parameters like temperature, partial pressures and material supply have to be adjusted such that the doping profile is preserved. Due to the two-dimensional character of the doping layer it is often referred to as doping sheet.

The profile of delta doping is described by two parameters: The location of the dopand sheet z_{2D} and the areal density of doping atoms in the sheet n_{2D} or p_{2D}. To be specific we consider donor dopants and a complete ionization. The 2D donor density n_{2D} may be estimated from growth parameters used to obtain a 3D bulk carrier concentration n by scaling with the thickness $r \times t$ of the doping sheet,

$$n_{2D} = nrt.$$

Here r is the growth rate and t the duration of the doping-sheet deposition. The relation applies well if the incorporation efficiency of dopants is not affected by the preceding growth interruption. The effective 3D concentration of donors n_{3D} is obtained from n_{2D} of a homogeneously doped sheet by considering the mean distance between donors in the sheet $(n_{2D})^{-1/2}$ and assuming the same mean distance in 3D bulk, yielding

$$n_{3D} = (n_{2D})^{3/2}. \tag{10.23}$$

In semiconductors with delta doping the concentration of dopants varies strongly over short distances. The free carrier concentration is then spread much further than the profile of doping atoms. This feature is different to a slowly varying doping concentration, where the free-carrier profile follows the doping profile. Let us assume in a first approximation that all (ionized) doping atoms are located within a single atomic layer in the xy plane located at $z = z_{2D}$ with a density n_{2D}. The doping profile is then described by $n(z) = n_{2D}\delta(z - z_{2D})$, δ being the delta function. The two-dimensional charge $e \times n_{2D}$ of the doping sheet creates a one-dimensional electric potential $V(z)$ which is calculated using Poisson's equation,

$$\frac{\partial^2 V}{\partial z^2} = -\frac{en(z)}{\varepsilon}. \tag{10.24}$$

$\varepsilon = \varepsilon_r \varepsilon_0$ is the permittivity of the semiconductor. Twofold integration yields the potential

$$V(z) = \frac{en(z)}{2\varepsilon}|z - z_{2D}|. \tag{10.25}$$

We note that V is a linear and symmetric function of z and is V-shaped with $V(z_{2D}) = 0$. For an interaction with *negative* free carriers $V \leq 0$. The slope of the potential $V(z)$ represents the electric field created by the ionized impurity atoms in the doping sheet, $E = -\partial V/\partial z = \text{const}$. It depends on the doping density and the permittivity, yielding typical values exceeding 10^6 V/m. Such strong fields lead to a strong attractive interaction with free carriers: The potential $V(z)$ forms a narrow well with a width on the order of the de Broglie wavelength. Similar to the narrow square potential treated in Sect. 4.3 we obtain size quantization with discrete energy levels. We should note that the shape of the potential actually deviates from a simple V. The delta-doped semiconductor is neutral, because the charge of the doping sheet is balanced by the opposite charge of the released free carriers. The electric field therefore approaches 0 at some distance from the sheet. Figure 10.13 depicts a self-consistent solution of an effective-mass calculation of carriers confined in a potential which is created by delta doping [35]. The calculation assumes data of GaAs with an electron mass $m^* = 0.067m_0$.

The donor-doping profile $n(z)$ illustrated in Fig. 10.13a creates a local bending of the conduction-band edge E_c that confines free electrons to discrete energies E_i. For the parameters assumed in the calculation four energy levels E_0 to E_3 are occupied with fractions of 61%, 24%, 11%, and 4%, respectively. We note a steep and nearly

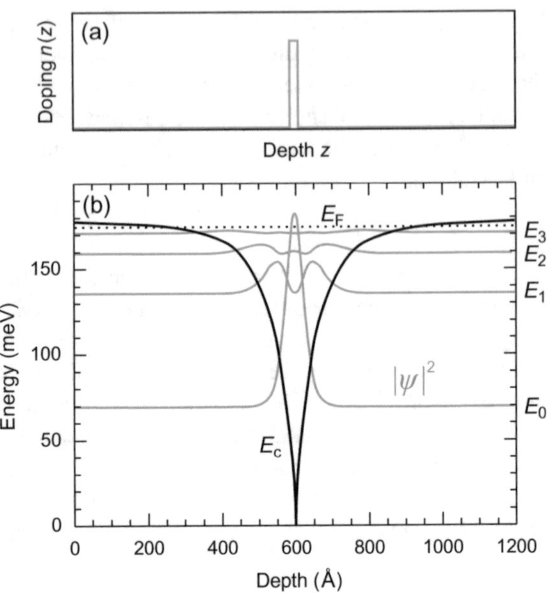

Fig. 10.13 **a** Schematic of delta-doping profile. **b** Calculated effective-field potential well (E_c) and confined carrier distribution $|\psi|^2$ of an n-type delta-doping sheet with 2 Å thickness and 5×10^{12} cm^{-2} density. E_F denotes the Fermi energy. Reproduced with permission from [35], © 1990 AIP

constant slope of the confining potential in a close vicinity to the doping sheet, and an approach to a constant value at some distance. The spatial extent of the carrier wave-function is much wider than the thickness of the doping sheet ($\sim$50 Å in the ground state vs. 2 Å, even more in the excited states). This finding leads to a more precise condition for delta doping: The distribution of doping atoms must be significantly more narrow than the spatial extent of carriers confined in the ground state. Obviously a factor 10 thicker doping sheet will lead to quite similar results in the present example.

The effect of delta doping on the Hall mobility of free carriers is illustrated in Fig. 10.14 [36]. The mobility μ of a carrier with drift velocity $\mathbf{v}$ (at low electric field $\mathbf{E}$) is generally given by $\mathbf{v} = \mu\, \mathbf{E}$, and is limited by scattering processes (see, e.g., [37]). The 2D mobility was measured using the two-dimensional electron gas created in the channel of a GaAs field-effect transistor [38]. Bulk values of the donor concentration n_{3D} corresponding to the 2D values of the delta-doping sheet were calculated from (10.23) and used in the abscissa of Fig. 10.14. The 2D mobility μ_{2D} is related to the doping-dependent mobility of bulk GaAs which is taken from the empirical expression $\mu_{3D} = \mu_L(1 + (n_{3D}/10^{17}\ \text{cm}^{-3})^{1/2})^{-1}$, $\mu_L = 10^4\ \text{cm}^2/(\text{Vs})$ being the mobility limit due to lattice scattering [42].

Figure 10.14 shows that the mobility of free carriers in the two-dimensional electron gas significantly exceeds that of bulk free carriers at high doping concentrations n_{3D}. The increase of mobility is assigned to various features of the 2D carriers. From an only weak dependence of the mobility μ_{2D} on temperature a reduced scattering at ionized impurities was concluded [36]. The carriers in the 2D electron gas have a high degeneracy on their energy levels E_i, connected to high kinetic energies parallel to the

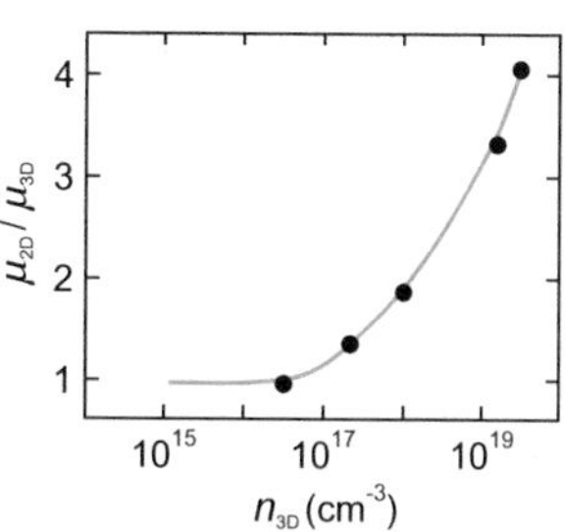

Fig. 10.14 Relative mobility enhancement of delta-doped n-type GaAs with respect to homogeneously doped n-type GaAs measured at 300 K. Data from [36]

plane of the doping sheet. Carriers with energies close to the Fermi surface are most sensitive to scattering. Their energies $E_F - E_i$ exceed well the thermal energy $k_B T$, leading to a reduced temperature-dependent impurity scattering. A further contribution originates from a decreased overlap of odd-numbered carrier wave-functions with the delta-doping plane of impurities. Wave functions $\psi_1, \psi_3, \ldots$ have a node at the position of the doping sheet (cf. Fig. 10.13b) and hence experience much less impurity scattering. Also the overlap of symmetric even-numbered excited wave functions $\psi_2, \psi_4, \ldots$ is much smaller than that of the ground state ψ_0, and even the spatial extent of the ground state is significantly larger than the thickness of the doping sheet.

A drawback of the delta-doping technique discussed above is that the mobile electrons are bound to the doping plane, where the statistically distributed donors provide a fluctuating potential; the electrons hence experience substantial impurity scattering. This disadvantage can be avoided if the donors are separated from the mobile electrons by confining them in a *separate* potential. This is achieved by introducing a heterointerface to confine the electrons, and placing the donor sheet at some distance away by inserting an undoped spacer layer; using such a spacer layer also a *homogeneous* doping can be applied, yielding a *modulation-doped* structure [39]. Band bending at the heterointerface creates a triangular confinement potential with discrete energies (along growth direction z) in the semiconductor with smaller bandgap as illustrated in Fig. 10.15 [43]. A similar potential can also be created at the oxide/semiconductor interface of a MOSFET device [40]. The electrons provided by the remote doping are free to move in the lateral (x, y) directions and form a two-dimensional electron gas (2DEG) with a high mobility, particularly at low temperature [41].

The low impurity scattering is utilized in modulation-doped field-effect transistors (MODFET, also termed HEMT, high electron-mobility transistor); it can be further reduced by delta doping. In a conventional MODFET a homogeneously doped wide-bandgap layer (usually AlGaAs) provides free carriers, which are trapped at the heterojunction to an adjacent undoped layer with a smaller bandgap (GaAs), see Fig. 10.15. Employing delta doping instead of homogeneous doping allows for a large and well-defined separation between free carriers and doping impurities [2].

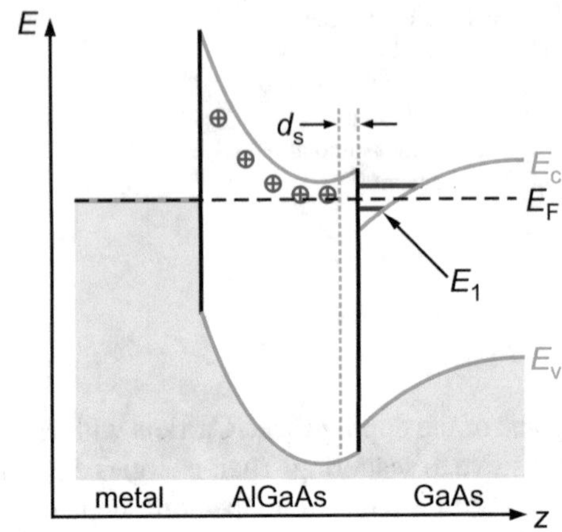

Fig. 10.15 Energy-band diagram of a modulation-doped AlGaAs/GaAs heterostructure with a metal gate contact, creating a 2D electron gas located at the interface; occupied energy level E_1 and an undoped spacer layer of thickness d_s are indicated

10.2 Diffusion

Atoms in a crystal may not stay fixed on their site. In presence of a concentration gradient and at sufficiently high temperature atoms redistribute by diffusion, thereby changing the interface abruptness in heterostructures and the spatial distribution of defects. The effect is utilized, e.g., for the indiffusion of dopants into semiconductors via the surface, but may also be detrimental for epitaxial nanostructures. Atoms diffuse via different mechanisms, which are controlled by temperature, partial pressure, and material composition. Since various diffusion paths usually act simultaneously and diffusivity may depend on defect concentration, diffusion is a complex subject. We will outline basic phenomena and illustrate some examples.

10.2.1 Diffusion Equations

The diffusion of an atom incorporated in a crystal is described similar to the surface diffusion of an adatom considered in Sect. 7.2.2. Each jump of the atom to an adjacent site is the result of an attempt to leave the actual site and of the success to surmount the potential barrier ΔE which tends to keep the atom at its site. The *diffusion coefficient* D is likewise given by the product of the mean square value of the displacement λ^2 and the rate of successful jumps ν, yielding

$$D = \lambda^2 \nu = \lambda^2 \nu_0 \exp\left(-\Delta E/(k_B T)\right) = D_0 \exp\left(-\Delta E/(k_B T)\right). \quad (10.26)$$

The *diffusion constant* $D_0 = \lambda^2 \nu_0$ has usually an only small temperature dependence compared to the exponential term and is often assumed constant with respect to temperature. Note that D_0 may still vary spatially in an inhomogeneous solid.

To obtain an expression for the motion of a diffusing atom we first consider a macroscopic view of the diffusion of impurities in a solid; an atomistic approach is discussed below in Sect. 10.2.2. We assume an impurity concentration c which varies only along a single coordinate z and a random diffusion of an impurity which is not affected by other impurities. In this case the net flux of impurities j per unit area and unit time also varies only along z and is given by *Fick's first law*,

$$j(z) = -D \frac{d}{dz} c(z). \tag{10.27}$$

The flux of impurities $j(z)$ is proportional to the gradient of the impurity concentration $c(z)$ and directed toward the low-concentration region (cf. the negative sign). Since the atoms move in such a way as to even the gradient up, the one-dimensional form of Fick's law is sufficient to describe the flux. We note that there is no net flux if the impurity concentration is constant. The diffusion coefficient D depends in a simple case not on the impurity concentration c and also not on the spatial position z (but still on the temperature). This case describes a *linear* Fick diffusion. A dependence $D(c, z)$ due to experimental conditions leads to a *non-linear* Fick diffusion also described by (10.27).

The impurity flux must comply with the *continuity condition*. The net flow into any volume element equals the increase of impurity concentration per unit time in this volume element,

$$\frac{\partial}{\partial z} j(z) = -\frac{\partial}{\partial t} c(z, t). \tag{10.28}$$

Inserting (10.27) into the one-dimensional continuity equation (10.28) yields the diffusion equation known as *Fick's second law*,

$$\frac{\partial}{\partial z}\left(D \frac{\partial c(z, t)}{\partial z} \right) = \frac{\partial c(z, t)}{\partial t}. \tag{10.29}$$

If the diffusion coefficient D is constant with respect to z, then (10.29) simplifies to

$$D \frac{\partial^2 c(z, t)}{\partial z^2} = \frac{\partial c(z, t)}{\partial t}. \tag{10.30}$$

The solution to Fick's second law depends on the boundary conditions which are given by the experiment.

We consider some solutions of the one-dimensional diffusion equation in the form of (10.30) for different experimental conditions; a more comprehensive treatment is given in [44]. The first example is the indiffusion of impurities from the surface with a surface concentration kept constant at a value c_0. Such a condition may be given by applying an external constant vapor pressure of impurity atoms on the surface. The impurity concentration in the solid depends only on the distance z from the surface, which is assumed to be located at $z = 0$. The resulting boundary conditions read

$$c(z = 0, t) = c_0 \quad \text{for all } t, \quad \text{and} \quad c(z, t) = 0 \quad \text{for } z > 0 \text{ and } t = 0.$$

The solution of the diffusion equation (10.30) for these conditions is

$$c(z, t) = c_0 \,\mathrm{erfc}\left(\frac{z}{2\sqrt{Dt}}\right),$$

where erfc is the complementary error function. The diffusion profile $c(z, t)$ depends on the diffusion coefficient $D(T)$ and the duration of the indiffusion process t. Profiles for varied values of the product Dt are shown in Fig. 10.16a. The concentration profiles $c(z)$ show a progressive indiffusion of impurities for increasing process time t. The material is constantly supplied from the external vapor phase via the surface, such that $c(z = 0, t) = c_0$ during the entire process.

In our second example the source of impurities is assumed to be located within the solid. We consider a solid being doped with impurities with an (initially) constant concentration c_d in a range $z < 0$ up to an interface located at $z = 0$. Beyond the interface at $z > 0$ the solid is assumed to be undoped. The boundary conditions are now

$$c(z) = c_d \quad \text{for } z < 0 \quad \text{and} \quad c(z) = 0 \quad \text{for } z > 0 \text{ for } t = 0.$$

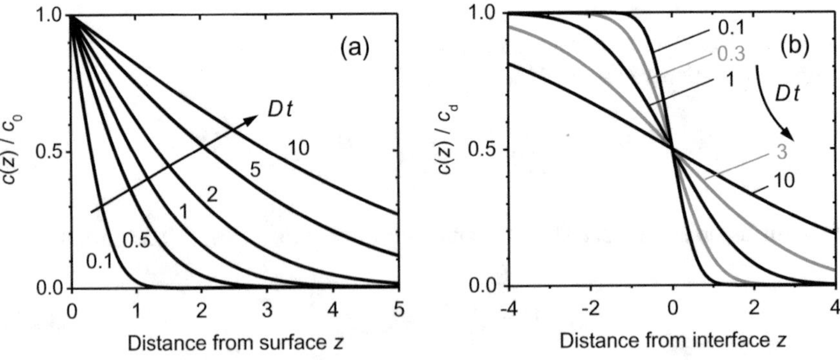

Fig. 10.16 Solutions of the one-dimensional diffusion equation for two different boundary conditions. Concentration profiles $c(z)$ are given for varied values of the quantity Dt as indicated on the curves. **a** Impurity distribution in an initially (at $Dt = 0$) undoped solid for a surface concentration kept constant at a value c_0. **b** Distribution for a solid being initially homogeneously doped with an impurity concentration c_d at $z < 0$ and undoped at $z > 0$

The solution of the diffusion equation (10.30) in this case is given by

$$c(z,t) = \frac{c_d}{2}\,\mathrm{erfc}\left(\frac{z}{2\sqrt{Dt}}\right).$$

The result is quite similar to the first case. In contrast to the first example the total amount of impurities diffusing in the solid now is constant. Impurities diffusing across the interface hence increase the concentration c at $z > 0$ on expense of c at $z < 0$. The initial step-function-like profile gradually smoothens and has the shape of the complementary error function as shown in Fig. 10.16b. The impurity concentration at the interface remains fixed at $c(z = 0, t) = c_d/2$. The example resembles the first one as long as the doped part can be considered semi-infinite thick with respect to the diffusion length $\sqrt{Dt}$.

In our third example a delta-doping sheet within an undoped solid is assumed to represent the source of impurities. The total impurity concentration c_{2D} is initially (at $t = 0$) located in the two-dimensional (xy)-plane at $z = 0$. The boundary conditions are in this case

$$c(z,t) = c_{2D}\delta(z) \quad \text{for } t = 0, \quad \text{and} \quad \int_{z=-\infty}^{\infty} c(z,t) = c_{2D} = \text{const} \quad \text{for all } t > 0.$$

The solution of the diffusion equation (10.30) for this geometry is given by

$$c(z,t) = \frac{c_{2D}}{2\sqrt{\pi Dt}}\exp\left(-\frac{z^2}{4Dt}\right).$$

The total amount of impurities diffusing in the solid is constant like in the previous example. The initial δ-function-like profile gradually broadens to a Gaussian distribution as illustrated in Fig. 10.17. The broadening is characterized by the increasing standard deviation of the distribution $\sigma = \sqrt{2Dt}$.

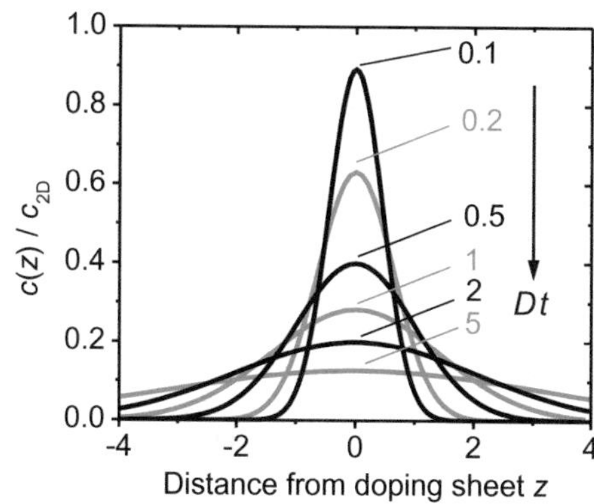

Fig. 10.17 Impurity concentration-profiles for a doping sheet located at $z = 0$ initially (at $Dt = 0$) described by a δ-function-like profile. Numbers on the curves signify respective values of Dt

10.2.2 Diffusion Mechanisms

The diffusion of atoms in a crystal depends on many experimental parameters in addition to the temperature, such as the material composition, the position of the Fermi level, or the concentration of point defects. An understanding of these dependences requires a microscopic view on the diffusion mechanisms. We consider some important mechanisms separately, even though usually a combination of these diffusion paths occurs in practice. Often the dominant mechanism provides a reasonable description of the diffusion process at least in some limited temperature range.

In a perfect crystal which does not contain point or line defects, the only mechanisms for atom diffusion on lattice sites are the exchange and ring mechanisms illustrated in Fig. 10.18a. Since many bonds need to be broken simultaneously in these collective mechanisms, they are associated with a very high activation energy for the migration and do not play a significant role in practice. Mechanisms involving defects are much more effective for both, diffusion of impurities and self-diffusion of crystal species. We focus on the effect of point defects and do not include phenomena in the presence of line defects.

Vacancy mechanism: Any crystal at finite temperature contains vacant lattice sites, which provide an efficient path of diffusion via substitutional sites. The elementary jump of an atom into a neighboring vacancy is depicted in Fig. 10.18b. Diffusion of substitutionally dissolved impurities or self-diffusion via this path is still slow compared to other mechanisms. Examples of slow impurity diffusors are the common Column III and Column V dopants in silicon as shown below in Fig. 10.19.

Interstitial mechanism: If atoms exist on interstitial sites, they can migrate by jumping from one interstitial site to another as illustrated in Fig. 10.18c. This mechanism is particularly favorable for small impurity atoms, which do not need to greatly displace crystal atoms from their regular lattice site. The mechanism is very efficient. Prominent fast diffusors are interstitially dissolved Cu, Li, H, or Fe in silicon [45]. Their large diffusivities are shown in Fig. 10.19.

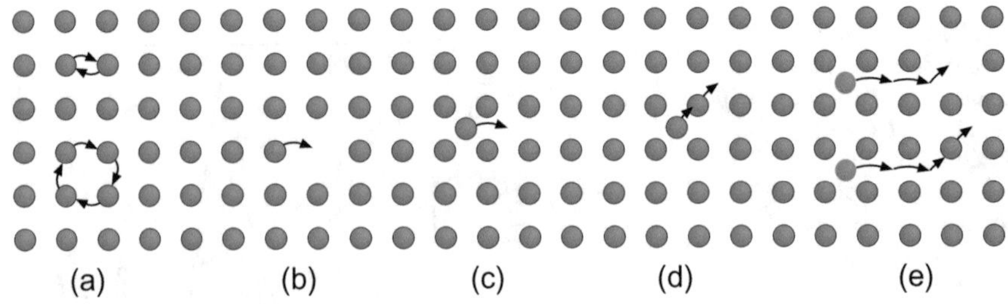

Fig. 10.18 Schematic illustration of diffusion mechanisms. **a** *Top*: exchange mechanism, *bottom*: ring mechanism. **b** Vacancy mechanism. **c** Interstitial mechanism. **d** Interstitialcy mechanism. **e** Substitutional-interstitial mechanisms, *top*: Frank-Turnbull mechanism, *bottom*: kick-out mechanism

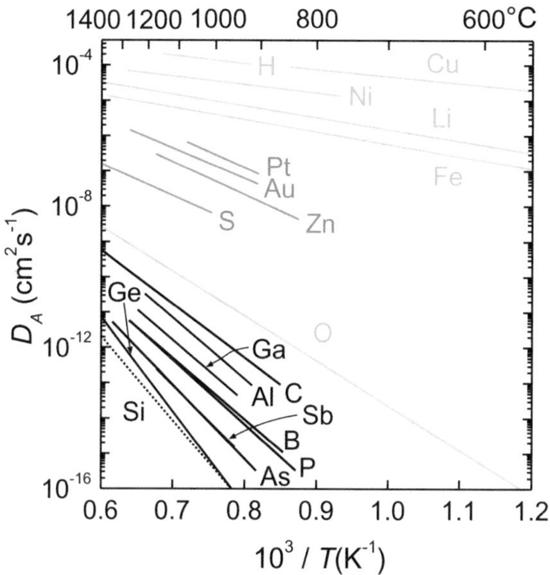

Fig. 10.19 Diffusion coefficients D_A of impurities A in Si. *Black lines* refer to elements which mainly dissolve substitutionally and diffuse via the vacancy or interstitialcy mechanism. *Gray lines* represent data of hybride elements which diffuse via the substitutional-interstitial mechanism. *Light gray lines* give data of elements diffusing via the direct interstitial mechanism. The *dotted line* represents Si self-diffusion. Data from [51], Li, Fe, and Ga from [52]

Interstitialcy or *indirect interstitial mechanism*: This path is more likely for impurities with a similar size as lattice atoms or for self-diffusion. The mechanism implies a cooperative motion of two atoms as shown in Fig. 10.18d. An interstitial atom moves into a lattice site by pushing an atom, which originally occupied this site, to a neighboring interstitial site.

Substitutional-interstitial mechanism: This effective diffusion mechanism in semiconductors may apply for impurity atoms A which can be incorporated on both a substitutional site A_s and an interstitial site A_i (hybrid solutes). Such impurities can diffuse via one of the two types of this mechanism depicted in Fig. 10.18e. The Frank-Turnbull (or, dissociative) mechanism involves vacancies V according to the reaction [46]

$$A_i + V \leftrightarrow A_s. \tag{10.31}$$

The kick-out mechanism involves self-interstitials I (interstitials of crystal species) according to the reaction [47, 48]

$$A_i \leftrightarrow A_s + I. \tag{10.32}$$

The diffusivity of A_i is generally much larger than that of A_s, while the solubility and hence also the equilibrium concentration c_i on interstitial sites is much less than on lattice sites c_s. An efficient incorporation of impurities A may then proceed by a fast interstitial diffusion for many consecutive jumps, eventually followed by the occupation of a regular lattice site. This process differs from the interstitialcy mechanism, where the atom in the interstitial position remains there for only a single step. The Frank-Turnbull and kick-out mechanisms are prevailing diffusion paths for many substitutionally dissolved fast diffusing elements in Column-IV and III–V semiconductors. Well established examples are the diffusion of Cu, Ag, and Au in Ge via the Frank-Turnbull mechanism and that of Au, Zn, Pt, and S in Si via the kick-out mechanism.

10.2.3 *Effective Diffusion Coefficients*

Usually several of the diffusion mechanisms outlined above act simultaneously. Since diffusion coefficients vary over many orders of magnitude depending on experimental conditions, the dominating path may already provide a reasonable description. We illustrate some dependences of the diffusivity for the important diffusion mechanisms with participating point defects. In compound semiconductors single point defects are vacancies, interstitials, and antisites. Their equilibrium concentrations depend on the parameters temperature, partial pressures, and composition, which are mutually related by the phase diagram. In this section we assume the concentrations C_{defect} as given quantities to point out their effect on diffusivity.

Vacancy mechanism: One diffusion jump of this mechanism requires the presence of one vacancy V. The diffusion coefficient D of the migrating species is therefore proportional to the concentration of vacancies C_V. An atom with no adjacent vacancy can also move, if first a vacancy diffuses to a neighboring lattice site. The diffusion coefficient D is therefore also proportional to the diffusion coefficient D_V of the vacancies, eventually yielding

$$D \propto C_V D_V \qquad \text{(vacancy mechanism)}.$$

Substitutional-interstitial mechanism: There are two types of this mechanism as outlined in the previous section. In the *Frank-Turnbull* mechanism (Fig. 10.18e, top) the diffusivity of the impurity A is essentially given by the (large) diffusivity D_i of the (small) fraction A_i dissolved on interstitial sites. The marginal contribution of substitutional impurities A_s, which either diffuse directly via the vacancy mechanism or interact with self-interstitials, is neglected. According to (10.31) the diffusion mechanism involves the three species A_i, A_s, and vacancies V. The diffusivity of A_s depends on the incorporation rate of impurities A on a lattice site. It is determined by the slower process of either supplying interstitials A_i or vacancies V from the sample surface. If the supply of interstitials limits the process due to experimental conditions (i.e., $C_i^{\text{eq}} D_i \ll C_V^{\text{eq}} D_V$), the effective diffusivity of the *Frank-Turnbull* mechanism is given by [45, 49]

$$D_{\text{eff}}^{\lim(\text{i})} = \left(\frac{C_{\text{i}}^{\text{eq}}}{C_{\text{i}}^{\text{eq}} + C_{\text{s}}^{\text{eq}}}\right) D_{\text{i}}$$
$$\approx \left(C_{\text{i}}^{\text{eq}}/C_{\text{s}}^{\text{eq}}\right) D_{\text{i}} \quad \text{(Frank-Turnbull, limited by interstitials).}$$

C_{i}^{eq} and C_{s}^{eq} are the local equilibrium concentrations of A_{i} and A_{s}, respectively. The index lim(i) denotes the limitation by the supply with interstitials. If the supply of vacancies limits the process (i.e., $C_V^{\text{eq}} D_V \ll C_{\text{i}}^{\text{eq}} D_{\text{i}}$), the local equilibrium concentration of vacancies C_V^{eq} controls the effective diffusivity according to

$$D_{\text{eff}}^{\lim(V)} = \left(C_V^{\text{eq}}/C_{\text{s}}^{\text{eq}}\right) D_V \quad \text{(Frank-Turnbull, vacancy-limited).}$$

The *kick-out* type of the substitutional-interstitial mechanism involves the species A_{i}, A_{s}, and self-interstitials I according to (10.32). Now the diffusivity of A_{s} depends on the rate of in-diffusion of A_{i} and the out-diffusion of self-interstitials. When a slow in-diffusion of A_{i} controls the process, $C_{\text{i}}^{\text{eq}} D_{\text{i}} \ll C_I^{\text{eq}} D_I$ holds and the diffusion coefficient is identical to $D_{\text{eff}}^{\lim(\text{i})}$ of the Frank-Turnbull mechanism noted above. If the out-diffusion of self-interstitials limits the process, the diffusivity becomes [45]

$$D_{\text{eff}}^{\lim(I)} \cong \left(C_I^{\text{eq}}/C_{\text{s}}^{\text{eq}}\right)\left(C_{\text{s}}^{\text{eq}}/C_{\text{s}}\right)^2 D_I \quad \text{(kick-out, self-interstitial-limited),}$$

where C_I^{eq} is the local equilibrium concentration of self-interstitials I and $C_{\text{s}} = C_{\text{s}}(\mathbf{r})$ is the actual local concentration of substitutional impurities A_{s}. A simultaneous operation of diffusion mediated by self-interstitials *and* vacancies leads for $C_{\text{i}}^{\text{eq}} D_i \gg (C_I^{\text{eq}} D_I + C_V^{\text{eq}} D_V)$ to a combined effective diffusivity composed of both parts, i.e. $D_{\text{eff}} = D_{\text{eff}}^{\lim(I)} + D_{\text{eff}}^{\lim(V)}$. Using such effective diffusivities the diffusion via the various substitutional-interstitial mechanisms can be described by a single effective process using Fick's diffusion equation (10.29).

Effect of charge: The diffusivity of a point defect D_{defect} depends strongly on its charge state. Since defects in semiconductors are usually charged, this property must be included into the terms discussed above. A stable charged point defect must possess an electronic level in the fundamental energy gap [48]. If a charged defect has several charge states, the thermal equilibrium concentrations of the differently charged defects depend on the position of the Fermi level. This feature is related to the dependence of the defect formation energy as illustrated in Fig. 10.6 for some native point defects in GaAs.

The substitutional-interstitial mechanism represents the diffusion of an impurity-defect complex. The interaction of the impurity and the defect depends on the charge state of the defect and the impurity, and so also the diffusivity of the complex. The Frank-Turnbull reaction (10.31) of an interstitial impurity A_{i}^{l+} with charge $l+$, an m-fold negatively charged vacancy V^{m-}, and a substitutional impurity A_{s}^{n-} with charge $n-$ reads [45]

$$A_{\text{i}}^{l+} + V^{m-} \leftrightarrow A_{\text{s}}^{n-} + (l - m + n)h.$$

h denotes the holes created or consumed in this reaction due to the charge balance condition, and l, m, n are integers. In compound semiconductors the vacancy is here assumed to be located on the same sublattice in which the substitutional impurity A_s is dissolved. Charged species were introduced in literature to describe the diffusion of Zn in GaAs according to the reaction $Zn_i^+ + V_{Ga} \leftrightarrow Zn_s^- + 2h^+$, and since then the Frank-Turnbull process involving charged species is also referred to *Longini mechanism* [50]. For the kick-out mechanism the relation is given by

$$A_i^{l+} \leftrightarrow A_s^{n-} + I^{m+} + (l - m + n)h.$$

Diffusivities following from the charge-including reactions depend on the supply of the charged defects similar to the reactions without charge discussed above. If the supply of interstitials A_i limits the process, we obtain [45]

$$D_{\text{eff}}^{\lim(\mathrm{i})} = (|n|+1)\left(\frac{C_i^{\text{eq}}(C_s^{\text{eq}})}{C_s^{\text{eq}}}\right)\left(\frac{C_s}{C_s^{\text{eq}}}\right)^{|n|\pm l} D_i$$

(Frank-Turnbull with charged defects, interstitial-limited).

A positive sign in the exponent applies for substitutional acceptors, a negative for substitutional donors. The reaction does not depend on the charge m of the vacancy. The equation accounts for a (possibly) locally varying electron or hole concentration, yielding an also locally varying concentration C_i^{eq} of charged interstitials A_i^{l+}. Similar to the uncharged case the same equation applies to the kick-out mechanism when a slow in-diffusion of A_i limits the incorporation rate of A_s. If the incorporation rate of A_s in the Frank-Turnbull process with charged defects is limited by the supply of vacancies, then the effective diffusivity of A_s impurities is described by [45]

$$D_{\text{eff}}^{\lim(V)} = (|n|+1)\left(\frac{C_V^{\text{eq}}(C_s^{\text{eq}})}{C_s^{\text{eq}}}\right)\left(\frac{C_s^{\text{eq}}}{C_s}\right)^{\pm m-|n|} D_V$$

(Frank-Turnbull with charged defects, vacancy-limited).

When in the kick-out mechanism with charged defects the incorporation of A_s is limited by the supply of self-interstitials, the effective diffusivity is analogously given by

$$D_{\text{eff}}^{\lim(I)} = (|n|+1)\left(\frac{C_I^{\text{eq}}(C_s^{\text{eq}})}{C_s^{\text{eq}}}\right)\left(\frac{C_s^{\text{eq}}}{C_s}\right)^{\pm m-|n|-2} D_I$$

(kick-out with charged defects, self-interstitial-limited).

If the diffusion coefficient of each charged complex is independent of that of other complexes, the resulting effective diffusion coefficient is given by a linear combination of the diffusivities of all complexes. This leads eventually to an effective diffusion coefficient, which in the general case is composed of the effective diffusion coefficients of all defect complexes with all occurring charge states. We abbreviate the notation for effective diffusivities of the two substitutional-interstitial mechanisms (10.31) and (10.32) by $D_{\text{eff}}(AV)$ for the Frank-Turnbull type and $D_{\text{eff}}(AI)$ for the kick-out type, and include the various charge states of vacancies $V^0, V^-, V^{2-}, \ldots,$ $V^+, V^{2+}, \ldots$ and the same for self-interstitials I. We then may write for the general case

$$D_{\text{eff}} = \sum_{z_{V,\min}}^{z_{V,\max}} D_{\text{eff}}\left(AV^{z_V}\right) + \sum_{z_{I,\min}}^{z_{I,\max}} D_{\text{eff}}\left(AI^{z_I}\right).$$

In the sums the charge states of the native defects characterized by the integers z_V and z_I start at the most negative values $z_{V,\min}$ and $z_{I,\min}$ occurring in the considered semiconductor, and end at the corresponding most positive values.

The relations discussed in this section point up the complexity of diffusion phenomena in semiconductors. A clear description is usually restricted to specific elements and a limited temperature range. Diffusivities of some impurity atoms in silicon are shown in Fig. 10.19.

The diffusion coefficients shown in the Arrhenius plot are described by the relation (10.26) and vary over many orders of magnitude. We note the very high diffusivity of the elements moving via the interstitial mechanism (light gray lines, except for O) and the slowly moving substitutional impurities diffusing via the vacancy mechanism or the interstitialcy mechanism (black lines). Such elements are favorable for fabricating stable doping profiles.

10.2.4 Disordering of Heterointerfaces

Epitaxial growth procedures have proved their ability to fabricate atomically sharp heterointerfaces. Such interfaces provide strong gradients in the distribution of different atom species and hence a driving force for disordering by diffusion. There may even exist growth conditions which counteract the formation of atomically sharp interfaces. Furthermore, also the preservation of sharp interfaces is important, e.g., during device processing of heterostructures. On the other hand, a well-directed locally enhanced layer disordering was employed for both studying diffusion mechanisms and defining lateral confinement for photons and charge carriers in optoelectronic devices. We consider disordering of heterointerfaces in more detail for $\text{Al}_x\text{Ga}_{1-x}\text{As}$-based heterostructures to illustrate the effect of experimental conditions on the diffusivity outlined in the previous section.

The self-diffusion of Ga in GaAs was found to be quite low under intrinsic conditions. A similar diffusivity was found for Ga self-diffusion $D_{\text{Ga}}(n_{\text{i}})$ and Ga-Al interdiffusion $D_{\text{Al-Ga}}(n_{\text{i}})$ obtained from AlGaAs/GaAs heterostructures, both being

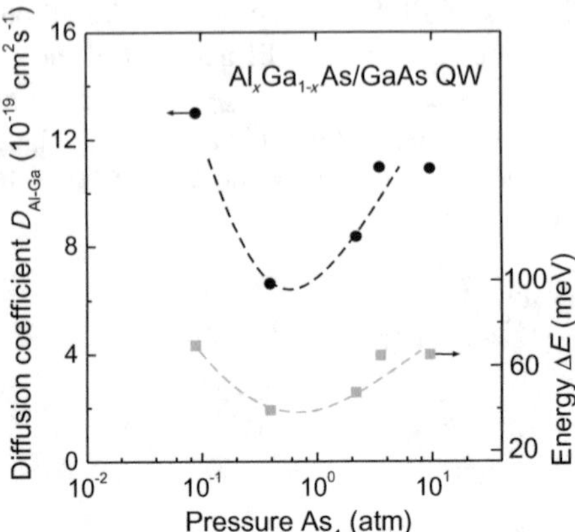

Fig. 10.20 Energy shift ΔE of the photoluminescence from an $Al_{0.25}Ga_{0.75}As/GaAs$ quantum well annealed for 25 h at 825 °C at various values of the As_4 ambient pressure (*gray squares*). The *black circles* are corresponding calculated diffusion coefficients of the Al-Ga interdiffusion. *Dashed lines* are guides to the eye. From [54]

well described by (10.26) with a diffusion constant $D_0 = 2.9 \times 10^8$ cm^2/s and a fairly high activation energy $\Delta E \cong 6$ eV [53]. Such values correspond to a diffusion length $\sqrt{Dt}$ of 1 nm for 30 h diffusion at 1100 K. Since the self- and interdiffusion on the Column III sublattice must proceed through native defects of the crystal, the diffusivity depends on the As pressure applied during the annealing at elevated temperature. Pressure-depending interdiffusion coefficients were derived from an evaluation of the photoluminescence energy-shift measured for annealed $Al_{0.25}Ga_{0.75}As/GaAs$ quantum well samples [54]. The diffusion of Al and Ga across the heterointerfaces of the 13 nm thick quantum well leads to a change of the confinement potential, resulting in a blue-shift of the energy levels.

Figure 10.20 shows that both, a high *and* a low As_4 ambient pressure applied during annealing leads to an enhanced $D_{\text{Al-Ga}}$ interdiffusion coefficient. The increase of $D_{\text{Al-Ga}}$ is generated by the change of the native defect concentration via the sample surface. Six single point-defect species may occur in GaAs. A shift of the crystal stoichiometry to the As-poor side favors the creation of As vacancies V_{As}, Ga interstitials I_{Ga}, and Ga antisites Ga_{As}. As-rich point defects are Ga vacancies V_{Ga}, As interstitials I_{As}, and As antisites As_{Ga}. Corresponding defects apply for Al which occupies the same sublattice as Ga, leading to the more general notation for Column III defects, e.g., Column III vacancies V_{III}.

A consistent description including also Fermi-level effects was obtained by considering the Column III self-diffusion as being due to neutral and singly ionized Column III vacancies V_{III} and Column III interstitials I_{III} [55]. A creation of such a defect pair can occur according to the reaction

$$0 \leftrightarrow I_{\text{III}} + V_{\text{III}} \quad \text{and} \quad C_{\text{I}_{\text{III}}} C_{\text{V}_{\text{III}}} = k_1 .$$

The second equation expresses the law of mass action with the concentrations C of the participating defects and a temperature-dependent constant k_1. Under As-rich conditions the crystal surface can act as a sink for the interstitials I_{III}, thereby increasing the concentration of vacancies $C_{V_{\mathrm{III}}}$. Under As-poor conditions, evaporation of As at the surface leads to a local excess of Ga atoms, which can diffuse into the crystal and increase the concentration of interstitials $C_{I_{\mathrm{III}}}$ according to the reaction

$$0 \leftrightarrow I_{\mathrm{III}} + \frac{1}{4}\mathrm{As}_4^{\mathrm{vapor}} \quad \text{and} \quad C_{I_{\mathrm{III}}} P_{\mathrm{As}_4}^{1/4} = k_2.$$

Both V_{III} and I_{III} can lead to Column III self-diffusion, yielding a diffusivity

$$D_{\mathrm{III}} = c_1 C_{V_{\mathrm{III}}} D_{V_{\mathrm{III}}} + c_2 C_{I_{\mathrm{III}}} D_{I_{\mathrm{III}}} = c_3 P_{\mathrm{As}_4}^{1/4} D_{V_{\mathrm{III}}} + c_4 P_{\mathrm{As}_4}^{-1/4} D_{I_{\mathrm{III}}}.$$

The c_i are constants containing the temperature-dependent constants k_j. In the second equation we used the law of mass action involving k_2. The equation describes the trend of the pressure-dependent diffusivity shown in Fig. 10.20. Under As-poor conditions the diffusivity is increased by Column III interstitials I_{III}, while under As-rich conditions Column III vacancies V_{III} enhance the diffusivity.

The diffusivity is also affected by doping. A shift of the Fermi level from the intrinsic position can increase the self-diffusion by orders of magnitude. Evidence for a combined effect of doping and ambient pressure on heterointerfaces applied during annealing is given in Fig. 10.21. The TEM images show cross sections of n-type (left, $n = 10^{18}\ \mathrm{cm}^{-3}$) and p-type (right, $p = 8 \times 10^{18}\ \mathrm{cm}^{-3}$) AlGaAs/GaAs superlattices, which were annealed for 10 h at elevated temperature [56]. The n-type structure shows some interdiffusion for an anneal under As-poor conditions, and a much stronger layer intermixing in As-rich ambient. In contrast, the p-type structure shows a complete intermixing for As-poor conditions, while the superlattice remains stable in As-rich ambient. Experiments with compensated doping (i.e., simultaneous doping with donors *and* acceptors) confirmed that the enhancement of the interdiffusivity is controlled by the position of the Fermi level and not by the presence of impurity atoms [57].

The experimental results point to the participation of charged defects in the effective diffusivity D_{III}. Their equilibrium concentration depends on their energy position in the bandgap with respect to the Fermi level. Considering only neutral and singly ionized vacancies V_{III} and interstitials I_{III} yields for the donor-like interstitials I_{III} the relation $C_{I_{\mathrm{III}}^+}/C_{I_{\mathrm{III}}^0} = \exp((E_{\mathrm{D}} - E_{\mathrm{F}})/(k_{\mathrm{B}}T))$, and for the acceptor-like vacancies V_{III} the ratio $C_{V_{\mathrm{III}}^-}/C_{V_{\mathrm{III}}^0} = \exp((E_{\mathrm{F}} - E_{\mathrm{A}})/(k_{\mathrm{B}}T))$ [55]. The difference of the ionization energy and the Fermi level in the exponential terms represents the energy gain when the native defect is charged, cf. Fig. 10.1 and Sect. 10.1.2. If the Fermi level is decreased by extrinsic p-type doping, the donor-like interstitials I_{III} are positively charged and attain an increased solubility and corresponding high concentration $C_{I_{\mathrm{III}}^+}$. In n-type heterostructures the solubility and thereby concentration of acceptor-like charged vacancies V_{III}^- is increased. Introducing respective diffusivities of the charged native defects, the effective diffusion constant for Column III self-diffusion noted above only for undoped defects is extended and reads [55]

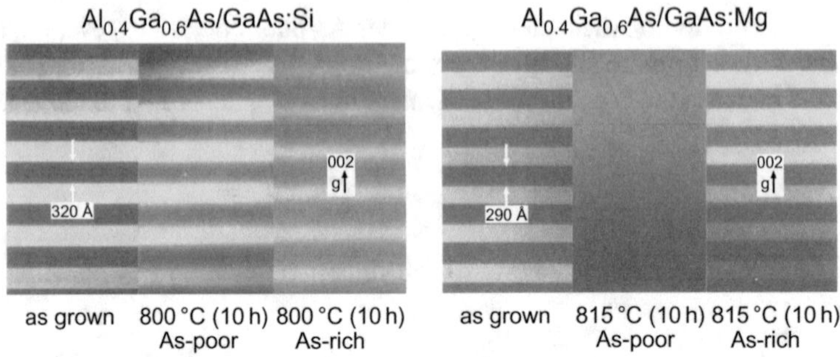

Fig. 10.21 Transmission-electron micrographs of cross sections of Si-doped (*left*) and a Mg-doped (*right*) $Al_{0.4}Ga_{0.6}As/GaAs$ superlattices. The images show the as-grown structures and the structures after annealing for 10 h under either As-poor or As-rich conditions. Reproduced with permission from [56], © 1988 MRS

$$D_{III} = c_5 P_{As_4}^{1/4}\big(D_{V_{III}^0} + D_{V_{III}^-}\exp\big((E_F - E_A)/(k_B T)\big)\big) + c_6 P_{As_4}^{-1/4}\big(D_{I_{III}^0} + D_{I_{III}^+}\exp\big((E_D - E_F)/(k_B T)\big)\big).$$

According to this relation the diffusivity increases when *p*-type AlGaAs heterostructures are annealed in an As-poor ambient, because both effects increase the concentration of interstitials I_{III}. The diffusivity increases also when *n*-type structures are annealed in an As-rich ambient. In this case both effects increase the concentration of vacancies V_{III}.

The effect of the Fermi-level position on the effective diffusivity is more pronounced for multiple charged defects. A cubic dependence of D_{III} on the electron concentration, e.g., points to the participation of a triply charged defect V_{III}^{3-} [53]. A definite assignment to a specific charged defect is difficult though, because both the enthalpy of formation and the enthalpy of migration enter the effective diffusivity, and *both* quantities change if the charge state of a defect changes due to a shift of the Fermi-level [55].

The spatially selective modification of heterostructure interfaces by layer disordering was also employed for device fabrication. By applying masks on the surface, a laterally selective intermixing of quantum-well heterostructures is obtained. The corresponding change of the bandgap and the refractive index with respect to the not intermixed regions can be used to locally confine charge carriers and photons. If, e.g., a sample containing a quantum well is exposed to intermixing conditions except along a stripe, the lateral regions adjacent to the stripe have a larger bandgap and a lower refractive index. These effects were used to fabricate, e.g., index-guided edge-emitting lasers of III–V [58, 59] and II–VI [60] semiconductors. Techniques employed for selective intermixing comprise impurity-induced layer disordering (IILD) by introducing dopants [61], impurity-free vacancy disordering (IFVD) by supplying vacancies [58], and ion-implantation-induced composition disordering (IID).

10.3 Metal-Semiconductor Contact

Any electronic semiconductor device requires an electric contact between the semiconductor and a metal. A metal-semiconductor junction may have either rectifying or ohmic characteristics, depending on the two materials which are brought into contact. The problem is closely related to the band alignment of a semiconductor heterostructure treated in Sect. 4.2. We will introduce the characteristics of a classical ideal metal-semiconductor contact, nonideal effects, and some approaches to fabricate ohmic junctions.

10.3.1 Ideal Schottky Contact

We consider a rectifying metal-semiconductor contact in the framework of the model introduced by Schottky [62] and Mott [63]. According to the simple Schottky-Mott model the properties of the junction are determined by the work function $e\phi_{\mathrm{m}}$ of the metal and that of the semiconductor $e\phi_{\mathrm{sc}}$. Since ϕ_{m} may be larger or smaller than ϕ_{sc} and the latter depends on the type of semiconductor (n or p), we have to distinguish four cases. We first consider a metal—n-type semiconductor junction with $\phi_{\mathrm{m}} > \phi_{\mathrm{sc}}$. Figure 10.22a shows the energy-band diagram of the two solids before contact, taking the vacuum level E_{vac} as reference energy. Since the work functions of the metal and the semiconductor differ, electrons will flow from the side with a high Fermi energy to that with a lower Fermi energy for equilibration if a contact is made. The difference $\phi_{\mathrm{m}} - \phi_{\mathrm{sc}}$ is referred to as *contact potential*. In the case depicted in Fig. 10.22a electrons are transferred to the metal, leaving positively charged donors without a balancing negative charge in the n-type semiconductor. A positive space charge is thereby built up in the interface-near region of the semiconductor, leading to a bending of the band edges over a width w, which is depleted from free electrons. Outside the depletion layer the semiconductor is neutral. The positive space charge in the semiconductor is balanced by an equal negative charge at the metal surface. The dipole potential equilibrates the contact potential. In thermal equilibrium these quantities are equal and the Fermi energy is constant through the junction. We note from Fig. 10.22b that now a barrier $e\phi_{\mathrm{Bn}}$ exists at the junction. The height of this Schottky barrier is given by the band bending eV_{bi}, the so-called built-in potential barrier, and the energy spacing between the Fermi level and the conduction-band edge in the semiconductor,

$$e\phi_{\mathrm{Bn}} = e(V_{\mathrm{bi}} + \phi_{\mathrm{sc}} - \chi) = e(\phi_{\mathrm{m}} - \chi), \qquad (10.33)$$

$e\chi$ being the electron affinity of the semiconductor as depicted in Fig. 10.22a. Equation (10.33) shows that the built-in potential equals the contact potential,

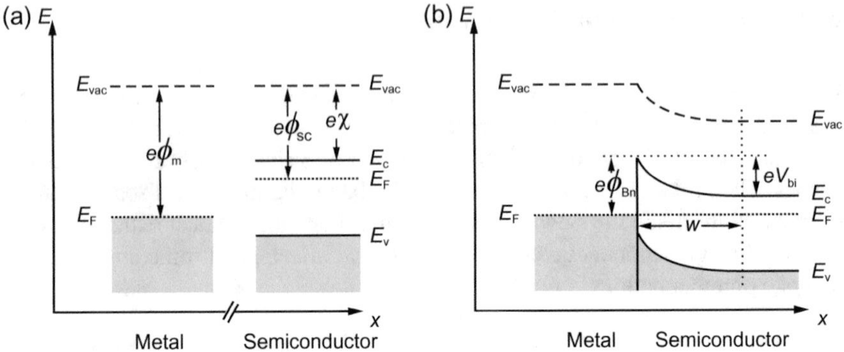

Fig. 10.22 Energy-band diagram of an ideal junction between a metal with a large work function $e\phi_{\mathrm{m}}$ and an n-type semiconductor with a smaller work function $e\phi_{\mathrm{sc}}$. **a** Before contact, **b** in contact without an external bias. $e\phi_{\mathrm{m}}$ and w denote the height of the Schottky barrier and the width of the space-charge region, respectively. *Gray shadings* indicate occupied bands

$V_{\mathrm{bi}} = \phi_{\mathrm{m}} - \phi_{\mathrm{sc}}$. Values of work functions and electron affinities for some solids are given in Table 10.2.

The width of the space-charge region w may be calculated in the Schottky-Mott model under the abrupt approximation, i.e., assuming that the charge density in this region is given by the constant concentration of the donors, $\rho = en_{\mathrm{D}}$. Outside the space-charge region the semiconductor is neutral and $\rho = 0$. The potential in the semiconductor is obtained from the one-dimensional Poisson equation

$$\frac{d^2V}{dx^2} = -\frac{\rho}{\varepsilon},$$

where $\varepsilon = \varepsilon_{\mathrm{r}} \times \varepsilon_0$ is the permittivity. Taking the location of the junction along x and E_{c} in the bulk of the semiconductor as origin, integration leads to

$$V(x) = \frac{en_{\mathrm{D}}}{\varepsilon}\left(wx - \frac{1}{2}x^2\right) - V_{\mathrm{bi}}. \tag{10.34}$$

The width of the depletion region w is obtained from the condition $V(x = w) = 0$, yielding

$$w = \left(\frac{2\varepsilon V_{\mathrm{bi}}}{en_{\mathrm{D}}}\right)^{1/2}. \tag{10.35}$$

In (10.35) we assumed a complete ionization of donors in the space-charge region and a zero bias across the junction. If the thermal distribution of the majority carriers is taken into account, the charge density $\rho = en_{\mathrm{D}}$ is replaced by $\rho =$

Table 10.2 Work functions ϕ_m of some metals and electron affinities χ of some semiconductors. Metal data from [64], semiconductors from [65]

Metal	$e\phi_{\rm m}$ (eV)	Semiconductor	$e\chi$ (eV)
Al	4.28	Ge	4.00
Au	5.10	Si	4.05
Ni	5.15	SiC (4H)	4.05
Pd	5.12	GaAs	4.07
Pt	5.65	InP	4.38
Ti	4.31	GaN	4.10
W	4.54	ZnSe	4.09

$en_{\rm D}(1 - \exp[eV(x)/(k_{\rm B}T)])$. $V_{\rm bi}$ in (10.35) is then replaced by $(V_{\rm bi} - k_{\rm B}T/e)$. If an external bias $V_{\rm ext}$ is applied to the metal-semiconductor junction, this quantity is also added. The more complete expression then reads

$$w = \left(\frac{2\varepsilon(V_{\rm bi} - V_{\rm ext} - k_{\rm B}T/e)}{en_{\rm D}}\right)^{1/2}$$
$$\text{(external bias and thermal carrier distribution included).} \qquad (10.36)$$

Application of an external bias $V_{\rm ext}$ such that the semiconductor is negative with respect to the metal (forward bias) lowers the barrier to $e(V_{\rm bi} - V_{\rm ext} - k_{\rm B}T/e)$, while a reverse bias leaves the barrier $e\phi_{\rm Bn}$, in this first-order model, unaffected. The Schottky barrier hence represents an asymmetric resistance for the current across the junction. The junction has a rectifying character, it represents a *Schottky diode*.

The treatment above refers to a metal—n-type semiconductor junction. We now compare the four basic cases of $\phi_{\rm m} > \phi_{\rm sc}$, $\phi_{\rm m} < \phi_{\rm sc}$, and contacts to either an n-type or a p-type semiconductor. If the junction to the n-type semiconductor is made using a metal with a *smaller* work function $\phi_{\rm m} < \phi_{\rm sc}$, electrons will flow from the metal to the semiconductor for equilibration. In thermal equilibrium (without external bias), when the Fermi energy is constant across the junction, a negative space charge in the semiconductor balances an equal positive charge at the junction in the metal. Figure 10.23b depicts the resulting band bending in the semiconductor. There exits no barrier for an electron flow from the n-type semiconductor to the metal, and a small barrier $e(\phi_{\rm sc} - \chi)$ for the reverse direction. Such a contact has nearly ohmic character.

A comparable case is the metal—p-type semiconductor junction for a *larger* work function of the metal $\phi_{\rm m} > \phi_{\rm sc}$ depicted in Fig. 10.23c. When the contact is made, electrons flow from the semiconductor to the metal to equilibrate the Fermi level $E_{\rm F}$. The band bending in the semiconductor originating from the space charge of the positively ionized donors is qualitatively similar to the case assumed in Fig. 10.23a, but now there exists no barrier for an electron flow from the metal to the p-type

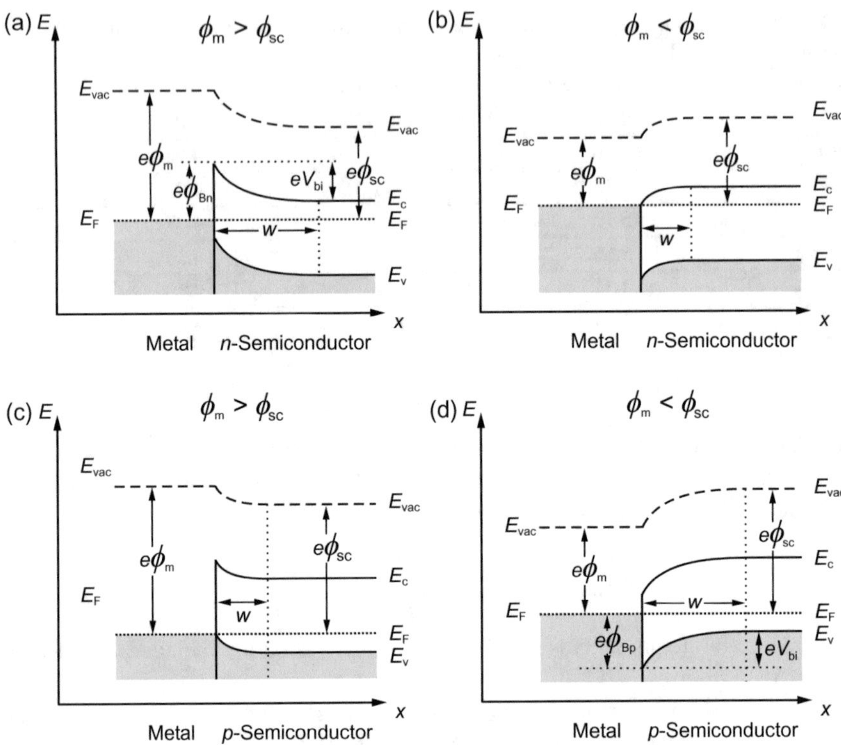

Fig. 10.23 Energy-band diagrams of ideal junctions between metal and an either n-type semiconductor (**a**, **b**) or a p-type semiconductor (**c**, **d**). Metal work functions $e\phi_{\mathrm{m}}$ being larger (**a**, **c**) or smaller (**b**, **d**) than the work functions of the semiconductors $e\phi_{\mathrm{sc}}$ are assumed

semiconductor, and an only small barrier $e(\chi - \phi_{\mathrm{sc}})$ exists for the reverse direction. This contact is nearly ohmic.

The last case of the metal—p-type semiconductor junction for a *smaller* work function of the metal $\phi_{\mathrm{m}} < \phi_{\mathrm{sc}}$ depicted in Fig. 10.23d has similarities to the first case shown in Figs. 10.23a and 10.22. The band bending in the semiconductor eV_{bi} due to an equilibrating electron flow from the metal to the semiconductor leads to the Schottky barrier $e\phi_{\mathrm{Bp}} = e(\chi - \phi_{\mathrm{m}})$. The barrier for the flow of holes is lowered to $e(V_{\mathrm{bi}} - V_{\mathrm{ext}})$ when an external bias V_{ext} is applied such that the semiconductor is positive with respect to the metal. For the reverse bias the barrier $e\phi_{\mathrm{Bp}}$ remains virtually unaffected, leading to a rectifying characteristics of this junction.

The height of the Schottky barriers in cases (a) and (d) above is somewhat reduced, if the *Schottky effect* is taken into account. A free carrier in the semiconductor with charge $-e$ experiences an image-charge effect near the junction to the metal, because the metal surface is an equipotential surface. This modifies the potential distribution similar to the case if an image charge $+e$ were equidistant to the junction on the metal side. The reduction of the Schottky barrier $\Delta\phi$ by this effect depends on the external bias and is given by the resulting electric field, $\Delta\phi = (e|\mathbf{E}|/(4\pi\,\varepsilon_r\varepsilon_0))^{1/2}$. The Schottky effect is hence diminished by the large relative permittivity ε_r of semiconductors and usually quite small (~10–20 meV). Since real metal-semiconductor junctions are dominated by other effects the Schottky effect is not detailed here.

10.3.2 Real Metal-Semiconductor Contact

The Schottky-Mott model pointed out above predicts the Schottky-barrier height $e\phi_B$ to be the difference between the semiconductor electron affinity $e\chi$ and the metal work function $e\phi_m$. Consequently the slope parameter $S = \delta\phi_B/\delta\phi_m$, which describes the variation of the barrier height of a given semiconductor with different metal contacts, is unity in this ideal case. Experimental Schottky barrier heights are, however, often only weakly dependent on the metal work-function. Some typical results are plotted in Fig. 10.24 for various semiconductors versus the work function of the applied metal. The steep dashed line indicates the expected ideal dependence of the Schottky barrier on the metal work function, i.e., $S = 1$. We note that the experimental slopes given in the figure are significantly smaller and differ among the semiconductors.

It is found that metal junctions to semiconductors with ionic bonding display a large dependence of the barrier height on the metal work-function with only little deviation from the ideal behavior ($S = 1$). On the other hand, metal junctions to semi-

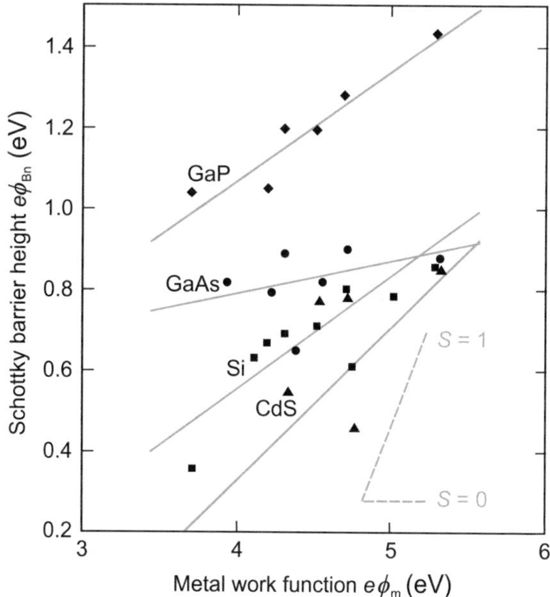

Fig. 10.24 Schottky barrier heights for various junctions between n-type semiconductor and metal. *Gray solid lines* are least-square fits to experimental data, *dashed lines* indicate the slope for slope parameters S equal 1 and 0. Data from [66]

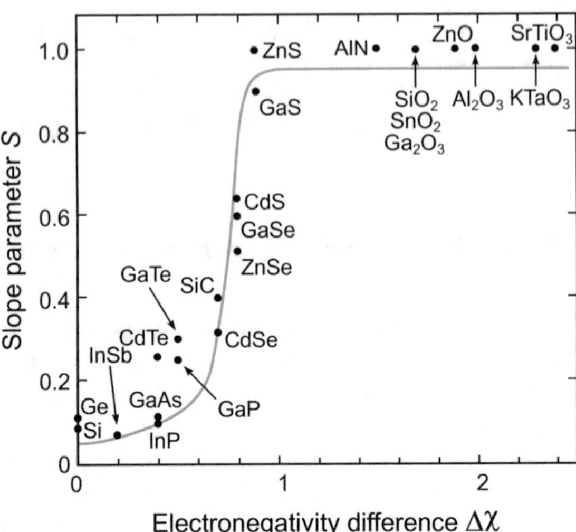

Fig. 10.25 Slope parameter S for various n-type semiconductor/metal junctions plotted versus the electronegativity difference of the constituents of the semiconductor. The *gray curve* is a guide to the eye. Adapted from [67]

conductors with a predominantly covalent bonding lead to an only weak dependence of the barrier height on the metal work function (S small). Experimental slope parameters S are given in Fig. 10.25 for a number of semiconductors. The ionicity of the bonding is expressed in terms of the difference in electronegativity $\Delta\chi = \chi_A - \chi_B$ of the atoms (or, more precisely, anions and cations) A, B of the semiconductor.

The low dependence of the Schottky-barrier height on the metal work-function is referred to as *Fermi-level pinning*. The phenomenon can be understood qualitatively in terms of a model based upon localized states located at the interface between semiconductor and metal. Let us consider such interface states (also termed surface states) with a distribution of electronic levels in the bandgap of the semiconductor. The distribution may be characterized by an energy $e\phi_0$, the so-called charge-neutrality level. States below this level are assumed neutral if they are filled with electrons, and states lying above are assumed neutral if they are empty. If the density of such interface states near $e\phi_0$ is large, then adding electrons to the semiconductor or extracting them from the semiconductor does not alter the position of the Fermi energy. The Fermi level is *pinned*. When the contact between the metal and the semiconductor is made, both, addition of electrons and extraction of electrons are accommodated by the interface states, leaving the Fermi level virtually unchanged. The Schottky-barrier height is then independent of the metal used, and given by

$$e\phi_B = E_g - e\phi_0,$$

$e\phi_0$ being the charge-neutrality level.

Two basically different models on the physical nature of the interface states have been proposed. The *model of virtual gap states* (ViGS, also referred to as metal induced gap states MIGS) assumes that the wave functions of the metal electrons have exponentially decaying tails into the semiconductor [68, 69]. These virtual states

Table 10.3 Schottky barrier height $e\phi_{Bn}^{theo}$ for an n-type semiconductor-Au metal junction and effective midgap point E_B calculated from the model of virtual gap states. Experimental barrier heights refer to Au metal. Data from [69], E_B from [70]

	E_g (eV)	E_B (eV)	$e\phi_{Bn}^{theo}$ (eV)	$e\phi_{Bn}^{exp}$ (Au) (eV)
Ge	0.66	0.18	0.48	0.59
Si	1.12	0.36	0.76	0.83
GaAs	1.56	0.70	0.74	0.94
ZnS	3.60		1.40	2.00

are located in the bandgap and decay on an atomic scale with a charge decay-length of some Å, making the first few layers of the semiconductor locally metallic: The local density-of-states in the semiconductor bandgap is filled with a smooth density of gap states. The gap states are related to those bands of the semiconductor that are nearest in energy. ViGS which are related to the valence band are then occupied, and those with conduction-band character are empty. At an effective midgap point E_B gap states change from primarily valence character to conduction character. The Fermi level is pinned at or near this energy E_B, yielding local charge neutrality. A relatively low number of ViGS (about one per 100 atoms at the interface) is required to produce the pinning effect. The strength of the ViGS model is its simplicity. Without adjustable parameters it could reasonably predict experimentally observed pinned Shottky barrier heights for a number of metal-semiconductor combinations and explain why more ionic semiconductors do not show a universal barrier height. Some results are listed in Table 10.3.

The ViGS model assumes a featureless interface between metal and semiconductor, neglecting structural details or the formation of strong local chemical bonds across the interface. Experiments on a variety of different junctions demonstrated, however, a systematic dependence of the barrier height on the chemical reactivity of the interface [71, 72]. Moreover, Fermi-level pinning was already obtained with a metal coverage much less than one monolayer. For reactive metals the position of the Fermi level was found to be largely independent of the metal used. Such findings lead to a different approach emphasizing the role of defects at the interface.

The *defect model* assumes that the Fermi-level pinning originates from localized electronic states originating from defects near the interface [72]. Such defects are, e.g., vacancies in the semiconductor. The energy for the formation of the defect can be created by the heat of condensation of surface adatoms or from the heat of formation of compounds made from metal and semiconductor atoms forming at the interface. A low number of defects (order of one per 100 interface atoms) is required to pin the Fermi energy, analogous to the ViGS model. There exist numerous experimental and theoretical studies on the microscopic nature of such defects and quite a number of related detailed models. Many chemical trends could be explained for specific junctions of semiconductors to metals. Sometimes both, ViGS model *and* defect model are needed to explain the data. No general model accounting for the rich variety of phenomena has been reached to date. The density of interface states cannot be predicted with any degree of certainty. Recent progress in computational approaches are encouraging [73]; still the Schottky barrier height must to date be considered a parameter which must experimentally be determined.

We should note that the fundamental mechanism for Fermi-level pinning is essentially the same for both, ViGS model and defect model. The concept is based upon energy levels in the gap of the semiconductor near the interface that can accommodate carriers flowing across the interface for equilibration of the electronegativity when the contact is made. The charge transferred between the metal and the semiconductor creates an interface dipole, which pins the Fermi level and hence controls the Shottky-barrier height.

10.3.3 Practical Ohmic Metal-Semiconductor Contact

Practical contacts should have a negligible resistance compared with that of the semiconductor device of which the contact forms part. Such contacts are often referred to as *ohmic contacts*, i.e., non-rectifying contacts. The *IV* characteristics of a device with low-resistance contacts is determined by the resistance of the semiconductor. The linearity of the contact is then actually not essential. The contact should, furthermore, not inject minority carriers.

The model of an ideal Schottky contact pointed out in Sect. 10.3.1 provides useful guidelines for fabricating a practical contact. The two basic approaches used to obtain a low contact resistance are depicted in Fig. 10.26. The first principle shown in Fig. 10.26a is based on the formation of a low Schottky-barrier height $e\phi_{\mathrm{Bn}}$. This requires—in absence of pinning effects—a metal with a very small work function for a junction to an n-type semiconductor, and a metal with a very large work function for a junction to an p-type semiconductor, cf. Fig. 10.23. The implementation of the concept is usually hampered by the problem of finding a metal with a suitable work function. The task is particularly difficult for p contacts to semiconductors with a wide bandgap.

The majority of practical contacts is therefore based on the approach of using a thin, heavily doped semiconductor layer adjacent to the metal as illustrated in Fig. 10.26b. The width of the depletion region of such a layer is very thin, as expressed by (10.35). The contact resistance is then dominated by the tunneling of carriers through the barrier.

The metal-semiconductor contact is characterized by the specific differential resistance R_{c} at zero bias,

$$R_{\mathrm{c}} = \left(\frac{dI}{dV} \right)^{-1}_{V=0}. \tag{10.37}$$

In the commonly applied approach depicted in Fig. 10.26b the tunneling current across the thin barrier comprises not only electrons with an energy close to the Fermi level (the so-called field emission dominating at very low temperatures), but also thermally excited electrons. The resulting current is known as *thermionic field*

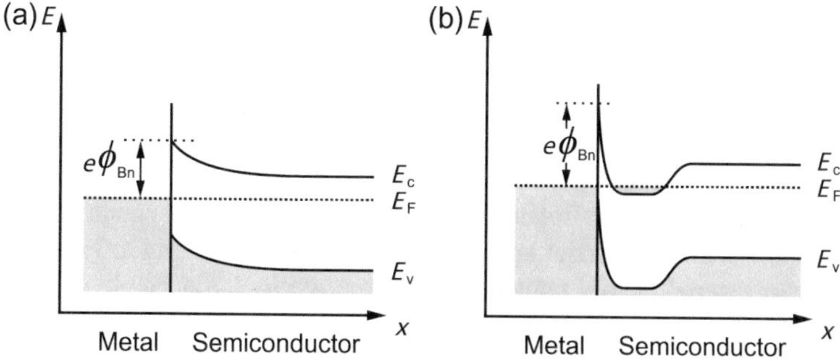

Fig. 10.26 Schematics of concepts for fabricating ohmic contacts: **a** formation of a low barrier height, **b** application of a thin layer with a high doping level

Table 10.4 Typical metallization layers applied for practical contacts to III–V semiconductors

Type of conductivity	Contact material	Composition (%)
n	AuGe	88 : 12
n	AuSn	95 : 5
p	AuBe	99.1 : 0.9
p	AuZn	95 : 5

emission. Its maximum passes the barrier at an energy E_m above the conduction band, where the tunneling probability is larger than at E_F due to a thinner barrier [74]. For this case the contact resistance is proportional to

$$R_c \sim \exp\left(\frac{2}{\hbar}\sqrt{\frac{\varepsilon m^*}{n_D}}\phi_{Bn}\right), \tag{10.38}$$

where ε is the permittivity of the semiconductor. We note the strong dependence on doping and on the height of the Schottky barrier.

Due to the large importance of ohmic contacts for semiconductor devices some standard technology has been developed. Generally a contact metal containing also some dopant material is evaporated onto the semiconductor surface like, e.g., AuZn on *p*-type GaAs. Table 10.4 summarizes some standard metallizations applied for contacts to III–V compound semiconductors. Often these contact layers are alloyed into the semiconductor at the respective eutectic temperature. In addition, further metal layers are usually added to improve the adhesion to the semiconductor and to lower the total contact resistance. An example is the layer sequence AuGe (100 nm)/Ni (50 nm)/Au (200 nm), where Ni provides good adhesion due to a low surface energy, and Au lowers the resistance.

10.3.4 Epitaxial Contact Structures

In some cases no suitable metallization for forming an ohmic contact can be found. The reason may be an unfavorable Fermi-level pinning or the lack of a metal with a suitable work function. A metal contact to, e.g., *p*-type ZnSe proved difficult, because no metal with a sufficiently large work function was found to avoid the formation of a Schottky barrier $e\phi_{\mathrm{Bp}}$ for hole injection according to Fig. 10.23d. In such cases a heteroepitaxial contact structure may provide a viable solution. The presented examples intend to introduce some basic concepts rather than representing general recipes.

The principle of an epitaxial contact structure is based on the introduction of a material between the metal and the semiconductor that accommodates the difference in metal and semiconductor work functions and avoids Fermi-level pinning. There exist various implementations of the idea. A simple approach is the application of a heavily doped thin semiconductor layer on top of the semiconductor used in the device. Such a structure was reported for a contact to *n*-type GaAs and is shown in Fig. 10.27 [75].

By inserting a heavily doped Ge layer between Au metal and GaAs, the large Schottky barrier forming at the metal/*n*-GaAs junction is split into two smaller barriers. The smaller metal/*n*-Ge barrier height is, furthermore, quite thin due to the heavy-doping ability of the narrow-gap semiconductor Ge ($n \geq 10^{20}\ \mathrm{cm}^{-3}$). A contact resistance below $10^{-7}\ \Omega\,\mathrm{cm}^2$ was reported for such a non-alloyed contact [75].

Another concept is the application of a semiconductor material which provides a good contact to a metal and can be alloyed during growth with the semiconductor used in the device. An alloy layer is then inserted the composition of which is graded from the semiconductor forming the interface with the metal to that used in the device. The principle of this approach is illustrated in Fig. 10.28.

The example depicted in Fig. 10.28 employs the property of InAs with an untypical Fermi-level pinning in the conduction band. Therefore a good ohmic contact with a small Schottky barrier $e\phi_{\mathrm{B}n} < 0$ can be made to *n*-type InAs as shown in Fig. 10.28b. The simple insertion of such a layer between the metal and *n*-type GaAs with a Fermi

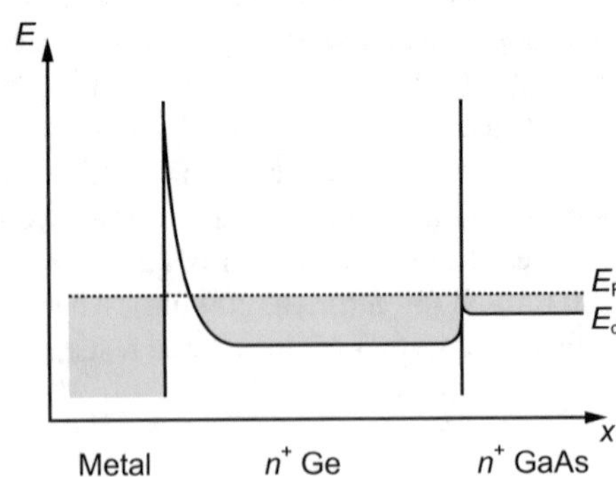

Fig. 10.27 Energy-band diagram of a heterojunction structure for a low-resistance ohmic contact

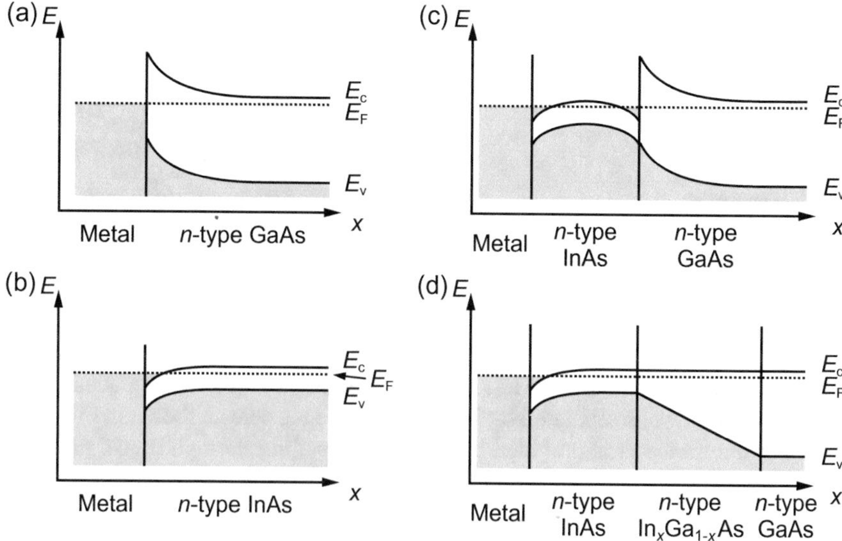

Fig. 10.28 Energy-band diagrams illustrating the concept of ohmic contact formation using a layer with a graded band gap: Metal contact to **a** n-type GaAs, **b** n-type InAs, **c** n-type GaAs with an inserted n-type InAs layer and abrupt interface, **d** n-type GaAs with an additional graded n-type $In_xGa_{1-x}As$ layer and non-abrupt interfaces

level pinned in the gap will not avoid the high Schottky barrier, which is then formed at the interface between the two semiconductors as illustrated in Fig. 10.28c. The problem may be solved by replacing the abrupt junction between the semiconductors by an n-type ternary $In_xGa_{1-x}As$ layer with a composition graded from $x = 1$ to $x = 0$ [76]. For the example given in Fig. 10.28 a contact resistance in the range $10^{-6}\ \Omega\,cm^2$ was reported. The principle was also applied to other materials. An ohmic p contact to ZnSe with a resistance in the mid $10^{-2}\ \Omega\,cm^2$ range was accomplished, e.g., by applying a Au metallization to$p$-type BeTe and using a pseudograded p-type BeTe/ZnSe superlattice [77]. Within the superlattice with 20 monolayer (ML) pseudoperiod the thickness of individual layers was varied in 1 ML steps starting at 19 ML BeTe + 1 ML ZnSe and ending at 1 ML BeTe + 19 ML ZnSe. The purpose of this superlattice was to mimic a random alloy for smoothly grading the valence-band offset.

An approach apparently similar to pseudograding but representing actually a different concept of an epitaxial contact structure is based on resonant tunneling within a multi-quantum well (MQW) structure. The principle of the approach is depicted for a p-type contact to ZnSe in Fig. 10.29. The metal-semiconductor junction is made to p-type ZnTe, which can be degenerately doped p-type and forms a good ohmic (tunnel) contact to Au for hole injection. A ZnTe/ZnSe MQW structure is placed between ZnTe and ZnSe to accommodate the large valence-band offset of about 0.5 eV between these binaries. The widths L_{QW} of the ZnTe quantum wells

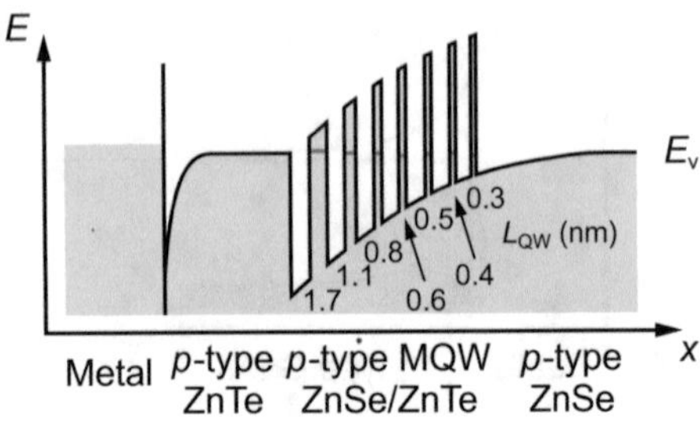

Fig. 10.29 Energy-band diagram depicting the concept of a resonant-tunneling contact using a multiple-quantum well (MQW) region. *Dark gray horizontal lines* represent the lowest quantized energy levels for holes

in this structure are designed such that the lowest hole levels align to the energies of the valence-band maxima of ZnTe and ZnSe, yielding a sequence of gradually narrowing quantum wells separated by 2 nm thick ZnSe barriers as shown in Fig. 10.29 [78]. Current transport is provided by resonant tunneling through the aligned QW levels. For the structure depicted in Fig. 10.29 also a contact resistance in the mid $10^{-2}\ \Omega\,\text{cm}^2$ range was obtained.

An issue of any contact is mechanical and thermal stability besides the band alignment and a low contact resistance as discussed above. The lattice mismatch of semiconductors applied in epitaxial contact structures must therefore also be considered. A high density of structural defects is usually detrimental to the contact lifetime. The InAs/GaAs and ZnTe/ZnSe junctions mentioned in the examples above introduce large strain due to ~7% misfit in both cases, while the Ge/GaAs and BeTe/ZnSe junctions are well lattice-matched and expected to contain a low density of defects.

10.4 Problems Chapter 10

10.1 (a) Find the intrinsic carrier concentrations in InAs and InP at 77 K and 300 K. Use band parameters given in Problem 2 of Chap. 4 and effective masses m_e, m_{hh}, and m_{lh} (all in units of the free-electron mass m_0) for InAs 0.02, 0.41, 0.03, and for InP 0.08, 0.60, 0.12, respectively.

(b) Calculate for both intrinsic semiconductors the temperature, where the Fermi energy deviates by +50 meV from the midgap position.

10.2 Compute the effective densities of states at the band edges, and estimate the maximum carrier concentration for electrons and holes in InP at 300 K using the Fermi-level stabilization model. Apply a Fermi-level stabilization energy of 0.76 eV above the valence-band edge, band parameters given in Problem 2 of Chap. 4, and the effective mass parameters given in the previous problem.

10.3 An undoped GaAs layer is grown at 700 °C on *p*-type GaAs bulk homogeneously doped with 1×10^{18} cm^3 Be atoms. The Be atoms redistribute due to the concentration gradient at the interface. Assume that the diffusion process is well described by a simple Fick diffusion with a single diffusion constant of 2×10^{-5} cm^2/s and an activation energy of 1.95 eV.

(a) How long does it take at the given temperature to obtain a softened Be concentration profile at the interface with a width (for drop from 90% to 10% of the initial concentration) of 200 nm? Approximate values of the complementary error function are (x, erfc(x): 0.0, 1.00; 0.2, 0.78; 0.4, 0.57; 0.6, 0.40; 0.8, 0.26; 1.0, 0.16; interpolate to obtain rough values in between).
(b) Diffusion slows down if the temperature is lowered. Which temperature is needed to double the time for obtaining the same softening of the concentration profile?
(c) Find the width of the drop of the softened Be concentration profile (drop from 90% to 10%) after an exposure of the interface (with initially step-like profile) at 700 °C for 24 h.
(d) Which Be concentration exists in the GaAs layer at a distance of 100 nm from the initial interface after the exposure at 700 °C for 24 h?

10.4 Consider a contact of platinum to n-type Si with 2×10^{16} cm^{-3} donors—all ionized at a temperature of 300 K—in the framework of the Schottky-Mott model. Si has a relative permittivity of 11.7 and—at the given temperature—an effective density of states of 2.8×10^{19} cm^{-3}.

(a) Calculate the barrier height, the energy difference between the conduction-band edge and the Fermi level in the bulk of Si, and the contact potential.
(b) What is the width of the depletion region without an externally applied voltage? What is the width of the space charge if a voltage of 1.0 V is applied in the forward direction? What is the width for a bias of 1.0 V in the reverse direction?
(c) What is the space-charge width without external bias at 400 K?

10.5 A contact resistance of 5×10^{-4} $\Omega \times$ cm^2 is measured for a junction between Al and n-Si highly doped to 2×10^{19} cm^{-3} at room temperature. Estimate the contact resistance if the doping level can be further raised to 1×10^{20} cm^{-3}, assuming no interface states occur. The effective mass of the density of states and the relative permittivity of Si are $1.2 \times m_0$ and 11.7, respectively.

10.5 General Reading Chapter 10

E.F. Schubert, *Doping in III–V Semiconductors* (Cambridge University Press, Cambridge, 1993)
D. Shaw (ed.), *Atomic Diffusion in Semiconductors* (Plenum Press, New York, 1973)
B. Tuck, *Atomic Diffusion in III–V Semiconductors* (Adam Hilger, Bristol, 1988)
P. Heitjans, J. Kärger (eds.), *Diffusion in Condensed Matter—Methods, Materials, Models* (Springer, Berlin, 2005)
H. Mehrer, *Diffusion in Solids—Fundamentals, Methods, Materials, Diffusion-Controlled Processes* (Springer, Berlin, 2007)
W. Mönch, *Semiconductor Surfaces and Interfaces*, 3rd edn. (Springer, Berlin, 2001)
S.L. Chuang, *Physics of Photonic Devices*, 2nd edn. (Wiley, Hoboken, 2009)

References

1. J.M. Luttinger, Quantum theory of cyclotron resonance in semiconductors: general theory. Phys. Rev. **102**, 1030 (1956)
2. E.F. Schubert, *Doping in III-V Semiconductors* (Cambridge University Press, Cambridge, 1993)
3. V.A. Karasyuk, D.G.S. Beckett, M.K. Nissen, A. Villemaire, T.W. Steiner, M.L.W. Thewalt, Fourier-transform magnetopholo♦luminescence spectroscopy of donor-bound excitons in GaAs. Phys. Rev. B **49**, 16381 (1994)
4. M. Grundmann, *The Physics of Semiconductors* (Springer, Berlin, 2006)
5. W. Götz, N.M. Johnson, C. Chen, H. Liu, C. Kuo, W. Imler, Activation energies of Si donors in GaN. Appl. Phys. Lett. **68**, 3144 (1996)
6. W. Götz, R.S. Kern, C.H. Chen, H. Liu, D.A. Steigerwald, R.M. Fletcher, Hall-effect characterization of III-V nitride semiconductors for high efficiency light emitting diodes. Mat. Sci. Engin. B **59**, 211 (1999)
7. M. Isshiki, T. Kyotani, K. Masumoto, W. Ichida, S. Suto, Emissions related to donor-bound excitons in highly purified zinc selenide crystals. Phys. Rev. B **36**, 2568 (1987)
8. H.E. Ruda, Theoretical study of hole transport in ZnSe. J. Appl. Phys. **59**, 3516 (1986)
9. R.N. Bhargava, R.J. Seymour, B.J. Fitzpatrick, S.P. Herko, Donor-acceptor pair bands in ZnSe. Phys. Rev. B **20**, 2407 (1979)
10. H. Reiss, Chemical effects due to the ionization of impurities in semiconductors. J. Chem. Phys. **21**, 1209 (1953)
11. D.B. Laks, C.G. Van de Walle, G.F. Neumark, P.E. Blöchl, S.T. Pantelides, Native defects and self-compensation in ZnSe. Phys. Rev. B **45**, 10965 (1992)
12. C.G. Van de Walle, D.B. Laks, G.F. Neumark, S.T. Pantelides, First-principles calculations of solubilities and doping limits: Li, Na, and N in ZnSe. Phys. Rev. B **47**, 9425 (1993)
13. Z. Zhu, H. Mori, M. Kawashima, T. Yao, Planar doping of p-type ZnSe layers with lithium grown by molecular beam epitaxy. J. Cryst. Growth **117**, 400 (1992)
14. M. Tao, A kinetic model for metalorganic chemical vapor deposition from trimethylgallium and arsine. J. Appl. Phys. **87**, 3554 (2000)
15. H. Künzel, J. Knecht, H. Jung, K. Wünstel, K. Ploog, The effect of arsenic vapour species on electrical and optical properties of GaAs grown by molecular beam epitaxy. Appl. Phys. A **28**, 167 (1982)
16. G. Mandel, Self-compensation limited conductivity in binary semiconductors. Phys. Rev. **134**, A1073 (1964)
17. G.A. Baraff, M. Schlüter, Electronic structure, total energies, and abundances of the elementary point defects in GaAs. Phys. Rev. Lett. **55**, 1327 (1985)
18. R.W. Jansen, O.F. Sankey, Theory of relative native- and impurity-defect abundances in compound semiconductors and the factors that influence them. Phys. Rev. B **39**, 3192 (1989)
19. F. El-Mellouhi, N. Mousseau, Self-vacancies in gallium aresenide: an ab initio calculation. Phys. Rev. B **71**, 125207 (2005)
20. S.B. Zhang, J.E. Northrup, Chemical potential dependence of defect formation energies in GaAs: application to Ga self-diffusion. Phys. Rev. Lett. **67**, 2339 (1991)
21. C.G. Van de Walle, J. Neugebauer, Hydrogen in semiconductors. Annu. Rev. Mater. Res. **36**, 179 (2006)
22. W. Götz, N.M. Johnson, J. Walker, D.P. Bour, Hydrogen passivation of Mg acceptors in GaN grown by metalorganic chemical vapor deposition. Appl. Phys. Lett. **67**, 2666 (1995)
23. N.M. Johnson, R.D. Burnham, R.A. Street, R.L. Thornton, Hydrogen passivation of shallow-acceptor impurities in p-type GaAs. Phys. Rev. B **33**, 1102 (1986)
24. K.J. Chang, D.J. Chadi, Theory of hydrogen passivation of shallow-level dopants in crystalline silicon. Phys. Rev. Lett. **60**, 1422 (1988)
25. F. Gendron, L.M. Porter, C. Porte, E. Bringuier, Hydrogen passivation of donors and acceptors in SiC. Appl. Phys. Lett. **67**, 1253 (1995)
26. P.M. Mooney, Deep donor levels (DX centers) in III-V semiconductors. J. Appl. Phys. **67**, R1 (1990)

27. D.J. Chadi, K.J. Chang, Energetics of DX-center formation in GaAs and $Al_xGa_{1-x}As$ alloys. Phys. Rev. B **39**, 10063 (1989)
28. D.J. Chadi, Doping in ZnSe, ZnTe, MgSe, and MgTe wide-band-gap semiconductors. Phys. Rev. Lett. **72**, 534 (1994)
29. W. Walukiewicz, Intrinsic limitations to the doping of wide-gap semiconductors. Physica B **302**, 123 (2001)
30. W. Walukiewicz, Fermi level dependent native defect formation: consequences for metal-semiconductor and semiconductor-semiconductor interfaces. J. Vac. Sci. Technol. B **6**, 1257 (1988)
31. E. Tokumitsu, Correlation between Fermi level stabilization positions and maximum free carrier concentrations in III-V compound semiconductors. Jpn. J. Appl. Phys. **29**, L698 (1990)
32. S.B. Zhang, The microscopic origin of the doping limits in semiconductors and widegap materials and recent developments in overcoming these limits: a review. J. Phys. Condens. Matter **14**, R881 (2002)
33. C.E.C. Wood, G. Metze, J. Berry, L.F. Eastman, Complex free-carrier profile synthesis by "atomic-plane" doping of MBE GaAs. J. Appl. Phys. **51**, 383 (1980)
34. E.F. Schubert, *Doping in III-V Semiconductors* (Cambridge University Press, Cambridge, 2005)
35. E.F. Schubert, Delta doping of III-V compound semiconductors: fundamentals and device applications. J. Vac. Sci. Technol. A **8**, 2980 (1990)
36. E.F. Schubert, J.E. Cunningham, W.T. Tsang, Electron-mobility enhancement and electron-concentration enhancement. Solid State Commun. **63**, 591 (1987)
37. K.W. Böer, U.W. Pohl, *Semiconductor Physics* (Springer, Cham, 2018)
38. E.F. Schubert, A. Fischer, K. Ploog, The delta-doped field-effect transistor. IEEE Trans. Electron Devices **33**, 625 (1986)
39. R. Dingle, H.L. Störmer, A.C. Gossard, W. Wiegmann, Electron mobilities in modulation-doped semiconductor heterojunction superlattices. Appl. Phys. Lett. **33**, 665 (1978)
40. T. Ando, A.B. Fowler, F. Stern, Electronic properties of two-dimensional systems. Rev. Mod. Phys. **54**, 437 (1982)
41. L.N. Pfeiffer, K.W. West, H.L. Störmer, K.W. Baldwin, Electron mobilities exceeding 10^7 cm^2/Vs in modulation-doped GaAs. Appl. Phys. Lett. **55**, 1888 (1989)
42. C. Hilsum, Simple empirical relationship between mobility and carrier concentration. Electron. Lett. **10**, 259 (1974)
43. W. Walukiewicz, H.E. Ruda, J. Lagowski, H.C. Gatos, Electron mobility in modulation-doped heterostructures. Phys. Rev. B **30**, 4571 (1984)
44. J. Crank, *The Mathematics of Diffusion*, 2nd edn. (Clarendon Press, Oxford, 1975)
45. U.M. Gösele, Fast diffusion in semiconductors. Annu. Rev. Mater. Sci. **18**, 257 (1988)
46. F.C. Frank, D. Turnbull, Mechanism of diffusion of copper in germanium. Phys. Rev. **104**, 617 (1956)
47. U. Gösele, W. Frank, A. Seeger, Mechanism and kinetics of the diffusion of gold in silicon. Appl. Phys. **23**, 361 (1980)
48. T.Y. Tan, U. Gösele, S. Yu, Point defects, diffusion mechanisms, and superlattice disordering in gallium arsenide-based materials. Crit. Rev. Solid State Mater. Sci. **17**, 47 (1991)
49. W.R. Wilcox, T.J. LaChapelle, Mechanism of gold diffusion in silicon. J. Appl. Phys. **35**, 240 (1964)
50. R.L. Longini, Rapid zinc diffusion in gallium arsenide. Solid-State Electron. **5**, 127 (1962)
51. H. Bracht, Diffusion mechanisms and intrinsic point-defect properties in silicon. Mater. Res. Soc. Bull. **25**(6), 22 (2000)
52. T.Y. Tan, U. Gösele, Diffusion in semiconductors, in *Diffusion in Condensed Matter*, ed. by P. Heitjans, J. Kärger (Springer, Berlin, 2005)
53. T.Y. Tan, U. Gösele, Mechanisms of doping-enhanced superlattice disordering and of gallium self-diffusion in GaAs. Appl. Phys. Lett. **52**, 1240 (1988)
54. L.J. Guido, N. Holonyak Jr., K.C. Hsieh, R.W. Kaliski, W.E. Plano, R.D. Burnham, R.L. Thornton, J.E. Epler, T.L. Paoli, Effects of dielectric encapsulation and As overpressure on Al-Ga interdiffusion in $Al_xGa_{1-x}As$-GaAs quantum-well heterostructures. J. Appl. Phys. **61**, 1372 (1987)

55. D.G. Deppe, N. Holonyak Jr., Atom diffusion and impurity-induced layer disordering in quantum-well III-V semiconductor heterostructures. J. Appl. Phys. **64**, R93 (1988)
56. D.G. Deppe, L.J. Guido, N. Holonyak, Impurity-induced layer disordering in $Al_xGa_{1-x}As$-GaAs quantum well heterostructures. Mater. Res. Soc. Symp. Proc. **126**, 31 (1988). Cf. also [55]
57. M. Kawabe, N. Shimizu, F. Hasegawa, Y. Nannichi, Effects of Be and Si on disordering of the AlAs/GaAs superlattice. Appl. Phys. Lett. **46**, 849 (1985)
58. D.G. Deppe, L.J. Guido, N. Holonyak Jr., K.C. Hsieh, R.D. Burnham, R.L. Thornton, T.L. Paoli, Stripe-geometry quantum well heterostructure $Al_xGa_{1-x}As$-GaAs lasers defined by defect diffusion. Appl. Phys. Lett. **49**, 510 (1986)
59. P.D. Floyd, C.P. Chao, K.-K. Law, J.L. Merz, Low-threshold lasers fabricated by alignment-free impurity-induced disordering. IEEE Photonics Technol. Lett. **5**, 1261 (1993)
60. M. Straßburg, O. Schulz, U.W. Pohl, D. Bimberg, S. Itoh, K. Nakano, A. Ishibashi, M. Klude, D. Hommel, A novel approach to improved green emitting II-VI lasers. IEEE J. Sel. Top. Quantum Electron. **7**, 371 (2001)
61. W.D. Laidig, N. Holonyak, M.D. Camras, H. Hess, J.J. Coleman, P.D. Dapkus, J. Bardeen, Disorder of an AlAsGaAs superlattice by impurity diffusion. Appl. Phys. Lett. **38**, 776 (1981)
62. W. Schottky, [De]Vereinfachte und erweiterte Theorie der Randschicht-Gleichrichter (Simplified and extended theory of boundary-layer rectifiers). Z. Phys. **118**, 539 (1942). (in German)
63. N.F. Mott, Note on the contact between a metal and an insulator or semiconductor. Math. Proc. Camb. Philos. Soc. **34**, 568 (1938)
64. H. Warlimont, W. Martienssen, *Springer Handbook of Condensed Matter and Materials Data*, 2nd edn. (Springer Nature, Cham, 2018)
65. Web-archive of semiconductor parameters of the Ioffe Physico-Technical Institute, St. Petersburg, Russian Federation. http://www.ioffe.rssi.ru/SVA/NSM/Semicond/
66. A.M. Cowley, S.M. Sze, Surface states and barrier height of metal-semiconductor systems. J. Appl. Phys. **36**, 3212 (1965)
67. S. Kurtin, T.C. McGill, C.A. Mead, Fundamental transition in the electronic nature of solids. Phys. Rev. Lett. **22**, 1433 (1969)
68. V. Heine, Theory of surface states. Phys. Rev. **138**, A1689 (1965)
69. J. Tersoff, Schottky barrier heights and the continuum of gap states. Phys. Rev. Lett. **52**, 465 (1984)
70. J. Tersoff, Recent models of Schottky barrier formation. J. Vac. Sci. Technol. B **3**, 1157 (1985)
71. L.J. Brillson, Transition in Schottky barrier formation with chemical reactivity. Phys. Rev. Lett. **40**, 260 (1978)
72. W.E. Spicer, P.W. Chye, P.R. Skeath, C.Y. Su, I. Lindau, New and unified model for Schottky barrier and III-V insulator interface states formation. J. Vac. Sci. Technol. **16**, 1422 (1979)
73. D. Stradi, U. Martinez, A. Blom, M. Brandbyge, K. Stokbro, General atomistic approach for modeling metal-semiconductor interfaces using density functional theory and nonequilibrium Green's function. Phys. Rev. B **93**, 155302 (2016)
74. F.A. Padovani, R. Stratton, Field and thermionic-field emission in Schottky barriers. Solid-State Electron. **9**, 695 (1966)
75. R. Stall, C.E.C. Wood, K. Board, L.F. Eastman, Ultra low resistance ohmic contacts to n-GaAs. Electron. Lett. **15**, 800 (1979)
76. J.M. Woodall, J.L. Freeouf, G.D. Pettit, T. Jackson, P. Kirchner, Ohmic contacts to n-GaAs using graded band gap layers of $Ga_{1-x}In_xAs$ grown by molecular beam epitaxy. J. Vac. Sci. Technol. **19**, 626 (1981)
77. F. Vigué, P. Brunet, P. Lorenzini, E. Tournié, J.P. Faurie, Ohmic contacts to p-type ZnSe using a ZnSe/BeTe superlattice. Appl. Phys. Lett. **75**, 3345 (1999)
78. F. Hiei, M. Ikeda, M. Ozawa, T. Miyajima, A. Ishibashi, K. Akimoto, Ohmic contacts to p-type ZnSe using ZnTe/ZnSe quantum wells. Electron. Lett. **29**, 878 (1993)

Chapter 11
Methods of Epitaxy

Abstract The fabrication of a semiconductor heterostructure with atomically sharp interfaces requires epitaxial growth. This chapter focuses on the widely applied growth techniques of liquid-phase epitaxy (LPE), metalorganic vapor-phase epitaxy (MOVPE), and molecular-beam epitaxy (MBE). The usually equilibrium-near LPE process is illustrated for different cooling procedures. For the growth employing MOVPE we consider properties of source precursors and processes of mass transport. The section on MBE concentrates particularly on vacuum requirements, the effusion of beam sources, and the uniformity of deposition.

The heterostructures discussed in the previous chapters represent the basis of advanced semiconductor devices. Such structures comprising quantum wells and superlattices can only be fabricated by epitaxial growth processes. Different methods for epitaxial growth have been established. They are basically named after the nutrition phase supplying the material for the growth of the solid phase. Prominent methods are techniques applying growth from the vapor phase. Vapor-phase epitaxy (VPE) is usually classified by the transport mechanism of the gaseous species: physical-vapor deposition (PVD), or chemical-vapor deposition (CVD). CVD is often further classified according to the chemistry of the source gases, such as metalorganic CVD (MOCVD, also termed metalorganic VPE, MOVPE, or organometallic VPE, OMVPE), chloride VPE (ClVPE), and hydride VPE (HVPE). PVD represents the vaporization of source material in vacuum. Again different methods are applied such as thermal evaporation, laser ablation, or sputtering. The most prominent PVD technique is molecular-beam epitaxy (MBE), where beams of species are provided by thermal heating in effusion cells.

Besides the rich variety of VPE techniques epitaxy is also performed from the liquid and even from the solid phase. In liquid-phase epitaxy (LPE) growth is performed from a liquid solution or a melt. Solid-phase epitaxy (SPE) is a transition between the solid amorphous and crystalline phases of a material. This kind of crystallization is primarily used for the annealing of crystal damage.

Each of the mentioned epitaxy methods has its strengths and weaknesses. We will focus on the widely applied techniques of liquid-phase epitaxy, metalorganic vapor-phase epitaxy, and molecular-beam epitaxy. Most electronic and optoelectronic

U. W. Pohl, *Epitaxy of Semiconductors*, Graduate Texts in Physics,
https://doi.org/10.1007/978-3-030-43869-2_11

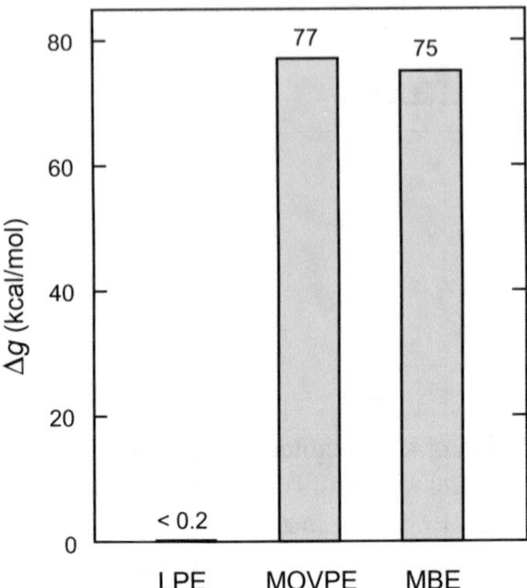

Fig. 11.1 Estimated Gibbs free energy differences between reactants and products for the epitaxy of GaAs at 1000 K using various growth methods. Calculation of the thermodynamic driving forces assumed for liquid-phase epitaxy (LPE) a supercooling below 10 K. In metalorganic vapor-phase epitaxy (MOVPE) growth with Me_3Ga and AsH_3 precursors, and in molecular-beam epitaxy (MBE) growth with Ga and As_4 species is assumed. Data from [1]

devices are fabricated using one of these methods. The principles applied in the three growth techniques are quite different. LPE operates usually close to thermodynamic equilibrium, while MOVPE and MBE occurs quite far from equilibrium. Figure 11.1 illustrates the apparent difference of the driving forces for the epitaxy of GaAs.

The equilibrium-near conditions in the LPE imply the possibility of reversible processes to occur at the growth interface. The LPE process is well described by thermodynamics. We note a much larger driving force in the MOVPE process. Here, growth is mostly not controlled by thermodynamics, but by the slow mass transport of reactants through the vapor to the growth interface. Similar conditions apply for the MBE process.

11.1 Liquid-Phase Epitaxy

In liquid-phase epitaxy (LPE) a crystalline layer usually grows from a supersaturated liquid solution on a substrate. The process has similarities to the seeded growth of bulk crystals from the melt, so that much experience could be transferred to LPE in the past. LPE is a mature technology which is widely used in industry, while it was largely replaced in universities by more flexible techniques like MBE or MOVPE.

LPE has a number of advantages over other growth techniques, the most unique features being:

- Layers with an extraordinary high structural perfection can be grown by LPE.
- High growth rates can be applied in the LPE growth process.

Since growth conditions of LPE processes are close to thermodynamic equilibrium, atoms can efficiently migrate to the growth interface and find energetically optimum positions for incorporation. This enables growth rates of one to two orders of magnitude above those of MOVPE or MBE, and may result in crystalline layers with a very low density of point and dislocation defects. Compound semiconductor layers grown by LPE exhibited point–defect densities orders of magnitudes lower than layers grown by other growth techniques. The carrier lifetime in such layers is consequently quite high. Red-emitting GaAs-based LEDs with highest efficiency and high-performance γ-ray and IR detectors based on HgCdTe are presently fabricated using LPE. The market share in the world's device production has recently given way in many areas to the more versatile techniques of MOVPE and MBE. Still LPE is a major growth method for fabricating magneto-optic layers. The technique was successfully applied to a large variety of materials, such as compound semiconductors (III–V, II–VI, IV–IV) and magnetic or superconducting oxides.

LPE is particularly useful for growing thick layers, also due to high deposition rates up to 1 μm/minute. On the other hand, LPE is less appropriate for fabricating nanostructures like quantum wells. The interface quality depends very strongly on crystallographic misorientation and lattice misfit. Rough interfaces are often also obtained due to back-dissolution. Another restriction of LPE is the limitation to materials which are miscible at growth temperature.

11.1.1 Growth Systems

There exist several methods to bring the substrate into contact with the growth solution prepared for epitaxial growth, and to separate them at the end of layer growth. They can be classified into the techniques of tipping, dipping, and sliding boat. They basically differ in the way to bring the substrate into contact with the solution. The material used for the fabrication of the crucible or the boat depends on the materials to be grown. Usually graphite is used for semiconductor growth systems, while platinum is applied for growing oxide materials like garnets. The schematics of the three mentioned techniques are illustrated below for epitaxy on single substrates. Today LPE is actually used for industrial mass production in upscaled systems employing multiple large-area substrates.

Tipping System

The tipping system was the earliest LPE system applied for growing III–V semiconductors [2]. The schematic of a tipping apparatus is given in Fig. 11.2. The set-up consists of a boat containing the source material solved in a saturated molten solution, a reactor tube allowing to control the gaseous ambient, and a furnace for precise temperature control. The furnace is tiltable such that the growth solution can either be separated from the substrate (position (a) in Fig. 11.2) or placed over the substrate (position (b)). The growth process starts by equilibrating the growth solution in position (a). After completing this step the temperature of the furnace is reduced and the furnace is tipped to position (b). The solution runs over the substrate and layer growth starts. After growing a layer of sufficient thickness, the furnace is tipped back to position (a) and the solution rolls off the layer. The removal of the solution is not quite reliable, a major drawback of the technique. Moreover, the system allows for growing only a single layer. The simple tipping technique was particularly useful for initial early experiments.

Dipping System

The principle of the dipping technique is illustrated in Fig. 11.3. The substrate is mounted on a substrate holder equipped with lift and rotation mechanisms in a horizontal or a vertical position. Layers are grown by dipping the substrate into the solution. The rotation mechanism is employed in the horizontal rotating-disk configuration to control the convection of the solute atoms in the liquid. It may also be used for removal of the solution by spinning off at the end of layer growth.

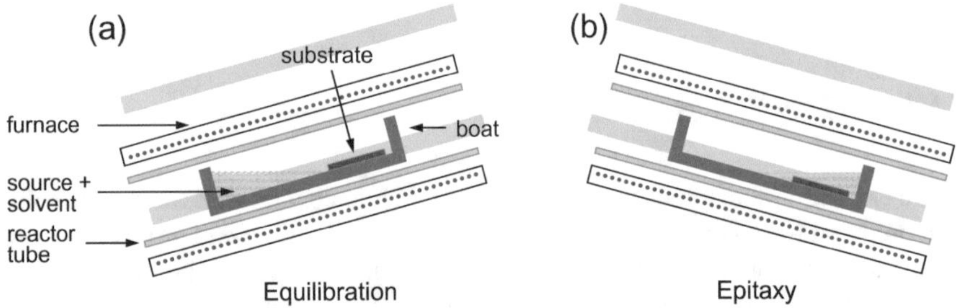

Fig. 11.2 Tipping apparatus for liquid-phase epitaxy. **a** Position for equilibrating the growth solution, **b** furnace tipped to the position for epitaxial growth on the substrate

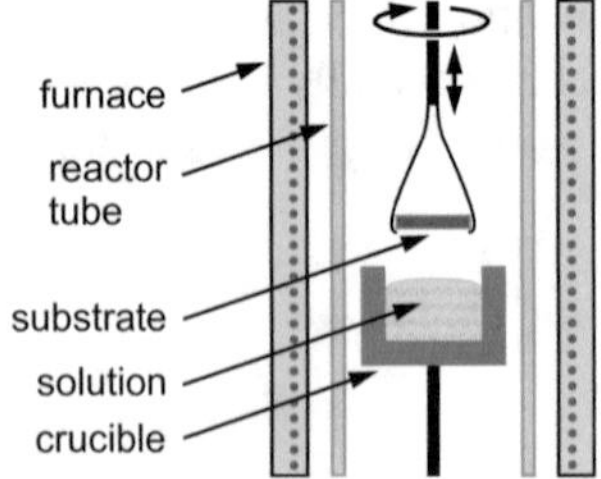

Fig. 11.3 Dipping LPE apparatus for liquid-phase epitaxy. In systems with horizontally mounted substrate rotation may be applied to set up convection in the solute

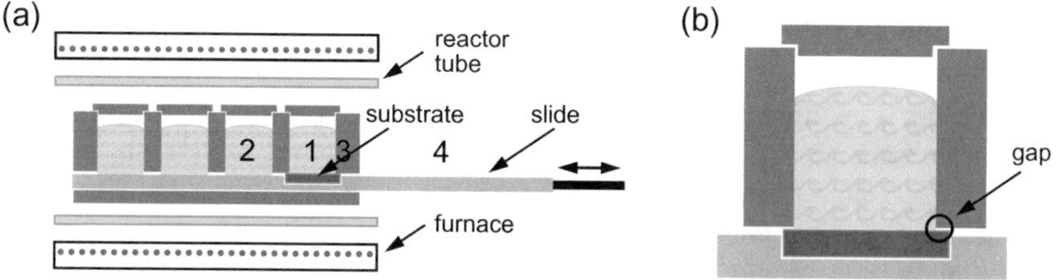

Fig. 11.4 **a** Sliding boat LPE apparatus. Numbers signify solutions with different compositions. **b** Location of the critical gap between the substrate and crucible containing the solution

Dipping systems are useful if thick layers are to be grown. A further advantage particularly for substrates with volatile components is that the substrate can be kept at a decreased temperature prior to growth. The dipping technique is widely applied for the LPE of oxide materials. The popular sliding-boat technique employed for growing semiconductors is not used to grow oxide layers, because platinum used as a boat material does not slide on platinum. Furthermore, no substrate rotation is applied for semiconductor growth due to the sensitivity of the solution surface to oxidation.

Sliding-Boat System

The sliding-boat technique is quite versatile and allows for multiple-layer growth. It is widely applied for the LPE of compound semiconductors. A schematic of a sliding-boat apparatus is given in Fig. 11.4a. The boat consists of two parts: a base, which carries the substrate in a recess, and wells in a block, which contain the solutions for growing successively different layers. Either the block with the wells or the base are movable, such that the solutions can be placed over the substrate. Similar to the previous systems the boat is located in a reactor tube for providing control of the gaseous ambient, and in a furnace for precise temperature control. The wells of the different solutions may be covered with caps as indicated in the figure.

Proper adjustment of the gap between the block containing the solutions and the substrate accommodated in the slide is a critical issue of the siding-boat technique. If the clearance depicted in Fig. 11.4b is too wide, mixing with an adjacent solution and aftergrowth effects may occur. If, on the other hand, the gap is too small, the grown layers are scratched when the slide is moved. This limits also the maximum total thickness of the epitaxial layer sequence which can be grown by the sliding-boat technique. Typical gaps are in the range of 25 to 100 μm.

11.1.2 Congruent Melting

We consider a melting process in a completely miscible two-component system to prepare the discussion of the LPE process. The phase diagram of the system composed of two components A and B is given in Fig. 11.5.

There are two liquidus curves: one separating the region of all liquid A and B at higher temperatures from a region at small concentrations of B (i.e., high concentrations of A) where a fraction of A is solid and in equilibrium with liquid A and liquid B. Another liquidus curve at higher concentration of B separates the all liquid region from a region with solid B and liquid A and B. At an even lower temperature T_{E} a solidus line bounds the regions with either only A or B solid from an all solid region. At the eutectic point denoted E the liquidus and solidus curves intersect. At this point all three phases (all) Liquid, Solid of A, and Solid of B coexist in equilibrium. The phase rule (6.10) states that this point is invariant at a selected fixed pressure: The number of independent variables N_{f} is determined by the number of components N_{c} and the number of coexisting phases N_{p}, $N_{\mathrm{f}} = N_{\mathrm{c}} - N_{\mathrm{p}} + 2 = 2 - 3 + 2 = 1$. If the temperature or the composition is changed at the given fixed pressure, then the number of phases N_{p} is reduced to 2.

At the composition $x_B = 0$ we obtain a one-component system of pure component A. This system melts at only one temperature, the melting temperature $T_{\mathrm{m}A}$. Correspondingly the pure one-component system B at $x_B = 1$ melts at $T_{\mathrm{m}B}$. For all compositions between these end points melting of the system begins at one temperature, the eutectic temperature T_{E}. Except for the eutectic composition melting occurs over a *range* of temperatures between T_{E} and the temperature of the liquidus curve at the given composition; the eutectic composition melts only at T_{E}. It is important to note that in a closed system in equilibrium the composition does not change during melting or, vice versa, during crystallization.

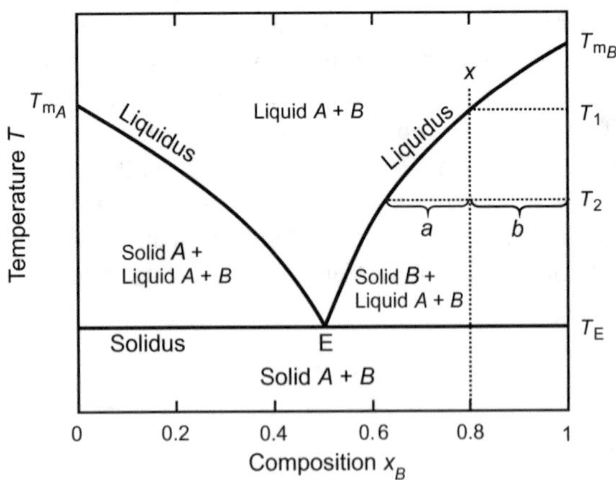

Fig. 11.5 Phase diagram of a two-component system at constant pressure. The abscissa values give the fraction of component B with respect to the sum $A + B$

To illustrate this rule we consider a crystallization of the system along the dotted line at $x_B = 0.8$ drawn in Fig. 11.5. Above the liquidus line the system is a liquid mixture of 80% of component B and 20% of component A. When the temperature is lowered, crystals of component B begin to form at T_1. The more T is lowered, the more of component B crystallizes. Since the overall composition is fixed at $x_B = 0.8$ and crystals only form from component B, the liquid becomes gradually more enriched in A due to the loss of liquid B. The 20% composition of B is hence composed of solid B and liquid B. The fraction of B in the solid state increases for decreasing temperature. At a temperature T_2 marked in Fig. 11.5 this fraction is given by the lever rule $x_{B,\text{solid}} = a/(a+b)$. The liquid fraction of B is correspondingly $x_{B,\text{liquid}} = b/(a+b)$, and A is all liquid. We note from Fig. 11.5 that section b remains constant and a increases as the temperature is lowered towards T_E. Thus, the proportional distance between the initial composition represented by the vertical dotted line and the liquidus curve gives the amount of solid B formed at a given temperature below T_1. The composition of the all Liquid region changes as given by the liquidus line. For decreasing temperature the Liquid contains gradually less B due to loss into the pure B solid. Still the overall composition $x_B = (B_\text{solid} + B_\text{liquid})/(A_\text{liquid}(+A_\text{solid} = 0))$ remains fixed at 0.8. At T_E also A begins to crystallize, keeping x_B constant (i.e., $A_\text{solid} \neq 0$). In a cooling process the temperature would remain constant at T_E until all of A is solid. The final solid is a mixture of 80% B_solid and 20% A_solid.

11.1.3 LPE Principle

The fundamental processes of liquid-phase epitaxy are similar to those of seeded growth of bulk crystals from a solution or a melt. There are major advantages of LPE growth from a solution rather than growing from a melt. Lower growth temperatures can be applied, leading to improved structural perfection and stoichiometry. In addition, the vapor pressure of volatile components in compound semiconductors is much reduced at temperatures far below melting, the interdiffusion of heterointerfaces is decreased, and detrimental effects of thermal expansion differences of substrate and epitaxial layer are reduced. Furthermore, solution growth allows for a more precise control of low growth rates for improved thickness adjustment, and unwanted spontaneous nucleation is reduced.

For a discussion of the basics we consider a part of the phase diagram of a congruently melting binary solid AB and the formation of a solid from a liquid solution of B atoms solved in a solvent A. Since the LPE process is usually controlled by varying the temperature, T is plotted as the independent variable. At a given temperature labeled T_{e2} in Fig. 11.6a the liquidus curve shows the equilibrium composition x_{e2} of B atoms in the saturated liquid at point L_e. The liquid composition x_{e2} is connected with a unique composition of the solid x_S. Points at the left of the liquidus, like L^+ above L_e, correspond to metastable states of a supersaturated or a supercooled liquid solution for the given temperature T_{e2}. Points at the right of the liquidus like

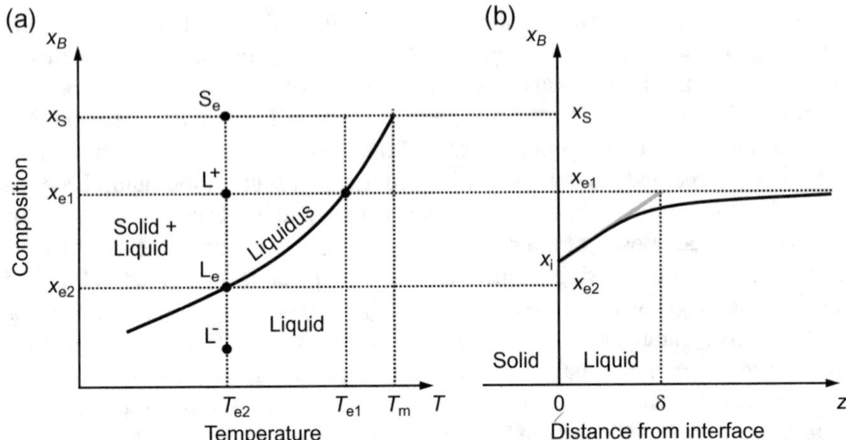

Fig. 11.6 **a** Phase diagram of a two-component system. Points L^+ and L^- signify a liquid in supersaturated and undersaturated states at temperature T_{e2}, respectively, x_{e2} is the equilibrium liquid composition corresponding to an equilibrium solid composition x_S. T_m is the melting temperature of the solid. **b** Concentration profile of composition x_B in the supersaturated liquid solution near the solid-liquid interface located at $z = 0$. δ denotes the diffusion boundary, the *gray line* represents the composition gradient at the interface

L^- correspond to undersaturated liquid solutions at T_{e2}. The deviation from thermal equilibrium can be expressed by the relative supersaturation

$$\sigma = \frac{x_{e1} - x_{e2}}{x_{e2}}, \tag{11.1}$$

where x_{e1} represents the actual concentration in the bulk of the liquid solution (i.e., far away from a phase boundary to a solid) being supersaturated at T_{e2}, and x_{e2} is the equilibrium concentration at T_{e2}. We discussed in Sect. 6.2 that nucleation of a solid in a homogeneous liquid phase requires a minimum driving force, or supersaturation, to take place. This quantity is the critical work ΔG_N^* for creating stable nuclei of critical size which can grow. In the phase diagram, the region between the corresponding critical supersaturation and the liquidus curve is a metastable region which can be used for heteronucleation, because the critical free energy of heteronucleation $\Delta G_{N\ \mathrm{hetero}}^*$ is significantly smaller than that of nucleation in a homogeneous (liquid) phase. Thus, if a supersaturated solution at point L^+ in Fig. 11.6a is brought into contact with a solid, e.g., a substrate, then the solid tends to grow. Vice versa, a contact to an undersaturated solution at point L^- tends to dissolve the solid. The width of the metastable region varies greatly for different solvent-solute systems, depending on the complexity of the crystallizing material, hydrodynamics, and other factors. While metallic semiconductor solutions support only a few degrees of undercooling $\Delta T_s = T_{e1} - T_{e2}$, undercoolings up to 150 °C are possible for oxides. For this reason LPE processes of these materials differ strongly.

We consider a solid B in contact with a supersaturated solution at temperature T_{e2}. Figure 11.6b shows the concentration $x_B(z)$ of component B in the liquid phase near the boundary to the solid. At the interface the concentration x_i is smaller than the concentration x_{e1} in the bulk of the liquid solution, because the solution depletes from B atoms due to incorporation into the solid. Usually kinetic limitations at the growing surface prevent x_i from attaining the equilibrium value x_{e2}. Away from the liquid-solid boundary the composition of the liquid approaches the prepared liquid bulk value x_{e1}.

In an LPE process the supersaturation is created either by cooling or by solvent evaporation of a saturated solution. During growth, a *diffusion boundary layer* of thickness δ is formed at the growth front, in which a concentration gradient and a temperature gradient exist. Growth species of the solute diffuse towards the liquid-solid interface, and solvent species diffuse contrariwise towards the bulk of the solution. Besides diffusion also hydrodynamic flow occurs near the interface. While hydrodynamic flow is more effective than diffusion in the bulk of the solution, its contribution decreases towards the interface and becomes negligible at the growth front. δ describes the concentration gradient in the solution at the interface, $(\partial x_B/\partial z)_{z=0} = (x_{e1} - x_i)/\delta$, cf. Fig. 11.6b. The thickness δ of the diffusion boundary layer is determined by the hydrodynamic flow, and the process within this layer is solely controlled by diffusion. Thus, δ eventually determines the growth rate.

The theory of diffusion-limited growth has been widely applied to describe the LPE process. It assumes that the rate-limiting step is mass transport, i.e., the diffusion of solute species to the growth interface. For simplicity we consider the growth of a one-component system like, e.g., growth from single solute species B in a solvent A. Such a description applies as well for the growth unit of a multicomponent system, if the growth is limited by low rate constants of one of the solute constituents in this unit. Under such conditions, the concentration of the growth unit in the liquid $x_L(z, t)$ (termed x_B for the system described in Fig. 11.6b) is described by the solution of the one-dimensional diffusion equation [3].

$$\frac{\partial x_L}{\partial t} = D\frac{\partial^2 x_L}{\partial z^2} + v\frac{\partial x_L}{\partial z}. \tag{11.2}$$

D is the diffusion coefficient of the solute species in the solvent and z is the distance to the solid-liquid interface as depicted in Fig. 11.6b. In a multicomponent system D refers to the limiting component of the growth unit. The second term on the right-hand side describes convection, where v is the growth velocity resulting from convection and the growth rate. In dipping processes convection is often enforced by substrate rotation to draw growth units in the normal direction towards the substrate and to control the thickness of the boundary layer by the rotation rate ω.

The simultaneously occurring diffusion of heat in the solid and liquid phases is described by an equation similar to (11.2),

$$\frac{\partial T}{\partial t} = K\frac{\partial^2 T}{\partial z^2} + v\frac{\partial T}{\partial z},$$

where T is the temperature and K is the thermal diffusivity. Thermal gradients due to the dissipation of the crystallization heat and to temperature differences in the setup have a major effect in growth *from the melt*. In solution growth usually isothermal conditions prevail, and heat diffusion may be neglected.

11.1.4 LPE Processes

We focus on the diffusion of growth species expressed by (11.2). If no convective field is set up by substrate rotation, the convection term is quite small. This is particularly fulfilled for small growth rates (order of 1 μm/h). We assume that convection is not enforced and neglect this term. We restrict our consideration to semi-infinite growth solutions, where the dimensions of the crucible or boat do not affect the process. This is a good approximation for growth times $t < l^2/D$, where l is the depth of the growth solution and D is the diffusion constant of the growth species. The concentration in the liquid $x_L(z,t)$ is then described by the solution of

$$\frac{\partial x_L}{\partial t} = D\frac{\partial^2 x_L}{\partial z^2}. \tag{11.3}$$

The solution of (11.3) depends on the boundary conditions of the growth process. Initially, the solute concentration is homogeneous, i.e., the concentration of the growth unit in the liquid $x_L(z, t=0) = x_{e1}$ for all z. We assume fast kinetics at the liquid-solid interface, such that the solute concentration at the interface x_i during growth equals the equilibrium concentration x_{e2} at any time, $x_L(z=0,t) = x_i(t) = x_{e2}(t)$.

Once the concentration profile x_L is known, the growth rate r is obtained by mass conservation, yielding

$$r = \frac{D}{(x_s - x_i)}\left(\frac{\partial x_L}{\partial z}\right)_{z=0}. \tag{11.4}$$

Here x_s is the concentration of growth units in the solid and $x_i \approx x_{e2}$ (for fast kinetics) is the interfacial concentration. Often $x_i \ll x_s$ applies, and x_i in the denominator is neglected. The thickness of the grown layer $d(t)$ results from integrating the growth rate,

$$d = \int_0^t r(t')\,dt'. \tag{11.5}$$

The temperature applied during growth may be a function of the growth time t, depending on the growth technique. There are various methods to adjust the temperature for producing the required supersaturation. Some temperature profiles are depicted in Fig. 11.7.

The process of liquid-phase epitaxy generally starts with heating the solution, followed by a period of homogenizing. The latter is of particular importance, since

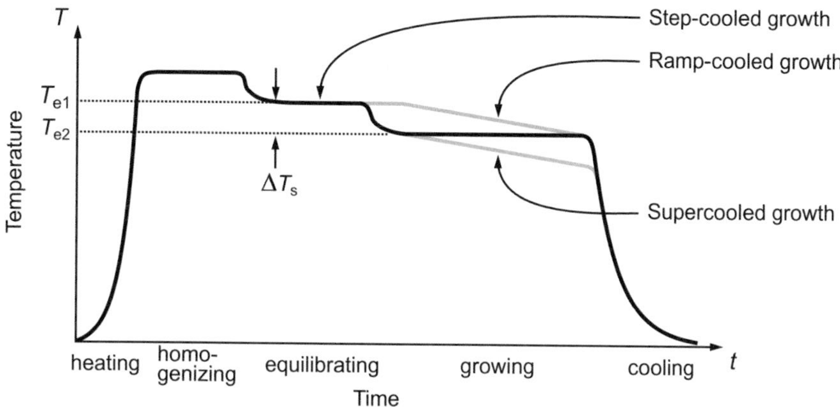

Fig. 11.7 Temperature profiles of various LPE growth processes. Profiles drawn in *gray* are alternatives to that applied for step-cooled growth drawn in *black*

volatile impurities are baked out in this time. Hereafter, the temperature of the solution is lowered for equilibration, and then growth is started. Major temperature profiles applied for LPE growth are pointed out in the following.

Step-cooling growth: During equilibration the solution attains a uniform solute concentration x_{e1} corresponding to a liquidus temperature T_{e1}, cf. Fig. 11.6. At growth start the temperature is lowered to $T_{e2} = T_{e1} - \Delta T_s$ as illustrated in Fig. 11.7. Thereby the equilibrium concentration changes to a value x_{e2} given by the liquidus curve at T_{e2}, and the solution, which is still at T_{e1}, is supersaturated. Inserting the slope of the liquidus curve $m = \partial T_e / \partial x_e$ and assuming a small ΔT_s such that the liquidus curve is approximately linear, we can express the equilibrium concentration on the substrate surface at T_{e2} by $x_L(z = 0, t) = x_{e2} = x_{e1} - \Delta T_s / m$. For this condition and the boundary condition $x_L(z, t = 0) = x_{e1}$ the solution $x_L(z, t)$ to (11.3) is given by the error function [4]

$$\frac{x_L - x_{e2}}{x_{e1} - x_{e2}} = \operatorname{erf}\left(\frac{z}{2\sqrt{Dt}}\right). \tag{11.6}$$

From (11.4) and (11.5) eventually the layer thickness is obtained as a function of the growth time, yielding [4]

$$d(t) = \frac{2\Delta T_s}{m x_s}\left(\frac{D}{\pi}\right)^{1/2} t^{1/2} \quad \text{(step cooling)}, \tag{11.7}$$

m being the slope of the liquidus curve and x_s is the concentration of growth units in the solid. We note that the layer thickness does not linearly increase with time, i.e., the growth rate is not constant. Experimental results for the LPE of GaAs using step cooling are given in Fig. 11.8a. GaAs was grown from a Ga-rich solution, i.e.,

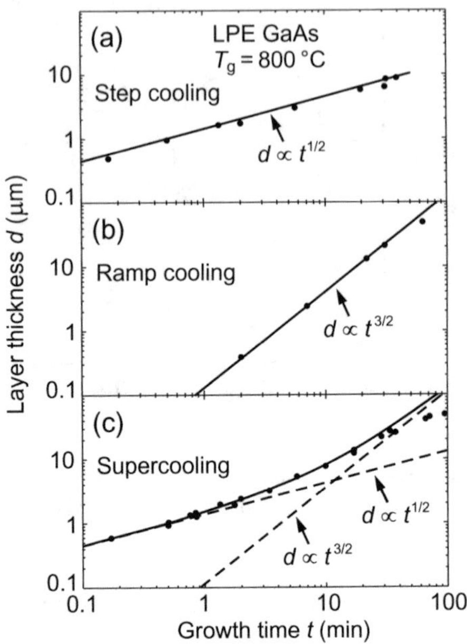

Fig. 11.8 Layer thickness d as a function of growth time t for the liquid-phase epitaxy of GaAs. **a** Layers grown by the step-cooling process. *Points* are measured data, the *line* is calculated from (11.7). **b** Layers grown by ramp cooling, the *line* is calculated from (11.9). **c** Layers grown by supercooling, the *continuous line* is calculated from (11.10). Reproduced with permission from [4], © 1974 Elsevier

Ga was also employed as solvent. The epitaxy was performed at 800 °C applying an undercooling $\Delta T_s = 5\,°\text{C}$. The straight line was calculated from (11.7) and describes well the measured data [4].

Ramp-cooling growth: The process starts with equilibration at a temperature T_{e1}, yielding $x_L(z, t = 0) = x_{e1}$. The temperature of the solution is then lowered at a linear rate α, such that $T(t) = T_{e1} - \alpha t$. If the total cooling interval is small, the slope of the liquidus curve $m = \partial T_e / \partial x_e$ is approximately constant and the equilibrium composition on the liquidus curve is a linear function of temperature. We then obtain the boundary condition $x_L(z = 0, t) = x_{e1} - (\alpha/m)t$. For these conditions the solution $x_L(z, t)$ to (11.3) is given by the complementary error function

$$x_L = x_{e1} - 4(\alpha t/m) i^2 \text{erfc}\left(\frac{z}{2\sqrt{Dt}}\right), \tag{11.8}$$

and by inserting into (11.4) and (11.5) the resulting layer thickness is [4]

$$d(t) = \frac{4\alpha}{3m x_s}\left(\frac{D}{\pi}\right)^{1/2} t^{3/2} \quad \text{(ramp cooling)}. \tag{11.9}$$

We note that the time dependence of $d(t)$ differs from the square-root dependence obtained for step cooling (11.7). Experimental data for ramp-cooling LPE of GaAs are given in Fig. 11.8b. A cooling rate $\alpha = 0.6$ °C/min was applied, and the given growth temperature T_g refers to the midpoint of the growth interval.

Supercooling growth: This process is a combination of step and ramp cooling processes. An undercooling temperature step of ΔT_s is introduced in addition to the temperature lowering at a linear rate α, yielding the boundary condition $x_L(z=0,t) = x_{e1} - \Delta T_s/m - (\alpha/m)t$. The solution $x_L(z,t)$ to (11.3) is accordingly a combination of both processes, yielding the time-dependent layer thickness [4]

$$d(t) = \frac{1}{m x_s}\left(\frac{D}{\pi}\right)^{1/2}\left(2\Delta T_s t^{1/2} + \frac{4}{3}\alpha t^{3/2}\right) \quad \text{(supercooling).} \tag{11.10}$$

LPE data of GaAs grown by supercooling given in Fig. 11.8c confirm the expected dependence. In the experiments temperature steps $\Delta T_s = 5$ °C and cooling rates $\alpha = 0.6$ °C/min were applied.

The chosen process depends on the requirements of the material to be grown, the phase diagram, the morphology of the layer structure, and other parameters. A survey of LPE growth of Column IV, III–V, II–VI semiconductors and other solids is given in [33].

The introduction outlined in this section illustrates that LPE relies on well-established data of the liquid-solid phase equilibrium. The technique is certainly challenging due to stringent demands of growth parameters, and the development of a growth process for a new layer structure requires much more time than for the more popular techniques of MOVPE and MBE pointed out in the next sections. On the other hand, liquid-phase epitaxy bares the potential for fabricating epitaxial layers with the highest possible structural perfection and homogeneity.

11.2 Metalorganic Vapor-Phase Epitaxy

Metalorganic vapor-phase epitaxy (MOVPE), also termed metalorganic chemical vapor deposition (MOCVD; sometimes O and M in the acronyms are exchanged), is the most frequently applied CVD technique for semiconductor device fabrication. Industrial large scale reactors presently have the capacity for a simultaneous deposition on thirty 4-inch wafers, and a majority of advanced semiconductor devices is produced using this technique. Applications of MOVPE are not restricted to semiconductors, but also include oxides, metals, and organic materials. The technique emerged in the 1960s [5–9], when epitaxy was dominated by liquid-phase epitaxy and chloride vapor-phase epitaxy, and molecular-beam epitaxy (Sect. 11.3) did not exist in its present form. Complex sample structures with abrupt interfaces down to the monolayer range and excellent uniformity may today be fabricated using either MOVPE or MBE, though application of MOVPE is advantageous in realizing graded layers or, e.g., in arsenide-phosphide alloys and nitride semiconductors.

Table 11.1 Dissociation energy E of the carbon-hydrogen bond for radicals R used in MOVPE source molecules

R	E (kJ/mol)	R	E (kJ/mol)
methyl (Me)	435	iso-propyl	398
ethyl (Et)	410	tert-butyl (^{t}Bu)	381
n-propyl	410	allyl	368

11.2.1 Metalorganic Precursors

A common feature of chemical vapor-phase techniques is the transport of the constituent elements in the gas phase to the vapor-solid interface in form of volatile molecules. In MOVPE these species consist of metalorganic compounds, and the transport is made by a carrier gas like hydrogen at typically 100 mbar total pressure. The gaseous species dissociate thermally at the growing surface of the heated substrate, thereby releasing the elements for layer growth. The dissociation at the surface is generally assisted by chemical reactions.

The net reaction for the MOVPE of GaAs using the standard source compounds trimethylgallium and arsine reads

$$\mathrm{Ga(CH_3)_3} + \mathrm{AsH_3} \rightarrow \mathrm{GaAs} + 3\mathrm{CH_4}\uparrow. \tag{11.11}$$

The reaction is actually much more complicate and comprises many successive steps and species in the chemistry of deposition [10] like, e.g., some steps of precursor decomposition

$$\mathrm{Ga(CH_3)_3} \rightarrow \mathrm{Ga(CH_3)_2} + \mathrm{CH_3} \rightarrow \mathrm{GaCH_3} + 2\mathrm{CH_3} \rightarrow \mathrm{Ga} + 3\mathrm{CH_3}.$$

Major species occurring in the gas phase near the substrate surface are indicated in Fig. 11.14. The source compounds employed for MOVPE must meet some basic requirements. Their stability is low to allow for decomposition in the process, but still sufficient for long-term storage. Furthermore the volatility should be high, and a liquid state is favorable to provide a steady state source flow. Most source molecules have the form MR_n, where M denotes the element used for MOVPE, and R are alkyls like methyl CH_3. By choosing a suitable organic ligand, the bond strength to a given element M can be selected to comply with the requirements of MOVPE for the solid to be grown. The metal-carbon bond strength depends on the electronegativity of the metal M and the size and configuration of the ligand R [11]. As a thumb rule the bond strength decreases as the number of carbon atoms bonded to the central carbon in the alkyl is increased. This trend is also reflected in the dissociation energy of the first carbon-hydrogen bond given in Table 11.1 [12].

Organic radicals R most frequently used for MOVPE precursors are depicted in Fig. 11.9a, some metalorganic precursors are given in Fig. 11.9b.

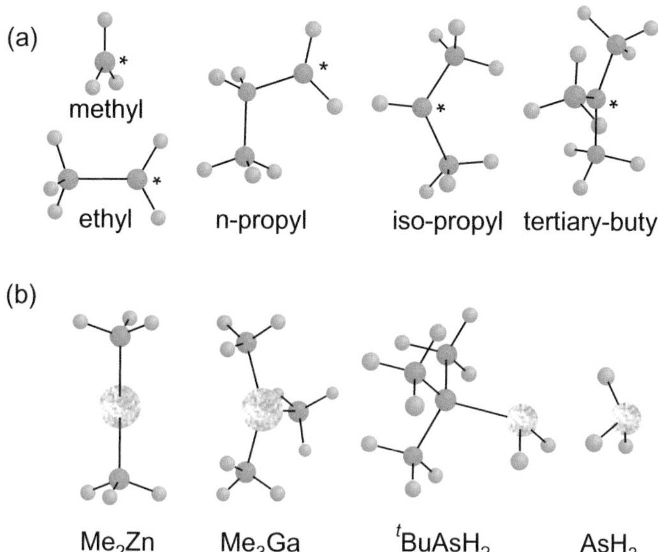

Fig. 11.9 **a** Alkyl radicals used as organic ligands in source molecules, and **b** some source molecules for metalorganic vapor phase epitaxy. *Red* and *blue* spheres represent carbon and hydrogen atoms, respectively, the location for a bond to an element *M* at the radicals is indicated by an *asterisk*

Besides metalorganic sources also hydrides like arsine are employed as precursors. Their use is interesting since they release hydrogen radicals under decomposition that can assist removal of carbon-containing radicals from the surface. A major obstacle is their high toxicity and their very high vapor pressure, requiring extensive safety precautions. To reduce the hazardous potential, hydrides are increasingly replaced by metalorganic alternatives, e.g., arsine by tertiarybutylarsine, where one of the three hydrogen radicals is replaced by a tertiarybutyl radical. Thereby the vapor pressure is strongly reduced, yielding usually liquids at ambient conditions. In addition the toxicity decreases significantly.

Partial pressures for some standard precursors used in the MOVPE of As-related III–V semiconductors are given in Table 11.2. The values are expressed in terms of the parameters a and b to account for the exponential temperature dependence of the vapor pressure according to

$$\log(P_{\text{eq MO}}) \cong a - b/T, \tag{11.12}$$

where the equilibrium pressure of the metalorganic source $P_{\text{eq MO}}$ and the temperature of the source T are given in units of mbar and K, respectively. Hydrides AsH_3 and PH_3 are stored at 20 °C as liquids under pressures of 15 bar and 40 bar, respectively, and introduced as gases to the MOVPE setup.

Precursor molecules may decompose by a number of pyrolytic mechanisms, the most simple being free radical homolysis, i.e., a simple bond cleavage. Since the

Table 11.2 Equilibrium vapor-pressure data of some metalorganic compounds used for III–V MOVPE (Vapor-pressure data taken from data sheets of several precursor suppliers)

Element	Precursor	Vapor pressure		
		a	b (K)	$P_{eq\,MO}$ (mbar) at 20 °C
Al	trimethylaluminum	8.349	2135	11.5
Ga	trimethylgallium	8.195	1703	241
	triethylgallium	8.208	2162	6.7
In	trimethylindium (solid)	10.645	3014	2.3
P	tertiarybutylphosphine	7.711	1539	187 (10 °C)
As	tertiarybutylarsine	7.368	1509	109 (10 °C)
Sb	trimethylantimony	7.833	1697	110
	triethylantimony	8.029	2183	3.8
N	dimethylhydrazine	8.771	1921	164

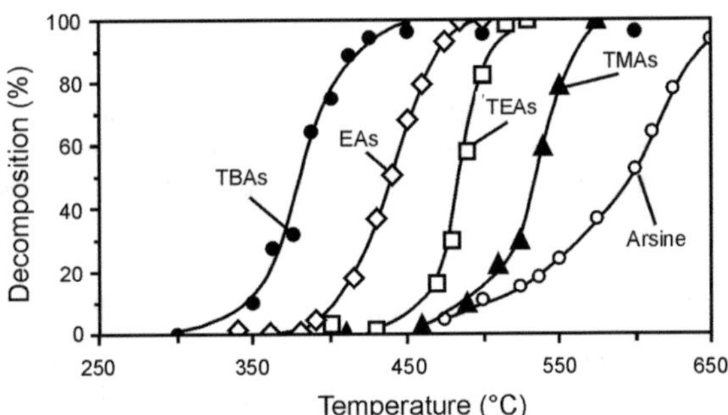

Fig. 11.10 Decomposition of As precursors, the labels *TBAs*, *EAs*, *TEAs* and *TMAs* denote tertiarybutyl-As, ethyl-As, triethyl-As, and trimethyl-As, respectively. Reproduced with permission from [1], © 1991 Elsevier

M-H bond is generally stronger than the M-C bond, metalorganic alternatives of the stable hydrides decompose at lower temperatures—a further incentive for their use. Results of pyrolysis studies for various As precursors, performed in an isothermal flow tube, are given in Fig. 11.10. The bond-strength thumb rule noted above is well reflected in these curves.

11.2.2 The Growth Process

Most metalorganic sources are liquids which are stored in bubblers. For transport to the reactor a carrier gas (usually hydrogen) with a flow Q_{MO} is introduced by a dip tube ending near the bottom, see Fig. 11.11. At a fixed temperature the metalorganic liquid forms an equilibrium vapor pressure $P_{eq\,MO}$ given by (11.12), and the bubbles of the carrier gas saturate with precursor molecules. At the outlet port of the bubbler a pressure controller is installed, which acts like a pressure-relief valve and allows to define a fixed pressure P_B ($>P_{eq\,MO}$) in the bubbler, thereby decoupling the bubbler pressure from the equilibrium vapor pressure of the MO source. Also the

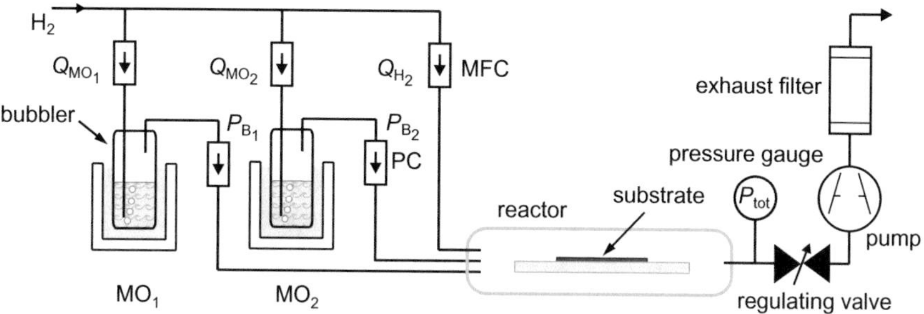

Fig. 11.11 Schematic diagram of a metalorganic vapor phase epitaxy apparatus. Hydrogen is used as carrier gas and introduced into the metalorganic sources MO_1 and MO_2. MFC and PC denote mass-flow and pressure controllers, respectively

total pressure P_{tot} in the reactor is controlled independently. The partial pressure of a metalorganic source in the reactor P_{MO} results from the mentioned parameters by

$$P_{\text{MO}} = \frac{Q_{\text{MO}}}{Q_{\text{tot}}} \times \frac{P_{\text{tot}}}{P_{\text{B}}} \times P_{\text{eq MO}}, \tag{11.13}$$

Q_{tot} denoting the total flow in the reactor. The two fractions in (11.13) are employed to control the partial pressure P_{MO} of the source in the reactor. For sources used as dopants or compounds with very high vapor pressures an additional dilution by mixing with a controlled flow of carrier gas is applied. The gaseous hydrides are directly controlled by their flow Q_{Hyd}, and (11.13) simplifies to

$$P_{\text{Hyd}} = \frac{Q_{\text{Hyd}}}{Q_{\text{tot}}} \times P_{\text{tot}}. \tag{11.14}$$

The total flow in the reactor Q_{tot} results from the sum of all component flows plus the flow of the carrier gas which is additionally introduced into the reactor by a separate mass-flow controller. This flow is generally much higher than that of all sources, and the sum of all source partial pressures P_{MO} and P_{Hyd} is consequently much smaller than the total pressure in the reactor P_{tot}. The reactor pressure P_{tot} is controlled as an independent parameter by a control valve attached to an exhaust pump behind the reactor (Fig. 11.11). The flow rate is usually specified in terms of a mass flow dm/dt (in units of g/min or mol/min) or volume flow dV/dt (standard cubic centimeters per minute, sccm), both defined at some standard conditions with respect to temperature and pressure.

The complete treatment of the MOVPE growth process involves numerous gas phase and surface reactions, in addition to hydrodynamic aspects. Such complex

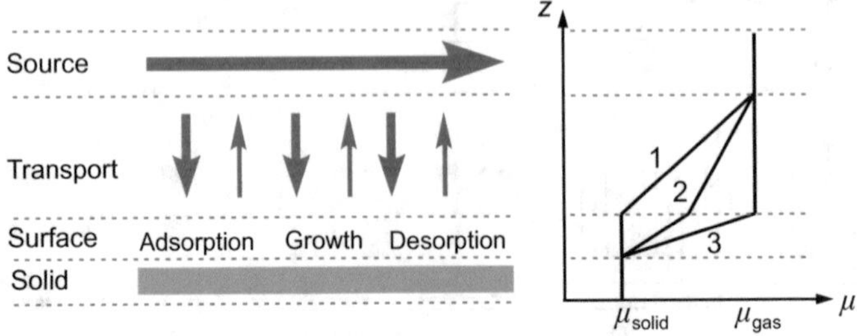

Fig. 11.12 Scheme of the chemical potential μ near the surface of the growing solid during MOVPE. Path *1* signifies growth controlled by mass transport, paths *3* and *2* denote growth being limited by interface reactions and the general case, respectively

studies require a numerical approach, and solutions were developed for specific processes like MOVPE of GaAs from trimethylgallium and arsine [10, 13, 14]. We will first draw a more general picture of the growth process.

Growth represents a nonequilibrium process. The driving force is given by a drop in the chemical potential μ from the input phase to the solid. For the discussion of the MOVPE process a description by consecutive steps as depicted in Fig. 11.12 is instructive. The reactants in the carrier gas represent the source. Near the solid surface a vertical diffusive transport component originates from reactions of source molecules and incorporation into the growing layer. All processes from adsorption at the surface to the incorporation are summarized to interface reactions. Finally excess reaction products desorb from the interface by diffusion.

The slowest process of the successive steps limits the growth rate. Without considering mechanisms of growth in detail, processes limited by either transport or kinetics can be well distinguished. Figure 11.13 shows on a logarithmic scale the dependence of the GaAs growth rate from the reciprocal substrate temperature. At low temperature experiment and simulation show an exponential relation, indicating that thermally activated processes limit the growth rate. Precursor decomposition and interface growth reactions lead to a pronounced temperature dependence; the slope $\propto -\Delta E/(k_B T)$ yields an activation energy ΔE near 19 kcal/mole for the given process. This regime is referred to as *kinetically limited growth*. The gas phase supplies precursors to the surface at a rate well exceeding the rate of growth reactions. As the temperature is increased, the growth rate becomes nearly independent on temperature. In this range precursor decomposition and surface reactions are much faster than mass transport from the source to the interface of the growing solid. Since diffusion in the gas phase depends only weakly on temperature, this process is called *transport-limited growth*. Mass transport in this regime depends on the geometry of the reactor, because flow field and temperature profile above the substrate affect cracking and arrival of precursors at the interface. This fact accounts for the difference in the maximum growth rates in Fig. 11.13. In the high-temperature range growth rates decrease due to enhanced desorption and parasitic deposition at the reactor walls, inducing a depletion of the gas phase.

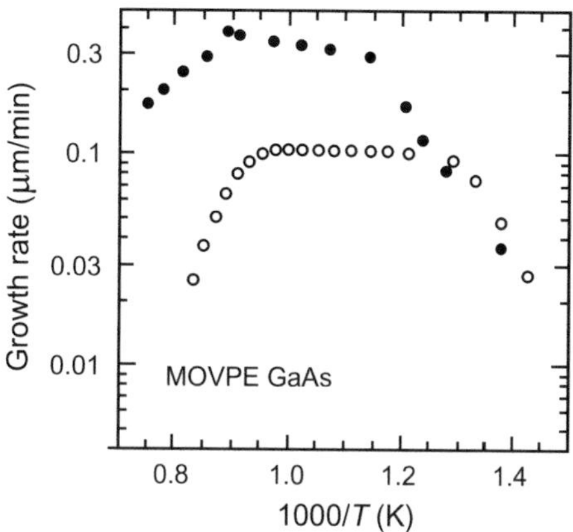

Fig. 11.13 Growth rate in the MOVPE of (001)-oriented homoepitaxial GaAs layers as a function of reciprocal temperature. Trimethylgallium and arsine are used as precursors. *Full* and *open circles* represent measured data from [15] and model predictions from [10], respectively

MOVPE is usually performed in the mid-temperature range of transport limited growth, where variations of the substrate temperature have only a minor effect on the growth rate, the composition of alloys, and on the doping level. For arsenide and phosphide semiconductors the range is typically between 500 and 800 °C, for nitrides above 1000 °C.

The decomposition of the precursors at the heated substrate and subsequent reactions leads to a large number of species in the gas phase and on the surface. We consider some results of a detailed study of the elementary processes for the homoepitaxial growth of GaAs on (111)A-oriented substrates [14]. The model for the MOVPE from trimethylgallium and arsine source compounds according to the net reaction (11.11) included 60 species and more than 200 reactions in the gas phase, and a total of 19 species and more than 100 processes at the surface. The model considered the flow, heat and mass transfer in a vertical reactor with forced convection like that shown in Fig. 11.15b. The gaseous reactants enter through a nozzle placed at right angles to the heated substrate. The gas flows toward the substrate and then flows radially outward. Gas inlet at 298 K is assumed, with inlet pressures of 3.4×10^{-4} bar for $Ga(CH_3)_3$ and 6.8×10^{-3} bar for AsH_3, a total flow rate of 1.5 standard liter per minute at 1 bar total reactor pressure, and a substrate temperature of 1000 K. Surface processes are considered for two different sites: the planar Ga-face of the (111)-oriented GaAs surface and ledge sites at monoatomic steps on the surface. Basic predictions of this model were verified experimentally.

The results demonstrate that the sites on the planar (111)A GaAs surface are basically occupied by AsH_n species with a decreasing fractional coverage for $n =$ 2, 3 at increased temperature, and a low, only slightly varying occupancy by AsH and As. $Ga(CH_3)_2$ and $GaCH_3$ species occupy ledge sites [14]. The rate-controlling steps for the modeled MOVPE of GaAs depend strongly on the substrate temperature. At low temperature the growth rate is controlled by the activation energy of surface kinetics involving $Ga(CH_3)_n$ species ($n = 0, 1$), and reactions in the gas phase are not important. The rate-limiting step at 773 K is the removal of CH_3 radicals from

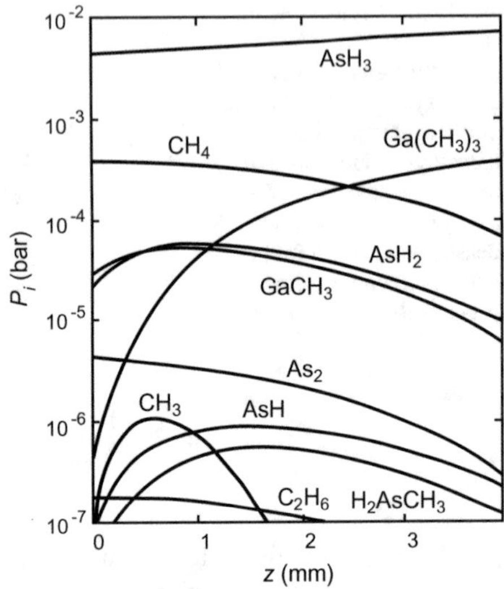

Fig. 11.14 Calculated partial pressures P_i of major species i in the gas phase for the metalorganic vapor-phase epitaxy of GaAs from $Ga(CH_3)_3$ and AsH_3 precursors on GaAs (111)A substrate. z designates the vertical distance from the growth surface, which is kept at 1000 K. Reproduced with permission from [14], © 1988 Elsevier

$Ga(CH_3)_2$ adsorbed at a ledge site. As the temperature is raised to intermediate values between 900 and 1000 K, surface reactions become fast and mass transport and gas-phase reactions become dominant. The reaction of AsH_3 with CH_3 radicals to AsH_2 and CH_4 in the gas phase enhances the decomposition of $Ga(CH_3)_3$ by removing the radicals. At high temperatures above 1000 K, deposition of arsenic in the form of As_2 and a reduced adsorption of $GaCH_3$ impose kinetic barriers.

The composition of the gas phase at a substrate temperature of 1000 K is given in Fig. 11.14. We note that the partial pressures of major species vary strongly in the vicinity of the substrate surface [14]. The amount of the readily decomposing Ga precursor $Ga(CH_3)_3$ drops to very small values, while the partial pressure of the stable arsine AsH_3 decreases only slightly. The concentration of CH_3 radicals remains low, mainly due to the consuming gas-phase reaction with AsH_3 noted above and the parallel reaction $CH_3 + H_2 \Leftrightarrow CH_4 + H$. GaAs growth at this high temperature occurs basically by adsorption of monomethylgallium $GaCH_3$ at ledge sites and subsequent surface reaction with AsH or with As.

A serious problem arises if the precursors of different components react already in the gas phase. Such parasitic gas-phase reactions may lead to the formation of stable particles, which disturb the epitaxial growth on the surface; in addition the gas phase depletes from source material, and the growth rate decreases and gets laterally inhomogeneous. Gas-phase reactions are known to affect the epitaxy of III-nitrides when the standard sources III-$(CH_3)_3$ and NH_3 are used. The issue is particularly critical in the epitaxy of AlN due to a strong Al-N coordination interaction favoring the formation of stable adducts and oligomers with simultaneous elimination of hydrocarbons [34, 35]. Special designs for the gas inlet into the reactor and operating conditions were developed to solve this issue [36]. Separate inlets for anions and

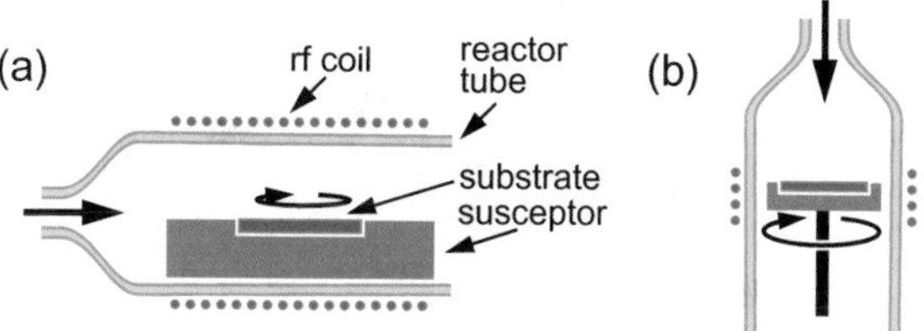

Fig. 11.15 Reactor types applied in the metalorganic vapor-phase epitaxy. **a** Horizontal reactor with inductive heating of the susceptor which carries the substrate. The *circular arrow* indicates the wafer rotation. **b** Schematic of a vertical reactor

cations are required, and a small spacing between inlet and substrate in vertical reactors is favorable [37].

11.2.3 Mass Transport

The access of supplied precursor molecules to the growth surface occurs by various transport processes. Diffusion and convection in the presence of large temperature and concentration gradients affect the growth rate and uniformity of layer composition. The transport processes depend strongly on the configuration of the reactor and the adjustment of operating parameters. The main geometries of MOVPE reactors are shown in Fig. 11.15. Horizontal and vertical reactors are used, and both designs are also upscaled for simultaneous multiwafer epitaxy. For simplicity we consider single-wafer setups. Reactor vessels are made of quartz glass or steel. The reactor walls are usually kept cold to minimize unwanted deposition at the walls. The wafer is placed on a (graphite) susceptor, which often is inductively heated by a radio-frequency (rf) generator via an rf coil for coupling. Usually wafer rotation (typ. 30 rpm) is applied during deposition to even out gas supply and heating nonuniformities for achieving a laterally uniform growth rate. In vertical reactors also rotation at high speed is used (>500 rpm) to emulate a rotating-disk flow.

Both reactor types depicted in Fig. 11.15 are designed for a laminar gas flow. Typical operating conditions with a total pressure P_{tot} of 10–100 mbar (low-pressure MOVPE) or 1000 mbar (atmospheric-pressure MOVPE) yield about 10 cm/s mean gas velocity (somewhat higher at low pressure, also depending on the total flow in the reactor Q_{tot}), leading to a flow well below the onset of turbolence.

The description of the mass transport in the gas phase consists of three-dimensional nonlinear, coupled partial differential equations. The conservation of momentum, energy, total mass and individual species is expressed by the Navier-Stokes, the energy transport, and mass continuity equations, respectively. In addition, the system is specified by the equation of state for the involved gases and suitable boundary conditions. The large temperature gradient created by the heated susceptor and the cold walls, and the gradient of species concentrations originating from decomposi-

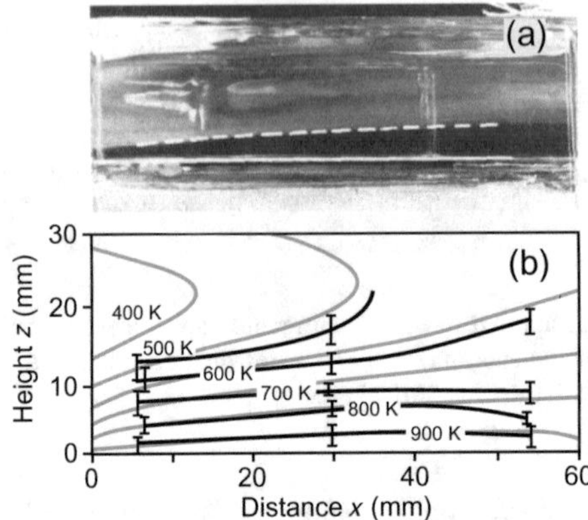

Fig. 11.16 Flow visualization and isotherms in MOVPE reactors. **a** Horizontal reactor with 8 l/min total flow of H_2 and TiO_2 particles above a susceptor kept at 1000 K. Reproduced with permission from [20], © 1986 Elsevier. **b** Measured (*black lines*) and calculated isotherms (*gray lines*) above the susceptor kept at 1000 K for 8 l/min inlet H_2 flow at 1000 mbar total pressure. Reproduced with permission from [16], © 1990 Elsevier

tion and chemical reactions lead to strong diffusion processes and a coupling of mass transport and thermal transport. A correct treatment of mass transport is very complex, because the coupled equations must be solved simultaneously. Furthermore, a complete description must also account for the chemical reactions illustrated in the previous section. An analytical solution of such a complex problem is not feasible. Experimental data have, however, been well described by extensive numerical modeling. Results of a comprehensive numerical treatment for some typical reactor geometries are given in [13, 38, 39]. Since general guidelines for the dependences of growth parameters are difficult to extract from numerical solutions, simplified analytical approaches are quite popular. Comparison with experimental and numerical data demonstrate that the results obtained from such models are often wrong and misleading. We focus on hydrodynamic aspects of the carrier gas and do not consider the complex chemistry. Some justification for this reasonable approach is given by the usually small partial pressures of the species.

The flow conditions in a horizontal reactor are visualized in Fig. 11.16a. The flow lines are traced by micron-sized TiO_2 particles, which are produced by leading H_2 carrier gas with $TiCl_4$ into a bubbler filled with H_2O. The image shows Mie scattering of light illuminating the glass reactor. The white line at the bottom indicates the top of the graphite susceptor, which is heated to 1000 K. The gas is introduced from the left at room temperature with a flow of 8 l/min at 1000 mbar total pressure. We note the dark region limited by the dashed line, indicating that the flow does not enter the hottest region above the substrate. This region is often termed boundary layer, because an apparently similar phenomenon of this name is observed in the isothermal parallel flow over a flat plate as illustrated in Fig. 11.17.

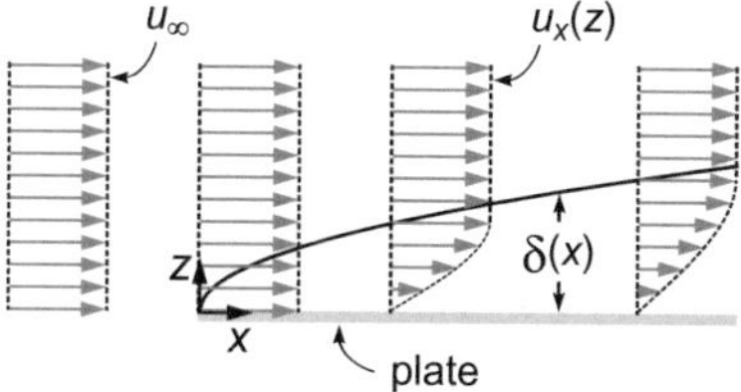

Fig. 11.17 Development of a boundary layer of thickness $\delta(x)$ above the surface of a plate of negligible thickness, placed in an isothermal laminar flow of constant velocity u_∞. The flow field below the plate is mirror inverted with respect to that above the plate and not drawn. The *arrows* give the distribution of the velocity u_x over the distance from the plate z at three different locations x

The temperature distribution in the horizontal MOVPE reactor was measured using Raman scattering from rotational transitions of the H_2 carrier gas [16]. We note in Fig. 11.16b the strong temperature gradient at the bottom above the susceptor. At the selected high flow rate the hight-T isotherms are fairly parallel to the susceptor, while a larger wedge-shaped dark region in flow visualization and accordingly more inclined isotherms are found at lower flow rate. The experimental data agree well with calculated results obtained using the finite elements method. Heat conduction in the reactor wall, heat transfer to the surroundings, and radiative heat transfer were included in the calculations to obtain such a close agreement.

We now take a closer look to the boundary layer which develops at a plate in an isothermal laminar flow. Figure 11.17 depicts the velocity field near the surface of the flat plate, which is placed in the homogeneous gas flow moving with a constant free-stream velocity u_∞. The boundary conditions for the solution of the (two-dimensional) hydrodynamic steady-state equation of continuity and the equation of motion are given by the general assumption, that the fluid velocity at solid-fluid interfaces equals the velocity with which the solid surface moves itself: the fluid clings to the solid surface. In our case the plate is fixed, hence the x component of the velocity at the surface (at $z = 0$) is $u_x = 0$. Moreover, far away from the plate (for all x at large z) and at the very leading edge of the plate (at $x = 0$) the velocity is unaffected, $u_x = u_\infty$.

In the hydrodynamic problem illustrated in Fig. 11.17 a boundary layer of thickness $\delta(x)$ develops between the surface of the plate and the constant flow field far away from the plate. $\delta(x)$ is conventionally defined by the distance from the surface at which the velocity component u_x becomes 99% of the free-stream velocity u_∞. Below $\delta(x)$, i.e., in the region of the boundary layer, the velocity u_x gradually decreases to 0 as postulated from the boundary condition for the solid-fluid interface. The velocity distribution $u_x(x, z)$ is approximately given by [17]

$$u_x \cong \left(\frac{3}{2}\left(\frac{z}{\delta(x)}\right) - \frac{1}{2}\left(\frac{z}{\delta(x)}\right)^3\right)u_\infty; \quad 0 < z < \delta(x) \qquad (11.15)$$

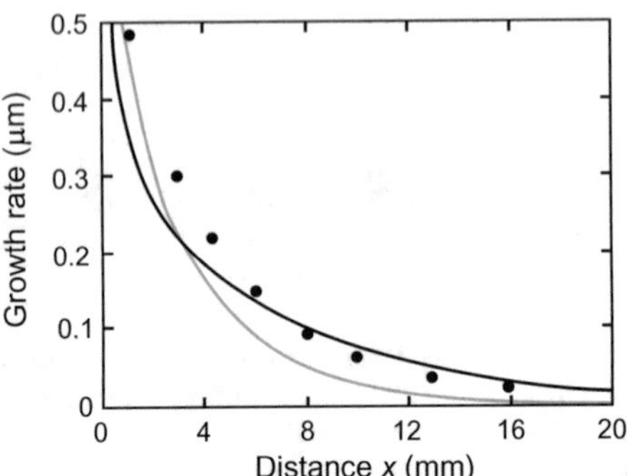

Fig. 11.18 Effect of thermal diffusion on the growth rate of GaAs in a horizontal MOVPE reactor. *Symbols* give experimental data, showing a decreasing growth rate due to a depletion of Ga species in the gas phase. The *black* and *gray lines* are numerically predicted growth rates with and without thermal diffusion, respectively. Reproduced with permission from [18], © 1988 Elsevier

where the thickness of the boundary layer is

$$\delta(x) \cong 5\sqrt{\frac{\eta}{\rho u_\infty}x}. \tag{11.16}$$

Here η and ρ are the dynamic viscosity and the mass density of the gas, respectively. Function (11.15) is drawn as dashed line in Fig. 11.17. The thickness of the boundary layer is proportional to the square root of the distance x down the plate along the direction of the flow, where x is measured from the leading edge of the plate. We note the similarity between the dark region in the flow above the hot susceptor marked in Fig. 11.16a and the boundary layer of the plate in the isothermal flow shown in Fig. 11.17. Calling the dark region in Fig. 11.16a boundary layer might meanwhile be misleading. Numerical simulations clearly demonstrate that the origin of the vertical flow component away from the hot susceptor is actually thermodiffusion (thermophoresis). The suddenly heated gas near the susceptor expands and becomes less dense. In the presence of gravity the gas rises due to buoyancy.

The effect of buoyancy-driven convection on the distribution of the GaAs growth rate along the flow direction in a horizontal reactor was evaluated by numerical modeling with and without the thermal-diffusion term included [18]. The growth rate under the selected experimental conditions was controlled by gas phase transport of the Ga component via diffusion and flow, and a significant gas-phase depletion was found [19]. The numerical results given in Fig. 11.18 show that thermal diffusion decreases the growth rate at the front edge of the susceptor, because the heavy reactant molecules are driven away due to buoyancy in the strong temperature gradient. The depletion of the gas phase from reactants is thereby diminished, leading to an increase of the growth rate in the downstream range. The influence of natural convection is reduced at lower total pressure.

Model of a Stagnant Boundary Layer

Extensive numerical treatment of the hydrodynamic conditions in the reactor yields a good description of experimental data, but it delivers only little insight into dependences of growth parameters. We consider the popular (stagnant) boundary-layer model to indicate qualitatively some tendencies. Since the assumptions of the simple model are too restrictive to yield reliable relations, we should consider the results merely as a rough guideline to understand some trends.

The model of a stagnant boundary layer interprets the region above the hot susceptor visualized in Fig. 11.16a as a stagnant layer, where mass transport occurs solely by diffusion. This means that the horizontal velocity component of the flow u_x is assumed to drop to zero below the upper bound of this layer as illustrated in Fig. 11.19.

The growth rate of the epilayer depends on the amount of source material supplied to the surface. The diffusive flux $\mathbf{j}_i$ of material component i is given by the diffusion along the partial pressure gradient $\partial P_i/\partial \mathbf{r}$ and the termodiffusion along the temperature gradient $\partial T/\partial \mathbf{r}$. The comparably small effect of thermodiffusion is neglected, and the partial pressures are assumed to drop over the stagnant layer of thickness d from their values P_i in the source to values $P_i^{\text{interface}}$ at the interface to the solid. A linear decrease of the partial pressure of component i is assumed, yielding a pressure gradient expressed by $(P_i - P_i^{\text{interface}})/d$. Only diffusion normal to the surface is considered. Ficks's first law then becomes a simple expression for the flux from the source above the stagnant layer to the surface [21],

$$j_i = \frac{D_i}{k_B T d}\left(P_i - P_i^{\text{interface}}\right). \tag{11.17}$$

D_i is the diffusion constant of species i in the carrier gas. The factor $D_i/(k_B T d)$ may be considered as an effective coefficient of mass transport for component i.

Due to the supersaturation set to induce growth, the partial pressures of the components at the inlet of the reactor P_i are much higher than the equilibrium-near

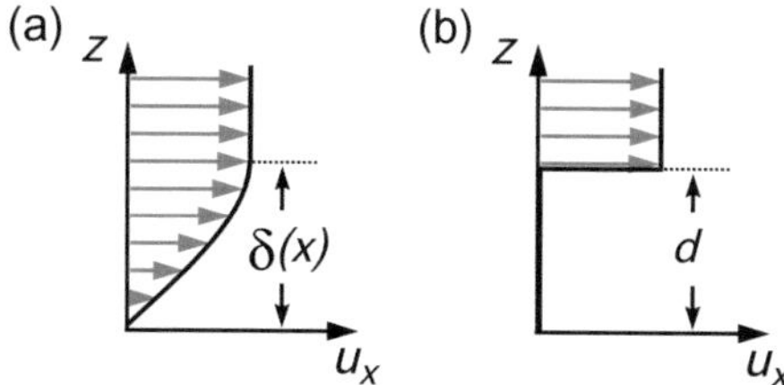

Fig. 11.19 Comparison of **a** the boundary layer of thickness δ of a flow above the surface of a fixed surface, and **b** the condition assumed in the boundary-layer model applied to describe the diffusion of source material from the flow over a stagnant layer of thickness d to the growing surface in the MOVPE. The *arrows* signify the flow velocity u_x

values at the interface to the solid $P_i^{\text{interface}}$. For III–V compounds like GaAs this means $P_{\text{III}} P_{\text{V}} \gg P_{\text{III}}^{\text{interface}} P_{\text{V}}^{\text{interface}}$. Furthermore, the Column-V precursors are far more volatile than the Column-III species (except for Sb-sources). III–V semiconductors are hence usually grown with a large excess of Column-V species, i.e., $P_{\text{V}}/P_{\text{III}} \gg 1$. At the interface to the solid the same number of Ga and As atoms is permanently removed from the gas phase due to the requirement of stoichiometric growth. Ga is therefore nearly depleted at the interface, while the arsenic partial pressure is only slightly reduced. These conditions lead to the relations of the partial pressures at the interface and the reactor inlet $P_{\text{III}}^{\text{interface}} \ll P_{\text{III}}$, and $P_{\text{V}}^{\text{interface}} \approx P_{\text{V}}$. The flux of Column-III species j_{III} arriving at the surface then reads

$$j_{\text{III}} = \frac{D_{\text{III}} P_{\text{III}}}{k_{\text{B}} T d}. \tag{11.18}$$

Since all Column-III species are incorporated into the solid, the growth rate r is controlled by the flux of Column-III species, $r \propto j_{\text{III}}$. Experimental evidence for the linear dependence is shown in Fig. 11.20.

The dependences of the growth rate on the total reactor pressure P_{tot} and on the total flow in the reactor Q_{tot} are estimated by substituting the thickness of the boundary layer δ in (11.16) for the stagnant layer thickness d in (11.18). The diffusion constant D in the gas phase is inverse to P_{tot}, and the density ρ is proportional to P_{tot}, yielding

$$r \propto P_{\text{III}} \sqrt{\frac{u_\infty}{P_{\text{tot}}}}. \tag{11.19}$$

Taking (11.13) into account we note that r is proportional to the flow of carrier gas through the bubbler Q_{MO} of the Column-III source. Equation (11.19) predicts a growth rate being *independent* on the total reactor pressure if all other parameters are kept constant, because both P_{III} ($\propto P_{\text{tot}}$, cf. (11.13)) and u_∞ (inverse to P_{tot}) are implicit functions of the reactor pressure. The dependence of r on the flow Q_{tot} is

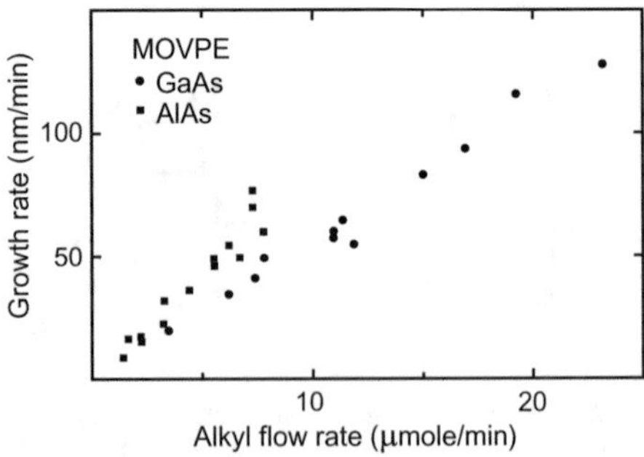

Fig. 11.20 Growth rate of GaAs (*circles*) and AlAs (*squares*) depending on the Column-III alkyl supply trimethylgallium or trimethylaluminum, respectively, at constant arsine supply. Reproduced with permission from [28], © 1984 Elsevier

also determined by the implicit dependences of P_{III} (inverse to Q_{tot}) and u_∞ ($\propto Q_{tot}$), yielding a decrease of the growth rate $\propto Q_{tot}^{-1/2}$ as the total flow in the reactor Q_{tot} is increased. We should not take these predictions literally and just regard them as some trends. Deviations are particularly expected at low pressures and low flow velocities, where the boundary-layer thickness is in the range of the reactor height.

11.3 Molecular Beam Epitaxy

Molecular beam epitaxy (MBE) is a physical-vapor deposition technique, which is widely applied in research labs and industrial production. The constituent elements of the crystalline solid are transported from the source(s) to the substrate using molecular beams. A molecular beam is a directed ray of neutral atoms or molecules in a vacuum chamber. In MBE the beams are usually thermally evaporated from solid or liquid elemental sources. Various names are used particularly for the epitaxy of compound semiconductors, if gas sources are employed as source materials: metalorganic MBE (MOMBE), if metalorganic compounds like those applied in MOVPE (Sect. 11.2.1) are used for metals and conventional sources for anions, gas-source MBE (GSMBE), if hydrides for anions and metals are employed, and chemical beam epitaxy (CBE) for the supply with all gas sources.

In early experiments during the 1950s and 60s various beam techniques were used for the crystalline and epitaxial deposition of II–VI [22], IV–VI [23], and III–V [24, 25] semiconductors. In the late 60s a study of the surface kinetics of Ga and As species in the epitaxy of GaAs provided a first insight into the growth mechanisms [26], and soon later epitaxial growth of GaAs layers with high quality was achieved [27].

11.3.1 MBE System and Vacuum Requirements

The characteristic feature of MBE is the mass transport in molecular or atomic beams. A vacuum environment is required to ensure that no significant collisions occur among the beam particles and between beam and background vapor. A schematic diagram of an MBE system is given in Fig. 11.21.

The vacuum is generated in a chamber by pumps and cryoshrouds. Usually effusion cells mounted opposite to the substrate produce beams of different species by evaporation. The duration of the exposure on the substrate is individually controlled by shutters for a rapid change of material composition or doping. The substrate is mounted on a heated holder and can be loaded and unloaded under vacuum conditions by a manipulating mechanism. A gauge can be placed at the position of the substrate to measure and calibrate the beam-equivalent pressure (BEP) produced by the individual sources. The vacuum environment maintained during epitaxy provides an excellent opportunity for in situ monitoring of the growth process. Virtually any MBE system is equipped with an electron-diffraction setup. Usually reflection high-energy

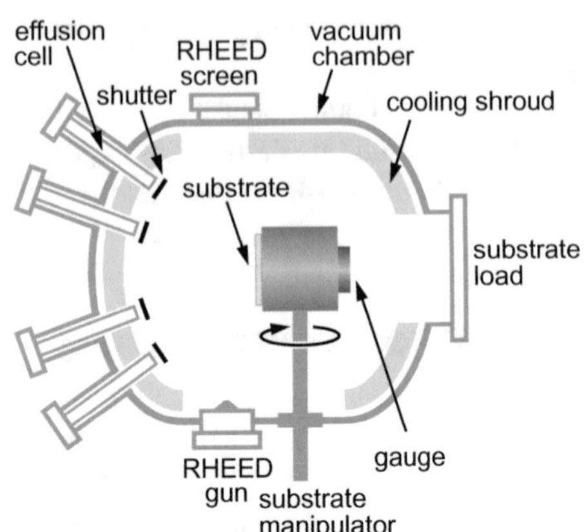

Fig. 11.21 Schematic representation of a molecular-beam epitaxy system. The *circular arrow* indicates the positioning of the gauge at the location of the substrate to calibrate the beam-equivalent pressure of the effusion cells which contain different source materials

electron diffraction (RHEED) with an electron beam nearly parallel to the growth surface is applied, yielding structural information on the surface crystallography during surface preparation and during epitaxy, see Sect. 8.3.1. The location of the electron gun and the monitoring screen is indicated in Fig. 11.21.

Molecular beam epitaxy is performed in ultra high vacuum (UHV), i.e., at a residual-gas pressure below 10^{-7} Pa (10^{-9} mbar). The need for such a low pressure originates from the required purity of epitaxial semiconductors. To obtain a relation for the maximum admissible pressure in the MBE chamber, we first consider the number of particles from the residual gas impinging on the substrate surface, and then relate this quantity to the particles of the molecular beams used to grow the epitaxial layer.

A number of N particles (molecules or atoms) that impinge on a surface with area A per time Δt produces a flux F given by

$$F = \frac{N}{A \Delta t}. \tag{11.20}$$

During the time interval Δt only particles with the velocity v_x and a maximum distance $\Delta x = v_x \Delta t$ can reach the surface, yielding the flux

$$F = \frac{N v_x}{A \Delta x} = \frac{N}{V} v_x. \tag{11.21}$$

According to (11.21) only particles contained in the volume V arrive on the surface during Δt. The velocity v_x of the ensemble of particles depends on the temperature T and is given by the Maxwell-Boltzmann-distribution

$$f(v_x)\, dv_x = \sqrt{\frac{2}{\pi}} \left(\frac{m}{k_B T} \right)^{3/2} v_x^2 e^{-m v_x^2/(2 k_B T)}\, dv_x, \tag{11.22}$$

where m is the mass of the particles and k_B is Boltzmann's constant. The area below the distribution function $f(v_x)$ over an interval dv_x denotes the fraction of particles with a velocity between v_x and $v_x + dv_x$, and $f(v_x)$ is normalized to yield all particles for an integration over all velocities, $\int_0^\infty f(v_x)\,dv_x = 1$. Using (11.22) we obtain for the flux

$$F = \int_0^\infty \left(\frac{N}{V}\right) v_x f(v_x)\,dv_x = \frac{N}{V}\left(\frac{k_B T}{2\pi m}\right)^{1/2}. \tag{11.23}$$

The flux F is proportional to the particle density N/V and the square root of T/m. The particle density may approximately be expressed in terms of the pressure P by using the state equation of the ideal gas $PV = Nk_BT$, eventually yielding

$$F = \frac{P}{\sqrt{2\pi k_B m T}} = 8.332 \times 10^{22} \times \frac{P}{\sqrt{MT}}\left(\frac{\text{particles}}{\text{m}^2\,\text{s}}\right). \tag{11.24}$$

M is the mole mass given by Avogadro's number N_A, $M = mN_A$. The second equation of (11.24) is given in SI units for all quantities.

For simplicity we assume only a single species producing the residual gas pressure, and we take oxygen O_2 with a mole mass $M = 32.0$ g/mol. At room temperature (300 K) we then obtain from (11.24) a pressure-dependent residual gas flux $F_{O_2} = 2.69 \times 10^{22} \times P_{O_2}$ (O_2 molecules m^{-2} s^{-1}), or twice this number for individual O species corresponding to a flux F_O. We now relate the flux F_O produced by the residual gas (assumed here to be given solely by O_2) to the flux F_{MBE} of the beam(s) in typical MBE conditions and the requirement for purity in the layer. A semiconductor has about mid 10^{22} atoms/cm^3. Let us assume that each residual gas atom arriving at the growing surface is incorporated into the epitaxial layer, and a maximum of mid 10^{17} impurities/cm^3 must not be exceeded. This leads to the requirement

$$\frac{F_O}{F_{MBE}} = 10^{-5} = \frac{5.38 \times 10^{22} P_{O_2}}{10^{19}} = 5.38 \times 10^3 P_{O_2}.$$

Resolving for P_{O_2}, the relation yields a maximum pressure 1.86×10^{-8} Pa. This is a pressure in the UHV regime. A range 10^{-8}–10^{-9} Pa corresponds to the typical residual gas pressure in an MBE chamber. The impurity level found in epitaxial layers grown under such conditions is actually significantly lower than the postulated mid 10^{17} cm^{-3}. The sticking coefficients of typical residual gas species are usually much less than the assumed unity.

The requirement of UHV conditions arising from the purity requirement also ensures the beam nature of the molecular sources. The condition of a sufficiently large mean free path of effused particles in the range of typical dimensions of the MBE setup (order of 10^{-1} m) is already fulfilled in a pressure range below 10^{-1} Pa.

To reach ultrahigh vacuum, all materials used in the vacuum chamber must have very low gas evolution and a high chemical stability. Tantalum and molybdenum are widely used for shutters, heaters, and other components. The entire MBE chamber is baked out typically at 200 °C for 24 h any time after having vented the system. Spurious fluxes of atoms and molecules from the walls of the chamber are minimized by a cryogenic cooling shroud chilled using liquid nitrogen as indicated in Fig. 11.21.

11.3.2 Beam Sources

The variety of source materials needed for MBE led to the development of different kind of sources with operation principles depending on the nature of the material. For the production of beams from solid or liquid materials usually Knudsen cells (K-cells) are employed. They are based on radiative heating and are limited to a maximum temperature of ~1300 °C for thermal evaporation. Sources for higher temperatures mostly use electron-beam evaporation, albeit also laser-induced evaporation and plasma ion sources were employed. Gaseous species are directly introduced or decomposed in gas sources. Sources for condensed and gaseous particles are considered in the following.

Sources for Condensed Materials

MBE sources for condensed (solid or liquid) source materials are usually based on thermal evaporation described by a—sometimes modified—Knudsen equation (11.27). The *ideal Knudsen cell* is an isothermal enclosure which contains the solid or liquid source material in thermodynamic equilibrium with its vapor at the pressure P_{eq}. Effusion occurs through a *small* orifice with an area much smaller than that of the evaporation surface of the source material, and the flux passing this aperture equals the flux of material which leaves the condensed phase to maintain the equilibrium pressure. The diameter of the aperture is also small compared to the mean free path of the particles in the gas phase at P_{eq}, and its wall thickness should be infinitely thin. All gaseous particles reaching the aperture from the inner side then escape to the vacuum chamber. The equilibrium pressure P_{eq} in the Knudsen cell depends on the temperature. For a pure source material in a closed enclosure with a temperature-independent evaporation enthalpy ΔH the dependence can be expressed by the Clausius–Clapeyron relation

$$P_{eq}(T) = P_0 \exp\left(-\frac{\Delta H}{k_B}\left(\frac{1}{T} - \frac{1}{T_0}\right)\right), \qquad (11.25)$$

where P_0 is the equilibrium pressure at some temperature T_0. The effusion rate Γ_{max} of the aperture is given by the product of the aperture area A and the flux F which passes the aperture. Using (11.24) we obtain

$$\Gamma_{max} = \frac{P_{eq} A}{\sqrt{2\pi k_B m T}} \left(\frac{\text{particles}}{\text{s}} \right). \tag{11.26}$$

Equation (11.26) represents the maximum effusion rate which can be achieved at a given cell temperature T. If the residual gas pressure P_{res} in the MBE chamber cannot be neglected, the equilibrium pressure in (11.26) is replaced by the difference pressure $(P_{eq} - P_{res})$ to express the effective effusion from the orifice area, yielding the *Knudsen equation*

$$\Gamma = \frac{(P_{eq} - P_{res}) A}{\sqrt{2\pi k_B m T}} \left(\frac{\text{particles}}{\text{s}} \right), \tag{11.27}$$

A being the aperture area. A real Knudsen cell may have a smaller effusion rate, and correction factors are introduced to account for the non-ideal behavior. A dimensionless evaporation coefficient a accounts for the microscopic condition of the evaporation surface and is taken as an additional experimentally determined correction factor to the effusion obtained from the Knudsen equation (11.27). Another limitation of a real cell is given by the infinitesimal thin orifice assumed for the ideal Knudsen cell. The orifice of a real Knudsen cell has a finite thickness, leading to diffuse scattering of particles at its side walls. This affects the angular distribution of the flux as illustrated in Fig. 11.22.

The ideal Knudsen cell with infinitely thin orifice wall ($L = 0$) effuses particles with a cosine angular dependence expressed by

$$\Gamma(\vartheta)/\omega = \Gamma(0) \cos \vartheta, \tag{11.28}$$

the angle ϑ referring to the direction normal to the orifice as shown in Fig. 11.22; ω is the unit solid angle comprising the considered flux. For a flux leaving the orifice of the cell at an angle ϑ the orifice area appears smaller by a factor $\cos \vartheta$, leading to this dependence. Equation (11.28) is referred to as Knudsen's cosine law of effusion. The constant $\Gamma(\vartheta = 0)$ is related to the total effusion rate Γ, yielding $\Gamma(0) = \Gamma/\pi$. An orifice with a *finite* thickness L and diameter d leads to some collimation of the particle beam. Consequently the angular distribution gradually narrows as the ratio L/d increases. The effect is shown in Fig. 11.22 for a ratio of unity.

For a given assembly of a substrate in front of an effusion cell the particle flux G per unit area of the substrate can be derived from (11.27), (11.28). If the substrate is placed at a distance l from the aperture directly in line (i.e., at $\vartheta = 0$), G is given by $G = \Gamma/(\pi l^2)$.

The beam flux of a Knudsen cell can be calculated from the Knudsen equation (11.27) and does not depend on the quantity of source material in the cell. Knudsen cells are hence also used for flux calibration. The schematic of a Knudsen cell is depicted in Fig. 11.23a. A Knudsen cell contains a crucible made of pyrolytic boron nitride pBN, graphite, quartz, or tungsten, equipped with a cap which can be removed

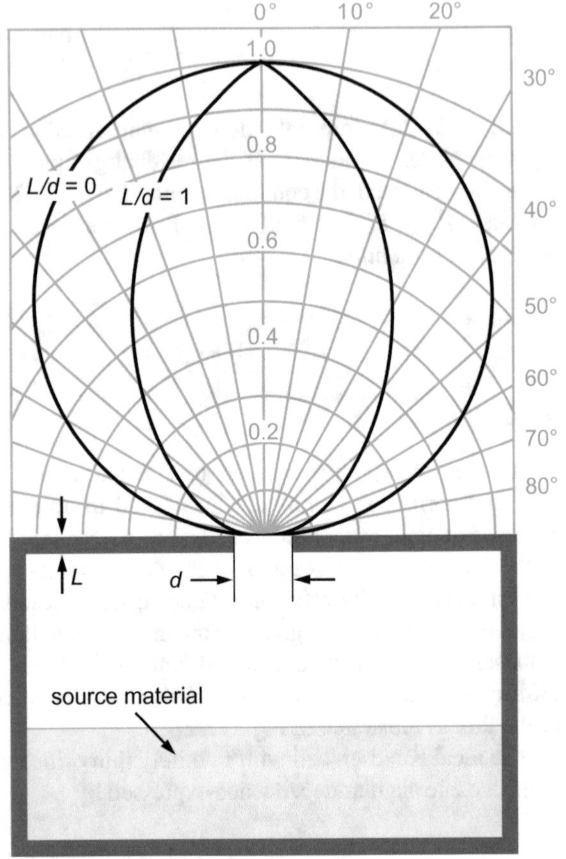

Fig. 11.22 Angular distribution of particles effused from a Knudsen cell with ideally thin orifice (wall thickness to diameter ratio $L/d = 0$) and a thick orifice wall with a ratio $L/d = 1$. Data from [29]

for filling the source material. Heating is provided by a filament coil or a heater foil (often made of metal tantalum, Ta). A radiation shield fabricated from multiple refractory metal foil (Ta or Mo) reduces heat losses. A water-cooling shield may be added to prevent heating of the cell environment. The temperature of the source material is controlled by a thermocouple, which is in intimate contact with the crucible of the cell and adjusts the heater power via a feed-back loop. Typically thermocouples of the thermally stable alloys W-Re (5 and 26%) are used to measure the temperature.

Knudsen cells are widely used in MBE as evaporators for elementary sources with relatively low partial pressure (e.g. Ga, Al, As, Hg), but they suffer from a number of limitations. We know from (11.27) that the effusion flux is proportional to the aperture area. Since the aperture of a Knudsen cell must be small, the flux intensity is quite restricted for most materials at moderate cell temperatures. At high cell temperature, however, excessive outgassing of cell materials degrades the purity of the source flux. Another problem is the loss of heat at the cap, leading to a temperature decrease at the orifice. Consequently evaporants tend to condensate at the orifice and change effusion conditions.

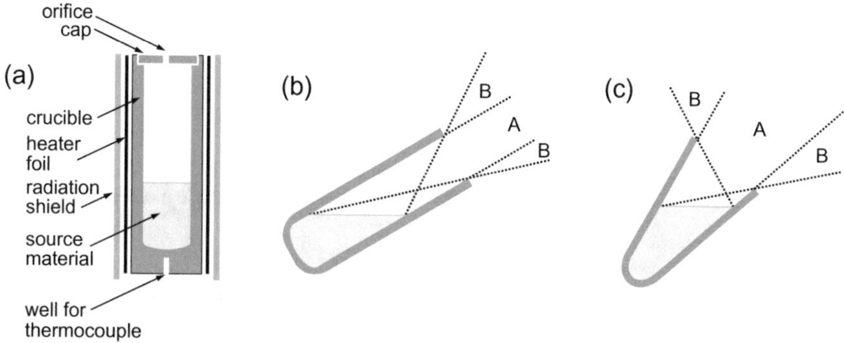

Fig. 11.23 **a** Schematic of a Knudsen cell used as MBE source for condensed materials. **b** Cylindrical and **c** conical crucibles with liquid charges and limits of flux areas indicated by *dotted lines*. *A* and *B* denote areas of full flux and partially shadowed flux, respectively

Improved cell designs do not use a cap, providing a large aperture for effusion. This allows for operation with a large flux at moderate cell temperature, leading to low flux contamination. The angular flux distribution of such a cell differs from that of a Knudsen cell. In particular, the flux distribution depends on the charging level of the cell. The beam is gradually collimated by the side walls of the crucible as the source material depletes. This effect is similar to that produced by a thick orifice wall shown in Fig. 11.22. Furthermore, source cells are usually mounted slantingly to allow for multiple-source arrangements. The surface normal of the source material in a cell is then inclined with respect to the crucible axis. Consequently the shadowing effects of the crucible walls are asymmetric and so is the angular flux distribution. The schematics of source cells with cylindrical or conical crucibles are depicted in Fig. 11.23b, c. Cylindrical crucibles allow for a larger charge of source material, and a better uniformity was reported for liquid charges such as Ga and Al [30]. The effect of a collimating shadowing is indicated by dotted lines in the figure. Area *B* denotes the penumbra, where only a part of the evaporant surface contributes to the flux. This area is particularly sensitive to the charge level. Conical crucibles have a large area *A* where the whole evaporant surface contributes to the flux.

Gaseous Sources

The use of gaseous source materials in MBE offers a couple of advantages. The lifetime of the source is not limited like that of condensed materials installed in the MBE system. Source gases are externally stored in cylinders and can readily be exchanged. Furthermore, the flux can be precisely controlled by pressure or mass-flow controllers which are also used in MOVPE setups. This allows also for a simple control of flux changes for, e.g., controlling alloy composition.

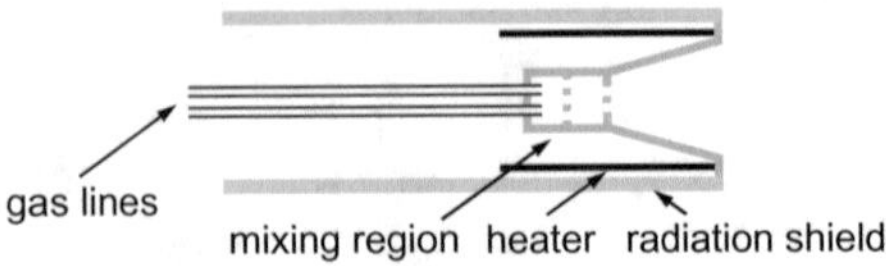

Fig. 11.24 Schematic of an inlet cell used for gaseous source materials in MBE. A heated region at gas mixing is used to prevent condensation or to decompose stable hydride molecules

A gas source consists of the gas-control system and a cell for the gas inlet into the MBE chamber. The gas-control system resembles that used in MOVPE or VPE systems. In the system either the mass-flow through the inlet tube or the pressure in the tube (which then has a fine nozzle) is controlled using mass-flow or pressure controllers, respectively. One inlet cell may comprise multiple gas lines of either hydrides (like AsH_3) or metalorganic compounds (like $Ga(CH_3)_3$). In a multiple gas cell gases are mixed and then introduced into the vacuum chamber. Metalorganic compounds are generally thermally less stable than hydrides and decompose at the heated substrate surface. The vacuum inlet cell for metalorganics is therefore a simple gas-feed nozzle. A heater in front of the nozzle provides moderate heating up to about 100 °C to prevent condensation of the gas in the cell. Hydrides usually require decomposition temperatures exceeding the substrate temperature at typical MBE conditions. They must therefore be decomposed in the vacuum inlet cell. Thermal dissociation is accomplished in a ceramic cracking stage of the cell by heating to high temperature (up to 1000 °C) or by catalytic decomposition on a metal surface at somewhat lower temperature. The assembly of such a stage is illustrated in Fig. 11.26. The gas outlet may be formed by a conical crucible defining the angle of the beam aperture. The schematic of an inlet cell for gaseous source materials is given in Fig. 11.24.

Dissociation Stage

Thermal dissociation as used in hydride cells may also be applied for decomposition of the evaporant molecules of condensed sources. In this case the source is equipped with an additional heated stage. Elemental sources for arsenic and phosphorous, e.g., produce a temperature-dependent mixture of dimers and tetramers. As_2 and P_2 dimer molecules are considered beneficial for the MBE of arsenides and phosphides. A cracker stage therefore dissociates As_4 and P_4 molecules and provides a pure beam of dimers. Typical temperatures for efficient cracking of both As_4 and P_4 are within the range 800–1000 °C. The thermal dissociation of As_4 tetramers to As_2 dimers in the cracker zone of an effusion cell kept at 327 °C is shown in Fig. 11.25.

The cracker region consists of a baffle assembly made of refractory metal (Ta), ceramic (pBN) or graphite, heated to a temperature much higher than that of the cell which evaporates the source material. The assembly provides multiple collision paths for the molecules. The schematic of the cracking region is depicted in Fig. 11.26.

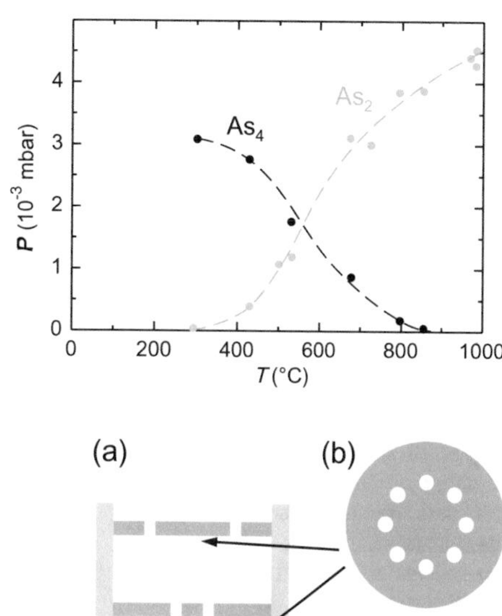

Fig. 11.25 Thermal dissociation of arsenic tetramers to dimers in a cracker stage. The *dashed curves* are guides to the eye. Data are from [31]

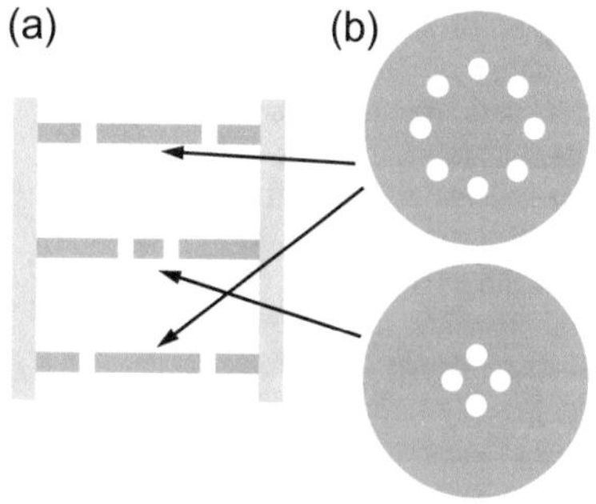

Fig. 11.26 Schematic of a thermal dissociation stage employed to decompose hydrides of a gas inlet cell or molecules effused from an attached evaporation source. **a** Side view, **b** plan view of baffles

11.3.3 Uniformity of Deposition

The uniformity of beam fluxes at the plane of the substrate surface is important to obtain a well-defined thickness, composition and doping of layers over the entire wafer of an epitaxial structure. Besides the angular flux distribution of a source treated above the uniformity of the deposition on the substrate depends on the geometry of the source-substrate assembly.

We illustrate the dependence for a point source with a cosine flux distribution as shown in Fig. 11.22. The flux distribution of such a source arriving at the plane of the substrate surface is depicted in Fig. 11.27. The off-center distance x indicated in the figure is a consequence of the usual multiple source arrangement of an MBE system outlined in Fig. 11.21. We note a strong variation of the flux intensity across the substrate area, depending also on the vertical distance z between substrate surface and orifice outlet of the source cell.

To obtain a largely homogeneous deposition on the substrate a rotation is applied as indicated by the circular arrow in the inset of Fig. 11.28. The flux is distributed rotationally symmetric on the substrate by averaging over the revolution around the axis. In principle a deposition uniformity better than 1% is then possible. The effect depends on the ratio of the source distance z versus the source displacement x as

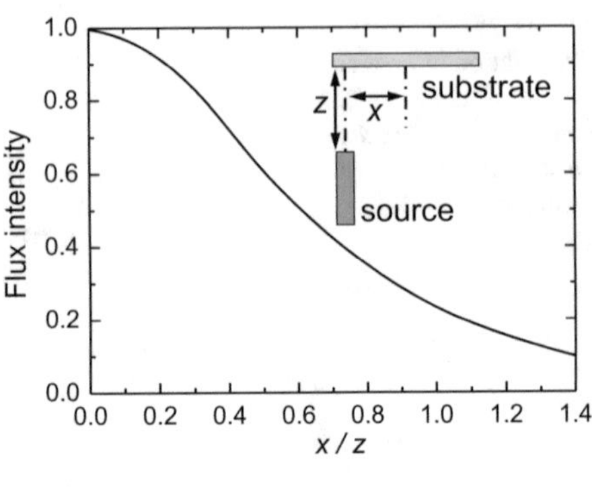

Fig. 11.27 Calculated normalized flux intensity of the source at the surface plane of the substrate, depending on the displacement x of the source axis with respect to the substrate center and the distance z of the source orifice to the surface plane. A source with a cosine angular flux distribution is assumed. After [32]

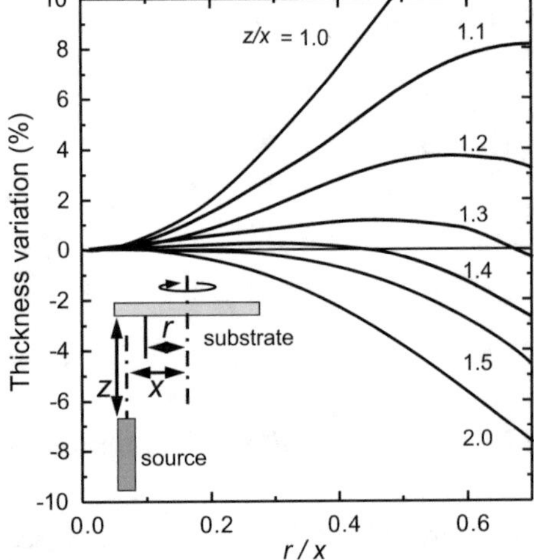

Fig. 11.28 Calculated deviation of the deposition thickness from the value at the substrate center a $x = 0$, depending on the radial distance r to the substrate axis and the distance z of the source orifice to the substrate surface. Values apply for substrates rotating about the axis as indicated in the *inset* and a source with a cosine angular flux distribution. After [32]

shown in Fig. 11.28. The ratio z/x of MBE setups is within the range 1.1 to 2. The calculated curves apply for point sources with a cosine flux distribution. Optimum condition for the assembly of multiple source cells is a symmetric arrangement of all cells with respect to the rotation axis.

The theoretical model considered above provides a reasonable rule-of-thumb for the design of MBE systems. In actual sources shadowing as illustrated in Fig. 11.23 and other effects affect the flux distribution. Source configurations are therefore also empirically optimized. Besides deposition uniformity, which improves as z is increased, also source yield, which improves for small z, and other factors are taken into account.

11.3.4 Adsorption of Impinging Particles

In the MBE growth process generally both, kinetic and thermodynamic processes are important. Since sources and substrate have different temperatures, no global equilibrium exists for the entire system. Particles effused from a source have an energy distribution according to the temperature of this specific source. If the particles impinge on the substrate surface they thermalize to the substrate temperature. Particles desorbing again from the surface were found to reflect an energy distribution according to the substrate temperature. This indicates that the time for thermalization is much less than the time required to grow one monolayer. Such a finding justifies to assume an at least partial or local equilibrium on a time scale relevant for the growth process. The relevant equilibrium temperature is that of the substrate.

In the description of kinetic steps discussed in Sect. 7.2 the re-evaporation of particles from the surface was taken into account. The fraction given by the number of particles sticking to the surface and being incorporated during epitaxy N_{sticking} with respect to the total number of impinging particles of a considered species N_{total} is referred to as the *sticking coefficient* s of this species,

$$s = \frac{N_{\text{sticking}}}{N_{\text{total}}}. \tag{11.29}$$

s may also depend on the flux of other species and may have any value between zero and unity. An example is the sticking coefficient of As_2 molecules on the growing GaAs surface, that sensitively depends on the Ga flux as outlined in Sect. 7.2.9.

The sticking of particles on the surface is described by an adsorption energy as applied in (7.6). Usually the terms *physisorption* and *chemisorption* are used to account for smaller and larger energies, respectively, although chemical interactions occur in both kind of adsorptions and the two terms are not well defined. The terms are still useful for an overall description of surface processes to express different mobilities of a considered adatom species diffusing on the surface. A more rigorous definition in surface science refers physisorption solely to Van-der-Waals interactions. For a simplified description of MBE growth a two-step condensation process of an impinging particle is assumed implying two sticking coefficients for a given species referring to the two adsorption energies. The high surface mobility often found for species arriving at the surface is then assigned to a physisorbed state with a larger desorption probability, and incorporation to the chemisorbed state.

The kinetic processes occurring during molecular-beam epitaxy strongly depend on the specific material grown. Details of these processes are quite complex and reasonably known for only few examples. The scenario of molecular beam epitaxy of GaAs treated in Sect. 7.2.9 indicates the challenge of the task, and also demonstrates the achievement accomplished by combining advanced experimental and theoretical techniques.

11.4 Problems Chapter 11

11.1 Liquid-phase epitaxy of GaAs is generally performed on the Ga-rich side with low concentrations x_{As} of As (as solute) dissoluted in Ga used as solvent. GaAs is grown in a step-cooling process with the solution at 805 °C (homogeneously in the bulk) and the solid-liquid interface at an equilibrium temperature of 800 °C. Assume that the liquidus curve $x(T)$ of the binary Ga-As system can be approximated by the empirical relation $x_{As} = 2352.8 \times \exp(-12404\ \mathrm{K}/T)$.

(a) What is the approximate concentration of As in the liquid directly at the interface to the solid GaAs? What is the excess As concentration in the bulk of the liquid due to the supercooling?
(b) What is the diffusion coefficient of As in liquid Ga, if the diffusion length, given by the solution thickness of 0.5 cm, is covered in a mean time of 104 min?
(c) Compute the thickness of the GaAs layer homoepitaxially grown after a time of 100 s and after a time of 200 s.

11.2 A GaAs layer is grown by metalorganic vapor-phase epitaxy using trimethylgallium (TMGa) and tertiarybutyl-arsenic (TBAs) precursors. The total flow rate in the reactor is 3 slm (standard liters per minute, i.e., 3000 sccm = 3000 cm^{-3}/min at 0 °C and 1013 mbar), the total pressure is 100 mbar.

(a) What is the flow rate required for the Ga source to obtain a partial pressure of 3 Pa in the reactor, if the bubbler is kept at −10 °C and 1 bar (10^5 Pa) pressure?
(b) Which mass has one molecule of TMGa? Which mass of TMGa is consumed during 1 hr of growth at the flow rate found in (a)? Assume ideal-gas behavior and a mole volume of 22.4 l at standard conditions; atomic masses of Ga, C and H are 69.7, 12.0, and 1.0 grams/mole, respectively.
(c) Which temperature is required for the tertiarybutyl-arsenic bubbler for a V/III ratio of 5 in the gas phase? A flow of 62.5 sccm and a bubbler pressure of 1500 mbar are applied.

11.3 Assume a total flow rate of 3 slm (standard liters per minute at 0 °C and 1013 mbar) hydrogen, homogeneously heated in a MOVPE reactor to 600 °C at 100 mbar total pressure.

(a) What is the free-stream velocity, if the cross-section area of the reactor is 40 cm^2 and surface effects are neglected? What would be the free-stream velocity at standard conditions?
(b) Calculate the approximate thickness of a boundary layer 10 cm behind the leading edge of a fixed plate placed in the homogeneous hydrogen flow of (a) at 600 °C. Assume a dynamic viscosity of 17 μPa s, ideal-gas behavior with a mole volume of 22.4 l at standard conditions, and 2.0 grams/mole atomic mass.

(c) Let the growth rate of a GaAs layer be controlled by the partial pressure of the Ga precursor. How will an increase of the temperature from 600 °C to 700 °C approximately change the growth rate in the regime limited by mass transport? Assume an empirical temperature dependence of the diffusion coefficient described by a factor $(T/T_0)^{1.8}$, T given in K.

11.4 A (001)-oriented Si layer is grown using molecular-beam epitaxy.

(a) What is the areal density of Si atoms on a (001) plane?
(b) Which flux per unit substrate-area is required to grow a 0.2 μm thick layer in 1 h? Assume a sticking coefficient of unity and ignore the formation of surface reconstructions which continuously reproduce during growth.
(c) Which effusion rate produces the flux density of (b)? Assume the substrate is mounted at a distance of 12 cm from the cell and is directly in line with the aperture.
(d) Estimate the substrate area with a deviation $\leq 5\%$ from the maximum deposition thickness, if a cosine law applies. Note that the substrate is planar: a beam inclined by an angle ϑ has a longer path and, furthermore, supplies atoms to a larger area for a considered cross-section.

11.5 A (001)-oriented GaAs layer is grown using MBE with a growth rate controlled by the Ga flux. The Ga Knudsen cell is kept at 960 °C, producing an equilibrium pressure of 2.5×10^{-3} mbar. The cell orifice has 0.8 cm diameter and the distance to the substrate is 13 cm.

(a) Calculate the areal flux density of Ga atoms at the substrate. Ga has 70 grams/mole atomic mass.
(b) Which growth rate of GaAs results from the areal flux density of (a)?
(c) Which cell temperature is required to double the growth rate of (b), if the enthalpy of Ga vaporization is 2.56×10^5 J/mol?

11.5 General Reading Chapter 11

M.A. Hermann, W. Richter, H. Sitter, *Epitaxy* (Springer, Berlin, 2004)
H. Asahi, Y. Horikoshi (Eds.), *Molecular Beam Epitaxy, Materials and Applications for Electronics and Optoelectronics* (Wiley & Sons, Chichester, 2019)
P. Capper, M. Mauk, *Liquid Phase Epitaxy of Electronic, Optical and Optoelectronic Materials* (Wiley, Chichester, 2007)
G.B. Stringfellow, *Organometallic Vapor-Phase Epitaxy*, 2nd edn. (Academic Press, New York, 1999)
J.E. Ayers, *Heteroepitaxy of Semiconductors: Theory, Growth, and Characterization* (CRC, Boca Raton, 2007)
A.C. Jones, P. O'Brien, *CVD of Compound Semiconductors* (VCH, Weinheim, 1997)
S.J.C. Irvine, P. Capper, *Metalorganic vapor phase epitaxy (MOVPE): Growth, materials properties and applications* (Wiley & Sons, Hoboken NJ, 2020)

References

1. G.B. Stringfellow, Fundamental aspects of vapor growth and epitaxy. J. Crystal Growth **115**, 1 (1991)
2. H. Nelson, Epitaxial growth from the liquid state and its application to the fabrication of tunnel and laser diodes. RCA Rev. **24**, 603 (1961)
3. E.A. Giess, R. Ghez, Liquid-phase epitaxy, in *Epitaxial growth part B*, ed. by J.W. Matthews (Academic Press, New York, 1975), pp. 183–213
4. J.J. Hsieh, Thickness and surface morphology of GaAs LPE layers grown by supercooling, step-cooling, equilibrium-cooling, and the two-phase solution techniques. J. Crystal Growth **27**, 49 (1974)
5. W. Miederer, G. Ziegler, R. Dötzer, Verfahren zum tiegelfreien Herstellen von Galliumarsenid-stäben aus Galliumalkylen und Arsenverbindungen bei niedrigen Temperaturen, German Patent 1,176,102, filed 25.9.1962
6. W. Miederer, G. Ziegler, R. Dötzer, Method of crucible-free production of gallium arsenide rods from alkyl galliums and arsenic compounds at low temperatures. U.S. Patent 3,226,270, filed 24.9.1963
7. H.M. Manasevit, W.I. Simpson, The use of metal-organics in the preparation of semiconductor materials on insulating substrates: I. Epitaxial III-V gallium compounds, J. Electrochem. Soc. **12**, 66C (1968)
8. H.M. Manasevit: The use of metalorganics in the preparation of semiconductor materials: Growth on insulating substrates, J. Crystal Growth **13/14**, 306 (1972)
9. R.W. Thomas, Growth of single crystal GaP from organometallic sources. J. Electrochem. Soc. **116**, 1449 (1969)
10. T.J. Mountziaris, K.F. Jensen, Gas-phase and surface reaction mechanisms in MOCVD of GaAs with trimethyl-gallium and arsine. J. Electrochem. Soc. **138**, 2426 (1991)
11. G.B. Stringfellow, *Organometallic Vapor-Phase Epitaxy*, 2nd edn. (Academic Press, New York, 1999)
12. R.T. Morrison, R.N. Boyd, *Organic Chemistry*, 5th ed. (Allyn & Bacon, New York, 1987)
13. K.F. Jensen, Transport phenomena in vapor phase epitaxy reactors. In: *Handbook of crystal growth*, ed. by D.R.T. Hurle (Elsevier, Amsterdam 1994), pp. 541–599
14. M. Tortowidjojo, R. Pollard, Elementary processes and rate-limiting factors in MOVPE of GaAs. J. Crystal Growth **93**, 108 (1988)
15. D.H. Reep, S.K. Ghandhi, Deposition of GaAs epitaxial layers by organometallic CVD. J. Electrochem. Soc. **130**, 675 (1983)
16. D.I. Fotiadis, M. Boekholt, K.F. Jensen, W. Richter, Flow and heat transfer in CVD reactors: Comparison of Raman temperature measurements and finite element model predictions. J. Crystal Growth **100**, 577 (1990)
17. R.B. Bird, W.E. Steward, E.N. Lightfood, *Transport phenomena* (Wiley, New York, 1960)
18. J. Ouazzani, K.-C. Chiu, F. Rosenberger, On the 2D modelling of horizontal CVD reactors and its limitations. J. Crystal Growth **91**, 497 (1988)
19. J. van de Ven, G.M.J. Rutten, M.J. Raaijmakers, L.J. Giling, Gas phase depletion and flow dynamics in horizontal MOCVD reactors. J. Crystal Growth **76**, 352 (1986)
20. L. Stock, W. Richter, Vertical versus horizontal reactor: an optical study of the gas phase in a MOCVD reactor. J. Crystal Growth **77**, 144 (1986)
21. R.B. Bird, W.E. Stewart, E.N. Lightfoot, *Transport phenomena* (Wiley, New York, 1962)
22. R.J. Miller, C.H. Bachmann, Production of cadmium sulfide crystals by coevaporation in a vacuum. J. Appl. Phys. **29**, 1277 (1958)
23. R.B. Schoolar, J.N. Zemel, Preparation of single-crystal films of PbS. J. Appl. Phys. **35**, 1848 (1964)
24. K.G.Günther, Aufdampfschichten aus halbleitenden III–V-Verbindungen. Z. Naturforschg. **13a**, 1081 (1958). (in german)
25. J.E. Davey, T. Pankey, Epitaxial GaAs films deposited by vacuum evaporation. J. Appl. Phys. **39**, 1941 (1968)

26. J.R. Arthur Jr., Interaction of Ga and As_2 molecular beams with GaAs surfaces. J. Appl. Phys. **39**, 4032 (1968)
27. A.Y. Cho, Film deposition by molecular-beam techniques. J. Vac. Sci. Technol. **8**, S31 (1971)
28. M. Mizuta, T. Iwamoto, F. Moriyama, S. Kawata, H. Kukimoto, AlGaAs growth using trimethyl and triethyl compound sources. J. Crystal Growth **68**, 142 (1984)
29. P. Clausing, Über die Strahlformung bei der Molekularströmung. Z. Physik **66**, 471 (1930). (in german)
30. T. Yamashita, T. Tomita, T. Sakurai, Calculations of molecular beam flux from liquid source. Jpn. J. Appl. Phys. **26**, 1192 (1987)
31. R.F.C. Farrow, P.W. Sullivan, G.M. Williams, C.R. Stanley, Proceedings 2nd international symposium molecular beam epitaxy and related clean techniques (Tokyo, 1982), p. 169
32. R.A. Kubiak, S.M. Newstead, P. Sullivan, Technology and design of molecular beam epitaxy systems. In: *Molecular beam epitaxy: Applications to key materials*, ed. by R.F.C. Farrow (Noyes Publications, Park Ridge, NJ, USA, 1995)
33. P. Capper, S. Irvine, T. Joyce, Epitaxial crystal growth: Methods and materials. In: *Springer Handbook of Electronic and Photonic Materials*, 2nd edn., ed. by S. Kasap, P. Capper (Springer Nature, Cham, 2017)
34. K. Nakamura, O. Makino, A. Tachibana, K. Matsumoto, Quantum chemical study of parasitic reaction in III–V nitride semiconductor crystal growth. J. Organomet. Chem. **611**, 514 (2000)
35. M.J. Almond, C.A. Jenkins, D.A. Rice, Organometallic precursors to the formation of GaN by MOCVD: structural characterisation of Me_3 Ga • NH_3 by gas-phase electron diffraction. J. Organomet. Chem. **439**, 251 (1992)
36. C. Theodoropoulos, T.J. Mountziaris, H.K. Moffat, J. Han, Design of gas inlets for the growth of gallium nitride by metalorganic vapor phase epitaxy. J. Crystal Growth **217**, 65 (2000)
37. C.-F. Tseng, T.-Y. Tsai, Y.-H. Huang, M.-T. Lee, R.-H. Horng, Transport phenomena and the effects of reactor geometry for epitaxial GaN growth in a vertical MOCVD reactor. J. Crystal Growth **432**, 54 (2015)
38. M. Dauelsberg, C. Martin, H. Protzmann, A.R. Boyd, E.J. Thrush, J. Käppeler, M. Heuken, R.A. Talalaev, E.V. Yakovlev, A.V. Kondratyev, Modeling and process design of III-nitride MOVPE at near-atmospheric pressure in close coupled showerhead and planetary reactors. J. Crystal Growth **298**, 418 (2007)
39. D. Sengupta, S. Mazumder, W. Kuykendall, S.A. Lowry, Combined ab initio quantum chemistry and computational fluid dynamics calculations for prediction of gallium nitride growth. J. Crystal Growth **279**, 369 (2005)

Chapter 12
Special Growth Techniques

Abstract A number of techniques applying MOVPE, MBE, or LPE have been developed to meet special requirements for the fabrication of semiconductor devices. Some popular methods are treated in this chapter. *Selective Area Growth* (SAG), also called *Selective Area Epitaxy* (SAE), is a widely applied technique to grow an epitaxial layer only in selected parts of a substrate; the areas are generally defined by windows cut through a mask layer. Selectivity is achieved best with the equilibrium-near growth conditions of liquid-phase epitaxy or hydride vapor-phase epitaxy, but SAG is also performed using MOVPE and even MBE. Epitaxy solely in the mask window leads to an enhanced growth rate due to additional material supply from the masked regions. The anisotropy of the growth rate along different crystallographic orientations is utilized to grow nanostructures such as ridges or nanowires. An important application is epitaxial lateral overgrowth of the mask to block or bend threading dislocations for fabricating layers with areas of low defect density. *Vapor–liquid–solid* (VLS) *growth* is a prominent method to fabricate semiconductor nanowires with high performance using CVD, MBE, or MOVPE. Usually a metal catalyst is employed to solute gaseous components, which precipitate at the liquid–solid interface and induce whisker growth. Controlled axial and radial growth enable the fabrication of axial and core–shell heterostructures, and elastic strain relaxation in these directions yield very large critical thicknesses for coherent lattice-mismatched semiconductors. The metal component of compound semiconductors allows for self-catalyzed nanowire growth without foreign catalyst. Nanowires of some semiconductors can also be grown self-induced without any catalyst. *Atomic Layer Epitaxy* (ALE) or *Deposition* (ALD), and the related *Migration Enhanced Epitaxy* (MEE) are techniques with an alternate pulsed supply of gaseous precursors. The MEE mode of MBE leads to an increased lifetime of cation adatoms, enabling the fabrication of particularly sharp interfaces and doping profiles. MEE also proved beneficial in selected area growth. ALD of amorphous or polycrystalline layers is widely-used in semiconductor technology due to an exceptional conformality and thickness control of layers on nonplanar surfaces. *Epitaxial deposition of organic crystals* is usually based on vapor-phase techniques of organic molecular beam deposition (OMBD) and organic vapor phase deposition (OVPD) using highly purified source materials. Basic differences between organic

U. W. Pohl, *Epitaxy of Semiconductors*, Graduate Texts in Physics,
https://doi.org/10.1007/978-3-030-43869-2_12

and inorganic layers with respect to bonding forces require comparably low deposition temperatures and careful substrate selection. Epitaxial relations may be established by suitable substrate-layer interaction, meeting demanding nucleation issues. Equilibrium-near van der Waals epitaxy for commensurate or coincident conditions, or quasiepitaxy for incommensurate conditions and more kinetic control are used.

12.1 Selective Area Growth

Selective area growth (SAG), also referred to as *selective area epitaxy* (*SAE*) or *selective epitaxy*, is a technique for growing an epitaxial layer solely in some selected part of a substrate; these areas are defined by windows cut through a mask layer. SAG is often combined with other techniques such as nanowire growth (Sect. 12.4) or anisotropic growth rates of different crystallographic faces to fabricate faceted surface structures or form defect-reduced layers by lateral overgrowth (Sects. 12.1.3 and 12.1.4).

12.1.1 Principle of Selective Area Growth

Selective area growth denotes the localized growth of an epitaxial material on predefined areas of the substrate. These locations are typically lithographically defined openings of a mask made of an oxide or another amorphous material. At appropriate conditions growth proceeds only in the mask windows, where the crystalline substrate surface is uncovered. The growth material is supplied by various paths. In addition to direct impingement of species on the windows area diffusion provides additional material as illustrated in Fig. 12.1. In MOVPE both vapor-phase diffusion and surface diffusion from masked areas contribute to growth, while in MBE only surface diffusion complements direct impingement. For a brief review see [1].

A basic requirement for selective epitaxy is the suppression of growth on the mask. Such conditions are favored by a large diffusion length and a small sticking coefficient of adatoms on the mask. Precursor molecules used in MOVPE have much smaller sticking coefficients than elemental particles used in MBE; the selectivity in

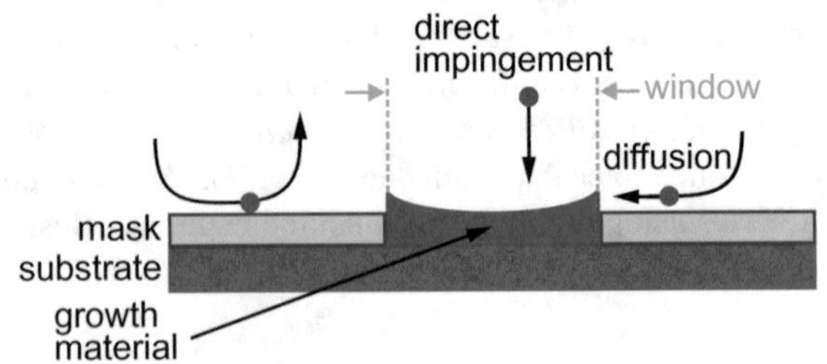

Fig. 12.1 Schematic illustrating the supply of growth material in selective area growth. Growth occurs in the window of the mask and is fed by both, direct impingement of particles and diffusion on or near the surface

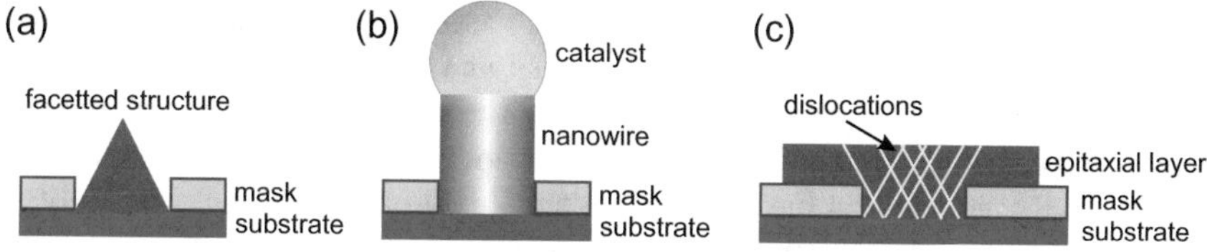

Fig. 12.2 Structures fabricated using selective area growth. **a** Growth of a faceted nanostructure by employing low grow rates of side facets, **b** growth of nanowires at predefined positions, **c** epitaxial lateral overgrowth of the mask performed for blocking dislocations to obtain layers with areas of low defect densities

MOVPE is hence much larger than in MBE, and usually polycrystalline growth on the mask is observed in MBE in addition to epitaxial growth in the windows.

Examples for applications of selective area growth are illustrated in Fig. 12.2. In panel *a* the dependence of the growth rate on the crystallographic orientation is applied for the self-organized formation of faceted structures of the epitaxial material; facets with slowest growth rate (usually low-index planes) eventually determine the shape of the structure, compare Fig. 6.18. Figure 12.2b depicts selective area growth for the position-controlled fabrication of nanowires, which otherwise nucleate at random locations on the wafer. In Fig. 12.2c the mask is used to block dislocations originating from the substrate or nucleating at the substrate interface; layers with low defect densities can then be fabricated by laterally overgrowing the mask.

12.1.2 Conditions for Selectivity

Any application of selective area growth requires a distinct selectivity of growth: epitaxy must only occur in the openings of the mask, while no nucleation and growth must occur on the mask layer. Such a selectivity is well achieved when growth takes place close to thermodynamic equilibrium. This requirement is fulfilled with liquid phase epitaxy (LPE) and chloride-based vapor-phase epitaxy [2]. The more versatile and widely applied techniques of MOVPE and MBE operate, however, far from equilibrium—see Fig. 11.1. Growth conditions and mask material must then appropriately be chosen to obtain the essential selectivity.

The *mask* is typically made of an amorphous dielectric. Usually a thin layer of SiO_x is used, and openings are cut out by conventional lithography and etching with fluoric acid; also SiN_x masks are applied, processed using reactive ion etching. To preserve selectivity any mask must not have large regions without openings: a large distance between openings are found to favor nucleation and correspondingly deposition also on the mask.

Growth conditions are selected for a maximum distance between mask openings while keeping the required selectivity. Results for MOVPE show that species of the growing material impinging on the solid surface desorb more easily from the mask than from the substrate in the openings. They are assumed to migrate in the surface-near gas phase due to collisions with gas particles, and reach a larger migration length at lower pressure and higher temperature [3, 4]. The selectivity is thereby connected to the distance between openings and increased at low reactor total pressure and lower particle fluxes; in addition the temperature should be preferably high. In MOVPE the selectivity is also enhanced by a solely *partial* precursor decomposition, since fully released metal components have large sticking coefficients also on the mask [4]; the selectivity is hence much smaller in MBE, where elemental source materials are used. The selectivity of compound semiconductors may be improved by atomic layer epitaxy (Sect. 12.3), where anion and cation components are supplied alternately [5].

A *growth rate enhancement* is usually observed in the mask openings compared to the growth rate on unmasked planar substrates. This effect originates from an additional flux of species provided by the masked nongrowth areas to the open areas. In MOVPE thereby a lateral concentration gradient of precursors exists near the openings in addition to the general vertical gradient, because consumption by growth only occurs in these windows. The enhancement of the growth rate in the windows is larger the smaller their area is compared to the masked area. While small windows have a nearly uniform enhancement, wide openings may have no enhancement in their center and hence a substantial non-uniform layer thickness [2].

The growth-rate enhancement *GRE* is given by

$$GRE = \frac{t_{\mathrm{SAG}}(x, y)}{t_{\mathrm{planar}}} = \frac{R_{\mathrm{SAG}}(x, y)}{R_{\mathrm{planar}}}, \tag{12.1}$$

where t denotes the layer thickness and R the growth rate observed in selective area growth (subscript SAG) compared to that on a planar substrate.

Selective Area Growth in MOVPE

A typical growth-rate enhancement observed in MOVPE SAG using dual stripe masks is shown in Fig. 12.3. The GaAs(001) epitaxy was performed at 650 °C and 150 mbar total pressure using TMGa and AsH_3 precursors; two parallel SiO_2 stripes aligned along [6] with an opening in between masked a part of the substrate, and the layer thickness was measured by optical interferometer microscopy [7]. The growth rate is clearly enhanced in the window between the stripes and also near the edges of the stripes. These measured growth-rate profiles were precisely described by a model assuming solely vapor-phase diffusion [8]; surface diffusion was neglected in the given geometry due to the very small diffusion length on the order of a few micrometers. For the applied growth conditions an effective diffusion length $(D/k_s)_{\mathrm{Ga}} = 85\ \mu\mathrm{m}$ was obtained for the Ga species, where D is the diffusion coefficient in the gas phase depending on pressure P, temperature T, and mass M by $D \propto T^{3/2}/\left(P \times M^{1/2}\right)$; k_s is a sticking-rate constant depending on the reactivity of the source molecules on the surface and on temperature by $k_s = k_0 \exp(-E_{\mathrm{act}}/kT)$ [1]. At the same growth conditions effective diffusion lengths of $(D/k_s)_{\mathrm{Al}} = 50\ \mu\mathrm{m}$ and $(D/k_s)_{\mathrm{In}} = 10\ \mu\mathrm{m}$ were obtained for the cation species in the selective MOVPE of AlAs and InP [7]. Data for varied temperature and pressure are reported in [9].

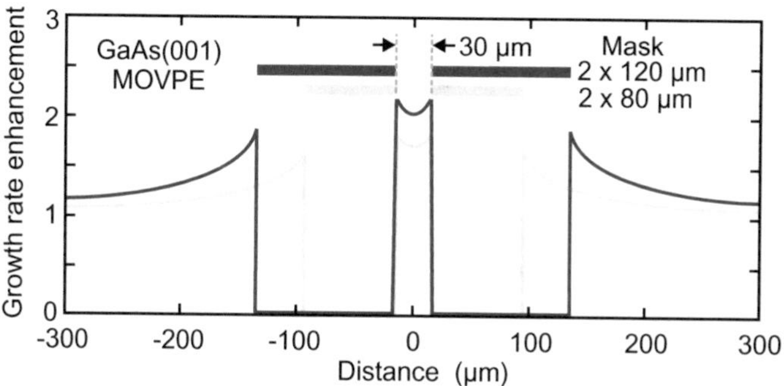

Fig. 12.3 Selective area growth of GaAs using SiO_2 masks with two long stripes (their width is shown in cross section) and 30 μm opening in between. *Blue and yellow curves* show the enhancement of the growth rate in the opening between the stripes at distance 0 and at the outer edges of the stripes. Data are from [7]

The example of Fig. 12.3 shows a substantial variation of the growth-rate enhancement in the window. Such a feature is usually detrimental for device fabrication. The effect basically disappears if the window is much smaller than the effective diffusion length of particles in the vapor phase. For typical stripe geometries applied in photonic devices with widths on the order of a few micrometers this is usually fulfilled in MOVPE.

The dependence of growth-rate enhancement on the width of the masked area is already seen in Fig. 12.3. The simultaneous dependence on width of the window is shown in Fig. 12.4 [10]. The SAG MOVPE of InP(001) with two parallel SiO_2 stripes aligned along [6] was performed at 630 °C and 101 mbar total pressure using TMIn and PH_3 precursors. The layer thickness was measured by a surface profiler in the center of the window; simulated curves were obtained from a vapor-phase diffusion model with an effective diffusion length of $(D/k_s)_{In} = 40\,\mu m$ fitted to the experimental data.

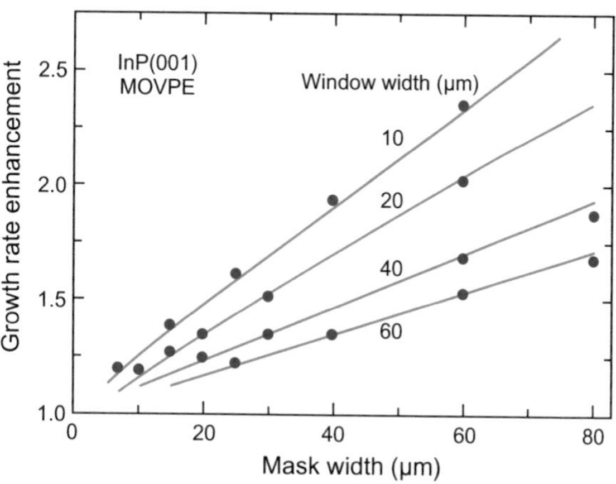

Fig. 12.4 Selective area growth of InP in dual stripe geometry. *Dots and curves* show experimental and calculated data, respectively. Reproduced with permission from [10], © 2002 Elsevier

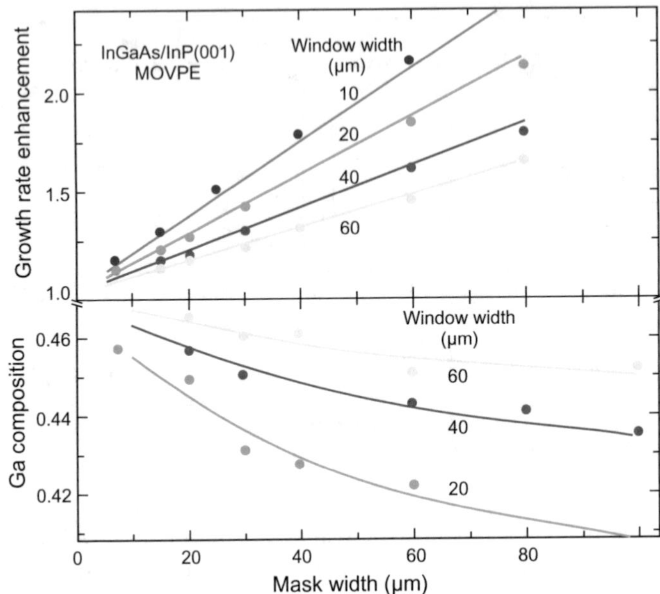

Fig. 12.5 Selective area growth of InP in dual stripe geometry. *Dots and curves* show experimental and calculated data, respectively. Reproduced with permission from [10], © 2002 Elsevier

In *ternary and quaternary compounds* the differences in diffusion length of various components may lead to a spatial variation of composition in the window region. Results for lattice-matched InGaAs/InP(001) layers grown by SAG MOVPE with similar conditions as above and TMIn, TEGa, and AsH_3 precursors are given in Fig. 12.5 The composition was determined using electron microprobe, yielding a significant dependence of the Ga concentration particularly for small windows. The growth-rate enhancement is largely comparable to that observed for the InP SAG shown in Fig. 12.4.

Selective Area Growth in MBE

The selectivity of growth in MBE is much lower than in MOVPE, and usually polycrystalline growth occurs on masked areas in addition to epitaxial growth in the windows. Since there is no vapor-phase diffusion in MBE, only surface diffusion with its inherent smaller diffusion lengths contributes to the selectivity. Furthermore, the sticking coefficient of the source material applied in MBE is generally higher than that of MOVPE precursors. Group-III particles adsorbed on the surface have a solely low desorption probability and hence contribute to parasitic growth on the mask.

A high growth temperature (compared to epitaxy on planar substrate) is used to achieve an optimum selectivity; at such conditions the re-evaporation of adatoms is enhanced. The difference in the cation desorption-rates between the crystalline surface and the amorphous mask determines the selectivity [11]. A study of selective GaAs epitaxy shows that the Ga flux must be kept low to avoid polycrystal growth on the mask, see Fig. 12.6 [12]. The growth rate is consequently low and decreases

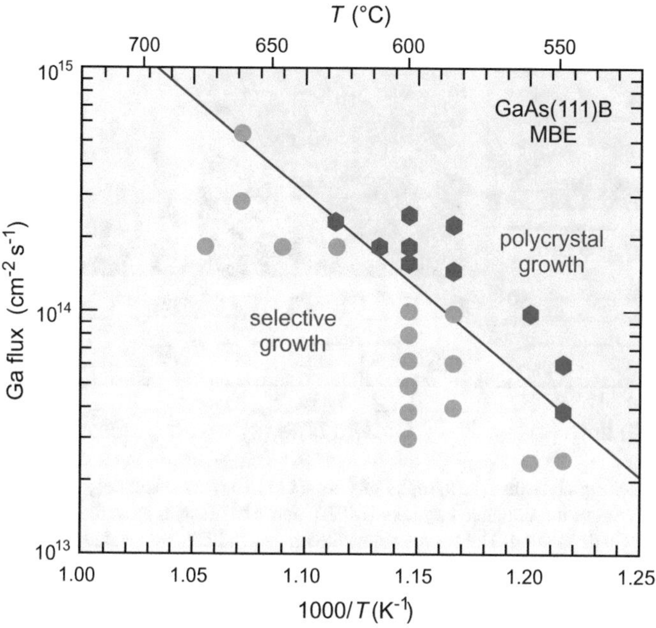

Fig. 12.6 Selectivity map for the selective area MBE of GaAs with masking SiO_2 stripes. After [12]

further with increased temperature due to the decreasing residence time of the growth-limiting Ga adatoms [13]. Smooth selective epitaxy without deposition on the mask was obtained above 550 °C, when the Ga flux was below the line indicated in Fig. 12.6; this critical flux is characterized by an Arrhenius dependence with 1.54 eV activation energy, which corresponds to the potential barrier for parasitic nucleation on the mask [12]. Growth was performed on (111)*B* oriented substrate, because the (001) surface is less stable and leads to a rough morphology at higher temperature due to Ga and also As loss. Furthermore, the top facet of the epitaxial ridge remains (111)*B*, while pyramidal cross sections with facets depending on the lateral ridge orientation are obtained for growth on (001) substrate [12].

On SiN_x surface a larger re-evaporation rate than on SiO_2 surface was observed for the selective MBE of GaAs; selective area growth of GaAs is hence more easily achieved using a SiN_x mask [14].

The growth-rate enhancement in MBE is difficult to estimate. In MOVPE the GRE is deduced from the vapor-phase diffusion and directly given by the concentration of the laterally diffusing rate-limiting species in the surface-near gas phase. In MBE the enhancement originates from surface diffusion, and randomly deposited polycrystalline material may act as sinks or diffusion barriers for adatoms, yielding a nonuniform layer thickness in the window.

A substantial improvement of selectivity is achieved by applying migration-enhanced epitaxy (MEE) to selective area growth [15]. The in the MEE technique the

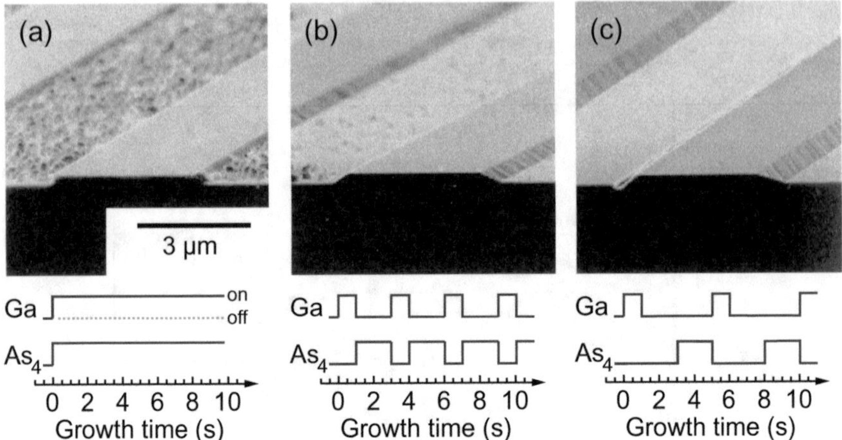

Fig. 12.7 Scanning electron micrographs of GaAs(111)*B* layers selectively grown by **a** conventional MBE, **b** migration-enhanced epitaxy (MEE), and **c** MEE with an additional brief annealing step after each Ga deposition. The curves below the images indicate the related switching sequences of the sources. Reproduced with permission from [15], © 1999 the Physical Society of Japan and the Japan Society of Applied Physics

anion and cation source materials are supplied alternately, see Sect. 12.2.2. Selective area growth of GaAs(111)*B* at 610 °C was performed using SiO_2 stripe masks oriented along $[\bar{2}11]$; the effect of non-simultaneous source fluxes is shown in Fig. 12.7. Panel *a* shows images of GaAs grown by conventional MBE in 3 μm wide windows bounded by 3 μm wide masks; the simultaneous supply of Ga and As_4 species clearly leads to substantial deposition of polycrystalline GaAs on the SiO_2 mask. Application of MEE at the same temperature and incident fluxes leads to a strong reduction of the deposition of the mask, see Fig. 12.7b. The selectivity map is quite similar to that of MBE shown in Fig. 12.6, but the boundary of critical Ga flux is shifted some tens of °C to lower temperature; raising the temperature above 650 °C yields rough surfaces due to high As evaporation at the given conditions. By applying an additional short annealing step ($\geq$2 s) after Ga deposition complete selectivity with no deposition on the mask is achieved, see Fig. 12.7c. No such improvement is observed when the annealing step is introduced after As_4 deposition.

12.1.3 Selective Area Growth of Faceted Structures

Differences in the growth velocity of different crystallographic surfaces can be employed to fabricate faceted structures by self-organized selective area growth. The principle outlined in Fig. 6.18 is used to grow, e.g., ridges for BH lasers, quantum wires, or quantum dots.

The energy of a crystallographic facet is given by the sum of broken bonds of this surface, see Sect. 6.2.4. Facets with low surface energy define the equilibrium crystal shape of a solid. This also applies for growing surfaces, even if their chemical potential differs from equilibrium as illustrated in Fig. 6.20. The surface energy depends on the chemical potential of the ambient (Fig. 6.17).

Facets with low surface energy have low growth rates; these are usually low-index crystal planes, such as {111}*A*, {111}*B*, {110} and {100} for zincblende semiconductors. These low-index facets often appear in selective area growth. The order of surface energies of different facets can to some extent be controlled by growth conditions due to the dependence on the ambient. In addition, facet formation is affected by direction and shape of the windows in the mask.

Low index facets which can appear for growth on (001) surfaces of zincblende semiconductors are shown in Fig. 12.8a. There are two {111}*A* facets and two {111}*B* facets; the {111}*A* facets are terminated by cations (group-III atoms for III–V semiconductors) and the {111}*B* facets by anions (group-V atoms). For selective area growth on a (111)*A* or (111)*B* substrate faceting shows a different geometry, see Fig. 12.8b,c.

{111}*A* facets are: (111), $(\bar{1}\bar{1}1)$, $(1\bar{1}\bar{1})$, $(\bar{1}1\bar{1})$,
{111}*B* facets are: $(\bar{1}\bar{1}\bar{1})$, $(\bar{1}11)$, $(1\bar{1}1)$, $(11\bar{1})$.

MOVPE on (001)-oriented substrates of zincblende semiconductors is usually performed in the transport-limited regime, where the growth rate is limited by diffusion in the gas phase and does not significantly depend on the temperature, see Sect. 11.2.2 (Fig. 11.13); also a variation of the anion partial pressure in a wide range does not change the growth rate. This does not necessarily also apply for MOVPE on other low-index surfaces. On a (111)*B* surface the growth rate was found to increase as either the growth temperature is increased or the anion partial pressure is decreased [16], while an inverse behavior was reported for growth on a (111)*A* surface [17]; near {011} planes a dependence on AsH_3 partial pressure was found for atmospheric-pressure MOVPE of GaAs [18].

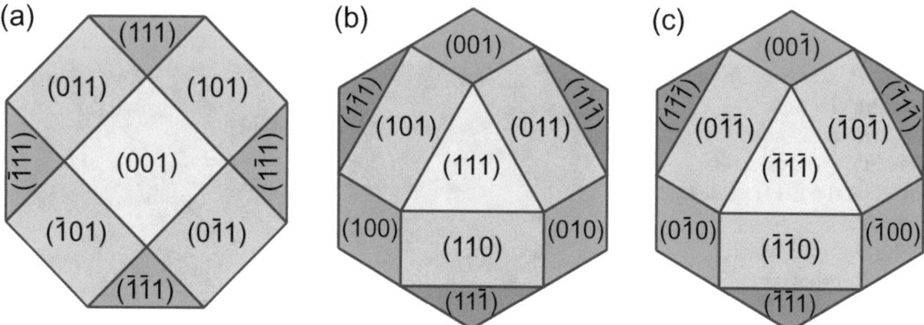

Fig. 12.8 Low-index facets of structures which may appear in selective area growth performed on a substrate with **a** (001) surface, **b** (111)*A* surface, and **c** (111)*B* surface. Darker shading indicates steeper facet

Which facets evolve from selective area growth can be estimated from the dependence of the growth rate from the crystallographic orientation. Once these growth rates $r(\mathbf{n})$ are known, the shape of locally grown crystals can be predicted by adopting the concept of the Wulff plot introduced in Sect. 6.2.4 [18, 19]. Wulff's theorem is a consequence of the *Gibbs-Curie relation*, which states that the shape of a crystal in thermal equilibrium with its ambient is given by minimizing the total surface energy,

$$G_{\text{surf}} = \oint_A \gamma(\mathbf{n}) dA \rightarrow min; \tag{12.2}$$

here $\mathbf{n}$ is the surface normal of the surface element dA with the (specific) surface energy γ. The geometrical construction of Wulff expresses (12.2) in terms of a polar diagram of surface energies. The validity of this approach is not restricted to thermal equilibrium; it may also be applied to growth or to dissolution (etching) of a crystal by replacing the specific surface-energy vectors by vectors expressing the localized growth rate $r(\mathbf{n})$ or the etch rate [19],

$$\oint_A r(\mathbf{n}) dA \rightarrow min. \tag{12.3}$$

Condition (12.3) applies for planar and convex surfaces. For concave surfaces such as a V groove the local growth rate at the common boundary of the facets tends to a maximum instead of a minimum [18].

Growth rates of GaAs using planar and non-planar substrates and various atmospheric-pressure MOVPE conditions were measured by scanning electron micrographs and optical interference-contrast microscopy. The cross-section SEM image Fig. 12.9 shows a facet formation at a [011] directed edge on a (001) surface [18]. The angle θ between the normal to two planes of a cubic lattice with Miller indices $h_i k_i l_i$ is given by

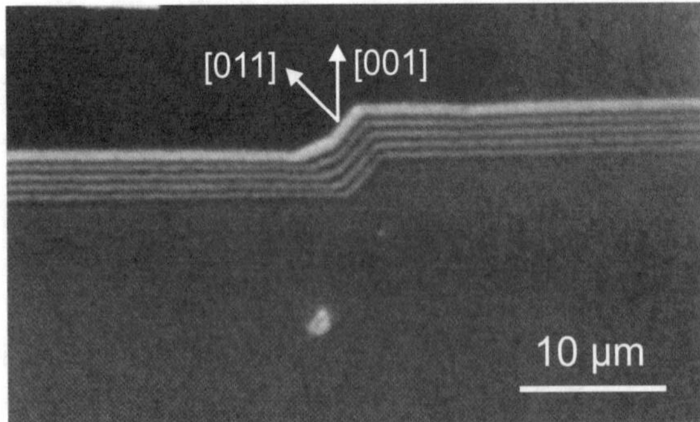

Fig. 12.9 Cross-sectional scanning electron micrograph of GaAs MOVPE layers on a (001) surface and a [011] directed edge. *Bright contrast* indicates doped marker layers. Reproduced with permission from [18], © 1991 Elsevier

$$\cos(\theta) = \frac{h_1 h_2 + k_1 k_2 + l_1 l_2}{\sqrt{h_1^2 + k_1^2 + l_1^2}\sqrt{h_2^2 + k_2^2 + l_2^2}}. \tag{12.4}$$

Gathering data for the {100}, {111}, {311}, and {511} surfaces, such measurements yield a general order of the GaAs growth rates of 20:15:9.7:2.5 corresponding to {100}:{111}Ga:{011}:{111}As.

A polar diagram is plotted from the experimental growth rates to obtain the modified Wulff diagram of the resulting crystal shape. Data points between the experimental points are obtained from an interpolation scheme given in more detail in [18]. In brief, all orientations are considered linear combinations of the {100}, {110}, and {111} orientations, with all surface steps being {111} microfacets; an average surface-reaction rate is assigned to the {100}, {110}, and {111} surfaces from the experimental growth rates, and for vicinal surfaces to these primary planes the growth rate is modified according to the density of steps. A polar diagram of GaAs growth rates obtained from this procedure is shown in Fig. 12.10a. This diagram refers to growth on a [20] surface.

The shape of a layer grown in the window area (or on non-planar substrates) can now be deduced from the orientation-dependent growth rate. The cross-sectional shape of a layer grown on a stripe window oriented along [011] is given in Fig. 12.10b. We note from panel (a) that the growth rate of {111}*B* facets (As faces) is smallest; these facets build the side walls of the ridge growing in the window area. Minima or cusps contribute most to the shape in Wulff constructions, but also regions adjacent to the minima may have a significant effect. The fast growing (100) top facet eventually disappears at continued growth. This shape is actually obtained in MOVPE for the given window orientation. Different facets are reported for a stripe window oriented

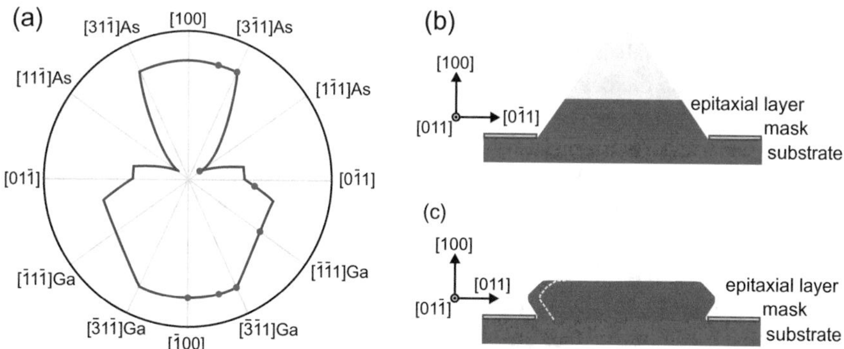

Fig. 12.10 **a** Semi-empirical polar diagram [(011) cross section] of the GaAs MOVPE growth rate on a (100) surface at 750 °C and 1013 mbar total pressure (*blue curve*); the *red dots* mark experimentally measured data. **b** Ridge shape resulting from a Wulff construction for growth on a (100) surface with stripe mask along [011] using the diagram of (**a**). Light shading indicates the shape for continued layer growth. **c** Wulff construction (*dahed yellow line*, shifted to right) and shape of a MOVPE ridge grown at 650 °C and 13 mbar for a stripe mask along [0$\bar{1}$1]. Data are taken from [18, 21]

along $[0\bar{1}1]$; the side walls are then bounded by $\{111\}A$ and $\{111\}B$ facets as shown in Fig. 12.10c [21]. Although this ridge was grown under different MOVPE conditions, the shape is reasonably well described by the diagram of panel (a) turned upside down, i.e., mirrored at the horizontal line connecting the $[01\bar{1}]$ and $[0\bar{1}1]$ labels; the corresponding Wulff construction is indicated by the dashed line in Fig. 12.10c.

If the selective epitaxy occurs in a relatively wide mask window, the polar diagram must be split into halves to construct the shapes at the left and right edges separately [18, 19]. It should be noted that the described study assumed uniform growth in the window region. If the window width exceeds twice the surface-assisted diffusion from the mask region nonuniform growth occurs as pointed out in Sect. 12.1.2. At high total pressure and a related short diffusion length a growth-rate enhancement was observed near the edges of a stripe otherwise similar to that of Fig. 12.10c [18].

12.1.4 Epitaxial Lateral Overgrowth (ELO)

Epitaxial lateral overgrowth (ELO, also referred to as lateral epitaxial overgrowth LEO) is a technique employing selective area epitaxy to fabricate areas of a layer with a strongly reduced dislocation density. In conventional epitaxy dislocations originating from the substrate and those generated by structural mismatch penetrate through the epitaxial layer, see Fig. 2.20. ELO is based on the finding that threading dislocations of the crystalline substrate cannot penetrate through an amorphous mask. This fact is used to selectively grow a layer in a "seed" window of a mask, and then overgrow the amorphous mask by a lateral extension of the seed region as illustrated in Fig. 12.2c. This leads to a dislocated layer above the opening of the mask, and a layer with a low dislocation density (reduced by orders of magnitude) above the mask—the wing regions. On these wings devices which are sensitive to threading dislocations can be fabricated.

The number of dislocations propagating through the seed region of the layer decreases as the width of the window in the mask decreases. A layer basically free of dislocations is hence expected if the window width in both lateral directions is significantly smaller than the average distance between dislocations in the substrate; a linear dislocation density of 10^5 cm^{-1} (corresponding to 10^{10} cm^{-2} areal density) then requires a window width well below 100 nm. For practical growth the width is generally larger and on the order of several micrometers.

Two requirements for ELO must be met by the epitaxial process: the lateral growth rate must be much larger than the vertical rate, and growth conditions must prevent nucleation on the mask material. These conditions were well achieved with MOVPE, LPE, and chloride or hydride VPE for many materials, such as Si [22], GaN [23], GaAs [24–26], and InP [27].

The equilibrium-near LPE is particularly well suited for epitaxial lateral overgrowth, and much early work was performed using this technique; a review on ELO using LPE is given in [25]. The ELO technique is of particular importance for the technology of GaN growth. Native substrates of GaN are exceedingly difficult to grow due to the high vapor pressure of nitrogen at melting point, and all devices are fabricated to date on foreign substrates (sapphire Al_2O_3, Si, or 6H SiC); structural and thermal mismatch then lead to a density of threading dislocations on the order of 10^{10} cm^{-2}. The seminal breakthrough for GaN-based lasers was achieved by dislocation blocking applying ELO, and growing devices on the wing regions [28, 29]. LPE is unfortunately not well applicable for GaN ELO due to the low solubility of nitrogen in liquid metals, and hydride VPE or MOVPE are used instead.

Epitaxial Lateral Overgrowth of GaN

ELO of GaN is performed using hydride VPE [30] or MOVPE [31, 32]. Any heteroepitaxy of GaN requires a series of process steps, starting with a nitridation of the substrate and (or) the deposition of a nucleation layer deposited at low temperature. There are two different approaches for the nucleation employing either a 2D or a 3D growth mode, and both nucleation layers need annealing at a high temperature (typ. 1050 °C) also used for subsequent MOVPE; details are reviewed in [23]. The 2D nucleation leads to GaN with typ. 10^{10} cm^{-2} dislocation density, while a density down to the 10^8 cm^{-2} range was achieved by 3D nucleation assisted by a Si/N treatment [33]. The reduced dislocation density is reflected in a strong increase of PL intensity, indicating reduced non-radiative recombination of carriers. Improvement of GaN-based devices such as lasers and LEDs require a further reduction of the dislocation density, which can be achieved by the ELO technique.

The ELO technique is usually applied to growth on basal-plane substrates, particularly to *c*-plane Al_2O_3. The examples outlined below refer to these structures. ELO was also successfully applied to growth along nonpolar and semipolar orientations, with stripe directions along the intersection of the basal plane with the growth plane; a review is given in [34].

One-step ELO The ELO process is usually performed on a thick (several μm) GaN buffer layer on either *c*-plane Al_2O_3 or 6H SiC, or on Si(111) substrate. The procedure implies either one or two steps as illustrated in Fig. 12.11. In one-step ELO parallel stripes of the dielectric mask are deposited and patterned using (PE)CVD and optical lithography, and regrowth starting in the window area eventually leads to lateral overgrowth and coalescence of adjacent ridges.

The epitaxial relation of the GaN buffer layer and a hexagonal Al_2O_3 substrate is given by $[0001]_{Al_2O_3} \| [0001]_{GaN}$ and $[10\bar{1}0]_{Al_2O_3} \| [11\bar{2}0]_{GaN}$. SiO_2 or SiN_x (typ. ≤ 0.1 μm thick) are used as mask dielectric, and typically 1–3 μm wide windows

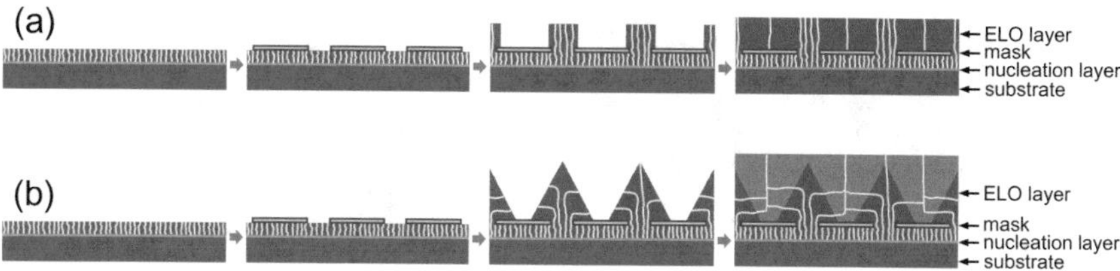

Fig. 12.11 Schematic of the **a** one-step and **b** two-step ELO process of GaN. *Yellow lines* indicate dislocation lines

separated by 4–7 μm wide mask stripes are defined on GaN. The stripes are usually aligned along a $\langle 1\bar{1}00\rangle$ direction of GaN, yielding for lower temperature and higher total pressure MOVPE (950 °C, 400 mbar) triangular-shaped ridges with (0001) top facet and inclined $\{11\bar{2}2\}$ side facets, and vertical $\{11\bar{2}0\}$ side facets at changed conditions (1000 °C, 100 mbar) [35]. After coalescence of the wing areas of adjacent seeds eventually a continuous GaN ELO layer with a smooth (0001) surface may be obtained.

In the final ELO layer three areas can be distinguished: the coherent area of the seed layer above the window, the wing area above the mask, and the coalescence area where wings of two adjacent seed layers meet—see Fig. 12.12a. The GaN buffer layer has a very high density of threading dislocations (basically pure edge and mixed edge/screw dislocations), which mostly propagate in [0001] direction. In the window region the dislocations propagate through the layer to the surface, keeping the high dislocation density. The wing region has a strongly reduced dislocation density; due to the creation of new dislocations this region is not entirely free of extended defects. The coalescence boundary in the middle of the wing region has again a high dislocation density, because the laterally growing facets converging in the middle of the window region are not in perfect registry: ELO GaN is composed of three materials with different thermal expansion (substrate, GaN, and mask) creating stress, and the GaN wings are usually slightly tilted. At coalescence dislocation half-loops and threading dislocations are generated.

In the coalescence area over the mask, the overgrown GaN does not adhere to the amorphous mask; this leads usually to the formation of a void between mask and the overgrown GaN. Such voids do not affect the quality of the final layer. The three regions of seed (above the window), wing (left L and right R), and coalescence area can be well distinguished in the TEM image shown in Fig. 12.13. The high dislocation density of the GaN seed layer between Al_2O_3 substrate and SiO_2 mask proceeds in the GaN layer between the SiO_2 stripes with a density of 10^8–10^9 cm^{-2}, while the GaN layer of the wings over the mask is almost free of dislocations; in the coalescence range where left and right wings meet a small void is observed above the mask [36].

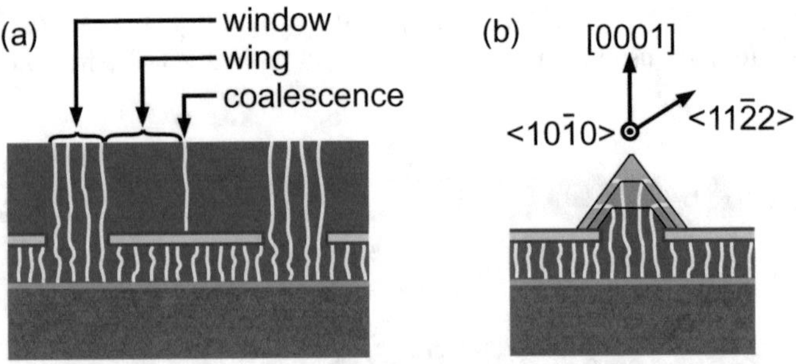

Fig. 12.12 **a** Regions of a one-step ELO GaN layer with different densities of threading dislocations. **b** Bending of dislocation lines in the first step of a two-step ELO process of GaN. Different *shading of the ridge* indicates subsequent stages of growth

Fig. 12.13 Transmission-electron micrograph of an ELO GaN/Al_2O_3(0001) layer grown with a one-step process. The *inset* at the bottom shows a small void at the coalescence front of the two meeting wings. Reproduced with permission from [36], © 2008 Springer Nature

Two-step ELO The ratio of lateral versus vertical growth rate in selective area growth depends on growth conditions. In the two-step ELO process the initial growth parameters favor vertical growth, but with inclined side walls of the ridges: the growth rate of the (0001) surface is significantly larger (by a factor of ~3) than that of the two side walls; the latter are inclined $(1\bar{2}12)$ and $(\bar{1}2\bar{1}2)$ facets for windows along the $[10\bar{1}0]$ direction of GaN; this yields a trapezoidal cross section of the ridge as indicated in Fig. 12.12b. When the fast growing (0001) top facet disappears the growth parameters are changed (step 2) to increase the lateral growth [37]. An increased lateral growth of GaN is obtained for appropriate facets at increased V/III ratio or increased temperature, or an N_2 admixture to the H_2 carrier gas or a surfactant such as Mg, or by other parameters. A schematic of the cross section of the ridge at the end of step 1 is illustrated in Fig. 12.12b.

The essential effect of the two-step ELO process is a bending of the dislocation lines. The first growth step is basically comparable to the one-step ELO process, reproducing above the seed window the high density of vertically threading dislocations of the GaN buffer beneath. When these dislocations meet the common edge of the inclined side walls and the (0001) top facet they bend by 90° as shown in Fig. 12.14, and continue within a basal plane of the lateral region grown at higher temperature. Thereby the dislocation density at the final GaN surface above the seed window is strongly reduced. The trenches between the triangular-shaped ridges are eventually filled by coalescence of neighboring, predominantly laterally growing ridges; the final GaN layer may be continuous with a smooth (0001) surface and a high dislocation density only at the coalescence boundaries.

Pendeo Epitaxy of GaN

Various approaches related to the ELO process were developed to further reduce the dislocation density of GaN layers. An interesting version is pendeo epitaxy (from Latin *to hang*) [39, 40]. The basic idea is avoiding growth through the windows on the (0001) surface; instead, a mask is used to pattern the GaN layer into deep ridges,

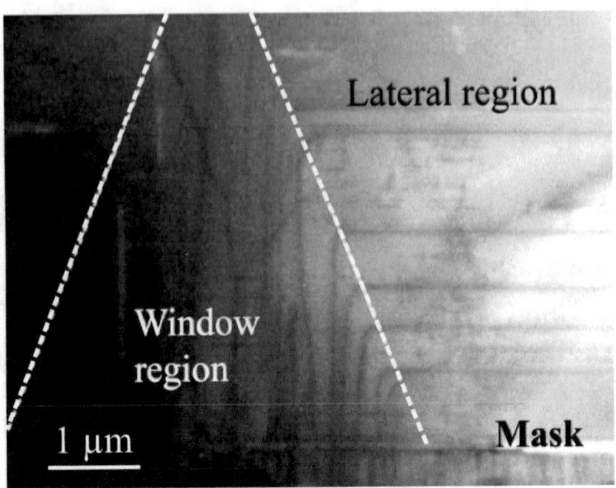

Fig. 12.14 Bright-field transmission-electron micrograph of an ELO GaN/Al_2O_3(0001) layer grown by MOVPE with a two-step process (cross section along [$10\bar{1}0$]). *Dark lines* are threading dislocations, the *dashed line* connects points where the dislocation lines bend. The *white region* at the mask originates from a void above the thin mask made of 3 nm SiN_x. Reproduced with permission from [38], © 2006 Elsevier

and subsequent growth begins laterally on the GaN *sidewalls* of the etched ridges. After coalescence of adjacent laterally growing (*hanging*) wings the layer grows vertically and eventually encloses the stripes of the mask as shown in Fig. 12.15a. The GaN(0001) pendeo layer was grown on a 500 nm GaN(0001) seed layer on a 100 nm AlN(0001) buffer layer deposited on a 5 µm thick cubic SiC(111) transition layer on Si(111), after defining a SiN_x stripe mask oriented along [$1\bar{1}00$] and etching ridges into the seed layer [41].

A drawback of the mask used in pendeo and ELO epitaxy is the introduction of thermally induced strain due to the thermal mismatch of the mask and the layer material. This strain leads to local tilt and the generation of dislocations in the layer [25, 40]. Since lateral growth of GaN(0001) at increased temperature is faster than vertical growth the mask is actually dispensable; in *maskless* pendeo epitaxy the mask is removed after etching the ridges into the seed layer, and after coalescence during regrowth a continuous layer with (0001) surface and no tilted regions forms. Unfortunately the threading dislocations blocked in masked pendeo epitaxy extend here into the subsequently grown layer, see Fig. 12.15b. The maskless GaN(0001) pendeo layer was grown on a GaN seed layer deposited on an AlN(0001) buffer layer on hexagonal 6H SiC(0001) substrate, with ridges along [$1\bar{1}00$]. The inset in Fig. 12.15b shows a high density of dislocations (10^9–10^{10} cm^{-2}) in the seed layer (below the dashed line), penetrating also the GaN layer above; the laterally grown pendeo layer contains a strongly reduced dislocation density of 10^4–10^5 cm^{-2} [39].

Epitaxy of GaN-based devices on patterned sapphire substrates (PSS) [42, 43] is of utmost importance for white LEDs (WLEDs) applied in solid-state lighting (SSL). WLEDs are presently generally replacing traditional incandescent and fluorescence lamps due to a superior luminous efficacy and device lifetime [44, 45].

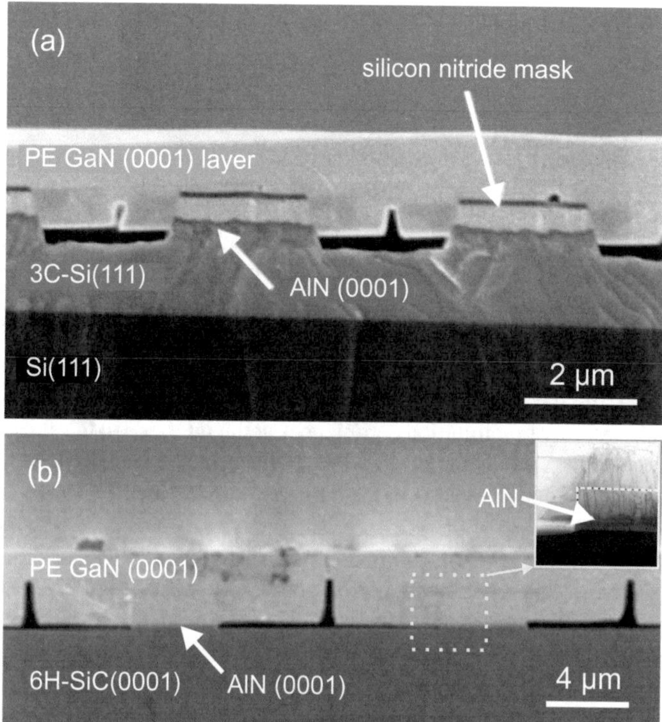

Fig. 12.15 Cross-sectional scanning electron micrograph of GaN grown using pendeo (PE) epitaxy (**a**) with SiN_x mask on AlN/SiC/Si(111) and (**b**) without mask on AlN/SiC(111). The *inset* in (**b**) shows a TEM micrograph corresponding to the *dotted area*. Reproduced respectively with permission from [39, 41], © 2000 and 1999 Springer Nature

In most WLEDs a GaN-based LED emitting in the blue is combined with a color-conversion layer consisting of an inorganic phosphor (an oxide or nitride) emitting in the yellow, usually cerium doped yttrium aluminum garnet (YAG:Ce^{3+})—sometimes complemented with a red-emitting phosphor [46, 47].

12.2 Vapor–Liquid–Solid Growth of Nanowires

One-dimensional (1D) semiconductors are interesting for devices exploiting ballistic transport, Coulomb blockade effects, and many-body phenomena. The fabrication of respective structures in the 10 nm scale faces technological challenges. Fluctuation of thickness or alloy composition leads to a large spread in electronic properties and usually to accidental zero-dimensional localization; this obstacle is very often encountered in epitaxial quantum wires pointed out in Sect. 7.3.3. The VLS growth offers the opportunity for reproducibly fabricating 1D semiconductors of high structural uniformity with high yield. Such nanowires comprise also structures with larger diameters in the μm range that are interesting for optoelectronic devices with small footprint.

12.2.1 Outline of the VLS Method

The *vapor–liquid–solid* (VLS) mechanism is an approach for the self-organized creation of 1D nanowires (NWs) [48]. Usually a metal catalyst (such as gold) is employed; at high temperature the catalyst forms liquid alloy droplets by absorbing gaseous components of the material to be grown as illustrated in Fig. 12.16. The continuous inclusion of the semiconductor material into the droplet leads to a super-saturation; the solute components then precipitate at the liquid–solid interface, which acts as a sink. Initially precipitation commences at the interface to the substrate; once nucleated, the solid semiconductor grows as a 1D whisker with lengths up to several tens of micrometers and diameters in the nanometer to micrometer range. The liquid droplet remains at the top of the growing nanowire.

The gaseous semiconductor components can be provided by various methods. Precursor generation by chemical vapor deposition (CVD or MOCVD/MOVPE) is most widely applied, but also molecular beam epitaxy (MBE) or pulsed laser ablation (PLA) are used. Semiconductor nanowires grown by the VLS method were fabricated for group IV (Si, SiC), III–V, and II–VI semiconductors. For Si the precursors are usually provided by CVD, while for compound semiconductors MOVPE, MBE, or PLA are more popular. Reviews are given in [49–51]. Details of the growth process of Si and GaAs-based NWs are discussed in Sects. 12.2.2 and 12.2.3.

The driving force for NW growth originates from the differences in chemical potentials of the phases. During steady-state VLS growth the chemical potentials of vapor, liquid, and solid are related by $\mu_{\text{vapor}} > \mu_{\text{liquid}} > \mu_{\text{solid}}$, see Fig. 12.17. Since a series of phase transitions occurs within a very small reaction space, also shape factors and non-equilibrium conditions may have a significant effect. Furthermore, μ_{liquid} is usually a complicated quantity due to interactions between the catalyst and reaction species.

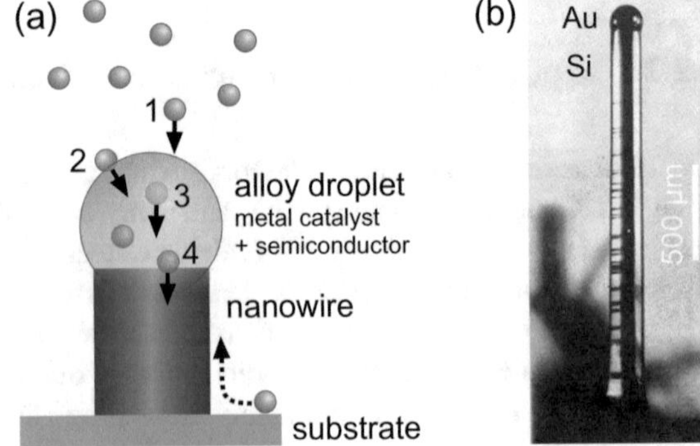

Fig. 12.16 **a** Successive steps in the catalyzed VLS mechanism applied to create a nanowire. **b** Si whisker with solidified Au–Si alloy catalyst on top, grown by the VLS mechanism on a {111} Si substrate; reproduced from [48], © 1964 AIP

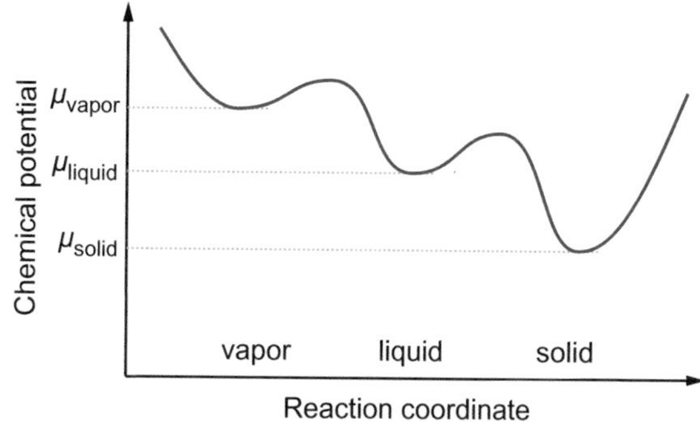

Fig. 12.17 Schematic of the chemical potential in vapor–liquid–solid phase transitions

Nanowire growth proceeds at *two competing interfaces*, the liquid/solid interface at the NW top and the gas/solid interface at the NW side walls. Precipitation at the liquid/solid interface leads to growth along the NW axis; such an axial VLS growth is usually desired to obtain a large length/diameter aspect ratio. Adsorption of gaseous components at the gas/solid interface leads to growth along the radial direction; this radial growth usually implies a dissociation of semiconductor reactants at the NW side wall. The ratio of the competing two growth rates is controlled by growth conditions such as temperature and partial pressures of gaseous species. For most applications growth of NWs with uniform diameters is intended. Tapered NWs can be fabricated by simultaneous growth in axial *and* radial directions.

Nanowire *heterostructures* can be fabricated by changing the composition of the gas phase during growth. Both radial (*core/shell*) and axial heterostructures (*quantum wells, quantum dots*) can be grown depending on the dominating growth direction. In addition branched NW structures (*nanotrees*) can be produced by interrupting NW growth, reintroducing metal-catalyst nanoclusters, and resuming NW growth.

Axial heterostructures are formed by a change of composition or doping along the wire axis. The elastic lateral strain relaxation at the free surfaces of axial heterostructures is very efficient. This leads to a strong increase of the critical layer thickness for lattice-mismatched heterostructures with layers stacked along the NW axis. Also *radial heterostructures* with interfaces parallel to the wire axis can be formed: after completing the growth of a nanowire core the growth conditions are altered to deposit a shell material. Multiple shell structures are produced by subsequent introduction of different materials.

Elastic strain relaxation via lateral relief in nanowires allows for much thicker coherent lattice-mismatched heterostructures than their counterparts in uniform tick layers [52]. Modelling the NW as a circular cylinder with a misfit layer on top in the framework of isotropic elasticity, the critical thickness t_c of the top layer can be expressed in terms of the NW radius [53]. The result is shown in Fig. 12.18 for plastic relaxation beyond t_c by 60° dislocations with $|\mathbf{b}| = 0.4$ nm Burgers vector and $\nu = 1/3$ Poisson ratio; the misfit f is defined here positive for laterally compressed layers according to (2.20b). The curves in Fig. 12.18a represent the boundary for coherent growth (to the left of the respective curve, i.e., at smaller radii) for given

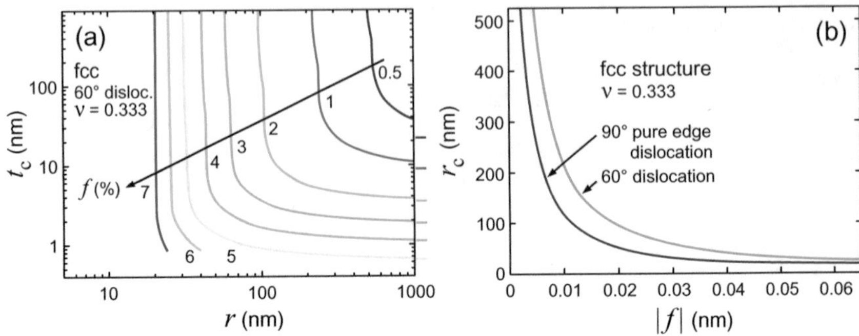

Fig. 12.18 **a** Calculated critical thickness for a layer with misfit f on top of a cylindrical nanowire with radius r. Face-centered cubic crystal structures and strain-reliving 60° dislocations are assumed. *Horizontal bars* at the right indicate curve asymptotes for $r \to \infty$. **b** Critical radius r_c below which a misfit top layer may grow coherently for infinite thickness; the limit depends on the strain-relaxing misfit dislocation as indicated at the curves. Reproduced with permission from [53], © 2006 APS

misfits as a parameter. We note that for a given misfit f the critical thickness for maximum coherent growth tends to become infinite below some critical radius of the nanowire. The dependence of this critical radius r_c on the absolute value of the misfit in given in Fig. 12.18b. We see that nanowires with very small radii allow for growth of thick layers even when a substantial misfit exists. Comparable results are obtained for a different equilibrium model based on the Matthews critical thickness [54] or an energy balance with finite elements [55].

12.2.2 Growth of Si Nanowires

The VLS mechanism was first proposed for whisker growth of Si seeded by Au particles [48]. The whiskers were grown by CVD using small gold particles placed on a Si(111) substrate, and $SiCl_4$ precursors in H_2 ambient. The VLS process can be described in the phase diagram of Au and Si given in Fig. 12.19. The binary eutectic Au–Si system has continuous solubility only in the liquid phase; in the solid phase the solubility is limited to less than 2% Si in Au and likewise for Au in Si. The phase diagram shows the eutectic melting point at 18.6 atomic % Si in Au at a temperature of 363 °C, much below the melting points of the pure elements. Above this temperature a liquid Au–Si alloy with a composition depending on the temperature and the total composition x_{Si} forms. Below 363 °C both Au and Si precipitate (get solid) with trace amounts of the respective other material incorporated.

Si in the vapor phase for VLS growth is usually supplied by CVD using either halide ($SiCl_4$) or the less stable hydride (SiH_4, Si_2H_6) precursors. *CVD at high temperature* (800 °C) using $SiCl_4$ leads to equilibrium-near growth well described by thermodynamics [56]. The conditions allow for high growth rates on the order of 1–10 μm/h with an Arrhenius-like temperature dependence, described by an activation

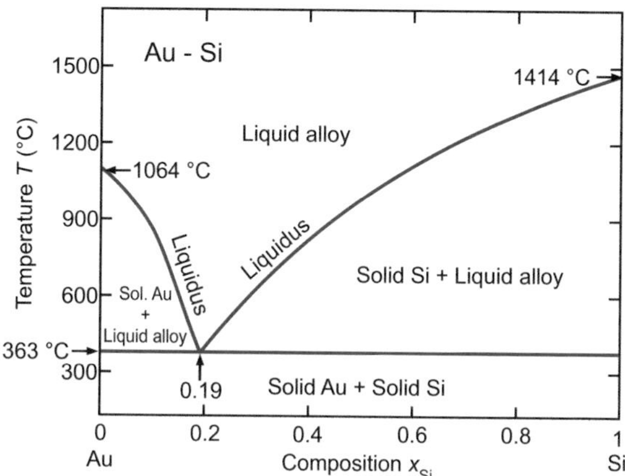

Fig. 12.19 Phase diagram of the binary Au–Si system

energy of 32 kcal/mol. The high temperature enables also the use of various catalyst materials, see Table 12.1 [57]. Au and Ag have a moderate solubility of Si in the liquid phase and are hence conductive to VLS growth. There are many other catalysts such as Cu, Ni, or Pd; they have high eutectic temperatures, a high solubility of Si in the liquid, and form several silicide phases, leading hence to complex phase diagrams. Such catalysts were also used for vapor–solid–solid (VSS) growth employing a solid silicide particle. *CVD at low temperature* (350–700 °C) is performed at low-pressure conditions using hydrides. The reduced metal diffusion at the decreased temperature leads to smaller NW diameters. Growth is controlled by both thermodynamic and kinetic effects. Besides Au also other metals with low eutectic temperature are used; the metals listed in Table 12.1 have an eutectic temperature close to their melting temperature and hence a very low solubility of Si in the liquid phase.

The *diameter* of the Si NW is controlled by the size of the liquid droplet on the substrate and its contact angle [58]. The diameter is largest at the base and decreases over a short height before reaching a steady state. The inclination angle of the side wall and the balance of the forces discussed in Sect. 6.2.2 determine the equilibrium shape of the droplet, yielding a NW shape depending on the surface tensions of the

Table 12.1 Metal catalysts applied for Si nanowire growth

Element	Melting temperature (°C)	Eutectic temperature (°C)	Eutectic composition (at.% Si)
Au	1064.2	363	18.6
Ag	961.8	835	11.0
In	156.6	156.6	0.004
Ga	29.8	29.8	0.0
Sn	231.9	231.9	0.0
Pb	327.5	327.5	0.0

liquid droplet and the solid Si, and on the interface tension between liquid and solid [58].

The *growth rate* of the NW depends on the wire diameter. A pronounced effect occurs at small diameters where the growth rate is strongly reduced. This effect is assigned to the change of chemical potential at highly curved surfaces, expressed by the *Gibbs–Thomson equation*

$$\mu(r) = \mu_\infty + \frac{2\Omega\sigma}{r}. \tag{12.5}$$

Here $\mu(r)$ is the chemical potential of the vapor in equilibrium with a droplet of radius r and μ_∞ is this quantity for an equilibrium with a liquid of the same material but a flat surface (i.e., $r \to \infty$); Ω is the volume per atom and σ is the surface free energy of the droplet. At small droplet radius the chemical potential is hence increased and consequently the supersaturation of Si in the droplet is decreased, leading to a decreased growth rate. Based on the Gibbs-Thomson equation the growth rate v can be described by the dependence [59]

$$v(r) = b\left(\frac{\Delta\mu_0 - 2\Omega\sigma/r}{kT}\right)^n, \tag{12.6}$$

where b and n ($\cong$ 2) are fit parameters, and $\Delta\mu_0$ is the difference between the chemical potential of Si in the vapor phase and Si in the solid with a flat boundary. The dependence of the growth rate on the diameter of Si nanowires grown from $SiCl_4$ in H_2 ambient on Au-coated Si (111) substrate is shown in Fig. 12.20.

There is often also a (slight) decrease of growth rate observed for Si nanowires with *increased* diameter. Various models were developed to account for this effect,

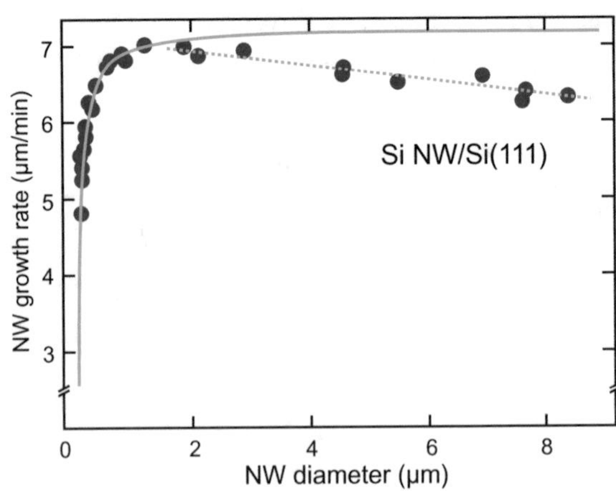

Fig. 12.20 Growth rate of Si nanowires seeded by Au particles on Si (111) substrate. The *solid line* is a fit to (12.6), the *dashed line* accounts for a delayed supersaturation in larger droplets. Reproduced with permission from [60], © 2011 Elsevier

considering limitations for incorporation and crystallization velocities, mass transport via diffusion along the NW side walls to the droplet, or delayed supersaturation in larger droplets [49].

The *growth direction* of Si NWs shows also some dependence on the diameter. The nanowires usually grow in $\langle 111 \rangle$ direction. The $\{111\}$ facets of Si have lowest surface energy, and growth with a liquid interface at this facet most favorable. The side walls of these nanowires are usually $\{112\}$ and $\{110\}$ faces, yielding a hexagonal cross section. Very *small nanowires* tend to grow along $\langle 110 \rangle$ direction; the cross-over radius for the change of growth direction occurs between 25 and 10 nm [61, 62]. These thin Si nanowires are typically terminated by two $\{100\}$ and four $\{111\}$ sidewall facets.

12.2.3 Growth of III–V Nanowires

VLS growth of nanowires made of compound semiconductors is more complex than that of Si NWs. The chemical potential of the liquid phase depends on both, cation and anion species with very different solubility [63], phase diagrams are more complicate and have high eutectic temperatures [57], and the structure of the solid may change between zincblende and wurtzite phase. The metal component of these semiconductors enables self-induced NW growth via a droplet of the native group-III liquid without the need for a foreign catalyst such as Au. We first address some general properties of III–V nanowires (which basically also apply for other compound semiconductors such as II–VI), and then focus on GaAs and GaN NWs.

Zincblende–Wurtzite Structure of Nanowires

Most semiconductors crystallize in the cubic zincblende structure; for III–V compounds these are the phosphides, arsenides, and antimonides, while nitrides prefer the hexagonal wurtzite structure, see Sect. 2.1 (Table 2.2). The anion-terminated $(111)B$ facets of zincblende semiconductors have lowest surface free energy; nanowires of these solids hence grow preferentially along the $\langle 111 \rangle B$ direction. The wurtzite nanowires grow along $\langle 0001 \rangle$ directions. Since usually vertically standing NWs are wanted, substrates of these orientations are used to prevent growth of slanted NWs.

The nanowires usually grow in the equilibrium structure of their solid bulk. However, the zincblende NWs often adopt the wurtzite phase or contain polytypes. Growth studies for GaAs NWs indicate that the wurtzite phase is favored at high supersaturation in the liquid droplet [64]. Formation of the wurtzite structure is also favored at small NW diameters; this effect indicates the importance of the surface/volume ratio. First-principles calculations of the dangling bonds at the side walls of small NWs indicate a lower energy of wurtzite phase, which has a smaller number of dangling bonds [65]. The study yields critical diameters for various III–V NWs, below which the wurtzite phase should prevail [66]. The computed values are, however, much too small to explain the experimentally observed data by solely this effect. The effects of both high supersaturation and energies at high surface/volume ratio were taken into account in a thermodynamic model; various strain-free III–V zincblende and polytype nanowires, grown along the $\langle 111\rangle B$ direction and bound by $\{110\}$ side faces for zincblende and along $\langle 0001\rangle$ direction with $\{1100\}$ or $\left\{11\bar{2}0\right\}$ side faces for wurtzite structure were considered. The study yields reasonable estimates for the radius, below which the zincblende structure of the NW becomes unfavorable [67]. This characteristic radius $r_{\mathrm{wz-zb}}$ is related to $\gamma_{\mathrm{zincblende}}^{\mathrm{side\ facet}}$ (the surface energy of side facets) and the difference $\Delta\mu_{\mathrm{wz-zb}} = \mu_{\mathrm{wurtzite}} - \mu_{\mathrm{zincblende}}$ of chemical potentials between the bulk wurtzite (wz) and zincblende (zb) phases by

$$r_{\mathrm{wz-zb}} = \frac{\gamma_{\mathrm{zincblende}}^{\mathrm{side\ facet}}}{\Delta\mu_{\mathrm{wz-zb}}}. \tag{12.7}$$

Values for the quantities in the fraction of (12.7) are fairly well known [66, 68]. Results of the characteristic radius $r_{\mathrm{wz-zb}}$ are listed in Table 12.2; at a NW radius $r \leq r_{\mathrm{wz-zb}}$ the hexagonal phase in the 4H polytype is expected to become prevalent, and at $r \leq r_{\mathrm{wz-zb}}/2r \leq r_{\mathrm{wz-zb}}/2$ the pure wurtzite phase [67]. Since kinetics effects are not included in the model, experimental values may deviate from the calculated data.

Growth of GaAs Nanowires

Growth of III–V NWs can be performed by applying metal catalysts similar to Si NW growth. Since the eutectic temperature of the GaAs pseudobinary system lies well above typical NW growth temperatures, no arsenic is expected to dissolve in the catalyst particle during growth; the mobility of atoms in small nanoparticles is high though, also evidenced by a decrease of melting point at particle diameters in the nm range. A corresponding fairly fast diffusion even in the solid [50] or along the catalyst/nanowire interface [69] suggests a possible NW growth by VSS (vapor–solid–solid) instead of VLS, proceeding in a similar manner.

We focus in the following on GaAs NWs grown on GaAs(111)B substrate. The interface at the *perimeter* of the base of the catalyst particle is a preferential site for nucleation. Surface diffusion of adatoms is an important process for material

Table 12.2 Calculated characteristic radius $r_{\mathrm{wz-zb}}$ for the transition from zincblende to the wurtzite phase of $\langle 111\rangle B$ oriented III–V nanowires according to (12.7). Data are from [67]

Semiconductor	(110) surface energy $\gamma_{\mathrm{zincblende}}^{\mathrm{side\ facet}}$ (J/m^2)	Characteristic radius $r_{\mathrm{wz-zb}}$ (nm)
InP	1.3	60.4
AlP	2.4	52.4
GaP	2	43.6
InAs	1	33.5
AlAs	1.8	30.9
GaAs	1.5	25.5
InSb	0.75	19.4
AlSb	1.3	24.3
GaSb	1.1	19.7

supply to the NW, in addition to direct attachment to the catalyst particle illustrated in Fig. 12.16 [70]. Kinetic processes also affect the crystal structure of the nanowire [71] and were accounted for in more recent simulations [72].

The *growth rate* of GaAs NWs may be a function of the diameter; however, results depend on growth conditions which may comprise competing processes. The nanowire is fed by various channels; in addition to direct impingement on the droplet, where precursor decomposition may have a strong effect, material transport from the substrate surface and side-facets are important. In addition, re-emission from these surfaces and subsequent impingement on the droplet occurs. In the early stages of NW growth, transport from the substrate within the diffusion length dominates; close standing nanowires with overlapping collect area compete for material and grow more slowly. The growth rate decreases with time and becomes constant, when the wire length exceeds the diffusion length of the farthest diffusing species (generally the group-III atoms). The growth rate then depends on the material impingement rate, and inversely on the diameter of the growing NW. Usually a relation similar to that shown for Si nanowires in Fig. 12.20 is obtained. Nanowires with a smaller diameter are hence longer than thicker ones, except for very thin NWs, where the Gibbs–Thomson effects leads to a strong decrease of the growth rate. Models for describing the processes are reported in [70, 73, 74].

The Ga component allows for *self-catalyzed NW growth*, i.e., using a Ga-rich droplet and no foreign catalyst. Such a III–V system is expected to be thermodynamically unstable, with the liquid phase tending to grow or shrink without bound; however, transport kinetics lead to conditions where the wire and droplet will both evolve toward a steady state [75]. Self-catalyzed VLS growth benefits from the thermodynamic properties of GaAs well known from LPE studies. Arsenic has a very low solubility in the liquid phase at typical NW growth temperature (on the order of 1%). The small As concentration controls the growth rate of GaAs NWs, in contrast to findings in the cation-limited epitaxy of layers; MBE studies proved that the growth rate depends linearly on the As beam flux, largely independent of the Ga flux [76, 77]. In situ TEM studies show that growth of GaAs NWs in the zincblende phase proceeds in a layer-by-layer mode by the successive addition of Ga–As bilayers, starting with a small 2D nucleus at the solid-liquid interface; the nucleus edge then extends quickly across the growth interface, and the nucleation of the next layer is slightly delayed [78]. By contrast, growth in the wurtzite phase proceeds by step flow across the droplet/nanowire interface, with each step flow representing the addition of one GaAs(0001) bilayer.

The numerous detailed data on self-catalyzed growth of zincblende GaAs NW enable a quantitative *modelling* of the process [79]. The model outlined here considers feeding of the droplet by direct impingement from an MBE As_4 beam and by As species re-emitted from the surface, and As-related emptying of the droplet by nucleation and evaporation. According to the classical nucleation theory the nucleation rate J_n reads (Sect. 6.2.5)

$$J_n = A(T) \times c_{\mathrm{As}} \left(\frac{\Delta\mu}{kT} \right)^{1/2} \times \exp\left(\frac{-\Delta G^*}{kT} \right), \tag{12.8}$$

where the prefactor $A(T)$ solely depends on temperature. $\Delta\mu(c_{As}, T)$ is the difference of the chemical potential per III–V pair between the Ga liquid with As composition c_{As} and the solid NW, and ΔG^* is the nucleation barrier given by

$$\Delta G^* = \chi \frac{a^4}{4\sqrt{3}} \frac{\gamma}{\Delta\mu} \tag{12.9}$$

for $\langle 111 \rangle$-oriented growth of zincblende GaAs with lattice parameter a; γ is the surface energy of the vertical edge of the 2D nuclei appearing during growth, and the shape factor χ equals π for circular and $3\sqrt{3}$ for triangle-shaped nuclei. The quantities of (12.8) and (12.9) lead to the growth rate

$$gr = \pi r^2 h J_n, \tag{12.10}$$

with the height $h = a/\sqrt{3}$ of one Ga–As monolayer and the radius r of the nanowire [79]. Putting the surface energy $\gamma = 0.123\,\mathrm{J/m^2}$ and the (over a wide range only slowly varying) prefactor $A(T) = A = 7.01 \times 10^{18}\,\mathrm{m^2/s}$ from fits to experimental data, the MBE growth of $\langle 111 \rangle$ oriented zincblende GaAs nanowires can be described. Results of the model with quantities difficult to access experimentally are given in Fig. 12.21.

Self-induced GaAs nanowires can also be grown in a *noncatalytic regime* by selective area epitaxy on SiO_2-masked Si(111) using MOVPE [80] or MBE [81]. This non-VLS growth appears for effective As/Ga ratio above unity and shows immediate onset of NW growth after source switching, which may be beneficial for obtaining sharp interfaces in NW heterostructures.

Growth of Nanowire Heterostructures

Axial heterostructures can be fabricated using the VLS mechanism by changing the vapor composition during NW growth; a review is given in [82]. There are basically two obstacles encountered concerning material combinations and interface sharpness. Due to generally different surface energies of two materials *A* and *B* (such as InAs and GaAs) only one of the two structures *A/B* or *B/A* is energetically favored; the respective considerations for layer epitaxy in Sects. 9.1.1 and 6.2.3 also apply to nanowire growth [83]. The favored sequence (e.g., *A* on *B*) then yields straight nanowires, while the wires with the invers sequence tend to grow with a kink along the axial growth direction [84]. By selecting appropriate growth conditions such a kinked or even downward growth may be suppressed kinetically to enable the fabrication of double heterostructures. Materials with similar surface energies such as GaAs and GaP or InAs and InP do not suffer from this obstacle and tend to grow straight. The other restraint refers to the delay occurring in VLS NW growth after changing the composition in the gas phase; the time required to obtain supersaturation with the altered composition in the catalyst leads to a gradual composition change along the NW axis. Structures for size quantization with short sequences (typically below 10 nm) may hence require a dedicated study [85]. The reservoir effect is less distinct when elements with a low solubility in the catalyst are switched, such as group-V

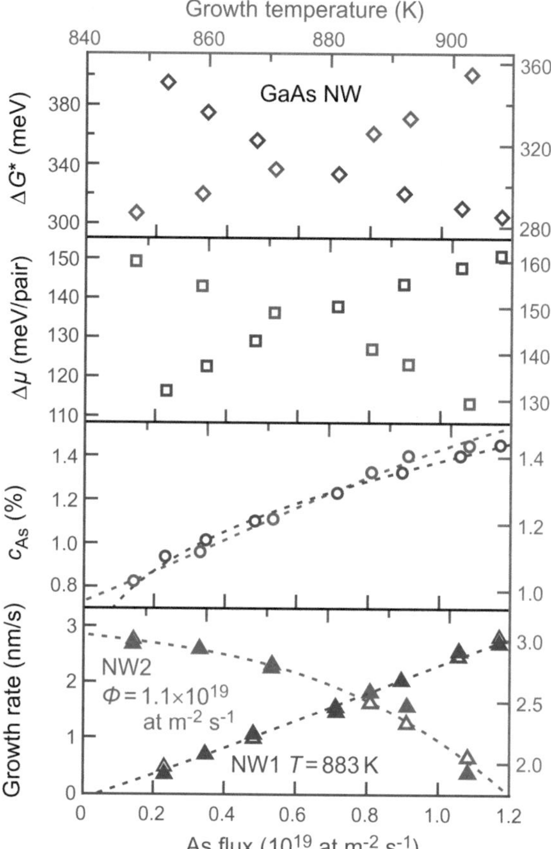

Fig. 12.21 Measured (*solid symbols*) and calculated (*open symbols*) data of self-induced ⟨111⟩ oriented growth of two GaAs nanowires. NW1 with 32 nm radius is grown at a fixed temperature of 883 K with varied As flux (*blue symbols*), NW2 with 37 nm radius is grown at a fixed As flux with varied growth temperature (*red symbols*). Reproduced with permission from [79], © 2013 APS

elements like As and P. Methods for the more difficult switching group-III elements are discussed in [77].

Core–shell heterostructures are grown after VLS growth of the homogeneous core nanowire by altering growth conditions to favor adsorption, precursor decomposition, and adatom incorporation at the NW side walls. The NW core is generally grown at a low temperature where growth on the uncatalyzed surfaces is kinetically hindered. For growth of the shell material the temperature is usually increased to remove the kinetic barriers, and the shell material is deposited at suitable partial pressure. The non-planar {110} side facets of III–V semiconductors have no surface reconstruction, and group V species desorb quite readily; lateral growth is hence promoted at increased V/III ratio [86].

The radial shell leads to a substantial improvement of electronic properties; surface states reduce the carrier lifetime and degrade the device performance due to the large surface-to-volume ratio of narrow nanowires. A shell of large-bandgap material around the nanowire confines the carriers to the core or to radial quantum wells so that they do not interact with the surface states. In electronic devices surface recombinations are hence reduced, and in photonics devices the emission efficiency is increased.

The orientation of NW side facets depends on the growth procedure. Zincblende NWs grown with Au catalyst have three $\{112\}A$ and three $\{112\}B$ faces. $\{112\}B$ facets grow slower at high V/III ratio; the $\{112\}A$ surfaces hence get smaller (compare Fig. 6.18), yielding a truncated triangular shape of the NW cross section. At low V/III ratio the conditions are reversed [87]. Self-catalyzed zincblende NWs have six $\{110\}$ side facets, and wurtzite NWs have either all $\left\{11\bar{2}0\right\}$ or all $\left\{1\bar{1}00\right\}$ side facets [88]; these non-polar facets usually yield a regular prismatic cross section of the nanowire.

The composition of ternary shell materials is usually not uniform. The corners of the side facets have an increased surface chemical potential due to the large curvature (Gibbs-Thomson effect), driving adatoms away [89]. Elements with larger bonding energy and consequently shorter diffusion length (such as Al with respect to Ga in AlGaAs) hence tend to accumulate at the corners. Figure 12.22 illustrates the enhancement of the Al concentration from 0.34 to 0.62 at the corner of the AlGaAs shell in a GaAs/AlGaAs nanowire. Similar effects are observed for epitaxy on non-planar surfaces, see Sect. 7.3.3.

Elastic strain relaxation in core–shell NWs is shared in axial and radial directions with a usually somewhat larger part in the radial directions, and a considerable part is relieved by the core; below some critical core radius there is even no limit for the layer thickness of coherent shell growth and no formation of defects expected [91]. Above this critical core thickness the limit for coherent shell growth is much larger than for 2D layer growth; e.g., for $In_{0.2}Ga_{0.8}As/GaAs$ with 2% misfit the 2D limit of about 10 nm is extended in 600 nm core–160 nm shell structures to this shell thickness [92].

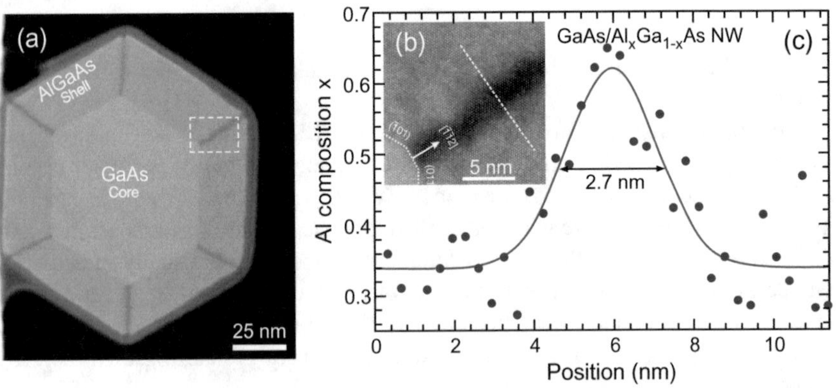

Fig. 12.22 **a** Cross-section scanning transmission-electron micrograph of a $GaAs/Al_{0.3}Ga_{0.7}As$ core–shell NW with Al-rich stripes (*dark contrast*) near the corner within the AlGaAs shell. **b** Detail of the region marked by the *dashed square* in (**a**). **c** Al concentration measured along the *dashed line* indicated in (**b**) with a Gaussian function fitted to the data. Reproduced with permission from [90], © 2013 ACS

Growth of GaN Nanowires

GaN nanowires are mainly grown by MBE and MOVPE on c-plane sapphire, on Si(111), and also on SiC(111) substrates. Both catalyst-assisted and catalyst-free growth are achieved. Au [93], Ni [94], and other metals are employed as catalysts. More recently self-catalyzed growth with Ga as a catalyst or spontaneous self-induced growth without any catalyst are widely applied.

Growth of GaN nanowires in wurtzite structure proceeds usually either along the Ga-polar [0001] ($+c$) direction or the N-polar $[000\bar{1}]$ ($-c$) direction; the polarity depends sensitively on the nucleation process outlined in more detail below. The sidewalls of the hexagonal wires are non-polar $\{1\bar{1}00\}$ m-planes. Usually the top facet of N-polar wires is a flat $(000\bar{1})$ facet [95], while that of Ga-polar wires is a hexagonal pyramid with $\{1\bar{1}02\}$ top facets [96]; the top facet can be reversibly altered to semipolar facets by a thermal treatment [97]. The morphology observed in GaN nanowire growth complies with the equilibrium crystal shape calculated for wurtzite crystals by an extended Wulff construction [98], indicating a significant thermodynamic influence on axial growth. The defect density in GaN nanowires is typically very low; most nanowires are fabricated applying some mask template, where a dislocation filtering occurs and threading dislocations are bend towards the sidewall facets [99, 100].

Self-catalyzed VLS growth of GaN nanowires may be achieved without applying a mask, and wire structures with good optical properties were reported [20, 101]. Since the random nucleation process of this approach leads to a distribution of morphology, sizes, and positions, selective area growth with lithographically defined mask openings are more widely applied.

Self-induced nanowire growth without self-catalyzed Ga droplets was found for III-nitrides under N-rich growth conditions and is used by many groups [102–104]. We consider the growth process of GaN nanowires in more detail. Growth conditions are typically N-rich, yielding an effective V/III ratio above unity, and the growth temperature is high (depending on the V/III ratio and growth method): in plasma-assisted MBE below ~780 °C continuous layers form and above ~830 °C no growth occurs; in MOVPE higher temperatures (~1000 °C) are used and higher growth rates are achieved.

The self-induced GaN nanowire formation proceeds by two distinct phases: nucleation and wire growth. Details of these processes depend strongly on substrates and growth conditions; still some general conclusions can be drawn. The nanowires are mostly grown on an AlN buffer layer or a on a Si_xN_y interlayer. *Nucleation* of GaN starts with an incubation period lasting in the range of seconds to minutes after supplying Ga and N species [105–107]; during this time the formation of stable nuclei is delayed by a high Ga desorption rate. In situ studies showed that initially metastable GaN clusters appear, which eventually lead to stable nuclei. Such a nucleus forming on an AlN/Si(111) buffer layer during plasma-assisted MBE is shown in Fig. 12.23a [108]. When growth proceeds the density of nuclei increases and they change their shape for elastically relieving some lattice-mismatch strain, eventually leading to

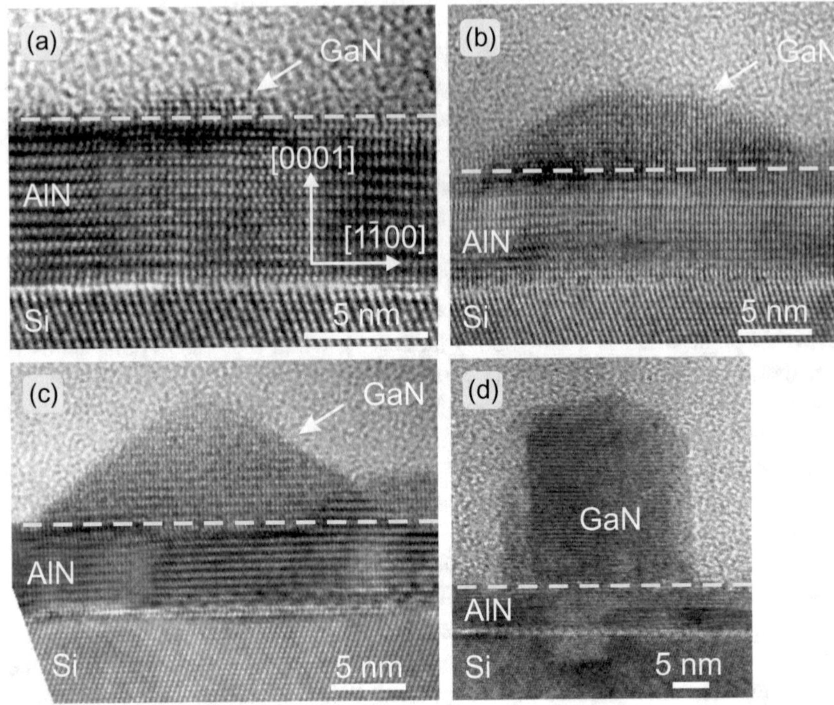

Fig. 12.23 High-resolution transmission-electron micrographs of different stages of GaN nanowire nucleation on AlN/Si(111). **a** GaN island formed at the onset of nucleation, **b** and **c** facetted coherent islands consecutively formed at later nucleation stages, **d** plastically relaxed hexahedral nanowire at beginning of axial growth. The *dashed lines* mark the surface of the AlN buffer layer. Reproduced with permission from [108], © 2010 APS

coherent islands with full pyramidal shape (panels *b* and *c*). Subsequently misfit dislocations are introduced at the edges to fully relax the strain, and the islands assume the form of hexahedral nanowires with *m*-plane side facets as shown in Fig. 12.23d.

Growth of GaN nanowires proceeds after nucleation with a diameter given by that of the full pyramidal islands after plastic strain relaxation [108]. It is generally accepted that GaN nanowires on Si_xN_y interlayers grow along the N-polar [000$\bar{1}$] direction, while NW polarity for growth on AlN buffers depends on the AlN polarity and on the surface morphology [6, 109]. In the beginning of *axial growth* the limiting Ga particles are supplied by both direct impingement and adatom diffusion from the substrate along the side walls; since the diffusion length on the *m*-plane side facets is quite small (~50 nm), the latter contribution ceases for longer wires [110]. The length of GaN nanowires is found to be invers to the diameter for a given growth time [111]. *Radial growth* occurs simultaneously to axial growth; axial growth can substantially be promoted by adding Si during growth [112].

Core–shell nanowires of nitride semiconductors are shown in Fig. 12.24, where five InGaN/GaN quantum wells were radially grown using MOVPE on the $\{1\bar{1}00\}$

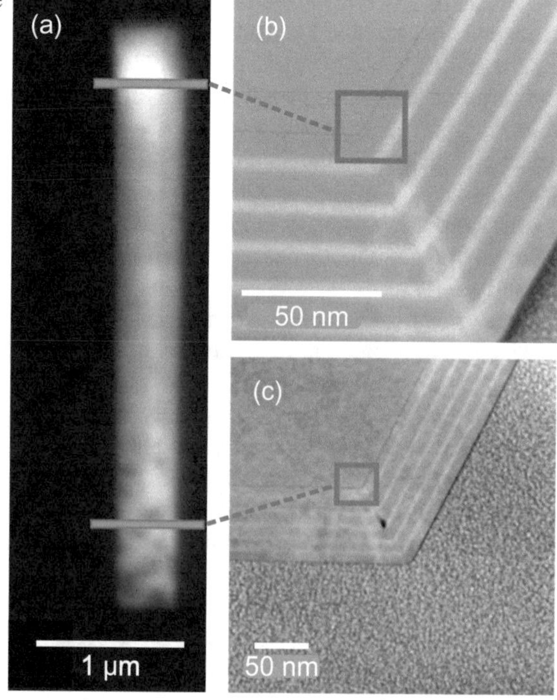

Fig. 12.24 Cathodoluminescence **a** and scanning-TEM cross-sectional images of the upper **b** and lower **c** part of a five-period InGaN/GaN core–shell nanowire grown on *c*-plane sapphire substrate. *Bright and dark areas* in the TEM images correspond to InGaN quantum wells and GaN barriers, respectively. Reproduced with permission from [113], © 2018 AIP

side facets of a nanowire [113]. The axial growth rate of the GaN core was enhanced by adding silane to the vapor phase, and the GaN barriers and InGaN wells were grown at different temperatures using N_2 carrier gas. Sample analysis showed a yet not optimum QW homogeneity; the QW width and In composition increase from the bottom to the top of the nanowire, leading to a red-shift of the detected cathodoluminescence along the wire axis (not shown in Fig. 12.24a).

12.3 Atomic Layer Epitaxy and Related Techniques

Atomic layer epitaxy (ALE) and related deposition methods are vapor-phase growth techniques with a pulsed supply of the gaseous precursors to the substrate. The techniques are suitable particularly for the growth of compounds such as III–V or I–V–VI_2 semiconductors, or metal oxides. Basic advantages of this techniques are a precise control of the layer thickness on the monolayer scale, sharp interfaces, and a very good uniformity of layers on large substrates; for polycrystalline or amorphous films excellent conformality for nonplanar surfaces comprising trenches or pores, good reproducibility, and a high quality at low deposition temperature are achieved. Completely smooth films are, however, only observed in epitaxial or amorphous growth; in polycrystalline growth nucleation stages similar to those of other deposition methods occur and lead to some roughening [114].

The growth process proceeds by cycles. During one cycle precursors of cations and those of anions are supplied in an alternating sequence. Thereby exactly one complete molecular layer is grown per cycle. In the ideal case the growth rate is hence proportional to the number of cycles, and not given by the material flux and growth time as in common deposition methods. The sequence of steps in one cycle is illustrated schematically in Fig. 12.25. A prerequisite of the technique is a self-saturating nature of each deposition, which leaves no more than one monolayer on the surface.

Precursors used in ALE or the more general atomic layer deposition (ALD) should be reactive, in contrast to requirements for sources used in CVD. A high reactivity yields a low threshold energy for surface reactions and a good monolayer coverage of the surface. No gas-phase reactions between different precursors occur, because they are supplied subsequently. After each step of material supply the surplus precursors and reaction products are removed from the reaction zone by a purge step.

Also metals or covalent semiconductors such as Si can be grown using ALE or ALD. In this case a compound precursor is required. A saturated monolayer of this reactant on the surface is then reduced to a monolayer of atoms of the covalent material or metal to be grown before the next reaction step; the reduction can be performed chemically with a reducing surface reaction or with a pulse of extra energy (e.g., by photons) to the surface [115].

The technique of atomic layer growth was introduced for the epitaxy of II–VI semiconductors and successfully applied to III–V epitaxy, mainly using MOVPE or halogen-based VPE [1]; it was named atomic layer epitaxy (ALE). In MBE the method is basically applied in the form of migration-enhanced epitaxy (MEE) [116]. In the early 2000s the growth technique was extended to the deposition of polycrystalline or amorphous materials such as insulating high-oxide dielectrics in CMOS circuits and then referred to as atomic layer deposition (ALD) [117, 118]. Today fabrication of precisely controlled deposition of amorphous and polycrystalline films is the main application. In the following the methods of ALE, MEE, and ALD are outlined.

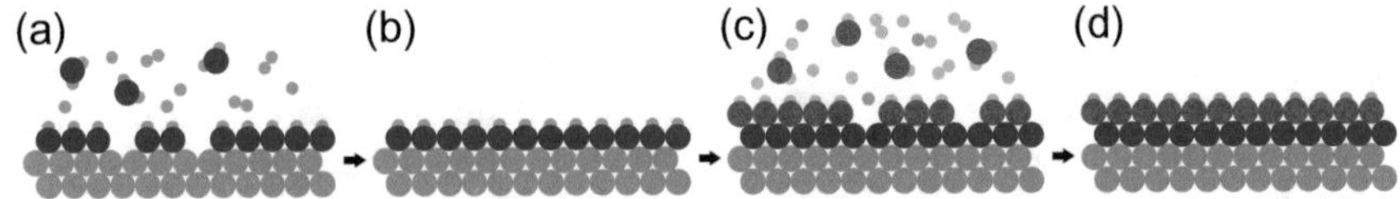

Fig. 12.25 Schematic of steps during one cycle in an ALE or ALD process. *Blue and red circles* indicate anions and cations, respectively, *light blue and red dots* symbolize ligands; the substrate is drawn in *gray*. **a** Supply of anion precursors and monolayer deposition on the surface, **b** removal of surplus precursors and reaction products, **c** and **d** similar as (**a**) and (**b**) with cation precursors

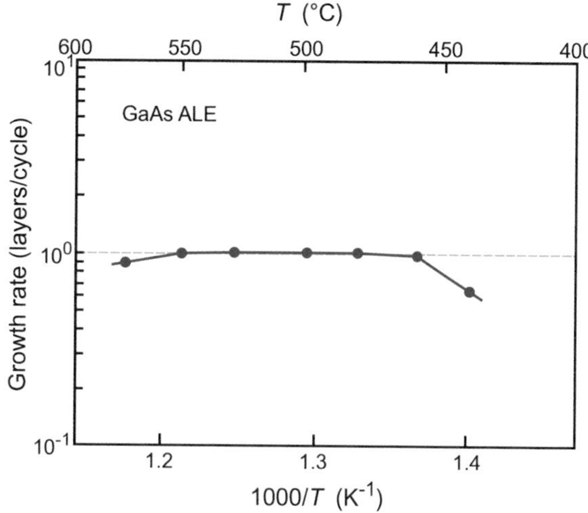

Fig. 12.26 Growth rate in units per cycle at varied temperature in the ALE of GaAs grown by MOVPE using TMGa and AsH_3 precursors; data are from [121]

12.3.1 *Atomic Layer Epitaxy (ALE)*

Atomic layer epitaxy of III–V arsenides and phosphides using MOVPE and halide VPE was widely investigated in the late 1980s [119]. A basic feature of the process is the temperature range where the criteria for ALE growth are met, the so-called processing window. This quantity denotes the amount of deposited material per cycle (*growth per cycle*, GPC, also referred to as *growth rate*). The temperature dependence of ALE (and ALD) is generally much lower than that of CVD, and the growth per cycle is ideally constant in the processing window. It proved difficult to find such a window for the ALE of GaAs; generally no saturation was observed with standard conditions, but using photoexcitation with above-bandgap energy at least a small processing window was found [120]. A wide process window in MOVPE was achieved by a short precursor injection followed by an effective purge with a fast hydrogen gas stream at low total pressure, see Fig. 12.26; this ensured a decomposition of the Ga precursor only at the substrate surface and not in the gas phase [121].

Good results were reported for the epitaxial growth of layers and superlattices applying ALE; there were, however, no significant benefits of ALE compared to well optimized epitaxy applying conventional MOVPE or MBE. The interest in ALE hence ceased in the 1990s.

12.3.2 *Migration-Enhanced Epitaxy (MEE)*

In MBE the application of ALE is affected by the poor desorption of elemental cations with low vapor pressure from the surface; this impedes a straightforward deposition of exactly one atomic layer. ALE can still be performed by a precise dose of cations to cover a complete monolayer, but the self-limiting saturation which enables an uncritical supply in MOVPE is lost [122].

The alternating supply of species in the MBE of compound semiconductors has still a very beneficial effect, as studied in detail for the epitaxy of GaAs and AlGaAs [116]. Ga atoms have a high sticking coefficient near unity, but a high mobility on the GaAs surface. The lifetime of a migrating Ga adatom is limited by the reaction with As; at common standard conditions with high V/III ratio thus two-dimensional nucleation easily occurs in addition to Ga incorporation at steps and kinks. Thereby the density of steps increases, leading to some surface roughness on an atomic scale. If Ga is supplied in an As-free ambient the migration length is substantially increased; Ga adatoms then incorporate at *existing* steps, leading eventually to atomically flat surfaces—even at very low substrate temperature. This is the principle of migration-enhanced epitaxy.

A flat growth surface and low growth temperature enable sharp interfaces for heterostructures and doping profiles, since bulk diffusion is largely suppressed. RHEED oscillation studies and PL measurements demonstrated good GaAs/AlGaAs quantum wells grown at 300 °C and GaAs layers even at 200 °C; the switching sequence required for the deposition of a complete cation layer (where the number of Ga atoms equals the number of surface sites) could be adjusted by a detailed study of the recovery of the RHEED intensity, but proved not very critical [123].

Besides for arsenides, MEE was also successfully applied to phosphides, nitrides, and other compound semiconductors. While the inherent low growth rate of this technique is less suitable for thick layers, it may be an interesting option for initial buffer growth or thin layers with sharp interfaces.

MEE proved also very beneficial for selective area growth described in Sect. 12.1: since cations and anions are alternately deposited on the substrate, cations cannot react with anions to nucleate and grow on areas covered with the amorphous mask.

12.3.3 Atomic Layer Deposition (ALD)

Many process steps in the fabrication of electronic devices do not require monocrystalline deposition. Progressing miniaturization of critical device dimensions and deposition on increasingly larger substrates create a need for a strict growth control also for amorphous or polycrystalline films. While most processes can be performed using CVD or PVD, the technique of atomic layer deposition offers superior control to meet particularly high demands. Basic advantages of ALD are an exceptional conformality, precise thickness control – also over large areas, and a good tuning ability for film composition; furthermore, many materials can be deposited at low temperature with a high quality and purity. These unique properties make ALD an important technique for fabricating advanced electronic and optoelectronic devices.

ALD is introduced in many reviews, e.g. [124–127], and books, e.g. [128–130]. Due to its close relation to ALE and its importance for semiconductor device fabrication some basic features are briefly outlined, albeit deposition of amorphous and polycrystalline films is beyond the scope of this book.

Conformality

Conformality designates the ability of a film-deposition process to coat the surfaces of a non-planar substrate uniformly. The top and side-wall surfaces are then all coated with the same film thickness, and even undercut recesses are well coated.

Conformality is important for many layers employed in semiconductor devices. For dielectric layers, e.g., a high dielectric strength at high electric fields is required; this property is related to the density of pinholes (keyholes) and the bulk density of the film. The same requirements apply for passivation layers.

The conformality of a film can be assessed by evaluating the ratio of film thicknesses at different locations on a nonplanar surface. Often structures with vertical trenches are applied, and the conformality can be defined by the aspect ratio of the thicknesses as indicated in Fig. 12.27a, i.e., $conformality = t_{side}/t_{top}$. An effective conformal coating yields a ratio of unity at all locations. Such a coating can even package microcontamination particles which may occur also under clean-room conditions and reduce the device yield; ALD coating of such particles is illustrated at the bottom of Fig. 12.27a. Evaporation and sputtering processes generally lead to a nonconformal coating as shown in Fig. 12.27b. The conformality then varies and depends on the surface structure. Here microcontamination particles can lead to uncoated locations and consequently to pinhole formation as illustrated at the bottom

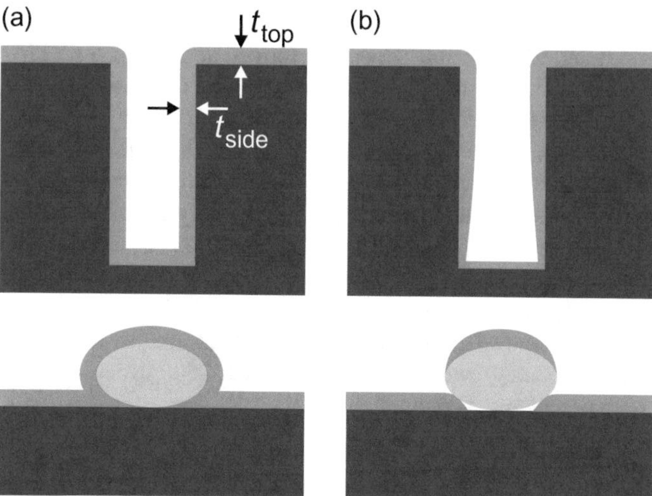

Fig. 12.27 **a** Schematic of the conformal film coating (*orange*) of a surface structure (*blue*) with a vertical trench shown in cross section; the bottom figure illustrates conformal packaging of a microcontamination particle. **b** Nonconformal coating of the structures shown in (**a**)

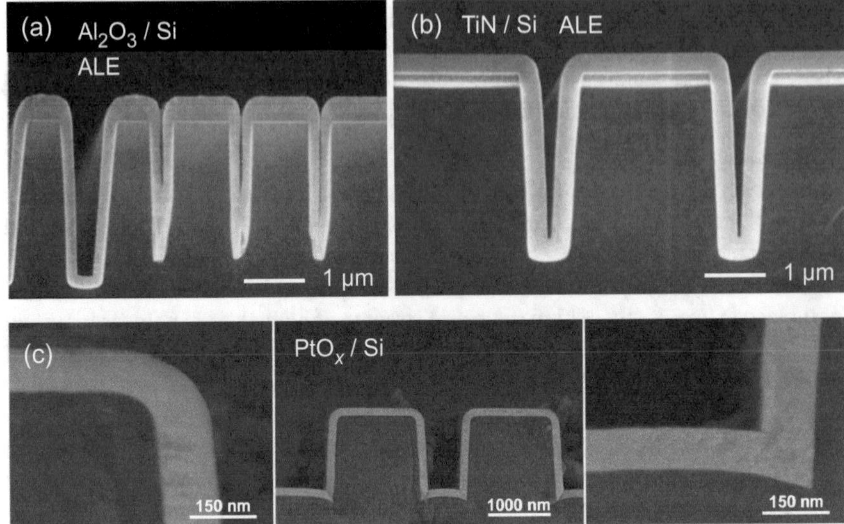

Fig. 12.28 **a** 300 nm Al_2O_3 film and **b** 160 nm TiN film deposited on trench-patterned Si substrate. Reproduced with permission from [131], © 1999 John Wiley & Sons. **c** Scanning electron micrographs of a PtO_x film deposited on patterned Si; reproduced with permission from [132], © 2008 ACS

of the figure. Even when narrow trenches become filled a smooth film may result without the formation of buried keyholes.

Examples for conformal coating of trench-patterned Si substrates using ALD are shown in Fig. 12.28. Deposition of an amorphous Al_2O_3 film from trimethylaluminum (TMAl) and water at 300 °C was performed applying 1.5 s TMAl and H_2O exposure times and 3.0 s purge periods; the film was deposited with 2730 cycles at a rate of about 1.1 Å/cycle using N_2 carrier gas at 10 mbar total pressure [131]. Taking a 100 Å thick native oxide on Si into account, a conformality of unity was achieved also in narrow trenches, see Fig. 12.28a; a merely small leakage current at high electric field was reported for such layers. The polycrystalline TiN film shown in Fig. 12.28b was deposited from $TiCl_4$, and NH_3 at 500 °C, with addition of some Zn as an additional reducing agent to decrease the film resistivity; exposure times were 0.2 s for $TiCl_4$ and Zn, and 0.5 s for NH_3, and all the purge times were 0.5 s [131]. Figure 12.28c shows a PtO_x film with good conformality deposited at 120 °C using Pt(acetylacetonato)$_2$ and ozone as precursors; 2500 cycles of 5 s pulses and purges for both precursors were applied [132].

Thickness Control

ALD enables ultimate thickness control on the atomic scale. Such a precision implies that the same amount of material is deposited in each cycle, which in turn requires the self-limiting nature of the precursor reactions. If this precondition is met and each monolayer sticks well during purge steps, the film thickness increases linearly with the number of ALD cycles.

There may occur an offset at initial growth on the substrate due to nucleation processes; precursor reactions with substrate material may differ from those with the film. The initial growth may then be accelerated or delayed for a few ALD cycles. Once steady-state conditions are established the growth per cycle remains constant within the processing window.

Uniformity of the film over substrates with a large area is another merit of ALD. If ternary films are deposited the requirements are even more strict than those for uniform thickness; this quantity also implies a film composition independent on the location.

Composition Control
For many ternary materials the stoichiometry can be well controlled in ALD. E.g., separate binary deposition cycles can be mixed to obtain the target composition. Usually the optimum process temperatures of the binary processes differ, yielding a narrow temperature window for ternary deposition. Since also the binary growth rates per cycle differ, dosing or purge times must be carefully adjusted. Often a nonlinear relation between the cycle ratio and the stoichiometry of the film complicates the composition control. The issues for quaternary films are generally more demanding.

12.4 Deposition of Organic Crystals

Organic semiconductors have lately found applications in increasingly demanding thin-film electronic and optical applications. They are inherently materials for low-temperature growth and processing due to the absence of strong bonding between molecules; typical temperatures are below 100–150 °C. This makes organic semiconductors compatible with direct-write printing-based manufacturing techniques and applications employing flexible plastic substrates or biological molecules. Beside the epitaxial techniques discussed in this section, techniques based on solution-processable methods are of particular interest for device production; a review is given in [133]. However, low-cost fabrication using spin-casting or printing generally yields highly defective semiconductors with low mobility and quantum yield. Improved device performance requires a high degree of crystallinity, which can be provided by organic epitaxy.

There is generally no lateral 1:1 lattice relation in organic epitaxy as most often required in the epitaxy of inorganic semiconductors (Sect. 3.2). Fabrication techniques hence rather use the term *deposition* instead of epitaxy. Two widely applied methods are discussed below, followed by the peculiarities of nucleation and growth of crystalline organic layers. We focus on small-molecule crystals of prominent species used for organic devices, which may establish an epitaxial relation to the substrate and usually show better performance than non-crystalline or polymer-based applications.

12.4.1 Methods of Organic Layer Deposition

Organic semiconductors have usually only little solubility in solvents. For the fabrication of organic semiconductors hence mostly vapor-phase techniques similar to those employed for inorganic semiconductors are used. The popular methods of molecular beam and of vapor phase depositions are discussed below.

Purification of Source Materials
Just as in the growth of inorganic semiconductors any deposition technique for organic layers requires the application of pure source materials. Commercial chemical compounds generally do not have sufficient purity for fabricating organic semiconductors. Atomic and molecular impurities have a substantial impact on the molecular adsorption on the growth surface and on the nucleation and geometry of an organic crystal layer. Means to reduce defect densities are both the crucial purification of the source material and the application of a preferably high deposition rate exceeding the adsorption rate of impurities. Impurities in organic compounds originate from the synthesis of the source material, but they can also be created from decomposition of the source material during evaporation. The latter may occur even below the sublimation point by catalysis involving contaminants like water [134].

Methods to remove impurities usually utilize differences in vapor pressure or distribution coefficients; gradient sublimation and zone refining are widely applied techniques. For *gradient sublimation* [135] the typically powdered starting material is loaded into a tube inserted in a furnace and evacuated to below 10^{-6} mbar. When the temperature is slowly raised to the sublimation point, purified material will condense in a zone of the tube kept at lower temperature, while more volatile impurities are removed by pumping and less volatile impurities remain in the zone of the starting material. The purified material can be further purified by a second cycle. *Zone refining* employs differences in the distribution coefficients [136]. Crystalline material is locally melted, and this zone is moved across the material. Multiple zones in the furnace can be used, and the process can be cycled. Limits are given for impurities with a distribution coefficient close to unity.

Organic Molecular Beam Deposition
Devices with organic semiconductors are often fabricated in high vacuum with a typical residual pressure of 10^{-6} mbar or in an inert gas atmosphere. In such conditions any surface is quickly covered with a monolayer of various species such as water, hydrocarbons, or oxygen—see Sect. 11.3.1. This surface coverage has a strong impact on the growth of a molecular layer, and also on the band alignment at interfaces as pointed out in Sect. 5.3. Results obtained in such an ambient therefor differ distinctly from those obtained in ultrahigh vacuum (UHV, residual pressure below 10^{-9} mbar); growth of PTCDA (Fig. 3f) on oriented graphite, e.g., was reported to yield monolayer films allowing for clear STM images when prepared in UHV, but only films clumped into several monolayer-thick islands when prepared in high vacuum [137].

Organic molecular beam deposition (OMBD), also referred to as OMB *epitaxy* (OMBE) when an epitaxial relation between substrate and layer exists, is a widely applied technique for preparing organic crystals under controlled conditions. The OMBE system is similar to that for MBE of inorganic semiconductors outlined in Sect. 11.3. It comprises the vacuum chamber with turbo and cryo pumps providing UHV conditions, effusion cells with shutters, and in situ characterization tools for growth control.

Molecular beams are provided by Knudsen cells filled with powdered, or sometimes liquid, source materials. The flux is controlled by the cell temperature kept above the sublimation temperature, but well below the decomposition temperature of the molecules. Typical evaporation temperatures are between 50 and 400 °C, aiming to produce about 10^{-4} mbar (= 0.75 Torr) vapor pressure for obtaining reasonable growth rates. Vapor pressures of many molecules are very well described in a limited temperature range by the Arrhenius dependence

$$\log(P/\text{mbar}) = a - b/T, \tag{12.11}$$

with P for a and b of Table 12.3 given in units of mbar (= 0.75 Torr = 100 Pa) and T in K; here $b = \Delta H_{\text{sub}}/R$, and the enthalpy of vaporization ΔH_{sub} is assumed constant in the considered temperature range, see (6.13). Examples for prominent organic semiconductors are given in Fig. 12.29 and Table 12.3. Straight curves in the data of Fig. 12.29 show that the assumption of constant enthalpy of sublimation is reasonable.

At the comparably low effusion temperatures many crucible materials can be used such as glasses and metals. Keeping the cell temperature steadily at elevated temperature in UHV for long periods and between growth runs proved beneficial for obtaining optimum purity of source materials.

Growth rates are typically ranging from 0.1 to 10 nm/s, translating to 0.2–30 ML/s (monolayers/s) for a thickness of 0.3–0.5 nm per ML. Optimum results are often obtained at medium growth rates due to enhanced impurity incorporation at very low rates and difficult control at very high rates. A drawback of OMBD is a substantial loss of evaporated material due to depositions at the cold walls of the chamber, and low throughput and high cost related to the UHV ambient.

Table 12.3 Vapor pressure data of prominent organic semiconductors in units of mbar according to (12.11)

Compound	Formula	Weight (g/mol)	Melting point (K)	T range (K)	a	b (K)
Anthracene	$C_{14}H_{10}$	178.24	491	318–363	12.454	5222
Tetracene	$C_{18}H_{12}$	228.29	630	386–472	12.590	6580
Pentacene	$C_{22}H_{14}$	278.35	>573	443–483	13.558	8194
Perylene	$C_{20}H_{12}$	252.3	551	391–424	15.505	6929
CuPc (β)	$CuN_8C_{32}H_{16}$	576.07	350	360–420	13.63	11025
6T	$C_{24}H_{14}S_6$	494.76	290	230–290	16.52	10817
Alq3	$Al(C_9H_6NO)_3$	459.43	>300	200–300	10.58	7192

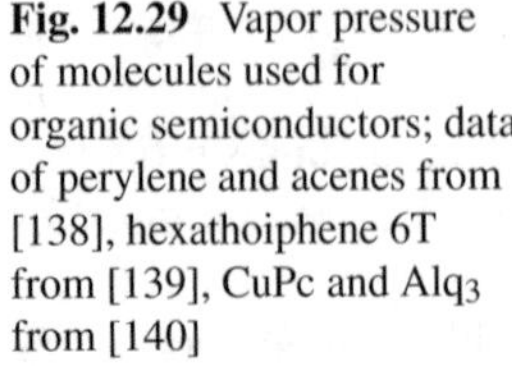

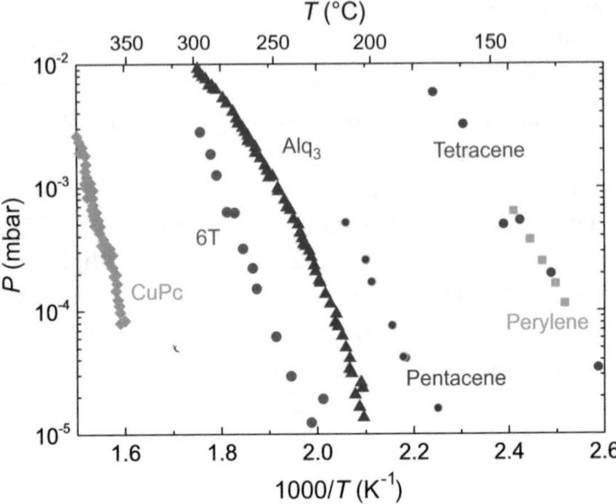

Fig. 12.29 Vapor pressure of molecules used for organic semiconductors; data of perylene and acenes from [138], hexathoiphene 6T from [139], CuPc and Alq_3 from [140]

Organic Vapor Phase Deposition

The demanding requirement of UHV conditions made technically more simple techniques attractive for mass production. *Organic Vapor Phase Deposition* (OVPD) is widely applied in manufacturing of devices based on organic semiconductors, allowing for high throughput and improved uniformity of film-thickness and dopant concentrations over large areas. The system is depicted in the schematic Fig. 12.30. The organic source material is sublimated in crucibles at the hot zone of a furnace and transported by a heated inert carrier gas (N_2) to the substrate kept at a low temperature. The deposition rate is controlled by the source temperatures and the mass

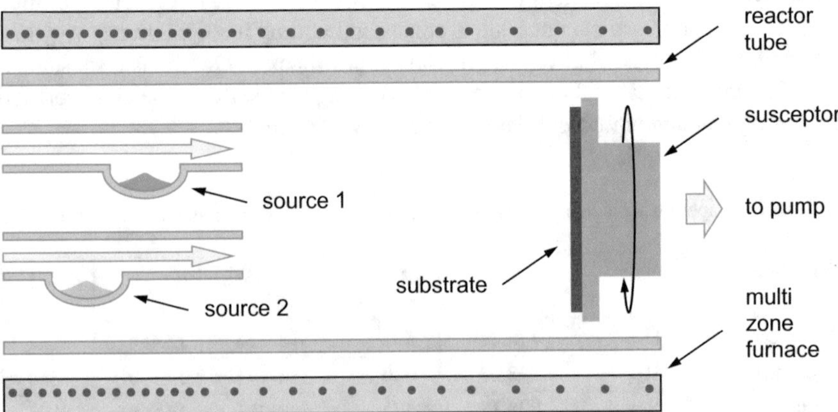

Fig. 12.30 Schematic diagram of an organic vapor phase deposition apparatus. The flow of carrier gas is set by mass-flow controllers

flow, comparable to the MOVPE process discussed in Sect. 11.2; the sticking coefficient of molecules is usually assumed close to unity. A convenient total pressure of typically 1 mbar is provided by a pumping unit. The reduced pressure increases the gas diffusivity and improves the mass transfer from the component streams to the substrate. The mass flow in the reactor can be engineered to achieve a uniform organic vapor distribution and consequently a uniform layer thickness. The setup is placed in a reactor tube with hot walls preventing parasitic deposition; an optimized reactor has minimum materials waste. A detailed introduction to the technique is given in [141–143].

The molar rate r of organic vapor provided by a source in an OVPD setup can be evaluated from the carrier-gas mass flow and the vapor pressure of the source. At a low flow rate of the carrier gas the vapor and the solid source material in the source region can equilibrate; the rate is then given by

$$r = \dot{V} \times \frac{P}{RT} \tag{12.12}$$

where $\dot{V}$ is the volume flow-rate of the carrier gas (usually measured in standard cubic centimeters per minute sccm), P is the equilibrium vapor pressure of the organic source material according to (12.11), and T is the source temperature. Under such low-flow conditions r is proportional to the flow rate providing good control ability; at high flow rate no equilibrium can be established and r depends more on the evaporation rate and less on the flow rate of the carrier gas [142].

Multiple sources allow for doping control and heterostructure deposition. Some systems use a shower head in front of the deposition region for homogeneous large-area film thickness, and a shutter for abrupt changes of composition.

12.4.2 Epitaxy of Organic Semiconductors

Differences Between Organic and Inorganic Layers

Kinetic processes of organic thin-film formation are much like those occurring in inorganic epitaxy discussed in Sect. 7.2. Still a number of basic differences strongly affects nucleation and growth of organic layers.

The growth units of inorganic crystals are atoms with usually spherical symmetry. Organic crystals are composed of extended molecules with a characteristic low symmetry (Sect. 3.1.2); various mutual orientations and displacements add additional degrees of freedom which gives rise to many different morphological structures in these crystals.

The weak intermolecular forces allow for accommodating considerable strain in organic crystals, but it also facilitates the formation of many polymorphs and structural defects. Variations in the conformation of individual molecules may add additional distortions.

A strong molecule adsorption on inorganic substrates often creates an organic wetting layer with a symmetry distinctly different from that of the organic bulk crystal. This layer forms a template for subsequently deposited layers, which sense gradually less bonding to the organic–inorganic interface with increasing thickness and more organic–organic bonding to neighboring molecules. The layer structure may consequently change completely, with a tendency to form the equilibrium bulk structure.

Impurities are mostly interstitial, because they typically do not share host bonds which create the electronic bands of van der Waals crystals. Excess carriers are then tightly localized at the defect and are usually not electrically active. Particularly molecular impurities may, however, affect nucleation and the stacking order of the organic crystal.

Due to the complex conditions mentioned above no strict rules can be given for the growth of organic crystals; there are still frequently observed phenomena, which allow to discuss some general aspects in the following.

Substrates for Organic Epitaxy

Usually best performance of organic semiconductors is achieved with highly crystalline material. Well-ordered crystallization of an organic layer on a substrate is generally a challenging task, and there are applications requiring substrate-layer combinations for which no single-crystal growth can be established. In any case the adsorption and initial orientation of the molecules depends sensitively on the chosen combination of the substrate and the molecular layer deposited on top.

Due to their limited stability organic layers are deposited at comparably low temperature, typically between 80 and 400 °C; the substrate temperature is frequently quoted rather in units of Kelvin than Celsius, because the substrate is often not heated but cooled. The low temperature allows to use many layer-substrate combinations. Also the relaxed requirement of lattice matching due to weak intermolecular forces enables a rich choice.

Inorganic substrates have much smaller unit cells than organic crystals, and it is difficult to find substrate-layer combinations with matching lattice symmetry and lattice constants. There is generally no 1:1 substrate-layer relation as known from inorganic epitaxy; instead, a large organic molecule (large compared to the small atoms of an inorganic substrate) attaches at chemically bonded docking groups to the substrate, or it is bonded by weaker electrostatic or van der Waals interactions.

One approach for establishing an epitaxial substrate-layer relation is the insertion of an interlayer to control the interaction strength between the π system of the molecules and the substrate surface: the interaction should be sufficiently strong to align the molecules according to an epitaxial relation—usually one of the coincidence relations discussed in Sect. 3.2.1; it should, however, be weak enough to allow the molecules to align properly, e.g., in a standing-up configuration. Such a control was found to be connected to the local density of states (LDOS) near the Fermi energy of the substrate at the surface [144]. A high LDOS leads to a strong interaction with the π electrons and consequently to a planar adsorption as usually obtained with metal substrates, while at a low LDOS the intermolecular van der Waals interaction may dominate and enable a standing-up alignment. Epitaxial stand-up configurations of pentacene were reported for the semimetallic Bi(0001)/Si(111)(7 × 7) and Si(111)(5 × 2)Au/Si(111)(7 × 7) surfaces [144], where an initial laying-down alignment is obtained without the interlayer. Further interlayers are mentioned in Sect. 3.2.1.

In some cases a substrate can be found providing a *commensurate* molecular arrangement with small integer numbers m_{ij} according to (3.4). An example is the use of graphite for the epitaxy of polycyclic aromatic hydrocarbons like pentacene discussed in Sect. 3.2.1. The basal plane of highly oriented pyrolytic graphite provides a hexagonal lattice nearly identical to the frame of the carbon backbone of hydrocarbon molecules; furthermore, an only weak adsorption energy allows for some degree of freedom in molecular alignment. Pentacene on a *smooth* graphite basal plane was reported to form a commensurate but not densely packed monolayer of laying down and slightly tilted molecules, while multilayer films evolve in islands with a nonplanar arrangement with (022) contact planes; on a *rough* graphite surface instead a standing-up orientation was observed in the first monolayer evolving to a (001) oriented Campbell bulk-phase [145].

Modes of Organic Epitaxy

The large variety of the alignment of molecules on a substrate and variations in conformations within individual molecules provide many degrees of freedom, giving rise to complex nucleation phenomena, to a large number of lattice distortions, and to many polymorphs in organic thin films.

The alignment of organic layer molecules with respect to the substrate is basically controlled by the interface potential discussed in Sect. 3.2.1. During nucleation often a preferred azimuthal orientation of the layer is found, which is generally an indication of an epitaxial relation and reflects a minimum of the interface potential. Still no smooth layers may form, and a coincident mode of the oblique organic unit cell may be difficult to identify. This led in practice to two widely used terms for the epitaxy of ordered organic crystals: *van der Waals epitaxy* and *quasiepitaxy*. These types of epitaxy are not well defined, but rather express some apparent features of studied layer/substrate combinations.

Van der Waals epitaxy refers to commensurate or coincident conditions with weak van der Waals bonds to the substrate and small elastic constants of the layer. This mode of epitaxy may occur at growth conditions near thermal equilibrium, i.e., at low growth rate and relative high substrate temperature; it is favored if interlayer bonds dominate over intralayer bonds (cf. Fig. 3.8).

Strain can be accommodated in case of some mismatch and may result in one of the three growth modes depicted in Fig. 6.12, depending on the intralayer forces and the adhesion to the substrate. The weak van der Waals bonds allow for substantial distortions with respect to the bulk structure within the first few monolayers and thus for larger strain than that observed in inorganic heteroepitaxy. With increasing layer thickness and hence decreasing influence of the substrate the mismatch relaxes toward the bulk structure by the creation of defects. The relaxation occurs typically within the first five monolayers and prevents the growth of thicker molecularly smooth films. Frequently layers are observed to grow in a Stranski-Krastanow-like mode, with a stable two-dimensional wetting layer and the formation of islands on top; such islands show preferred azimuthal orientations and often large lateral aspect ratios.

Quasiepitaxy (QE) may occur at incommensurate matching conditions (Fig. 3.7c) if intralayer bonds dominate. It is favored under kinetically controlled nonequilibrium conditions, i.e., at high deposition rate and low growth temperature.

Despite incommensurate conditions in QE the layer may show azimuthal order with respect to the substrate; often the layer is then not uniformly oriented but shows crystallites with different lateral alignments. QE layers may be strained; in contrast to layers grown by van der Waals epitaxy such a strain does not relax with increasing thickness. Instead, accommodation occurs within the unit cells by internal degrees of freedom, rendered possible by the weak intermolecular forces and the related small elastic constants.

Compared to inorganic heteroepitaxy with covalent or ionic bonding the binding energy of organic layers to the substrate is weak. For a given critical thickness t_c the amount of strain which can be accommodated in the layer while maintaining pseudomorphism is invers to the binding energy: at low values a larger strain is required for a constant t_c. This is expressed by the curve of Fig. 12.31, yielding a qualitative relation between both types of epitaxy [146]. We note that both van der Waals (vdW) epitaxy and quasiepitaxy (QE) may occur at a given strain, depending on growth conditions. Layers with strong bonds and large strain have values above the constant-t_c curve and relax or grow polycrystalline or amorphous.

Nucleation

Nucleation of crystalline organic layers depends critically on the interactions among molecules of the layer and the interaction to the substrate, as discussed above and in Sect. 3.2.1. The crucial nucleation step is therefor specific for any substrate-layer combination and additionally depends on growth conditions. We hence mention only a few examples.

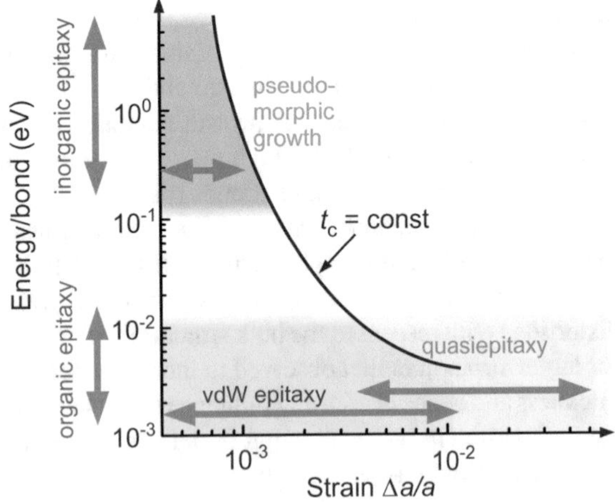

Fig. 12.31 Schematic roughly indicating ranges (*gray arrows*) for the epitaxy of strained layers. The *curve* represents the limit of coherent growth for an assumed constant critical thickness t_c

Epitaxial growth of phthalocyanines (Fig. 3.3c) and the perylene-derivative PTCDA (Fig. 3.3f) was reviewed in [137]. The molecule-substrate interaction of PbPc on alkali halides was reported to be dominated by electrostatic forces between substrate anions and the Pc metal. Growing at equilibrium-near conditions on (001) substrates, commensurate conditions with low strain on NaCl and with high strain on KCl were found, while on KBr coexistence of a commensurate layer lattice and an incommensurate fraction was assumed. Electrostatic interactions are also assumed to control the alignment of PTCDA on NaCl and KCl; commensurate conditions with two slightly different polymorphs with different lateral rotations of the unit cells were found on NaCl, while both polymorphs have the same preferred alignment on KCl. The monolayer thick structure on NaCl was found to be incompatible with a multilayer structure, triggering a dewetting and a mixture of epitaxial $p(3 \times 3)$ layer and bulk-like crystallites at higher coverage [147]. Such a Stranski–Krastanow-like transition beyond one monolayer toward a bulk-like herringbone arrangement was also observed on KCl [148].

Epitaxy of sexithiophene (6T, Fig. 3.3d) and *para*-hexaphenyl (*p*-6P) on mica(001), KCl(001), and other substrates was reviewed in [149]. For *p*-6P on a Cu–O(110) surface a similar dewetting on top of a stable monolayer is reported; accumulated strain destabilizes the second monolayer leading to stable nuclei with bulk-like arrangement at further deposition [150]. At low surface coverage on mica and KCl both *p*-6P and 6T tend to form needle-like nano-sized fibers with an orientation reflecting the symmetry of the substrate. For *p*-6P on muscovite mica (with no rotational symmetry of the surface unit cell) parallel needles are observed, while on phlogopite mica (three-fold symmetry) or KCl (four-fold symmetry) respective three or two orthogonal alignments are found [149]. Within a needle the rodlike molecules are stacked face-to-face, such that the long needle axis is perpendicular to the long molecular axis.

There are many nucleation phenomena discussed in detail in the literature. Reviews are given, e.g., in [137, 151, 152].

12.5 General Reading Chapter 12

(12.1) Z.R. Zytkiewicz, Epitaxial lateral overgrowth of semiconductors, in *Springer Handbook of Crystal Growth*, ed. by G. Dhanaraj, K. Byrappa, V. Prasad, M. Dudley (Springer, Berlin Heidelberg, 2010), pp. 999–1039.
(12.2) V.G. Dubrovskii, *Nucleation Theory and Growth of Nanostructures* (Springer, Berlin, 2014).
(12.3) H.C.M. Knoops, S.E. Potts, A.A. Bol, W.M.M. Kessels, Atomic layer deposition, in *Handbook of Crystal Growth*, 2nd edn, vol. III part B, ed. by T.F. Kuech (Elsevier, Amsterdam 2015), pp. 1101–1134.
(12.4) M. Shtein, Organic Vapor Phase Deposition, in *Organic Materials—Materials, Processing, Devices, and Applications*, ed. by F. So (CRC Press, Boca Raton, FL USA, 2010), pp. 27–57.

References

1. J.D. Kim, X. Chen, J.J. Coleman, Selective area masked growth (nano to micro), in *Handbook of Crystal Growth*, 2nd edn., vol. 3, Part A, ed. by T.F. Kuech (Elsevier, Amsterdam, 2015), pp. 441–481
2. R. Bhat, Current status of selective area epitaxy by OMCVD. J. Cryst. Growth **120**, 362 (1992)
3. E. Colas, A. Shahar, J.B.D. Soole, W.J. Tomlinson, J.R. Hayes, C. Caneau, R. Bhat, Lateral and longitudinal patterning of semiconductor structures by crystal growth on nonplanar and dielectric-masked GaAs substrates: application to thickness-modulated waveguide structures. J. Cryst. Growth **107**, 226 (1991)
4. O. Kayser, Selective growth of InP/GaInAs in LP-MOVPE and MOMBE/CBE. J. Cryst. Growth **107**, 989 (1991)
5. M. Ozeki, K. Mochizuki, K. Kodama, Growth of GaAs and AlAs thin films by a new atomic layer epitaxy technique. Thin Solid Films **174**, 63 (1989)
6. T. Auzelle, B. Haas, A. Minj, C. Bougerol, J.- Rouvière, A. Cros, J. Colchero, B. Daudin, The influence of AlN buffer over the polarity and the nucleation of self-organized GaN nanowires. J. Appl. Phys. **117**, 245303 (2015)
7. N. Dupuis, J. Décobert, P.-Y. Lagrée, N. Lagay, F. Poingt, C. Kazmierski, A. Ramdane, A. Ougazzaden, Mask pattern interference in AlGaInAs selective area metal-organic vapor-phase epitaxy: Experimental and modeling analysis. J. Appl. Phys. **103**, 113113 (2008)
8. M. Gibbon, J.P. Stagg, C.G. Cureton, E.J. Thrush, C.J. Jones, R.E. Mallard, R.E. Pritchard, N. Collis, A. Chew, Selective-area low-pressure MOCVD of GaInAsP and related materials on planar InP substrates. Semicond. Sci. Technol. **8**, 998 (1993)
9. H.-J. Oh, M. Sugiyama, Y. Nakano, Y. Shimogaki, Surface reaction kinetics in metalorganic vapor phase epitaxy of GaAs through analyses of growth rate profile in wide-gap selective-area growth. Jpn. J. Appl. Phys. **42**, 6284 (2003)
10. J.E. Greenspan, Alloy composition dependence in selective area epitaxy on InP substrates. J. Cryst. Growth **236**, 273 (2002)
11. P. Aseev, A. Fursina, F. Boekhout, F. Krizek, J.E. Sestoft, F. Borsoi, S. Heedt, G. Wang, L. Binci, S. Martí-Sánchez, T. Swoboda, R. Koops, E. Uccelli, J. Arbiol, P. Krogstrup, L.P. Kouwenhoven, P. Caroff, Selectivity map for molecular beam epitaxy of advanced III–V quantum nanowire networks. Nano Lett. **19**, 218 (2019)
12. F. Allegretti, M. Inoue, T. Nishinaga, In-situ observation of GaAs selective epitaxy on GaAs (111)B substrates. J. Cryst. Growth **146**, 354 (1995)
13. A. Okamoto, Selective epitaxial growth by molecular beam epitaxy. Semicond. Sci. Technol. **8**, 1011 (1993)
14. T. Sugaya, Y. Okada, M. Kawabe, Selective growth of GaAs by molecular beam epitaxy. Jpn. J. Appl. Phys. **31**, L713 (1992)
15. K. Suzuki, M. Ito, Y. Horikoshi, Selective growth of GaAs on GaAs (111)B substrates by migration-enhanced epitaxy. Jpn. J. Appl. Phys. **38**, 6197 (1999)
16. S.S. Chang, S. Ando, T. Fukui, Fabrication of high density ultrafine GaAs quantum wire array structures using selective metalorganic chemical vapor deposition. Surf. Sci. **267**, 214 (1992)
17. K. Tomoika, T. Fukui, Growth of semiconductor nanocrystals, in *Handbook of Crystal Growth*, vol. 1 part A, ed. by T. Nishinaga (Elsevier, Amsterdam, 2015), pp. 749–793
18. S.H. Jones, L.K. Seidel, K.M. Lau, Patterned substrate epitaxy surface shapes. J. Cryst. Growth **108**, 73 (1991)
19. D.W. Shaw, Morphology analysis in localized crystal growth and dissolution. J. Cryst. Growth **47**, 509 (1979)
20. C. Tessarek, M. Bashouti, M. Heilmann, C. Dieker, I. Knoke, E. Spiecker, S. Christiansen, Controlling morphology and optical properties of self-catalyzed, mask-free GaN rods and nanorods by metal-organic vapor phase epitaxy. J. Appl. Phys. **114**, 144304 (2013)
21. K. Kamon, S. Takagishi, H. Mori, Selective epitaxial growth of GaAs by low-pressure MOVPE. J. Cryst. Growth **73**, 73 (1985)

22. L. Jastrzebski, SOI by CVD: Epitaxial lateral overgrowth (ELO) process. J. Cryst. Growth **63**, 493 (1983)
23. P. Gibart, Metal organic vapour phase epitaxy of GaN and lateral overgrowth. Rep. Prog. Phys. **67**, 667 (2004)
24. R.P. Gale, R.W. McClelland, J.C.C. Fan, C.O. Bozler, Lateral epitaxial overgrowth of GaAs by organometallic chemical vapor deposition. Appl. Phys. Lett. **41**, 545 (1982)
25. Z.R. Zytkiewicz, Epitaxial lateral overgrowth of semiconductors, in *Springer Handbook of Crystal Growth*, ed. by G. Dhanaraj, K. Byrappa, V. Prasad, M. Dudley (Springer, Berlin Heidelberg, 2010), pp. 999–1039
26. D.J. Ironside, A.M. Skipper, T.A. Leonard, M. Radulaski, T. Sarmiento, P. Dhingra, M.L. Lee, J. Vučković, S.R. Bank, High-quality GaAs planar coalescence over embedded dielectric microstructures using an all-MBE approach. Cryst. Growth Des. **19**, 3085 (2019)
27. N. Julian, P. Mages, C. Zhang, J. Zhang, S. Kraemer, S. Stemmer, S. Denbaars, L. Coldren, P. Petroff, J. Bowers, Coalescence of InP epitaxial lateral overgrowth by MOVPE with V/III ratio variation. J. Electron. Mater. **41**, 845 (2012)
28. S. Nakamura, M. Senoh, S.-i. Nagahama, N. Iwasa, T. Yamada, T. Matsushita, H. Kiyoku, Y. Sugimoto, T. Kozaki, H. Umemoto, M. Sano, K. Chocho, Present status of InGaN/GaN/AlGaN-based laser diodes. J. Cryst. Growth **189/190**, 820 (1998)
29. S. Nakamura, M. Senoh, S.-I. Nagahama, N. Iwasa, T. Yamada, T. Matsushita, H. Kiyoku, Y. Sugimoto, T. Kozaki, H. Umemoto, M. Sano, K. Chocho, InGaN/GaN/AlGaN-based laser diodes with modulation-doped strained-layer superlattices. Jpn. J. Appl. Phys. **36**, L1568 (1997)
30. A. Usui, H. Sunakawa, A. Sakai, A.A. Yamaguchi, Thick GaN epitaxial growth with low dislocation density by hydride vapor phase epitaxy. Jpn. J. Appl. Phys. Lett. **36**, L899 (1997)
31. H.-O. Nam, M.D. Bremser, T.S. Zheleva, R.F. Davis, Lateral epitaxy of low defect density GaN layers via organometallic vapor phase epitaxy. Appl. Phys. Lett. **71**, 2638 (1997)
32. H. Marchand, X.H. Wu, J.P. Ibbetson, P.T. Fini, P. Kozodoy, S. Keller, J.S. Speck, S.P. DenBaars, U.K. Mishra, Microstructure of GaN laterally overgrown by metalorganic chemical vapor deposition. Appl. Phys. Lett. **73**, 747 (1998)
33. E. Frayssinet, B. Beaumont, J.P. Faurie, P. Gibart, Zs. Makkai, B. Pécz, P. Lefebvre, P. Valvin, Micro epitaxial lateral overgrowth of GaN/sapphire by metal organic vapour phase epitaxy, MRS Int. J. Nitride Semicond. Res. **7**, 8 (2002)
34. P. Vennéguès, Defect reduction methods for III-nitride heteroepitaxial films grown along nonpolar and semipolar orientations. Semicond. Sci. Technol. **27**, 024004 (2012)
35. K. Hiramatsu, K. Nishiyama, A. Motogaito, H. Miyake, Y. Iyechika, T. Maeda, Recent progress in selective area growth and epitaxial lateral overgrowth of III-nitrides: effects of reactor pressure in MOVPE growth. Phys. Stat. Sol. A **176**, 535 (1999)
36. Z. Liliental-Weber, X. Ni, H. Morkoc, Structural perfection of laterally overgrown GaN layers grown in polar- and non-polar directions. J. Mater. Sci.: Mater. Electron. **19**, 815 (2008)
37. P. Vennéguès, B. Beaumont, V. Bousquet, M. Vaille, P. Gibart, Reduction mechanisms for defect densities in GaN using one- or two-step epitaxial lateral overgrowth methods. J. Appl. Phys. **87**, 4175 (2000)
38. D. Cai, J. Kanga, S. Ito, Dislocations formed under longitudinal stress field in epitaxial-lateral-overgrowth GaN. Mat. Sci. Semicond. Process. **9**, 15 (2006)
39. T.S. Zheleva, S.A. Smith, D.B. Thomson, K.J. Linthicum, P. Rajagopal, R.F. Davis, Pendeo-epitaxy: a new approach for lateral growth of gallium nitride films. J. Electron. Mater. **28**, L5 (1999)
40. R.F. Davis, T. Gehrke, K.J. Linthicum, T.S. Zheleva, E.A. Preble, P. Rajagopal, C.A. Zorman, M. Mehregany, Pendeo-epitaxial growth of thin films of gallium nitride and related materials and their characterization. J. Cryst. Growth **225**, 134 (2001)
41. T. Gehrke, K.J. Linthicum, E. Preble, P. Rajagopal, C. Ronning, C. Zorman, M. Mehregany, R.F. Davis, Pendeo-epitaxial growth of gallium nitride on silicon substrates. J. Electron. Mater. **29**, 306 (2000)

42. K. Tadatomo, H. Okagawa, Y. Ohuchi, T. Tsunekawa, Y. Imada, M. Kato, T. Taguchi, High output power InGaN ultraviolet light-emitting diodes fabricated on patterned substrates using metalorganic vapor phase epitaxy. Jpn. J. Appl. Phys. **40**, L583 (2001)
43. M. Yamada, T. Mitani, Y. Narukawa, S. Shioji, I. Niki, S. Sonobe, K. Teguchi, M. Sano, T. Mukai, InGaN-Based near-ultraviolet and blue-light-emitting diodes with high external quantum efficiency using a patterned sapphire substrate and a mesh electrode. Jpn. J. Appl. Phys. **41**, L1431 (2002)
44. R. Haitz, J.Y. Tsao, Solid-state lighting: 'the case' 10 years after and future prospects. Phys. Stat. Sol. A **208**, 17 (2011)
45. J. Cho, J.H. Park, J.K. Kim, E.F. Schubert, White light-emitting diodes: history, progress, and future. Laser Photonics Rev. **11**, 1600147 (2017)
46. P. Pust, P.J. Schmidt, W. Schnick, A revolution in lighting. Nature Mat. **14**, 454 (2015)
47. T. Guner, M.M. Demir, A Review on halide perovskites as color conversion layers in white light emitting diode applications. Phys. Stat. Sol. A **215**, 1800120 (2018)
48. R.S. Wagner, W.C. Ellis, Vapor-liquid-solid mechanism of single crystal growth. Appl. Phys. Lett. **4**, 89 (1964)
49. J.M. Redwing, X. Miao, X. Li, Vapor–liquid–solid growth of semiconductor nanowires, in *Handbook of Crystal Growth, Thin Films and Epitaxy,* vol. 3 part A, 2nd edn., ed. by T.F. Kuech (Elsevier, Amsterdam, 2015), pp. 399–439
50. K.A. Dick, A review of nanowire growth promoted by alloys and non-alloying elements with emphasis on Au-assisted III-V nanowires. Prog. Cryst. Growth Character. Mater. **54**, 138 (2008)
51. Y. Zhang, J. Wu, M. Aagesen, H. Liu, III–V nanowires and nanowire optoelectronic devices. J. Phys. D **48**, 463001 (2015)
52. G.E. Cirlin, V.G. Dubrovskii, I.P. Soshnikov, N.V. Sibirev, Y.B. Samsonenko, A.D. Bouravleuv, J.C. Harmand, F. Glas, Critical diameters and temperature domains for MBE growth of III–V nanowires on lattice mismatched substrates. Phys. Stat. Sol. (RRL) **3**, 112 (2009)
53. F. Glas, Critical dimensions for the plastic relaxation of strained axial heterostructures in free-standing nanowires. Phys. Rev. B **74**, 121302 (2006)
54. E. Ertekin, P.A. Greaney, D.C. Chrzan, Equilibrium limits of coherency in strained nanowire heterostructures. J. Appl. Phys. **97**, 114325 (2005)
55. H. Ye, P. Lu, Z. Yu, Y. Song, D. Wang, S. Wang, Critical thickness and radius for axial heterostructure nanowires using finite-element method. Nano Lett. **9**, 1921 (2009)
56. J. Bloem, Y.S. Oei, H.H.C. Demoor, J.H.L. Hanssen, L.J. Giling, Near equilibrium growth of silicon by CVD I, the Si–Cl–H system. J. Cryst. Growth **65**, 399 (1983)
57. P. Villars, H. Okamoto, K. Cenzual (eds.), *ASM alloy phase diagrams database* (ASM International, Materials Park, 2006)
58. V. Schmidt, S. Senz, U. Gösele, The shape of epitaxially grown silicon nanowires and the influence of line tension. Appl. Phys. A **80**, 445 (2005)
59. E.I. Givargizov, Fundamental aspects of VLS growth. J. Crystal Growth **31**, 20 (1975)
60. C.E. Kendrick, J.M. Redwing, The effect of pattern density and wire diameter on the growth rate of micron diameter silicon wires. J. Cryst. Growth **337**, 1 (2011)
61. V. Schmidt, S. Senz, U. Gösele, Diameter-dependent growth direction of epitaxial silicon nanowires. Nano Lett. **5**, 931 (2005)
62. C.X. Wang, M. Hirano, H. Hosono, Origin of diameter-dependent growth direction of silicon nanowires. Nano Lett. **6**, 1552 (2006)
63. F. Glas, Chemical potentials for Au-assisted vapor-liquid-solid growth of III–V nanowires. J. Appl. Phys. **108**, 073506 (2010)
64. F. Glas, J.-C. Harmand, G. Patriarche, Why does wurtzite form in nanowires of III–V zinc blende semiconductors? Phys. Rev. Lett. **99**, 146101 (2007)
65. T. Akiyama, K. Nakamura, T. Ito, Structural stability and electronic structures of InP nanowires: Role of surface dangling bonds on nanowire facets. Phys. Rev. B **73**, 235308 (2006)

66. T. Akiyama, K. Sano, K. Nakamura, T. Ito, An empirical potential approach to wurtzite-zinc-blende polytypism in group III–V semiconductor nanowires. Jpn. J. Appl. Phys. **45**, L275 (2006)
67. V.G. Dubrovskii, N.V. Sibirev, Growth thermodynamics of nanowires and its application to polytypism of zinc blende III-V nanowires. Phys. Rev. B **77**, 035414 (2008)
68. *Group IV Elements, IV-IV and III-V Compounds*, ed. by U. Rössler, Landolt-Börnstein, New Series, Group III, vol. 41, Part A (Springer, Berlin, 2006)
69. A.A. Koryakin, S.A. Kukushkin, N.V. Sibirev, On the mechanism of the vapor–solid–solid growth of Au-catalyzed GaAs nanowires. Semiconductors **53**, 350 (2019)
70. V.G. Dubrovskii, G.E. Cirlin, I.P. Soshnikov, A.A. Tonkikh, N.V. Sibirev, Y. Samsonenko, V.M. Ustinov, Diffusion-induced growth of GaAs nanowhiskers during molecular beam epitaxy: theory and experiment. Phys. Rev. B **71**, 205325 (2005)
71. V.G. Dubrovskii, N.V. Sibirev, J.C. Harmand, F. Glas, Growth kinetics and crystal structure of semiconductor nanowires. Phys. Rev. B **78**, 235301 (2008)
72. E.K. Mårtensson, S. Lehmann, K.A. Dick, J. Johansson, Simulation of GaAs nanowire growth and crystal structure. Nano Lett. **19**, 1197 (2019)
73. J. Bauer, V. Gottschalch, H. Paetzelt, G. Wagner, B. Fuhrmann, H.S. Leipner, MOVPE growth and real structure of vertical-aligned GaAs nanowires. J. Cryst. Growth **298**, 625 (2007)
74. J. Johansson, B.A. Wacaser, K.A. Dick, W. Seifert, Growth related aspects of epitaxial nanowires. Nanotechnol. **17**, S355 (2006)
75. J. Tersoff, Stable self-catalyzed growth of III–V nanowires. Nano Lett. **15**, 6609 (2015)
76. C. Colombo, D. Spirkoska, M. Frimmer, G. Abstreiter, A. Fontcuberta i Morral, Ga-assisted catalyst-free growth mechanism of GaAs nanowires by molecular beam epitaxy. Phys. Rev. B **77**, 155326 (2008)
77. F. Glas, Comparison of modeling strategies for the growth of heterostructures in III–V nanowires. Cryst. Growth Des. **17**, 4785 (2017)
78. D. Jacobsson, F. Panciera, J. Tersoff, M.C. Reuter, S. Lehmann, S. Hofmann, K.A. Dick, F.M. Ross, Interface dynamics and crystal phase switching in GaAs nanowires. Nature **531**, 317 (2016)
79. F. Glas, M. Reda Ramdani, G. Patriarche, J.-C. Harmand, Predictive modeling of self-catalyzed III–V nanowire growth. Phys. Rev. B **88**, 195304 (2013)
80. K. Tomioka, Y. Kobayashi, J. Motohisa, S. Hara, T. Fukui, Selective-area growth of vertically aligned GaAs and GaAs/AlGaAs core–shell nanowires on Si(111) substrate. Nanotechnol. **20**, 145302 (2009)
81. D. Rudolph, S. Hertenberger, S. Bolte, W. Paosangthong, D. Spirkoska, M. Döblinger, M. Bichler, J.J. Finley, G. Abstreiter, G. Koblmüller, Direct observation of a noncatalytic growth regime for GaAs nanowires. Nano Lett. **11**, 3848 (2011)
82. M. Royo, M. De Luca, R. Rurali, I. Zardo, A review on III–V core–multishell nanowires: growth, properties, and applications. J. Phys. D **50**, 143001 (2017)
83. K.A. Dick, S. Kodambaka, M.C. Reuter, K. Deppert, L. Samuelson, W. Seifert, F.M. Ross, The morphology of axial and branched nanowire heterostructures. Nano Lett. **7**, 1817 (2007)
84. M. Paladugu, J. Zou, Y.-N. Guo, G.J. Auchterlonie, H.J. Joyce, Q. Guo, H.H. Tan, C. Jagadish, Y. Kim, Novel growth phenomena observed in axial InAs/GaAs nanowire heterostructures. Small **3**, 1873 (2007)
85. M.T. Borgström, M.A. Verheijen, G. Immink, T. de Smet, E.P.A.M. Bakkers, Interface study on heterostructured GaP–GaAs nanowires. Nanotechnology **17**, 4010 (2006)
86. M. Heigoldt, J. Arbiol, D. Spirkoska, J.M. Rebled, S. Conesa-Boj, G. Abstreiter, F. Peiró, J.R. Morantece, A. Fontcuberta i Morral, Long range epitaxial growth of prismatic heterostructures on the facets of catalyst-free GaAs nanowires. J. Mater. Chem. **19**, 840 (2009)
87. J. Zou, M. Paladugu, H. Wang, G.J. Auchterlonie, Y.N. Guo, Y. Kim, Q. Gao, H.J. Joyce, H.H. Tan, C. Jagadish, Growth mechanism of truncated triangular III–V nanowires. Small **3**, 389 (2007)
88. H.J. Joyce, Q. Gao, H.H. Tan, C. Jagadish, Y. Kim, J. Zou, L.M. Smith, H.E. Jackson, J.M. Yarrison-Rice, P. Parkinson, M.B. Johnston, III–V semiconductor nanowires for optoelectronic device applications. Prog. Quantum Electron. **35**, 23 (2011)

89. J.B. Wagner, N. Sköld, L.R. Wallenberg, L. Samuelson, Growth and segregation of GaAs–$Al_xIn_{1-x}P$ core-shell nanowires. J. Cryst. Growth **312**, 1755 (2010)
90. D. Rudolph, S. Funk, M. Döblinger, S. Morkötter, S. Morkötter, S. Hertenberger, L. Schweickert, J. Becker, S. Matich, M. Bichler, D. Spirkoska,†I. Zardo, J.J. Finley, G. Abstreiter, G. Koblmüller, Spontaneous alloy composition ordering in GaAs-AlGaAs core–shell nanowires. Nano Lett. **13**, 1522 (2013)
91. S. Raychaudhuri, E.T. Yu, Calculation of critical dimensions for wurtzite and cubic zinc blende coaxial nanowire heterostructures. J. Vac. Sci. Technol., B **24**, 2053 (2006)
92. M.V. Nazarenko, N.V. Sibirev, K.W. Ng, F. Ren, W.S. Ko, V.G. Dubrovskii, C. Chang-Hasnain, Elastic energy relaxation and critical thickness for plastic deformation in the core-shell InGaAs/GaAs nanopillars. J. Appl. Phys. **113**, 104311 (2013)
93. J.-P. Ahl, H. Behmenburg, C. Giesen, I. Regolin, W. Prost, F.J. Tegude, G.Z. Radnoczi, B. Pécz, H. Kalisch, R.H. Jansen, M. Heuken, Gold catalyst initiated growth of GaN nanowires by MOCVD. Phys. Stat. Sol. C **8**, 2315 (2011)
94. F. Qian, Y. Li, S. Gradečak, D. Wang, C.J. Barrelet, C.M. Lieber, Gallium nitride-based nanowire radial heterostructures for nanophotonics. Nano Lett. **4**, 1975 (2004)
95. M.D. Brubaker, S.M. Duff, T.E. Harvey, P.T. Blanchard, A. Roshko, A.W. Sanders, N.A. Sanford, K.A. Bertness, Polarity-controlled GaN/AlN nucleation layers for selective-area growth of GaN nanowire arrays on Si(111) substrates by molecular beam epitaxy. Cryst. Growth Des. **16**, 596 (2016)
96. A. Urban, J. Malindretos, J.-H. Klein-Wiele, P. Simon, A. Rizzi, Ga-polar GaN nanocolumn arrays with semipolar faceted tips. New J. Phys. **15**, 053045 (2013)
97. T. Auzelle, G. Calabrese, S. Fernández-Garrido, Tuning the orientation of the top-facets of GaN nanowires in molecular beam epitaxy by thermal decomposition. Phys. Rev. Mater. **3**, 013402 (2019)
98. H. Li, L. Geelhaar, H. Riechert, C. Draxl, Computing equilibrium shapes of wurtzite crystals: the example of GaN. Phys. Rev. Lett. **115**, 085503 (2015)
99. R. Colby, Z. Liang, I.H. Wildeson, D.A. Ewoldt, T.D. Sands, R.E. Garcia, E.A. Stach, Dislocation filtering in GaN nanostructures. Nano Lett. **10**, 1568 (2010)
100. S.D. Hersee, A.K. Rishinaramangalam, M.N. Fairchild, L. Zhang, P. Varangis, Threading defect elimination in GaN nanowires. J. Mater. Res. **26**, 2293 (2011)
101. S.D. Carnevale, J. Yang, P.J. Phillips, M.J. Mills, R.C. Myers, Three-dimensional GaN/AlN nanowire heterostructures by separating nucleation and growth processes. Nano Lett. **11**, 866 (2011)
102. V. Consonni, Self-induced growth of GaN nanowires by molecular beam epitaxy: a critical review of the formation mechanisms. Phys. Stat. Sol. RRL **7**, 699 (2013)
103. H. Sekiguchi, T. Nakazato, A. Kikuchi, K. Kishino, Structural and optical properties of GaN nanocolumns grown on (0 0 0 1) sapphire substrates by rf-plasma-assisted molecular-beam epitaxy. J. Cryst. Growth **300**, 259 (2007)
104. R. Koester, J.S. Hwang, C. Durand, D. Le Si Dang, J. Eymery, Self-assembled growth of catalyst-free GaN wires by metal–organic vapour phase epitaxy. Nanotechnol. **21**, 015602 (2010)
105. V. Consonni, M. Hanke, M. Knelangen, L. Geelhaar, A. Trampert, H. Riechert, Nucleation mechanisms of self-induced GaN nanowires grown on an amorphous interlayer. Phys. Rev. B **83**, 035310 (2011)
106. K. Hestroffer, C. Leclere, V. Cantelli, C. Bougerol, H. Renevier, B. Daudin, In situ study of self-assembled GaN nanowires nucleation on Si(111) by plasma-assisted molecular beam epitaxy. Appl. Phys. Lett. **100**, 212107 (2012)
107. M. Sobanska, V.G. Dubrovskii, G. Tchutchulashvili, K. Klosek, Z.R. Zytkiewicz, Analysis of incubation times for the self-induced formation of GaN nanowires: Influence of the substrate on the nucleation mechanism. Cryst. Growth Des. **16**, 7205 (2016)
108. V. Consonni, M. Knelangen, L. Geelhaar, A. Trampert, H. Riechert, Nucleation mechanisms of epitaxial GaN nanowires: origin of their self-induced formation and initial radius. Phys. Rev. B **81**, 085310 (2010)

109. S. Fernández-Garrido, X. Kong, T. Gotschke, R. Calarco, L. Geelhaar, A. Trampert, O. Brandt, Spontaneous nucleation and growth of GaN nanowires: the fundamental role of crystal polarity. Nano Lett. **12**, 6119 (2012)
110. V. Consonni, V.G. Dubrovskii, A. Trampert, L. Geelhaar, H. Riechert, Quantitative description for the growth rate of self-induced GaN nanowire. Phys. Rev. B **85**, 155313 (2012)
111. R.K. Debnath, R. Meijers, T. Richter, T. Stoica, R. Calarco, H. Lüth, Mechanism of molecular beam epitaxy growth of GaN nanowires on Si(111). Appl. Phys. Lett. **90**, 123117 (2007)
112. C. Tessarek, M. Heilmann, E. Butzen, A. Haab, H. Hardtdegen, C. Dieker, E. Spiecker, S. Christiansen, The role of Si during the growth of GaN micro- and nanorods. Cryst. Growth Des. **14**, 1486 (2014)
113. C.X. Ren, F. Tang, R.A. Oliver, T. Zhu, Nanoscopic insights into the effect of silicon on core-shell InGaN/GaN nanorods: luminescence, composition, and structure. J. Appl. Phys. **123**, 045103 (2018)
114. M. Ritala, M. Leskelä, Atomic layer epitaxy—a valuable tool for nanotechnology? Nanotechnology **10**, 19 (1999)
115. T. Suntola, Atomic layer epitaxy. Mater. Sci. Rep. **4**, 261 (1989)
116. Y. Horikoshi, Migration-enhanced epitaxy of GaAs and AlGaAs. Semicond. Sci. Technol. **8**, 1032 (1993)
117. M. Leskelä, M. Ritala, Atomic layer deposition (ALD): from precursors to thin film structures. Thin Solid Films **409**, 138 (2002)
118. H.C.M. Knoops, S.E. Potts, A.A. Bol, W.M.M. Kessels, Atomic layer deposition, in *Handbook of Crystal Growth*, 2nd edn, vol. III part B, ed. by T.F. Kuech (Elsevier, Amsterdam, 2015), pp. 1101–1134
119. M. Ozeki, Atomic layer epitaxy of III–V compounds using metalorganic and hydride sources. Mater. Sci. Rep. **8**, 97 (1992)
120. A. Doi, Y. Aoyagi, S. Namba, Stepwise monolayer growth of GaAs by switched laser metalorganic vapor phase epitaxy. Appl. Phys. Lett. **49**, 785 (1986)
121. M. Ozeki, New approach to the atomic layer epitaxy of GaAs using a fast gas stream. Appl. Phys. Lett. **53**, 1509 (1988)
122. T. Suntola, Atomic layer epitaxy. Thin Solid Films **216**, 84 (1992)
123. Y. Horikoshi, M. Kawashima, H. Yamaguchi, Migration-enhanced epitaxy of GaAs and AlGaAs. Jpn. J. Appl. Phys. **27**, 169 (1988)
124. R.W. Johnson, A. Hultqvist, S.F. Bent, A brief review of atomic layer deposition: from fundamentals to applications. Mater. Today **17**(5), 236 (2014)
125. S.M. George, Atomic layer deposition: an overview. Chem. Rev. **110**, 111 (2010)
126. V. Miikkulainen, M. Leskelä, M. Ritala, R.L. Puurunen, Crystallinity of inorganic films grown by atomic layer deposition: overview and general trends. J. Appl. Phys. **113**, 021301 (2013)
127. J. Sheng, J.-H. Lee, W.-H. Choi, T.H. Hong, M.J. Kim, J.-S. Park, Atomic layer deposition for oxide semiconductor thin film transistors: advances in research and development. J. Vac. Sci. Technol., A **36**, 060801 (2018)
128. C. S. Hwang (ed.), *Atomic Layer Deposition for Semiconductors* (Springer, New York, 2014)
129. J. Bachmann (ed.), *Atomic Layer Deposition in Energy Conversion Applications* (Wiley-VCH, Weinheim, 2017)
130. T. Kääriäinen, D. Cameron, M.-L. Kääriäinen, A. Sherman, *Atomic Layer Deposition: Principles, Characteristics, and Nanotechnology*, 2nd edn. (Scrivener Publishing, Salem, MA, 2013)
131. M. Ritala, M. Leskelä, J.-P. Dekker, C. Mutsaers, P.J. Soininen, J. Skarp, Perfectly conformal TiN and Al_2O_3 films deposited by atomic layer deposition. Chem. Vap. Deposition **5**, 7 (1999)
132. J. Hämäläinen, F. Munnik, M. Ritala, M. Leskela, Atomic layer deposition of platinum oxide and metallic platinum thin films from $Pt(acac)_2$ and ozone. Chem. Mater. **20**, 6840 (2008)
133. Z. He, J. Chen, D. Li, Crystal alignment for high performance organic electronics devices. J. Vac. Sci. Technol. A **37**, 040801 (2019)
134. T.U. Kampen, *Low Molecular Weight Organic Semiconductors* (Wiley-VCH, Weinheim, 2010)

135. F. Gutmann, L.B. Lyon, *Organic Semiconductors* (R.E. Krieger Publishing, Malabar, FL, 1981)
136. N. Karl, J. Marktanner, R. Stehle, W. Warta, High-field saturation of charge carrier drift velocities in ultrapurified organic photoconductors. Synth. Met. **42**, 2473 (1991)
137. S.R. Forrest, Ultrathin organic films grown by organic molecular beam deposition and related techniques. Chem. Rev. **97**, 1793 (1997)
138. V. Oja, E.M. Suuberg, Vapor pressures and enthalpies of sublimation of polycyclic aromatic hydrocarbons and their derivatives. J. Chem. Eng. Data **43**, 486 (1998)
139. C. Kloc, R.A. Laudise, Vapor pressures of organic semiconductors: α-hexathiophene and α-quarter-thiophene. J. Cryst. Growth **193**, 563 (1998)
140. K. Yase, Y. Takahashi, N. Ara-Kato, A. Kawazu, Evaporation rate and saturated vapor pressure of functional organic materials. Jpn. J. Appl. Phys. **34**, 636 (1995)
141. M. Baldo, M. Deutsch, P. Burrows, H. Gossenberger, M. Gerstenberg, V. Ban, S. Forrest, Organic vapor phase deposition. Adv. Mater. **10**, 1505 (1998)
142. M. Shtein, Organic Vapor Phase Deposition, in *Organic Materials—Materials, Processing, Devices, and Applications*, ed. by F. So (CRC Press, Boca Raton, FL, USA, 2010), pp. 27–57
143. M. Heuken, N. Meyer, Organic Vapor Phase Deposition, in *Organic Electronics—Materials, Processing, Manufacturing and Applications*, ed. by H. Klauk (Wiley-VCH, Weinheim, 2006), pp. 203–232
144. G.E. Thayer, J.T. Sadowski, F. Meyer zu Heringdorf, T. Sakurai, R.M. Tromp, Role of surface electronic structure in thin film molecular ordering. Phys. Rev. Lett. **95**, 256106 (2005)
145. J. Götzen, D. Käfer, C. Wöll, G. Witte, Growth and structure of pentacene films on graphite: weak adhesion as a key for epitaxial film growth. Phys. Rev. B **81**, 085440 (2010)
146. S.R. Forrest, P.E. Burrows, Growth modes of organic semiconductor thin films using organic molecular beam deposition: epitaxy, van der Waals epitaxy, and quasi-epitaxy. Supramol. Sci. **4**, 127 (1997)
147. S.A. Burke, W. Ji, J.M. Mativetsky, J.M. Topple, S. Fostner, H.-J. Gao, H. Guo, P. Grütter, Strain induced dewetting of a molecular system: bimodal growth of PTCDA on NaCl. Phys. Rev. Lett. **100**, 186104 (2008)
148. T. Dienel, C. Loppacher, S.C.B. Mannsfeld, R. Forker, T. Fritz, Growth-mode-induced narrowing of optical spectra of an organic adlayer. Adv. Mater. **20**, 959 (2008)
149. C. Simbrunner, H. Sitter, Organic van der Waals epitaxy versus templated growth by organic-organic heteroepitaxy, in *Handbook of Crystal Growth of Thin Films and Epitaxy: Basic Techniques*, vol. 3 part A, 2nd edn., ed. by T.F. Kuech (Elsevier, Amsterdam, 2015), pp. 483–508
150. A.J. Fleming, F.P. Netzer, M.G. Ramsey, Nucleation and 3D growth of para-sexiphenyl nanostructures from an oriented 2D liquid layer investigated by photoemission electron microscopy. J. Phys. Condens. Matter. **21**, 445003 (2009)
151. A.A. Virkar, S. Mannsfeld, Z. Bao, N. Stingelin, Organic semiconductor growth and morphology considerations for organic thin-film transistor. Adv. Mater. **22**, 3857 (2010)
152. P.G. Evans, J.W. Spalenka, Epitaxy of small organic molecules, in *Handbook of Crystal Growth—Thin Films and Epitaxy: Basic Techniques*, vol. 3 part A, 2nd edn., ed. by T.F. Kuech (Elsevier, Amsterdam, 2015), pp. 509–554

Appendix

Answers to Problems

2.1 (a) $f = 24.8\%$
(b) $4a_{\text{GaAs}} \approx 5\, a_{\text{GaN}}, f = –0.2\%$
(c) $a_{\text{wurtzite}} = 4.0$ Å, use $a_{\text{hex}} = a_{\text{cub}}/\sqrt{2}$

2.2 (a) $f = +49.1\%, f_{\text{alternative1}} = -32.9\%$
(b) $f = -14.0\%$ ($f_{\text{alternative1}} = 16.2\%$),
(c) $f_1 = -1.8\%, f_2 = -0.5\%$
(d) $f_a = -3.5\%, f_c = -3\%$

2.3 (a) $x = 0.475–1.016y$
(b) $y_{\max} = 0.468$, $x_{\max} = 0.475$
(c) $(Al_{0.475}In_{0.525}As)_z(Ga_{0.468}In_{0.532}As)_{1-z}$

2.4 (a) $f = 0.24\%$
(b) $x = 0.5\%$
(c) $\varepsilon_{\parallel} = +0.15\%$
(d) $x = 8.41\%, f = -0.15\%$

2.5 (a) $\varepsilon_{\perp} = +0.15\%$
(b) $\Delta V/V = -2.1 \times 10^{-3}$
(c) $d_{\text{unstrained}} = 3.272$ Å, $d_{\text{strained}} = 3.270$ Å, The diagonal of the zincblende unit cell of $a_0\sqrt{3}$ length comprises 3 anion-cation (111) layers in *ABC* sequence, the nearest (111) layer distance is thus $1/3 \times \sqrt{3} \times a_0 = a_0/\sqrt{3}$
(d) $n = 100.3$
(e) $(E/A)_1 = 0.131$ J/m^2, $(E/A)_2 = 2.10$ J/m^2
(f) $E/V_{\text{EZ}} = 7.36 \times 10^{-11}$ J/m^3

2.6 (a) $t = 11.52$ Å $= 4$ ML
(b) $y = 0.3755$

2.7 (a) $E_{\text{screw}}/L = 7.10 \times 10^{-6}$ J/m
(b) $E_{\text{edge}}/L = 3.03 \times 10^{-5}$ J/m
(c) $E_{60}/L = 1.38 \times 10^{-5}$ J/m

2.8 (a) From Fig. 2.15: $t_c \approx 6 \times 10^2 \times a_s = 330$ nm
(b) $t_{c2}/t_{c1} = 0.90$

U. W. Pohl, *Epitaxy of Semiconductors*, Graduate Texts in Physics,
https://doi.org/10.1007/978-3-030-43869-2

(c) $f = 2 \times 10^{-4}$

2.9 (a) $\Delta\Theta_{\text{relaxed}} = -342$ s, $\Delta\Theta_{\text{strained}} = -744$ s
(b) $\Delta\Theta_{\text{relaxed}} = -528$ s, $\Delta\Theta_{\text{strained}} = -1100$ s,
(c) $I(004)/I(115) \approx 3.6$

2.10 (a) $x_1 = 18.5\%$
(b) $x_2 = 25.1\%$, $x_3 = 22.6\%$

3.1 (a) $V = -2.1$ meV
(b) $\mathrm{d}V/\mathrm{d}r = 6\alpha r^{-7} - \beta\gamma \exp(-\gamma r)$;
2.8 Å: $\mathrm{d}V/\mathrm{d}r = -23.9$ meV/Å (steep negative slope of $V(r)$, repulsion)
3.3 Å: $\mathrm{d}V/\mathrm{d}r = 0.038$ meV/Å (0 with more precise equilibrium near $r = 3.3032$ Å)
3.8 Å: $\mathrm{d}V/\mathrm{d}r = +1.62$ meV/Å (shallow positive slope of $V(r)$, attraction)

3.2 (a) $V_{\text{molecule}} = 257$ Å^3, density = 1.34 g/cm^3
(b) $\Delta V/V = +3.7\%$, density = 1.29 g/cm^3

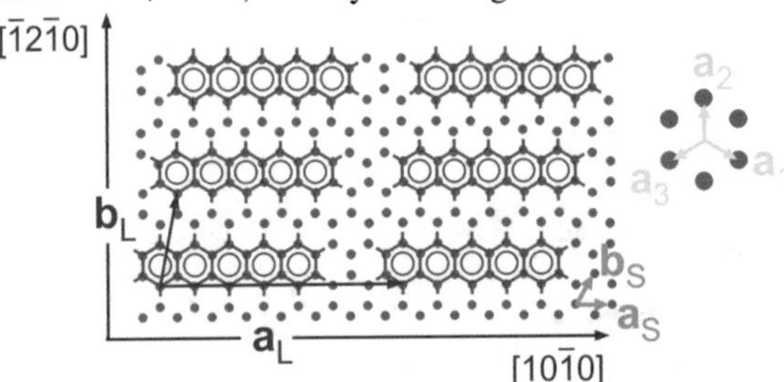

3.3 Coordinates:
(a) 12.3 Å, orientation $\mathbf{a}_\mathrm{L} \| \mathbf{a}_\mathrm{S} \| \langle 1\ 0\ \bar{1}0 \rangle$; a total of 6 orientations, but only 3 can be distinguished.
(b) $a_\mathrm{L} = 17.2$ Å, $b_\mathrm{L} = 6.5$ Å, angle 79.1°
(c) $d_{\min}(\mathrm{C}_{\text{mol1}} - \mathrm{C}_{\text{mol2}}) = 3.8$ Å

3.4 (a) 4.01 Å, 4.19 Å
(b) $V_{4.10\,\text{Å}} = -2.39$ eV, $V_{3.98\,\text{Å}} = -2.55$ eV

4.1 (a) $x = 48.5\%$, $E^\Gamma = 1.92$ eV (direct), order (increasing E) $E^\Gamma, E^\mathrm{L}, E^\mathrm{X}$, VCA: $E^\Gamma, E^\mathrm{X}, E^\mathrm{L}$
(b) $x = 30.5\%$, crossing E^Γ and E^X, $E = 2.20$ eV

4.2 $\Delta E_\mathrm{c}(77\text{ K}) = 0.25$ eV, $\Delta E_\mathrm{c}(300\text{ K}) = 0.26$ eV

4.3 (a) $n = 4.45 \times 10^{12}$ cm^{-2}
(b) $\varepsilon_{\|} = -2.62\%$, $n_{\text{tot}} = 4.41 \times 10^{12}$ cm^{-2}

4.4 (a) $\Delta E_\mathrm{v} = -0.24$ eV, $\Delta E_\mathrm{c} = -0.13$ eV
(b) $\Delta E_\mathrm{v} = +0.24$ eV, $\Delta E_\mathrm{c} = -0.01$ eV,
(c) using $b^{\text{theo}} = -1.90$ eV: $E_{\mathrm{v,lh}} - E_{\mathrm{v,hh}} = 0.35$ eV, splitting exceeds calculated offsets significantly—the strain is very large, $E^\mathrm{X}_{\mathrm{lh}} < E^\mathrm{X}_{\mathrm{hh}}$, $\Delta E_\mathrm{v} = 0.45$ eV, ΔE_c is unchanged.

4.5 (a) $E_\mathrm{g}(77\text{ K}) = 1.51$ eV, $E_\mathrm{g}(300\text{ K}) = 1.42$ eV
(b) $E_{\mathrm{g,\,eff}} = 1.55$ eV

(c) $\Delta E_{g,\,eff} = \pm 8$ meV

5.1 (a) 2.23 eV

(b) $E_g^{opt} = 3.25$ eV, $E_X = 0.29$ eV

5.2 $\beta = -2.5$, μ (0 °C) = 1.28 cm²/(Vs)

5.3 (a) $E(\text{HOMO}_{\text{BCP}}) = E(\text{HOMO}_{\text{Alq3}}) + 0.2$ eV, $E(\text{LUMO}_{\text{BCP}}) = E(\text{LUMO}_{\text{Alq3}}) - 0.6$ eV

(b) $E_{vac}(\text{Alq}_3) = E_{vac}(\text{PTCDA}) - 0.5$ eV, $E(\text{HOMO}_{\text{Alq3}}) = E(\text{HOMO}_{\text{PTCDA}}) + 0.4$ eV

(c) $E_{vac}(\alpha\text{-NPD}) = E_{vac}(\text{PTCDA}) - 0.2$ eV

5.4 (a) $E_F(\alpha\text{-NPD/Au}) = 0.2$ eV, $E_F(\alpha\text{-NPD/Mg}) = 1.6$ eV

(b) $e\Delta\phi_{\text{Au}} = 1.2$ eV, $e\Delta\phi_{\text{Mg}} = 0.4$ eV, $S = 0.57$

6.1 (a) $\Delta\mu = 94$ J/K

(b) $\Delta\mu = 6.1$ J/K

6.2 (a) Δg_m (800 °C) = 0.66 kJ, Δg_m (1200 °C) = −1.00 kJ

(b) $T = 959$ °C

(c) $x_1 = 0.17$, $x_2 = 0.83$

6.3 Elements of the plot: axes [100], [011], and [111] lie in a common $(1\bar{1}0)$ plane considered here. Angles from [100] are 54.7° to [111] and 90° to [011], distances from the origin to respective Wulff planes are $r_{100} = 1.16 \times r_{011}$ and $r_{111} = 1.05 \times r_{011}$. The facets meet the condition, {110}: 0.83 J/m² $< \sqrt{2} \times$ 0.96 J/m² = 1.36 J/m², {111}: 0.87 J/m² $< \sqrt{3} \times$ 0.96 J/m² = 1.66 J/m²

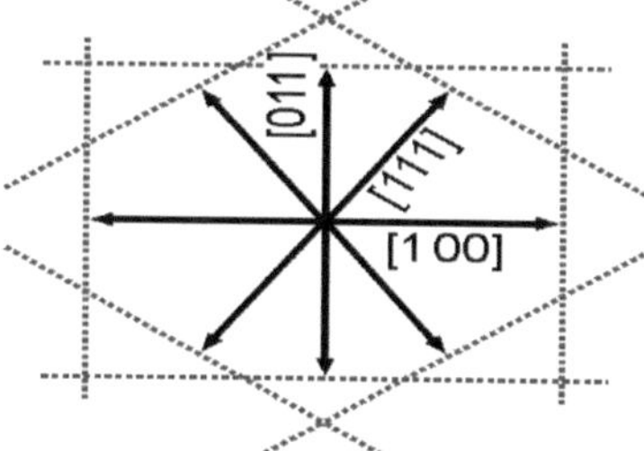

6.4 (a) $\Delta G_\gamma^{2D} = 2.0$ eV, for $\gamma_{S_B} = \gamma_{S_A}$ is $\Delta G_\gamma^{2D} = 0.9$ eV

(b) $N = 362$, $\Delta\mu \approx 1.1$ kJ/mol

6.5 (a) $R \approx 2$ ML/s (1.98 ML/s)

(b) $v = 61$ nm/s, $l = 251$ nm

7.1 (a) bulk $(n-2)^3$, faces $6 \times (n-2)^2$, edges $12 \times (n-2)$, corners 12×1, sum 1000 for $n = 10$, 10^9 for $n = 1000$

(b) bulk $3E_b$, face $2.5E_b$, edge $2E_b$, corner $1.5E_b$, fractions 4.34×10^{-1} for $n = 10^1$, 4.93×10^{-2} for 10^2, 4.99×10^{-3} for 10^3, 5.00×10^{-4} for 10^4; virtually unchanged without edge and corner atoms, e.g., 4.985×10^{-3} instead of 4.993×10^{-3} for 10^3, i.e., the small fraction of edge and corner atoms gets negligible.

7.2 The figure shows a top view of all atoms in the first 4 layers of a unit cell for the two reconstructions (no atom hidden), the largest in the 1st layer, open and filled symbols for either As and Ga, respectively, (in β2(2 × 4)), or vice versa (in β2(4 × 2)); dimers are encircled. Electron numbers see list, showing both reconstructions comply with the ECR.

Layer	Atoms	Electrons for β2(2 × 4)		Electrons for β2(4 × 2)	
		Required	Available	Required	Available
1	2 dimers	28	20	20	12
2	6 atoms	24	18	32	30
3	6 atoms + 1 dimer	24 + 14	30 + 10	24 + 10	18 + 6
4	8 atoms		12		20
Sum		90	90	86	86

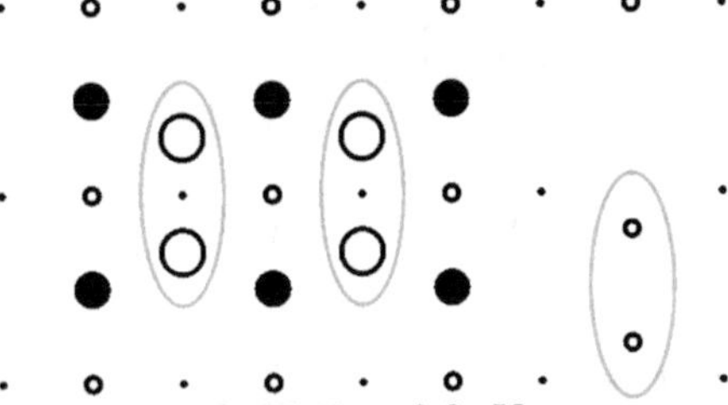

7.3 (a) $E_d = 1.2$ eV, $E_a = 2.0$ eV
(b) 1041 K and 759 K

7.4 (a) $\tau_a = 8.2 \times 10^{-6}$ s, $\lambda_2 = 1.9 \times 10^{-4}$ cm
(b) $\Theta = 0.039°$, $r_{step} = 0.4$ mm/s

7.5 (a) $n_{In} = n_{As} = 2937$ atoms
(b) $d = 1.62$ ML
(c) $n_{In} = n_{As} = 396$ atoms, $d = 1.43$ ML

8.1 (a) $E_d = 3.18$ eV
(b) $\phi_{800°C} = 12.5$ ML/min
(c) 7.41×10^{15}

8.2 (a) $k_{rel}/k_{nonrel} = 1.0049$ (10 kV), $k_{rel}/k_{nonrel} = 1.0241$ (50 kV)
(b) $k_{10\,kV}/g_{GaAs} = 291$

8.3 (a) $s_{800\,°C/0.89} = 44.0$ nA, $T_{44\,nA/0.85} = 777.4$ °C, $s_{800\,°C/0.85} = 60.0$ nA
(b) $T = 990$ °C

8.4 (a) $f_{therm} = 1.60 \times 10^{-3}$
(b) $M_L = 478.5$ GPa, $\sigma_L = -0.764$ GPa
(c) $\kappa = 74.7$ km^{-1}, $R_c = 13.4$ m

8.5 (a) $R(0) = 0.600$, $R(150\text{ s}) = 0.249$, $R(600\text{ s}) = 0.453$
(b) $T = 314$ s, $t = 1279$ s
(c) $n = 4.40$, $\kappa = 0.40$
(d) $g = 0.30$ nm/s

9.1 (a) Applying definition of (2.20a): $f_{Ge/Si} = -4.005\%$, $f_{24Ge/25Si} = -0.005\%$
(b) Visible D1 $\frac{1}{6}[11\bar{2}]$, D1 $\frac{1}{6}[1\bar{2}1]$, D2 $\frac{1}{6}[\bar{1}2\bar{1}]$, D2 $\frac{1}{6}[\bar{1}\bar{1}2]$, not visible D1 $\frac{1}{6}[\bar{2}11]$, D2 $\frac{1}{6}[2\bar{1}\bar{1}]$ 81 Å apart: set of D1 $\frac{1}{6}[1\bar{2}1]$ and set of D2 $\frac{1}{6}[\bar{1}\bar{1}2]$, 40.5 Å apart: D1 $\frac{1}{6}[11\bar{2}]$, D2 $\frac{1}{6}[\bar{1}2\bar{1}]$

9.2 (a) $a_{hop} = 3.84$ Å, $D_0 = 1.47 \times 10^{-2}$ cm^2/s, $n_{hop} = 1341$, $d_{10ns} = 140$ Å
(b) $\lambda = 186$ Å, $n_{hop} = 2347$
(c) $D_{eff}/D_{bare} = 5.66 \times 10^{-3}$
(d) $E_{de\text{-}exc} = 7.34 \times 10^{-1}$ eV

9.3 (a) $E_{Si/Si} \cong 0.53$ eV, $E_{Si/Sb/Si} \cong 1.53$ eV
(b) $R_{Si/Si} \cong 1450$ Å, $g \cong 4.6 \times 10^{-2}$, $T_{100nm,Si/Si} \cong 807$ K, $T_{100nm,Si/Sb/Si} \cong$ 977 K

9.4 (a) $E_{Si-Sb} = 1.07$ eV, $E_{Si,unsat} - E_{Si,sat} = 0.72$ eV
(b) $E_{\gamma,i^*=1} = -1.49$ eV, $E_{\gamma,i^*=2} = -2.23$ eV

9.5 $a_{InAs,400\,°C}/a_{InAs,300K} = 1.0017$, $(a_{InAs} - a_{GaAs}/a_{GaAs})|_{400\,°C} = 7.11\%(a_{InAs_\parallel} - a_{GaAs})/a_{GaAs} \cong 6.3\%, \varepsilon = -0.002)(a_{InAs_\parallel/Te} - a_{GaAs})/a_{GaAs} \cong 4.9\%, \varepsilon = -0.021)$

10.1 (a) n_{InAs} (77 K) $= 4.3 \times 10^3$ cm^{-3}, n_{InAs} (300 K) $= 7.5 \times 10^{14}$ cm^{-3}, n_{InP} (77K) $= 2 \times 10^{-17}$ cm$^{-3} \approx 0$ cm^{-3}, n_{InP} (300 K) $= 1.2 \times 10^7$ cm^{-3}
(b) T_{InAs} 255 K (−18 °C), T_{InP} 373 K (100 °C)

10.2 $D_c^{eff} = 5.8 \times 10^{17}$ cm^{-3}, $D_v^{eff} = 1.3 \times 10^{19}$ cm^{-3}, $n_{max} = 6 \times 10^{19}$ cm^{-3}, $p_{max} = 5 \times 10^{19}$ cm^{-3}

10.3 (a) $t = 1.82 \times 10^4$ s ≈ 5 h
(b) $T_2 = 672$ °C
(c) $w = 435$ nm
(d) $c = 2.8 \times 10^{17}$ cm^{-3}

10.4 (a) $e\phi_{Bn} = 1.60$ eV, $E_c - E_F \approx 0.19$ eV, $eV_{bi} = 1.41$ eV
(b) $w_0 = 3.0 \times 10^{-7}$ m ≈ 300 nm, $w_{forward} \approx 1.6 \times 10^{-7}$ m, $w_{reverse} \approx 3.9 \times 10^{-7}$ m
(c) Slightly reduced by 2%

10.5 $R_{c2} = 1.6 \times 10^{-6}$ Ω cm^2

11.1 (a) $x_{interface} \approx 0.0224 = 2.24\%$, $x_{excess} \approx 0.0012 = 0.12\%$
(b) $D = 4 \times 10^{-5}$ cm^{-2}/s
(c) $d_{100s} \approx 1.7$ μm, $d_{200s} \approx 2.4$ μm

11.2 (a) $Q_{TMGa} = 17.2$ sccm
(b) $m_{TMGa\ molecule} = 1.9 \times 10^{-22}$ g, $m_{TMGa} = 0.28$ g consumption in 1 h
(c) $T_{TBAs} = 10$ °C

11.3 (a) $u \approx 40$ cm/s, $u_{standard} = 1.25$ cm/s
(b) $\delta \approx 20$ cm
(c) $r_2/r_1 \approx 1.09$

11.4 (a) $n_{Si} = 6.8 \times 10^{14}$ cm^{-2}
(b) $G = 2.8 \times 10^{14}$ cm^{-2} s^{-1}
(c) $\Gamma = 2.0 \times 10^{17}$ s^{-1}
(d) $A \approx 18$ cm^2

11.5 (a) $G = 8.5 \times 10^{14}$ cm^{-2} s^{-1}
(b) $r = 1.4$ μm/h
(c) Double the Ga equilibrium pressure, $T_{Ga} = 995$ °C

Index

U. W. Pohl, *Epitaxy of Semiconductors*, Graduate Texts in Physics,
https://doi.org/10.1007/978-3-030-43869-2

T

Fundamental Physical Constants

Quantity	Symbol	Value	SI Unit
Avogadro constant	N_A	6.02214×10^{23}	mol^{-1}
Bohr radius	a_B	5.29177×10^{-11}	m
Boltzmann constant	k_B	1.38065×10^{-23}	J/K
Elementary charge	e	1.60218×10^{-19}	As
Electron mass	m_0	9.10938×10^{-31}	kg
Molar gas constant	$R = k_B \times N_A$	8.31446	J/(mol K)
Permeability in vacuum	$\mu_0 = 1/(\varepsilon_0 c^2)$	1.25664×10^{-6}	Vs/(Am)
Permittivity in vacuum	ε_0	8.85419 10^{-12}	As/(Vm)
Planck constant	h	6.62607×10^{-34}	Js
	$\hbar = h/2\pi$	1.05457×10^{-34}	Js
Proton mass	m_p	1.67262×10^{-27}	kg
Rydberg energy	$hcR_\infty = m_0 e^4/(2\hbar^2)$	2.17987×10^{-18}	J
Speed of light in vacuum	c	2.99792×10^{8}	m/s
Unified atomic mass unit	$u = \frac{1}{12} m(C_6^{12})$	1.66054×10^{-27}	kg
1 Electron volt	eV	1.60218×10^{-19}	J
1 Angstrom	Å	1.00000×10^{-10}	m

Source: CODATA internationally recommended values of the fundamental physical constants, https://physics.nist.gov/cuu/Constants/index.html.

U. W. Pohl, *Epitaxy of Semiconductors*, Graduate Texts in Physics,
https://doi.org/10.1007/978-3-030-43869-2

Printed in the United States
by Baker & Taylor Publisher Services